FOUNDATIONS OF INFORMATION TECHNOLOGY IN THE ERA OF NETWORK AND MOBILE COMPUTING

IFIP - The International Federation for Information Processing

IFIP was founded in 1960 under the auspices of UNESCO, following the First World Computer Congress held in Paris the previous year. An umbrella organization for societies working in information processing, IFIP's aim is two-fold: to support information processing within its member countries and to encourage technology transfer to developing nations. As its mission statement clearly states,

IFIP's mission is to be the leading, truly international, apolitical organization which encourages and assists in the development, exploitation and application of information technology for the benefit of all people.

IFIP is a non-profitmaking organization, run almost solely by 2500 volunteers. It operates through a number of technical committees, which organize events and publications. IFIP's events range from an international congress to local seminars, but the most important are:

- The IFIP World Computer Congress, held every second year;
- open conferences;
- working conferences.

The flagship event is the IFIP World Computer Congress, at which both invited and contributed papers are presented. Contributed papers are rigorously refereed and the rejection rate is high.

As with the Congress, participation in the open conferences is open to all and papers may be invited or submitted. Again, submitted papers are stringently refereed.

The working conferences are structured differently. They are usually run by a working group and attendance is small and by invitation only. Their purpose is to create an atmosphere conducive to innovation and development. Refereeing is less rigorous and papers are subjected to extensive group discussion.

Publications arising from IFIP events vary. The papers presented at the IFIP World Computer Congress and at open conferences are published as conference proceedings, while the results of the working conferences are often published as collections of selected and edited papers.

Any national society whose primary activity is in information may apply to become a full member of IFIP, although full membership is restricted to one society per country. Full members are entitled to vote at the annual General Assembly, National societies preferring a less committed involvement may apply for associate or corresponding membership. Associate members enjoy the same benefits as full members, but without voting rights. Corresponding members are not represented in IFIP bodies. Affiliated membership is open to non-national societies, and individual and honorary membership schemes are also offered.

FOUNDATIONS OF INFORMATION TECHNOLOGY IN THE ERA OF NETWORK AND MOBILE COMPUTING

IFIP 17th World Computer Congress —
TC1 Stream / 2nd IFIP International Conference on Theoretical Computer Science (TCS 2002)
August 25-30, 2002, Montréal, Québec, Canada

Edited by

Ricardo Baeza-Yates
Universidad de Chile
Chile

Ugo Montanari
Università di Pisa
Italy

Nicola Santoro
Carleton University
Canada

KLUWER ACADEMIC PUBLISHERS
BOSTON / DORDRECHT / LONDON

Distributors for North, Central and South America:
Kluwer Academic Publishers
101 Philip Drive
Assinippi Park
Norwell, Massachusetts 02061 USA
Telephone (781) 871-6600
Fax (781) 681-9045
E-Mail <kluwer@wkap.com>

Distributors for all other countries:
Kluwer Academic Publishers Group
Post Office Box 322
3300 AH Dordrecht, THE NETHERLANDS
Telephone 31 786 576 000
Fax 31 786 576 474
E-Mail <services@wkap.nl>

Electronic Services <http://www.wkap.nl>

Library of Congress Cataloging-in-Publication Data

IFIP World Computer Congress (17th : 2002 : Montréal, Québec)
Foundations of information technology in the era of network and mobile computing: IFIP 17th World Computer Congress—TC1 stream/2nd IFIP International Conference on Theoretical Computer Science (TCS 2002), August 25-30, 2002, Montréal, Québec, Canada / edited by Ricardo Baeza-Yates, Ugo Montanari, Nicola Santoro.
p. cm. — (International Federation for Information Processing ; 96)
Includes bibliographical references and index.
ISBN 1-4020-7181-7
1. Information technology—Congresses. 2. Computer networks—Congresses. 3. Mobile computing—Congresses. I. Baeza-Yates, R. (Ricardo). II. Montanari, U. (Ugo). III. Santoro, N. (Nicola), 1951- . IV. TCS 2002 (2002 : Montréal, Québec). V. Title. VI. International Federation for Information Processing (Series) ; 96.
T58.5 I357 2002

2002026736

Printed on acid-free paper.

Printed in Great Britain by IBT Global, London

Contents

Preface

This volume contains the papers presented at the *2nd IFIP International Conference on Theoretical Computer Science* (TCS 2002), held in Montreal, Quebec, Canada, on August 26-29, 2002.

The International Conference on Theoretical Computer Science is sponsored by the *IFIP Technical Committee on Foundations of Computer Science* (IFIP TC1), in co-operation with the *European Association for Theoretical Computer Science* (EATCS) and the *ACM Special Interest Group on Automata and Computing* (ACM SIGACT). For the first time, the Conference has been held as part of the *IFIP World Computer Congress.*

The special focus of the conference has been *Foundations of Information Technology in the Era of Network and Mobile Computing.* In recent years, IT application scenarios have evolved in very innovative ways. Highly distributed networks have now become a common platform for large-scale distributed programming, high bandwidth communications are inexpensive and widespread, and most of our work tools are equipped with processors enabling us to perform a multitude of tasks. In addition, mobile computing (referring specifically to wireless devices and, more broadly, to dynamically configured systems) has made it possible to exploit interaction in novel ways. To harness the flexibility and power of these rapidly evolving, interactive systems, there is need of radically new foundational ideas and principles; there is need to develop the theoretical foundations required to design these systems and to cope with the many complex issues involved in their construction; there is need to develop effective principles for building and analyzing such systems.

Original and significant contributions on the special focus and on foundational questions have been sought from all areas of theoretical computer science.

Reflecting the diverse and wide spectrum of topics and interests within the theoretical computer science community, the areas have been divided in two distinct, but interrelated tracks:

Track 1, focusing on Algorithms, Complexity and Models of Computation,
Track 2, focusing on Logic, Semantics, Specification and Verification.

Following the Call for Papers, there have been 121 submissions, out of which 45 papers have been selected for presentation at the Conference. The international Program Committee consisted of:

Track (1)	Track (2)
Eric Allender	*Gabriel Baum*
Jos Balcazar	*Luca Cardelli*
Andrej Brodnik	*Frank DeBoer*
Volker Diekert	*Ursula Goltz*
David Fernandez-Baca	*Roberto Gorrieri*
Kazuo Iwama	*Jieh Hsiang*
Jan van Leeuwen	*Takayasu Ito*
Xuemin Lin	*Alexander Letichevsky*
Alberto Marchetti-Spaccamela	*Jean-Jacques Levy*
David Peleg	*Huimin Lin*
Prabhakar Raghavan	*Kim Marriott*
Venkatesh Raman	*Narciso Marti-Oliet*
Siang Song	*John Mitchell*
Paul Spirakis	*Luis Monteiro*
Luca Trevisan	*Peter Mosses*
Brigitte Valle	*Prakash Panangaden*
Alfredo Viola	*Benjamin Pierce*
Manfred Warmuth	*Amir Pnueli*
Sue Whitesides	*Leila Ribeiro*
Peter Widmayer	*Gheorghe Stefanescu*
Jiri Wiedermann	*Andrzej Tarlecki*
	P.S. Thiagarajan

To all of them our sincere thanks for contributing to the high scientific quality of this volume.

The final program, and the content of this volume, comprises the revised versions of those accepted papers, as well as *invited* talks by four leading scientists who gracefully accepted our invitation:

Andy Gordon, Jozef Gruska, Carl Gunter, Jon Kleinberg

There are many other individuals who have contributed to the success of TCS 2002. In particular, we would like to thank the TCS 2002 Organizing Committee:

Michel Barbeau, Amiya Nayak, Giuseppe Prencipe,

and the Steering Committee of IFIP TC1:

Giorgio Ausiello, Wilfried Brauer, Takayasu Ito,
Michael O. Rabin, and *Joseph Traub.*

For their organizational help and support, we would like to thank: the WCC 2002 International Organizing Committee, chaired by *George Boyton*; the International Program Committee of IFIP WCC 2002, chaired by *Jan Wibe*; Kluwer's IFIP Editor, *Yana Lambert.* We also like to thank *Luz Adriana Jaramillo* and *Fabio Gadducci* for their help.

A special thank goes to *Giorgio Ausiello* for his helpful advice and encouragment.

Last but not least we would like to thank the authors who responded to our Call for Papers and have contributed, through this volume, to developing the new foundations of Information Technology.

Ricardo Baeza-Yates, Ugo Montanari, Nicola Santoro
Chairs

External Referees

Andrea Asperti
James Bailey
Markus Blaeser
Matthias Blume
Chiara Bodei
Hans-J. Boehm
Marcello Bonsangue
Michele Boreale
Andrzej Borzyszkowski
Paolo Bottoni
Victor Braberman
Mario Bravetti
Gerhard Buntrock
Marzia Buscemi
Nadia Busi
Ed Coffman
Patrick Dehornoy
Mariangiola Dezani
Vladimir Estivill-Castro
Thomas Firley
Cedric Fournet
Marcelo Frias
Yuxi Fu
Silvia Ghilezan
Rob van Glabbeek
Andy Gordon
Xudong Guan
Joshua D. Guttman
Annegret Habel
Ulrich Hertrampf
Fritz Hohl
Michael Houle
Yu-Ping Hsieh
Atsushi Igarashi
Mark Jones
Idit Keidar
Lefteris Kirousis
Naoki Kobayashi
Beata Konikowska
Yassine Lakhnech
Klaus-Jorn Lange
Kim G. Larsen
Weifa Liang
Zsuzsanna Lipt'ak
Xinxin Liu
Meena Mahajan
Pablo E. Martinez Lopez
Ralph Matthes
Massimo Merro
Eugenio Moggi
Till Mossakowski
Uwe Nestmann
Peter Niebert
Flemming Nielson
Alfredo Olivero
Victor Pan
Ioannis Papoutsakis
Wiesław Pawłowski
Paolo Penna
J. M. Piquer

Marco Pistore
Michel Pocchiola
Claudia Pons
Rajeev Raman
Arend Rensink
Gianluca Rossi
Luca Roversi
Jan Rutten
Ken Satoh
Aleksy Schubert
Peter Sewell
Riccardo Silvestri
Mark-Oliver Stehr
Paul Strooper
Werner Struckmann
C.R. Subramanian
Martin Sulzmann
Eijiro Sumii
Mario Szegedy
Kenjiro Taura
Phil Trinder
Athanassios Tsakalidis
Emilio Tuosto
Irek Ulidowski
Jrme Vouillon
Klaus Wagner
Nobuo Yamashita
Naoki Yonezaki
Nobuko Yoshida
Albert Zuendorf

Track 1: Algorithms, Complexity and Models of Computation

POWER OF QUANTUM ENTANGLEMENT

Jozef Gruska
*Faculty of Informatics, Masaryk University, Botanická 68a, Brno, Czech Republik**

Hiroshi Imai and Keiji Matsumoto
ERATO Quantum Computation and Information Project, Hongo 5-28-3, Bunkyo-ku, Tokyo 113-0033, Japan

Progress in theory is sometimes done by pessimists.
Progress in technology is always done by optimists.

—

Abstract Quantum entanglement is arguably the most inherently quantum feature of quantum information, computation and communication – a feature that is at heart of quantum physics. Quantum entanglement is also increasingly often considered as being behind new and surprising power quantum computations and communications exhibit – comparing to the classical computation and communication.

Quantum entanglement used to be seen, practically until 1993, especially due to its accompanying non-locality impacts, as being behind various mysteriously looking and weird phenomena of quantum world, and of interest mainly to the philosophers of science. Since then our perception of entanglement has changed much. Currently, quantum entanglement is increasingly believed to be a resource that can be exploited to implement various quantum information processing tasks, at spatially separated locations, and to be behind new gold mine for science and technology to which the outcomes of the research in quantum information science and quantum information technology seem to pave the road.

*Support of GAČR grant 201/01/0413 is highly appreciated.

Quantum entanglement implications are also a deep reason to attempt to develop new, quantum information processing based, foundations of quantum mechanics. To help to do that might be one of big challenges for Informatics.

1. Introduction

In the last ten years, enormous progress has been made in quantum information processing and communication. In 1992, only few very theoretical results were available, with almost no indication that a rapid development and important outcomes for science and technology could follow soon. Landauer's thesis (**Information is physical**) was already known. However, on a more specific level, we had only Deutsch's (not yet fully satisfactory) model of (universal) quantum Turing machine, a very simple Deutsch's quantum algorithm, a cryptography experiment to the distance of 32cm, and with hardly anybody who could imagine to have soon implementations of quantum gates. After 10 years, we have a rapidly developing area, with about 300 papers per month. We have flourishing and increasingly broad theory, rich on deep and lasting results. We know a variety of techniques and technologies allowing some elementary experiments to be performed (and ambitious goals to be put on the agenda). We have quantum cryptography, aiming at unconditionally secure communications (with perfect detection of eavesdropping), approaching already a developing phase. (Recently, transmissions of photons through optical fibers to distance 67 km have been reported by N. Gisin and through the open air in night time to distance 23.4 km by H. Weinfurter.) Moreover, projects for ground-satellite or plane-satellite cryptography communications are under way. NMR (nuclear magnetic resonance) and quantum optics have been technologies used so far for most of the experiments. However, future seems to be in solid-state technologies, where recently quite a bit of progress has been reported in developing ways how to store, process and transform quantum information in a robust and reliable way. Shor's and Grover's algorithms made revolution in algorithmic design, and threatened to break main current cryptosystems, provided quantum computers would be available. Moreover, in the last few years, the whole field of quantum information processing and communication (QIPC) science and technology has been changing, from the theory driven to the experiments driven field. This can be seen as a positive feature, especially from the point of view of developing new quantum information technologies.

There have been several other important impacts of the developments in QIPC. Informaticians have learned that their complexity theory views of computational problems may need a change whenever a new physical theory is developed. Physicists have learned that tools of informatics, especially computational complexity concepts, methods, paradigms and results, are of the large importance for formulation and evaluation of the research goals, methods, paradigms and results, as well as for solution of physical problems. Shortly, (some) physicists have learned that (theoretical) informatics is a much needed and useful body of concepts, models, knowledge and methods that can guide their thinking and help to formulate and to solve their fundamental problems.[1] Finally, a fundamentally new view of physics started to be developed. Namely, that **physics is informational**. That is, that investigations of information processing laws and limitations of Nature may be an important new (or even the key) way to understand Nature (and, perhaps, also to develop a theory of Nature that could be used to understand both physical and biological worlds).

In general, QIPC is a rapidly developing interdisciplinary research area that encompasses many disciplines, such as physics, computer science, engineering, mathematics, chemistry and so on. It has clearly potential to revolutionize information technology, to develop Quantum Information Technology, and by that to have very broad impact on science, technology and society in large.

Quantum entanglement is often characterized and quantified as a feature of composed quantum systems that causes non-local effects, leads to pseudo-telepathy, and cannot be created through local quantum operations and classical communications among the parties.

Since the famous EPR article [EPR35], in which Einstein, Podolsky and Rosen pointed out non-locality implications of quantum entanglement, until 1993, **entanglement was considered as a strange quantum feature**, that is behind various quantum mysteries and of interest/importance mainly only for quantum theory people and especially for philosophers of quantum physics (science). (These aspects of quantum entanglement have been well expressed by A. Peres (see [Bru01]): *Quantum entanglement allows quantum magicians to produce phenomena that cannot be imitated by classical magicians.*)

Since the discovery of the surprising power of quantum computation and of quantum teleportation, in 1993-96, **quantum entanglement is increasingly being considered and explored as a new and important physical resource** of (quantum) communication and computation.

Currently, quantum entanglement is considered to be of large importance for the theory and practice of quantum information processing because it allows:

- to perform tasks that are not possible otherwise;
- to speed-up much some computations and to economize (even exponentially) some communications;
- to increase capacity of (quantum) communication channels;
- to implement perfectly secure information transmissions;
- to develop more general and more powerful theories of computations and communications than the framework of classical physics allows;
- to develop a new, better, information based, understanding of the key quantum phenomena and of Nature.

For more details see paper [Gru00] or the book [Gru02], and its web-updatings. For more about puzzling aspects of quantum entanglement see [GI01]

2. Basics of quantum entanglement

The concept of quantum entanglement of pure states goes back to Erwin Schrödinger [Sch35], and for mixed states to Werner [Wer89].

2.1. Basic concepts

A pure state $|\phi\rangle$ of a bipartite quantum system $A \otimes B$ is **entangled** if $|\phi\rangle$ is not a tensor product of a pure state from A and a pure state from B. An example of such a state is so called *EPR state* of the Hilbert space $H_2 \otimes H_2$

$$|\mathrm{EPR}\rangle = \frac{1}{\sqrt{2}}(|00\rangle + |11\rangle).$$

The case of entanglement of mixed states of bipartite systems $A \otimes B$ seems to be slightly less natural. A mixed state ρ is **entangled** if ρ cannot be written as a convex combination of the tensor products of mixed states

$$\rho = \sum_{i=1}^{k} p_i \rho_i^A \otimes \rho_i^B,$$

where ρ_i^A (ρ_i^B) are mixed states of the quantum system A (B) and $\sum_{i=1}^{k} p_i = 1$.

Operationally, the same idea is behind both definitions: a state is entangled if it cannot be created by two parties if they perform quantum operations only on their subsystems and they communicate only classically.

Both definitions generalize naturally to the case of multipartite systems. However, it has turned out that it is useful to consider many different types of multipartite entanglement. For example, an entangled state $|\phi\rangle$ of an m-partite quantum system $S_1 \otimes S_2 \otimes \ldots \otimes S_m$ is called $(\mathcal{M}_1 : \mathcal{M}_2 : \ldots : \mathcal{M}_k)$-separable, where sets $\mathcal{M}_1, \mathcal{M}_2, \ldots, \mathcal{M}_k$ form a partition of the set $\{1, 2, \ldots, m\}$, if the state $|\phi\rangle$ is separable with respect to the tensor product of the quantum system $S_{\mathcal{M}_i}$ that are themselves tensor products of quantum systems $\{S_j \mid j \in \mathcal{M}_i\}$.

Design of bipartite entangled states on demand is still experimentally a (very) difficult task, and so is any design of 3- and more-partite entangled states.

3. Quantum entanglement and quantum non-locality

As already indicated, quantum entanglement is being increasingly considered, especially due to its non-locality implications, to be the key resource for computation and communication.[2]

3.1. From Einstein's doubts through Bell's inequalities to quantum teleportation

If two particles are in the EPR state $\frac{1}{\sqrt{2}}(|00\rangle + |11\rangle)$, then, as theory says, and ever more perfect experiments strongly confirm, they can stay in that state even if they are taken much apart. Moreover, a measurement of one of the particles, with respect to the standard basis, causes that the overall state of the particles collapses, immediately and randomly, into one of the states $|00\rangle$ or $|11\rangle$. Consequently, the result of the measurement of the first (second) particle uniquely determines the result of the measurement of the second (first) particle.

In a slightly different form was this fact first observed in [EPR35]. Einstein and his colleagues did not want to accept the existence of non-local phenomena and argued that the description of the physical reality, which quantum mechanics provides, is incomplete. They suggested that such strange phenomena disappear once some hidden variables are used to complete the description of the physical reality that is missing in the Hilbert space view of states.[3]

Einstein's discovery, as well as their suggestions, caused first an uproar in physics community, but, after a while, they stopped to bother too much "working physicists" because the phenomenon they pointed out seemed to be on the philosophical level, without any visible effect on the more pragmatically oriented research in quantum mechanics, dominating at that time.

An important change in the situation came when Bell, in 1964, suggested an experimental setup that could be used to verify whether a theory of hidden variables without non-locality effects can exist. Bell derived, for results of the measurements of a Gedanken experiment, certain inequalities, which should be satisfied by physical reality provided a theory of hidden variables without non-locality is valid, and that should be violated if quantum mechanics, with non-local phenomena, is valid. In this way, the existence of the non-local phenomena stopped to be the issue only for the philosophers of quantum mechanics. Experimentalists were to resolve the dilemma.

Another important step was made by in 1993 [BBC+93]. It was shown, in a way that could also be experimentally verified, that quantum entanglement can be an important resource, to teleport unknown quantum states.

3.2. From Aspect to Zeilinger

It took quite a while until two important experimental confirmation of nonlocality were made in quite a convincing, even not yet absolutely perfect, way.

From several attempts to show that Bell's inequalities[4] can be violated by bipartite entangled states, Aspect was the first to come, in 1982, with convincing results. Since then, many other experiments have been performed and a consensus has emerged that non-locality has been demonstrated even between particles quite far apart. (Experiments with more-partite states could provide even stronger tests of non-locality.) A consensus has also emerged that any local hidden variable theory that could exploit some loopholes in a way that has not been demonstrated in experiments yet, would have to be "so conspiratorial as to be almost incredible" [Ken02].

Until 1989 it was widely believed that a quantum state is entangled if and only if it violates some Bell's inequality. However, this is not true. A state violating a Bell's inequality has to be entangled, but not vice versa, as shown first in [Wer89], for mixed states, and in [VW01], for multipartite pure states.

The next step was to demonstrate experimentally that quantum entanglement is indeed a resource that can be used to do what is impossible without it. Zeilinger's group was first to publish, in 1997, outcomes of the experiments demonstrating that quantum teleportation is possible [BPM+97].

3.3. Non-locality – optimists versus pessimists

Acceptance of non-locality is such a strong departure from the former basic beliefs about Nature that it is natural, and perhaps also fortunate, that there are still pessimists who do not believe that quantum non-locality has been demonstrated beyond reasonable doubts. They try to point out, on one side, loopholes in the experiments, and/or in the conclusions made by experimentalists, or even in the hidden variable model itself. In addition, some try to offer new physical theories explaining without non-locality phenomena for which current quantum physics needs non-locality.

Experiments confirming non-locality have clearly some loopholes. Well known is the detection loophole (not sufficiently perfect detectors were used in experiments) and also the locality loophole (experiments were performed with not sufficiently far away particles). Moreover, no one seems to see a way how to close (soon) all loopholes. Detection loophole was recently closed and, almost, also locality loophole (that cannot be, in principle fully closed).

However, new loopholes are being discovered and explored again and again. For example, the *collapse locality loophole*, [Ken02], due to the fact that there may be some time between a particle enters a measuring device and the state collapse, due to the measurement, takes place. In addition, *memory loophole*, in the hidden variable model itself. Moreover, experiments are always based on some *common sense assumptions* and experimentalists from them conclude validity of some *non-common sense conclusions*, as that of non-locality. Some complain to such reasoning [Mar02], saying that correct conclusion should be to cast doubts on the assumptions.

In addition to non-locality, there are several other reasons why some believe that current quantum theory might be wrong and there might be a need to look for alternative theories: problems to understand fully quantum measurement, difficulties in matching general relativity with quantum theory and so on. For example, in so called *causal quantum theory*, measurement events that are space-like correlated do not have non-local correlations. Some versions of such theories have not been refuted yet by (Bell) experiments (due to a possibility to exploit collapse locality loophole) [Ken02].

Moreover, outcomes of QIPC seem to suggest a very new way, quantum information processing based, to build foundations of quantum mechanics, again without non-locality puzzles [Fuc02]. In any case, it seems that one of the main challenges of theoretical informatics is to help to develop such new foundations.

3.4. From particles to macroscopic objects

It used to be said that one of the puzzling facts about Nature is that two key features of the microscopic quantum world, superposition and entanglement, have not been (much) seen (yet) in the macroscopic world.

One of the main task of the current experimental research in QIPC is to demonstrate that both superposition and entanglement can be witnessed not only on particles. Zeilinger's group (see [BHU+02]) demonstrated superposition for special molecules, and several other experiments of this type have already been performed. It is an important open question to determine for how large molecules such superpositions can be demonstrated. Is there an upper bound? (It does not seem to be.)

Most of the experiments demonstrating entanglement did that for states of light. For storage and processing of quantum information, entanglement of material particles seems to be of larger importance. Polzik's group (see [JKP01]), has demonstrated (robust with respect to quantum standards) entanglement of two objects consisting of about 10^{12} atoms.

4. Quantum teleportation

Discovery of quantum teleportation has been so far one of the major demonstrations of the power of entanglement as an information transmission resource.

Indeed, let us assume that two parties, called in quantum computing usually Alice and Bob, share two particles in the EPR state, and Alice gets another particle, in an unknown state $|\phi\rangle = \alpha|0\rangle + \beta|1\rangle$, to teleport. The total state of the system is then, after a proper rearrangement of terms,

$$\begin{aligned}|\phi\rangle|EPR\rangle &= \frac{1}{2}|\Phi^+\rangle(\alpha|0\rangle + \beta|1\rangle) + \frac{1}{2}|\Psi^+\rangle(\beta|1\rangle + \alpha|0\rangle) \\ &\quad + \frac{1}{2}|\Phi^-\rangle(\alpha|0\rangle - \beta|1\rangle) + \frac{1}{2}|\Psi^-\rangle(-\beta|0\rangle + \alpha|1\rangle),\end{aligned}$$

where

$$|\Phi^\pm\rangle = \frac{1}{\sqrt{2}}(|00\rangle \pm |11\rangle), \quad |\Psi^\pm\rangle = \frac{1}{\sqrt{2}}(|01\rangle \pm |10\rangle),$$

are so called Bell states (that form the (Bell) basis in H_4). If Alice performs a measurement on her two particles with respect to the Bell basis, then Bob's particle state collapses, with equal probability, into one of the states

$$\alpha|0\rangle + \beta|1\rangle, \quad \alpha|1\rangle + \beta|0\rangle, \quad \alpha|0\rangle - \beta|1\rangle, \quad -\alpha|1\rangle + \beta|0\rangle$$

and Alice gets information, 2 bits, which of the four potential outcomes of the measurement has been obtained. If Alice then sends these two bits, using a public classical channel, to Bob, then he can achieve that his particle gets into the state that was teleported, by just performing on his particle, depending on two bits received, one of the following Pauli operators

$$I = \begin{pmatrix} 1 & 0 \\ 0 & 1 \end{pmatrix}, \sigma_x = \begin{pmatrix} 0 & 1 \\ 1 & 0 \end{pmatrix}, \sigma_z = \begin{pmatrix} 1 & 0 \\ 0 & -1 \end{pmatrix}, i\sigma_y = \begin{pmatrix} 0 & 1 \\ -1 & 0 \end{pmatrix}.$$

An inverse process, in a sense, to quantum teleportation, in which one qubit is used to send two bits, is called **superdense quantum coding**.

Quantum teleportation is still one of the most studied subjects of QIPC. On the theoretical level, various its generalizations are being explored and also their relations to other problems [Wer00]. On the experimental level, a big challenge is to increase reliability and distance at teleportation.

5. Power of entanglement

There are several ways power of entanglement has been demonstrated.

5.1. Speeding-up quantum computations

It is intuitively clear that entanglement plays an important role in many quantum algorithms that exhibit much better performance than known classical algorithms for the same tasks. However, it is less clear how to demonstrate such intuition formally.

In case of pure states, it is already known that without processing with increasingly growing multipartite entanglement, we cannot have an exponential speed-up, compared with the classical case. This follows from the results in [JL02], presented bellow.

Definition For a given integer p, a pure state of n qubits is p-blocked, if none of its $p+1$ qubits are entangled (after tracing out the remaining qubits).

Theorem Consider any quantum computational process on *pure* states with an increasing input size. Suppose there is a p such that at every stage of the computation the states produced by the process are p-blocked. Then such a computation process can be classically simulated in polynomial time.

However, it is not clear whether a necessity of increasingly growing entanglement can be demonstrated also for computations with mixed states.

Let us now mention some of the main success in the design of efficient quantum algorithms

- Shor designed, in 1994, polynomial time quantum algorithm to factorize integers [Sho97]— which would allow to break the RSA cryptosystem – no classical polynomial time algorithm for factorization is known. (Shor and other showed that quantum computers could break also other well known public-key cryptosystems.)

- Grover [Gro97] showed that quantum search in an unordered database of n elements can be done with $O(\sqrt{n})$ queries — that would allow to break the DSA cryptosystem – while classically n queries may be needed.

- Shor's result has been generalized to show that polynomial time algorithms exist for all *Hidden subgroups problem* for Abelian groups.[5]

- For several other computational and simulation problems, for which no classical polynomial time algorithm is known, polynomial time quantum algorithms exist. For example, for solving so called Pell equation, $x^2 - dy^2 = 1$, with d as the parameter (Hallgren in 2002).

There are several other ways entanglement is of importance for quantum information processing.

- It was shown [GC99] that entanglement is a computational primitive. Indeed, it is possible to realize any quantum computation by starting with some GHZ states $\frac{1}{\sqrt{2}}(|000\rangle + |111\rangle)$, and then performing one qubit operations and Bell measurements.

  Moreover, it was shown [RB00] that universal quantum computation is possible by initializing a proper multipartite entangled state (a sort of a computational substrate) and then performing only one qubit measurements.

- Entanglement can also serve as a catalyst. Indeed, it was shown [JP99] that there are pairs of pure states $(|\phi_1\rangle, |\phi_2\rangle)$ such that, using local quantum operations and classical communications (LOCC), one cannot transform $|\phi_1\rangle$ into $|\phi_2\rangle$, but with the assistance of an appropriate entangled state $|\psi\rangle$, as a catalyst, one can transfer $|\phi_1\rangle$ into $|\phi_2\rangle$, using LOCC, in such a way that the state $|\psi\rangle$ is not changed in the process.

  Moreover, as shown in [BR01], entanglement can serve as a **super-catalyst** that not only allows to perform operations otherwise impossible, but during such a process the catalyst can even increase its entanglement.

5.2. Making communication more efficient

There are five basic ways entanglement can provide a new quality for communication. One of them, quantum teleportation, has already been discussed. Three other are analyzed in this section. The last one, concerning security of communication, is discussed in Section 10.

Decreasing communication complexity: While in quantum computation we merely believe that quantum mechanics allows exponential speed-up for some computational tasks, in quantum communication we can prove that quantum tools can provide exponential savings. And not only that. It can be shown that

- Entangled parties can make a better use of the classical communication channels than non-entangled parties.
- Entangled parties can benefit from their entanglement even if they are not allowed any form of direct (classical or quantum) communication.

In the so called *entanglement-enhanced quantum communication model*, only classical bits are communicated, but communication is facilitated by an a priori distribution of entangled qubits among the communicating parties.

An exponential gap between the bounded-error classical and quantum entan-
glement-enhanced communication complexity has been shown in [Raz99], for a *promise problem*. This seems to be so far the strongest separation result for communication complexity

Proving lower bounds is notoriously hard. One of the main recent results in this direction has been a complete characterization, up to a logarithmic factor, of the bound-error quantum communication complexity for every symmetric predicate $f(x,y), x, y \in \{0,1\}^n$, depending only on $|x \cap y|$; that is if $f(x,y) = D(|x \cap y|)$ for some $D \in \{1,2,\ldots,n\} \to \{0,1\}$. Lower bound was shown in [Raz02], even for the model with preshared entanglement, in the form $\sqrt{nl_0(D,n)} + l_1(D,n)$ for certain functions l_0 and l_1. In two special cases, $D(s) \equiv (s = 0)$ and $D(s) \equiv s \pmod 2$, one gets lower bounds for much studied communication problems of disjointness and of inner product.

Concerning "communication without communication", we say, see [BCT99], that a *spooky communication* or *pseudo-telepathy* takes place, if a task cannot be done using classical communication only, but it can be done once parties share entanglement. **Spooky communication complexity** of the task is then defined as the amount of entanglement required to perform the task.

It has been shown, for a certain relation, that the number of bits to compute the relation classically is exponentially larger than the number of entangled pairs shared by the parties at the spooky communication (and therefore that spooky communication can be exponentially more efficient).

Increasing capacity of quantum channels: If two communicating parties share entangled states, then classical communication capacity of their noisy quantum channel can be increased, with respect to the best achievable capacity without preshared entanglement, by arbitrarily large constant factor [BSST99]. Surprisingly, from several capacities that have been defined for quantum channels with quantum inputs and outputs, entanglement-assisted capacity of noisy quantum channels is the only one that we already know how to compute.

Fighting decoherence: Until 1995, there was strong pessimism whether meaningful quantum information processing would eventually be possible. The main reason behind was **quantum decoherence** – the fact that due to the unavoidable entanglement of any computational quantum system with its environment, fragile quantum superpositions, that are behind powerful quantum parallelism, can get exponentially fast destroyed. In addition, it had been believed that efficient quantum error-correcting codes cannot exist because: (a) number of quantum errors seemed to be infinite; (b) quantum copying, needed to create redundancy, so vital for classical error-correcting codes, is impossible; (c) measurement of an erroneous state could, in general, irreversibly destroy the state to be corrected. However, Shor [Sho96] showed, that not only quantum error-correcting codes, but also quantum fault-tolerant computations, are possible. (The main new and ingenious idea was to use multipartite entanglement to fight, in polynomial time, exponentially fast growing decoherence (caused, actually, by entanglement itself).)

6. Basic approaches to study entanglement

Two main approaches to develop qualitative and quantitative theory of entanglement have been pursued so far. They are related to the fact that in order to understand entanglement as a resource, we need to understand laws and methods how entanglement can be transformed, from one form to another, and also how to quantify entanglement.

6.1. Investigation of (reversible) transformations of states

Entanglement may appear in a form not suitable for a specific application. Therefore, it is of importance to understand how and when one can transform entanglement from one form into another form. The basic problem is to determine, when we can transfer one given state (or several copies of it), into some other given state (or several copies of it), using certain quantum and classical tools. Of special importance are the cases when the transformations allowed are from the following three classes.

LOCC This stands for the case that parties can perform only local quantum operations and classical communication.

SLOCC Operations are as in the LOCC case, but this time it is enough if the result is obtained with some non-zero probability.

EALOCC As above, but communicating parties are allowed to share some entanglement – these are *entanglement-assisted LOCC.*

Of a special importance are the following problems.

Entanglement concentration: How to obtain, from n copies of a non-maximally entangled pure state $|\phi\rangle$, using LOCC, as many as possible (m) copies of a maximally entangled state, and how large the ratio $\frac{m}{n}$ can be. (A variety of methods have been developed to do that – see [Gru02] for an overview.) Surprisingly, there is also a way to do entanglement concentration using only local quantum operations [HM01].

Entanglement purification – distillation: How to obtain, from n copies of a given mixed state ρ, as many as possible copies (say m) of a maximally entangled pure state, and how large the ratio $\frac{m}{n}$ can be – see [Gru02] for an overview.

Discovery, due to Horodeckis family, [HHH98b], that not all entangled mixed states are distillable, has been one of the big surprises at the development of QIPC science. Such states are called **bound entangled** and will be discussed in more details later. Moreover, it has been shown [VC01], that there are mixed states from which one can distill some entanglement, but less than it is needed to create those states.

Transformations of one state into another:
Of large importance is to determine, when it is possible to transform one given state into another given one, exactly or asymptotically (having enough copies), deterministically (with probability one) or stochastically (with non-zero probability), and without or with a catalyst assistance, using operations of a certain type $\mathcal{O}$ (for example, LOCC). The related problem is to find out when are two states equivalent (mutually/reversibly transformable) using certain operations and modes of transformations. As expected, asymptotic and stochastic transformations yield usually simpler classifications. This problem is already quite well understood, see [Gru02] for overview, but still needs a lot of attention, especially for multipartite states, because it is so basic for classifications of entangled states.

MREGS (Minimal Reversible Entanglement Generating Sets):
They are, for a given m, sets of states of minimal cardinality sufficient to generate all m-partite pure states by asymptotically reversible LOCC transformations. For bipartite states, the set containing one EPR state is a MREGS set. It is an open problem whether also for $m > 2$ there are finite MREGS (or whether there are infinitely many inequivalent types of entanglement), though some lower bounds for size of MREGS, depending on m, are known.

6.2. Quantification of entanglement

One of the key difficulties at the study of entanglement is that it is not clear how to quantify entanglement – especially of mixed and multipartite states.

In case of pure states of a bipartite system $A \otimes B$, a reasonable measure of entanglement of a state $|\phi\rangle$ is von Neumann entropy of the reduced density matrix, that is $E(|\phi\rangle) = -Tr\rho_A \lg \rho_A = -Tr\rho_B \lg \rho_B$, where $\rho = |\phi\rangle\langle\phi|$.

In the case of mixed bipartite states, the most natural – physically well motivated – measures of entanglement are entanglement of formation and distillation.

Entanglement of formation, $E_f(\rho)$, is defined as

$$E_f(\rho) = \inf \sum_j p_j E_f(\phi_j),$$

where infimum is taken over all pure-state decompositions of $\rho = \sum_j p_j|\phi_j\rangle\langle\phi_j|$. *Entanglement of distillation*, $E_d(\rho)$, is the average amount of maximally entangled states that can be distilled from several copies of ρ.

Another class of important measures of bipartite entanglement are that of *entanglement of relative entropies*, $E_r^S(\rho)$, that equal to the minimal "distance" of ρ to a set S of states (that are separable, or not-distillable, or have a similar property). These measures have interesting properties and serve as upper (lower) bounds for entanglement of distillation (asymptotic version of entanglement of formation).

For all basic measures of entanglement E, of importance is to consider their asymptotic (or regularized) versions $E^\infty(\rho) = \lim_{n\to\infty} \frac{E(\rho^{\otimes n})}{n}$. (Observe that entanglement of distillation is already the asymptotic measure.)

For any entanglement measure $\bar{E}$, it holds $\bar{E} = \bar{E}^\infty$ if $\bar{E}$ additive. It is a major open problem whether entanglement of formation is additive. A method has been developed, [MSW02], how to transfer additivity results for the Holevo capacity of quantum channels (another major open problem) to the additivity results for entanglement of formation of certain (sets of) states.

The problem with all above measures of entanglement is that they are difficult to compute because they require to solve an extremization problem. Only for 2×2 and 3×2 dimensional systems, we have an easily computable measure of entanglement, so called **concurrence** $C(\rho)$ [DW01], that can also be used to determine entanglement of formation because it holds $E_f(\rho) = H\left(\frac{1+\sqrt{1-(C(\rho))^2}}{2}\right)$, where $H(x)$ is the binary entropy function. The concurrence is defined as $C(\rho) = \max\{0, \lambda_1-\lambda_2-\lambda_3-\lambda_4\}$, where λ_i are, in the descending order, eigenvalues of the matrix $\rho\hat{\rho}$, where σ_y is the Pauli matrix and $\hat{\rho} = (\sigma_y\otimes\sigma_y)\rho^*(\sigma_y\otimes\sigma_y)$. This strangely looking measure has turned out to be surprisingly useful.

For bipartite states, both basic measures of entanglement can be seen as being defined in terms of such a "standard currency" as Bell states. They can be seen as the average amount of Bell states, needed to create a given state or to get distilled from it. This approach does not seem to be possible in the case of multipartite systems.

There have been various attempts to develop measures of entanglement for multipartite states. Most of them are attempts to generalize entanglement of relative entropy or concurrence – see [Gru02] for an overview.

Several of such generalizations are special cases of the hierarchy of measures of entanglement based on concurrence that was defined in [FMI02]. The k-th level of concurrence, for $k \leq d$, is defined for d-dimensional systems, by

$$C_k(|\phi\rangle) = \sum_{0\leq i_0<i_1<\ldots<i_k\leq d-1} \lambda_{i_0}^{\downarrow}\lambda_{i_1}^{\downarrow}\ldots\lambda_{i_k}^{\downarrow},$$

where $\lambda_i^{\downarrow}$ is the ith largest eigenvalue of the reduced density matrix.

In general, it is already understood that a multicomponent entanglement measure is needed to quantify multipartite states, with each component representing a different type of entanglement.

7. Bound entanglement

In case of multipartite systems, bound entangled states (BE-states, for short), are such states that some initial entanglement is needed to create them, using LOCC, but no entanglement can be distilled from them by LOCC.

A simple example of bound entangled state is the state [Smo00]

$$\rho_s = \frac{1}{4}\sum_{i=1}^{4} |\Phi_i\rangle\langle\Phi_i| \otimes |\Phi_i\rangle\langle\Phi_i|, \tag{1}$$

where $|\Phi_i\rangle, i = 1, 2, 3, 4$, are all Bell states.

The existence of bound entangled states was considered to be a puzzling phenomenon. Such states cannot be used at many important applications of entanglement, such as teleportation, where usually maximally entangled pure states are needed. Of interest and importance has therefore been to find out whether we can make any use at all of bound entangled states. There are now several results showing that this is indeed the case.

First of all, in case of multipartite systems, a bound entangled state can be distillable if some groups of parties "get together". Indeed, if we denote parties as A, B, C, D, then the state (1) is $\{A, B\} : \{C, D\}-$, $\{A, C\} : \{B, D\}-$ and $\{A, D\} : \{B, C\}$-separable.

However, quite surprisingly, bound entangled states are of some use. They can be *activated* [HHH98a], and *super-activated* [SST00]. In addition, a protocol has been described, so called *remote information concentration protocol* [MV00], that uses BE-states.

Activation of bound entanglement, or so called *quasi-distillation*, refers to the process in which a finite number of free entangled mixed states are distilled with the assistance of a large number of BE-states, but without such an assistance no useful entanglement can be distilled from these states. In a *super-activation*, two BE-states are combined (tensored) to get a state which is not bound entangled. In other words, in this case BE-states are activated by BE-states — and therefore one can distill entanglement out of them. Hence, distillable entanglement is superadditive.

An intriguing problem is to find out which BE-states do not violate any Bell inequality. By [Dür01], there are multipartite BE-states that violate a Bell inequality, and by [KZG02], there is a bipartite BE-state that does not violate any Bell inequality.

8. ZOO of multipartite entanglement

As already mentioned, we still do not know whether for m-partite states, with $m > 2$, there are finite MREGS. This indicates that we can expect to have (very) many different types of entanglement.

Classification of multipartite states with respect to reversible transformations is, of course, not the only reasonable way to classify quantum states. Another important way is to consider two states as (stochastically) equivalent if each of them can be obtained from the other one by SLOCC – what actually means that two such states contain the same amount of entanglement.

For a special case of 3-partite and 4-partite qubit systems, quite a bit is already known about such different types of entanglement. For example, in 3-qubit systems [DVC00], we have four different types of pure states: (a) **Separable states** that are tensor products of three qubit states; (b) **Biseparable states,** that are not $A : B : C$ separable, but they are $AB : C$ or $AC : B$ or $A : BC$-separable; (c) So called **W-states,** with genuine entanglement of all three parties. They are the states that can be transformed, in a reversible way, by SLOCC, to the state $W = \frac{1}{\sqrt{3}}(|001\rangle + |010\rangle + |100\rangle)$; (d) So called **GHZ-states,** again with genuine entanglement of all three parties together, but with no two of them entangled separatedly. They are states that can be transformed, in a reversible way, by SLOCC, to the GHZ state.

For the case of four-qubit systems, there are 9 such different types of entangled pure states [VDMV01].

Four different types of three-qubit mixed states, that parallel the above classification of pure three-qubit states, have been shown in [ABLS01].

9. Entanglement sharing

Being a resource, it is intuitively clear that there have to be some restrictions on how entanglement can be shared. Some of the basic related questions are

- To which extend does entanglement between two objects restrict their entanglement with other objects?
- What are the general laws and limitations of entanglement sharing?
- Does the entanglement sharing potential grow with the dimension of particles?

For example, if two qubit-particles are maximally entangled, then they cannot be entangled with other particle. (This property of entanglement is usually called *monogamy*).

In this context, a key question is how are related multipartite and bipartite entanglement. Namely, which pairs of parties are entangled in a given multipartite system, if other parties are traced out.

Of interest is also to study the function $E(d, n)$ that denotes, for the case of states of n-partite d-dimensional systems, the maximum of the minimal entanglement of formation in any two subsystems [DW01].

One way to characterize entanglement of multipartite states is to specify how entangled become different subparties if the rest of parties is traced out. Results in [KZM02] show that for a "typical" pure state of n qudits (states in d-dimensional systems) all subsystems of less than $\frac{n}{3}$ parties are either separable or bound entangled and that probability of finding an n-qudit entangled state having some $\frac{n}{3}$ qubits entangled falls exponentially with the dimension of the Hilbert space. This means that most of the states are highly entangled, but entanglement is quite spread out and not shared by small group of parties.

10. Entanglement in quantum cryptography

There are two ways entanglement plays an important role in quantum cryptography: a positive one and a negative one.

Positive is the fact that quantum entanglement allows perfectly secure transmission of information and unconditionally secure generation of perfectly secret random binary keys. Indeed, in case two parties share enough EPR states, they can encode a state to be transmitted through a sequence of qubits and then to teleport these qubits. This is an absolutely secure way of transmission, because no physical systems are transmitted.

Moreover, by sharing n pairs of particles in the EPR state, both parties can implement quantum one-time pad cryptosystem without a need to share a classical key [Leu00]. This is again an absolutely secure way of transmission.

There are also several ways how entanglement can be used to generate shared and perfectly secret binary key. This is of importance for classical secret key cryptography which is so secure how secure is the key distribution.

For example, let Alice and Bob share n pairs of particles in the EPR state. If both parties measure their particles in the standard basis (and it does not matter in which order), they receive, as the result of their measurement, the same random binary string of length n. This way of binary key generation is again absolutely secure, because no information is transmitted.

Negative impact has entanglement on security of such basic quantum protocols as is bit commitment. It can be shown, due to the fact that using entanglement one party can always cheat, that no unconditional secure bit commitment is possible (in non-relativistic physical setting).

11. Frequency and robustness of entanglement

There are many basic questions concerning frequency and robustness of entanglement. It is of large importance to answer them for getting a more clear picture concerning the role entanglement can play in theory and especially in practice of quantum information processing and in quantum physics in general. Some of these questions to be asked for any type $\mathcal{T}$ of entangled states are:

- Given any state $|\phi\rangle$ of type $\mathcal{T}$, is there always a ball, in some reasonable distance measure, around $|\phi\rangle$, such that all states in that ball are of the type $\mathcal{T}$?

- Are there states of type $\mathcal{T}$ such that some ball of non-zero radius contains only states of the type $\mathcal{T}$?

For example, it has been shown [ZHSL98], that there are separable, entangled and also bound entangled states, such that a ball around them contains the same type of states.

Numerical results, reported in [ZHSL98], showed that the ratio of the volume of separable states, and also of bound entangled states, to the volume of all states goes down exponentially with the dimension of the system.

Results obtained so far also show that a pure state is more likely to be entangled, but a mixed state is more likely to be separable.

12. Challenges

Let us try to summarize some of the main challenges research in quantum entanglement is to deal with.

(a) To develop a comprehensive theory of all correlations – quantum and classical. (b) To explore entanglement capabilities of such physical processes as Hamiltonian interactions, unitary operations, ... Which physical interactions can create entanglement? How much? How to use them optimally? (c) To demonstrate experimentally entanglement for increasingly larger distances and to find out how much entanglement extend into the macroscopic world to systems of increasing complexity. (d) To search for entanglement in Nature and to explore how robust it is.[6] (e) To clear-up the role of mixed-states entanglement for quantum computations. (f) To qualify and quantify multipartite entanglement. (g) To discover laws and limitations of entanglement sharing.

Notes

1. In this context, it is interesting to ask: ***Why von Neumann, who played such important role in the development of both quantum mechanics and computing, did not come up with the idea of quantum information processing***, or, even better, ***whether somebody could come up at all, at his time, with such an idea***? (It seems safe to say that the main reason was the fact that at his time there was no computation complexity theory and therefore there was no way to see that the idea of quantum information processing, clearly extremely complex from the technology point of view, could pay off at all.)

2. However, one should note that the term ***quantum non-locality*** is quite confusing. Indeed, the best quantum theories we have – quantum field theories – satisfy the following locality condition: ***operators with support in space-like separated regions commute.***

3. The term ***hidden variable*** is quite confusing. Such variables should not be hidden from us – they are hidden only with respect to the current formalism of quantum theory.

4. The term ***Bell's inequalities*** is nowadays used in a wide sense, to denote a whole set of inequalities, between average values of correlations of some quantum experiments, that can be used to demonstrate that no local hidden-variable model of the reality can derive all predictions of quantum mechanics.

5. An important open problem is whether this is true for all non-Abelian groups. A positive answer would imply the existence of polynomial time quantum algorithm for graph isomorphism.

6. For example, every Bose-Einstein condensate is in a highly entangled state.

References

[ABLS01] A. Acin, D. Bruß, M. Lewenstein, and A. Sanpera. Classification of mixed three-qubit states. quant-ph/0103025, 2001.

[BBC+93] Ch. H. Bennett, G. Brassard, C. Crépeau, R. Jozsa, A. Peres, and W. K. Wootters Teleporting an unknown quantum state via dual classical and Einstein-Podolsky-Rosen channels. *Physical Review Letters*, 70:1895–1899, 1993.

[BCT99] G. Brassard, R. Cleve, and A. Tapp. Cost of exactly simulating quantum entanglement with classical communication. *Physical Review letters*, 83(9):1874–1787, 1999. quant-ph/9901035.

[BHU+02] B. Brezger, L. Hackermüler, S. Uttenhalter, J. Petachinka, and A. Zeilinger. Matter-wave interferometer for large molecules. quant-ph/0202158, 2002.

[BPM+97] D. Bouwmeester, J-W. Pan, K. Mattle, M.Eibl, H. Weinfurter, and A. Zeilinger. Experimental quantum teleportation. *Nature*, 390:575–579, 1997.

[BR01] S. Bandyopadhyay and V. Roychowdhury. Supercatalysis. quant-ph/0107103, 2001.

[Bru01] D. Bruß. Characterizing entanglement. quant-ph/0110078, 2001.

[BSST99] Ch. H. Bennett, P. W. Shor, J. A. Smolin, and A. V. Thapliyal. Entanglement-assisted classical capacity of noisy quantum channels. quant-ph/9904023, 1999.

[Dür01] W. Dür. Multipartite bound entangled states that do not violate Bell's inequality. quant-ph/0107050, 2001.

[DVC00] W. Dür, G. Vidal, and J. I. Cirac. Three qubits can be entangled in two inequivalent ways. quant-ph/0005115, 2000.

[DW01] K. Dennison and W. K. Wootters. Entanglement sharing among qudits. quant-ph/0106058, 2001.

[EPR35] A. Einstein, B. Podolsky, and N. Rosen. Can quantum mechanical description of physics reality be considered complete? *Physical Review*, 47:777–780, 1935.

[FMI02] H. Fan, K. Matsumoto, and H. Imai. Quantifying entanglement by concurrence hierarchy. quant-ph/0204041, 2002.

[Fuc02] Ch. A. Fuchs. Quantum mechanics as quantum information. q-p/0205039, 2002.

[GC99] D. Gottesman and I. L. Chuang. Quantum teleportation is a universal computational primitive. quant-ph/9908010, 1999.

[GI01] J. Gruska and H. Imai. Puzzles, mysteries and power of quantum entanglement. In *Proceedings of MCU'01, Cisenau, LNCS 2055*, pages 25–69, 2001.

[Gro97] L. K. Grover Quantum mechanics helps in searching for a needle in a haystack. *Physical Review Letters*, 78:325–328, 1997.

[Gru00] J. Gruska. *Mathematics unlimited, 2001 and beyond*, chapter Quantum computing challenges, pages 529–564. Springer-Verlag, 2000.

[Gru02] J. Gruska. *Quantum computing.* McGraw-Hill, 1999-2002. See also additions and updatings of the book on http://www.mcgraw-hill.co.uk/gruska.

[HHH98a] M. Horodecki, P. Horodecki, and R. Horodecki. Bound entanglement can be activated. quant-ph/9806058, 1998.

[HHH98b] M. Horodecki, P. Horodecki, and R. Horodecki. Mixed-state entanglement and distillation: is there a "bound" entanglement in nature? quant-ph/9801069, 1998.

[HM01] M. Hayashi and K. Matsumoto. Variable length universal entanglement concentration by local operations and its application to teleportation and dense coding. quant-ph/0109028, 2001.

[JKP01] B. Julsgaard, A. Kozhekin, and E. S. Polzik. Experimental long-lived entanglement of two macroscopic objects. quant-ph/0106057, 2001.

[JL02] R. Jozsa and N. Linden. On the role of entanglement in quantum computational speed-up. quant-ph/0201143, 2002.

[JP99] I. D. Jonathan and M. B. Plenio. Entanglement-assisted local manipulation of pure quantum states. quant-ph/9905071, 1999.

[Ken02] A. Kent. Causal quantum theory and the collapse locality loophole. quant-ph/0204104, 2002.

[KZG02] D. Kaszilowski, M. Zukowski, and P. Gnacinski. Bound entanglement and local realism. *Physical Review A*, 65:032107, 2002.

[KZM02] V. M. Kendom, K. Zyckowski, and W. J. Munro. Bounds on entanglement in qudit systems. quant-ph/0203037, 2002.

[Leu00] D. W. Leung. Quantum Vernam cipher. quant-ph/0012077, 2000.

[Mar02] T. W. Marshall. Nonlocality – the party may be over. quant-ph/0203042, 2002.

[MSW02] K. Matsumoto, T. Shimono, and A. Winter. Additivity of the Holevo channel capacity and of the entanglement of formation. In preparation, 2002.

[MV00] M. Murao and V. Vedral. Remote information concentration using a bound entangled state. quant-ph/0008078, 2000.

[Raz99] R. Raz. Exponential separation of quantum and classical communication complexity. In *Proceedings of* 31*st ACM STOC*, pages 358–367, 1999.

[Raz02] A. A. Razborov. Quantum communication complexity of symmetric predicates. quant-ph/0204025, 2002.

[RB00] R. Raussendorf and H. J. Briegel. Quantum computing with measurement only. quant-ph/0010033, 2000.

[Sch35] E. Schrödinger. Die gegenwartige Situation in der Quanenmechanik. *Natürwissenschaften*, 23:807–812, 823–828, 844–849, 1935.

[Sho96] P. W. Shor. Fault-tolerant quantum computation. In *Proceedings of* 37*th IEEE FOCS*, pages 56–65, 1996.

[Sho97] P. W. Shor Polynomial time algorithms for prime factorization and discrete logarithms on quantum computer. *SIAM J. on Computing*, 26(5):1484–1509, 1997.

[Smo00] J. A. Smolin. A four-party unlockable bound-entangled state. quant-ph/0001001, 2000.

[SST00] P. W. Shor, J. Smolin, and A. Thapliyal. Superactivation of bound entanglement. quant-ph/0005117, 2000.

[VC01] Guifré Vidal and J. I. Cirac. When only two thirds of the entanglement can be distilled. quant-ph/0107051, 2001.

[VDMV01] F. Verstraete, J. Dehaene, B. De Moor, and H. Verschede. Four qubits can be entangled in nine different ways. quant-ph/0109033, 2001.

[VW01] F. Verstraete and M. M. Wolf. Entanglement versus Bell violations under local filtering operations. quant-ph/0112012, 2001.

[Wer89] R. F. Werner. Quantum states with Einstein-Podolsky-Rosen correlations admitting a hidden-variable model. *Phys. Review A*, 40:4277–4281, 1989.

[Wer00] R. F. Werner. All teleportation and dense coding schemes. quant-ph/0003070, 2000.

[ZHSL98] K. Zyczkowski, P. Horodecki, A. Sampera, and M. Lewenstein. On the volume of mixed entangled states. quant-ph/9804024, 1998.

INFORMATION NETWORKS, LINK ANALYSIS, AND TEMPORAL DYNAMICS

(Summary of Invited Paper)

Jon Kleinberg
Cornell University

The Internet has given rise to two widespread communication media: the World Wide Web, and electronic mail. Both are sources of fearsome complexity, though in quite different ways.

Unlike other great networks of the past century — the electric power grid, the telephone system, or the highway and rail systems — the Web is not fundamentally an engineered artifact; its growth has been sudden, populist, and anarchic. The emergence of the Web has crystallized a view of large networks not just as technological creations, but as complex phenomena to be studied on their own terms. We are discovering that the Web and related information networks exhibit a characteristic 'geography'; they share a number of fundamental structural properties that presumably reflect the forces driving their growth and evolution [7, 19, 20, 27]. The study of these systems has led to methods for organizing the content of on-line document collections through analysis of their underlying link structures [6, 8, 16], and it has suggested research directions in models for large graphs [1, 3, 13, 21, 24], as well as computational perspectives on social network analysis [17, 30, 31].

E-mail has forced on us a different spectrum of problems — the personal complexity of managing a message stream that can reach a hundred pieces of mail per day, and organizing personal archives of correspondence that can easily grow to hundreds of megabytes in size. And at a still larger scale, e-mail has become the raw material for legal proceedings and historical investigation [22]. How can an algorithmic perspective suggest organizing principles for message streams of this magnitude? There has been research aimed at structuring e-mail archives by topic classification and keyword indexing [5, 9, 11, 12, 26]. A promising approach, complementary to these methods, is to make use of the tight relationship between topics and temporal dynamics — as time progresses, topics of interest are signaled by 'bursts of activity' in the stream. Using a concrete computational model for such 'bursts,' one can begin to structure the underlying content around them [18]. The resulting set of issues has interesting connections to research in topic detection and tracking [2, 4, 28, 29], as well as to probabilistic models from queueing theory [14] and temporal data mining [10, 15, 23, 25].

References

[1] W. Aiello, F. Chung, L. Lu. "Random evolution of massive graphs," *Proc. 42nd IEEE Symposium on Foundations of Computer Science*, 2001.

[2] J. Allan, J.G. Carbonell, G. Doddington, J. Yamron, Y. Yang, "Topic Detection and Tracking Pilot Study: Final Report," *Proc. DARPA Broadcast News Transcription and Understanding Workshop*, Feb. 1998.

[3] A.-L. Barabasi, R. Albert. "Emergence of scaling in random networks," *Science*, 286(509), 1999.

[4] D. Beeferman, A. Berger, J. Lafferty, "Statistical Models for Text Segmentation," *Machine Learning* 34(1999), pp. 177-210.

[5] A. Birrell, S. Perl, M. Schroeder, T. Wobber, *The Pachyderm E-mail System*, 1997, at http://www.research.compaq.com/SRC/pachyderm/.

[6] S. Brin, L. Page, "Anatomy of a Large-Scale Hypertextual Web Search Engine," *Proc. 7th International World Wide Web Conference*, 1998.

[7] A. Broder, R. Kumar, F. Maghoul, P. Raghavan, S. Rajagopalan, R. Stata, A. Tomkins, J. Wiener. "Graph structure in the Web," *Proc. 9th International World Wide Web Conference*, 2000.

[8] S. Chakrabarti, B. Dom, D. Gibson, J. Kleinberg, S.R. Kumar, P. Raghavan, S. Rajagopalan, A. Tomkins, "Mining the link structure of the World Wide Web," *IEEE Computer*, August 1999.

[9] W. Cohen. "Learning rules that classify e-mail." *Proc. AAAI Spring Symp. Machine Learning and Information Access*, 1996.

[10] D. Hand, H. Mannila, P. Smyth, *Principles of Data Mining*, MIT Press, 2001.

[11] J. Helfman, C. Isbell, "Ishmail: Immediate identification of important information," AT&T Labs Technical Report, 1995.

[12] E. Horvitz, "Principles of Mixed-Initiative User Interfaces," *Proc. ACM Conf. Human Factors in Computing Systems*, 1999.

[13] B. Huberman, L. Adamic, "Growth dynamics of the World Wide Web," *Nature* 401(1999).

[14] F.P. Kelly, "Notes on effective bandwidths," in *Stochastic Networks: Theory and Applications*, (F.P. Kelly, S. Zachary, I. Ziedins, eds.) Oxford Univ. Press, 1996.

[15] E. Keogh, P. Smyth, "A probabilistic approach to fast pattern matching in time series databases," *Proc. Intl. Conf. on Knowledge Discovery and Data Mining*, 1997.

[16] J. Kleinberg. "Authoritative sources in a hyperlinked environment." *Proc. 9th ACM-SIAM Symposium on Discrete Algorithms*, 1998. Extended version in *Journal of the ACM* 46(1999).

[17] J. Kleinberg. "Navigation in a Small World." *Nature* 406(2000).

[18] J. Kleinberg, "Bursty and Hierarchical Structure in Streams," *Proc. 8th ACM SIGKDD Intl. Conf. on Knowledge Discovery and Data Mining*, 2002.

[19] J. Kleinberg, S.R. Kumar, P. Raghavan, S. Rajagopalan, A. Tomkins. "The Web as a graph: Measurements, models and methods." *Proc. Intl. Conf. on Combinatorics and Computing*, 1999.

[20] J. Kleinberg, S. Lawrence, "The Structure of the Web," *Science* 294(2001).

[21] R. Kumar, P. Raghavan, S. Rajagopalan, A. Tomkins. "Stochastic models for the Web graph," *Proc. 41st IEEE Symposium on Foundations of Computer Science*, 2000.

[22] S.S. Lukesh, "E-mail and potential loss to future archives and scholarship, or, The dog that didn't bark," *First Monday* 4(9) (September 1999), at http://firstmonday.org.

[23] H. Mannila, M. Salmenkivi, "Finding simple intensity descriptions from event sequence data," *Proc. 7th ACM SIGKDD Intl. Conf. on Knowledge Discovery and Data Mining*, 2001.

[24] D. Pennock, G. Flake, S. Lawrence, E. Glover, C.L. Giles, "Winners don't take all: Characterizing the competition for links on the Web," *Proc. Natl. Acad. Sci.* 99(2002).

[25] L. Rabiner, "A tutorial on hidden Markov models and selected applications in speech recognition," *Proc. IEEE* 77(1989).

[26] R. Segal, J. Kephart. "Incremental Learning in SwiftFile," *Proc. Intl. Conf. on Machine Learning*, 2000.

[27] S. Strogatz, "Exploring complex networks," *Nature* 410(2001).

[28] R. Swan, J. Allan, "Automatic generation of overview timelines," *Proc. SIGIR Intl. Conf. on Research and Development in Information Retrieval*, 2000.

[29] R. Swan, D. Jensen, "TimeMines: Constructing Timelines with Statistical Models of Word Usage," *KDD-2000 Workshop on Text Mining*, 2000.

[30] D. Watts, *Small Worlds*, Princeton University Press, 1999.

[31] D. Watts, S. Strogatz, "Collective dynamics of small-world networks," *Nature* 393(1998).

GEOMETRIC SEPARATION AND EXACT SOLUTIONS FOR THE PARAMETERIZED INDEPENDENT SET PROBLEM ON DISK GRAPHS

(EXTENDED ABSTRACT)

Jochen Alber*
Universität Tübingen, Wilhelm-Schickard-Institut für Informatik, Sand 13, D-72076 Tübingen, Germany.
alber@informatik.uni-tuebingen.de

Jiří Fiala†
Charles University, KAM, DIMATIA and ITI‡, Faculty of Mathematics and Physics, Malostranské nám. 2/25, 118 00 Prague, Czech Republic.
fiala@kam.mff.cuni.cz

Abstract We consider the parameterized problem, whether a given set of n disks (of bounded radius) in the Euclidean plane contains k non-intersecting disks. We expose an algorithm running in time $n^{O(\sqrt{k})}$, that is—to our knowledge—the first algorithm for this problem with running time bounded by an exponential with a sublinear exponent.

The results are based on a new "geometric $\sqrt{\cdot}$-separator theorem" which holds for all disk graphs of bounded radius. The presented algorithm then performs, in a first step, a "geometric problem kernelization" and, in a second step, uses divide-and-conquer based on our geometric separator theorem.

Keywords: Disk graphs, independent set, fixed parameter tractable problems.

*Supported by the Deutsche Forschungsgemeinschaft (DFG), research project PEAL (Parameterized complexity and Exact Algorithms), NI 369/1-1, 1-2.
†Research of this author supported in part by Czech research grant GAČR 201/99/0242 and by NATO science fellowship, administered through Norwegian Research Council project no. 143192/410 during his postdoctoral fellowship at the University of Bergen.
‡Supported by the Ministry of Education of the Czech Republic as project LN00A056.

1. Introduction

The problem and its motivation. In this paper, we study the parameterized INDEPENDENT SET problem on disk graphs, which takes as an input a set $\mathcal{D}$ of disks in the plane and an integer k and the task is to determine whether there are k mutually disjoint disks in $\mathcal{D}$. The problem is motivated by various applications, among which we want to mention the area of frequency assignment problems in cellular networks.

Previous work. It is known [4] that the problem is NP-hard even for unit disk graphs. A way to cope with this hardness was proposed by approximation theory [8, 10]. Very recently, Erlebach *et al.* [8] gave a PTAS for INDEPENDENT SET on disk graphs. In this paper, however, we are interested in *exact* solutions for the given problem. We want to briefly summarize various results on the parameterized and classical complexity of the INDEPENDENT SET problem for general graphs and for planar graphs (the latter are equivalent to the class of coin graphs [12], i.e., disk graphs where disks are not allowed to overlap).

In parameterized complexity theory, it is known [5] that INDEPENDENT SET is complete for the class $W[1]$, which captures *intractable* parameterized problems (see [6] for details). However, restricted to planar graphs, INDEPENDENT SET is fixed parameter tractable, i.e., in the class FPT (see [6] for a precise definition). For the (asymptotically) best known algorithm we get running time $O(c^{\sqrt{k}}+n)$ being sublinear in k (see [2]). Moreover, very recently, Cai and Juedes [3] showed that this is best possible, in the sense that a running time of the form $O(c^{o(\sqrt{k})}n^{O(1)})$ cannot be achieved unless 3SAT$\in DTIME(2^{o(n)})$, which is considered to be very unlikely.

In the classical (one-dimensional) complexity study, the best known algorithm running in time $2^{e(n)}$ with $e(n) = 0.276n$ is due to Robson [15]. Moreover, $e(n) \in o(n)$ is impossible unless 3SAT$\in DTIME(2^{o(n)})$ (see [11]). If restricted to planar graphs, Lipton and Tarjan applied their well-known planar separator theorem [13] to get an algorithm with $e(n) = O(\sqrt{n})$. This is the best possible asymptotic behavior for e one can hope for, since otherwise an algorithm with $e(n) \in o(\sqrt{n})$ in combination with a known linear problem kernel would lead to an algorithm for the parameterized problem better than the relative lower bound shown by Cai and Juedes.

Main results and methods used. In this paper, for the case of disk graphs, we pursue the strategy of combining a geometric version of reduction to problem kernel with a divide-and-conquer approach based on an appropriate separator theorem. However, for disk graphs, so far such separator theorems are known only for so-called intersection graphs of τ-neighborhood systems [7, 14], which are closely related to (unit) disk graphs with λ-precision, where all centers are at mutual distance of at least $\lambda > 0$. With respect to general unit disk graphs, we quote from the introduction of Hunt *et al.* [10]:

> *"The [...] drawback is that problems such as maximum independent set [...]* **cannot** *be solved at all by the separator approach. This is because an arbitrary (unit) disk graph of n vertices can have a clique of size n."*

Table 1. Relating our results on INDEPENDENT SET on disk graphs to known results for general graphs and for planar graphs. (Lower bounds are under the assumption that 3SAT $\notin DTIME(2^{o(n)})$.)

graph class	(classical) complexity	parameterized complexity
general graphs	$2^{0.276\,n}$ [15] lower bound: $2^{\Omega(n)}$ [11]	W[1]-complete [5]
disk graphs DG_σ	$2^{O(\sqrt{n}\log(n))}$ [Rem.16]	$2^{O(\sqrt{k}\log(n))}$ [Thm.15], open: FPT or W[1]-h ?
disk graphs $DG_{\sigma,\lambda}$ (with λ-precision)	$2^{O(\sqrt{n})}$ [Rem.12]	$O(2^{O(\sqrt{k}\log(k))}+n)$ [Cor.17], hence: FPT
planar graphs	$2^{O(\sqrt{n})}$ [13] lower bound: $2^{\Omega(\sqrt{n})}$ [3]	$O(2^{O(\sqrt{k})}+n)$[2], hence: FPT

The key result in this paper is to show a way out of this dilemma by proving a new type of "geometric separator theorem" which holds even for the more general class of disk graphs with bounded radius ratio. Our geometric separator theorem can be seen as a generalization of (classical) separator theorems, where the guarantee is not on the size of the separator in terms of its number of vertices, but in terms of the space occupied by its disks.

This result is used to optimally solve the parameterized INDEPENDENT SET problem on disk graphs of bounded radius ratio in time $n^{O(\sqrt{k})}$ which is—to our knowledge—the first algorithm for this problem with running time bounded by a function with an exponent sublinear in k. In the worst case (i.e., when $k = n$) this turns into an algorithm of running time $2^{e(n)}$ with the sublinear term $e(n) = \sqrt{n}\log(n)$; a running time which cannot be achieved for general graphs (unless 3SAT $\in DTIME(2^{o(n)})$) [11].

In addition, in the case of disk graphs with λ-precision, we can show that the INDEPENDENT SET problem is in *FPT*. The results are summarized in Table 1.

Various proofs are omitted in this version due to space restrictions.

2. Preliminaries and Notation

If $\mathcal{S} = \{S_1, \ldots, S_n\}$, $S_i \subseteq \mathbb{R}^2$ is a collection of geometric objects, we denote by $\bigcup\mathcal{S} = \bigcup_{i=1}^{n} S_i$ the union of $\mathcal{S}$. For a collection $\mathcal{S}$, let $G_\mathcal{S} = (V_\mathcal{S}, E_\mathcal{S})$ denote the ***intersection graph*** of $\mathcal{S}$, i.e., $V_\mathcal{S} = \{v_1, \ldots, v_n\}$ and $E_\mathcal{S} = \{(v_i, v_j) \mid S_i \cap S_j \neq \emptyset\}$. The collection $\mathcal{S}$ is called the ***representation*** of $G_\mathcal{S}$. Moreover, for a subset $\mathcal{S}' \subseteq \mathcal{S}$, we denote by $V_{\mathcal{S}'} \subseteq V_\mathcal{S}$ the subset of vertices induced by $\mathcal{S}'$, i.e., $V_{\mathcal{S}'} = \{v_i \mid S_i \in \mathcal{S}'\}$. In this setting $G_{\mathcal{S}'} = G_\mathcal{S}[V_{\mathcal{S}'}]$ is the subgraph of $G_\mathcal{S}$ induced by the set of vertices $V_{\mathcal{S}'}$.

Disk graphs. A disk $D \subsetneq \mathbb{R}^2$ is specified by a triple $(r, x, y) \in \mathbb{R}^3$, where (x, y) are coordinates of the center of the disk in the Euclidean plane and r is its radius. The class of *disk graphs*, denoted by DG, is the set of all graphs G, for which we find a collection of disks $\mathcal{D} = \{D_1, \dots, D_n\}$ such that $G = G_{\mathcal{D}}$. Note that for a collection $\mathcal{D}$, the graph $G_{\mathcal{D}}$ has a natural embedding in the plane, where v_i sits in the position of the center of D_i.

The class of disk graphs of *bounded radius ratio* σ is the subclass $DG_\sigma \subset DG$ of all graphs $G \in DG$ which admit a representation $\mathcal{D} = \{D_1, \dots, D_n\}$, such that $(\max_{i=1,\dots n} r_i)/(\min_{i=1,\dots n} r_i) \leq \sigma$, where r_i denotes the radius of disk D_i. The parameter σ is called *radius ratio*. By a rescaling argument, for a graph $G \in DG_\sigma$ with representation $\mathcal{D}$, we can always achieve, that the smallest disk in $\mathcal{D}$ has radius one and, hence, all radii being upper bounded by σ.

Finally, a collection $\mathcal{D}$ is said to have λ-precision if all centers of disks are pairwise at least λ apart (see [10, Definition 3.2]). By a rescaling argument, all disk graphs have a representation with λ-precision, however only some graphs of DG_σ, allow a representation with radii in $[1, \sigma]$ and precision λ. We denote this class of graphs by $DG_{\sigma,\lambda}$.

Throughout the paper, we assume that a disk graph G is given together with its representation witnessing its membership in DG, DG_σ, or $DG_{\sigma,\lambda}$, respectively.

Grid graphs. Fix an arbitrary constant $\delta > 0$, and consider the infinite grid of span δ as the planar graph $H^\delta = (W^\delta, E^\delta)$ with vertices $W^\delta = \{w_{i,j} \mid i, j \in \mathbb{Z}\}$ and edges $E^\delta = \{(w_{i,j}, w_{k,l}) \mid |i - j| + |k - l| = 1\}$. The canonical (straight-line) embedding of H^δ in the plane is given by putting vertex $w_{i,j}$ at the coordinates $(i\delta, j\delta)$. The set of faces $\mathcal{F}^\delta$ of H^δ contains all closed squares $F^\delta_{i,j} = [i\delta, (i+1)\delta] \times [j\delta, (j+1)\delta] \subseteq \mathbb{R}^2$. For a grid vertex $w \in W^\delta$, we define the *face neighborhood* $\hat{N}(w) := \{F \in \mathcal{F}^\delta \mid w \in F\}$. Similarly for $W' \subseteq W^\delta$, $\hat{N}(W') = \bigcup_{w \in W'} \hat{N}(w)$.

Definition 1 *For a collection of disks $\mathcal{D} = \{D_1, \dots, D_n\}$, we define $H^\delta_{\mathcal{D}}$ to be the smallest subgraph of the infinite grid H^δ induced by a set of grid points which completely covers all disks in $\mathcal{D}$. We call H^δ the* covering grid (of span δ) for $\mathcal{D}$. *In other words, if we define the set of faces hit by $\mathcal{D}$ as $\mathcal{F}^\delta_{\mathcal{D}} = \{F \in \mathcal{F}^\delta \mid F \cap \bigcup \mathcal{D} \neq \emptyset\}$, and if $W^\delta_{\mathcal{D}}$ is the set of all grid points of $\mathcal{F}^\delta_{\mathcal{D}}$, then the covering grid $H^\delta_{\mathcal{D}}$ is the subgraph of H^δ induced by $W^\delta_{\mathcal{D}}$.*

An example which illustrates the construction of $H^\delta_{\mathcal{D}}$ is given in Figure 1.

Finally, for a collection $\mathcal{D}$ of disks and a set $S \subseteq \mathbb{R}^2$ (e.g., a set of grid vertices or a set of faces), we call $\mathcal{D}[S] := \{D \in \mathcal{D} \mid D \cap S \neq \emptyset\}$ the set of disks induced by S.

Measures. We use the standard Lebesgue measure μ in $\mathbb{R}^2$ as follows: For a Lebesgue measurable set $S \subseteq \mathbb{R}^2$, $\mu(S)$ denotes the *size* of S, i.e., the space in $\mathbb{R}^2$ occupied by S. In particular, for a collection of disks $\mathcal{D} = \{D_1, \dots, D_n\}$ let $\mu(\mathcal{D}) = \mu(\bigcup \mathcal{D})$ be the space covered by the union of disks $D_1, \dots, D_n$.

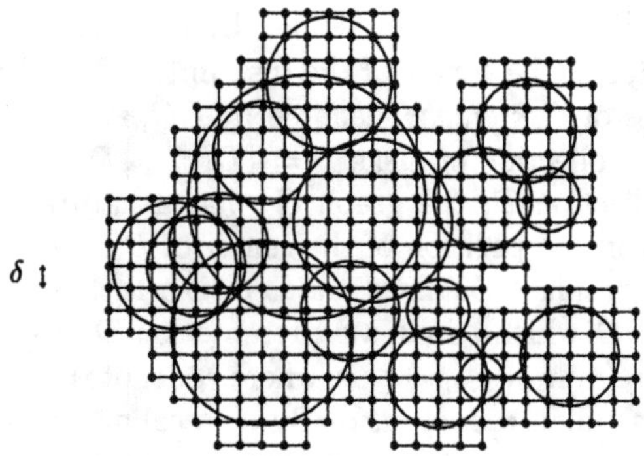

Figure 1. Constructing the grid graph $H^{\delta}_{\mathcal{D}}$ induced by a collection of disks $\mathcal{D}$.

Definition 2 *Let $\mathbb{G}$ be a graph class closed under taking subgraphs. A function $\xi : \mathbb{G} \to \mathbb{R}^+$, that is monotonous with respect to the subgraph ordering, i.e., $\xi(G) \leq \xi(G')$ if $G \subseteq G'$, and for which $\xi((\emptyset, \emptyset)) = 0$, is called a* graph measure.

Example 3 *We specify two graph measures which play a decisive role throughout the paper.*

1 *The usual* counting measure $|\cdot|$ *which assigns to any graph G the size of its vertex set $|V_G|$ is clearly a graph measure.*

2 *The* Lebesgue measure $\mu(\cdot)$ *assigning to a disk graph $G_{\mathcal{D}}$ with representation $\mathcal{D}$ the value $\mu(G_{\mathcal{D}}) = \mu(\mathcal{D})$ is a graph measure for DG, when we restrict the subgraph ordering to $G_{\mathcal{D}} \subseteq G_{\mathcal{D}'} \Leftrightarrow \mathcal{D} \subseteq \mathcal{D}'$.*

3. A Geometric Problem Kernelization

Reduction to problem kernel is a core technique in the design of fixed parameter algorithms.

Definition 4 *Let $\mathcal{L}$ be a parameterized problem, i.e., $\mathcal{L}$ consists of pairs (I, k), where problem instance I has a solution of size k (the parameter).* Reduction to problem kernel, *then, means to replace problem (I, k) by a "reduced" problem (I', k') (called* problem kernel*) such that*

$$k' \leq c \cdot k, \quad |I'| \leq p(k), \quad \text{and} \quad (I,k) \in \mathcal{L} \text{ iff } (I',k') \in \mathcal{L}, \tag{1}$$

where c is a constant. Furthermore, we require that the reduction from (I, k) to (I', k') is computable in time $T_K(|I|, k)$, which is a polynomial.

It is well-known that a parameterized problem is fixed parameter tractable if and only if it admits a reduction to a problem kernel.

For disk graphs, we can prove a geometric version of a problem kernel , where the size of the reduced instance is upper bounded by $O(k)$, when measured by the (Lebesgue) measure $\mu(\cdot)$ instead of the counting measure $|\cdot|$.

kernelize (disk graph $G_{\mathcal{D}}$, integer k)
- scale $\mathcal{D}$ such that the smallest disk has unit radius.
- set $\mathcal{F}_\bullet = \mathcal{F}^\bullet = \emptyset$, $\delta = \frac{1}{20}$
- *for each* $D \in \mathcal{D}$ *do*
  $\mathcal{F}_\bullet := \mathcal{F}_\bullet \cup \{F \in \mathcal{F}^\delta \mid F \subseteq D\}$
  $\mathcal{F}^\bullet := \mathcal{F}^\bullet \cup \{F \in \mathcal{F}^\delta \mid F \cap D \neq \emptyset\}$
- *if* $|\mathcal{F}^\bullet|\delta^2 > 9\pi\sigma^2 k$ *then* return "YES"
  else if $|\mathcal{F}_\bullet|\delta^2 < \pi k$ *then* return "NO"
  else return $(G_{\mathcal{D}}, k)$

Figure 2. Geometric problem kernel reduction.

Proposition 5 *For the parameterized* INDEPENDENT SET *(IS) problem on DG_σ there exists a "geometric" problem kernel, i.e., there is a procedure , that transforms an instance $(G_{\mathcal{D}}, k)$ to an instance $(G_{\mathcal{D}'}, k)$, such that $(G_{\mathcal{D}}, k) \in IS$ iff $(G_{\mathcal{D}'}, k) \in IS$ and*

$$\pi k \;\leq\; \mu(G_{\mathcal{D}'}) \;\leq\; 9\pi\sigma^2 k.$$

Proof. By our assumptions all disks in $\mathcal{D}$ have radius in the range $[1, \sigma]$.

Observe first, that $\mu(G_{\mathcal{D}}) > 9\pi\sigma^2 k$ implies that $(G_{\mathcal{D}}, k) \in IS$. We use the fact that $\mu(\mathcal{D}[N(v)]) \leq (3\sigma)^2\pi$ for any vertex $v \in V$, i.e., that the neighborhood of any vertex may occupy the space at most $9\pi\sigma^2$. And, secondly, if $\mu(G_{\mathcal{D}}) < \pi k$, then $(G_{\mathcal{D}}, k) \notin IS$, since the representation of any independent set of k vertices needs space at least πk. The procedure which in linear time transforms $(G_{\mathcal{D}}, k)$ to $(G_{\mathcal{D}'}, k)$ with $\mu(G_{\mathcal{D}'}) \leq 9\pi\sigma^2 k$ is given in Figure 2. □

Note that this is not a problem kernel according to Definition 4, since the size of $G_{\mathcal{D}}$ is measured by the (Lebesgue) measure $\mu(\cdot)$, which, in general, is not related to the (input) size of G. For disk graphs with λ-precision, however, we can prove an upper bound the counting measure by the Lebesgue measure.

Lemma 6 *Let $G_{\mathcal{D}} = (V, E) \in DG_{\sigma,\lambda}$ be a graph with and representation $\mathcal{D}$. Then, $|V| \leq 4\pi^{-1}\lambda^{-2}\mu(G_{\mathcal{D}})$.*

Corollary 7 *The parameterized* INDEPENDENT SET *problem on disk graphs $DG_{\sigma,\lambda}$ (with λ-precision) admits a linear problem kernel (in terms of the counting measure) of size $c = 36(\frac{\sigma}{\lambda})^2 k$, which can be computed in linear time.*

4. A Geometric Separator Theorem

In the following, we prove our key result—a geometric $\sqrt{\cdot}$-separator theorem—that makes our divide-and-conquer strategy work.

4.1. Classical $\sqrt{\cdot}$-separator theorems

We start with a somewhat generalized notion of separator theorems.

Definition 8 *Let* $G = (V, E)$ *be an undirected graph. A* separator $V_S \subseteq V$ *of* G *partitions* V *into two* parts V_A *and* V_B *such that*

- $V_A \dot{\cup} V_S \dot{\cup} V_B = V$*, and*
- *no edge joins a vertex of* V_A *to* V_B*.*

The triple (V_A, V_S, V_B) *is also called a* separation *of* G*.*

In order to provide a quantitative approach to separators, we need the notion of "measure" as introduced in Section 2.

Definition 9 *Let* ξ *be a graph measure. An* $f(\cdot)$-separator theorem for the measure ξ (and constants $\alpha < 1$, $\beta > 0$) *on a class of graphs* $\mathbb{G}$ *which is closed under taking subgraphs is a theorem of the following form:*
For any $G \in \mathbb{G}$ *there exists a separation* (V_A, V_S, V_B) *of* G *such that*

1 $\xi(G[V_S]) \leq \beta \cdot f(\xi(G))$

2 $\xi(G[V_A]), \xi(G[V_B]) \leq \alpha \cdot \xi(G)$

$\sqrt{\cdot}$-separator theorems on planar graphs. In this framework, the planar separator theorem due to Lipton and Tarjan [13] can be formulated as follows.

Theorem 10 *On the class of planar graphs, there exists a* $\sqrt{\cdot}$*-separator theorem for the counting measure* $|\cdot|$ *with constants* $\alpha = 2/3$ *and* $\beta = 2\sqrt{2}$*. Moreover, the separation can be found in linear time.*

$\sqrt{\cdot}$-separator theorems on disk graphs with λ-precision. In terms of geometric graphs, a $\sqrt{\cdot}$-separator theorem for the counting measure was proven on the class of intersection graphs of so-called τ-neighborhood systems (see [10]). Here, a τ-neighborhood system is a collection $\mathcal{B} = \{B_1, \ldots, B_n\}$ of balls in a space of arbitrary fixed dimension, such that the intersection of any $(\tau + 1)$ distinct balls in $\mathcal{B}$ is empty. It can be verified that every unit disk graph with λ-precision is an intersection graph of a τ-neighborhood system (τ depending on λ) and, vice versa, every intersection graph of a τ-neighborhood system in $\mathbb{R}^2$ is λ-precision disk graph (λ being the minimum distance between the centers of any two disks), see [10]. In the two-dimensional case the corresponding separator theorem reads as follows (see [14, Theorem 2.5] and [7, Theorem 5.1]):

Theorem 11 *On the class of intersection graphs of* τ*-neighborhood systems, there exists a* $\sqrt{\cdot}$*-separator theorem for the measure* $|\cdot|$ *with constants* $\alpha = 3/4$ *and* $\beta = O(\sqrt{\lambda})$*. Moreover, the separation can be found in linear time.*

Remark 12 *As exhibited, e.g., in [1, Section 4.1], a divide-and-conquer approach yields that* INDEPENDENT SET *on intersection graphs of* τ*-neighborhood systems, and hence, on unit disk graphs with* λ*-precision as well as for graphs from* $DG_{\sigma,\lambda}$*, can be solved in time* $2^{O(\sqrt{n})}$*.*

4.2. A new geometric $\sqrt{\cdot}$-separator theorem

In this subsection, we prove an analogue to the classical $\sqrt{\cdot}$-separator theorems for the class DG_σ of disk graphs with bounded radius ratio. Note that graphs in DG_σ may contain arbitrary large cliques, which means that a $\sqrt{\cdot}$-separator theorem does not hold for the counting measure. Using the Lebesgue measure instead, we get the following result.

Theorem 13 *On the class DG_σ of disk graphs with bounded radius ratio, there exists a $\sqrt{\cdot}$-separator theorem for the Lebesgue graph measure $\mu(\cdot)$.*

More precisely there exist constants $\alpha < 1$ and β such that, for every graph $G_\mathcal{D} \in DG_\sigma$ with representation $\mathcal{D}$, we find three sets $\mathcal{D}_A, \mathcal{D}_S, \mathcal{D}_B \subseteq \mathcal{D}$, such that $(V_{\mathcal{D}_A}, V_{\mathcal{D}_S}, V_{\mathcal{D}_B})$ is a separation for $G_\mathcal{D}$, satisfying

1 $\mu(\mathcal{D}_S) \leq \sigma^2 \beta \sqrt{\mu(\mathcal{D})}$,

2 $\mu(\mathcal{D}_A), \mu(\mathcal{D}_B) \leq \alpha\mu(\mathcal{D})$.

Moreover, this separation can be found in time linear in $|\mathcal{D}|$.

The idea for the proof of Theorem 13 is to construct the covering grid $H_\mathcal{D}^\delta$ (of suitable span) for a given collection $\mathcal{D}$ of disks. Then, in a second step, one applies a planar $\sqrt{\cdot}$-separator theorem on $H_\mathcal{D}^\delta$ from which, in a suitable manner, the three sets $\mathcal{D}_A, \mathcal{D}_S, \mathcal{D}_B \subseteq \mathcal{D}$ will be constructed. We need the following key result interrelating the space covered by $\mathcal{D}$ with the size of the covering grid.

Proposition 14 *For any ε there exists a δ such that for any set $\mathcal{D}$ of disks of radius at least one, we have $|W_\mathcal{D}| \leq (1+\varepsilon)\,\delta^{-2}\mu(\mathcal{D})$.*

Proof. (of Theorem 13) The sets $\mathcal{D}_S$, $\mathcal{D}_A$ and $\mathcal{D}_B$ will be determined according to the procedure given in Figure 3. We now prove that, indeed $(V_{\mathcal{D}_A}, V_{\mathcal{D}_S}, V_{\mathcal{D}_B})$ is a separation of G and that properties (1.) and (2.) of the theorem hold for the computed sets $\mathcal{D}_S, \mathcal{D}_A$, and $\mathcal{D}_B$:

$(V_{\mathcal{D}_A}, V_{\mathcal{D}_S}, V_{\mathcal{D}_B})$ is a separation of G: Showing that $(V_{\mathcal{D}_A}, V_{\mathcal{D}_S}, V_{\mathcal{D}_B})$ is a separation of $G_\mathcal{D}$ is equivalent to proving that $\bigcup \mathcal{D}_A \cap \bigcup \mathcal{D}_B = \emptyset$. Recall that (W_A, W_S, W_B) is the separation of the covering grid $H_\mathcal{D}^\delta$ obtained by the algorithm of Lipton and Tarjan. First of all we claim that

$$\bigcup \mathcal{D}_A \cap W_\mathcal{D}^\delta \subseteq W_A, \quad \text{and} \quad \bigcup \mathcal{D}_B \cap W_\mathcal{D}^\delta \subseteq W_B. \tag{2}$$

To see this, note that $\bigcup \mathcal{D}_A \cap W_S = \emptyset$, since if there is a disk $D \in \mathcal{D}_A$ containing a point $w \in W_S$—by construction of $\mathcal{D}_S$—we had $D \in \mathcal{D}_S$. Suppose now that there is a vertex $w_B \in W_B$ which lies in a disk $D \in \mathcal{D}_A$. By definition of the set $\mathcal{D}_A$, we find a vertex $w_A \in W_A$ inside D as well. Then, there exists a path P in $H_\mathcal{D}^\delta$ which connects w_A and w_B and which is completely located inside the disk D. Since (W_A, W_S, W_B) is a separation of $H_\mathcal{D}^\delta$, there exists a vertex of W_S on P contradicting $\bigcup \mathcal{D}_A \cap W_S = \emptyset$. This also implies that

geometric_separator(disk graph $G_{\mathcal{D}}$)
- scale $\mathcal{D}$ such that the smallest disk has unit radius.
- fix any $\varepsilon < \frac{1}{2}$ and select δ according to Proposition 14
- run the algorithm of Lipton and Tarjan (see Theorem 10) on $H^\delta_{\mathcal{D}}$ to obtain a separation (W_A, W_S, W_B) with
  $|W_S| \leq 2\sqrt{2|W^\delta_{\mathcal{D}}|}$ and
  $|W_A|, |W_B| \leq \frac{2}{3}|W^\delta_{\mathcal{D}}|$,
- return the three sets
  $\mathcal{D}_S := \mathcal{D}[\hat{N}(W_S)],\ \mathcal{D}_A := \mathcal{D}[W_A] \setminus \mathcal{D}_S,\ \ \mathcal{D}_B := \mathcal{D}[W_B] \setminus \mathcal{D}_S.$

Figure 3. Separator algorithm corresponding to Theorem 13.

$\bigcup \mathcal{D}_A \cap W_B = \emptyset$. Since $W^\delta_{\mathcal{D}} = W_A \cup W_S \cup W_B$, we get $\bigcup \mathcal{D}_A \cap W^\delta_{\mathcal{D}} \subseteq W_A$. The property $\bigcup \mathcal{D}_B \cap W^\delta_{\mathcal{D}} \subseteq W_B$ follows similarly.

Assume now, for contradiction that vertices $v_A \in V_{\mathcal{D}_A}$ and $v_B \in V_{\mathcal{D}_B}$ form an edge of $E_{\mathcal{D}}$, that means there exists some point z in $D_A \cap D_B$. Let $F_z \in \mathcal{F}^\delta$ be any of the (at most four) squares containing z, and let W_z be the four grid vertices adjacent to F_z in $H^\delta_{\mathcal{D}}$.

We first consider the case when D_B does not intersect W_z. Then it intersects one side of F_z, and since $\delta \ll 1$ it contains two grid points of the square sharing this side. Note that we use equation (2) in the sense that all grid points intersected by D_B belong to W_B. Then as D_A must contain at least one point of W_z (if not, either there is no possible place for z or $W_A \cap W_B \neq \emptyset$).

In this case, or when we symmetrically exchange subscripts A and B, and also when both D_A and D_B intersect W_z, there are two grid points $w_A \in D_A$, $w_B \in D_B$, that are in $H^\delta_{\mathcal{D}}$ at distance at most two. Since w_A and w_B must be separated by W_S, and at the same time they belong to $\hat{N}(W_S)$, that is in contradiction with the definition of sets $\mathcal{D}_A$ and $\mathcal{D}_B$.

ad property (1.): Consider a vertex $w \in W_S$ and the neighborhood $\hat{N}(w)$. All disks in $\mathcal{D}$ which intersect $\hat{N}(w)$ must lie inside a cycle of radius $(2\sigma + \sqrt{2}\delta)$ centered at w. This is clear, since disks in $\mathcal{D}$ have radius bounded by σ and since the grid has span δ. More formally, we get

$$\mu(\mathcal{D}[\hat{N}(w)]) \leq (2\sigma + \sqrt{2}\delta)^2\pi.$$

Moreover, this implies that

$$\begin{aligned} \mu(\mathcal{D}_S) &= \mu(\mathcal{D}[\hat{N}(W_S)]) = \mu\Big(\bigcup_{w\in W_S} \mathcal{D}[\hat{N}(w)]\Big) \\ &\leq \sum_{w\in W_S} \mu(\mathcal{D}[\hat{N}(w)]) \leq \Big((2\sigma + \sqrt{2}\delta)^2\pi\Big)|W_S| \leq 5\sigma^2\pi|W_S|. \end{aligned} \tag{3}$$

Since, by our choice, $\varepsilon < \frac{1}{2}$ we may use in the last step $\sqrt{2}\delta < \frac{1}{6} \leq \frac{\sigma}{6}$. Using property (a) of step (3.) of the algorithm in Figure 3 and Proposition 14, we have $|W_S| \leq \beta' \sqrt{|W_\mathcal{D}|} \leq \beta' \sqrt{\frac{1+\varepsilon}{\delta^2} \mu(\mathcal{D})}$ which together with the estimate (3) establishes

$$\mu(\mathcal{D}_S) \leq \sigma^2 \beta \sqrt{\mu(\mathcal{D})} \quad \text{for} \quad \beta = \frac{5\pi\beta'}{\delta} \sqrt{(1+\varepsilon)}. \tag{4}$$

ad property (2.): First of all, observe that the set $\bigcup \mathcal{D}_A$ is completely covered by the square faces of the subgraph $H^\delta_\mathcal{D}[W_A]$, induced by the vertices of W_A. To see this, suppose there is a point $z \in D$ (for some $D \in \mathcal{D}_A$) which lies in some square $F_z \in \mathcal{F}^\delta$ of $H^\delta_\mathcal{D}$ but not of $H^\delta_\mathcal{D}[W_A]$. As above, if the four vertices adjacent to F_z host a vertex of W_B or W_S, we get $D \cap \hat{N}(W_S) \neq \emptyset$, a contradiction. By this observation and by the fact that $|W_A| \leq \alpha' |W_\mathcal{D}|$, we get

$$\mu(\mathcal{D}_A) \;\leq\; \mu(H^\delta_\mathcal{D}[W_A]) \;\leq\; \delta^2 |W_A| \;\leq\; \delta^2 \alpha' |W^\delta_\mathcal{D}| \;\leq\; \alpha'(1+\varepsilon)\,\mu(\mathcal{D}),$$

where Proposition 14 was used in the last step.

Similarly, one proves $\mu(\mathcal{D}_B) \;\leq\; \alpha'(1+\varepsilon)\,\mu(\mathcal{D})$. □

We note here that by our choice of $\varepsilon = \frac{1}{4}$, we get $\alpha = \alpha'(1+\varepsilon) = \frac{5}{6}$; but α can be arbitrarily close to α' by a sufficiently small choice of ε. Then, however, we have to consider the tradeoff of getting a small β according to Equation (4) on the one hand, and enlarging $|W^\delta_\mathcal{D}|$ on the other hand.

5. The Algorithm and its Analysis

We use the geometric kernelization of Section 3 and a divide-and-conquer approach based on the geometric separator theorem from Section 4.2 to derive an algorithm for the INDEPENDENT SET problem on disk graphs DG_σ.

Theorem 15 *Let $\mathcal{D}$ be a collection of n disks with $G_\mathcal{D} \in DG_\sigma$. Then, there is an algorithm running in time $n^{O(\sqrt{k})}$ which decides if $G_\mathcal{D}$ admits an independent set of size at least k, and if so constructs one.*

Proof. On input instance $(G_\mathcal{D}, k)$, in a first step, perform the geometric kernelization explained in Section 3. After this step, without loss of generality, we may assume that $\mu(\mathcal{D}) \leq ck$ for the constant c given in Proposition 5. In a second step, the divide-and-conquer procedure **indep_set** shown in Figure 4 is applied to the instance $(\mathcal{D}, ck)$.

Denote by $T(n, s)$ the time needed to execute **indep_set**$(\mathcal{D},s)$ on a collection of n disks $\mathcal{D}$ with $\mu(\mathcal{D}) \leq s$. Let $p(|\mathcal{D}|)$ be the polynomial time needed to compute the sets $\mathcal{D}_S, \mathcal{D}_A$, and $\mathcal{D}_B$ according to Theorem 13, and $q(|\mathcal{D}|)$ be the polynomial time needed to perform constructions of $\mathcal{D}'_A$ and $\mathcal{D}'_B$. Note that in $\mathcal{D}_S$ at most $\lfloor \frac{\beta\sqrt{s}}{\pi} \rfloor$ many disks can be independent, since $\mu(\mathcal{D}_S) \leq \beta\sqrt{s}$ and every disk has radius at least one. Hence, the total number of independent sets in $G_{\mathcal{D}_S}$ is upper bounded by

$$\sum_{i=0}^{\lfloor \frac{\beta\sqrt{s}}{\pi} \rfloor} n()i \leq n^{\hat{\beta}\sqrt{s}},$$

disks **indep_set**(disks $\mathcal{D}$, space s)
- *if* $(\mathcal{D} = \emptyset)$ *then* return $\emptyset$; *else*
- compute $\mathcal{D}_S$, $\mathcal{D}_A$, and $\mathcal{D}_B$ according to Theorem 13
- *for all* independent sets V_{IS} of $G_{\mathcal{D}_S}$ *do*
  $\mathcal{D}'_A := \mathcal{D}_A \setminus \{D \in \mathcal{D}_A \mid \exists D' \in \mathcal{D}[V_{\text{IS}}] : D \cap D' \neq \emptyset\}$
  $\mathcal{D}'_B := \mathcal{D}_B \setminus \{D \in \mathcal{D}_B \mid \exists D' \in \mathcal{D}[V_{\text{IS}}] : D \cap D' \neq \emptyset\}$
  $\text{rst}_{V_{\text{IS}}} := V_{\text{IS}} \cup$ **indep_set**$(\mathcal{D}'_A, \alpha s) \cup$ **indep_set**$(\mathcal{D}'_B, \alpha s)$
- return $\text{rst}_{V_{\text{IS}_0}}$, for which $|\,\text{rst}_{V_{\text{IS}_0}}\,| = \min\{|\,\text{rst}_{V_{\text{IS}}}\,|\}$

Figure 4. Divide-and-conquer algorithm for INDEPENDENT SET on disk graphs of bounded radius ratio based on our geometric separator theorem.

where $\widehat{\beta}$ is some constant. Then, the recursion we have to solve in order to compute an upper bound on $T(n,s)$ reads as follows:

$$T(n,s) \leq p(n) \;+\; n^{\widehat{\beta}\sqrt{s}} \cdot q(n) \cdot 2\,T(n, \alpha s).$$

Hence, for n large enough, and a suitable constant $\widetilde{\beta}$ we have

$$\begin{aligned} T(n,s) \;&\leq\; n^{\widetilde{\beta}\sqrt{s}} \cdot T(n,\alpha s) \;\leq\; \prod_{i=0}^{\log_{1/\alpha}(s)} n^{\widetilde{\beta}\sqrt{\alpha^i s}} \cdot T(n,1) \\ &\leq\; n^{\widetilde{\beta}\sqrt{s}(\sum_{i=0}^{\infty}(\sqrt{\alpha})^i)} \cdot T(n,1) \;=\; n^{\frac{\widetilde{\beta}\sqrt{s}}{1-\sqrt{\alpha}}} \cdot T(n,1). \end{aligned}$$

Note that $T(n,1)$ is constant, since $\mu(\mathcal{D}) \leq 1$ implies $\mathcal{D} = \emptyset$, because for every disk D we have $\mu(D) \geq \pi$. By plugging in the values $n = |\mathcal{D}|$ and $s = ck$, we obtain the running time as we have claimed. □

Remark 16 *In the worst case $k = n$, we have an $O(2^{O(\sqrt{n}\log(n))})$-algorithm, a running time with a sublinear exponent that cannot be achieved for general graphs (unless 3SAT$\in DTIME(2^{o(n)})$) [11].*

Corollary 17 *By Corollary 7, it follows that the parameterized* INDEPENDENT SET *problem on disk graphs $DG_{\sigma,\lambda}$ (with λ-precision) can be solved in time $O(k^{O(\sqrt{k})} + n)$, hence, the problem is fixed parameter tractable.*

We leave it as an open problem, whether INDEPENDENT SET on disk graphs is in *FPT* or complete for the classes $W[1]$ or $W[2]$, respectively. We want to emphasize, that we are not aware of a (non-artificial) $W[1]$-hard problem which can be solved in time bounded by an exponential with a sublinear exponent. Similarly, its W[1]-hardness would expose the first example of a fixed parameter intractable problem that simultaneously allows a PTAS.

Acknowledgments. We thank Thomas Erlebach and Rolf Niedermeier for introducing us to the problem and for fruitful discussion and valuable comments during this project. We thank also Jan Arne Telle for his hospitality and encouragement during our visit at University of Bergen. Finally, we thank Jiří Matoušek for pointing us to an elegant tool which was used in the proof of Lemma 6.

References

[1] Alber, J., Fernau, H., and Niedermeier., R. (2001a). Graph separators: a parameterized view. In *Proc. 7th COCOON 2001*, pages 318–327. Springer-Verlag LNCS 2108.

[2] Alber, J., Fernau, H., and Niedermeier., R. (2001b). Parameterized complexity: exponential speed-up for planar graph problems. In *Proc. 28th ICALP 2001*, pages 261–272. Springer-Verlag LNCS 2076.

[3] Cai, L. and Juedes, D. W. (2001). On the existence of subexponential parameterized algorithms. revised version of: Subexponential parameterized algorithms collapse the W-hierarchy. In *Proc. 28th ICALP 2001*, pages 273–284. Springer-Verlag LNCS 2076.

[4] Clark, B. N., Colbourn, C. J., and Johnson, D. S. (1990). Unit disk graphs. *Discrete Math.*, 86(1–3):165–177.

[5] Downey, R. G. and Fellows, M. R. (1995). Fixed-parameter tractability and completeness II: On completeness for W[1]. *Theor. Comput. Sci.*, 141:109–131.

[6] Downey, R. G. and Fellows, M. R. (1999). *Parameterized Complexity*. Texts and Monographs in Computer Science. Springer-Verlag.

[7] Eppstein, D., Miller, G. L., and Teng, S.-H. (1993). A deterministic linear time algorithm for geometric separators and its applications. In *Proc. of 9th ACM Symposium on Computational Geometry*, pages 99–108.

[8] Erlebach, T., Jansen, K., and Seidel, E. (2001). Polynomial-time approximation schemes for geometric graphs. In *Proc. 12th ACM-SIAM SODA*, pages 671–679.

[9] Hochbaum, D. S. and Maass, W. (1985). Approximation schemes for covering and packing problems in image processing and VLSI. *Journal of the ACM*, 32(1):130–136.

[10] Hunt, H. B., Marathe, M. V., Radhakrishnan, V., Ravi, S., Rosenkrantz, D. J., and Stearns, R. E. (1998). NC-approximation schemes for NP- and PSPACE-hard problems for geometric graphs. *Journal of Algorithms*, 26(2):238–274.

[11] Impagliazzo, R., Paturi, R., and Zane, F. (1998). Which problems have strongly exponential complexity? In *Proc. of the 39th IEEE FOCS*, pages 653–664.

[12] Koebe, P. (1936). Kontaktprobleme der konformen Abbildung. *Berichte über die Verhandlungen d. Sächs. Akad. d. Wiss.* Math-Phys. Klasse, 88:141–164.

[13] Lipton, R. J. and Tarjan, R. E. (1980). Applications of a planar separator theorem. *SIAM Journal on Computing*, 9:615–627.

[14] Miller, G. L., Teng, S.-H., and Vavasis, S. A. (1991). A unified geometric approach to graph separators. In *Proc. of the 32nd IEEE FOCS*, pages 538–547.

[15] Robson, J. M. (1986). Algorithms for maximum independent sets. *Journal of Algorithms*, 7:425–440.

BIN-PACKING WITH FRAGILE OBJECTS

Nikhil Bansal
Computer Science Department
Carnegie Mellon University
Pittsburgh, PA 15213, USA
nikhil@cs.cmu.edu

Zhen Liu
IBM T. J. Watson Research Center
PO Box 704
Yorktown Heights, NY 10598, USA
zhenl@us.ibm.com

Arvind Sankar
Department of Mathematics
Massachusetts Institute of Technology
Cambridge, MA 02139-4307, USA
arvinds@mit.edu

Abstract We consider an extension of the classical bin packing problem, motivated by a frequency allocation problem arising in cellular networks. The problem is as follows: Each object has two attributes, weight and fragility. The goal is to pack objects into bins such that, for every bin, the sum of weights of objects in that bin is no more than the fragility of the most fragile object in that bin.

We look for approximation algorithms for this problem. We provide a 2-approximation to the problem of minimizing the number of bins. We also show a lower bound of 3/2. Unlike in traditional bin packing, this bound holds in the asymptotic case. We then consider the approximation with respect to fragility and provide a 2-approximation algorithm. Our algorithm uses the same number of bins as the optimum but the weight of objects in a bin can exceed the fragility by a factor of 2.

1. Introduction

We consider a generalization of the classical bin-packing problem. In the traditional bin-packing problem, we are given a collection of n objects, where each object has an arbitrary non-negative weight no more than 1. These objects are to be placed in bins, such that the total weight of the objects in each bin is at most 1, and the total number of bins used is minimized. In our problem the objects are fragile, that is, in addition to its weight, each object has a fragility associated with it. The more fragile an object, the lower is its fragility value. In this model, an object breaks down if the total weight of objects in the bin in which it is placed exceeds the fragility of this object. Thus, we seek to place the objects in bins such that for every bin, the sum of the weights of objects in that bin is no more than the fragility of the most fragile object in that bin.

Clearly, if we set the fragility of each object to 1, this places an upper bound of 1 on the total weight in each bin, and hence our problem reduces to the traditional bin-packing problem. As classical bin packing is known to be NP-hard, we look for efficient approximation algorithms for our fragile bin packing problem. There are two natural notions of approximation for the problem. The first (as in the classical bin packing problem) is the number of bins used. We give a factor 2 approximation with respect to the number of bins. We also show that, unlike the traditional bin packing problem which admits an asymptotic PTAS [de la Vega and Lueker, 1981, Karmarkar and Karp, 1982], our problem cannot be approximated by a polynomial time algorithm to a factor better than 3/2, unless P=NP. Second, we can also consider an approximation with respect to fragility. Specifically, we can imagine an algorithm which produces an assignment of objects to bins that uses the same number of bins as the optimal algorithm, however we might violate the fragility upto some factor greater than 1. That is, the sum of weights of objects in a bin can be upto some factor times more than the fragility of the most fragile object in that bin. We will give an algorithm which achieves an approximation ratio of 2 with respect to this measure.

The bin-packing problem and its various variants have been extensively studied, in the context of approximation algorithms [de la Vega and Lueker, 1981, Karmarkar and Karp, 1982], online algorithms [Garey et al., 1972, Yao, 1980, van Vliet, 1996] and average case analysis [Shor, 1984, Coffman and Lueker, 1991]. A survey article by Coffman et. al. [Coffman et al., 1997] and numerous references therein provide a comprehensive list of the results in this area. While our variant of the problem is fairly natural, we do not know of any previous work on this problem. Our variant is motivated by the following problem arising in frequency allocation in cellular networks.

Consider a base station in a cellular network which has various users communicating with it on various frequency channels. In CDMA (a commonly used technology in wireless systems), a wide channel with capacity much larger than an individual user's information rate is allocated and multiple users share a single channel. To utilize the bandwidth efficiently it is necessary to assign as many users as possible to a single channel. However, the tradeoff in assigning many users to a particular channel is that there is loss of quality (given by signal to noise ratio (SNR), usually denoted by β), due to interference among users sharing the channel. The goal of the channel assignment scheme is to minimize the number of channels used while guaranteeing that each user achieves a minimum SNR of β.

To be more specific, consider n users communicating with a central base station. If user i transmits with power p_i, then the signal received by the base station is $s_i = p_i g_i$, where g_i is the channel gain for user i. Let $U(t)$ be the set of users transmitting to the base station at time t, then transmission from user $i \in U(t)$ is successful if and only if

$$\frac{s_i}{N_0 + \sum_{j \in U(t), j \neq i} s_j} \geq \beta$$

where β is the signal-to-noise ratio (SNR) requirement for successful communication and N_0 is the background noise power. While we consider only single cell networks in this paper, N_0 can be used to model the interference due to thermal noise and the interference due to the transmission of the neighboring channels in a multicellular network.

Thus, we can model the frequency channels as bins, the users are objects with weights equal to the amount of power received at the base station. Since the users can tolerate a power of up to $1/\beta$ times their power from other users on the same frequency channel, we set their fragilities to be $1 + 1/\beta$ times their weight. Thus the frequency allocation problem is a special case of our general bin-packing problem with fragile objects.

We note that a widely studied solution (assuming certain hardware capabilities) to the frequency allocation problem is *power control* [Hanly and Tse, 1999, Yates, 1995, Yates and Huang, 1995, Zander, 1993, Viterbi, 1995]. Here, the users control their transmission power levels such that the power received at the base station (the weight in our bin packing problem) is almost equal for all users. However, power control involves expensive and relatively sophisticated hardware. For example, most of the wireless Ethernet cards available presently cannot adjust their transmission power [Feeney and Nilsson, 2001]. Thus, our work applies to scenarios where power control is not available.

2. Problem Formulation

We are given n objects $1, \ldots, n$ with weights $w_1, \ldots, w_n$ and fragility $f_1, \ldots, f_n$ respectively. We have bins (of potentially infinite capacity), in which we want to place the objects. Suppose objects $j_1, j_2, \ldots, j_{n_i}$ are assigned to bin B_i. Then the assignment to bin B_i is feasible for bin B_i if $w_{j_1} + \cdots + w_{j_{n_i}} \leq \min\{f_{j_1}, f_{j_2}, \ldots, f_{j_{n_i}}\}$. In words, the total weight of the objects assigned to bin B_i does not exceed the fragility of any object in the bin.

Let A be an assignment of objects to bins. We say that A is feasible if each object is assigned to some bin, and each bin is feasible. The cost of the assignment A is the number of bins used by the assignment. Let OPT be the assignment which uses the minimum number of bins. We also abuse the notation to let OPT also denote the number of bins used the the optimum solution.

We consider two measures of approximation. In the first, we measure the approximation ratio by the number of bins used by our algorithm against that used by the optimum. We will denote this as "approximation with respect to the number of bins".

For the second measure, we define an assignment of objects to bin B_i to be c-feasible if $w_{j_1} + \cdots + w_{j_{n_i}} \leq c \min\{f_{j_1}, f_{j_2}, \ldots, f_{j_{n_i}}\}$. That is, the bin is allowed to deviate from its fragility constraint by a factor of up to c. An assignment is c-feasible if each bin is c-feasible. According to this measure, our algorithm has an approximation ratio of c if it finds an assignment which uses at most OPT bins, but only guarantees that each bin is c-feasible (as opposed to being 1-feasible as in the optimum solution). We will denote this as "approximation with respect to fragility".

3. Approximation with Respect to the Number of Bins

3.1. Algorithm

Before we describe the algorithm, we first fix some notation and labeling conventions.

We will label the users according to the non-decreasing order of their fragility values, thus $f_n \geq f_{n-1} \geq \ldots \geq f_2 \geq f_1$. We will say that i is to the left of j if $f_i > f_j$, in the above ordering. Thus 1 is the rightmost object. Consider an assignment A of objects into bins. For a bin B, let $r(B)$ denote the index of the object with the least fragility assigned to bin B. Suppose A uses m bins, we will label the bins $B_1, \ldots, B_m$ such that $r(B_1) < r(B_2) < \ldots < r(B_m)$. Thus, the bin containing user 1 is bin 1, the bin containing the rightmost object other than the ones in bin 1 will be denoted by bin 2 and so on.

To avoid trivialities, we will assume that $w_i \leq f_i$ for each object i. Observe that if $w_i > f_i$ for some i, then the object cannot be placed in any bin, hence there is no feasible solution.

Consider the following greedy algorithm:

1 Sort and label all the objects according to non-decreasing f_i, i.e. $f_n \geq \ldots \geq f_1$.

2 Initialize $i \leftarrow 1$, $j \leftarrow 1$ and $w \leftarrow 0$, $f \leftarrow f_i$.

3 While $i \leq n$
  If $B_j \cup \{i\}$ is feasible $(w + w_i \leq \min\{f, f_i\})$
    $B_j \leftarrow B_j \cup \{i\}$, $w \leftarrow w + w_i$, $f \leftarrow \min\{f, f_i\}$
  Else
    $j \leftarrow j + 1$
    $B_j \leftarrow \{i\}$, $w \leftarrow w_i$, $f \leftarrow f_i$
  $i \leftarrow i + 1$

Observe that the algorithm above fills the bins greedily starting from object 1. It adds objects to a bin, until adding the next object makes the bin infeasible. At this point, the algorithm creates a new bin and continues to add objects into it.

We will say that a solution is *banded* if each bin consists only of consecutive objects. Formally, if $i > j$, the for any k, l such that $k \in B_i$ and $l \in B_j$, we have that $k > l$. It is easy to see that the greedy algorithm above produces the optimum banded solution, and requires time $O(n \log n)$.

We will show that the greedy algorithm above requires at most twice the number of bins required by the optimum algorithm. The idea of the proof is that, given an optimum solution using OPT bins, we first produce a *fractional banded solution* (defined below) which consists of OPT bins. We then round this fractional solution to produce a banded solution with at most $2 \cdot OPT$ bins.

3.2. Worst-Case Analysis

Consider an optimum assignment. We will abuse notation and let $B_1, \ldots, B_{OPT}$ denote the sets of objects assigned to bins $B_1, \ldots, B_{OPT}$ respectively. Define W_i to be the sum of all the weights of objects in B_i, thus $W_i = \sum_{j \in B_i} w_j$. From the optimum solution construct a fractional banded solution as follows:

Greedily assign objects, possibly fractionally, to bins B'_i such that sum of the weights of objects in bin B'_i is exactly W_i. That is, start from the rightmost object, put it in bin B'_1. Let k be such that $w_1 + \ldots + w_k \leq W_1$ and $w_1 + \ldots + w_{k+1} > W_1$. In this case, bin B'_1 will comprise of objects $1, 2, ..., k$ and a fraction x of the object $k+1$ such that $w_1 + \ldots + w_k + xw_{k+1} = W_1$. Continue packing the remaining objects in bin B'_2 and so on.

Informally, if we think of the objects laid out (in the order of their fragility) along the line as segments of length equal to their respective weight, then what we are doing is marking intervals of length equal to the weight of each bin of the optimal solution. In the fractional banded solution, these intervals are interpreted as the bins, and each object is put into the bin corresponding to the interval in which it falls, fractionally if it overlaps two of the intervals.

We now define some notation to make this precise. Let l'_k and r'_k be the index of the leftmost and the rightmost object having a non-zero fraction in B'_k. Let $x_{l'_k}$ and $x_{r'_k}$ denote the fraction of l'_k and r'_k in B'_k, hence $0 \leq x_{l'_k}, x_{r'_k} \leq 1$. Then, $\forall k$, $r'_k \leq l'_{k-1}$ and $x_{l'_k} w_{l'_k} + \sum_{i, l'_k < i < r'_k} w_i + x_{r'_k} w_{r'_k} = W_k$

Note that B'_k completely contains all objects i such that $l'_k < i < r'_k$. In the case when $l'_k = r'_k$, then B'_k contains at most one object, moreover, since $w_i \leq f_i$ for all i, we know that B'_k contains exactly one object and $x_{l'_k} = x_{r'_k} = 1$.

We now observe a few properties of this assignment.

Property 1 *The number of bins B' formed thus will be exactly OPT.*

Proof: This follows directly from the procedure to construct the fractional bins. ■

Property 2 *Let r'_i and r_i be the index of the object with the least fragility (hence rightmost) in bin B'_i and B_i respectively. Then $r_i \leq r'_i$.*

Proof: In OPT, all objects to the right of r_i will lie in bins B_j such that $j < i$. Hence the sum of the weights of the objects to the right of r_i will be at most $\sum_{j=1}^{i-1} W(j)$. Since the sum of the objects to the right of r'_i in the fractional solution is exactly $\sum_{j=1}^{i-1} W(j)$, if follows that the rightmost object of B'_i is to the left of the rightmost object of bin B_i in OPT. ∎

We will now convert this fractional banded solution into a banded solution which uses at most $2 \cdot OPT$ bins.

Rounding Procedure:

1 If an object i has a fraction strictly less than 1 in some bin B'_j, then remove i from all bins B'_j, which contain some fraction of i. Put object i into a new bin by itself. Call these bins of *type H*.

2 Repeat Step 1 until all the bins B'_k contain a 0 or 1 fraction of every object. Now, each B'_k consists of objects in the range $[l''_k, \ldots, r''_k]$ Denote these new bins by B''_k.

Clearly the number of bins B''_k will be at most OPT. Also, each bin of type H is feasible since it contains exactly one object. The next lemma shows that the number of bins of type H is also bounded, and all the bins B'' produced are feasible.

Lemma 3 *The above rounding procedure produces a* banded *assignment of objects to bins, the number of bins used is at most $2 \cdot OPT$ and each bin is feasible.*

Proof: We first bound the number of bins of type H. For any bin B'_j at most one object has the property that it lies in both B'_j and B'_{j+1}. Since the number of such bins B' is at most OPT, it follows that the number of objects covered fractionally is at most OPT.

We now show that the bins B''_k are feasible. The sum of the weights of the objects in bin B''_k is clearly no more than that of B'_k and hence less than W_k. Also, the index of the rightmost object of B''_k is at least as large as that of B'_k, which by Property 2 is at least as large as that of B_k. Thus, as the fragility of the least fragile object of B''_k is not less than that of B_k and the weight of objects in B''_k is not greater than that of B_k, it follows that B''_k is feasible. Thus the result follows. ∎

Since the greedy algorithm produces the optimum banded solution, Lemma 3 gives,

Theorem 1 *The number of bins used by the greedy algorithm is at most twice the optimum, and it runs in time $O(n \log n)$.*

3.3. Lower Bound

In this section we show that no algorithm can achieve an approximation ratio of less than $\frac{3}{2}$ unless $P = NP$. This lower bound holds even in the asymptotic case, unlike the traditional bin packing which has a PTAS in the asymptotic case.

We reduce the partition problem to our packing problem. Given a set A with each object having a size $s(a) \in Z^+$, the partition problem is to decide if $\exists A' \subset A$, such that $\sum_{a \in A'} s(a) = \sum_{a \in A-A'} s(a)$ [Garey and Johnson, 1979]. We can easily transform this into our problem. Let s be the sum of objects in A. We form an instance of our problem, where we have n objects with weights $s(a)$ and all fragilities set to $s/2$.

Clearly, there exists a solution to the partition problem iff the objects can be packed in 2 bins. To show a lower bound for the asymptotic case, we construct m copies, $A_1, \ldots, A_m$ of the set A. The set A_i is obtained from A by scaling both the weight and the fragility of the objects by s^{i-1} (i.e. in A_i the weight of object a is $s^{i-1}s(a)$ and the fragility is $s/2 \cdot s^{i-1}$).

We solve the bin packing problem on the instance $A_1 \cup \ldots \cup A_m$. Observe that, if $i \neq j$ no two objects from A_i and A_j can lie in the same bin, because one of the objects will always have more weight than the fragility of the other object. Thus, if we obtain a solution which uses less than $3m$ bins we know that the partition problem has a solution. Hence the bin packing problem has an optimum solution which either uses only $2m$ bins or at least $3m$ depending on whether the partition problem has a solution. Thus we have shown that

Theorem 2 *The number of bins used the bin packing problem cannot be approximated to within a factor of $\frac{3}{2}$ by a polynomial time algorithm, unless $P = NP$.*

4. Approximation with Respect to Fragility

In this section we will construct a solution where the number of bins used is OPT, but each of the bins is only guaranteed to be 2-feasible rather than 1-feasible as in the optimum solution.

Given some optimum solution OPT, we consider the fractional solution using OPT bins as in Section 3. However we modify our rounding procedure.

Rounding Procedure:

1 Given a bin B_i' in the fractional solution, form a new bin B_i'' by including the rightmost object of B_i' completely in B_i''.

2 If $x_{l_i'} < 1$, (that is if the leftmost object of B_i' occurs fractionally in B_i') then remove it from B_i''.

Lemma 4 *The rounding procedure gives an assignment of objects to bins which is banded, has number of bins at most OPT, and each bin is 2-feasible.*

Proof: In the rounding procedure, observe that the rightmost object of the bin B_i'' remains the same as the rightmost object of B_i'. Moreover, if B_i' contains an object which is not contained by B_i'', we know that the object is l_i' and $x_{l_i'} < 1$. Hence $x_{r_{i+1}'} < 1$ and l_i' will be included in B_{i+1}''. Thus every object (except possibly n) lies in some B_i'' for some i. To see that n also lies in B_{OPT}'', observe that $l_n' = 1$, since $w_n \leq f_n$. Hence the number of bins B_i'' is exactly OPT, and they contain all the objects.

We now show that each bin B_i'' is 2-feasible. Consider the bin B_i'', we know that its rightmost object is r_i', and by Property (2), $r_i' \geq r_i$. Moreover we know that the sum of weights of objects in B_i'' is at most $W_i + w_{r_i}$.

Now since bin B_i was feasible, $W_i \leq f_{r_i}$. As $w_{r_i} \leq f_{r_i}$ it follows that $W_{r_i} + w_{r_i} \leq 2f_{r_i} \leq 2f_{r_i'}$. ∎

Thus there exists a banded solution where the bins are 2-feasible and the number of bins used is at most OPT. So our algorithm will be as follows.

1 Sort and label all the objects according to non-decreasing f_i, i.e. $f_n \geq \ldots \geq f_1$.

2 Initialize $i \leftarrow 1$, $j \leftarrow 1$ and $w \leftarrow 0$, $f \leftarrow f_i$.

3 While $i \leq n$
    If $B_j \cup \{i\}$ is 2-feasible (i.e. $w + w_i \leq 2\min\{f, f_i\}$)
        $B_j \leftarrow B_j \cup \{i\}$, $w \leftarrow w + w_i$, $f \leftarrow \min\{f, f_i\}$
    Else
        $j \leftarrow j + 1$
        $B_j \leftarrow \{i\}$, $w \leftarrow w_i$, $f \leftarrow f_i$
    $i \leftarrow i + 1$

The above algorithm produces an optimal banded solution which is 2-feasible. It follows from Lemma 4 that

Theorem 3 *The algorithm produces a solution where each bin is 2-feasible and the number of bins used is at most that used by the optimum 1-feasible solution. Moreover the algorithm runs in time $O(n \log n)$.*

5. Conclusion

In this paper we considered a variant of the bin packing problem arising in frequency allocation of cellular networks. We showed that two simple heuristics for assigning objects to bins based on their fragility are provably good. We believe that the results can be extended in various directions along the lines of other results for the traditional bin-packing problem. For example, an average case analysis for the case when the weights and fragilities come from some distribution would be interesting. It would also be interesting to study the online case, and the case when the objects are present for a temporary duration (this would correspond to users entering and leaving the cellular network). Also the problem of d-dimensional bin packing with fragile objects would be interesting.

Acknowledgments

The authors are grateful to Professors Avrim Blum and Volker Diekert for useful comments which helped improving the presentation of the paper.

References

[Coffman et al., 1997] Coffman, E., Garey, M., and Johnson, D. (1997). Approximation algorithms for bin packing: a survey. In *D. S. Hochbaum, ed. Approximation Algorithms for NP-hard Problems*, pages 46–93. PWS Publishing, Boston.

[Coffman and Lueker, 1991] Coffman, E. G. and Lueker, G. S. (1991). *Probabilistic Analysis of Packing and Partitioning Algorithms.* Wiley-Interscience.

[de la Vega and Lueker, 1981] de la Vega, W. F. and Lueker, V. (1981). Bin packing can be solved within $1 + \epsilon$ in linear time. *Combinatorica*, pages 349–355.

[Feeney and Nilsson, 2001] Feeney, L. and Nilsson, M. (2001). Investigating the energy consumption of a wireless network interface in an ad hoc networking environment. In *Proceedings of IEEE INFOCOM.*

[Garey and Johnson, 1979] Garey, M. and Johnson, D. (1979). *Computers and Intractability: A Guide to the Theory of NP Completeness.* W.H. Freeman and Company, New York.

[Garey et al., 1972] Garey, M. R., Graham, R. L., and Ullman, J. D. (1972). Worst-case analysis of memory allocation algorithms. In *ACM Symposium on Theory of Computing*, pages 143–150.

[Hanly and Tse, 1999] Hanly, S. and Tse, D. (1999). Power control and capacity of spread-spectrum wireless networks. *Automatica*, 35(12):1987–2012.

[Karmarkar and Karp, 1982] Karmarkar, N. and Karp, R. M. (1982). An efficient approximation scheme for the one–dimensional bin–packing problem. In *Proc. 23rd Ann. Symp. on Foundations of Computer Science.*

[Shor, 1984] Shor, P. W. (1984). The average-case analysis of some on-line algorithms for bin packing. In *IEEE Symposium on Foundations of Computer Science*, pages 193–200.

[van Vliet, 1996] van Vliet, A. (1996). On the asymptotic worst case behavior of harmonic fit. *J. Algorithms*, 20:113–136.

[Viterbi, 1995] Viterbi, A. (1995). *CDMA - Principles of Spread Spectrum Communication.* Addison–Wesley.

[Yao, 1980] Yao, A. C.-C. (1980). New algorithms for bin packing. *Journal of the ACM*, 27:207–227.

[Yates, 1995] Yates, R. (1995). A framework for uplink power control in cellular radio systems. *IEEE Jour. Sel. Areas Communication*, 13(7):1341–1348.

[Yates and Huang, 1995] Yates, R. and Huang, C. (1995). Integrated power control and base station assignment. *IEEE Transactions on Vehicular Technology*, 44(3):638–644.

[Zander, 1993] Zander, J. (1993). Transmitter power control for co-channel interference management in cellular radio systems. In *Proc. WINLAB Workshop.*

LOWER AND UPPER BOUNDS FOR TRACKING MOBILE USERS

Sergei Bespamyatnikh
Department of Computer Science, Duke University
Box 90129, Durham, NC 27708, USA
besp@cs.duke.edu

Binay Bhattacharya
School of Computing Science, Simon Fraser University
Burnaby, B.C. Canada, V5A 1S6
binay@cs.sfu.ca

David Kirkpatrick
Department of Computer Science, University of British Columbia
Vancouver, B.C. Canada V6T 1Z4
kirk@cs.ubc.ca

Michael Segal
Communication Systems Engineering Department, Ben-Gurion University of the Negev
Beer-Sheva 84105, Israel
segal@cse.bgu.ac.il

Abstract In this paper we prove improved lower and upper bounds for the location of mobile facilities (in L_∞ and L_2 metrics) under the motion of clients when facility moves faster than clients. This paper continues the research started in our joint paper where we present lower bounds and efficient algorithms for exact and approximate maintenance of the 1-center for a set of moving points in the plane. Our algorithms are based on the *kinetic* framework introduced by Basch, Guibas and Hershberger.

1. Introduction

The goal of this paper is to study problems related to the location of mobile facilities serving a set of customers. The notion of mobile servers has been well studied in connection with stationary but transitory customer sets [24]. Here we deal with the problems related to the location of mobile facilities, where customers are modelled by points moving continuously in d-dimensional space, $d \geq 1$. Specific example of such problems is the maintenance of the 1-center for moving points under the L_p metric. This has applications, for example in mobile wireless communication networks when the broadcast range should contain all the customers so as to provide service to the cellular phones. There are likely to be many applications of mobile facilities, particularly with respect to tactical mobile packet-radio networks. Servers such as battlefield map servers, or even servers whose sole purpose is to provide support for network control (such as those that maintain entity-to-address information) could be considered to be mobile facilities. Another application of these problems is locating a welding robot in an automobile manufacturing plant. Both problems are interesting from both theoretical and practical points of view.

Facility location is a classical problem of operations research that has also been examined in the computational geometry community. Most of the problems described in the facility location literature are concerned with finding a "desirable" facility location: the goal is to *minimize* a distance function between the facility (e.g., a service) and the sites (e.g., the customers). As was pointed out in [6], the data structures and algorithms that have been developed for the *static* problems (i.e., customers are not moving) are not directly applicable to the setting of moving points when the motion of the facilities must satisfy natural constraints. In this paper we continue the study of mobile version of the classical facility location problem:

- **k-center:** Given a set S of n demand points in d-dimensional space ($d \geq 1$), find a set P of k supply points so that the *radius* defined as the maximum L_p distance between a demand point and its nearest supply point in P is minimized. If P is a subset of S then P called a set of *discrete* centers. Note that for some metrics such a set P is not necessarily unique even for $k = 1$ and $d = 2$.

This problem has been well studied in both the exact [2, 7, 8, 9, 10, 11, 13, 21, 25] and approximate [10, 12, 18] versions. In approximate versions a set P provides $(1 + \varepsilon)$-approximate k-center if the associated radius is at most $(1 + \varepsilon)$ times the optimal radius, for any $\varepsilon > 0$. Facility location problems for time varying networks (when edge distances satisfy the triangle inequality) also have been studied, see [16, 19]. Let us define the *mobile k-center* problem as follows. Given a set $S = \{p_1, p_2, \ldots, p_n\}$ of n continuously moving points specified by a piecewise differentiable functions $\{g_1, g_2, \ldots, g_n\}$, where $g_i, 1 \leq i \leq n$ maps time interval $[0, T]$ to $\mathbb{R}^d$, we want to determine whether there exist k continuous functions $f_1, \ldots f_k$; $f_i : [0, T] \to \mathbb{R}^d$ such that at any given moment $t \in [0, T]$, the points $f_1(t), \ldots, f_k(t)$ form a k-center for the points at locations $g_1(t), \ldots, g_n(t)$, and, if so, find $f_1, \ldots, f_k$. In this paper we concentrate mainly on the instances of the mobile 1-center problem, when $p = \{1, 2, \infty\}$ and assume that $d = 2$ if the dimension is not specified. We present improved lower and upper bounds for mobile rectilinear and Euclidean 1-center when facility is allowed to move faster than the clients.

The mobile approximate 1-center and related problems are studied by Agarwal and Har-Peled [1] very recently. For the Euclidean 1-center in the plane, they presented an approximation algorithm based on a notion of *extent*. Their algorithm aims to maintain an approximate 1-center with a few events only. However, the velocity of the facility can be arbitrary large (Lemma 4). We describe improved strategies that guarantee both an approximation factor and a bounded velocity of the facility (Theorem 5 and Lemma 7).

Gao et al. [14] proposed a randomized algorithm for maintaining a set of clusters among moving points in the plane. They presented algorithms with expected approximation factor (on the optimal number of centers) of $c \log n$ for intervals and of $c\sqrt{n} \log n$ for squares. The probability that there are more than expected number of centers is $O(1/n^{\Theta(c^2)})$ for the case of intervals and $O(1/n^{c^2 \log n})$ for the case of squares. An extension of this basic algorithm led to a hierarchical algorithm, also presented in [14], based on kinetic data structures [5] with expected constant approximation factor on the number of discrete centers, where the approximation factor also depends linearly on the constant c. The dependency of the approximation factor and the probability that the algorithm chooses more than a constant times the optimal number of centers is similar to that of the non-hierarchical algorithm for the squares case. The constant c, which has not been explicitly determined in [14], can be shown to be very large. Their algorithm has an expected update time of $O(\log^{3.6} n)$, the number of levels used in the obtained hierarchy is $O(\log \log n)$ with $O(n \log n \log \log n)$ space. Har-Peled [17] found a scheme of determining centers in advance, i.e. if in the optimal solution the number of centers is k and r is the optimal radius for the points moving with the degree of motion l, then his scheme guarantees 2^{l+1}-approximation of the radius with k^{l+1} centers chosen from the set of input points before the points start to move. Huang et al. [20] present fully distributed (decentralized) constant-factor approximation algorithms for k-center problem in the L_1- and L_∞ norms for any d-dimensional space $\mathbb{R}^d$, $d \geq 1$, and for the L_2 norm in $\mathbb{R}^2$ and $\mathbb{R}^3$, which either match or improve the best known constant approximation factors of a centralized algorithm, or have (asymptotically) optimal update costs improving [14].

Our algorithms for mobile 1-center problem are based on a relatively new class of data structures (used also in [6, 14]) — *kinetic data structures (KDS)* — aimed at keeping track of attributes of interest in systems of moving objects [5, 15]. In the kinetic setting, a set of points is assumed to be continuously changing, or moving. Each point follows a posted *flight plan*, but a plan can change at any moment through a *flight plan update*. A KDS maintains a *configuration function* of interest (which in our case will be the set P) by watching for critical events as the objects move. A KDS for computing a particular attribute for a set of points in motion maintains a set of *certificates*. A certificate based on a tuple of points is a continuous function that associates a real number with each configuration of these points. For example, the certificates for a convex hull KDS are a collection of the triple of points, each with a particular orientation. At any one time, the conjunction of all the certificates being maintained by the kinetic data structure proves the combinatorial correctness of its output. As the points move, some certificates may become invalid, e.g., a triple of points changes orientation. When a certificate fails, the proof structure needs to be modified and the combinatorial description of the configuration function may need to be updated. Certificates are stored in a priority queue, ordered by failure time, and processed in order as they fail. A good kinetic data structure will take advantage of the continuity of the point motions to select certificate structures that are easy to update at these critical events; a structure satisfying this condition is called *responsive*. Other criteria for a KDS are *efficiency*, i.e., the number of events processed by KDS is not much greater than the number of combinatorial changes in the configuration function itself, *compactness*, i.e., the number of active certificates at any one time is roughly linear in the complexity of the moving points, and *locality*, i.e., a flight plan update for any one point affects only a small number of certificates.

2. Mobile 1-center problems

In this section we present new lower and upper bounds for mobile 1-center problems under the L_∞ and L_2 metrics for facility moving faster than the clients and also extend a lower bound proof for unit velocity of the facility presented in [6] to work in any metric and any dimension.

2.1. Rectilinear 1-center

We consider the case when the velocity of the facility belongs to the range $[1, \sqrt{2}]$, define the *mixing strategy* and analyze it. Then we show the lower bound on the approximation factor for this case. At the end of this section we show an extended lower bound presented in [6] for facility moving with velocity 1 that works for any dimension and any metric.

Motion with bounded velocity. Suppose now, that the velocity of the facility is bounded by $v_{max} \in [1, \sqrt{2}]$. We notice that for values of $v_{max} = 1$ and $\sqrt{2}$ the best approximation factors are 2 and 1, respectively (see [6]). We *mix* our two strategies — the center of mass of the points and the center of the bounding box — in the following way. Let (f_1, v_1) be the location (vector in the plane) and velocity of the center of mass of the points. Similarly, (f_2, v_2) are the location and velocity of the center of the bounding box. (Bespamyatnikh et al. [6] explain how to maintain the center of bounding box and the center of mass of points using kinetic data structures.) The mixing strategy is to maintain the *mixing center* (f, v) defined as $(\alpha f_1 + (1-\alpha)f_2, \alpha v_1 + (1-\alpha)v_2)$, where $\alpha = (\sqrt{2} - v_{max})(\sqrt{2}-1)$. We analyze the mixing strategy and show how it can be improved in the following theorem.

Theorem 1 (Bounded Velocity) *Suppose that the facility is allowed to move with velocity at most* $v_{max} \in [1, \sqrt{2}]$. *Let* $a_1(v)$ *be the linear function defined by* $a_1(1) = 2$ *and* $a_1(v') = a'$ *where* $v' = \sqrt{2}\cos(\pi/8)$ *and* $a' = 1.25$. *Let* $a_2(v)$ *be the linear function defined by* $a_2(v') = a'$ *and* $a_2(\sqrt{2}) = 1$. *There is a strategy of the facility that guarantees an approximation factor* $max(a_1(v_{max}), a_2(v_{max}))$.

Proof. First we show that the velocity of the mixing center is bounded by v_{max}. We want to prove the inequality $\alpha v_1 + (1-\alpha)v_2 \leq v_{max}$. In order to do this, we replace v_1 by 1 and v_2 by $\sqrt{2}$. It can easily be seen that the value of the expression $\alpha + (1-\alpha)\sqrt{2}$ when $\alpha = (\sqrt{2} - v_{max})(\sqrt{2}-1)$ is equal to v_{max}.

Regarding the approximation factor we consider the case in which the center of mass is at largest L_∞ distance from the center of the bounding box. This distance is $1 - 2/n$ if the radius that is determined by the center of the bounding box is 1. Then the L_∞ distance between the mixing center and the center of the bounding box is $\frac{\sqrt{2}-v_{max}}{\sqrt{2}-1}(1-\frac{2}{n})$. The approximation factor bound is

$$1 + \frac{\sqrt{2} - v_{max}}{\sqrt{2}-1}\left(1 - \frac{2}{n}\right).$$

The smallest bound that suits all n is the linear function $1 + (\sqrt{2} - v_{max})(\sqrt{2}-1)$.

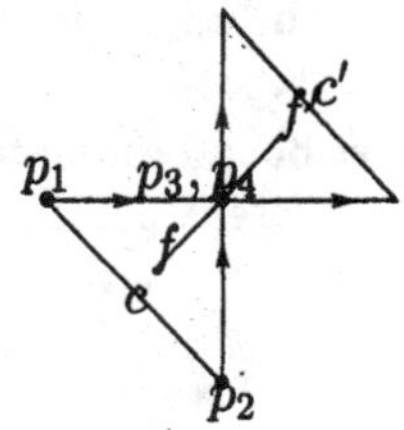

Figure 1. Movement of the center.

It turns out that the mixing strategy is not easy to improve. The reason is that the worst case velocities of two strategies, the center of mass (of all the points or a linear combination of a few points from S) and the center of bounding box, can be in the same direction as shown in Figure 1.

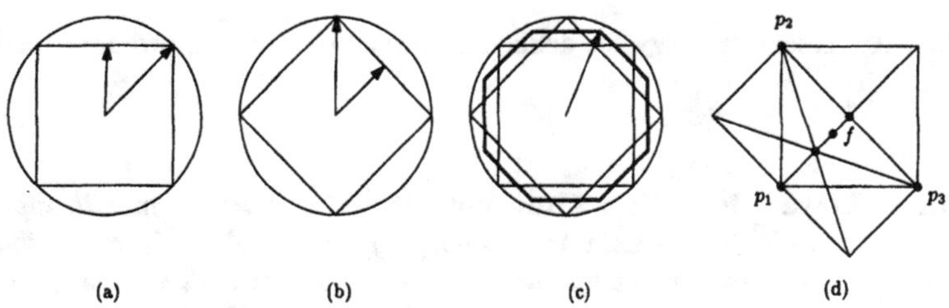

Figure 2. 8-gonstrategy.

In this case the best motion of the facility is to move in this direction within the maximum allowed velocity which corresponds to mixing strategy. The idea now is to interpolate two strategies that have the worst case velocities in different directions. Note that the maximum velocity of the center of bounding box depends on the direction and varies from 1 to $\sqrt{2}$, see Fig. 2(a). A good candidate for the interpolation is the strategy of the center of bounding box aligned to the coordinate axis rotated by 45°. The range of the maximum velocity is depicted in Fig. 2(b). Consider the midpoint of two centers. We call this *8-gon strategy* since the range of its maximum velocity forms the regular 8-gon, see Fig. 2(c). The maximum velocity of the facility following the 8-gon strategy is $v' = \sqrt{2}\cos(\pi/8)$ and the approximation in worst case is $a' = 1.25$ for three points $S = \{p_1, p_2, p_3\}$, see Fig. 2(d). One can show that the approximation factor of the 8-gon strategy is smaller than one of the mixing strategy for $v_{max} = v'$. In order to improve the approximation for other values of v_{max} we

(i) interpolate (or mix) the strategy of the center of mass and the 8-gon strategy for v_{max} in the range $[1, v']$, and

(ii) interpolate the strategy of the center of the bounding box and the 8-gon strategy for $v_{max} \in [v', \sqrt{2}]$.

The theorem follows. ∎

Lower bound for an approximation factor when $v_{max} \in [1, \sqrt{2}]$. In fact we can show two different lower bound estimates. The first bound is based on the example depicted in Figure 1 when $|ff'| = v_{max}$ and $|p_1p_4| = |p_2p_4| = 1$. In this case the exact rectilinear radius is $\frac{1}{2}$ and the approximate radius is $1 - \frac{v_{max}}{2\sqrt{2}}$. Thus, the approximation factor $\beta(v_{max}) = 2 - \frac{v_{max}}{\sqrt{2}}$.

In order to prove the second lower bound estimate, we first consider the case when $v_{max} = 1$. The original proof for this case is presented in [6]. We extend the proof to work for any dimension and any metric.

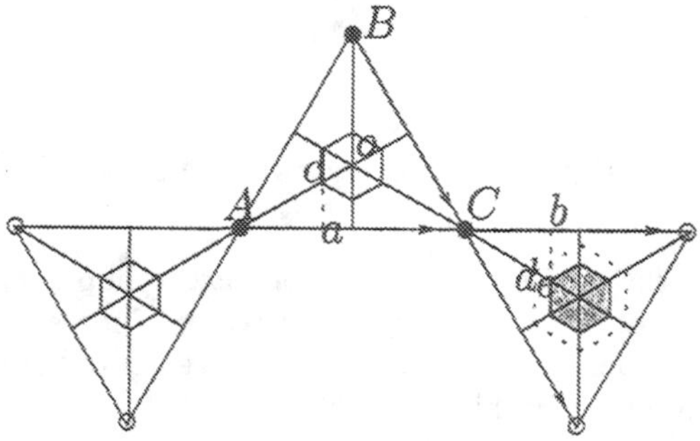

Figure 3. Adversary strategy for achieving lower bound for approximation factor.

Theorem 2 (Lower Bound for Unit Velocity) *Consider the mobile approximate 1-center problem in $\mathbb{R}^d, d \geq 2$ with any underlying metric. Any algorithm that starts with the facility inside the convex hull and moves the facility with at most unit velocity achieves an approximation factor of at least 2 (asymptotically) for the exact 1-center in the worst case.*

Due to lack of space the proof is omitted.

Now we are ready to prove the second lower bound estimate for an approximation factor when $v_{max} \in [1, \sqrt{2}]$. It is based on the proof of the theorem above. In fact, a similar proof can be applied for $v_{max} \in [1, \frac{2}{\sqrt{3}}]$. If $v_{max} \geq \frac{2}{\sqrt{3}}$ the facility can follow the center o of the triangle. For $v_{max} \in [1, \frac{2}{\sqrt{3}}]$ the facility can always be inside a hexagon of some fixed size. Let $x = |Aa|$ and $|ce| = 2v$ on Figure 3. One can show that $x = \sqrt{3v_{max}^2 - 3}$. The rectilinear distance from c to B is $\sqrt{3} - \frac{x}{\sqrt{3}}$ and the distance from c to C is $2 - x$. The exact radius is equal to 1. Therefore, the approximation factor is $\gamma(v_{max}) = \max(2 - x, \sqrt{3} - \frac{x}{\sqrt{3}})$, for $x = \sqrt{3v_{max}^2 - 3}$. One can show that $\beta(v_{max}) > \gamma(v_{max})$ if $v_{max} > 2/\sqrt{3}$. Combining all the results together we obtain

Lemma 3 *If the facility's maximal allowed velocity is $v_{max} \in [1, \sqrt{2}]$, then the worst-case approximation factor of* any *algorithm is at least* $\max(\beta(v_{max}), \gamma(v_{max}))$.

2.2. Euclidean 1-center

Bespamyatnikh et al. [6] provided an example with three points showing that the Euclidean 1-center may move with unbounded velocity. See Figure 5(a). We found another example with four points (see Figure 5(b)) that can be used for the same purpose and at the same time provides a better bound of the velocity of the exact Euclidean 1-center.

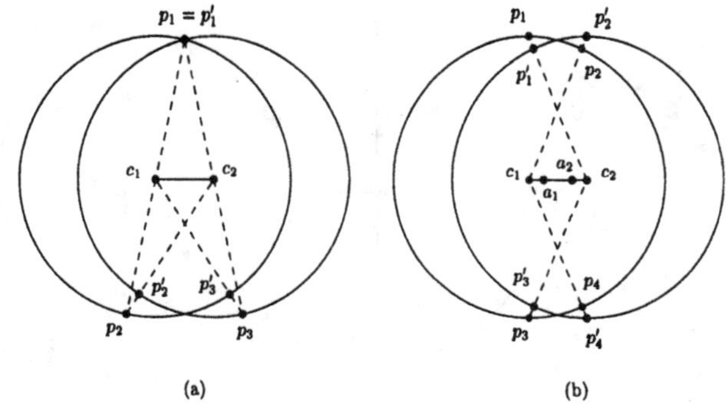

Figure 4. Euclidean 1-center may move with unbounded velocity.

To see this we observe that points $p_1, \ldots, p_4$ are located on the unit circle centered at c_1. A point $p_i, i = 1, \ldots, 4$ moves toward the point p_i'. The points p_i make the same length paths since $c_1c_2p_2'p_1$ and $c_1c_2p_4'p_3$ are rectangles. Let $x = |c_1c_2|$ be the length of the path made by the 1-center and let $y = |p_1p_1'|$ be the length of the path made by the point p_1. It suffices to show that y/x tends to 0 if x tends to 0. Indeed, $1 + x^2 = (1 + y)^2$ since the triangle $c_1c_2p_1$ is right. It implies $x^2 = 2y + y^2$ and $2y/x = x - y^2/x \leq x$.

Next, we show that velocity of an approximate 1-center must be high.

Lemma 4 (Lower Bound) *For every $\varepsilon > 0$, any $(1+\varepsilon)$-approximate mobile Euclidean 1-center has velocity at least $1/4(\sqrt{2\varepsilon} + \varepsilon) = 1/(4\sqrt{2\varepsilon}) - O(1)$ in worst case.*

Due to lack of space the proof is omitted.

We show that the lower bound of the velocity needed to approximate mobile Euclidean 1-center established in Lemma 4 is optimal up to a constant factor.

Theorem 5 (Upper Bound) *For any $\varepsilon > 0$ there is a strategy for mobile approximate Euclidean 1-center that guarantees the approximation factor $1 + \varepsilon$ using velocity of the facility $4/\sqrt{\varepsilon} + o(1/\sqrt{\varepsilon})$ in the worst case.*

Proof. We apply the following strategy for the mobile facility. Let V be the maximum velocity of the facility that will be specified later. Consider the initial configuration. We assume that in the beginning the facility f is located so that the radius of the smallest circle enclosing all the customer points is at most $(1 + \delta)$ times the optimal radius where $\delta \leq \varepsilon$ and will be specified later. The goal of the facility now is to reach the position of the exact Euclidean 1-center at this moment. The facility heads to the target using the velocity V. We can compute the time t needed for this. After reaching the target we repeat the procedure. It is interesting that the motion of the facility is independent on the motion of customers during the time when facility moves from one position to another. It is also interesting that the exact 1-center can achieve any velocity sometimes and the approximate facility just ignores any acceleration of the exact 1-center.

Let c be the position of the exact Euclidean 1-center at the initial time t_0 and let a be the position of the facility (the approximate 1-center) at the same time. Let t_1 be the time when the facility reaches c. Let r denote the radius associated with c at the time t_0. Without loss of generality we can assume that $r = 1$.

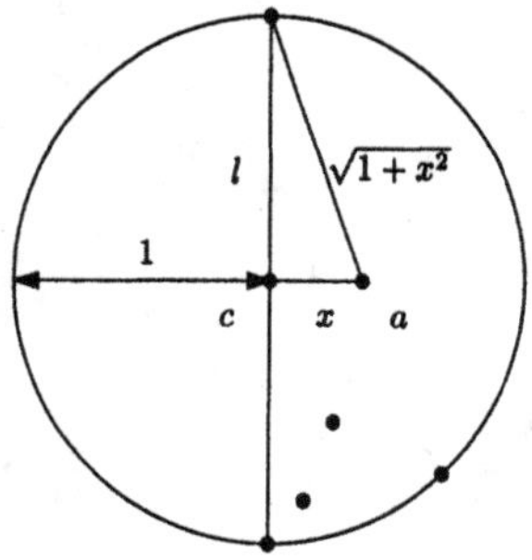

Figure 5. $(1+\varepsilon)$-approximation of Euclidean 1-center.

First we estimate the distance between the exact and approximate centers at the time t_0. Denote $x = |ac|$. The line l orthogonal to the line ac passing through c partition the unit circle centered at c into 2 arcs, see Fig. 5. Each arc contains at least one customer point since the exact radius of the 1-center disk is 1. The minimum distance from a to the arc beyond l is $\sqrt{1+x^2}$. Clearly, $\sqrt{1+x^2} \le 1+\delta$. Therefore $x = |ac| \le \sqrt{2\delta+\delta^2}$.

Let t be the time the facility need to get the exact center c, i.e., $t \le |ac|/V$. Consider the moment t_1. The radius associated with the approximate 1-center c is at most $1+t$ since the velocities of the customers are bounded by 1. Let c' be the location of the exact 1-center. We show that the associated radius is at least $1-t$. Construct the line passing through c and orthogonal to cc'. The arc of the unit circle centered at c from which the exact center moves toward c' contains at least one customer point at the moment t_0. This customer is at distance at least $1-t$ from c' at the moment t_1. Thus, the approximation ratio at the moment t_1 is at least

$$\frac{1+t}{1-t} = 1 + \frac{2t}{1-t}.$$

We want to show that this is at most $1+\delta$. It follows from

$$\frac{2t}{1-t} \le \delta \text{ or, equivalently, } \quad t \le \frac{\delta}{2+\delta}. \tag{1}$$

We also want the approximation factor to be at most $1+\varepsilon$ for any moment $t_0+\Delta \in [t_0, t_1]$. The radius associated with the approximate 1-center is at most $1+\delta+\Delta$. The radius is at least $1-\Delta$. Their ratio is at most

$$\frac{1+\delta+\Delta}{1-\Delta} = 1 + \frac{2\Delta+\delta}{1-\Delta}.$$

The largest value of the upper bound is achieved when $\delta = t$. It suffices to have

$$\frac{2t+\delta}{1-\delta} \le \varepsilon \text{ or, equivalently, } \quad t \le \frac{\varepsilon-\delta}{2+\varepsilon}. \tag{2}$$

Making the right sides of the equations 1 and 2 we obtain $\delta^2 + 4\delta - 2\varepsilon = 0$ and $\delta = \sqrt{4+2\varepsilon} - 2 \approx \varepsilon/2$. By the equation 1 we can set $t = \delta/(2+\delta) \approx \varepsilon/4$. Therefore, the velocity of the facility can be $V = \sqrt{2\delta+\delta^2}/t \approx 4/\sqrt{\varepsilon}$. The theorem follows. ∎

We also can achieve reasonably good approximation factor for Euclidean 1-center when we trace the center of the bounding box of the points.

Theorem 6 (Bounding Box Strategy) *The strategy of tracing the center of the bounding box of $n \geq 3$ points has an approximation factor of $(1+\sqrt{2})/2$ for the mobile Euclidean 1-center and this bound is tight.*

Proof. Let $s = (1+\sqrt{2})/2$. Without less of generality we can assume that the bounding box is $B = [0,2] \times [0,2t]$ for some $t \geq 1$, see Fig. 6 (a). Let S be any configuration of $n \geq 3$ points in B such that each side of B contains at least one point of S. Let r be the approximate radius, i.e., $r_a = \max_{p \in S}\{|po|\}$ where $c_a = (1,t)$ is the center of B. Let r be the exact radius of 1-center. We want to prove that $r_a \leq sr$.

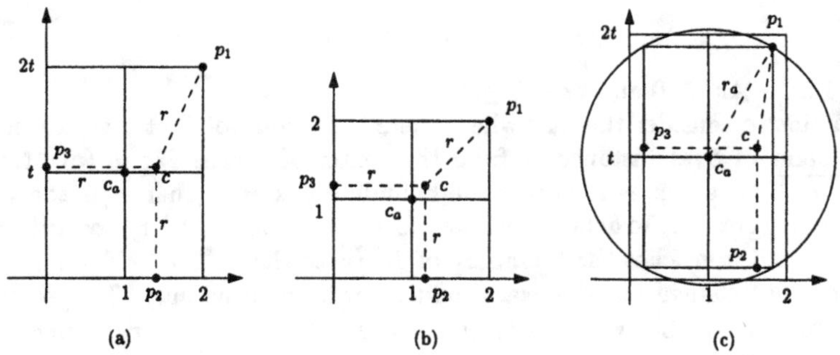

Figure 6. Bounding box strategy for approximating Euclidean 1-center.

We first consider the case where one of the points of S is located at a vertex of B, say $p_1 = (2, 2t)$. It implies $r_a = \sqrt{1+t^2}$. Two sides of B contain p_1, and there are at least two points p_2 and p_3 lying on the sides of B along x- and y-axis, respectively. We can assume that $p_2 \neq p_3$, otherwise $p_2 = p_3 = (0,0)$ and $r = r_a$. If $t > 2$ then $r \geq |p_1p_2| \geq t$ and one can check that

$$\frac{r_a}{r} \leq \frac{\sqrt{1+t^2}}{t} = \sqrt{\frac{1}{t^2}+1} \leq \sqrt{1\frac{1}{4}} < s.$$

We assume that $t \in [1,2]$. The smallest value of r is achieved when p_2 is strictly below the 1-center and p_3 is to the left of it, see Fig. 6 (a). Thus the exact 1-center has coordinates $c = (r,r)$. The 1-center is at distance r from p_1,

$$(2-r)^2 + (2t-r)^2 = r^2.$$

This can be transformed to the quadratic equation

$$r^2 - 4(t+1)r + 4(t^2+1) = 0.$$

$r = 2(t+1-\sqrt{2t})$ since the root with "+" is greater than $2t$ which is impossible.

r_a and r can be viewed as functions of t. In order to prove that $r_a/r \leq s$ for $t \in [1,2]$ we show that (i) $r_a(1)/r(1) = s$, and (ii) $(r_a/r)' \leq 0$ for any $t \in [1,2]$. If $t = 1$ then $r_a = \sqrt{2}$ and $r = 2(2-\sqrt{2})$. One can check that $r_a/r = s$. Note that this gives an example with tight bound, see Fig. 6 (b).

To prove the second condition we differentiate r and r_a

$$r' = 2(1 - 1/\sqrt{2t}) \quad \text{and} \quad r'_a = t/\sqrt{1+t^2}.$$

We want to prove that $r'_a r - r_a r' \leq 0$ or, equivalently,

$$\frac{2t}{\sqrt{1+t^2}}\left(t + 1 - \sqrt{2t}\right) \leq 2\sqrt{1+t^2}\left(1 - \frac{1}{\sqrt{2t}}\right).$$

Let $x = \sqrt{t/2}$. Then $t = 2x^2$ for $x \in [\sqrt{2}/2, 1]$ and the inequality can be written as

$$2x^2(2x^2 + 1 - 2x) \leq (1 + 4x^4)\left(1 - \frac{1}{2x}\right).$$

Multiplying by $2x$ we obtain

$$(12x^4 - 4x^3) + (2x - 1) \geq 0.$$

Clearly, $12x^4 - 4x^3 \geq 0$ and $2x - 1 \geq 0$.

It remains to consider the case where there is no point of S at a vertex of B. Let p_1 be a point of S at distance r_a from the center of B, see Fig. 6 (c). Note that $r_a < \sqrt{1+t^2}$. The idea is to shrink the bounding box B so that p_1 is the vertex of new bounding box B'. To achieve this we move every point of $S \setminus B'$ toward the exact 1-center c until it reaches the boundary of B' (note that $c \in B$). The approximate radius does not change but the exact radius can be reduced only. The exact radius can be bounded from below as described above and the inequality $r_a \leq sr$ holds. The theorem follows. ∎

Lemma 7 (Mixing Strategy) *If the facility is allowed to move with velocity* $v_{max} \in [1, \sqrt{2}]$, *the approximation factor*

$$\alpha\frac{1+\sqrt{2}}{2} + (1-\alpha)\left(2 - \frac{2}{n}\right), \quad \textit{where} \quad \alpha = \frac{\sqrt{2} - v_{max}}{\sqrt{2} - 1} \qquad (3)$$

is achievable.

Proof. The same mixing strategy described for rectilinear 1-center can be applied to the Euclidean 1-center problem for the facility moving with velocity $v_{max} \in [1, \sqrt{2}]$. Let c_m be the center of mass and let c_b be the center of the bounding box. Let r be the exact radius of 1-center and let r_m, r_b be the radii determined by c_m and c_b, respectively. The mixing center is defined as $c_{mix} = \alpha c_m + (1-\alpha)c_b$.

Using the result obtained in [6] which says that the center of mass of points guarantees a $(2 - 2/n)$-approximation of the rectilinear 1-center of S and using Theorem 6 we have $r_m \leq (2 - 2/n)r$ and $r_b \leq (\sqrt{2}+1)r/2$. Let p be any point in S. It suffices to prove that $|pc_{mix}| \leq Ar$, where A is the required approximation factor from (3). The mixing center has a property that $pc_{mix} = \alpha pc_m + (1-\alpha)pc_b$. Therefore

$$\begin{aligned} |pc_{mix}|^2 &= \alpha^2|pc_m|^2 + (1-\alpha)^2|pc_b|^2 + 2\alpha(1-\alpha)pc_m \cdot pc_b \\ &\leq \alpha^2|pc_m|^2 + (1-\alpha)^2|pc_b|^2 + 2\alpha(1-\alpha)|pc_m||pc_b| \\ &= (\alpha^2|pc_m| + (1-\alpha)|pc_b|)^2|. \end{aligned}$$

Thus, $|pc_{mix}| \leq \alpha|pc_m| + (1-\alpha)|pc_b| \leq \alpha(2 - 2/n)r + (1-\alpha)(\sqrt{2}+1)r/2 = Ar$ and we are done. ∎

3. Conclusion

In this paper we investigated mobile version of 1-center and problem in the plane. We provided several lower and upper bounds for the cases when facility is allowed to move with unit or bounded velocity. The interesting open question is to narrow the gap between upper and lower bounds for facility moving with velocity in the range $[1, \sqrt{2}]$ for the case of 1-center. Another possible direction is to generalize the results presented in this paper to the mobile k-center problem for $k \geq 2$ and in dimensions $d > 2$.

References

[1] P. K. Agarwal and S. Har-Peled. "Maintaining the Approximate Extent Measures of Moving Points", in *Proc. 12th ACM-SIAM Sympos. Discrete Algorithms*, pp. 148–157, 2001.

[2] P. Agarwal and M. Sharir. "Planar geometric location problems", *Algorithmica*, 11, pp. 185–195, 1994.

[3] C. Bajaj "Geometric optimization and computational complexity", Ph.D. thesis. Tech. Report TR-84-629. Cornell University (1984).

[4] J. Basch, "Kinetic Data Structures", Ph.D. thesis, Stanford University, USA, 1999.

[5] J. Basch, L. Guibas and J. Hershberger "Data structures for mobile data", *Journal of Algorithms*, 31(1), pp. 1–28, 1999.

[6] S. Bespamyatnikh, B. Bhattacharya, D. Kirkpatrick and M. Segal "Mobile facility location", *4th International ACM Workshop on Discrete Algorithms and Methods for Mobile Computing and Communications (DIAL M for Mobility'00)*, pp. 46–53, 2000.

[7] S. Bespamyatnikh, K. Kedem and M. Segal "Optimal facility location under various distance functions", *Workshop on Algorithms and Data Structures* Lecture Notes in Computer Science 1663, Springer-Verlag, pp. 318–329, 1999.

[8] S. Bespamyatnikh and M. Segal "Rectilinear static and dynamic center problems", *Workshop on Algorithms and Data Structures* Lecture Notes in Computer Science 1663, Springer-Verlag, pp. 276–287, 1999.

[9] J. Brimberg and A. Mehrez "Multi-facility location using a maximin criterion and rectangular distances", *Location Science* 2, pp. 11–19, 1994.

[10] Z. Drezner "The *p*-center problem: heuristic and optimal algorithms", *Journal of Operational Research Society*, 35, pp. 741–748, 1984.

[11] Z. Drezner "On the rectangular p-center problem", *Naval Res. Logist. Q.*, 34, pp. 229–234, 1987.

[12] M. Dyer and M. Frieze "A simple heuristic for the *p*-center problem", *Oper. Res. Lett.*, 3, pp. 285–288,1985.

[13] D. Eppstein "Faster construction of planar two-centers", *Proc. 8th ACM-SIAM Symp. on Discrete Algorithms*, pp. 131–138, 1997.

[14] J. Gao and L. Guibas and J. Hershberger and L. Zhang and A. Zhu "Discrete mobile centers", In *17th ACM Symposium on Computational Geometry*, 2001.

[15] L. Guibas "Kinetic data structures: A state of the art report", In *Proc. 1998 Workshop Algorithmic Found. Robot.*, pp. 191–209, 1998.

[16] S. L. Hakimi, M. Labbe and E. Schmeichel "Locations on Time-Varying Networks", Networks, Vol. 34(4), pp. 250–257, 1999.

[17] Sariel Har-Peled "Clustering motion", *FOCS'2001*.

[18] D. Hochbaum and D. Shmoys "A best possible approximation algorithm for the k-center problem", *Math. Oper. Res.*, 10, pp. 180–184, 1985.

[19] D. Hochbaum and A. Pathria "Locating Centers in a Dynamically Changing Network and Related Problems", Location Science, 6, pp. 243–256, 1998.

[20] Hai Huang and Andrea Richa and Michael Segal "Approximation algorithms for mobile piercing set problem with applications to clustering", Technical report, Arizona State University, 2001.

[21] J. Jaromczyk and M. Kowaluk, "An efficient algorithm for the Euclidean two-center problem", in *Proc. 10th ACM Sympos. Comput. Geom.*, pp. 303-311, 1994.

[22] H. Kuhn, "A note on Fermat's problem", *Mathematical Programming*, 4, pp. 98–107, 1973.

[23] N. Megiddo "Linear time algorithms for linear programming in R^3 and related problems", *SIAM J. Comput.*, 12, pp. 759–776, 1983.

[24] M. Manasse, L. McGeoch and D. Sleator "Competitive algorithms for server problems", *Journal of Algorithms*, 11, pp. 208–230, 1990.

[25] N. Megiddo and A. Tamir "New results on the complexity of p-center problems", *SIAM J. Comput.*, 12(4), pp. 751–758, 1983.

ON THE ENUMERABILITY OF THE DETERMINANT AND THE RANK

Alina Beygelzimer, Mitsunori Ogihara*
Department of Computer Science, University of Rochester, Rochester, NY 14627
{beygel,ogihara}@cs.rochester.edu

Abstract We investigate the complexity of enumerative approximation of two elementary problems in linear algebra, computing the rank and the determinant of a matrix. In particular, we show that if there exists an enumerator that, given a matrix, outputs a list of constantly many numbers, one of which is guaranteed to be the rank of the matrix, then it can be determined in AC^0 (with oracle access to the enumerator) which of these numbers is the rank. Thus, for example, if the enumerator is an FL function, then the problem of computing the rank is in FL. The result holds for matrices over any commutative ring whose size grows at most polynomially with the size of the matrix. The existence of such an enumerator also implies a slightly stronger collapse of the exact counting logspace hierarchy. For the determinant function we establish the following two results: (1) If the determinant is poly-enumerable in logspace, then it can be computed exactly in FL. (2) For any prime p, if computing the determinant modulo p is $(p-1)$-enumerable in FL, then computing the determinant modulo p can be done in FL. This gives a new perspective on the approximability of many elementary linear algebra problems equivalent to computing the rank or the determinant.

Introduction

Valiant (Valiant, 1979) proved that the permanent of integer matrices characterizes #P, the class of functions definable as the number of accepting computations of a nondeterministic polynomial-time Turing machine. A similar connection has been shown between the complexity of computing the determinant and #L, the logspace analog of #P (Toda, 1991; Valiant, 1992; Vinay, 1991; Damm, 1991). (Since the determinant of integer matrices can take on negative values, the determinant is in fact complete for GapL, the class of functions that can be expressed as the difference of two #L functions.) Toda's

*This work is supported in part by NSF grants EIA-0080124 and DUE-9980943, and in part by NIH grants RO1-AG18231 and P30-AG18254.

result (Toda, 1989) showing the surprising power of counting in the context of nondeterministic polynomial-time, namely that the polynomial hierarchy is contained in $\mathrm{P}^{\#\mathrm{P}}$, naturally raises the question of whether #P functions are at least easy to approximate. In the standard sense of coming close to the correct value (i.e., within a multiplicative factor), approximating #P functions is in Δ_3^p. Moreover, any technique for showing that it has complexity lower than Δ_2^p, would have to be non-relativizable. (See (Stockmeyer, 1985).) In search of a better answer, Cai and Hemachandra (Cai and Hemachandra, 1989) proposed an alternative notion of approximation, called enumerative counting. Instead of restricting the range of a function value to an interval, they consider enumerating a short list of (not necessarily consecutive) values, guaranteed to contain the correct one. Which of these approximation tasks is more natural depends on the function one is trying to approximate: enumerative counting is more suitable in cases when there is no natural total ordering on the range of the function; or when the range is either exponentially large (as in the case of the determinant), or some intervals of the range are substantially denser than others, and it is preferred to obtain a fixed number of candidates for every input of the same length, instead of dealing with an interval that, though bounded, may contain vastly different numbers of possible values for different inputs (depending *not* on the input itself, but on the value of the function on this input).

More formally, a function f is said to be *$g(n)$-enumerable* if there exists a function that, on input x, outputs a list of at most $g(|x|)$ values that is guaranteed to contain $f(x)$. Cai and Hemachandra (Cai and Hemachandra, 1991), and also Amir, Beigel, and Gasarch (Amir et al., 1990), showed that if the permanent function is poly-enumerable in polynomial time, then $\mathrm{P} = \mathrm{P}^{\#\mathrm{P}}$. (Their result is actually stated for #SAT, another classical #P-complete function.) Thus if we are given an $n \times n$ integer matrix A, and instead of $n!(2^n)^n$ values that perm(A) can possibly take, we get a restricted list of polynomially-many values, guaranteed to contain perm(A), then we can easily (in polynomial time) determine which one is the correct one. This is certainly a *non*-enumerability result, as it says that a very hard to compute function (hard for the whole polynomial hierarchy) can easily bootstrap its exact value if it is left with polynomially many candidates.

This paper investigates the enumerability of functions complete for logspace counting classes. In particular, it is interesting whether enumerability implies a similar decrease in the complexity of the determinant; and if so, to what extent. Since #L functions have significantly less computational power (they are contained in the NC^2, or even TC^1), the enumerability properties of logspace counting analogs can be very different from those of #P-complete functions. This is even more interesting, because, as far as we know, there are no results on approximating the determinant (or the rank) in the standard sense. Another purpose of such an investigation is to get a better understanding of the relationships among the complexity classes sandwiched between NL and uniform TC^1.

We show that if there exists an enumerator that, given a matrix, outputs a list of constantly many numbers, one of which is guaranteed to be the rank of the matrix, then it can be determined in AC^0 (with oracle access to the enumerator) which of these numbers is the rank. Thus, for example, if the enumerator runs in logspace, then the problem of computing the rank is in logspace. The result holds for any commutative ring with identity whose size grows polynomially with the size of the matrix. The existence of such an enumerator implies a slightly stronger collapse of the exact counting logspace hierarchy; namely, it shows that the the hierarchy collapses to the closure of its base class $C_{=}L$ under $\leq^{AC^0}_{O(1)\text{-tt}}$ reductions. We also consider a related problem of computing the number of dependent vectors in a given set, and prove it to be $\leq^{AC^0}_{O(1)\text{-tt}}$-equivalent to computing the rank. For the determinant function we show that if the determinant is poly-enumerable in logspace, then it can be computed exactly in logspace. We also establish a similar result for computing the determinant modulo any prime.

1. Preliminaries

We will be concerned with the complexity of computing the following functions (with input and output in $\{0,1\}^*$):

- DET$_F$: Given $A \in F^{n\times n}$, compute the determinant of A.
- RANK$_F$: Given $A \in F^{n\times n}$, compute the rank of A.
- SINGULARITY$_F$: Given $A \in F^{n\times n}$, determine whether A is singular.
- INDEPENDENCE$_F$: Given a set of vectors in F^n, determine whether they are linearly independent.

Here F is any commutative ring with unity. When $F = \mathbb{Z}$, the ring of integers, F is dropped. With some abuse of notation, we will consider these functions as sets by associating the function $f : \{0,1\}^* \to \{0,1\}^*$ with the set $\{(x,i) \mid \text{the } i\text{-th bit of } f(x) \text{ is } 1\}$. Notice that a function is in NC^k if and only if its associated set (its characteristic function) is in NC^k and is polynomially bounded; hence there will be no confusion in viewing circuit classes as functional, sets being identified with their characteristic functions.

Uniformity. In order to make circuit classes comparable to traditional classes defined by time and space, we need to place uniformity restrictions on circuit families. For our purposes it will be sufficient to use logspace uniformity, meaning that there exists a logspace machine that, on input 1^n, generates a description of the circuit for n inputs. For a detailed treatment of uniformity and a discussion of other uniformity conditions, see (Ruzzo, 1981; Barrington and Immerman, 1997).

Reductions. We use Wilson's model (Wilson, 1985) of oracle circuits to define the reductions. A function f is AC^0-*reducible* to function g, if there is a logspace uniform AC^0 family of circuits that computes f, where in addition to the usual gates, oracle gates for g are allowed. Similarly we define NC^1-*reducibility* except that now the circuits have a bounded fan-in, and thus an oracle gate with fan-in m has to count as depth $\log m$. For a circuit class $\mathcal{C}$, we write $\mathcal{C}(f)$ to denote the class of functions $\mathcal{C}$-reducible to f. For a function class $\mathcal{F}$, $\mathcal{C}(\mathcal{F})$ denotes the class of functions $\mathcal{C}$-reducible to some function in $\mathcal{F}$. A function f is AC^0 ***many-one reducible*** to a function g (written as $f \leq_m^{\mathrm{AC}^0} g$) is there exists an AC^0 family of circuits $\langle C_n \rangle$ such that for every x of length n, we have $f(x) = g(C_n(x))$.

Logspace counting classes and the complexity of problems in linear algebra

Many basic linear algebra problems are known to be in NC^2. In order to classify and capture the exact complexity of these problems, Cook (Cook, 1985) defined the class of problems NC^1-reducible to the determinant of integer matrices, and showed that most linear algebra problems with fast parallel algorithms are in this class. Many are in fact complete for this class; others were shown to be complete for the (potentially smaller) class of problems reducible to computing the rank (von zur Gathen, 1993; Santha and Tan, 1998). Santha and Tan (Santha and Tan, 1998) defined a more refined hierarchy of problems that reflects the computational difference between the functional and the verification versions of the problems under AC^0-Turing and AC^0-many-one reductions. Toda (Toda, 1991) gave many examples of graph-theoretic problems that are (under appropriate reductions) equivalent to computing the determinant. (Although some of these problems still have natural matrix interpretations when graphs are identified with their adjacency matrices.)

Allender and Ogihara (Allender and Ogihara, 1996) observed that, even though for most natural problems the closures under AC^0 and NC^1-reductions coincide, this does not seem to be apparent for the determinant. This motivated the definition of the following hierarchies (defined using the "Ruzzo-Simon-Tompa" oracle access model (Ruzzo et al., 1984), which is standard for defining Turing reductions for space-bounded nondeterministic machines, see (Allender and Ogihara, 1996)):

- The exact counting logspace hierarchy $\mathrm{C_=L} \cup \mathrm{C_=L}^{\mathrm{C_=L}} \cup \mathrm{C_=L}^{\mathrm{C_=L}^{\mathrm{C_=L}}} \cdots = \mathrm{AC}^0(\mathrm{C_=L})$

  The class C$_=$L is defined as the class of languages, for which there exists a GapL function f such that for every x, x is in the language if and only if $f(x) = 0$. If follows immediately that the set of singular matrices is complete for C$_=$L. Allender, Beals, and Ogihara (Allender et al., 1999) showed that AC^0 and NC^1 reducibilities coincide on C$_=$L; furthermore, the hierarchy collapses to $\mathrm{L}^{\mathrm{C_=L}}$. We show that, if Rank is $O(1)$-enumerable in logspace, then it collapses to the $\leq_{O(1)\text{-tt}}^{\mathrm{AC}^0}$-closure of C$_=$L.

- The PL hierarchy $\mathrm{PL} \cup \mathrm{PL}^{\mathrm{PL}} \cup \mathrm{PL}^{\mathrm{PL}^{\mathrm{PL}}} \cup \cdots = \mathrm{AC}^0(\mathrm{PL})$

Ogihara (Ogihara, 1998) showed that the PL hierarchy collapses to PL under AC^0 (in fact TC^0) reductions, which was improved to NC^1 reducibility by Beigel and Fu (Beigel and Fu, 1997). A problem, easily seen to be complete for PL, is checking whether the determinant of integer matrices is positive.

- The #L hierarchy $L \cup L^{\#L} \cup L^{\#L^{\#L}} \cdots = AC^0(\#L) = AC^0(\text{DET})$

  It is not known whether the #L hierarchy collapses, or whether $AC^0(\#L) = NC^1(\#L)$. The latter would imply the collapse (Allender, 1997).

Allender et al. (Allender et al., 1999) showed that the problems of computing the rank of integer matrices, determining whether the rank is odd, and determining the solvability of a system of linear equations, are all complete for $AC^0(C_=L)$. Clearly, the problems of computing and verifying the rank of a matrix are AC^0-equivalent (since there are just $n+1$ possibilities for the rank). However, Allender et al. classified the complexity of verifying the rank exactly, showing that it is complete for the second level of the Boolean Hierarchy above $C_=L$ (i.e. the class of sets expressible as an intersection of a $C_=L$ and a co-$C_=L$ set).

Our results. We show that if there exists a logspace computable $O(1)$-enumerator for RANK, then RANK is $\leq^{AC^0}_{O(1)\text{-tt}}$-reducible to INDEPENDENCE, and thus to SINGULARITY. (The reduction holds for arbitrary rings.) SINGULARITY is complete for $C_=L$, and thus the existence of the enumerator implies that $AC^0(C_=L)$ coincides with the closure of $C_=L$ under $O(1)$-tt-reductions, a slight improvement over $O(\text{poly}(n))$-tt that follows from (Allender et al., 1999) (unconditionally). We also show that if RANK_F is $O(1)$-enumerable in logspace, then $\text{RANK}_F \in \text{FL}$, where F is any commutative ring with identity whose size grows at most polynomially with the size of the input matrix. Finally, we consider a related problem of computing the number of dependent vectors in a given set (i.e., vectors involved in some non-trivial linear dependencies with other vectors in the set), and show it to be $\leq^{AC^0}_{O(1)\text{-tt}}$-equivalent to computing the rank. For the determinant function DET we establish the following two results: (1) If DET is poly-enumerable in logspace, then DET is in FL. (2) For any prime p, if the determinant-modulo-p function is $(p-1)$-enumerable in FL, then it can be computed exactly in FL.

Organization of the paper. All the results pertaining to the rank and the determinant are collected in Sections 3 and 4, respectively. Section 5 concludes with a discussion.

2. Enumerability of the Rank

Recall that a function f is logspace $g(n)$-enumerable if there exists a logspace computable function that, on input x, outputs a list of at most $g(|x|)$ values, one of which is $f(x)$. The following lemma shows how to combine several matrices into a single matrix such that the ranks of the original matrices can be read off the rank of the combined matrix.

Lemma 2.1 [Block diagonal construction] There exists a logspace computable function S that given an ordered list $\mathcal{Q} = \langle A_1, \ldots, A_q \rangle$ of $n \times n$ matrices, outputs a single matrix $S(\mathcal{Q})$ of dimension $O(n^q)$ such that a logspace procedure can uniquely decode the sequence of ranks $\langle \text{rank}(A_1), \ldots, \text{rank}(A_q) \rangle$ from the value of $\text{rank}(S(\mathcal{Q}))$. Moreover, both procedures can be implemented by uniform AC^0 circuit families.

Proof: Consider the following combining construction. On input $\mathcal{Q} = \langle A_1, \ldots, A_q \rangle$, S outputs a block diagonal matrix (i.e. a matrix of $n \times n$ blocks sitting on the main diagonal) with the following block structure. The first block of $S(\mathcal{Q})$ corresponds to A_1, the next $(n+1)$ blocks correspond to A_2, and so on, until we get to $n^q + n^{q-1} + \cdots + 1$ blocks of A_q. The multiplicity of A_i as a block is $\sum_{j=1}^{i} n^{j-1}$, thereby the dimension of $S(\mathcal{Q})$ is $\sum_{i=1}^{q} n^{i+1}(q - i + 1) = O(n^q)$. The rank of $S(\mathcal{Q})$ is the sum of all block ranks, and since the rank of each block is at most n, the original sequence of ranks $\langle \text{rank}(A_1), \ldots, \text{rank}(A_q) \rangle$ can be read off from the value of $\text{rank}(S(\mathcal{Q}))$. It is easy to see that both the construction and the decoding can be done in uniform AC^0. ∎

The combining construction above allows one to eliminate many candidate rank sequences. For example, if we were to feed each of q matrices to an r-enumerator separately, we would get r^q purported rank sequences, whereas combining the matrices into a single query reduces the number of candidates to r. In order for the dimension of $S(\mathcal{Q})$ to be polynomial in n, the number of matrices, q, has to be constant. A simple information-theoretic argument shows that this is the best possible. Indeed, the dimension of a matrix whose rank can encode $(n+1)^q$ possible rank sequences must be at least $(n+1)^q$. Notice that combining matrices into a single query to an r-enumerator allows one to link matrices in the following sense.

Definition 1 Two r-element sequences $\{p_1, \ldots, p_r\}$ and $\{q_1, \ldots, q_r\}$ are said to be *linked* if $p_i = p_j$ if and only if $q_i = q_j$, for all $1 \leq i < j \leq r$.

In other words, two matrices are linked – relative to an enumerator – if there is a direct correspondence between the values on their claimed lists of ranks; hence knowing the rank of one immediately gives the rank of the other.

Claim 2.1 There exists r_0 such that for any $r > r_0$, any set of $(\frac{r}{\ln r})^r$ r-element sequences, contains at least one linked pair.

Proof: The number of r-element sequences sufficient to guarantee the existence of a linked pair is precisely one more than the number of partitions of an r-element set into non-empty subsets. The latter is known as the rth Bell number, B_r. De Bruijn (de Bruijn, 1970) gave the asymptotic formula

$$\ln B_r r = \ln r - \ln\ln r - 1 + \ln\ln r \ln r + 1 \ln r + 12 \left(\ln\ln r \ln r\right)^2 + O\left(\ln\ln r (\ln r)^2\right),$$

immediately yielding the claim. Other (less explicit) asymptotic approximations for B_r are known (see, for example, (Lovász, 1993)). ∎

Let $\kappa(r)$ be the minimum number of matrices that are guaranteed to contain a linked pair. By Claim 2.1, $\kappa \stackrel{\text{def}}{=} \kappa(r) \leq (\frac{r}{\ln r})^r$. Given an $n \times n$ matrix A, let A_i (for $1 \leq i \leq n$) denote the $n \times n$ matrix with the first i rows of A and 0s elsewhere, so that $A_n = A$. An r-enumerator for the rank function defines the *equivalence graph* of A, a labeled graph on $[n] \stackrel{\text{def}}{=} \{1, \dots, n\}$ with the set of nodes corresponding to $\mathcal{A} = \{A_1, \cdots, A_n\}$, and an edge between nodes i and j if and only if there is a set of $\kappa - 2$ matrices in $\mathcal{A} - \{A_i, A_j\}$ certifying the equivalence between A_i and A_j (i.e. witnessing that A_i and A_j are linked). (We can assume without loss of generality that the enumerator is deterministic and the combining encoding of queries in Lemma 2.1 is symmetric.) The label of an edge is defined by the equivalence (i.e. direct correspondence between the r claimed values for rank(A_i) and the r claimed values for rank(A_j)) given by the lexicographically smallest κ-tuple linking A_i and A_j.

Notice that by definition every subset of κ nodes in the equivalence graph induces at least one edge. Hence the number of connected components in the equivalence graph is at most $\kappa - 1$. The following proposition shows that in this case every pair of nodes is connected by a short path (where the length of a path is the number of edges it contains). The proof can be found in (Beygelzimer and Ogihara, 2002).

Proposition 2.1 *Any pair of nodes in the equivalence graph is connected by a path of length at most* $2\kappa - 3$.

We will use equivalence graphs in the proof of the theorem below.

Theorem 1 If, for some integer r, there exists an r-enumerator for RANK_F, then RANK_F is computable in AC^0 with oracle calls to the enumerator. Here F is any commutative ring with identity whose size grows polynomially with the size of the input matrix.

Proof: Given an enumerator and a set of matrices, the equivalence graph is uniquely defined. Recall that the number of equivalence classes (i.e. the number of connected components in the graph) is at most $\kappa - 1$. Consider guessing the number of equivalence classes, a representative matrix from each equivalence class, and, finally, the ranks of the representatives chosen. Since there are at most $(\kappa - 1)n()\kappa - 1r^{\kappa - 1}$ possibilities total,which is polynomial in n when r is constant, we have no problem checking them all in parallel; thus we will concentrate on a single guess.

Once we have guessed the number of equivalence classes and their representatives, we can check whether every node is reachable from at least one representative, and whether the representatives are not reachable from each other. Recall that we only need to check all paths of length at most $2\kappa - 3$ from every representative node. If at least one of these conditions is not satisfied, we reject; otherwise we proceed to checking the consistency of ranks, as we do next.

Let $R_1, \ldots, R_k \in \mathcal{A}$ be the representative matrices, and $\tau_1, \ldots, \tau_k$ be the corresponding ranks, where $1 \leq k < \kappa$ is the number of equivalence classes. Note that $\tau_1, \ldots, \tau_k$ uniquely define the ranks of all matrices in $\mathcal{A}$. Of course, we do not know $\tau_1, \ldots, \tau_k$. Instead, we will use the block diagonal construction in Lemma 2.1 to pack $R_1, \ldots, R_k$ into a single matrix, which we can then feed to the enumerator to get a list of r sequences of ranks, one of which is $\langle \tau_1, \ldots, \tau_k \rangle$. Denote the sequences by $\langle \tau_1^1, \ldots, \tau_k^1 \rangle$, ..., $\langle \tau_1^r, \ldots, \tau_k^r \rangle$. Each $\langle \tau_1^i, \ldots, \tau_k^i \rangle$ uniquely defines the rank sequence $\langle v_1^i, \ldots, v_n^i \rangle$ claimed to be $\langle \text{rank}(A_1), \ldots, \text{rank}(A_n) \rangle$. Let $U_i = \{1 \leq j \leq n \mid v_j^i = v_{j-1}^i + 1\}$, where we define $v_0^i = 0$ for all $1 \leq i \leq r$; thus each of the $U_1, \cdots, U_r$ claims to be a maximal set of linearly independent rows of A. Our goal is to test whether each U_i is indeed maximal, i.e. whether every remaining row of A is a linear combination of the rows in U_i. As there are only constantly many U_i's, testing them in parallel causes no problem. The rank of A is given by the size of the smallest U_i that passes the maximality test. (This number can be found as the corresponding v_n^i.)

Remark 1 Alternatively, we could have obtained the (alleged) maximally independent sets of columns $V_1, \ldots, V_r$ (using the same procedure as for the rows). The square submatrices indexed by $U_1 \times V_1$, $\cdots$, $U_r \times V_r$ all claim to be non-singular. (We can obviously discard all candidate sequences with $|U_i| \neq |V_i|$.) Now instead of verifying the maximality claim, we can test, in parallel, which submatrices are indeed non-singular, and then take the maximum over all that pass the test.

Notice that the discussion above is valid for arbitrary matrices. Now we show how to test the maximality of U_i's for matrices with entries from any commutative ring whose size does not grow more than polynomially with the size of the input matrix.

Testing maximality:. Given row vectors $v_1, \ldots, v_q, w \in F^n$, verify that w is in subspace generated by $v_1, \ldots, v_q$.

Let $F = \{a_1, \cdots, a_m\}$ be the ground field. If $v_1, \ldots, v_q$ are linearly independent, w is dependent on $v_1, \ldots, v_q$, and $w \neq 0^n$ (where 0 is the null element of F), then there must exist unique coefficients $c_1, \ldots, c_q \in F$ such that $c_1 v_1 + \cdots + c_q v_q + w = 0$. For each i and j, $1 \leq i \leq m$, $1 \leq j \leq q$, define the matrix

$$M_i^j = \begin{pmatrix} v_1 \\ \vdots \\ a_i v_j + w \\ \vdots \\ v_q \end{pmatrix}.$$

If the above conditions hold, then for each j, there is a unique i such that $\text{rank}(M_i^j) = q-1$; namely, $\text{rank}(M_i^j) = q-1$ iff $a_i = c_j$; otherwise $\text{rank}(M_i^j) = q$. We have $qm < nm$ matrices[1], which is polynomial in n, provided that m grows at most polynomially in n. We want to reduce the number of possibilities for the ranks of these matrices to a constant, and we already know how to force the enumerator to do this for us: Recall that there is a constant $\kappa = \kappa(r)$ such that combining any κ matrices into a single query witnesses at least one equivalence relation between a pair of claimed lists of ranks. Furthermore, there are at most $\kappa - 1$ equivalence classes, and thus only polynomially (in nm) many choices of how to partition M_i^js into the equivalence classes. All choices can be verified in parallel, each one gives only a constant number of possible values for the ranks of M_i^js. For each candidate sequence of ranks we collect the coefficients c_is, assuming that this sequence is correct (i.e. for each j there is a unique i such that $\text{rank}(M_i^j) = q-1$, and $\text{rank}(M_l^j) = q$ for all $l \neq i$); then we verify (column-wise in parallel) that the equality $c_1 v_1 + \cdots + c_q v_q + w = 0$ holds. Notice that the maximality test can be run in parallel not only for all of $U_1, \cdots, U_r$, but also for *all* rows w claimed to be linear combinations of the rows in U_h (for each h, $1 \le h \le r$). The number of matrices that we are dealing with for each h is less than $n^2 m$, so if m is polynomial in n, testing them in parallel causes no problem. It is easy to see that the entire computation can be done in AC^0 with oracle gates for the enumerator, since we are dealing with fields of polynomial size (and thus elements of logarithmic length).

Now, to extend this method to commutative rings of polynomial size, we just have to test whether, for each $d \in F$, there exist coefficients $c_1, \ldots, c_q \in F$ satisfying $c_1 v_1 + \cdots + c_q v_q + dw = 0$, which can be done in parallel for each possible d. For some i there can be more than one value of j for which $\text{rank}(M_i^j) = q-1$, but since the underlying computation is AC^0 we have only to select the smallest such j. ∎

Corollary 2.1 Let F be any commutative ring with identity whose size grows polynomially with the size of the input matrix. If there exists a $O(1)$-enumerable for RANK_F that runs in logspace, then RANK_F is computable in logspace.

[1]If $q = n$ (i.e. if the enumerator has claimed that A has full rank), we can verify whether each row is a linear combination of the remaining rows. Since there are just n rows, the verification can be done in parallel. The sequence passes the test if and only if it passes all of the n tests. Thus we may assume that $q < n$.

Computing the number of dependent rows

Consider the following related problem: Given a set of row vectors $v_1, \ldots, v_n \in F^n$, determine how many of them are *dependent*, i.e., are involved in some non-trivial linear dependency with other vectors in the set. Define the problem DEP_F: Given a matrix $A \in F^{n\times n}$, compute the number of dependent rows of A, written as $\text{dep}(A)$.

Proposition 2.2 If, for some integer r, there exists an r-enumerator for DEP_F, then DEP_F is computable in TC^0 with oracle calls to the enumerator. Here F is any commutative ring of polynomial size.

Proposition 2.3 For an arbitrary ring F, DEP_F is $\leq^{\text{TC}^0}_{O(n)\text{-tt}}$-equivalent to RANK_F.

The proofs are not included due to space limitations; they can be found in (Beygelzimer and Ogihara, 2002).

3. Enumerability of the Determinant

Theorem 2 If, for some k, DET is n^k-enumerable in logspace, then $\text{DET} \in \text{FL}$.

Proof: Let G be the configuration graph of a nondeterministic logspace machine on some input x. Let $\#\text{path}_G(s,t)$ denote the number of directed paths from node s to node t in G. Define $f(G)$ as the function, whose value (written in binary) consists of a sequence of n^2 blocks of length $s = 2n\lceil\log n\rceil$, the $(n(i-1)+j)$th segment corresponding to $\#\text{path}_G(i,j)$, where $1 \leq i,j \leq n$; thus

$$f(G) = \sum_{i,j=1}^{n} 2^{(n(i-1)+j)s} \#\text{path}_G(i,j).$$

It is easy to see that f is a #L function; this can be done by exhibiting an NL machine N whose number of accepting computation paths on G is $f(G)$. The machine N, on input G, nondeterministically guesses a number $p = n(i-1)+j$, $1 \leq p \leq n^2$, after which it guesses q, $1 \leq q \leq 2^{ps}$, followed by a guess of path from node i to node j in G. If the guess is correct, N accepts; otherwise, it rejects. It is easy to see that N is a nondeterministic logspace machine that has the required number of accepting paths. Hence $f \in \text{GapL}$, and a function is in GapL if and only if it is logspace many-one reducible to the determinant. Thus there must exist a logspace function g such that for all G, $f(G) = \text{DET}(g(G))$. We shall use g to transform G into a matrix $M = g(G)$, and then run the n^k-enumerator on M to obtain a list of n^k values, one of which is the determinant of M. Using the equality in the above reduction, we convert this list to a list of n^k candidates for $f(G)$ (each of which, if correct, certifies that the corresponding claimed value for $\det(M)$ is correct). Since we have only n^k candidates for $f(G)$, they can be checked in parallel. Given a purported $f(G)$, we can uniquely read off $\#\text{path}_G(i,j)$ for each pair (i,j), $1 \leq i,j \leq n$. These values can then be locally checked using the self-reducibility of $\#\text{path}_G$. ∎

A natural question is whether the same theorem holds for finite fields. We can only show a similar result for the determinant of integer matrices modulo some integer p. The problem of computing the determinant mod p is $\leq_m^{\log}$-complete for $\mathrm{Mod}_p\mathrm{L}$, defined in (Buntrock et al., 1992). $\mathrm{Mod}_p\mathrm{L}$ is the class of sets A for which there exists $f \in \#\mathrm{L}$ such that for all x, $x \in A$ iff $f(x) \not\equiv 0 \bmod p$. We will also use the notion of membership comparability, due to Ogihara (Ogihara, 1995).

Definition 2 A set S is *$g(n)$-membership comparable*, written as $S \in \text{L-mc}(g(n))$, if there exists a function $f \in \mathrm{FL}$ such that for any set of $g(n)$ inputs $x_1, \ldots, x_{g(n)}$, each of length at most n, f excludes one of $2^{g(n)}$ candidates for the characteristic sequence $\chi_S(x_1, \ldots, x_{g(n)})$.

We will also use the predicate version of the problem, namely DET-mod-p = $\{(A, i) \mid \det(A) \equiv i \text{ (modulo } p)\}$.

Theorem 3 Let p be any prime. If there exists a logspace computable $(p-1)$-enumerator for DET-mod-p, then DET-mod-$p \in \mathrm{FL}$.

Proof: Omitted due to space limitations; see (Beygelzimer and Ogihara, 2002). ∎

4. Open questions

A natural question is whether the rank being, say, $O(\log n)$-enumerable implies that the rank is in logspace. As we mentioned, it does not seem possible to combine more than a constant number of queries into a single query to the enumerator. Another improvement would be to show that Proposition 2.2 holds for AC^0 in place of TC^0, the question being whether counting the number of dependent basis vectors can be avoided in this context.

References

Allender, E. (1997). A clarification concerning the #L hierarchy. Available at http://www.cs.rutgers.edu/~allender/.

Allender, E., Beals, R., and Ogihara, M. (1999). The complexity of matrix rank and feasible systems of linear equations. *Computational Complexity*, 8:99–126.

Allender, E. and Ogihara, M. (1996). Relationships among PL, #L, and the determinant. *Theoretical Informatics and Applications*, 30(1):1–21.

Amir, A., Beigel, R., and Gasarch, W. (1990). Some connections between bounded query classes and nonuniform complexity. In *5th Structure in Complexity Theory Conference*, pages 232–243.

Barrington, D. and Immerman, N. (1997). Time, hardware, and uniformity. In Hemaspaandra, L. and Selman, A., editors, *Complexity Theory Restrospective II*, pages 1–22. Springer-Verlag.

Beigel, R. and Fu, B. (1997). Circuits over PP and PL. In *12st IEEE Conference on Computational Complexity*, pages 24–35.

Beygelzimer, A. and Ogihara, M. (2002). On the enumerability of the determinant and the rank. Electronic Colloquium on Computational Complexity TR02-016.

Buntrock, G., Damm, C., Hertrampf, U., and Meinel, C. (1992). Structure and importance of Logspace-MOD class. *Mathematical Systems Theory*, 25(3):223–237.

Cai, J. and Hemachandra, L. (1989). Enumerative counting is hard. *Information and Computation*, 82(1):34–44.

Cai, J. and Hemachandra, L. (1991). A note on enumerative counting. *Information Processing Letters*, 38:215–219.

Cook, S. (1985). A taxonomy of problems with fast parallel algorithms. *Information and Control*, 64:2–22.

Damm, C. (1991). DET = $L^{\#L}$? Informatik-Preprint 8, Fachbereich Informatik der Humboldt-Universität zu Berlin.

de Bruijn, N. G. (1970). *Asymptotic methods in analysis*. North-Holland, Amsterdam.

Lovász, L. (1993). *Combinatorial problems and exercises*. North-Holland, 2nd edition.

Ogihara, M. (1995). Polynomial-time membership comparable sets. *SIAM Journal on Computing*, 24(5):1168–1181.

Ogihara, M. (1998). The PL hierarchy collapses. *SIAM Journal on Computing*, 27:1430–1437.

Ogihara, M. and Tantau, T. (2001). On the reducibility of sets inside NP to sets with low information content. Preprint.

Ruzzo, W. (1981). On uniform circuit complexity. *Journal of Computer and System Sciences*, 22(3):365–383.

Ruzzo, W., Simon, J., and Tompa, M. (1984). Space-bounded hierarchies and probabilistic computations. *Journal of Computer and System Sciences*, 28:216–230.

Santha, M. and Tan, S. (1998). Verifying the determinant in parallel. *Computational Complexity*, 7:128–151.

Stockmeyer, L. (1985). On approximation algorithms for #P. *SIAM Journal on Computing*, 14(4):849–861.

Toda, S. (1989). On the computational power of PP and $\oplus$P. In *30th IEEE Symposium on Foundations of Computer Science*, pages 514–519.

Toda, S. (1991). Counting problems computationally equivalent to computing the determinant. Technical Report CSIM 91-07, Department of Computer Science, University of Electro-Communications, Tokyo, Japan.

Valiant, L. (1979). The complexity of computing the permanent. *Theoretical Computer Science*, 8:189–201.

Valiant, L. (1992). Why is boolean complexity theory difficult. In Paterson, M., editor, *Boolean Function Complexity*, London Mathematical Society Lecture Notes Series 169, pages 84–94. Cambridge University Press.

Vinay, V. (1991). Counting auxiliary pushdown automata and semi-unbounded arithmetic circuits. In *6th IEEE Structure in Complexity Theory Conference*, pages 270–284.

von zur Gathen, J. (1993). Parallel linear algebra. In Reif, J., editor, *Synthesis of Parallel Algorithms*, pages 574–615. Morgan Kaufmann.

Wilson, C. (1985). Relatizived circuit complexity. *Journal of Computer and System Sciences*, 31:169–181.

ON THE SYMMETRIC RANGE ASSIGNMENT PROBLEM IN WIRELESS AD HOC NETWORKS

Douglas M. Blough[1], Mauro Leoncini[2], Giovanni Resta[3], Paolo Santi[4]
[1] *School of ECE, Georgia Institute of Technology, Atlanta GA.*
[2] *Dip. di Dipartimento di Scienze Sociali, Cognitive e Quantitative, Università di Modena e Reggio Emilia, Italy.*
[3,4] *Istituto di Informatica e Telematica del CNR, Area della Ricerca di Pisa, Italy.*
[1]doug.bloug@ece.gatech.edu, [2]leoncini@unimo.it, [3]p.santi@iit.cnr.it, [4]g.resta@iit.cnr.it

Abstract In this paper we consider a constrained version of the range assignment problem for wireless ad hoc networks, where the value the node transmitting ranges must be assigned in such a way that the resulting communication graph is strongly connected and the energy cost is minimum. We impose the further requirement of symmetry on the resulting communication graph. We also consider a weaker notion of symmetry, in which only the existence of a set of symmetric edges that renders the communication graph connected is required. Our interest in these problems is motivated by the fact that a (weakly) symmetric range assignment can be more easily integrated with existing higher and lower-level protocols for ad hoc networks, which assume that all the nodes have the same transmitting range. We show that imposing symmetry does not change the complexity of the problem, which remains NP-hard in two and three-dimensional networks. We also show that a weakly symmetric range assignment can reduce the energy cost considerably with respect to the homogeneous case, in which all the nodes have the same transmitting range, and that no further (asymptotic) benefit is expected from the asymmetric range assignment. Hence, the results presented in this paper indicate that weak symmetry is a desirable property of the range assignment.

Introduction

Recent emergence of affordable, portable, wireless communication and computation devices has resulted in the rapid growth of mobile wireless networks. Among these, *ad hoc networks*, i.e. networks of mobile, untethered units communicating with each other via radio transceivers, are receiving increasing attention in the scientific community. Ad hoc networks can be used wherever a wired backbone is not viable, e.g. in mobile computing applications in areas where other types of infrastructures are unavailable, to provide communications during emergencies, or to monitor remote geographical regions.

In ad hoc networks, every node u is characterized by a transmitting range r_u: when u sends a message, all the nodes at distance at most r_u from u can potentially receive the message. If the recipient is not an immediate neighbor of u, the message must be routed to the destination through a multi-hop path. For this reason, ad hoc networks are also called *multi-hop packet radio networks.*

One of the major concerns in wireless ad hoc networks is reducing node power consumption. In fact, nodes are usually equipped with a limited capacity battery, and battery recharge and/or replacement is very difficult or even impossible in many application scenarios (e.g., wireless sensor networks [23]). Hence, reducing power consumption is often the only way to extend network lifetime. It is known that one of the main sources of power consumption in a wireless node is communication, and that the power p_u required by node u to transmit data is related to its transmitting range r_u [10]. Thus, node transmitting ranges should be set as small as possible consistently with some requirement (e.g., strong connectivity) on the resulting network topology.

Given a transmitting range assignment (range assignment for short) for each node in the network, the *communication graph* G is defined, where directed edge (u, v) exists in G if and only if v is at distance at most r_u from u. Although current transceivers and communication protocols are designed for a fixed transmitting range (e.g., 250 meters in the widely used IEEE 802.11 standard [1]), a scenario in which the transmitting range is not fixed is fully compatible with current technology. The range assignment could be decided prior to node deployment in the case of stationary networks when information on the physical node placement are available, or it can be varied dynamically in presence of mobility or when the physical node placement is unknown. Distributed topology control protocols aimed at dynamically changing the transmitting range assignment in order to keep the network connected and minimize energy consumption have been recently presented in [17, 24, 26].

The problem of assigning transmitting range to nodes in such a way that the resulting communication graph is strongly connected and the energy cost is minimized is called the *range assignment problem* (RA), and was first studied in [18]. In [18], it is shown that RA for one-dimensional networks (i.e., nodes in a line) is in P, while it is NP-hard in the case of three-dimensional networks. For two-dimensional networks, the problem remains NP-hard [7].

Results concerning a variant of RA in which the range assignment induces a communication graph of diameter at most h, for some constant h, were also derived in [7, 8, 9, 18]. However, we believe this version of the problem is less interesting from a practical point of view. In fact, imposing a topology which is "too connected" would often cause communication interference to occur even between nodes that are far apart, thus decreasing the network capacity. This phenomenon is confirmed by theoretical as well as experimental results [13, 15, 16], which show that the communication graph in wireless ad hoc networks should be as sparse as possible, while preserving connectivity.

A simpler version of RA, in which all the units must have the same transmitting range r, was also investigated. We call this problem the ***homogeneous range assignment problem*** (HRA). The value of r ensuring connectivity with high probability when nodes are distributed in a given region according to some probability distribution was derived in [4, 14, 20, 21, 22, 27].

In this paper, we consider RA with the constraint that the range assignment be symmetric, i.e. such that edge (u, v) is in G if and only if (v, u) is in G. We call this problem the ***symmetric range assignment problem*** (SRA). We will also investigate a weaker version of the problem, called ***weak symmetric range assignment*** (WSRA), in which the requirement for symmetry applies only to a well defined subset of the edges. We are aware of only one paper addressing the symmetric range assignment problem [6], where the authors present a $(1 + \ln 2 + \epsilon)$ approximation algorithm for a problem equivalent to WSRA. Our interest in studying SRA and WSRA will be clearly motivated in the next section.

First, we show that SRA (and, hence, WSRA) remains NP-hard in two and three-dimensional networks. Hence, imposing (weak) symmetry does not change the complexity of the problem. Then, we investigate the asymptotic cost of the solution of WSRA. We prove that the solutions to RA and WSRA have the same asymptotic cost, thus showing that imposing weak symmetry on the communication graph has only a marginal influence on its energy cost. Finally, we determine bounds on the magnitude of this cost for two typical instances of the problem, i.e. the ***random instance***, in which nodes are distributed uniformly at random in $[0, 1]^d$, and the (Δ, δ)***-instance***, in which the maximum and minimum mutual distances between nodes are Δ and δ, respectively. The bounds presented in this paper can be compared to similar bounds for different variants of RA obtained in [4, 7, 8, 9, 18].

1. Motivation

While transceivers with dynamically changing transmitting range are compatible with current technology, most of the existing wireless devices, which are commonly based on either the IEEE 802.11 or the Bluetooth standard, have a fixed transmitting range. As a consequence, most of existing work on routing, clustering and broadcasting protocols for ad hoc networks assume that all the nodes have the same transmitting range [3, 11, 25].

Observe that routing, broadcasting and clustering protocols are not concerned with the network topology, but they simply assume that transmitting ranges are set in such a way that the resulting communication graph is connected. This means that the presence of an intermediate-level topology control service that dynamically changes node transmitting ranges in order to maintain connectivity while reducing power consumption can be considered. However, due to the homogeneous range assignment assumption, most of the protocols rely (either implicitly or explicitly) on the fact that whenever u sends a message to v, v is capable of communicating directly with u, e.g. to acknowledge the message reception. Hence, a topology control mechanism which returns a symmetric range assignment is transparent to higher level protocols, which can continue to operate as the range assignment was homogeneous. Further motivations for studying SRA can be found in the full version of the paper [5].

From the discussion above it is clear that the symmetry of the range assignment is a useful property in the design of protocols for wireless ad hoc networks. However, it should be noted that what is really important is the existence of a set of symmetric edges that connect all the nodes in the network. In other words, there could exist further edges for which symmetry is not guaranteed, but removing these edges from the communication graph does not cause disconnection. We call a range assignment with this property *weakly symmetric*.

It is important to note that while imposing weak symmetry does not impair connectivity, it has a beneficial effect on both node power consumption and network capacity. For a given set of nodes, it can be easily seen that the cost of the optimal symmetric range assignment is higher than that of the weakly symmetric one. Although quantifying the relation between these costs is not immediate, examples can be found in which the cost reduction is considerable. For instance, consider the node placement depicted in Figure 1. For the network to be connected, nodes u and v_1 must have a transmitting range of at least $n\delta$. If the range assignment must be symmetric, the transmitting range of v_{n+1} must be $n\delta$ too. Hence, all the v_i's, for $i = 2, \ldots, n+1$ must have a transmitting range that enables them to reach the farther between node v_1 and v_{n+1}, which is $\Theta(n\delta)$. Assuming that the energy cost is proportional to the square of the transmitting range (see the next section for a definition of the energy cost), we have that the cost of the symmetric range assignment must be $\Theta(n^3\delta^2)$. However, if only weak symmetry is required, all the v_i's, for $i = 2, \ldots, n+1$, have a transmitting range of δ, and the total cost is $\Theta(n^2\delta^2)$.Thus, the saving with respect to the symmetric case is of a factor $\Theta(n)$.

Summarizing, we can say that considering WSRA instead of SRA reflects the requirement for a connected but as sparse as possible communication graph, and it is then fully consistent with the philosophy of wireless ad hoc networks.

2. Preliminaries

Let $V = \{v_1, \ldots, v_n\}$ be a set of points in the d-dimensional Euclidean space. The set V represents the nodes of the network. For any two points v_i, v_j in V, $d(v_i, v_j)$ denotes the Euclidean distance between them.

A range assignment for V is a function $RA : V \to \mathbb{R}^+$. Given any range assignment RA for V, the communication graph induced by RA is the directed graph $G = (V, E)$, where edge $(v_i, v_j) \in E$ if and only if $d(v_i, v_j) \leq RA(v_i)$. A range assignment is said to be *connecting* if the resulting communication graph is strongly connected, and it is said to be *symmetric* if $(v_i, v_j) \in E$ if and only if $(v_j, v_i) \in E$. If the range assignment is symmetric, the communication graph can be regarded as undirected, and we are interested in characterizing connectivity instead of strong connectivity. A particular case of symmetric range assignment is the r-homogeneous range assignment, defined as $RA(v_i) = r$ for

$i = 1, ..., n$, where r is a positive constant. Given a range assignment RA and the corresponding communication graph G, we define the ***symmetric restriction*** of G as the subgraph G_S of G obtained by deleting non-symmetric edges, i.e. $G_S = (V, E_S)$, with $E_S = \{(v_i, v_j) | (d(v_i, v_j) \leq RA(v_i)) \wedge (d(v_i, v_j) \leq RA(v_j))\}$. A range assignment such that the symmetric restriction of its communication graph is strongly connected is said to be ***weakly symmetric***.

It is known [19] that the power p_i required by node v_i to correctly transmit data to node v_j must satisfy inequality $p_i / d(v_i, v_j)^\alpha \geq \beta$, where $\alpha \geq 1$ is the ***distance-power gradient*** and $\beta \geq 1$ is the ***transmission quality*** parameter. In ideal conditions we have $\alpha = 2$; however, in general it is $1 \leq \alpha \leq 6$ depending on environmental conditions.

Setting $\beta = 1$, we can define the cost of a range assignment RA as $c(RA) = \sum_{v_i \in V} (RA(v_i))^\alpha$. We are now ready to formally define the range assignment problems considered in this paper:

Definition 1 *Let $V = \{v_1, ..., v_n\}$ be a set of point in the d-dimensional space:*

- *RA* — *Determine a connecting range assignment RA such that c(RA) is minimum.*
- *WSRA* — *Determine a weakly symmetric range assignment RA such that c(RA) is minimum.*
- *SRA* — *Determine a connecting symmetric range assignment RA such that c(RA) is minimum.*
- *HRA* — *Determine the minimum value of r such that the r-homogeneous range assignment is connecting.*

In the following, the cost of the solutions of RA, WSRA, SRA and HRA will be denoted c_R, c_{WS}, c_S and c_{HR}, respectively. By definition, it is immediate that $c_R \leq c_{WS} \leq c_S \leq c_{HR}$.

3. Complexity of SRA

Due to space limitations, the proof that SRA (and, consequently, WSRA) for two and three-dimensional networks is NP-hard is not reported. See [5] for details.

4. Bounds on c_{WS}

In this section we investigate the cost of the optimal solution of WSRA and its relation with the cost of the solutions of other versions of the problem. We start by showing that c_R and c_{WS} have the same magnitude.

Given a set $V = \{v_1, ..., v_n\}$ of points in the d-dimensional space, denote with c_{MST} the cost of a minimum spanning tree on the same set of points, where edge (v_i, v_j) has cost $d(v_i, v_j)^\alpha$. The following theorem is a straightforward consequence of Theorem 3.2 of [18]. For the sake of completeness, we report part of its proof.

Theorem 1 *Let $V = \{v_1, ..., v_n\}$ be a set of points in the d-dimensional space, for $d = 2,3$. For every $\alpha \geq 1$ we have $c_{MST} < c_R \leq c_{WS} \leq 2c_{MST}$.*

Proof.
1) $c_{WS} \leq 2c_{MST}$. Given the MST T for V, we construct a range assignment RA' assigning to each node u a transmitting range equal to the α-th root of the maximum edge weight among the tree edges incident in u. Denote with G and G_S the communication graph and the symmetric restriction of the communication graph induced by RA'. It can be easily seen that every edge of T corresponds to a pair of symmetric edges in G_S. Hence, G_S is connected and RA' is a weakly symmetric range assignment. Considering that during the construction of RA' each edge of T can be chosen as the "longest" edge, i.e., as transmitting range, by at most two nodes, we have $c_{WS} \leq c(RA') \leq 2c_{MST}$.
2) $c_{MST} < c_R$. See proof of Theorem 3.2 in [18].

Theorem 1 proves that the solutions of RA and WSRA have the same asymptotic cost. Hence, the requirement for weak symmetry has only a marginal effect on the energy cost, while it eases significantly the integration of topology control mechanisms with existing higher and lower-level protocols.

Observe that Theorem 1 states that c_{MST}, c_R and c_{WS} have the same magnitude, but gives no clue on how large this magnitude actually is. In the following sub-sections we evaluate the magnitude of c_{WS} (hence, of c_R) for two typical instances of the range assignment problem.

4.1. The cost of the random instance

In the random instance, the network nodes are distributed uniformly at random in $[0,1]^d$. In this case, the magnitude of the cost of the Euclidean MST has been evaluated. The following theorem summarizes some results presented in [2, 28, 29].

Theorem 2 *Let X_i, $i \geq 1$, be i.i.d. random variables with values in $[0,1]^d$, $d \geq 2$, and let $M^\alpha(X_1, \ldots, X_n)$ be the cost of the* MST *of the Euclidean graph whose vertices are identified with $X_1, \ldots, X_n$, and such that for each $1 \leq i \neq j \leq n$ the weight of the edge e_{ij} is $d(X_i, X_j)^\alpha$. Then for all $d \geq 2$ and $\alpha \geq 1$,*

$$\lim_{n\to\infty} \frac{M^\alpha(X_1, \ldots, X_n)}{n^{1-\alpha/d}} = C(\alpha, d)$$

and

$$\left| \frac{M^\alpha(X_1, \ldots, X_n)}{n^{1-\alpha/d}} - C(\alpha, d) \right| \leq \frac{C}{n^{1/d}},$$

where $C(\alpha, d)$ denotes a positive constant depending only on α and d and C is a constant.

Combining theorems 1 and 2 we obtain the following tight bound for c_{WS} (and, consequently, for c_R) when nodes are distributed uniformly at random in $[0,1]^d$.

Theorem 3 *Let $V = \{v_1, ..., v_n\}$ be a set of points chosen uniformly and independently at random in $[0,1]^d$, for $d = 2,3$, and assume n is sufficiently large. Then, for every $\alpha \geq 1$ we have $c_{WS} = \Theta(n^{1-\frac{\alpha}{d}})$.*

The bound stated in Theorem 3 can be used to compare the magnitude of c_{WS} with the optimal cost of different versions of the range assignment problem in case of random instance. In the following discussion we assume $\alpha = d = 2$, since most of existing bounds are for this case. In [20] it is shown that for homogeneous range assignments, the communication graph is connected with high probability if and only if $r = \Omega(\sqrt{\frac{\log n}{n}})$. Hence, $c_{HR} = \Theta(\log n)$. When the diameter of the communication graph must be at most h, for some positive constant h, the cost of the optimal (asymmetric) range assignment is $\Theta(n^{1/h})$ [8]. Thus, while imposing weak symmetry on the range assignment increases its cost of at most a constant factor, the stronger constrains of either homogeneity or small diameter increases the cost significantly, namely of a factor at most $\log n$ in the first case and at most $n^{1/h}$ in the second. On one hand, these results encourage the utilization of a topology control mechanism to reduce power consumption. On the other hand, they discourage the stronger requirement of small diameter of the communication graph, which causes an increased energy cost and reduces (as discussed in the introduction) the network capacity.

4.2. The cost of the (Δ, δ)-instance

In this sub-section we consider the (Δ, δ)-instance of WSRA, in which the maximum and minimum mutual distances between nodes in V are Δ and δ, respectively. Since no results on the cost of the Euclidean MST are known in this case, we recur to a simple recursive construction technique that allows to obtain upper bounds to c_{WS} for $d = 2,3$ and $\alpha \geq 1$. These bounds are shown to be tight in some cases.

For the sake of simplicity, we describe the construction for $d = 2$. The construction for $d = 3$ is an easy modification. Observe that, without loss of generality, we can assume that all the nodes are placed in a square region S whose diagonal is Δ.

Let us begin our construction by dividing S into 4 quadrants. For each quadrant that contains at least one node, we choose one of them as representative. We set the transmitting ranges of these $p_1 \leq 4$ representatives in such a way that each of them can communicate directly with all the others. For the remaining $n - p_1$ nodes, we set the transmitting range to a value sufficient to communicate with the representative in their quadrant. Observe that this construction is weakly symmetric, its cost is at most $p_1\Delta^\alpha + (n - p_1)(\Delta/2)^\alpha$, and each node is at most 3 hops away from every other.

Let us proceed a step further, subdividing each non-empty quadrant in 4 subquadrants (see Figure 2). Again, in each of the non-empty subquadrants (excluding those containing the p_1 nodes chosen as representatives in the previous step) we select a node as representative. These $p_2 \leq 3 \cdot 4$ nodes will have a transmitting range large enough to communicate with their representative, i.e.,

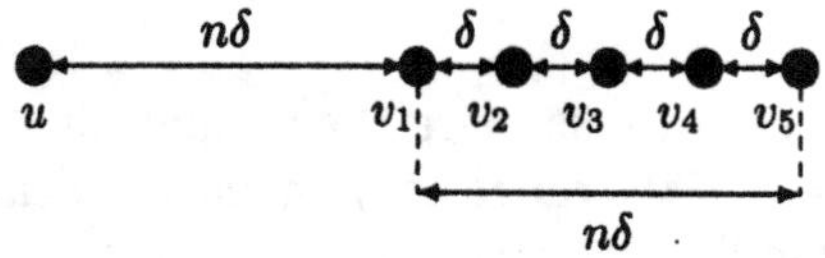

Figure 1. Node placement with consistent cost reduction from the symmetric to the weakly symmetric range assignment. The figure refers to the case of $n = 4$.

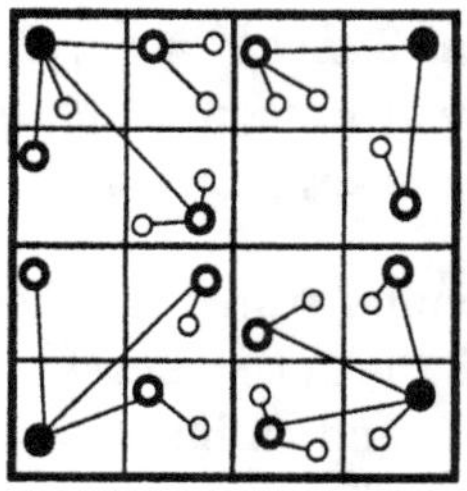

Figure 2. The second step of the construction. For simplicity, we omitted the connections between the first 4 representatives (the black full dots).

at most $\Delta/2$, while the remaining $(n - p_1 - p_2)$ nodes can communicate with the representative in their subquadrant with a transmitting range not exceeding $\Delta/4$. Again, the construction is weakly symmetric, its cost is at most

$$p_1\Delta^\alpha + p_2\left(\frac{\Delta}{2}\right)^\alpha + (n - p_1 - p_2)\left(\frac{\Delta}{4}\right)^\alpha,$$

and each station is at most 5 hops away from every other.

Repeating this construction for k steps, we obtain a weakly symmetric range assignment RA_k with cost

$$c(RA_k) \leq \sum_{i=1}^{k} p_i\left(\frac{\Delta}{2^{i-1}}\right)^\alpha + \left(n - \sum_{i=1}^{k} p_i\right)\left(\frac{\Delta}{2^k}\right)^\alpha,$$

and each station is at most $1+2k$ hops away from every other. This construction can be easily applied to the case $d = 3$, by means of recursive subdivisions in $2^3 = 8$ subcubes instead of $2^2 = 4$ subquadrants.

In general we have $p_1 \leq 2^d$, and $p_i \leq (2^d - 1)\cdot(2^d)^{i-1}$ for every $i > 1$, hence, by simple manipulation we obtain

$$c(RA_k) \leq \Delta^\alpha + (2^d - 1)\Delta^\alpha \sum_{i=0}^{k-1} \frac{2^{di}}{2^{\alpha i}} + \left(n - \sum_{i=1}^{k} p_i\right)\left(\frac{\Delta}{2^k}\right)^\alpha. \quad (1)$$

Let t be an integer such that $2^t > \Delta/\delta$. At step t each subquadrant has thus a diagonal smaller than δ and contains at most one node. Hence, at step $t+1$ all the nodes will be chosen as representative, i.e., $n = \sum_{i=1}^{t+1} p_i$, and the last term of (1) will vanish. So, letting $k' = \lceil \log_2(\Delta/\delta)\rceil + 1$ we have

$$c(RA_{k'}) \leq \Delta^\alpha + (2^d - 1)\Delta^\alpha \sum_{i=0}^{k'-1} \frac{2^{di}}{2^{\alpha i}}.$$

In general the ratio Δ/δ, and thus the number of recursive steps, can be arbitrarily large. However, it is easily seen that if $\Delta/\delta > n^{1/d}$, then a weakly symmetric range assignment with a smaller cost can be obtained by stopping the construction at step $k'' = (\log_2 n)/d$. In fact, by $(n - \sum_i p_i) \le n$, we obtain:

$$c(RA''_k) \le \Delta^\alpha + (2^d - 1)\Delta^\alpha \sum_{i=0}^{k''-1} \frac{2^{di}}{2^{\alpha i}} + n\left(\frac{\Delta}{2^{k''}}\right)^\alpha .$$

Considering the asymptotic values of $c(RA'_k)$ and $C(RA''_k)$ in the three cases $\alpha < d$, $\alpha = d$ and $\alpha > d$ we obtain:

$$c_{WS} = \begin{cases} O(\Delta^\alpha) & \text{if} \quad d < \alpha \\ O(\Delta^\alpha \min\{\log(\Delta/\delta), \log n\}) & \text{if} \quad d = \alpha \\ O(\Delta^\alpha \min\{(\Delta/\delta)^{d-\alpha}, n^{1-\frac{\alpha}{d}}\}) & \text{if} \quad d > \alpha \end{cases}$$

Observing that in general $\Delta/\delta = \Omega(n^{1/d})$ holds, the bounds on c_{WS} can be rewritten as

$$c_{WS} = \begin{cases} O(\Delta^\alpha) & \text{if} \quad d < \alpha \\ O(\Delta^\alpha \log n) & \text{if} \quad d = \alpha \\ O(\Delta^\alpha n^{1-\frac{\alpha}{d}}) & \text{if} \quad d > \alpha \end{cases} \tag{2}$$

The upper bounds established in equation (2) can be compared with the trivial lower bound of $\Omega(n\delta^\alpha)$ on the cost of any connecting range assignment for the (Δ, δ)-instance of the problem. Observe that this trivial lower bound cannot be improved, since it can actually be achieved when points are located in a lattice of step $n^{1/d}$. Hence, the construction presented above in general is not optimal. However, in the case of *well spread* instances, i.e. when $\Delta/\delta = \Theta(n^{1/d})$, our construction is optimal when $d > \alpha$ (for example, when $d = 3$ and $1 \le \alpha < 3$). In fact, in this case the lower bound can be rewritten as $\Omega(\Delta^\alpha n^{1-\frac{\alpha}{d}})$, which matches the cost of our construction when $d > \alpha$. It should also be observed that when the instance is well spread and $\alpha = d = 2$, an optimal asymptotic cost can be achieved using the construction based on the MST described in the proof of Theorem 1. This follows by results presented in [12, 28], where it is proved that given n points in $[0,1]^2$, there exists a spanning tree T such that $\sum_{(v_i,v_j)\in T}(d(v_i,v_j))^2 \le 2\sqrt{2}$. Scaling by a factor Δ, we have that the weakly symmetric range assignment obtained as described in the proof of Theorem 1 has cost $O(\Delta^2)$, which matches the lower bound. Unfortunately, the methods used in [12, 28] to derive these results depend heavily on geometric properties on the plane, hence they do not extend immediately to the case $\alpha = d = 3$. Our discussion can be summarized in the following theorem.

Theorem 4 *Let $V = \{v_1, ..., v_n\}$ be a set of points in the d-dimensional space, for $d = 2, 3$, corresponding to a well spread (Δ, δ)-instance of WSRA. Then*

$$c_{WS} = \begin{cases} \Theta(\Delta^\alpha) & \text{if} \quad d = \alpha = 2 \\ \Theta(\Delta^\alpha n^{1-\frac{\alpha}{d}}) & \text{if} \quad d > \alpha \end{cases}$$

Consider now the (Δ, δ)-instance of WSRA in which the nodes are concentrated at opposite corners of a cube of diagonal Δ. It is immediate that connectivity is achieved if and only if at least two nodes have a $\Theta(\Delta)$ transmitting range. Hence, we have $c_{WS} = \Omega(\Delta^\alpha)$, and the bound (2) is optimal in the worst-case when $d < \alpha$.

In a certain sense, the bounds for c_{WS} presented here extends those presented in [8] for the version of RA in which the diameter of the communication graph must be at most h, denoted RA_h in the following. The authors only considered the case $\alpha = d = 2$, and they show that the cost c_{R_h} of the solution of RA_h for a (Δ, δ)-instance is $\Omega(\delta^2 n^{1+1/h})$, for any positive constant h. They also present a construction which yields a solution of cost $O(\Delta^2 n^{1/h})$, which is optimal in the case of well-spread instances. Observe that our construction indeed yields a communication graph of diameter $O(\log n)$, hence it can be seen as a generalization of the construction of [8] to the case of $h = O(\log n)$. Furthermore, we cover also the case $d = 3$ and different combinations of the values α and d. When $\alpha = d = 2$, our construction yields a solution of cost $O(\Delta^2 \log n)$, which is smaller (as expected) than the cost $O(\Delta^2 n^{1/h})$ needed when the diameter must be constant. This indicates that also in the case of the (Δ, δ)-instance the diameter of the communication graph can be traded off with the energy cost.

Finally, we observe that when $d > \alpha$ (as it is likely to be for three-dimensional networks) and the instance is well spread, c_{WS} and c_{R_h} have the same magnitude for $h = O(\log n)$. This follows easily by the fact that our construction, which is optimal for well spread instances of WSRA when $d > \alpha$, produces a range assignment of diameter $O(\log n)$. Hence, for well spread instances (and when $d > \alpha$), a $O(\log n)$ diameter (instead of $O(n)$) comes with no additional (asymptotic) cost.

5. Conclusions

In this paper we have studied the impact of imposing the (weak) symmetry constraint to the range assignment problem for wireless ad hoc networks. We have shown that the requirement for symmetry (hence, for weak symmetry) does not change the complexity of the problem, which remains NP-hard for two and three-dimensional networks. We have also shown that the solutions of RA and WSRA have the same asymptotic cost. This means that the requirement for weak symmetry have small impact on the energy cost of the optimal solution. We have also determined bounds on the magnitude of the solution of WSRA for two typical instances of the problem, i.e. the random instance, that accounts for those situations in which node positions are not known in advance or may change with time, and the (Δ, δ)-instance, that accounts for the case in which at least partial information on node positions are available.

In summary, the results presented in this paper have shown that a weakly symmetric range assignment can reduce the energy cost considerably with respect to the homogeneous case, and that no further (asymptotic) benefit is expected from the asymmetric range assignment. Thus, the goal of a "good" topology control mechanism should be to provide a weakly symmetric range assignment, rather than an asymmetric range assignment as in the case of existing protocols. On the other hand, a stronger requirement on the diameter (constant or logarithmic in n) of the communication graph would increase the energy cost significantly while reducing the network capacity. However, when $d > \alpha$ a communication graph with diameter $O(\log n)$ is achievable with no additional (asymptotic) cost if the instance is well spread.

Observe that, due to the complexity of the problem, only heuristic approaches can be considered. In this perspective, the relation between WSRA and MST can be very useful in driving the design of a topology control protocol that returns a good approximation of the optimal solution. The design of a *distributed* weakly symmetric topology control mechanism is one of the most important problems left open.

Finally, establishing the relation between the cost of the solution to WSRA and SRA remains open.

References

[1] Wireless LAN Medium Access Control and Physical Layer Specifications, Aug. 1999. IEEE 802.11 Standard (IEEE Computer Society LAN MAN Standards Committee).

[2] D. Aldous, J.M. Steele, "Asymptotics for Euclidean Minimal Spanning Trees on Random Points", *Probab. Theory Relat. Fields*, Vol. 92, pp. 247–258, 1992.

[3] S. Basagni, D. Bruschi, I. Chlamtac, "A Mobility-Transparent Deterministic Broadcast Mechanism for Ad Hoc Networks", *IEEE Transactions on Networking*, Vol. 7, n. 6, pp. 799–807, 1999.

[4] D.M. Blough, P. Santi, "The Random Point Graph Model for Ad Hoc Networks and its Application to the Range Assignment Problem", *Tech. Rep. IMC-B4-01-05*, Istituto di Matematica Computazionale del CNR, Pisa - Italy, Dec. 2001.

[5] D.M. Blough, M. Leoncini, G. Resta, P. Santi, "On the Symmetric Range Assignment Problem in Wireless Ad Hoc Networks", *Tech. Rep. IMC-B4-01-07*, Istituto di Matematica Computazionale del CNR, Pisa - Italy, Nov. 2001.

[6] G. Calinescu, I.I. Mandoiu, A. Zelikovsky, "Symmetric Connectivity with Minimum Power Consumption in Radio Networks", *to appear in Proc. 2nd IFIP Conf. on Theoretical Computer Science*, Montreal, August 2002.

[7] A.E.F. Clementi, P. Penna, R. Silvestri, "Hardness Results for the Power Range Assignment Problem in Packet Radio Networks", *Proc. 2nd International Workshop on Approximation Algorithms for Combinatorial Optimization Problems (RANDOM/APPROX '99)*, LNCS (1671), pp. 197–208, 1999.

[8] A.E.F. Clementi, P. Penna, R. Silvestri, "The Power Range Assignment Problem in Radio Networks on the Plane", *Proc. XVII Symposium on Theoretical Aspects of Computer Science (STACS 00)*, LNCS (1770), pp. 651–660, 2000.

[9] A.E.F. Clementi, A. Ferreira, P. Penna, S. Perennes, R. Silvestri, "The Minimum Range Assignment Problem on Linear Radio Networks", *Proc. 8th European Symposium on Algorithms (ESA 2000)*, LNCS (1879), pp. 143–154, 2000.

[10] L.M. Feeney, M. Nilson, "Investigating the Energy Consumption of a Wireless Network Interface in an Ad Hoc Networking Environment", *Proc. IEEE INFOCOM 2001*, pp. 1548–1557, 2001.

[11] M. Gerla, J. Tzu-Chen Tsai, "Multicluster, Mobile, Multimedia Radio Network", *Wireless Networks*, Vol. 1, pp. 255–265, 1995.

[12] E.N. Gilbert, H.O. Pollak, "Steiner Minimal Trees", *SIAM J. Appl. Math.*, Vol. 16, pp. 1–29, 1968.

[13] M. Grossglauser, D. Tse, "Mobility Increases the Capacity of Ad Hoc Wireless Networks", *Proc. IEEE INFOCOM 2001*, pp. 1360–1369, 2001.

[14] P. Gupta, P.R. Kumar, "Critical Power for Asymptotic Connectivity in Wireless Networks", *Stochastic Analysis, Control, Optimization and Applications*, Birkhauser, Boston, pp. 547–566, 1998.

[15] P. Gupta, P.R. Kumar, "The Capacity of Wireless Networks", *IEEE Trans. Information Theory*, Vol. 46, n. 2, pp. 388–404, 2000.

[16] J. Li, C. Blake, D.S.J. De Couto, H. Imm Lee, R. Morris, "Capacity of Ad Hoc Wireless Networks", *Proc. ACM MOBICOM 2001*, pp. 61–69, 2001.

[17] L. Li, J.H. Halpern, P. Bahl, Y. Wang, R. Wattenhofer, "Analysis of a Cone-Based Distributed Topology Control Algorithm for Wireless Multi-hop Networks", *Proc. ACM PODC 2001*, 2001.

[18] L.M. Kirousis, E. Kranakis, D. Krizanc, A. Pelc, "Power Consumption in Packet Radio Networks", *Theoretical Computer Science*, Vol. 243, pp. 289–305, 2000.

[19] K. Pahlavan, A. Levesque, *Wireless Information Networks*, John Wiley and Sons, New York, 1995.

[20] P. Panchapakesan, D. Manjunath, "On the Transmission Range in Dense Ad Hoc Radio Networks", *Proc. IEEE SPCOM 2001*, 2001.

[21] T.K. Philips, S.S. Panwar, A.N. Tantawi, "Connectivity Properties of a Packet Radio Network Model", *IEEE Trans. Information Theory*,Vol. 35, n. 5, pp. 1044–1047, 1989.

[22] P. Piret, "On the Connectivity of Radio Networks", *IEEE Trans. Information Theory*,Vol. 37, n. 5, pp. 1490–1492, 1991.

[23] G.J. Pottie, W.J. Kaiser, "Wireless Integrated Network Sensors", *Communications of the ACM*, Vol. 43, n. 5, pp. 51– 58, 2000.

[24] R. Ramanathan, R. Rosales-Hain, "Topology Control of Multihop Wireless Networks using Transmit Power Adjustment", *Proc. IEEE Infocom 2000*, pp. 404–413, 2000.

[25] S. Ramanathan, M. Steenstrup, "A Survey of Routing Techniques for Mobile Communication Networks", *Mobile Networks and Applications*, Vol. 1, n. 2, pp. 89–104, 1996.

[26] V. Rodoplu, T.H. Meng, "Minimum Energy Mobile Wireless Networks", *IEEE Journal Selected Areas in Comm.*, Vol. 17, n. 8, pp. 1333–1344, 1999.

[27] P. Santi, D.M. Blough, F. Vainstein, "A Probabilistic Analysis for the Range Assignment Problem in Ad Hoc Networks", *Proc. ACM MobiHoc01*, pp. 212–220, 2001.

[28] J.M. Steele, "Growth Rates of Euclidean Minimal Spanning Trees with Power Weighted Edges", *Annals of Probability*, Vol. 16, pp. 1767–1787, 1988.

[29] J.E. Yukich, "Asymptotics for Weighted Minimal Spanning Trees on Random Points", *Stochastic Processes and their Appl.*, Vol. 85, pp. 123–128, 2000.

PARITY GRAPH-DRIVEN READ-ONCE BRANCHING PROGRAMS AND AN EXPONENTIAL LOWER BOUND FOR INTEGER MULTIPLICATION

(Extended Abstract)

Beate Bollig*

FB Informatik, LS2, Univ. Dortmund, Germany

bollig@ls2.cs.uni-dortmund.de

Stephan Waack

Institut für Numerische und Angewandte Mathematik

Georg-August-Universität Göttingen, Germany

waack@math.uni-goettingen.de

Philipp Woelfel*

FB Informatik, LS2, Univ. Dortmund, Germany

woelfel@ls2.cs.uni-dortmund.de

Abstract Branching programs are a well-established computation model for boolean functions, especially read-once branching programs have been studied intensively. Exponential lower bounds for deterministic and nondeterministic read-once branching programs are known for a long time. On the other hand, the problem of proving superpolynomial lower bounds for parity read-once branching programs is still open. In this paper restricted parity read-once branching programs are considered and an exponential lower bound on the size of well-structured parity graph-driven read-once branching programs for integer multiplication is proven. This is the first strongly exponential lower bound on the size of a nonoblivious parity read-once branching program model for an explicitly defined boolean function. In addition, more insight into the structure of integer multiplication is yielded.

*Supported in part by DFG grant WE 1066.

1. Introduction

Branching programs (BPs) or Binary Decision Diagrams (BDDs) are a well-established representation type or computation model for boolean functions.

Definition 1 A *branching program* (BP) or binary decision diagram (BDD) on the variable set $X_n = \{x_1, \ldots, x_n\}$ is a directed acyclic graph with one source and two sinks labeled by the constants 0 and 1. Each non-sink node (or internal node) is labeled by a boolean variable and has two outgoing edges, one labeled by 0 and the other by 1. A *nondeterministic branching program* is a generalized branching program where the number of edges leaving an internal node is not restricted.

An input $a \in \{0,1\}^n$ activates all edges consistent with a, i.e., the edges labeled by a_i which leave nodes labeled by x_i. A computation path for an input a in a BP G is a path of edges activated by a that leads from the source to a sink. A computation path for an input a which leads to the 1-sink is called accepting path for a.

The output for an input a is 1 iff there is an accepting path for a. A *parity branching program* is a nondeterministic branching program with the parity acceptance mode, i.e., an input is accepted iff the number of its accepting paths is odd.

The size of a branching program G is the number of its nodes and is denoted by $|G|$. The branching program size of a boolean function f is the size of the smallest BP representing f. The length of a branching program is the maximum length of a path.

The branching program size of a boolean function f is known to be a measure for the space complexity of nonuniform Turing machines and known to lie between the circuit size of f and its $\{\wedge, \vee, \neg\}$-formula size (see, e.g., [19]). Hence, one is interested in exponential lower bounds for more and more general types of BPs (for the latest breakthrough for semantic super-linear length BPs see [1], [3] and [4]). In order to develop and strengthen lower bound techniques one considers restricted computation models.

Definition 2 i) A branching program is called (syntactically) *read k times* (BPk) if each variable is tested on each path at most k times.

ii) A BP is called *s-oblivious*, for a sequence of variables $s = (s_1, \ldots, s_l)$, $s_i \in X_n$, if the set of its internal nodes can be partitioned into disjoint sets V_i, $1 \le i \le l$, such that all nodes from V_i are labeled by s_i and the edges which leave V_i-nodes reach a sink or a V_j-node, $j > i$.

Bryant [9] has introduced ordered binary decision diagrams (OBDDs) which are up to now the most popular representation for formal circuit verification. OBDDs are oblivious BP1s, where on each path from the source to a sink the variables are tested accoding to a *variable ordering* given by a permutation π on the variable set. Unfortunately, several important and also quite simple functions have exponential OBDD size. Therefore, Gergov and Meinel [12] and Sieling and Wegener [17] have generalized independently the concept of variable orderings.

Definition 3 A *graph ordering* is a branching program with a single sink, where on each path from the source to the sink all variables appear exactly once. A (parity) *graph-driven* BP1 with respect to a graph ordering G_0, (parity) G_0-BP1 for short, is a (parity) BP1 with the following additional property. If for an input a, a variable x_i appears on the unique computation path of a in G_0 before the variable x_j, then x_i also appears on all computation paths of a in G before x_j.

(Note that the size of a (parity) G_0-BP1 G is the number of nodes in G and not in G and G_0.)

For many restricted (nondeterministic) variants of branching programs exponential lower bounds are known (for a survey see e.g. [15]). Moreover, Thathachar [18] has been able to prove an exponential gap between the size of nondeterministic BPks and deterministic BP$(k+1)$s for an explicitly defined boolean function. His results have demonstrated that the lower bound techniques for these models are highly developed. Nevertheless, the problem of proving superpolynomial lower bounds for parity read-once branching programs is still open. Krause [13] has proved the first exponential lower bounds for oblivious parity branching programs with bounded length. Later, Savický and Sieling [16] have presented exponential lower bounds for restricted parity read-once branching programs. In their model only at the top of the read-once branching program parity nodes are allowed. Recently, Brosenne, Homeister, and Waack [8] have proved the first (not strongly) exponential lower bound on the size of restricted parity graph-driven BP1s representing the characteristic function of linear codes.

Motivated by applications the analysis of *natural* functions like the basic arithmetic functions is of interest.

Definition 4 *Integer multiplication* MUL_n maps two n-bit integers $x = x_{n-1}\dots x_0$ and $y = y_{n-1}\dots y_0$ to their product $x \cdot y = z = z_{2n-1}\dots z_0$. $\mathrm{MUL}_{i,n}$ denotes the boolean function defined as the ith bit of MUL_n.

The middle bit of multiplication ($\mathrm{MUL}_{n-1,n}$) is known to be the *hardest* bit. Hence, in the following we only consider the function $\mathrm{MUL}_n := \mathrm{MUL}_{n-1,n}$. For OBDDs Bryant [10] has presented an exponential lower bound of size $2^{n/8}$ for MUL_n. Incorporating Ramsey theoretic arguments of Alon and Maass [2] and using the rank method of communication complexity, Gergov [11] has extended the lower bound to arbitrary nondeterministic linear-length oblivious BPs. Recently, Woelfel [21] has improved Bryant's lower bound up to $\Omega(2^{n/2})$. The first exponential lower bound on the size of deterministic BP1s has been proven by Ponzio [14]. His lower bound is of order $2^{\Omega(n^{1/2})}$ and has been improved by Bollig and Woelfel [7] to the first strongly exponential lower bound of size $\Omega(2^{n/4})$ for MUL_n. Bollig [5] has presented the first (not strongly) exponential lower bound on the size of MUL_n for so-called nondeterministic tree-driven BP1s. Her result also holds for parity tree-driven BP1s. Until now exponential lower bounds on the size of MUL_n for general nondeterministic BP1s or BPks with $k \geq 2$ are unknown. Here we present an exponential lower bound on the size of restricted parity graph-driven BP1s for MUL_n. This is the first strongly exponential lower bound for this branching program model. In addition, we yield more insight into the structure of integer multiplication.

Due to the lack of space we have to omit some of the proofs. For a full version of the paper see [6].

2. The Lower Bound Criterion

In [17] a restricted variant of graph-driven BP1s has been investigated.

Definition 5 A graph-driven BP1 $G = (V, E)$ with respect to a graph ordering $G_0 = (V_0, E_0)$ is called *well-structured* if there exists a representation function $\alpha : V \to V_0$ with the following properties. The nodes v and $\alpha(v)$ are labeled by the same variable and for all inputs a such that v lies on the computation path for the input a the node $\alpha(v)$ lies on the path in G_0 which is activated by a.

Similar to the deterministic case well-structured parity G_0-BP1s are defined. The difference between graph-driven and well-structured graph-driven BP1s is the following one. In the general graph-driven model it is possible that two different inputs reach in G the same node labeled by x_i, whereas they reach in the graph-ordering G_0 different nodes labeled by x_i. This is not allowed in the well-structured case.

Brosenne, Homeister, and Waack [8] have realized how this restriction can be used to determine the number of nodes that is necessary to represent a boolean function f in a well-structured parity graph-driven BP1. A further observation which turns out to be very helpful in order to prove exponential lower bounds is the following one. The size of a well-structured parity graph-driven BP1 G and the size of a graph ordering G_0 of minimal size such that G is G_0-driven are polynomially related. First, we need the following lemma which is a slight generalization of a result from [17].

Lemma 1 ([8]) *Let G_0 be a graph ordering, v a node in a well-structured parity G_0-BP1 G, α the representation function, and $c \in \{0, 1\}$. If w is one of the c-successors of v in G then all paths to the sink in G_0 which leave $\alpha(v)$ via the c-edge pass through $\alpha(w)$.*

Proposition 1 *Let G be a well-structured parity graph driven* BP1 *on n variables. There exists a graph ordering G_0 such that G is G_0-driven and $|G_0| \leq 2n|G|$.*

Proof. Let G_0' be a graph ordering such that G is G_0'-driven and let $\mathcal{N}_v(G)$ be the set of nodes u in G such that $\alpha(u) = v$. First, we mark all nodes v in G_0' for which $\mathcal{N}_v(G)$ is not empty. Afterwards we eliminate all nodes which have not been marked in G_0'. An edge leading to one of theses nodes v is redirected to the first successor of v which has been marked. Because of Lemma 1 this node is uniquely determined. The resulting graph is a read-once branching program with one sink and at most $|G|$ nodes. Finally, we use the usual algorithm (see also [20]) to insert nodes such that on each path from the source to the sink there exist for each variable x_i exactly one node labeled by x_i. According to a topological ordering of the nodes, for each node v the set $V(v)$ of variables tested on some path from the source to v excluding the label of v is computed. Afterwards on each edge (v, w) dummy tests of the variables in $V(w) \setminus V(v)$ excluding the variable tested at v are added. A dummy test is a node where the 0- and the 1-edge lead to the same node.

The resulting graph ordering G_0 consists of at most $2n|G|$ nodes. It is easy to see that G is G_0-driven. □

The proof of Proposition 1 cannot be generalized in a straightforward way for (general) parity graph-driven BP1s because the existence of the α-function is an essential part of the proof. Until now exponential lower bounds on the size of general parity graph-driven BP1s are unknown.

In the following, we consider the representation of a boolean function f by its value table as an element of $(\mathbb{Z}_2)^{2^n}$. This set is a $\mathbb{Z}_2$ vector space where addition is component-wise parity and scalar multiplication by 0 or 1 is defined in the obvious way. Before we state our lower bound criterion, we have to introduce some notations. Let v be a node in the graph ordering G_0, G a well-structured parity G_0-driven BP1, $\mathcal{N}_v(G)$ the set of nodes u in G such that $\alpha(u) = v$, and f a boolean function. On all paths from the source to v the same set of variables has to be tested. W.l.o.g. let $x_1, \dots, x_{i-1}$ be the previously tested variables and let v be labeled by x_i. Let $A(v) \subseteq \{0,1\}^{i-1}$ be the set of vectors $(a_1, \dots, a_{i-1})$ such that v is reached for all inputs a starting with $(a_1, \dots, a_{i-1})$. We define $\mathcal{F}_v := \{f_{|x_1=a_1,\dots,x_{i-1}=a_{i-1}} | (a_1, \dots, a_{i-1}) \in A(v)\}$. The functions of $\mathcal{F}_v$ depend syntactically on all variables $x_1, \dots, x_n$ but they do not depend essentially on $x_1, \dots, x_{i-1}$. (A function g essentially depends on a variable x_j iff $g_{|x_j=0} \neq g_{|x_j=1}$.) Now let $\mathcal{P}_v$ be the set of all nodes that lie on a path leaving v in G_0 including v. Then we define $B_{f,v}^{G_0}$ as the boolean vector space spanned by all functions in $\bigcup_{w \in \mathcal{P}_v} \mathcal{F}_w$.

Let V be a vector space and V_1, V_2 be sub vector spaces of V. V_1 is said to be *linearly independent modulo* V_2, if $V_1 \cap V_2 = \{o\}$, i.e., $\dim V_1 + \dim V_2 = \dim(V_1 + V_2)$.

Lemma 2 *Let $A'(v)$ be a subset of $A(v)$ such that the subfunctions $f_{|x_1=a_1,\dots,x_{i-1}=a_{i-1}}$, $(a_1, \dots, a_{i-1}) \in A'(v)$, are linearly independent, and let $B_{f,A'}^{G_0}$ be the vector space spanned by these subfunctions. If $B_{f,A'}^{G_0}$ is linearly independent modulo the vector space of all subfunctions in $B_{f,v}^{G_0}$ not essentially depending on x_i, then $|\mathcal{N}_v(G)| \geq |A'(v)|$.*

3. Integer Multiplication and the Matrix Game

We start our investigations with two technical lemmas which provide important properties of the function MUL_n.

In the rest of the paper $[x]_{n-k}^{n-1}$ denotes the bits at position $n-1$ to $n-k$ in the binary representation of the integer x. Using universal hashing Bollig and Woelfel [7, proof of Lemma 5] have shown the following.

Lemma 3 (Covering Lemma) *Let* $X \subseteq \mathbb{Z}_{2^n}$ *and* $Y \subseteq \mathbb{Z}_{2^n}^* := \{1, 3, \ldots, 2^n - 1\}$. *If* $|X| \cdot |Y| \geq 2^{n+2k+1}$, $k \geq 0$, *then there exists an element* $y^* \in Y$ *such that*

$$\forall z \in \{0, \ldots, 2^k - 1\}\ \exists x \in X : \ [xy^*]_{n-k}^{n-1} = z.$$

The lemma states that if X and Y are large enough sets of (odd) n-bit integers, then by choosing an appropriate $y \in Y$, the possible outcomes in the bits $n-1, \ldots, n-k$ of the products xy for $x \in X$ cover all possible k-bit values. (Note that Bollig and Woelfel [7] have proved this statement only implicitly in a non-parameterized form.) We now state another important lemma about integer multiplication, which is a generalization of Lemma 6 from [7].

Lemma 4 (Distance Lemma) *Let* $Y \subseteq \mathbb{Z}_{2^{n-1}}^*$, $1 \leq k \leq n-3$, *and* $(z_i, z_i') \in \mathbb{Z}_{2^{n-1}} \times \mathbb{Z}_{2^{n-1}}$, *where* $z_i \neq z_i'$, $1 \leq i \leq t$. *Then there exists a subset* $Y' \subseteq Y$ *with*

$$\forall y \in Y' : 4 \cdot 2^{n-k-1} \leq ((z_i - z_i')y) \bmod 2^{n-1} \leq 2^{n-1} - 4 \cdot 2^{n-k-1}$$

such that $|Y'| \geq |Y| - t \cdot 2^{n-k+1}$.

Proof. Let $\delta_i := (z_i - z_i') \bmod 2^{n-1}$, $1 \leq i \leq t$, and

$$\begin{aligned} M' &:= \{0, \ldots, 4 \cdot 2^{n-k-1} - 1\} \text{ and} \\ M'' &:= \{2^{n-1} - 4 \cdot 2^{n-k-1} + 1, \ldots, 2^{n-1} - 1\}. \end{aligned}$$

Let Y' be the set of all $y \in Y$ where $(y\delta_i) \bmod 2^{n-1} \notin M' \cup M''$ for all $i \in \{1, \ldots, t\}$. Bollig and Woelfel [7, proof of Lemma 4] have shown that the number of $y \in Y$ with $(y\delta_i) \bmod 2^{n-1} \in M' \cup M''$ for a fixed $i \in \{1, \ldots, t\}$ is bounded above by 2^{n-k+1}. Therefore, for at most $t \cdot 2^{n-k+1}$ elements $y \in Y$ there exists at least one element $i \in \{1, \ldots, t\}$ such that $(y\delta_i) \bmod 2^{n-1} \in M' \cup M''$. Altogether, we have proved that the size of Y' is at least $|Y| - t \cdot 2^{n-k+1}$. □

Before we state our main lemma about properties of integer multiplication, we motivate our investigations. Let G_0 be a graph ordering which is not too large. Then we can prove that there exists a node v such that w.l.o.g. at least as many x- as y-variables have been tested from the source to v, v is labeled by a variable x_i, and there is a partial assignment a^* to the y-variables tested on the paths to v such that many paths which agree for the tested y-variables with a^* lead to v. Let $A'(v)$ be the set of these assignments. Now our aim is to prove that the boolean vector space spanned by the subfunctions of MUL_n according to $A'(v)$ is linearly independent modulo the vector space spanned by all subfunctions not essentially depending on V^*, where V^* contains x_i and the variables which have been tested on the paths to v. Then we can conclude using Lemma 2 that the size of well-structured parity G_0-BP1s representing MUL_n is large.

In the following, we investigate integer multiplication of two binary numbers $x = (x_{n-1}, \ldots, x_0)$ and $y = (y_{n-1}, \ldots, y_0)$, where $x_{n-1} = y_{n-1} = 0$ and $x_0 = y_0 = 1$. Let $V_x = \{x_1, \ldots, x_{n-2}\}$ and $V_y = \{y_1, \ldots, y_{n-2}\}$. Furthermore, let $V'_x \subseteq V_x$ ($V'_y \subseteq V_y$) be a set of m x-variables (y-variables), where $m \leq \lfloor (n-17)/6 \rfloor$. We fix an arbitrary assignment of the V'_y-variables. Now we consider a $2^m \times 2^{2n-2m-4}$ matrix M. Each row is associated with one assignment of the V'_x-variables and each column with an assignment of the variables from $V_x \setminus V'_x$ and $V_y \setminus V'_y$. Together with the fixed assignment of the V'_y-variables, $x_{n-1} = y_{n-1} = 0$ and $x_0 = y_0 = 1$, we obtain two well-defined n-bit numbers $x_{r,c}$ and y_c for each pair (r, c) of a row and a column. We define $M_{r,c}$ as $\mathrm{MUL}_n(x_{r,c}, y_c)$. Finally, we define for an arbitrary fixed variable $x_i \in V_x \setminus V'_x$ and a column c the column c' as the one which only differs from c by the assignment to the variable x_i.

Now our aim is to show that for an arbitrary choice of different rows $r^1, \ldots, r^l$, there exists a column c such that

$$\bigoplus_{j=1}^{l} M_{r^j,c} \neq \bigoplus_{j=1}^{l} M_{r^j,c'}. \tag{1}$$

Before we show (1) we illustrate how this property can be used to prove lower bounds using Lemma 2. The set of all possible assignments of the V'_x- and V'_y-variables is a superset of the set $A(v)$. By fixing the V'_y-variables by an arbitrary assignment, we obtain a set $A^*(v)$ which determines the matrix M. The number of a row of M identifies an assignment α determined by an element in $A^*(v)$ and the row itself represents the function vector of the subfunction $\mathrm{MUL}_{|\alpha}$. In this setting, (1) is the following. If we take an arbitrary linear combination of subfunctions (represented by the rows $r^1, \ldots, r^l$), then there exist two assignments to the variables in $(V_x \setminus V'_x) \cup (V_y \setminus V'_y)$ differing only in their setting to x_i such that the function value of the linear combination is different for both assignments. Hence, no subfunction not essentially depending on the V'_x- and V'_y-variables and x_i can be represented as a linear combination of the subfunctions determined by $A^*(v)$. By Lemma 2, this allows the conclusion that $|\mathcal{N}_v(G)| \geq |A'(v)|$, where $A'(v) \subseteq A^*(v)$.

We return to the proof of (1). Let $x_{r,c}$ be the number in $\mathbb{Z}^*_{2^n-1}$ defined by the choice of a row r and a column c and y_c the number in $\mathbb{Z}^*_{2^n-1}$ defined by the choice of the column c and the fixed assignment of the V'_y-variables. Therefore,

$$M_{r,c} = [x_{r,c} \cdot y_c]_{n-1}.$$

The number $x_{r,c}$ can be written as the sum of two components $x_r^{row} + x_c^{col}$, where x_r^{row} is the number defined by the partial assignment of the V_x'-variables given by the row r and the 0-assignment of the variables from $V_x \setminus V_x'$ and x_c^{col} is the number defined by the partial assignment of the variables from $V_x \setminus V_x'$, $x_0 = 1$, and the 0-assignment of the V_x'-variables. It follows that $M_{r,c} = [(x_r^{row} + x_c^{col}) \cdot y_c]_{n-1}$.

We take a look at the columns where for an arbitrary i the variable x_i is set to 0. Obviously the set of all pairs (x_c^{col}, y_c) of theses columns c corresponds to a set $X \times Y$ where $X, Y \subseteq \mathbb{Z}^*_{2^{n-1}}$, $|X| = 2^{n-m-3}$, and $|Y| = 2^{n-m-2}$. Furthermore, $x_{c'}^{row} - x_c^{row} = 2^i$. Finally, the choice of l rows $r^1, \ldots, r^l$ corresponds to the numbers $x_{r^1}^{row}, \ldots, x_{r^l}^{row}$. For the ease of description we denote these numbers by $x^1, \ldots, x^l$.

Summarizing, our aim is to prove that, under the assumption discussed above, for arbitrarily chosen $x^1, \ldots, x^l$ there exists a pair $(x, y) \in X \times Y$ such that the number of indices $j \in \{1, \ldots, l\}$ for which

$$[(x^j + x)y]_{n-1} \neq [(x^j + x + 2^i)y]_{n-1}$$

is odd. Formally this leads to the statement of Lemma 5.

Lemma 5 *Let $m \le \lfloor (n-17)/6 \rfloor$, $1 \le l \le 2^m$, $X, Y \subseteq \mathbb{Z}^*_{2^{n-1}}$, $d \neq 0$, and let $x^1, \ldots x^l$ be elements from $\mathbb{Z}_{2^{n-1}}$ with the following properties:*

i) $|X| \ge 2^{n-m-3}$ *and* $|Y| \ge 2^{n-m-2}$,

ii) $\forall 2 \le j \le l: \ x^1 \neq x^j$ *and* $\forall 1 \le j \le l: \ x^1 \neq x^j + d$,

iii) for all $x \in X$ *and all* $1 \le j \le l$: $x + x^j + d < 2^{n-1}$.

Let $(x, y) \in X \times Y$ and let $\sigma(x, y)$ be the number of indices $j \in \{1, \ldots, l\}$ where $[(x^j + x)y]_{n-1} \neq [(x^j + x + d)y]_{n-1}$. Then there exists a pair $(x, y) \in X \times Y$ such that $\sigma(x, y)$ is odd.

Obviously, the conditions of Lemma 5 are fulfilled for $d = 2^i$ and our choice of $x^1, \ldots, x^l$ and X and Y as described above. (Note, that we have achieved (iii) by setting $x_{n-1} = y_{n-1} = 0$.)

Proof. Let $k = 2m + 5$ and $X' := \{x^1 + x \mid x \in X\}$. Clearly $|X'| = |X| \ge 2^{n-m-3}$. Because of condition (iii), X' is a subset of $\mathbb{Z}_{2^{n-1}}$. First, we consider the $2l - 1$ pairs (x^1, z) where $z \in Z := \{x^2, \ldots, x^l\} \cup \{x^1 + d, \ldots, x^l + d\}$. Because of condition (iii), all $z \in Z$ are elements of $\mathbb{Z}_{2^{n-1}}$ and, because of condition (ii), they are all different from x^1. Let Y' be the set of all $y \in Y$ such that for all pairs (x^1, z), $z \in Z$,

$$4 \cdot 2^{n-k-1} \le ((z - x^1)y) \bmod 2^{n-1} \le 2^{n-1} - 4 \cdot 2^{n-k-1}. \tag{2}$$

According to Lemma 4

$$\begin{aligned} |Y'| &\ge |Y| - (2l-1)2^{n-k+1} > |Y| - 2^{m+1+n-k+1} \\ &\ge 2^{n-m-2} - 2^{n-m-3} = 2^{n-m-3}. \end{aligned}$$

Here we have used the fact that $2l \le 2^{m+1}$. Using $m \le \lfloor (n-17)/6 \rfloor$ we can conclude that

$$|X'| \cdot |Y'| \ \ge \ 2^{2n-2m-6} \ \ge \ 2^{2n-n/3+17/3-6} \ = \ 2^{n+(2/3)n-1/3}.$$

Since $k = 2m+5$, it follows that

$$2^{n+2k+1} \ = \ 2^{n+4m+11} \ \le \ 2^{n+(2/3)n-34/3+11} \ = \ 2^{n+(2/3)n-1/3}$$

such that we obtain $|X'| \cdot |Y'| \ \ge \ 2^{n+2k+1}$. Now we can apply Lemma 3. According to this there exist an element $y^* \in Y'$ and $x^*, x^{**} \in X'$ such that

$$[x^* y^*]_{n-k}^{n-1} \ = \ 2^{k-1} - 1 \quad \text{and} \quad [x^{**} y^*]_{n-k}^{n-1} \ = \ 2^{k-1}.$$

Let $y = y^*$. According to the definition of X' we can write x^* as $x^1 + x$ and x^{**} as $x^1 + x'$ for two elements $x, x' \in X$ such that

$$[(x^1 + x)y]_{n-k}^{n-1} \ = \ 2^{k-1} - 1 \quad \text{and} \quad [(x^1 + x')y]_{n-k}^{n-1} \ = \ 2^{k-1}. \tag{3}$$

Next we prove the following claims for x and x':

(C1) $[(x^1 + x)y]_{n-1} \neq [(x^1 + x')y]_{n-1}$.

(C2) For all $2 \le i \le l$: $[(x^i + x)y]_{n-1} = [(x^i + x')y]_{n-1}$.

(C3) For all $1 \le i \le l$: $[(x^i + x + d)y]_{n-1} = [(x^i + x' + d)y]_{n-1}$.

Using these claims we can prove in the following way that either $\sigma(x,y) = \sigma(x',y) - 1$ or $\sigma(x,y) = \sigma(x',y) + 1$. From (C1) and (C3) for $i = 1$ we can conclude that

$$[(x^1 + x)y]_{n-1} = [(x^1 + x + d)y]_{n-1} \ \Leftrightarrow \ [(x^1 + x')y]_{n-1} \neq [(x^1 + x' + d)y]_{n-1},$$

and from (C2) and (C3) that

$$[(x^i + x)y]_{n-1} = [(x^i + x + d)y]_{n-1} \ \Leftrightarrow \ [(x^i + x')y]_{n-1} = [(x^i + x' + d)y]_{n-1}$$

for $i = 2, \ldots, l$.

Therefore, exactly one of the values $\sigma(x,y)$ or $\sigma(x',y)$ is odd and we can complete our proof by proving (C1)-(C3). (C1) follows immediately from equation (3). To prove (C2) and (C3) we reconsider the pairs (x^1, z), $z \in Z = \{x^2, \ldots, x^l, x^1 + d, \ldots, x^l + d\}$. Obviously, it is sufficient to prove that $[(z + x)y]_{n-1} = [(z + x')y]_{n-1}$, for all $z \in Z$. We assume that this is not the case, w.l.o.g. $[(z + x)y]_{n-1} = 0$ and $[(z + x')y]_{n-1} = 1$ (the other case follows similarly).

According to equation (3) it follows that

$$2^{n-1} - 2^{n-k} \ \le \ ((x^1 + x)y) \bmod 2^n \ < \ 2^{n-1} \quad \text{and} \tag{4}$$

$$2^{n-1} \ \le \ ((x^1 + x')y) \bmod 2^n \ < \ 2^{n-1} + 2^{n-k}. \tag{5}$$

From this it follows that

$$1 \le ((x' - x)y) \bmod 2^n < 2 \cdot 2^{n-k}. \qquad (6)$$

From our assumption $[(z+x)y]_{n-1} = 0$ and $[(z+x')y]_{n-1} = 1$ we know that

$$((z+x)y) \bmod 2^n < 2^{n-1} \le ((z+x')y) \bmod 2^n.$$

Since $((z+x')y) \bmod 2^n - ((z+x)y) \bmod 2^n = ((x'-x)y) \bmod 2^n$, we can conclude using inequality (6)

$$2^{n-1} - 2 \cdot 2^{n-k} \le ((z+x)y) \bmod 2^n < 2^{n-1}.$$

Together with inequality (4) we obtain

$$-2 \cdot 2^{n-k} < ((z+x)y) \bmod 2^n - ((x^1+x)y) \bmod 2^n < 2^{n-k}.$$

Considering all terms in this inequality modulo 2^{n-1} it follows that

$$((z-x^1)y) \bmod 2^{n-1} < 2^{n-k} \quad \text{or} \quad ((z-x^1)y) \bmod 2^{n-1} > 2^{n-1} - 2 \cdot 2^{n-k}.$$

But this is a contradiction to inequality (2) and we are done. □

Altogether, we have proved that the vector space spanned by all subfunctions of MUL_n according to all assignments of the m V'_x-variables and an arbitrary assignment a^* of the m V'_y-variables is linearly independent modulo the vector space spanned by all subfunctions of MUL_n according to all assignments of the V'_x- and V'_y-variables not essentially depending on a variable x_i from $V_x \setminus V'_x$.

4. A Strongly Exponential Lower Bound for Integer Multiplication

Combining the new lower bound technique for well-structured parity graph-driven BP1s with Lemma 5 we prove the first strongly exponential lower bound on the size of a nonoblivious parity branching program model.

Theorem 1 *The size of well-structured parity graph-driven BP1s representing* MUL_n *is bounded below by* $2^{(n-46)/12}/n$.

Proof. Let G be a well-structured parity graph-driven BP1 representing MUL_n and G_0 be a graph ordering of minimal size such that G is G_0-driven. We may assume that the size of G_0 is at most $2^{1/2\lfloor (n-17)/6 \rfloor}$, because otherwise using Proposition 1 we can conclude that the size of parity graph-driven BP1s representing MUL_n is bounded below by

$$2^{1/2\lfloor (n-17)/6 \rfloor}/(4n) \ge 2^{(1/2)\cdot(n-22)/6}/(4n) = 2^{(n-46)/12}/n.$$

Let $m := \lfloor (n-17)/6 \rfloor$, $V_x = \{x_1, \ldots, x_{n-2}\}$, and $V_y = \{y_1, \ldots, y_{n-2}\}$. Since on all paths in G_0 all variables have to be tested, it is obvious that on all paths from the source to a node v the same set of variables is tested. In the following we only investigate paths where $x_0 = y_0 = 1$ and $x_{n-1} = y_{n-1} = 0$. We define a cut in the graph ordering G_0 in the following way. The cut consists of all nodes v where v is labeled by a V_x-variable and on all paths to v exactly m V_x-variables and at most m V_y-variables have been tested (or vice versa). On each path in G_0 there is exactly one node of the cut. Using the pigeonhole principle there exists one node v which lies on at least $2^{2n-4}/|G_0|$ paths from the source to the sink. W.l.o.g. v is labeled by x_i, and m V_x-variables and m' V_y-variables, $m' \leq m$, have been tested. Using the pigeonhole principle again there exists one partial assignment a^* to the V_y-variables tested on the paths from the source to v such that there are at least $2^m/|G_0|$ paths to v which agree for the V_y-variables with the partial assignment a^*. Let $A'(v)$ be the set of all assignments associated with these paths, V'_x (V'_y) be the set of the x-variables (y-variables) which have been tested, and let v be labeled by x_i. Clearly the requirements from Lemma 5 are fulfilled and we can conclude that the vector space spanned by all subfunctions according to $A'(v)$ is linearly independent modulo the vector space of all subfunctions not essentially depending on the V'_x- and the V'_y-variables and x_i. Therefore, we obtain the result

$$|\mathcal{N}_v(G)| \geq |A'(v)| \geq 2^{1/2\lfloor (n-17)/6 \rfloor}.$$

Altogether, we have proved a lower bound of $2^{1/2\lfloor (n-17)/6 \rfloor}/4n$,which is at least $2^{(n-46)/12}/n$, on the size of well-structured parity graph-driven BP1s representing MUL_n. □

Acknowledgments

We would like to thank Stefan Droste and Ingo Wegener for proofreading and fruitful discussions on the subject of the paper.

References

[1] M. Ajtai. A non-linear time lower bound for boolean branching programs. In *Proceedings of the 40th Annual IEEE Symposium on Fountations of Computer Science*, pp. 60–70. 1999.

[2] N. Alon and W. Maass. Meanders and their applications in lower bounds arguments. *Journal of Computer and System Sciences*, 37:118–129, 1988.

[3] P. Beame, M. Saks, X. Sun, and E. Vee. Super-linear time-space tradeoff lower bounds for randomized computation. In *Proceedings of the 41st Annual IEEE Symposium on Fountations of Computer Science*, pp. 169–179. 2000.

[4] P. Beame and E. Vee. Time-space tradeoffs, multiparty communication complexity, and nearest-neighbor problems. In *Proceedings of the 34th Annual ACM Symposium on Theory of Computing*. 2002. To appear.

[5] B. Bollig. Restricted nondeterministic read-once branching programs and an exponential lower bound for integer multiplication. *RAIRO Theoretical Informatics and Applications*, 35:149–162, 2001.

[6] B. Bollig, S. Waack, and P. Woelfel. Parity graph-driven read-once branching programs and an exponential lower bound for integer multiplication. Technical Report TR01-73, Electronic Colloquium on Computational Complexity, 2001.

[7] B. Bollig and P. Woelfel. A read-once branching program lower bound of $\Omega(2^{n/4})$ for integer multiplication using universal hashing. In *Proceedings of the 33rd Annual ACM Symposium on Theory of Computing*, pp. 419–424. 2001.

[8] H. Brosenne, M. Homeister, and S. Waack. Graph-driven free parity BDDs: Algorithms and lower bounds. In *Mathematical Foundations of Computer Science: 26th International Symposium*, volume 2136 of *Lecture Notes in Computer Science*, pp. 212–223. 2001.

[9] R. E. Bryant. Graph-based algorithms for boolean function manipulation. *IEEE Transactions on Computers*, C-35:677–691, 1986.

[10] R. E. Bryant. On the complexity of VLSI implementations and graph representations of boolean functions with applications to integer multiplication. *IEEE Transactions on Computers*, 40:205–213, 1991.

[11] J. Gergov. Time-space tradeoffs for integer multiplication on various types of input oblivious sequential machines. *Information Processing Letters*, 51:265–269, 1994.

[12] J. Gergov and C. Meinel. Efficient analysis and manipulation of OBDDs can be extended to FBDDs. *IEEE Transactions on Computers*, 43:1197–1209, 1994.

[13] M. Krause. Separating ⊕L from L, NL, co-NL, and AL (=P) for oblivious turing machines of linear access time. *RAIRO Theoretical Informatics and Applications*, 26:507–522, 1992.

[14] S. Ponzio. A lower bound for integer multiplication with read-once branching programs. *SIAM Journal on Computing*, 28:798–815, 1998.

[15] A. Razborov. Lower bounds for deterministic and nondeterministic branching programs. In *Proc. of Fundamentals in Computation Theory*, volume 529 of *Lecture Notes in Computer Science*, pp. 47–60. 1991.

[16] P. Savický and D. Sieling. A hierarchy result for read-once branching programs with restricted parity nondeterminism. In *Mathematical Foundations of Computer Science: 25th International Symposium*, volume 1893 of *Lecture Notes in Computer Science*, pp. 650–659. 2000.

[17] D. Sieling and I. Wegener. Graph driven BDDs – a new data structure for Boolean functions. *Theoretical Computer Science*, 141:283–310, 1995.

[18] J. S. Thathachar. On separating the read-k-times branching program hierarchy. In *Proceedings of the 30th Annual ACM Symposium on Theory of Computing*, pp. 653–662. 1998.

[19] I. Wegener. *The Complexity of Boolean Functions*. Wiley-Teubner, 1987.

[20] I. Wegener. *Branching Programs and Binary Decision Diagrams - Theory and Applications*. SIAM, first edition, 2000.

[21] P. Woelfel. New bounds on the OBDD-size of integer multiplication via universal hashing. In *Proceedings of the 18th Annual Symposium on Theoretical Aspects of Computer Science*, volume 2010 of *Lecture Notes in Computer Science*, pp. 563–574. 2001.

COMPUTABILITY OF LINEAR EQUATIONS

Vasco Brattka*
Theoretische Informatik 1, FernUniversität Hagen
58084 Hagen, Germany
Vasco.Brattka@FernUni-Hagen.de

Martin Ziegler**
Heinz Nixdorf Institute, University of Paderborn
33095 Paderborn, Germany
ziegler@uni-paderborn.de

Abstract Do the solutions of linear equations depend computably on their coefficients? Implicitly, this has been one of the central questions in linear algebra since the very beginning of the subject and the famous Gauß algorithm is one of its numerical answers. Today there exists a tremendous number of algorithms which solve this problem for different types of linear equations. However, actual implementations in floating point arithmetic keep exhibiting numerical instabilities for ill-conditioned inputs. This situation raises the question which of these instabilities are intrinsic, thus caused by the very nature of the problem, and which are just side effects of specific algorithms. To approach this principle question we revisit linear equations from the rigorous point of view of computability. Therefore we apply methods of computable analysis, which is the Turing machine based theory of computable real number functions. It turns out that, given the coefficients of a system of linear equations, we can compute the space of solutions, if and only if the dimension of the solution space is known in advance. Especially, this explains why there cannot exist any stable algorithms under weaker assumptions.

Track 1: Algorithms, Complexity and Models of Computation.

Keywords: Computable Analysis, Linear Equations.

*Work partially supported by DFG Grant BR 1807/4-1
**Work partially supported by DFG Grant Me 872/7-3

1. Introduction

In this paper we want to study computability properties of systems of linear equations $Ax = b$, where $A \in \mathbb{R}^{m \times n}$ is a real matrix, $b \in \mathbb{R}^m$ is a real vector and $x \in \mathbb{R}^n$ is a variable. Solving systems of linear equations *effectively*, that is, by means of some sort of computing machinery, has been a particularly prominent subject in both mathematics and computer science. Gauß, when inventing his now famous algorithm, considered real numbers as entities which, in one step, can be operated on exactly. This idea has been captured, in order to analyze the complexity of other numerical algorithms as well, by Blum, Cucker, Shub, and Smale's model of real number computation [1].

In contrast, numerical mathematics takes into account properties of floating point numbers by tracing propagation of rounding errors throughout the computational process. There, the input is presumed to be *exact* (a rational number of for example type `float`) and one asks for the required accuracy (e.g., `double`) for intermediate computations in order to obtain the result with desired precision (say, `float` again). Notice that, strictly speaking, no reals but only rational numbers are involved for input, processing, and output.

Numerical analysis provides a large number of algorithms to solve linear equations, such as Gauß' elimination algorithm, Cholesky's decomposition algorithm and many others. Unfortunately, the applicability of these algorithms is limited by their numerical instability and error analysis is a non-trivial topic of research (cf. [12] as a standard text on this topic).

In the present work, we consider a truly *real* model of computation based on approximation by sequences of rational numbers. This model takes into account that, in practice, input data like $1.6 \cdot 10^{-19}$ (negative charge, in Coulomb, of an electron) in fact is *not* exact but has been obtained, for instance, by a physical measurement (here: by R.A. Millican, 1911) together with some error bound (say, $\pm 0.1 \cdot 10^{-19}$). Improved experimental technique later yielded better approximations like $(1.6022 \pm 0.0001) \cdot 10^{-19}$; a further increase in precision can be expected for the future.

This observation is captured in a model of real number computation introduced by Alan Turing [9] and Grzegorczyk [4] which nowadays forms the basis of *Computable Analysis* [8, 6, 10]: A real number $x \in \mathbb{R}$ is fed into a computer by means of an infinite sequence of rationals q_n and corresponding error bounds ε_n such that $|x - q_n| \leq \varepsilon_n \to 0$. Upon this input, it *computes* result $y = f(x)$ if it outputs two corresponding rational sequences (p_n) and (δ_n) for $y \in \mathbb{R}$.

One advantage of this approach is that it allows to unify the development of algorithms and their stability analysis in a single model. And although real numbers are represented by approximations in this model, one can still consider them as entities to operate on algorithmically: The *feasible real RAM* model, introduced in [2], characterizes the Turing machine approach in terms of real RAMs.

Since Turing's seminal work, a vast number of publications have investigated of various problems over real numbers from areas ranging from Classical Analysis [8] over Geometry, Theory of Fractals [7, 5] to Physics [8, 11]. In fact, several of them turned out to be *un*computable with respect to this model and thus explained for numerical instabilities experienced in actual software implementations.

In the present work, we address one of the very basic problems in Linear Algebra: Solving systems of linear equations $Ax = b$. The set of solutions of such equations are exactly the *affine* subsets $L \subseteq \mathbb{R}^n$ and any such subset can equivalently be characterized by an affine basis. But are these characterizations also *computably* equivalent? In other words: Can some Turing Machine effectively convert between the following representations for an affine subspace $L \subseteq \mathbb{R}^n$:

a) matrix $A \in \mathbb{R}^{m \times n}$ and vector $b \in \mathbb{R}^m$, given by rational approximations and error bounds such that $L = \{x \in \mathbb{R}^n : Ax = b\}$;

b) affine basis $x_1, ..., x_d \in \mathbb{R}^n$ and $x_0 \in \mathbb{R}^n$, given by rational approx. and error bounds, such that $L = \{x_0 + \lambda_1 x_1 + ... + \lambda_d x_d : \lambda_i \in \mathbb{R}\}$;

c) distance function $d_L : \mathbb{R}^n \to \mathbb{R}, y \mapsto \inf\{\|x - y\| : x \in L\}$, given by a "program" of d_L (or, equivalently, by approximating rational Weierstraß polynomials).

Converting from a) to b) or c) means finding all solutions x to $Ax = b$ in a more ore less explicit way, that is, '*solving*' the equation. Our main result shows that this conversion is possible, if and only if the dimension of the solution space is known in advance (that is, if $\text{rank}(A, b) = \text{rank}(A)$ is given as additional input information).

Since by Church's thesis the Turing machine model characterizes those functions which are realizable on physical machines, our results imply certain intrinsic limitations of algorithmic solutions of systems linear equations: there is no general algorithm which could be performed on a physical machine and which solves systems of linear equations without knowing the dimension of the solution space in advance. And actually, our conclusions are in best conformity with the practical knowledge in numerical analysis. But since numerical analysis does not use any formal model of computation, it was not before such a theoretical study that this heuristic knowledge on principal limitations could be expressed in form of concise theorems.

In a previous paper [13] we have started to link linear algebra to computable analysis and we have investigated the question in which sense the dimension of a linear subspace can be computed. The present article continues along this line. The following section contains a short introduction to computable analysis and our previous results. Section 3 contains the technical main part of the paper and discusses how certain types of information on linear subspaces can be computably translated into each other. Finally, Section 4 applies the results to linear equations in order to study their computability properties.

2. Computable Analysis and Linear Algebra

In this section we briefly present some basic notions from computable analysis and some direct consequences of well-known facts. We will use Weihrauch's representation based approach to computable analysis, the so-called *Type-2-Theory of Effectivity*, since it allows to express computations with real numbers, continuous functions and subsets in a highly uniform way. For a precise and comprehensive reference we refer the reader to [10]. Roughly speaking, a partial real number function $f :\subseteq \mathbb{R}^n \to \mathbb{R}$ is computable, if there exists a Turing machine which transfers each sequence $p \in \Sigma^\omega$ that represents some input $x \in \mathbb{R}^n$ into some sequence $F_M(p)$ which represents the output $f(x)$. Since the set of real numbers has continuum cardinality, real numbers can only be represented by infinite sequences $p \in \Sigma^\omega$ (over some finite alphabet Σ) and thus, such a Turing machine M has to compute infinitely long, but eventually it transfers each input sequence p into an appropriate output sequence $F_M(p)$. It is reasonable to allow only one-way output tapes for infinite computations since otherwise the output after finite time would give no information on the final result (because it could possibly be replaced later by the machine). It is straightforward how this notion of computability can be generalized to other sets X with a corresponding ***representation***, that is a surjective partial mapping $\delta :\subseteq \Sigma^\omega \to X$.

Definition 1 (Computable functions) Let δ, δ' be representations of X, Y, respectively. A function $f :\subseteq X \to Y$ is called (δ, δ')*-computable*, if there exists some Turing machine M such that $\delta' F_M(p) = f\delta(p)$ for all $p \in \mathrm{dom}(f\delta)$.

Here, $F_M :\subseteq \Sigma^\omega \to \Sigma^\omega$ denotes the partial function, computed by the Turing machine M. It is straightforward how to generalize this definition to functions with several inputs and it can even be generalized to multi-valued operations $f :\subseteq X \rightrightarrows Y$, where $f(x)$ is a subset of Y instead of a single value. In this case we replace the condition in the definition above by $\delta' F_M(p) \in f\delta(p)$. We can also define the notion of (δ, δ')*-continuity* by replacing F_M by a continuous function $F :\subseteq \Sigma^\omega \to \Sigma^\omega$ (with respect to the Cantor topology on Σ^ω).

Already in case of the real numbers it appears that the defined notion of computability sensitively relies on the chosen representation of the real numbers. The theory of ***admissible*** representations completely answers the question how to find "reasonable" representations of topological spaces [10]. We will make no formal use of admissibility, and hence the reader may ignore all corresponding information. Let us just mention that for admissible representations δ, δ' each (δ, δ')-computable function is necessarily continuous (with respect to the final topologies of δ, δ').

An example of an admissible representation of the real numbers is the so-called *Cauchy representation* $\rho :\subseteq \Sigma^\omega \to \mathbb{R}$, where roughly speaking, $\rho(p) = x$ if p is an (appropriately encoded) sequence of rational numbers $(q_i)_{i\in\mathbb{N}}$ which converges rapidly to x, i.e. $|q_i - q_k| \leq 2^{-k}$ for all $i > k$. By standard coding techniques this representation can easily be generalized to a representation of the n-dimensional Euclidean space $\rho^n :\subseteq \Sigma^\omega \to \mathbb{R}^n$ and to a representation of $m \times n$ matrices $\rho^{m\times n} :\subseteq \Sigma^\omega \to \mathbb{R}^{m\times n}$. A vector $x \in \mathbb{R}^n$ or a matrix $A \in \mathbb{R}^{m\times n}$ will be called *computable*, if it has a computable ρ^n–, $\rho^{m\times n}$*–name*, i.e. if there exists a computable $p \in \Sigma^\omega$ such that $x = \rho^n(p)$ or $A = \rho^{m\times n}(p)$, respectively. A function $f :\subseteq \mathbb{R}^n \to \mathbb{R}$ is just called *computable*, if it is (ρ^n, ρ)–computable.

If δ and δ' are admissible representations of topological spaces X and Y, respectively, then there exists a canonical representation $[\delta,\delta'] :\subseteq \Sigma^\omega \to X \times Y$ of the product space $X \times Y$ and a canonical function space representation $[\delta \to \delta'] :\subseteq \Sigma^\omega \to C(X,Y)$ of the set $C(X,Y)$ of total continuous functions $f : X \to Y$. We mention that these representations allow evaluation and type conversion. Evaluation means that the evaluation function $C(X,Y) \times X \to Y$, $(f,x) \mapsto f(x)$ is $([[\delta \to \delta'],\delta],\delta')$–computable and type conversion means that a function $f : Z \times X \to Y$ is $([\delta'',\delta],\delta')$–computable, if and only if the canonically associated function $f' : Z \to C(X,Y)$ with $f'(z)(x) := f(z,x)$ is $(\delta'',[\delta \to \delta'])$–computable. As a direct consequence we obtain that matrices $A \in \mathbb{R}^{m\times n}$ can effectively be identified with linear mappings $f \in \text{Lin}(\mathbb{R}^n,\mathbb{R}^m)$, see Proposition 2.1 and 2.2 below. Especially, a matrix A is computable, if and only if the corresponding linear mapping is a computable function.

To express weaker computability properties, we will use two further representations $\rho_<,\rho_> :\subseteq \Sigma^\omega \to \mathbb{R}$. Roughly speaking, $\rho_<(p) = x$ if p is an (appropriately encoded) list of all rational numbers $q < x$. (Analogously, $\rho_>$ is defined with $q > x$.) It is known that a mapping $f :\subseteq X \to \mathbb{R}$ is (δ,ρ)–computable, if and only if it is $(\delta,\rho_<)$– and $(\delta,\rho_>)$–computable [10]. The $(\rho^n,\rho_<)$–, and the $(\rho^n,\rho_>)$–computable functions $f : \mathbb{R}^n \to \mathbb{R}$ are called *lower* and *upper semi-computable*, respectively.

Moreover, we will also need a representation of the space $\mathcal{L}^n$ of linear subspaces $V \subseteq \mathbb{R}^n$. Since all linear subspaces are non-empty closed spaces, we can use well-known representations of the hyperspace $\mathcal{A}^n$ of all closed non-empty subsets $A \subseteq \mathbb{R}^n$ (cf. [3, 10]). One way to represent such spaces is via the distance function $d_A : \mathbb{R}^n \to \mathbb{R}$, defined by $d_A(x) := \inf_{a\in A} d(x,a)$, where $d : \mathbb{R}^n \times \mathbb{R}^n \to \mathbb{R}$ denotes the Euclidean metric of $\mathbb{R}^n$. Altogether, we define three representations $\psi^n,\psi^n_<,\psi^n_> :\subseteq \Sigma^\omega \to \mathcal{A}^n$. We let $\psi^n(p) = A$, if and only if $[\rho^n \to \rho](p) = d_A$. In other words, p encodes a set A w.r.t. ψ^n, if it encodes the distance function d_A w.r.t. $[\rho^n \to \rho]$. Analogously, let $\psi^n_<(p) = A$, if and only if $[\rho^n \to \rho_>](p) = d_A$ and let $\psi^n_>(p) = A$, if and only if $[\rho^n \to \rho_<](p) = d_A$. One can prove that $\psi^n_<$ encodes "positive" information about the set A (all open rational balls $B(q,r) := \{x \in \mathbb{R}^n : d(x,q) < r\}$ which intersect A can be enumerated), and $\psi^n_>$ encodes "negative" information about A (all closed rational balls $\overline{B}(q,r)$ which do not intersect A can be enumerated). The final topology induced by ψ^n on $\mathcal{A}^n$ is the so-called *Fell topology*. It is a known fact that a mapping $f :\subseteq X \to \mathcal{A}^n$ is (δ,ψ^n)–computable, if and only if it is $(\delta,\psi^n_<)$– and $(\delta,\psi^n_>)$–computable [10].

A closed set $A \subseteq \mathbb{R}^n$ is called *r.e., co-r.e.* or *recursive*, if it is empty or if there is a computable $p \in \Sigma^\omega$ such that $A = \psi^n_<(p)$, $A = \psi^n_>(p)$, $A = \psi^n(p)$, respectively. Thus, the non-empty r.e., co-r.e. or recursive subsets $A \subseteq \mathbb{R}^n$ are exactly those with upper, lower semi-computable or computable distance function $d_A : \mathbb{R}^n \to \mathbb{R}$, respectively and a closed set is recursive, if and only if it is r.e. and co-r.e. By duality, an open subset $U \subseteq \mathbb{R}^n$ is called *r.e., co-r.e.* or *recursive*, if and only if its complement $\mathbb{R}^n \setminus U$ is co-r.e., r.e. or recursive. Given a representation δ of X, we will say more generally that a subset $U \subseteq Y \subseteq X$

is *δ–r.e. open* in Y, if $\delta^{-1}(U)$ is r.e. open in $\delta^{-1}(Y)$. Here a set $A \subseteq B \subseteq \Sigma^\omega$ is called *r.e. open* in B, if there exists some computable function $f :\subseteq \Sigma^\omega \to \Sigma^*$ with $\mathrm{dom}(f) \cap B = A$. Intuitively, a set U is δ–r.e. open in Y, if and only if there exists a Turing machine which halts for an input $x \in Y$ given w.r.t. δ, if and only if $x \in U$. It is known that a set $U \subseteq \mathbb{R}^n$ is ρ^n–r.e. open in $\mathbb{R}^n$, if and only if it is r.e. open. If a set $U \subseteq X$ is δ–r.e. open in X, then we will say for short that it is *δ–r.e. open.*

We close this section with a short survey on computability results in linear algebra which have been established in our previous paper [13]:

Proposition 2 *Consider the following canonical mappings from linear algebra:*

1. $\mathrm{Lin}(\mathbb{R}^n, \mathbb{R}^m) \to \mathbb{R}^{m\times n}$ *is* $([\rho^n \to \rho^m], \rho^{m\times n})$*-computable,*
2. $\mathbb{R}^{m\times n} \to \mathrm{Lin}(\mathbb{R}^n, \mathbb{R}^m)$ *is* $(\rho^{m\times n}, [\rho^n \to \rho^m])$*-computable,*
3. $\ker : \mathbb{R}^{m\times n} \to \mathcal{A}^n$ *is* $(\rho^{m\times n}, \psi^n_>)$*-computable,*
4. $\mathrm{span} : \mathbb{R}^{m\times n} \to \mathcal{A}^m$ *is* $(\rho^{m\times n}, \psi^m_<)$*-computable,*
   but neither $(\rho^{m\times n}, \psi^m_>)$*-computable, nor -continuous,*
5. $\det : \mathbb{R}^{n\times n} \to \mathbb{R}$ *is* $(\rho^{n\times n}, \rho)$*-computable,*
6. $\mathrm{rank} : \mathbb{R}^{m\times n} \to \mathbb{R}$ *is* $(\rho^{m\times n}, \rho_<)$*-computable,*
   but neither $(\rho^{m\times n}, \rho_>)$*-computable, nor -continuous,*
7. $\dim :\subseteq \mathcal{A}^n \to \mathbb{R}$ *is* $(\psi^n_<, \rho_<)$*- and* $(\psi^n_>, \rho_>)$*-computable.*

3. Linear Subspaces and their Dimension

Considering the computability results about linear algebra known so far from Proposition 2, what can be said about linear equations? If we consider only homogeneous equations $Ax = 0$ in the first step, then we obtain the solution space $L = \ker(A)$ and we can deduce from Proposition 2.3 that there exists a Turing machine which takes A as input with respect to $\rho^{m\times n}$ and which computes the space of solutions with respect to $\psi^n_>$. Unfortunately, this type of "negative" information about the space of solutions is not very helpful; in general it does not even suffice to find a single point of the corresponding space (cf. [10]). Thus, it is desirable to obtain the "positive" information (i.e. a $\psi^n_<$–name) about the space of solutions too. On the other hand we can deduce from $\mathrm{rank}(A) = n - \dim\ker(A)$ and Proposition 2.6 and 2.7 that $\ker : \mathbb{R}^{m\times n} \to \mathcal{A}^n$ is not $(\rho^{m\times n}, \psi^n_<)$–continuous. In other words: without any additional input information, positive information about the solution space is not available in principle.

What kind of additional information could suffice to obtain positive information about the solution space? We will show that it is sufficient to know the dimension of the solution space, i.e. $\mathrm{codim}(A) = \dim\ker(A)$ in advance. More precisely, the following theorem states that given a linear subspace $V \subseteq \mathbb{R}^n$ with respect to $\psi^n_>$ and given its dimension $\dim(V)$, we can effectively find a $\psi^n_<$–name of V. The remaining part of this section will be devoted to the proof of the following theorem, separated in several lemmas.

Theorem 3 *There exists a Turing machine which on input of a linear subspace $V \subseteq \mathbb{R}^n$ and $d = \dim(V)$ with respect to $\psi^n_>$ and ρ, respectively, outputs V with respect to $\psi^n_<$, more precisely, the function*

$$f :\subseteq \mathcal{A}^n \times \mathbb{R} \to \mathcal{A}^n, (V, d) \mapsto V$$

with $\mathrm{dom}(f) := \{(V, d) \in \mathcal{A}^n \times \mathbb{R} : V \in \mathcal{L}^n$ *and* $d = \dim(V)\}$ *is* $([\psi^n_>, \rho], \psi^n_<)$-*computable.*

The main technical tool for the proof of this theorem is given in the following definition. Here and in the following $|x| := \sqrt{\sum_{i=1}^n |x_i|^2}$ denotes the *Euclidean norm* of $x = (x_1, ..., x_n) \in \mathbb{R}^n$.

Definition 4 *Let $W \subseteq \mathbb{R}^n$ be a linear subspace and $\varepsilon > 0$. Denote by*

$$W_\varepsilon := \bigcup_{w \in W} B(w, \varepsilon|w|) = \{x \in \mathbb{R}^n : (\exists w \in W)\, |x - w| < \varepsilon|w|\}$$

the relative blow-up *of W by factor ε with respect to Euclidean norm.*

The following Figure 1 shows the blow-up W_ε of a one-dimensional subspace $W \subseteq \mathbb{R}^3$ by factor $\varepsilon = 1/4$ together with a one-dimensional subspace $V \subseteq W_\varepsilon \cup \{0\}$. The first useful property of the blow-up is given in the following

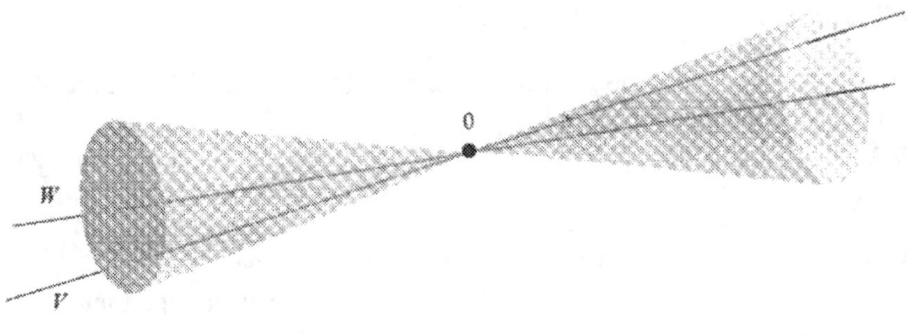

Figure 1. The blow-up W_ε of a linear subspace

lemma, which roughly speaking states that each linear subspace is contained in an arbitrarily small blow-up of a linear subspace of the same dimension but with rational basis.

Lemma 5 *Let $V \subseteq \mathbb{R}^n$ be a linear subspace of dimension d and $\varepsilon > 0$. Then there are $w_1, ..., w_d \in \mathbb{Q}^n$ such that $V \subseteq W_\varepsilon \cup \{0\}$, where $W :=$* $\mathrm{span}(w_1, ..., w_d)$.

We leave the proof of this and the following three lemmas to the reader. Before we formulate the next property of the blow-up, we state an intermediate lemma about linear independence.

Lemma 6 *For each $n \geq 1$ there exists a constant $\Delta > 0$ such that, whenever $b_1, ..., b_d \in \mathbb{R}^n$ are pairwise orthogonal normed vectors and $x_1, ..., x_d \in \mathbb{R}^n$ with $|b_i - x_i| < \Delta$ for $i = 1, ..., d$, then $(x_1, ..., x_d)$ is linearly independent.*

From now on we assume without further mentioning that $\Delta < 1$ is a fixed rational constant as in the previous lemma (where we consider $n \geq 1$ to be arbitrary but fixed). The next lemma formulates another property of the blow-up which roughly speaking states that if a linear subspace V is contained in a sufficiently small blow-up of a linear subspace W of the same dimension, then this blow-up already approximates V quite well.

Lemma 7 *Let $V, W \subseteq \mathbb{R}^n$ be linear subspaces of equal dimension d and let $\varepsilon > 0$ with $\delta := 2\sqrt{d} \cdot \varepsilon/(1-\varepsilon) < \Delta$. If $V \subseteq W_\varepsilon \cup \{0\}$, then $B(w, \delta|w|)$ intersects V for any $w \in W \setminus \{0\}$.*

Now we formulate the last lemma of this section which states an effectivity property of the blow-up. Roughly speaking, the property $V \subseteq W_\varepsilon \cup \{0\}$ can be recognized by a Turing machine in a certain sense.

Lemma 8 *There exists a Turing machine which, on input of linear subspaces $V, W \subseteq \mathbb{R}^n$ with respect to representations $\psi^n_>$ and $\psi^n_<$ and $\varepsilon > 0$ halts, if and only if $V \subseteq W_\varepsilon \cup \{0\}$, more precisely*

$$\{(V, W, \varepsilon) \in \mathcal{A}^n \times \mathcal{A}^n \times \mathbb{R} : V \subseteq W_\varepsilon \cup \{0\} \text{ and } \varepsilon > 0\}$$

is $[\psi^n_>, \psi^n_<, \rho]$-r.e. open in $\mathcal{L}^n \times \mathcal{L}^n \times \mathbb{R}$.

Finally, we can combine Lemma 5, 7 and 8 to a proof of Theorem 3.

Proof of Theorem 3. Let $V \subseteq \mathbb{R}^n$ be a linear subspace and let $d = \dim(V) > 0$. We claim

$$B(q,r) \cap V \neq \emptyset \iff \begin{array}{l} (\exists w_1, ..., w_d \in \mathbb{Q}^n)(\exists \lambda_1, ..., \lambda_d \in \mathbb{Q})(\exists \varepsilon > 0) \\ \delta < \Delta,\ (w_1, ..., w_d) \text{ is linearly independent,} \\ V \subseteq W_\varepsilon \cup \{0\} \text{ and } B(w, \delta|w|) \subsetneq B(q,r), \\ \text{where } W := \mathrm{span}(w_1, ..., w_d),\ w := \sum_{i=1}^d \lambda_i w_i \neq 0 \\ \text{and } \delta := 2\sqrt{d} \cdot \varepsilon/(1-\varepsilon) \end{array}$$

for all $q \in \mathbb{Q}^n$ and $r \in \mathbb{Q}$ with $r > 0$. By Lemma 7 it is clear that "$\Leftarrow$" holds. Let on the other hand $B(q,r) \cap V \neq \emptyset$ with $q \in \mathbb{Q}^n$ and $r \in \mathbb{Q}$ with $r > 0$. Then there exists some $v \in V \cap B(q,r)$, $v \neq 0$. Let $\delta(\varepsilon) := 2\sqrt{d} \cdot \varepsilon/(1-\varepsilon)$ for all $\varepsilon > 0$. Since $|q - v| < r$ there is some ε with $0 < \varepsilon < 1$ such that

$$\left(1 + \tfrac{\varepsilon + \delta(\varepsilon)}{1-\varepsilon}\right)|q - v| + \tfrac{\varepsilon+\delta(\varepsilon)}{1-\varepsilon}|q| < r.$$

Let $\delta := \delta(\varepsilon)$. By Lemma 5 there exist $w_1, ..., w_d \in \mathbb{Q}^n$ such that $V \subseteq W_\varepsilon \cup \{0\}$ with $W := \mathrm{span}(w_1, ..., w_d)$. Thus, there is some $w \in W \setminus \{0\}$ with $|v - w| < \varepsilon|w|$

and without loss of generality we can even assume that there are $\lambda_1, ..., \lambda_d \in \mathbb{Q}$ with $w = \sum_{i=1}^{d} \lambda_i w_i$. We obtain $|q-w| \le |q-v| + |v-w| < |q-v| + \varepsilon|w|$ and $|w| \le |q-w| + |q| \le |q-v| + \varepsilon|w| + |q|$, and hence $|w| \le 1/(1-\varepsilon)(|q-v| + |q|)$ and thus

$$|q-w| + \delta|w| < |q-v| + (\varepsilon+\delta)|w| \le \left(1 + \tfrac{\varepsilon+\delta}{1-\varepsilon}\right)|q-v| + \tfrac{\varepsilon+\delta}{1-\varepsilon}|q| < r,$$

i.e. $B(w, \delta|w|) \subsetneqq B(q, r)$. Thus, "$\Rightarrow$" holds too and the above equivalence is proved.

Thus, given V by $\psi^n_>$ and $d = \dim(V)$ by ρ, we can recursively enumerate all $q \in \mathbb{Q}^n$, $r \in \mathbb{Q}$ with $r > 0$ such that $B(q, r) \cap V \neq \emptyset$ by virtue of Lemma 8. In this way we obtain a $\psi^n_<$–name of V. □

Using Theorem 3 we can improve the statement of Corollary 1 in [13] in the following way.

Corollary 9 *The multi-valued mapping* basis $:\subseteq \mathcal{A}^n \times \mathbb{R} \rightrightarrows \mathcal{A}^n$,

$$(V, d) \mapsto \{\{b_1, ..., b_d\} \subseteq \mathbb{R}^n : (b_1, ..., b_d) \text{ is a basis of } V\}$$

with $\text{dom(basis)} := \{(V, d) : d = \dim(V)\}$ *is* $([\psi^n_<, \rho], \psi^n)$*-computable and* $([\psi^n_>, \rho], \psi^n)$*-computable.*

Here, the $([\psi^n_<, \rho], \psi^n)$–computability of basis has been proved in [13] and the $([\psi^n_>, \rho], \psi^n)$–computability follows with Theorem 3. Roughly speaking, we can deduce that the following equivalences hold for different types of information about linear subspaces:

positive + dimension ≡ negative + dimension ≡ positive + negative ≡ basis

These equivalences could be made precise by defining corresponding representations of $\mathcal{L}^n$ and by proving their equivalence, but we are not going to discuss this here. Instead of that, we mention that for single linear subspaces one obtains the following less uniform corollary.

Corollary 10 *A linear subspace* $V \subseteq \mathbb{R}^n$ *is r.e., if and only if it is co-r.e., if and only if it is recursive, if and only if it admits a computable basis.*

Since the dimension is always a computable number, the proof of this corollary follows directly from the previous corollary and the fact that the mapping span $:\subseteq \mathbb{R}^{n \times d} \to \mathcal{A}^n$, restricted to linear independent inputs $(b_1, ..., b_d)$, is $(\rho^{n \times d}, \psi^n)$–computable, which has been proved in [13].

4. Linear Equations

In this section we want to apply the results of the previous section to solve linear equations $Ax = b$. It is a well-known and obvious fact from linear algebra that such a linear equation is solvable, if and only if $\text{rank}(A) = \text{rank}(A, b)$. The following theorem is the main result of this paper. It states that the solution operator of solvable linear equations is computable, provided that the rank of the linear equation is given as additional input.

Theorem 11 *There exists a Turing machine which takes a solvable linear equation $Ax = b$ together with $d = \mathrm{rank}(A, b)$ as input and which computes the space of solutions $L = \{x : Ax = b\}$. More precisely, the function*

$$\mathrm{solve} :\subseteq \mathbb{R}^{m\times n} \times \mathbb{R}^m \times \mathbb{R} \to \mathcal{A}^n, (A, b, d) \mapsto L = \{x \in \mathbb{R}^n : Ax = b\}$$

with $\mathrm{dom(solve)} := \{(A, b, d) \in \mathbb{R}^{m\times n} \times \mathbb{R}^m \times \mathbb{R} : \mathrm{rank}(A) = \mathrm{rank}(A, b) = d\}$ *is* $([\rho^{m\times n}, \rho^m, \rho], \psi^n)$*-computable.*

Proof. Notice that $x \in L$, if and only if, in homogeneous coordinates, x is a solution to $(A, b) \cdot {}^t(x, -1) = 0$. We therefore may determine the kernel of $(A, b) \in \mathbb{R}^{m\times(n+1)}$ and scale the results x such that $x_{n+1} = -1$.

To realize this idea precisely, we perform several steps: let $A \in \mathbb{R}^{m\times n}$ be given by $\rho^{m\times n}$, let $b \in \mathbb{R}^m$ be given by ρ^m and let $d = \mathrm{rank}(A) = \mathrm{rank}(A, b)$ be given by ρ. First, we determine $\ker(A, b)$ w.r.t. $\psi^{n+1}_{>}$, which is possible by Proposition 2.3. Then we can use Theorem 3 and the formula $\dim\ker(A, b) = n + 1 - d$ to determine a $\psi^{n+1}_{<}$–name of $\ker(A, b)$. Especially, this name allows to find effectively a point $z = (z_1, ..., z_{n+1}) \in \ker(A, b)$ w.r.t. ρ^{n+1} such that $z_{n+1} < 0$. Let $c_i := z_i/|z_{n+1}|$ for $i = 1, ..., n$. Then $c := (c_1, ..., c_n)$ is a solution of $Ax = b$ and $L = \{x : Ax = b\} = c + \ker(A)$. Since $\dim\ker(A) = n - d$ we can compute a ψ^n–name of $\ker(A)$ by Proposition 2.3 and Theorem 3. Finally, we note that the function $\mathbb{R}^n \times \mathcal{A}^n \to \mathcal{A}^n, (x, A) \mapsto x + A := \{x + a \in \mathbb{R}^n : a \in A\}$ is $([\rho^n, \psi^n], \psi^n)$–computable. Altogether, this allows us to compute a ψ^n–name of L. □

Regarding the proof and Corollary 9 we can even conclude the following corollary, which states that given a solvable linear equation together with its rank we can effectively find a specific solution and a basis for the homogeneous equation.

Corollary 12 *The map* $s :\subseteq \mathbb{R}^{m\times n} \times \mathbb{R}^m \times \mathbb{R} \rightrightarrows \mathbb{R}^n \times \mathcal{A}^n, (A, b, d) \mapsto S$*, where* S *is the set*

$$\{(c, \{b_1, ..., b_{n-d}\}) \in \mathbb{R}^n \times \mathcal{A}^n : c + \mathrm{span}(b_1, ..., b_{n-d}) = \{x : Ax = b\}\},$$

and $\mathrm{dom}(s) := \{(A, b, d) \in \mathbb{R}^{m\times n} \times \mathbb{R}^m \times \mathbb{R} : \mathrm{rank}(A) = \mathrm{rank}(A, b) = d < n\}$*, is* $([\rho^{m\times n}, \rho^m, \rho], [\rho^n, \psi^n])$*-computable.*

Moreover, the previous theorem allows to deduce an immediate consequence about single linear equations.

Corollary 13 *If* $A \in \mathbb{R}^{m\times n}$ *is a computable matrix and* $b \in \mathbb{R}^m$ *a computable vector, then* $L = \{x \in \mathbb{R}^n : Ax = b\}$ *is a recursive set. If, additionally,* $Ax = b$ *has a unique solution* $x \in \mathbb{R}^n$*, then this solution is computable.*

It is interesting to note that our results also allow to handle the problem which is inverse to solving a linear equation: given an affine subspace, we can find a linear equation with this affine subspace as solution space.

Theorem 14 *There exists a Turing machine which takes an affine space L as input and computes a linear equation $Ax = b$ with $L = \{x : Ax = b\}$ as set of solutions. More precisely, the function* solve *admits a multi-valued right inverse $r :\subseteq \mathcal{A}^n \rightrightarrows \mathbb{R}^{m\times n} \times \mathbb{R}^m \times \mathbb{R}$ which is $(\psi^n, [\rho^{m\times n}, \rho^m, \rho])$-computable, for any $m \geq n$.*

Proof. Let L be given w.r.t. ψ^n. Then we can effectively find some point $c \in L$ w.r.t. ρ^n. As in the proof of Theorem 11 we can compute $L - c$ w.r.t. ψ^n. By Corollary 1 from [13] we can find a basis $(b_1, ..., b_k) \in \mathbb{R}^{n\times k}$ of $L-c$ w.r.t. $\rho^{n\times k}$. If $d := n - k = 0$, then $A = 0$ and $b = 0$ defines a linear equation with $L = \mathbb{R}^n$. Otherwise, apply the Gram-Schmidt orthogonalization process to determine an orthogonal basis $(o_1, ..., o_k)$ of $L - c$ w.r.t. $\rho^{n\times k}$, i.e.

$$o_1 := b_1,\ o_{j+1} := b_{j+1} - \sum_{i=1}^{j} \frac{b_{j+1} \cdot o_i}{|o_i|^2} o_i$$

for $j = 1, ..., k - 1$. Then, find some vectors vectors $b_{k+1}, ..., b_n \in \mathbb{R}^n$ w.r.t. ρ^n such that $(o_1, ..., o_k, b_{k+1}, ..., b_n)$ is linear independent, which is possible by Lemma 4 in [13]. Then, apply the Gram-Schmidt orthogonalization process again to determine vectors $o_{k+1}, ..., o_n$ w.r.t. ρ^n such that $(o_1, ..., o_n)$ is an orthogonal basis of $\mathbb{R}^n$. Thus, $(o_{k+1}, ..., o_n)$ is an orthogonal basis of the orthogonal complement of $L - c$. Now, we can compute $A := {}^t(o_{k+1}, ..., o_n, 0, ..., 0) \in \mathbb{R}^{m\times n}$ w.r.t. $\rho^{m\times n}$ and $b := Ac$ w.r.t. ρ^m. Then we obtain $\ker(A) = L - c$ and $L = \{x : Ax = b\}$. Altogether, the procedure describes how to compute a right inverse r of the function solve. □

Again we can deduce a simple fact about single spaces and equations.

Corollary 15 *If $L \subseteq \mathbb{R}^n$ is a recursive non-empty affine subspace, then there exists a computable matrix $A \in \mathbb{R}^{m\times n}$ and a computable vector $b \in \mathbb{R}^m$ such that $L = \{x \in \mathbb{R}^n : Ax = b\}$ for any $m \geq n$.*

5. Conclusion

In this paper we have continued our project to investigate computability properties in linear algebra with rigorous methods from computable analysis. This project has been started with [13] and could be continued along several different lines. On the one hand, it would be interesting to extend the investigation to complexity questions. Surely, this is possible as long as one considers some non-uniform versions of our results. However, the fully uniform versions would require complexity measures for asymmetric hyperspaces which are beyond the state of the art. On the other hand, it is a promising topic to study other parts of linear algebra such as spectral theory or linear inequalities. Some steps in this direction have been presented in [14, 15].

Last but not least, our results give further ground to the hope that computable analysis can help to explain fundamental limitations of real number computations. Many practical observations of numerical analysis, e.g. the fact that numerical differentiation is much more difficult than numerical integration, already found natural explanations in computable analysis (see [10]). We have tried to extend these applications of computable analysis to linear algebra topics.

References

[1] L. Blum, F. Cucker, M. Shub, and S. Smale, *Complexity and Real Computation*, Springer, New York 1998.

[2] V. Brattka and P. Hertling, Feasible real random access machines, *J. Complexity* **14** (1998) 490–526.

[3] V. Brattka and K. Weihrauch, Computability on subsets of Euclidean space I: Closed and compact subsets, *Theoret. Comp. Sci.* **219** (1999) 65–93.

[4] A. Grzegorczyk, On the definitions of computable real continuous functions, *Fund. Math.* **44** (1957) 61–71.

[5] H. Kamo, K. Kawamura, and I. Takeuti, Computational complexity of fractal sets, *Real Analysis Exchange* **26** (2000/01) 773–793.

[6] K.-I. Ko, *Complexity Theory of Real Functions*, Birkhäuser, Boston 1991.

[7] K.-I. Ko, On the computability of fractal dimensions and Hausdorff measure, *Ann. Pure Appl. Logic* **93** (1998) 195–216.

[8] M. B. Pour-El and J. I. Richards, *Computability in Analysis and Physics*, Springer, Berlin 1989.

[9] A. M. Turing, On computable numbers, with an application to the "Entscheidungsproblem", *Proc. London Math. Soc.* **42** (1936) 230–265.

[10] K. Weihrauch, *Computable Analysis*, Springer, Berlin 2000.

[11] K. Weihrauch and N. Zhong, Turing computability of a nonlinear Schrödinger propagator, in: J. Wang (ed.), *Computing and Combinatorics*, vol. 2108 of *Lect. Not. Comp. Sci.*, Springer, Berlin 2001, 596–599.

[12] J. H. Wilkinson, *The Algebraic Eigenvalue Problem*, Oxford University Press, Oxford 1965.

[13] M. Ziegler and V. Brattka, Computing the dimension of linear subspaces, in: V. Hlaváč, K. G. Jeffery, and J. Wiedermann (eds.), *SOFSEM 2000: Theory and Practice of Informatics*, vol. 1963 of *Lect. Not. Comp. Sci.*, Springer, Berlin 2000, 450–458.

[14] M. Ziegler and V. Brattka, A computable spectral theorem, in: J. Blanck, V. Brattka, and P. Hertling (eds.), *Computability and Complexity in Analysis*, vol. 2064 of *Lect. Not. Comp. Sci.*, Springer, Berlin 2001, 378–388.

[15] M. Ziegler and V. Brattka, Turing computability of (non-)linear optimization, in: T. Biedl (ed.), *Thirteenth Canadian Conference on Computational Geometry*, University of Waterloo 2001, 181–184.

HIERARCHY AMONG AUTOMATA ON LINEAR ORDERINGS

Véronique Bruyère
Institut d'Informatique, Université de Mons-Hainaut,
Le Pentagone, 6 avenue du Champ de Mars,
B-7000 Mons, Belgium
Veronique.Bruyere@umh.ac.be

Olivier Carton
Institut Gaspard Monge, Université de Marne-la-Vallée,
5 boulevard Descartes,
F-77454 Marne-la-Vallée Cedex 2, France
Olivier.Carton@univ-mlv.fr

Abstract In a preceding paper, automata and rational expressions have been introduced for words indexed by linear orderings, together with a Kleene-like theorem. We here pursue this work by proposing a hierarchy among the rational sets. Each class of the hierarchy is defined by a subset of the rational operations that can be used. We then characterize any class by an appropriate class of automata, leading to a Kleene theorem inside the class. A characterization by particular classes of orderings is also given.

1. Introduction

The first result in automata theory and formal languages is the Kleene theorem which establishes the equivalence between sets of words accepted by automata and sets of words described by rational expressions. Since the seminal paper of Kleene [8], this equivalence has been extended to many kinds of structures: infinite words, bi-infinite words, finite and infinite trees, finite and infinite traces, pictures, *etc.*

In [3], we have considered linear structures in a general framework, *i.e.*, words indexed by a linear ordering. This approach allows to treat in the same way finite words, left- and right-infinite words, bi-infinite words, ordinal words which are studied separately in the literature. We have introduced a new notion of automaton accepting words on linear orderings, which is simple, natural and includes previously defined automata. We have also defined rational expressions for such words. We have proved the related Kleene-like theorem when the orderings are restricted to countable scattered linear orderings. This result extends Kleene's theorem for finite words [8], infinite words [4, 10], bi-infinite words [7, 11] and ordinal words [5, 6, 16].

Another jewel of formal languages is the characterization of star-free languages by first-order logic [9] or by group-free semigroups [14]. A set of finite words is star-free if it can be described by a rational expression using concatenation, union and complementation only. The class of star-free sets is thus obtained by restricting the rational operations. The star iteration is replaced by complementation which is weaker when union and concatenation are already allowed.

In this paper, we propose a hierarchy among rational sets of words on linear orderings. As for star-free sets, this hierarchy is obtained by restricting the rational operations that can be used. Each class contains the rational sets that can be described by a given subset of the rational operations.

The rational operations introduced in [3] include the usual operations of union, concatenation and star iteration. They also include the omega iteration usually used to construct infinite words and the ordinal iteration introduced by Wojciechowski [16] for ordinal words. Three new operations are added: the backwards omega iteration, the backwards ordinal iteration and a last operation which is a kind of iteration for all countable scattered linear orderings. The lowest class of the hierarchy contains sets that can be described by rational expressions using union, concatenation and star iteration. This is of course the class of rational sets of finite words. The greatest class contains sets that can be described by rational expressions using all rational operations introduced in [3]. It contains all rational sets of words on scattered linear orderings. Some other classes of words already studied in the literature appear naturally in our framework. Sets of words on ordinals introduced by Büchi [5] or sets of words on ordinals smaller that ω^ω studied by Choueka [6] form two classes of our hierarchy.

We give a characterization of each class of the hierarchy by a corresponding class of automata. A set of words belongs to the given class if and only if it is recognized by an automaton of the corresponding class. Each of these characterizations is thus a Kleene theorem which holds for that class. For well-known classes, these Kleene theorems were already proved by Wojciechowski [16] for words on ordinals or by Choueka [6] for words on ordinals smaller than ω^ω. In each case, the corresponding class of automata is obtained naturally by restricting the kind of transitions that can be used. For instance, the automata for words on ordinals do have left limit transitions but no right limit transitions as there were defined by Büchi [5].

The last rational operation defined in [3] works like an iteration for all countable scattered linear orderings. It is binary. In this paper, we consider a simpler definition of this iteration as a unary operation. This simplified definition seems to be more natural but it turns out to be weaker. The results of this paper show that the binary operation is really needed to obtain the Kleene theorem of [3]. This question was actually the original motivation of our work.

We also give a characterization of each class of the hierarchy by a corresponding class of orderings. A set of words belongs to the given class if and only if the length of each of its words belongs to the corresponding class of orderings. For some classes as the class of sets of words on ordinals, this characterization is straightforward. However for some other classes, suitable classes of orderings have to be defined. These definitions are inspired by the characterization of countable scattered orderings due to Hausdorff.

To summarize, the results of the paper establish a hierarchy among rational sets of words on linear orderings, with connections between natural classes of orderings, rational operations and the types of transitions in automata.

The paper is organized as follows. In Sections 2, 3 and 4, we briefly recall the new notions introduced in [3]: words on linear orderings, automata and rational expressions. We refer the reader to [13] for a complete introduction to linear orderings. The different classes of the hierarchy are described in Section 5. This hierarchy is summarized on Figure 6 and illustrated by some examples. The proofs are given in the Appendix.

2. Orderings and Words

A *linear ordering* J is an ordering $<$ which is total, that is, for any $j \neq k$ in J, either $j < k$ or $k < j$ holds. Given a finite alphabet A, a *word* $(a_j)_{j\in J}$ is a function from J to A which maps any element j of J to a letter a_j of A. We say that J is the *length* $|x|$ of the word x. For instance, the *empty word* ε is indexed by the empty linear ordering $J = \varnothing$. Usual finite words are the words indexed by finite orderings $J = \{1, 2, \dots, n\}$, $n \geq 0$. A word of length $J = \omega$ is a word usually called an ω-word or an infinite word. A word of length $J = \zeta$ is a sequence $\dots a_{-2}a_{-1}a_0a_1a_2\dots$ of letters which is usually called a bi-infinite word.

Given a linear ordering J, we denote by $-J$ the *backwards* linear ordering obtained by reversing the ordering relation. For instance, $-\omega$ is the backwards linear ordering of ω which is used to indexed the so-called left-infinite words. For a class $\mathcal{V}$ of linear orderings, we denote by $-\mathcal{V}$ the class $\{-J \mid J \in \mathcal{V}\}$.

Given two linear orderings J and K, the linear ordering $J + K$ is obtained by juxtaposition of J and K, i.e., it is the linear ordering on the disjoint union $J \cup K$ extended with $j < k$ for any $j \in J$ and any $k \in K$. For instance, the linear ordering ζ can be obtained as the sum $-\omega + \omega$. More generally, let J and K_j for $j \in J$, be linear orderings. The linear ordering $\sum_{j\in J} K_j$ is obtained by juxtaposition of the orderings K_j in respect of J. More formally, the *sum* $\sum_{j\in J} K_j$ is the set L of all pairs (k, j) such that $k \in K_j$. The relation $(k_1, j_1) < (k_2, j_2)$ holds iff $j_1 < j_2$ or $j_1 = j_2$ and $k_1 < k_2$ in K_{j_1}.

The sum operation on linear orderings leads to a notion of product of words as follows. Let J and K_j for $j \in J$, be linear orderings. Let $x_j = (a_{k,j})_{k\in K_j}$ be a word of length K_j, for any $j \in J$. The *product* $\prod_{j\in J} x_j$ is the word z of length $L = \sum_{j\in J} K_j$ equal to $(a_{k,j})_{(k,j)\in L}$. For instance, the word $a^\zeta = a^{-\omega} \cdot a^\omega$ of length $\zeta = -\omega + \omega$ is the product of the two words $a^{-\omega}$ and a^{ω} of length $-\omega$ and ω respectively.

In this paper as in [3], we only consider linear orderings which are *countable* and *scattered*, i.e., without any dense subordering. This class is denoted by $\mathcal{S}$ and its elements are shortly called *orderings*. We use notation $\mathcal{N}$ for the subclass of $\mathcal{S}$ of finite linear orderings and $\mathcal{O}$ for the subclass of countable ordinals. Recall that an *ordinal* is a linear ordering which is well-ordered, that is, without the subordering $-\omega$.

The following characterization of the class $\mathcal{S}$ is due to Hausdorff [13]. Notation 1 is used for the finite ordering with one element and notation $\beta < \alpha$ means the usual ordering on ordinals.

Theorem 1 (Hausdorff) $\mathcal{S} = \bigcup_{\alpha \in \mathcal{O}} U_\alpha$ *where the classes* U_α *are inductively defined by*

1 $U_0 = \{\varnothing, 1\}$;

2 $U_\alpha = \{\sum_{j \in J} K_j \mid J \in \mathcal{N} \cup \{\omega, -\omega, \zeta\}$ *and* $K_j \in \bigcup_{\beta < \alpha} U_\beta\}$.

For instance, the ordinal ω belongs to U_1 because $\omega = \sum_{j \in \omega} K_j$ with $K_j = 1 \in U_0$. More generally one can check that the ordinal ω^n belongs to $U_n \setminus U_{n-1}$, for all $n \geq 1$. Finally, ω^ω belongs to U_ω since it equals $\sum_{j \in \omega} K_j$ with $K_j = \omega^j \in U_j$, but it belongs to no U_n (see Chapter 5 of [13]).

We propose two new families of classes to characterize $\mathcal{S}$. The class $\mathcal{S}$ is equal to $\bigcup_{\alpha \in \mathcal{O}} V_\alpha$ where the classes V_α are inductively defined by

1 $V_0 = \{\varnothing, 1\}$;

2 $V_\alpha = \{\sum_{j \in J} K_j \mid J \in \mathcal{O} \cup \{-\omega, \zeta\}$ and $K_j \in \bigcup_{\beta < \alpha} V_\beta\}$.

It is also equal to $\bigcup_{\alpha \in \mathcal{O}} W_\alpha$ where the classes W_α are inductively defined by

1 $W_0 = \{\varnothing, 1\}$;

2 $W_\alpha = \{\sum_{j \in J} K_j \mid J \in \mathcal{O} \cup -\mathcal{O}$ and $K_j \in \bigcup_{\beta < \alpha} W_\beta\}$.

The ordinal ω^ω belongs to V_1 and W_1. The backwards ordinal $-\omega^\omega$ belongs to V_ω and to W_1. More generally, $\mathcal{O} \subseteq V_1$ and $\mathcal{O} \cup -\mathcal{O} \subseteq W_1$. It can be proved that the ordering ζ^ω (see [13] for a precise definition) belongs to W_ω but not to $\bigcup_{n < \omega} W_n$.

3. Automata

Automata accepting words on linear orderings are a natural extension of finite automata. As above they are defined as $\mathcal{A} = (Q, A, E, I, F)$. The set E is composed with three types of transitions: the usual *successor* transitions in $Q \times A \times E$, the *left limit* transitions which belong to $\mathcal{P}(Q) \times Q$ and the *right limit* transitions which belong to $Q \times \mathcal{P}(Q)$.

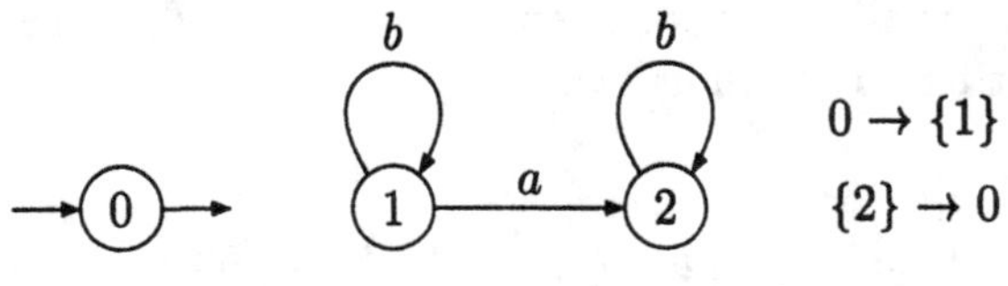

Figure 1. An automaton on linear orderings

Example 1 *The automaton depicted in Figure 1 has one left limit transition* $\{2\} \to 0$ *and one right limit transition* $0 \to \{1\}$.

The notion of cut is needed to define a path in such an automaton. A *cut* of an ordering J is a pair (K, L) of intervals such that $J = K \cup L$ and for any $k \in K$ and $l \in L$, $k < l$. The two subsets must be disjoint and they form a partition of the set J. The set of all cuts of the ordering J is denoted by $\hat{J}$. The set $\hat{J}$ can be linearly ordered as follows. For any cuts $c_1 = (K_1, L_1)$ and $c_2 = (K_2, L_2)$, define the relation $c_1 < c_2$ iff $K_1 \subsetneq K_2$. Note that $\hat{J}$ has always a least cut $(\varnothing, J)$ denoted $c_{\min}$ and a greatest cut $(J, \varnothing)$ denoted $c_{\max}$.

See Figure 2 where each element of J is represented by a bullet, and each cut by a vertical bar.

$$|\cdots|\bullet|\bullet|\bullet|\cdots|\cdots|\bullet|\bullet|\bullet|\cdots|$$

Figure 2. Ordering $J \cup \hat{J}$ for $J = \zeta + \zeta$.

A word $x = (a_j)_{j \in J}$ of length J is accepted by $\mathcal{A}$ if it is the label of a successful path. A *path* γ is a sequence of states $\gamma = (q_c)_{c \in \hat{J}}$ of length $\hat{J}$ verifying the following conditions. For two consecutive states in γ, there must be a successor transition labeled by the letter in between. For a state $q \in \gamma$ which has no predecessor on γ, there must be a left limit transition $P \to q$ where P is the limit set of γ on the left of q. Right limit transitions are used similarly when q has no successor on γ. A path is *successful* if its first state $q_{c_{\min}}$ is initial and its last state $q_{c_{\max}}$ is final.

More precisely, for any cut $c \in \hat{J}$, define the sets $\lim_{c^-} \gamma$ and $\lim_{c^+} \gamma$ as follows:

$$\lim_{c^-} \gamma = \{q \in Q \mid \forall c' < c \, \exists k \quad c' < k < c \text{ and } q = q_k\},$$

$$\lim_{c^+} \gamma = \{q \in Q \mid \forall c < c' \, \exists k \quad c < k < c' \text{ and } q = q_k\}.$$

For any consecutive cuts c_j^- and c_j^+ of $\hat{J}$, $q_{c_j^-} \xrightarrow{a_j} q_{c_j^+}$ must be a successor transition. For any cut $c \neq c_{\min}$ in $\hat{J}$ which has no predecessor, $\lim_{c^-} \gamma \to q_c$

must be a left limit transition. For any cut $c \neq c_{\min}$ in $\hat{J}$ which has no successor, $q_c \to \lim_{c^+} \gamma$ must be a right limit transition.

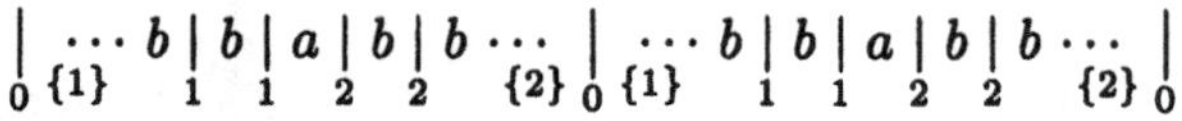

Figure 3. The word $(b^{-\omega}ab^{\omega})^2$ is accepted

The notion of path γ we have introduced for words on orderings coincide with the usual notion of paths considered in the literature for finite words [12], ω-words [15] and ordinal words [2]. For a Muller automaton accepting ω-words, a left limit set P is computed at the end of the path. It is nothing else than the states appearing infinitely often along the path. In our context, the path then ends with an additional left limit transition to a state q which is final.

4. Rational Expressions

We now introduce the notion of rational sets of words on linear orderings. The rational operations of course include the usual Kleene operations for finite words which are the union $+$, the concatenation $\cdot$ and the star operation $*$. They also include the omega iteration ω usually used to construct ω-words and the ordinal iteration $\sharp$ introduced by Wojciechowski [16] for ordinal words. Three new operations are also needed: the backwards omega iteration $-\omega$, the backwards ordinal iteration $-\sharp$ and a last binary operation denoted $\diamond$ which is a kind of iteration for all orderings.

Given two sets X and Y of words, we define

$$
\begin{aligned}
X + Y &= \{z \mid z \in X \cup Y\},\\
X \cdot Y &= \{x \cdot y \mid x \in X, y \in Y\},\\
X^* &= \{\textstyle\prod_{j\in\{1,\dots,n\}} x_j \mid n \in \mathcal{N}, x_j \in X\},\\
X^\omega &= \{\textstyle\prod_{j\in\omega} x_j \mid x_j \in X\},\\
X^{-\omega} &= \{\textstyle\prod_{j\in-\omega} x_j \mid x_j \in X\},\\
X^\sharp &= \{\textstyle\prod_{j\in\alpha} x_j \mid \alpha \in \mathcal{O}, x_j \in X\},\\
X^{-\sharp} &= \{\textstyle\prod_{j\in-\alpha} x_j \mid \alpha \in \mathcal{O}, x_j \in X\},\\
X \diamond Y &= \{\textstyle\prod_{j\in J\cup\hat{J}^*} z_j \mid J \in \mathcal{S}, z_j \in X \text{ if } j \in J \text{ and } z_j \in Y \text{ if } j \in \hat{J}^*\}.
\end{aligned}
$$

The last operation needs some explanation. Notation $\hat{J}^*$ is used for the set $\hat{J} \setminus \{(\varnothing, J), (J, \varnothing)\}$. A word x belongs to $X \diamond Y$ iff there is an ordering J such that x is the product indexed by the ordering $J \cup \hat{J}^*$ of words where each word indexed by an element of J belongs to X and each word indexed by a cut in $\hat{J}^*$ belongs to Y (see Figure 4). We use notation $X^\diamond$ for the set $X \diamond \varepsilon$. When operation $\diamond$ is used as a unary operation, we also use notation $\diamond_1$. When it is used as a binary operation, we use notation $\diamond_2$.

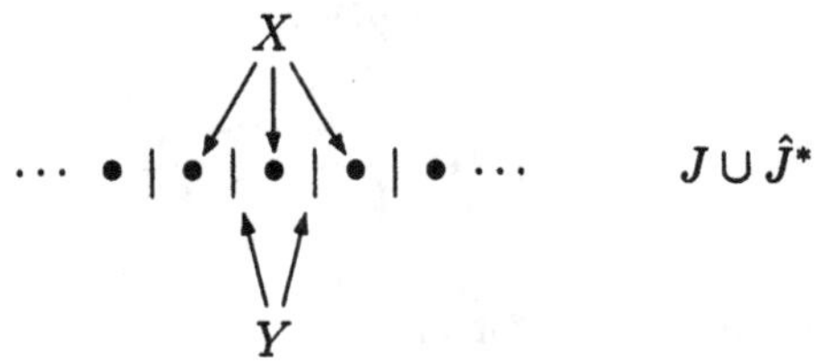

Figure 4. The operation $X \diamond Y$

Note that the definitions for the operation $*$ and $\sharp$ are closed to each other. The only difference is that the products are over any $J \in \mathcal{N}$ for operation $*$ whereas they are over any $J \in \mathcal{O}$ for operation $\sharp$.

Example 2 *The set of all words over the alphabet A is the rational set $A^\diamond = A \diamond \varepsilon$. The set of words accepted by the automaton of Figure 1 is the rational set $(b^{-\omega}ab^{\omega})^*$.*

In [3], we show that a set of words on countable scattered linear orderings is accepted by an automaton iff it can be described by a rational expression. This result extends the well-known Kleene theorem on finite words, its extension to ω-words [4] and to ordinal words [16].

Theorem 2 ([3]) *Over countable scattered linear orderings, $X \subseteq A^\diamond$ is recognizable iff it is rational.*

The proof that any rational set of words is recognizable is by induction on the rational expression denoting the set by giving the corresponding construction for the automaton. The constructions for the union, the concatenation and the star iteration are very similar to the classical ones for automata on finite words [12]. The proof that any set of words accepted by an automaton is rational is a generalization of McNaughton and Yamada algorithm. It is based on a induction on the number of states of the automaton and the type of transition that is used. The base of the induction is Kleene's theorem on finite and ω-words. This part of the proof is the most difficult.

Example 3 *The automaton pictured in Figure 5 accepts the set denoted by the rational expression $a^\zeta \diamond b$. The part of the automaton given by state 2 and the two limit transitions $0 \to \{2\}$ and $\{2\} \to 1$ accepts the word a^ζ whereas the part given by the successor transition from state 1 to state 0 accepts the word b. Any occurrence of a^ζ is preceded and followed by an occurrence of b in the automaton. Thanks to the limit transitions $0 \to \{0,1,2\}$ and $\{0,1,2\} \to 1$, the occurrences of a^ζ are indexed by an ordering J, the occurrences of b are indexed by the ordering $\hat{J}^*$ and they are interleaved according to the ordering $J \cup \hat{J}^*$.*

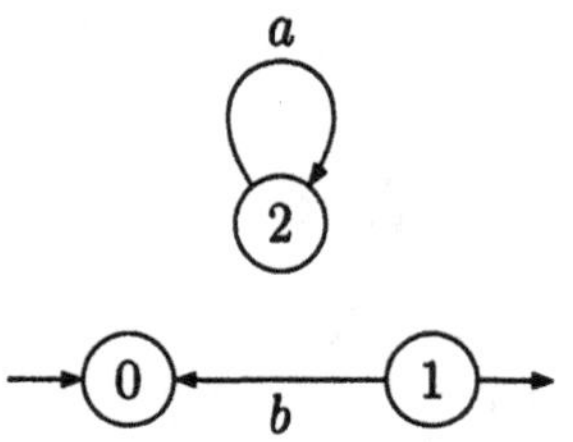

$0 \to \{2\}$

$0 \to \{0,1,2\}$

$\{2\} \to 1$

$\{0,1,2\} \to 1$

Figure 5. Automaton accepting the set $a^{\zeta} \diamond b$

5. Hierarchy

We now come to the main result of this paper. We introduce a hierarchy among rational sets of words on countable scattered linear orderings. Each class of this hierarchy is obtained by restricting the rational operations that can be used. It turns out that each of these classes can be characterized by automata with a particular form. These classes of automata are defined by restricting the limit transitions of the automata. In other words, a Kleene-like theorem holds for each class of the hierarchy. Finally, each class of rational sets of our hierarchy can be characterized by a class of orderings. For any rational set of a given class, the lengths of all its words belong to the corresponding class of orderings.

For instance, rational sets with operations restricted to $+$, $\cdot$ and $*$ correspond to automata without any limit transitions. The corresponding orderings are the finite ones.

The different classes of the hierarchy are summarized on Figure 6. Let us give a precise description of each class.

0 This class corresponds to Kleene theorem [8] on rational sets of finite words. The rational operations are the usual Kleene operations: union, concatenation and star iteration. The automata are the usual automata on finite words with no limit transitions. The orderings are the finite ones.

1 This class corresponds to Choueka theorem [6] on rational sets of words of length an ordinal smaller than ω^{ω}. The rational operations are the Kleene ones and the omega iteration. Automata considered by Choueka are a special kind of automata on ordinals introduced by Büchi. It can be shown [1] that these automata are equivalent to automata with no right limit transitions and such that the left limit transitions are of the form $P \to q$ with $q \notin P$. The corresponding orderings are the ordinals smaller than ω^{ω}.

2 The rational operations of Class 2 are the Kleene ones, the omega iteration and the backwards omega iteration. The automata have left limit transitions $P \to q$ with $q \notin P$ and right limit transitions $q' \to P'$

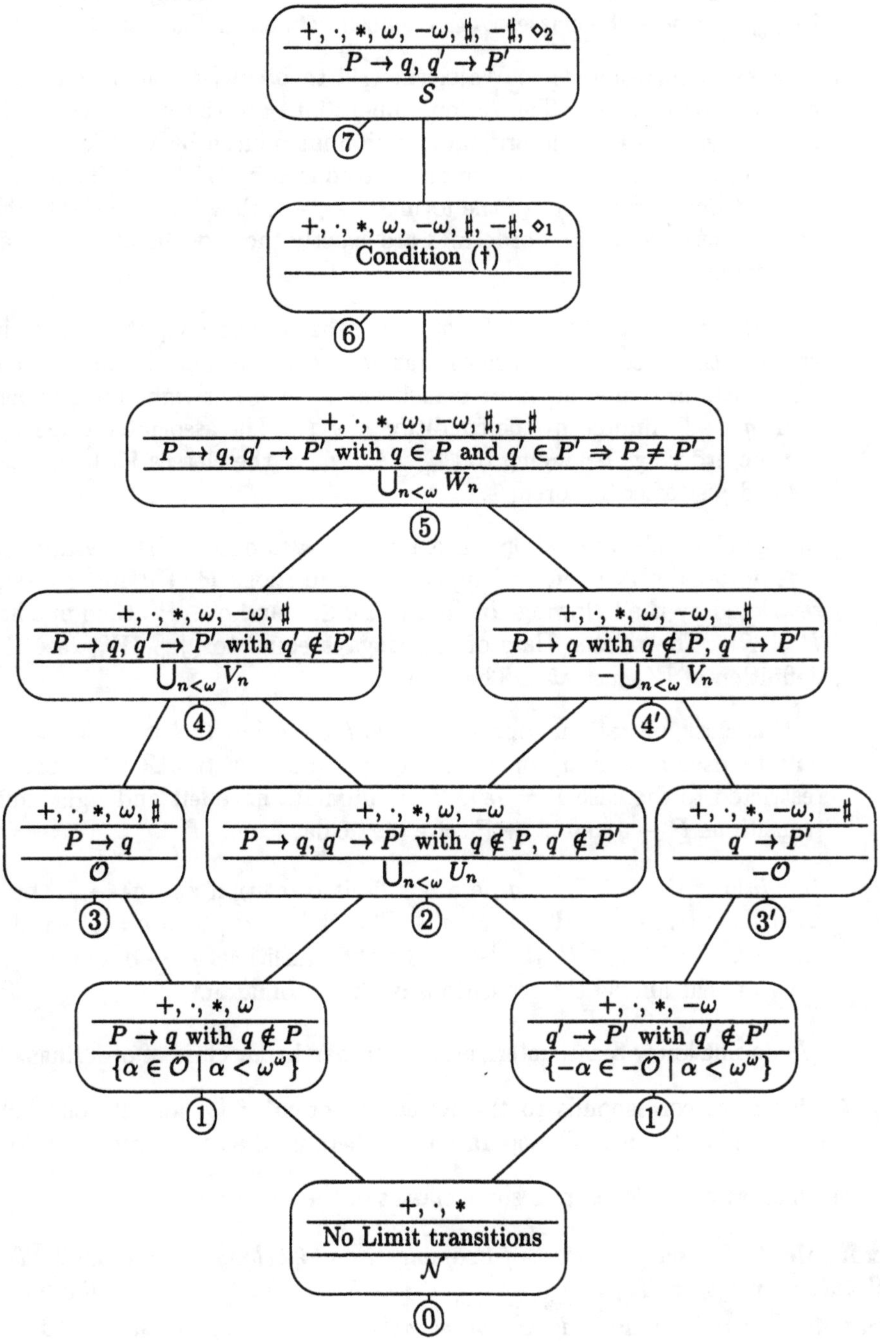

Figure 6. Classes of the hierarchy

with $q' \notin P'$. The associated orderings are the orderings belonging to $\bigcup_{n<\omega} U_n$ where the classes U_n have been defined in Theorem 1.

3 This class corresponds to Wojciechowski theorem [16] on rational sets of words on ordinals. The rational operations are the Kleene ones, the omega iteration and the ordinal iteration introduced by Wojciechowski. The automata are those on ordinals introduced by Büchi [5]. They have any left limit transitions of the form $P \to q$ but they have no right limit transitions. The related orderings are exactly those of the class $\mathcal{O}$ of all countable ordinals.

4 The rational operations of Class 4 are the Kleene one, the omega iteration, the backwards omega iteration and the ordinal iteration. The automata have any left limit transitions $P \to q$ but right limit transitions $q' \to P'$ limited by the condition $q' \notin P'$. The associated orderings are the orderings belonging to $\bigcup_{n<\omega} V_n$ where the classes V_n have been defined just after Theorem 1.

5 In this class, all rational operations are allowed except the operation $\diamond$. The automata have left and right limit transitions $P \to q$ and $P' \to q'$ restricted by the following condition: if $q \in P$ and $q' \in P'$, then one has $P \neq P'$. The related class of orderings is equal to $\bigcup_{n<\omega} W_n$ (see the definition of W_n just after Theorem 1).

6 In this class, all rational operations are allowed but the operation $\diamond$ can only be used as a unary operation, that is the operation $X \diamond Y$ must be restricted to the case $Y = \{\varepsilon\}$. The automata have left and right limit transitions $P \to q$ and $q' \to P'$ of a particular form.

Condition (†). Let $P \to q$ be a left limit transition and $q' \to P'$ be a right limit transition. If $q \in P$, $q' \in P'$ and $P = P'$, then $q = q'$ and for any $R \subseteq P$ with $q \in R$, the left and right transitions $R \to q$ and $q \to R$ must appear among the transitions of the automaton.

We do not know a characterization by a particular class of orderings.

7 This class corresponds to the Kleene theorem of [3] for all countable scattered orderings. The operation $\diamond$ is here used as a binary operation.

Let us illustrate by examples some classes of the hierarchy.

Example 4 *The set $(b^{-\omega}ab^{\omega})^*$ of Examples 1 and 2 belong to the Class 2. The related orderings are $K_n = \sum_{j\in\{1,\dots,n\}} \zeta$ which belong to U_2. Since the linear ordering ζ is neither an ordinal nor a backwards ordinal, the set $(b^{-\omega}ab^{\omega})^*$ cannot belong to a lower class.*

Example 5 *The automaton of Figure 7 accepts the rational set $(a+bb)^{\diamond}$. The operation $\diamond$ is unary and the automaton satisfies Condition (†). Hence it is*

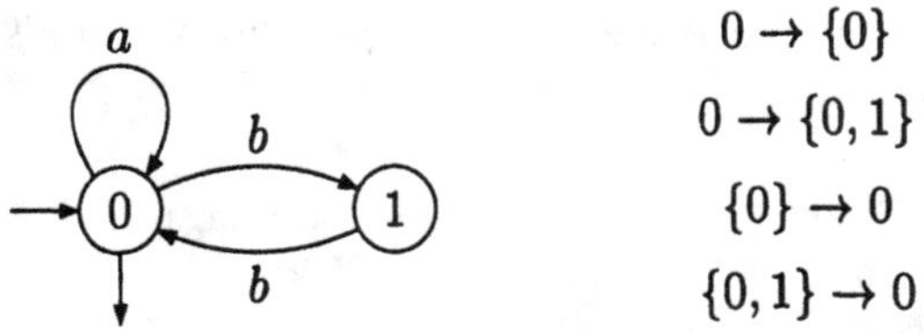

Figure 7. Automaton accepting the set $(a+bb)^\diamond$

an example of Class 6. Note that it does not belong to Class 5. Indeed, the condition on the limit transitions of the automaton is not respected. Moreover, the set $(a+bb)^\diamond$ *contains the word* $\prod_{j\in J} a$ *of length* $J=\zeta^\omega$ *and we have seen in Section 2 that* J *belongs to* W_ω *but not to* $\bigcup_{n<\omega} W_n$.

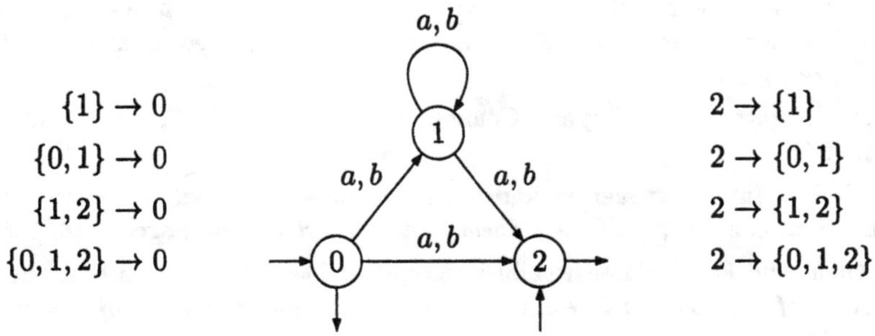

Figure 8. Automaton accepting the set $\varepsilon \diamond A$

Example 6 *The automaton pictured in Figure 8 accepts the set denoted by the rational expression* $\varepsilon \diamond A$ *where* $A=\{a,b\}$. *Recall that a linear ordering is* complete *if any subset which is upper bounded has a least upper bound (or equivalently if any subset which is lower bounded has a greatest lower bound). It can be shown that a scattered linear ordering* J *is complete iff there is a scattered linear ordering* K *such that* $J=\hat{K}$. *Therefore, the automaton of Figure 8 accepts words the length of which is a complete ordering.*

This automaton does not satisfy Condition (†). Indeed there exist left and right limit transitions $P\to q$ *and* $P'\to q'$ *such that* $q\in P$, $q'\in P'$ *and* $P=P'$ *but* $q\neq q'$. *Take* $P=P'=\{0,1,2\}$, $q=2$ *and* $q'=0$. *It can be shown that any rational expression with a unary* $\diamond$ *denotes a set which contains words with a non complete length. Therefore the set* $\varepsilon\diamond A$ *does not belong to Class 6.*

References

[1] N. Bedon. *Langages reconnaissables de mots indéxés par des ordinaux.* Thèse de doctorat, Université de Marne-la-Vallée, 1998.

[2] N. Bedon and O. Carton. An Eilenberg theorem for words on countable ordinals. In Cláudio L. Lucchesi and Arnaldo V. Moura, editors, *Latin'98: Theoretical Informatics*, volume 1380 of *Lect. Notes in Comput. Sci.*, pages 53–64. Springer-Verlag, 1998.

[3] V. Bruyère and O. Carton. Automata on linear orderings. In J. Sgall, A. Pultr, and P. Kolman, editors, *MFCS'2001*, volume 2136 of *Lect. Notes in Comput. Sci.*, pages 236–247, 2001. IGM report 2001-12.

[4] J. R. Büchi. Weak second-order arithmetic and finite automata. *Z. Math. Logik und grundl. Math.*, 6:66–92, 1960.

[5] J. R. Büchi. Transfinite automata recursions and weak second order theory of ordinals. In *Proc. Int. Congress Logic, Methodology, and Philosophy of Science, Jerusalem 1964*, pages 2–23. North Holland, 1965.

[6] Y. Choueka. Finite automata, definable sets, and regular expressions over ω^n-tapes. *J. Comput. System Sci.*, 17(1):81–97, 1978.

[7] D. Girault-Beauquier. Bilimites de langages reconnaissables. *Theoret. Comput. Sci.*, 33(2–3):335–342, 1984.

[8] S. C. Kleene. Representation of events in nerve nets and finite automata. In C.E. Shannon, editor, *Automata studies*, pages 3–41. Princeton university Press, Princeton, 1956.

[9] R. McNaughton and S. Papert. *Counter free automata.* MIT Press, Cambridge, MA, 1971.

[10] D. Muller. Infinite sequences and finite machines. In Proc. of Fourth Annual IEEE Symp., editor, *Switching Theory and Logical Design*, pages 3–16, 1963.

[11] M. Nivat and D. Perrin. Ensembles reconnaissables de mots bi-infinis. In *Proceedings of the Fourteenth Annual ACM Symposium on Theory of Computing*, pages 47–59, 1982.

[12] D. Perrin. Finite automata. In J. van Leeuwen, editor, *Handbook of Theoretical Computer Science*, volume B, chapter 1, pages 1–57. Elsevier, 1990.

[13] J. G. Rosenstein. *Linear ordering.* Academic Press, New York, 1982.

[14] M.-P. Schützenberger. On finite monoids having only trivial subgroups. *Inform. Control*, 8:190–194, 1965.

[15] W. Thomas. Automata on infinite objects. In J. van Leeuwen, editor, *Handbook of Theoretical Computer Science*, volume B, chapter 4, pages 133–191. Elsevier, 1990.

[16] J. Wojciechowski. Finite automata on transfinite sequences and regular expressions. *Fundamenta informaticæ*, 8(3-4):379–396, 1985.

SYMMETRIC CONNECTIVITY WITH MINIMUM POWER CONSUMPTION IN RADIO NETWORKS

G. Călinescu*, I.I. Măndoiu†, and A. Zelikovsky‡

**Department of Computer Science, Illinois Institute of Technology, Chicago, IL 60616. Research partially done while visiting the Department of Combinatorial Optimization of University of Waterloo, where supported by a NSERC grant.*

† *Department of Computer Science & Engineering, UC San Diego, La Jolla, CA 92093*

‡ *Department of Computer Science, Georgia State University, Atlanta, GA 30303. Research partially supported by NSF Grant CCR-9988331, MRDA and CRDF Award No. MM2-3018 and by the State of Georgia's Yamacraw Initiative.*

calinesc@cs.iit.edu, mandoiu@cs.ucsd.edu, alexz@cs.gsu.edu

Abstract

We study the problem of assigning transmission ranges to the nodes of a multi-hop packet radio network (also known as static ad hoc wireless network) so as to minimize the total power consumed under the constraint that enough power is provided to the nodes to ensure that the network is connected. Precisely, we require that the bidirectional links established by the transmission range of every node form a connected graph. We call this problem MIN-POWER SYMMETRIC CONNECTIVITY.

Implicit in previous work on transmission range assignment under asymmetric connectivity requirements is the proof that MIN-POWER SYMMETRIC CONNECTIVITY is NP-hard and that a simple algorithm has approximation ratio of 2. In this paper we establish the similarity between MIN-POWER SYMMETRIC CONNECTIVITY and the classic STEINER TREE problem in graphs, give a polynomial-time approximation scheme with performance ratio approaching $1 + \ln 2 \approx 1.69$, and present a more practical 15/8 approximation algorithm. We also show that the related MIN-POWER SYMMETRIC UNICAST problem can be solved efficiently by a shortest-path computation in an appropriately constructed graph.

1. Introduction

Ad hoc wireless networks have received significant attention in recent years due to their potential applications in battlefield, emergency disaster relief, and other application scenarios (see, e.g., [1, 4, 5, 7, 9, 11, 14, 17, 16]). Unlike wired networks or cellular networks, no wired backbone infrastructure is installed in ad hoc wireless networks. A communication session is achieved either through single-hop transmission if the recipient is within the transmission range of the source node, or by relaying through intermediate nodes otherwise. We assume that omnidirectional antennas are used by all nodes to transmit and receive signals. Thus, a transmission made by a node can be received by all nodes within its transmission range. This feature is extremely useful for energy-efficient multicast and broadcast communications.

For the purpose of energy conservation, each node can (possibly dynamically) adjust its transmitting power, based on the distance to the receiving node and the background noise. In the most common power-attenuation model [12], the signal power falls as $\frac{1}{r^\kappa}$ where r is the distance from the transmitter antenna and κ is a real *constant* dependent on the wireless environment, typically between 2 and 4. Assume that all receivers have the same power threshold for signal detection, which is typically normalized to one. With this assumption, the power required to support a link between two nodes separated by a distance r is r^κ. A crucial issue is how to find a route with minimum total energy consumption for a given communication session. This problem is referred to as *Minimum-Energy Routing* in [14, 17]. Having every link established in both directions simplifies the one-hop transmission protocols by allowing acknowledgement messages to be sent back for every packet (see, for example [15]). This motivates the study of the MIN-POWER SYMMETRIC CONNECTIVITY problem, where a link is established only if both nodes have transmission range at least as big as the distance between them, and the goal is to ensure that the network is connected.

Formally, given a set of points V (representing the nodes in the network) in E^2 (the two-dimensional Euclidean space) or in E^3 (the three-dimensional Euclidean space), a *transmission range assignment* (or *range assignment*, for short) is a function $r : V \to R_+$. A *unidirectional link* from node u to node v is established under the range assignment r if $r(u) \geq \|uv\|$, where $\|uv\|$ denotes the Euclidean distance between u and v. A *bidirectional link* uv is established under the range assignment r if $r(u) \geq \|uv\|$ and $r(v) \geq \|uv\|$. Let $B(r)$ denote the set of all bidirectional links established between pairs of nodes in V under the transmission range r. In this paper we study the following problem:

MIN-POWER SYMMETRIC CONNECTIVITY: Given a set of nodes V and $\kappa \geq 1$, find a transmission range assignment $r : V \to R_+$ minimizing $\sum_{v \in V} r(v)^\kappa$ subject to the constraint that the graph $(V, B(r))$ is connected.

Implicit in the work of Clementi, Penna, and Silvestri [5] is a proof that MIN-POWER SYMMETRIC CONNECTIVITY in E^2 is NP-Hard (radio "bridges" in canonical form gadgets, see Definition 3 on page 10 of [5], can be made to be bidirectional links). Therefore, we search for polynomial-time approximation algorithms for this problem. The *performance ratio* of an approximation algorithm A for a minimization problem is the supremum, over all possible instances I, of the ratio between the cost of the output of A when running on I and the cost of an optimal solution for I (the smaller the performance ratio, the better). We say that A is an *α-approximation algorithm* if its performance ratio is at most α. A *fully polynomial α-approximation scheme* is a family of algorithms A_ε such that, for every $\varepsilon > 0$, A_ε (1) has performance ratio at most $\alpha + \varepsilon$, and (2) runs in time polynomial in the size of the instance and $1/\varepsilon$.

Kirousis, Kranakis, Krizanc, and Pelc [7] give a minimum spanning tree (MST) based 2-approximation algorithm for MIN-POWER SYMMETRIC CONNECTIVITY (their algorithm is actually designed for a related problem, which we discuss in Section 1.1). We improve the approximation ratio under 2 by exploiting similarities between MIN-POWER SYMMETRIC CONNECTIVITY and the classic STEINER TREE problem: given an edge-weighted graph $G = (V, E, w)$ and a set $T \subseteq V$ of *terminals*, find a minimum weight *Steiner tree* for T, i.e., a minimum weight connected subgraph of G which contains T. Computing an MST in the complete graph on T with edge-weights equal to the minimum distance in G between corresponding terminals gives a 2-approximation algorithm for STEINER TREE [3, 8]. Zelikovsky [18] gave the first algorithm with approximation ratio less than 2: he used 3-restricted Steiner trees and the concept of *gain* to obtain an approximation ratio of 11/6. Promel and Steger [10] extend the results of Camerini, Galbiati, and Maffioli [2] and give a polynomial time 5/3-approximation scheme for STEINER TREE, by finding almost optimal 3-restricted Steiner tree. Zelikovsky [19] gives a polynomial time $(1 + \ln 2)$-approximation scheme.

In Section 3 we show that similar concepts can be used for approximating MIN-POWER SYMMETRIC CONNECTIVITY. In particular, we show that the algorithms of [10], [18], and [19] can be modified to give similar approximation ratios for MIN-POWER SYMMETRIC CONNECTIVITY. Our main results are a fully polynomial $1 + \ln 2$ approximation scheme based on [19] and a more practical 15/8 approximation algorithm based on [18].

Our algorithms have the same approximation guarantees when network nodes are located in E^3. In fact, since they work on a graph model of the network, our algorithms can be directly applied to more general problem formulations, e.g., observing given upper-bounds on the transmission range of each node and/or taking into account obstacles that completely block the communication between certain pair of nodes.

In Section 4 we address the related MIN-POWER SYMMETRIC UNICAST problem, which, for given source and destination nodes, $s, t \in V$, asks for a sequence $v_0 = s, v_1, \ldots, v_k = t$ of nodes and transmission ranges $r(v_i)$, $i = 0, \ldots, k$, under which all bidirectional links $v_i v_{i+1}$ are established. We show that MIN-POWER SYMMETRIC UNICAST can be solved efficiently by a shortest-path computation in an appropriately constructed graph.

1.1. Related Work

Previous research on symmetric connectivity has addressed only the objective of minimizing the maximum node power [9, 11]. The objective of minimizing the total power $\sum_{v \in V} r^\kappa(v)$ has been addressed under the related *asymmetric* connectivity model, in which unidirectional links give raise to a directed graph on V. Four problems have been studied under this model.

First, the ASYMMETRIC UNICAST problem requires establishing a minimum power directed path from a source s to a destination t, and is easily solved in polynomial time by shortest-path algorithms.

Second, the ASYMMETRIC BROADCAST problem [14, 17] requires establishing a minimum power arborescence rooted at a given vertex s. Clementi et al. [4] prove that ASYMMETRIC BROADCAST is NP-Hard when the nodes are in E^2. The best known approximation algorithm for ASYMMETRIC BROADCAST [16], based on computing a minimum spanning tree, has performance ratio of at most 12 when the nodes are in E^2.

Third, in ASYMMETRIC MULTICAST, one is given a root s and a set of terminals T, and the goal is to establish a minimum-power branching rooted at s which reaches all vertices of T. As a generalization of ASYMMETRIC BROADCAST, ASYMMETRIC MULTICAST is also NP-Hard, and based on the work of [16], it is immediate that a minimum Steiner tree would give an approximation ratio of 12ρ, where ρ is the approximation for Steiner tree in graphs (the best result known at this moment, given in [13], is $\rho = 1 + \frac{1}{2}\ln 3 + \varepsilon$).

Fourth, in the COMPLETE RANGE ASSIGNMENT problem the objective is establishing a strongly connected subgraph of V. Kirousis, Kranakis, Krizanc, and Pelc [7] prove that COMPLETE RANGE ASSIGNMENT in E^3 is NP-Hard and, based on the minimum spanning tree, give a 2-approximation algorithm. As opposed to the ASYMMETRIC BROADCAST approximation of [16], the COMPLETE RANGE ASSIGNMENT approximation of [7] is valid in arbitrary graphs (that is, the distance between two points could be arbitrary, not necessarily Euclidean). Clementi, Penna, and Silvestri [5] give an elaborate reduction proving that COMPLETE RANGE ASSIGNMENT in E^2 is also NP-Hard.

The power for the asymmetric COMPLETE RANGE ASSIGNMENT can be twice less than the power for MIN-POWER SYMMETRIC CONNECTIVITY as illustrated by the following example in which $\kappa = 2$. The terminal set (see Figure 1) consists of n groups of $n+1$ points each, located on the sides of a regular $2n$-gon. Each group has 2 terminals in distance 1 of each other (represented as thick circles in Figure 1) and $n-1$ equally spaced points (dashes in Figure 1) on the line segment between them. It is easy to see that the minimum range assignment ensuring asymmetric connectivity assigns power of 1 to the one thick terminal in each group and power of $\varepsilon^2 = (1/n)^2$ to all other points in the group. The total power then equals $n+1$. For symmetric connectivity it is necessary to assign power of 1 to all but two thick points, and of ε^2 to the remaining points, which results in total power of $2n - 1 - 1/n + 2/n^2$.

2. Preliminaries

We begin by formulating the graph-weighted extension of MIN-POWER SYMMETRIC CONNECTIVITY. For completeness, we then show that computing an MST gives a 2-approximation for this extension; this result is implicit in the work of Kirousis, Kranakis, Krizanc, and Pelc [7].

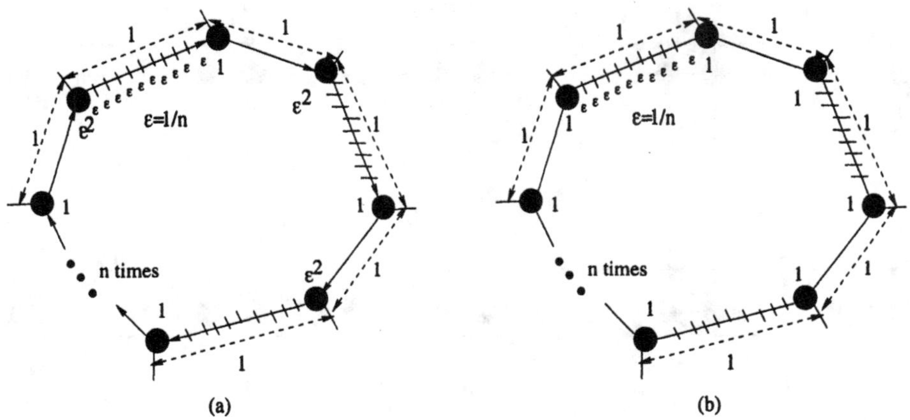

Figure 1. Total power for the asymmetric COMPLETE RANGE ASSIGNMENT can be twice less than the total power for MIN-POWER SYMMETRIC CONNECTIVITY ($\kappa = 2$). (a) Minimum range assignment ensuring asymmetric connectivity has total power $n + n^2\varepsilon^2 = n + n^2\frac{1}{n^2} = n + 1$. (b) Minimum range assignment ensuring symmetric connectivity has the total power $(2n-2) + (n^2 - n + 2)\varepsilon^2 = 2n - 1 - \frac{1}{n} + \frac{2}{n^2}$.

Let $G = (V, E, c)$ be an edge-weighted graph. For a spanning tree $T = (V, F)$ of G, let $\overline{r_T}(v) = \max_{u|uv \in F} c(uv)$. Define the *power-cost of T* by

$$p(T) = \sum_{v \in V} \overline{r_T}(v)$$

Since any connected graph contains a tree, an equivalent formulation of MIN-POWER SYMMETRIC CONNECTIVITY is to ask for a spanning tree with minimum power-cost in the complete graph on V with edge costs given by $c(uv) = \|uv\|^\kappa$. Thus, MIN-POWER SYMMETRIC CONNECTIVITY is a special case of the following problem:

MINIMUM POWER-COST SPANNING TREE: Given a connected edge-weighted graph $G = (V, E, c)$, find a spanning tree T of G with minimum power-cost.

All our algorithms work for the this graph-weighted extension of MIN-POWER SYMMETRIC CONNECTIVITY. From now on, we only use this formulation.

Theorem 1 *Computing an MST with respect to c gives a 2-approximation for* MIN-POWER SYMMETRIC CONNECTIVITY.

Proof: Let $c(T) = \sum_{uv \in F} c(uv)$. Claim 2 of Theorem 3.2 of [7] is equivalent to

$$p(T) = \sum_{v \in V} \max_{u|uv \in F} c(uv) \leq \sum_{v \in V} \sum_{u|uv \in F} c(uv) = 2c(T) \tag{1}$$

Let u be a vertex incident to an edge of maximum cost. If we root the tree T at u, and use v' to denote the parent of v in T, since $\overline{r_T}(v) \geq c(vv')$ we

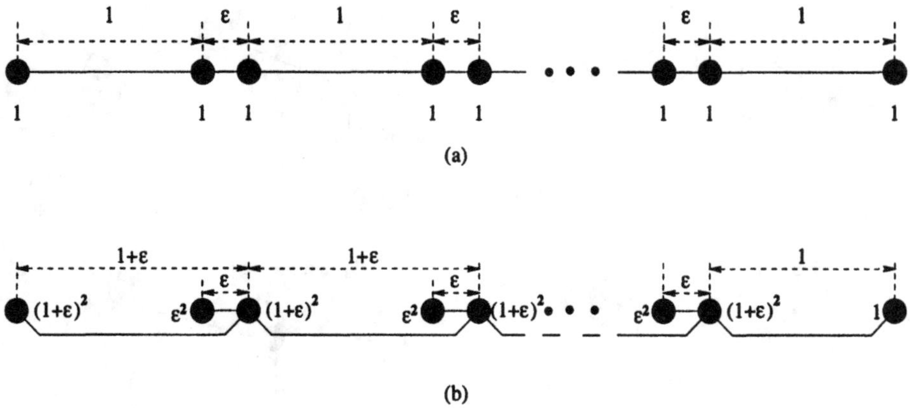

Figure 2. Tight example for the performance ratio of the MST algorithm ($\kappa = 2$). (a) The MST-based range assignment needs total power $2n$. (b) Optimum range assignment has total power $n(1+\varepsilon)^2 + (n-1)\varepsilon^2 + 1 \to n+1$.

conclude that $p(T) \geq c(T)$. Therefore, if MST is the minimum spanning tree with respect to c and OPT is the tree with minimum power-cost, we have

$$p(MST) \leq 2c(MST) \leq 2c(OPT) \leq 2p(OPT)$$

■

The following example shows that the ratio of 2 given in Theorem 1 is tight. Consider $2n$ points located on a single line such that the distance between consecutive points alternates between 1 and $\varepsilon < 1$ (see Figure 2) and let $\kappa = 2$. Then the minimum spanning tree MST connects consecutive neighbors and has power-cost $p(MST) = 2n$. On the other hand, the tree T with edges connecting each other node (see Figure 2(b)) has power-cost equal $p(T) = n(1+\varepsilon)^2 + (n-1)\varepsilon^2 + 1$. When $n \to \infty$ and $\varepsilon \to 0$, we obtain that $p(MST)/p(T) \to 2$.

3. k-Restricted Approach to Symmetric Min-Power Connectivity

We first give definitions of k-restricted decompositions and prove an upper bound on the power-cost of such decompositions. Then we will describe approximation algorithms whose approximation ratios follow from the approximation ratios of Steiner tree algorithms in graphs.

3.1. k-Restricted Decompositions

A k-restricted decomposition Q of the tree T is a partition of T into subtrees $T_1, T_2, \ldots, T_p$ each containing at most k vertices such that each edge of T belongs to exactly one subtree T_i. The power-cost $p(Q)$ of Q defined to be the sum of power-costs of all its elements:

$$p(Q) = \sum_{T_i \in Q} p(T_i)$$

Theorem 2 *For any weighted tree T and any $k \geq 1$, there is a 2^k-restricted decomposition Q of T such that $p(Q) \leq (1 + 1/k)p(T)$.*

Proof: Without loss of generality we can assume that all edge costs are different. Let the endpoints r and s of the heaviest edge h of T be the *roots* of T, which means that two subtrees of $T - \{h\}$ are rooted at r and s, respectively. Then each vertex v of T, except r and s, has a unique parent. We call the vertices adjacent to v, other than the parent of v (if defined), the children of v. For each vertex v of T, we sort the edges connecting v to its children in increasing order of their cost. For the most costly such edge e, we define $next(e) = f$, where f is the edge connecting v to its parent (if v has a parent), or $f = h$ if v does not have a parent. For some other edge e from v to one of its children, we define $next(e) = e'$, where e' is the next edge (in the sorted order above) from v to one of its children.

We now construct a rooted directed binary (with arcs going toward the root) tree B as follows. The vertices of B are the edges of T and the root of B is h, the heaviest edge of T. The arcs of B consists of arcs $(e, next(e))$ for each edge e of T. It is immediate that every vertex $e = uv$ of B has at most two incoming arcs. Indeed, if $e = rs$, then only the most costly edge of $T \setminus \{e\}$ incident to r and the most costly edge of $T \setminus \{e\}$ incident to s, have e as a parent. The other edges of T $e = uv$ have v as the parent of u (the other case being symmetric), and the arcs coming into e are only the most costly edge of $T \setminus \{e\}$ incident to u and the edge in between v and another child of v which precedes e in the sorted order above. Note that each vertex of B has cost since it is an edge of T.

Let B_i be the set of vertices of B in distance i from the root h. There is an integer $0 \leq l < k$ such that $\sum_{j \mid j \equiv l \pmod k} c(B_j) \leq \frac{1}{k}c(B) = \frac{1}{k}c(T)$, and let $\overline{B} = \cup_{j \mid j \equiv l \pmod k} B_j$. Removal of every edge outgoing from $\overline{B}$ decomposes B into subtrees Q_i corresponding to subtrees T_i of T. The number of vertices in Q_i is at most $2^k - 1$ since Q_i is a binary tree of height at most $k - 1$. Therefore, each T_i has at most 2^k vertices. We denote by Q the 2^k-restricted decomposition of T into T_i's.

Let $e_i = (v_i, u_i)$ be the root of Q_i (note that $e_i \in \overline{B}$) and, if $e_i \neq (r, s)$, rename v_i and u_i such that u_i is the parent of v_i in T. By the construction of B, we have that $\max_{u \mid uu_i \in E(T_i)} c(uu_i) = c(e_i)$. Then we have:

$$p(T_i) \leq c(e_i) + \sum_{v \in V(T_i) \setminus \{u_i\}} \max_{(v,u) \in E(T)} c(v, u).$$

For $i \neq j$, the sets $V(T_i) \setminus \{u_i\}$ and $V(T_j) \setminus \{u_j\}$ are disjoint. We conclude that

$$\begin{aligned} p(Q) &= \sum_i p(T_i) \\ &\le \sum_{v\in V(T)} \max_{(v,u)\in E(T)} c(v,u) + \sum_i c(e_i) \\ &\le p(T) + c(\overline{B}) \\ &\le p(T) + \frac{1}{k}c(T) \\ &\le (1+\frac{1}{k})p(T). \end{aligned}$$

■

A subtree of T with exactly three edges and two edges is called a *fork*. So a 3-restricted decomposition Q of T consists of forks and individual edges. We have:

Theorem 3 *For any weighted tree T, there is a 3-restricted decomposition Q of T such that $p(Q) \le \frac{7}{4}p(T)$.*

Proof: By induction on n, the number of vertices of the tree, with the base case obvious. Assume uv is the minimum cost edge in the tree (with $c(uv) = \varepsilon$), and let T_1 and T_2 be the two trees obtained from T after removing the edge uv. Let $c'(xy) = c(xy) - c(uv)$, for any edge xy in T_1 or in T_2.

By induction, there are 3-restricted decompositions Q_1 of T_1 and Q_2 of T_2 such that $p'(Q_1) \le \frac{7}{4}p'(T_1)$ and $p'(Q_2) \le \frac{7}{4}p'(T_2)$. We also assume that Q_1 and Q_2 are "fork-maximal", in the sense that no two edges can be merged in fork (such a merging decreases the power-cost). We count the number of forks in Q_1 and in Q_2. Assume T_i has n_i vertices, and Q_i has k_i forks and s_i single edges. Then $n_i = 2k_i + s_i + 1$. As Q_i is fork-maximal, we also have that $s_i \le 2k_i + 1$. Also, $n = n_1 + n_2$. Let Q be the partition of T obtained from Q_1, Q_2, and uv. Then we have:

$$p(Q) = p'(Q_1) + p'(Q_2) + \varepsilon(3k_1 + 2s_1 + 3k_2 + 2s_2 + 2),$$

while

$$p(T) = p'(T_1) + p'(T_2) + \varepsilon n.$$

So it suffices to show that $3k_1 + 2s_1 + 3k_2 + 2s_2 + 2 \le \frac{7}{4}n$, which follows immediately from the the equations on n, n_i, and s_i above. ■

3.2. Approximation Algorithms

Based on the results of [2] or [10] and Theorem 3, it is immediate there exists an algorithm with the approximation ratio $\rho_3 < \frac{7}{4} + \epsilon$.

Let ρ_k be the supremum, over all trees T, of the ratio of the power-cost of the minimum power-cost k-restricted decompositions to the power-cost of T.

Greedy Algorithm
Input: A complete graph $G = (V, E, cost)$ with edge costs
Output: A 3-restricted decomposition connecting all vertices in S
$T \leftarrow MST(G)$
$H \leftarrow G$
Repeat forever
  Find a triple K with the maximum $g = gain_T(K)$
  If $g \leq 0$ **then exit repeat**
  $H \leftarrow H \cup K$
  $V \leftarrow V/K$
Output H

Figure 3. The Greedy Algorithm

Theorem 2 implies that $\rho_k \leq 1 + \frac{1}{\lfloor \lg k \rfloor}$, Theorem 3 implies that $\rho_3 \leq 7/4$, and (1) together with the example in Figure 2 imply that $\rho_2 = 2$. In order to obtain better approximation ratios we can approximate the minimum power-cost k-restricted decompositions.

In Steiner trees, k-restricted decompositions correspond to k-restricted Steiner trees. Below we translate the Steiner tree terms into the language of decompositions.

For a set of vertices V, denote by $mst(V)$ the minimum cost of a spanning tree. For a tree H connecting some vertices from V, we denote V/H the set of vertices V after contracting of H, i.e., collapsing all vertices of H into a single vertex. Let *gain* of a subtree H, $gain(H)$, be

$$gain(H) = 2mst(V) - 2mst(V/H) - p(H)$$

It has been proved in [18] that the Greedy Algorithm (see Figure 3) has approximation ratio at most arithmetic mean of ρ_2 and ρ_3. Thus we have:

Theorem 4 *The Greedy Algorithm for* MIN-POWER SYMMETRIC CONNECTIVITY *(see Fig. 3) has approximation ratio of* $15/8$.

It has been shown in [19] that the k-restricted Relative Greedy Algorithm (see Figure 4) has approximation ratio at most $1 + \ln 2 + \varepsilon$ if $\lim_{k\to\infty} \rho_k = 1$.

Theorem 5 *The k-restricted Relative Greedy Algorithm for* MIN-POWER SYMMETRIC CONNECTIVITY *(see Fig. 4) has approximation ratio of* $1 + \ln 2 + \varepsilon$.

4. Minimum Power Symmetric Unicast

In this section we reformulate MIN-POWER SYMMETRIC UNICAST as a graph problem, and then reduce the latter problem to a single-source single-sink shortest-path computation in an appropriately constructed graph.

---

k-restricted Relative Greedy Algorithm (k-RGA)
Input: A complete graph $G = (V, E, cost)$ with edge costs and an integer $k \leq |V|$
Output: A k-restricted decomposition connecting all vertices in S
$T \leftarrow MST(G)$
$H \leftarrow G$
Repeat forever
  Find a k-restricted tree K with the minimum $r = \frac{p(H)}{2mst(V)-2mst(V/H)}$
  If $r \geq 1$ then exit repeat
  $H \leftarrow H \cup K$
  $V \leftarrow V/K$
Output H

Figure 4. The k-restricted Relative Greedy Algorithm (k-RGA).

MIN-POWER SYMMETRIC PATH IN GRAPHS: Given a graph $G = (V, E, c)$ with costs on edges a source $s \in V$ and a destination $t \in V$, find an $s - t$ path in G of the minimum power-cost.

The following example in the Euclidean plane shows that a straightforward application of Dijkstra's algorithm does not work, i.e., a minimum cost $s - t$ path does not always have minimum power-cost. Consider a network consisting of three nodes, $s = (0, 3)$, $t = (4, 0)$, and $x = (0, 0)$, and assume $\kappa = 2$. Then the two $s - t$ paths, namely, (s, t) and (s, v, t), have the same cost of 25 but different power-costs: the power-cost of (s, t) is 25+25=50 while the power-cost of (s, v, t) is 9+16+16=41.

Our solution of MIN-POWER SYMMETRIC UNICAST first modifies the given graph $G = (V, E, c)$ and then applies Dijkstra's algorithm to the resulted directed graph G'. We now describe the construction of the directed graph $G' = (V', E', c')$.

For any $u \in V$, we sort all adjacent vertices $\{v_1, \ldots, v_k\}$ in ascending order of costs of edges connecting them to u, i.e., $c(u, v_i) \leq c(u, v_{i+1})$. The vertex v is replaced with a *gadget* (see Figure 5(a)) as follows:

(i) each edge (u, v_i) is subdivided by a vertex $[u, v_i]$

(ii) for each u, we connect all vertices $[u, v_i]$'s by two directed paths:
$P_1 = (u, [u, v_1], \ldots, [u, v_{k-1}], [u, v_{k-1}])$ and
$P_2 = ([u, v_{k-1}], [u, v_{k-1}], \ldots, [u, v_1], u)$.

(iii) the costs of the arcs on path P_1 are $c(u, v_1)$, $c(u, v_2) - c(u, v_1)$, ..., $c(u, v_k) - c(u, v_{k-1})$, respectively, and the cost of all arcs on the path P_2 is zero.

Finally, each edge (u, v) of G is replaced in G' with the two arcs $([u, v], [v, u])$ and $([v, u], [v, u])$, both of cost $c(u, v)$. Figure 5(b) shows the graph G' for the

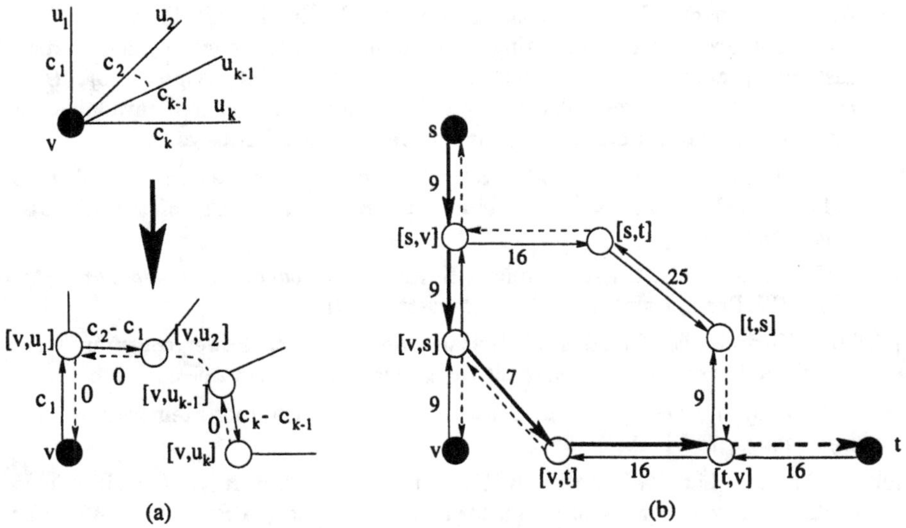

Figure 5. (a) A vertex v adjacent to k vertices $u_1, \ldots, u_k$ via edges of cost $c_1, c_2, \ldots, c_k$ and a gadget replacing v with a bidirectional path. The solid edges of the path $(v, [v, u_2])$, $([v, u_2]), [v, u_3], \ldots$, $([v, u_{k-1}], [v, u_k]$ have cost c_1, $c_2 - c_1$, ..., $c_k - c_{k-1}$, respectively. The dashed edges have zero cost. (b) The graph G' for the example below. Thick edges belong to the shortest path corresponding to the path (s, v, t) in G.

three-node example $s = (0, 3)$, $t = (4, 0)$, and $x = (0, 0)$, with $\kappa = 2$. It is easy to see that a shortest s–t path in G' corresponds to a minimum power-cost s–t path.

Using the Fibonnaci heaps implementation of Dijkstra's algorithm [6] to compute the shortest s–t path in G', and observing that $|V'| = O(|E|)$ and $|E'| = O(|E|)$, we obtain the following

Theorem 6 SYMMETRIC UNICAST *is solvable in time* $O(|E| \log |V|)$.

5. Acknowledgements

The first author thanks Joseph Cheryan, Francisco Zaragoza, and Bhaskar DasGupta for useful discussions in the early stage of this project.

References

[1] D.M. Blough, M. Leoncini, G. Resta, and P. Santi. On the symmetric range assignment problem in wireless ad hoc networks. In *Proc. 2nd IFIP International Conference on Theoretical Computer Science*, page (to appear), 2002.

[2] P.M. Camerini, G. Galbiati, and F. Maffioli. Random pseudo-polynomial algorithms for exact matroid problems. *Journal of Algorithms*, 13:258–273, 1992.

[3] E.-A. Choukhmane. Une heuristique pour le probleme de l'arbre de Steiner. *RAIRO Rech. Oper.*, 12:207–212, 1978.

[4] A.E.F. Clementi, P. Crescenzi, P. Penna, G. Rossi, and P. Vocca. On the complexity of computing minimum energy consumption broadcast subgraphs. In *Symposium on Theoretical Aspects of Computer Science*, pages 121–131, 2001, extended version available at http://www.mat.uniroma2.it/~penna/papers/stacs01-TR.ps.gz.

[5] A.E.F. Clementi, P. Penna, and R. Silvestri. On the power assignment problem in radio networks. *Electronic Colloquium on Computational Complexity (ECCC)*, (054), 2000.

[6] T.H. Cormen, C.E. Leiserson, and R.L. Rivest. *Introduction to algorithms (2nd ed.)*. MIT Press, Cambridge, Massachusetts, 2001.

[7] L.M. Kirousis, E. Kranakis, D. Krizanc, and A. Pelc. Power consumption in packet radio networks. *Theoretical Computer Science*, 243:289–305, 2000.

[8] L. Kou, G. Markowsky, and L. Berman. A fast algorithm for Steiner trees. *Acta Informatica*, 15:141–145, 1981.

[9] E. Lloyd, R. Liu, M. Marathe, R. Ramanathan, and S.S. Ravi. Algorithmic aspects of topology control problems for ad hoc networks. In *Proc. ACM MobiHoc*, page (to appear), 2002.

[10] H.J. Promel and A. Steger. A new approximation algorithm for the Steiner tree problem with performance ratio 5/3. *Journal of Algorithms*, 36:89–101, 2000.

[11] Ram Ramanathan and Regina Hain. Topology control of multihop wireless networks using transmit power adjustment. In *INFOCOM (2)*, pages 404–413, 2000.

[12] T.S. Rappaport. *Wireless Communications: Principles and Practices.* Prentice Hall, 1996.

[13] G. Robins and A. Zelikovsky. Improved Steiner tree approximation in graphs. In *Proceedings of the 11th ACM-SIAM Annual Symposium on Discrete Algorithms*, pages 770–779, 2000.

[14] S. Singh, C.S. Raghavendra, and J. Stepanek. Power-aware broadcasting in mobile ad hoc networks. In *Proceedings of IEEE PIMRC*, 1999.

[15] A.S. Tanembaum. *Computer Networks (3rd edition)*. Prentice Hall, 1996.

[16] P.-J. Wan, G. Calinescu, X.-Y. Li, and O. Frieder. Minimum energy broadcast routing in static ad hoc wireless networks. In *Proc. IEEE INFOCOM, vol. 2*, pages 1162–1171, 2001.

[17] J.E. Wieselthier, G.D. Nguyen, and A. Ephremides. On the construction of energy-efficient broadcast and multicast trees in wireless networks. In *Proc. IEEE INFOCOM*, pages 585–594, 2000.

[18] A. Zelikovsky. An 11/6-approximation algorithm for the network Steiner problem. *Algorithmica*, 9:463–470, 1993.

[19] A. Zelikovsky. Better approximation bounds for the network and Euclidean Steiner tree problems. Technical Report CS-96-06, Department of Computer Science, University of Virginia, 1996.

A MODIFIED RECURSIVE TRIANGULAR FACTORIZATION FOR CAUCHY-LIKE SYSTEMS*

Zhao Chen
Mathematics Department
New York City Technical College
300 Jay Street
Brooklyn, NY 11201
zchen@nyctc.cuny.edu

Abstract We derive new algorithms for solving strongly nonsingular Cauchy-like systems of linear equations $C\tilde{x} = \tilde{v}$ in $O(n \log^2 n)$ running time, where $\mathbf{F}$ is a field and $\tilde{v} \in \mathbf{F}^{n\times 1}$ is a vector, $C \in \mathbf{F}^{n\times n}$ is a strongly nonsingular Cauchy-like matrix. Morf, Bitmead and Anderson presented the efficient algorithms to solve strongly nonsingular Toeplitz-like equations of linear systems by using the Recursive Triangular Factorization in 1980. Recently, Pan and Zheng extended the Recursive Triangular Factorization to solve Cauchy-like systems with the complexity of $O(n \log^3 n)$ operations. This is the best known complexity bound by using the direct approach of Recursive Triangular Factorization in Cauchy-like cases. However, these algorithms are still slower than the well known algorithms with the asymptotic bound of $O(n \log^2 n)$ operations, which have been proposed by the means of reducing Cauchy-like matrices into Toeplitz-like matrices. In our present paper, we will modify the Recursive Triangular Factorization so that the complexity bound of the direct recursive approach can be decreased to $O(n \log^2 n)$ operations. This matches the asymptotic bound without transforming to Toeplitz-like matrices. Our improvement of the direct recursive approach is by a factor off $\log n$ due to changing the original vectors which expressed in the given Cauchy-like matrix into the special vectors, where the entries are unit roots. The applications of structured matrices include Nevanlinna-Pick tangential interpolation problems.

*Supported by PSC CUNY Awards 62835-00-31, NYCTC[1] and NYCLSAMP.

1. Introduction

Computations with displacement structured matrices such as Cauchy (or-like), Vandermonde (or-like) and Toeplitz (or-like) have a variety of applications to science and engineering, for example, conformal mapping, tangential Nevanlinna-Pick interpolation, solution of integral equations, rational interpolation, polynomial interpolation and evaluation (cf. [25], [24], [23], [4], [19], [22], [18], [14], [12], [16] and [15]).

By exploiting the displacement structure, we can save running time and memory spaces dramatically in solving Cauchy-like systems of linear equations. Instead of working on the entries of a matrix itself, we use its generators in our computations. Many matrices can be re-constructed through their generators with the displacement operators [11], [1], [10] and [9]. The displacement representation provides many opportunities to accelerate our computations.

Let $\mathbf{F}$ be a field and $\tilde{s}$ denotes a vector. We also denote $A_{M\tilde{v}}(n)$ as the complexity of the product of a matrix $M \in \mathbf{F}^{n\times n}$ by a vector $\tilde{v} \in \mathbf{F}^{n\times 1}$. Moreover, Let us work over a field $\mathbf{F}$ which supports Fast Fourier Transforms of length n.

To solve $n \times n$ strongly nonsingular Toeplitz-like equations of linear systems $T\tilde{x} = \tilde{v}$, where $T \in \mathbf{F}^{n\times n}$ is a Toeplitz-like matrix, Morf, Bitmead and Anderson (see e.g. [13] and [2]) applied the Recursive Triangular Factorization to design fast algorithms with complexity of $O(n \log^2 n)$ arithmetic operations (hereafter, we refer to arithmetic operations as *ao*). These algorithms are faster than Gaussian Elimination algorithms, which run $O(n^3)$ times.

The efficient techniques of the Recursive Triangular Factorization have been extended recently to $n \times n$ Cauchy-like matrices by Pan and Zheng [20] with complexity of

$$A_{C_{like}^{-1}\tilde{v}}(n) = O(A_{C\tilde{v}}(n) \log n) = O(n \log^3 n), \tag{1.1}$$

where C_{like}^{-1} denotes as the inverse of a Cauchy-like matrix and $A_{C\tilde{v}}(n) = O(n \log^2 n)$ denotes as the complexity of multiplication of an $n \times n$ Cauchy matrix by a vector. $O(n \log^3 n)$ is the best known bound by using direct recursive approach to solve Cauchy-like systems. However, these algorithms are still slower than the well known algorithms with the asymptotic bound of $O(n \log^2 n)$ operations, which have been proposed by the means of reducing Cauchy-like matrices into Toeplitz-like matrices. Due to the difference structure between the Cauchy-like matrix and Toeplitz-like matrix, transitions from Cauchy-like matrices into Toeplitz-like matrices will lose the Cauchy-like extension of the Recursive Triangular Factorization from the Toeplitz-like case. Moreover, the transitions which used Vandermonde matrices may produce errors in numerical computations because of the ill-condition of the Vandermonde matrices [5] (except the Fourier matrices). In our present paper, we will modify the Recursive Triangular Factorization so that the complexity bound of the direct recursive approach can be decreased to $O(n \log^2 n)$ operations. This matches the asymptotic bound without transforming to Toeplitz-like matrices.

We propose the new Cauchy-like extension of the Recursive Triangular Factorization, which is inferior to the best known direct recursive approach.

By observing from (1.1), the complexity of the Recursive Triangular Factorization depends on the complexity of the products of a Cauchy matrix by a vector. Let $\left(\frac{1}{e_i - f_j}\right)_{i,j=0}^{n-1}$ be a Cauchy matrix. It is well known that a Cauchy matrix times a vector can be computed in $O(n \log^2 n)$ *ao* (see [8]). If the entries of the vectors $\tilde{e} = e_0, \cdots, e_{n-1}$ and $\tilde{f} = f_0, \cdots, f_{n-1}$ are n-th roots of 1 and -1, it costs only $O(n \log n)$ *ao* to compute their product. This property motivates us to change the original vectors which expressed in generalized Cauchy-like matrices into the special vectors, where the entries the vectors are unit roots.

Due to the new matrix preserves the same Cauchy-like structure, the computations of finding the inverse of the new matrix is faster than the computations of finding the inverse of the original given matrix by using the Recursive Triangular Factorization. This new process of modifying the Recursive Triangular Factorization can be used to solve strongly nonsingular Cauchy-like systems of linear equations in the complexity of

$$A_{C_{like}^{-1}\tilde{v}}(n) = O(A_{C_s\tilde{v}}(n) \log\ n) = O(n\ \log^2 n), \tag{1.2}$$

where $A_{C_s\tilde{v}}(n) = O(n\ \log n)$ denotes the complexity of multiplication of an $n \times n$ special Cauchy matrix by a vector. (we refer a Cauchy matrix $\left(\frac{1}{e_i - f_j}\right)_{i,j=0}^{n-1}$ as a special Cauchy matrix if the entries of the vectors $\tilde{e}$, $\tilde{f}$ are roots of unity). Our improvement for the direct recursive approach is by a factor off $\log n$, versus running time of the best known algorithms for the direct recursive approach. In fact, Heinig [10] has proposed to change the vectors into unit roots by applying Vandermonde linear solver. Computations with Vandermonde matrices will create the numerical instabilities due to the ill-conditions of Vandermonde matrices (cf. [5]). Our algorithms avoid the computations of the solutions of Vandermonde linear systems. The complexity bound of our algorithms reaches to the asymptotic bound.

We organize our paper as follow. In section 2, we present our new representation of the inverse of Cauchy-like matrices. In section 3, we recall the basic algorithms of the Recursive Triangular Factorization. In our final section, we modify the Recursive Triangular Factorization to solve Cauchy-like linear systems of equations. Let us work over a field **F** which supports FFT and follow some lines of [20] and [21].

2. New Expression of the Inverse of a Cauchy-like matrix

It is fast for us to work with the Cauchy-like matrices with special vectors because of lowering our computation complexity. We will show that the vectors in Cauchy-like matrices can be reduced to special vectors. The new representation of Cauchy-like matrices will be given in this section. Let us review some well known definitions.

Definition 1 *Let* $\alpha = e^{-\frac{2\Pi\sqrt{-1}}{n}}$, $\beta = e^{\frac{\Pi\sqrt{-1}}{n}}$. *The elements* $\alpha_i = \alpha^i$, $\beta_j = \beta^j$, *for* $i, j = 0, \ldots, n-1$ *are n-roots of* 1 *and* -1 *respectively.*

Definition 2 *(cf.[17]) We denote* W^T *as the transpose of a matrix or a vector* W. *The inverse of a matrix* M *is denoted as* M^{-1}. *Let* $D(\tilde{x}) \in \mathbf{F}^{n\times n}$ *be a diagonal matrix with the diagonal entries* $\tilde{x} = x_0, \ldots, x_{n-1}$. *Let* $c(\tilde{t}, \tilde{v}) = \left(\frac{1}{t_i - v_j}\right)_{i,j=0}^{n-1} \in \mathbf{F}^{n\times n}$ *denoted as a Cauchy matrix.*

Let us recall some well known results as follows: (cf. [8])

Lemma 1 *Let* $\tilde{w}$, $\tilde{u}$ *and* $\tilde{v}$ *be triple vectors, where the elements* $u_i \in \tilde{u}$ *and* $w_j \in \tilde{w}$, *are not equal in a field* **F**, *for* $i, j = 0, 1, \ldots n-1$. *Let* $C(\tilde{u}, \tilde{w})$ *be an* $n \times n$ *Cauchy matrix. Then the computations of the product* $\tilde{x} = C(\tilde{u}, \tilde{w})\tilde{v}$ *cost*

$$A_{C\tilde{v}}(n) = O(n \log^2 n) \tag{2.1}$$

ao. *If the elements of the vectors* $\tilde{u}$ *and* $\tilde{w}$ *are roots of unity, then the complexity can be reduced to*

$$A_{C_s\tilde{v}}(n) = O(n \log n) \tag{2.2}$$

ao.

By following [6], and [9], we have the following definitions.

Definition 3 *A linear operator* $\Delta^{[E,F]}(\cdot) : \mathbf{F}^{n\times n} \to \mathbf{F}^{n\times n}$ *is defined to map each matrix* $C \in \mathbf{F}^{n\times n}$ *to its displacement* $EC - CF$, *where* $E \in \mathbf{F}^{n\times n}$, $F \in \mathbf{F}^{n\times n}$ *are given matrices. The operator* $\Delta^{[E,F]}(\cdot)$ *is* *called the symmetric Sylvester operator. The rank of the image,* $r = \text{rank}(EC - CF)$ *is called* $[E, F]$*-displacement rank of the matrix* C *for* $r << n$. *Let* $E = D(\tilde{a}) \in \mathbf{F}^{n \times n}$, $F = D(\tilde{b}) \in \mathbf{F}^{n \times n}$ *and* $G_r, H_r \in \mathbf{F}^{n \times r}$ *be matrices. A matrix* $C \in \mathbf{F}^{n \times n}$ *is called a Cauchy-like matrix if it satisfies*

$$\Delta^{[D(\tilde{a}),D(\tilde{b})]}(C) = D(\tilde{a})C - CD(\tilde{b}) = G_r H_r^T, \tag{2.3}$$

where the pair of the matrices G_r, $H_r \in \mathbf{F}^{n \times r}$ *are called the* $\Delta^{[D(\tilde{a}),D(\tilde{b})]}$*-generators of* C *with length* r. *We denote this matrix as* $C_{G_r H_r}(\tilde{a}, \tilde{b})$ *or* C_{like}.

It is easy to realize the following well known relationships of the Cauchy and Cauchy-like matrices (cf. [9]). We observe that $\Delta^{[D(\tilde{s}),D(\tilde{u})]}$ $C(\tilde{s}, \tilde{u}) = D(\tilde{s})C(\tilde{s}, \tilde{u}) - C(\tilde{s}, \tilde{u})D(\tilde{u}) = \tilde{1}\tilde{1}^T$, where $\tilde{1}^T = (1, \ldots, 1)$. Thus it follows.

Lemma 2 *let* $c(\tilde{s}, \tilde{u})$ *be an* $n \times n$ *Cauchy matrix with the vectors* $\tilde{s}$ *and* $\tilde{u}$ *having* $2n$ *distinct values in* $\mathbf{F}$, *then it has the* $[D(\tilde{s}), D(\tilde{u})]$*-displacement rank equal to 1.*

Let the Sylvester operator, $\Delta^{[E,F]}(\cdot) : \mathbf{F}^{n \times n} \to \mathbf{F}^{n \times n}$, act on the linear space $\mathbf{F}^{n \times n}$ with the empty kernel. The Cauchy-like matrices can be completely reconstructed as follows:

Proposition 1 *([7]) Let* $C_{G_k H_k}(\tilde{v}, \tilde{w}) \in \mathbf{F}^{n \times n}$ *be a Cauchy-like matrix associated with the Sylvester operator* $\Delta^{[D(\tilde{v}),D(\tilde{w})]}(\cdot) : \mathbf{F}^{n \times n} \to \mathbf{F}^{n \times n}$ *and generators* $G_k = [\tilde{g}_1, \ldots, \tilde{g}_k]$, $H_k = [\tilde{h}_1, \ldots, \tilde{h}_k]$ *as defined in Definition 3. Then the Cauchy-like matrix* $C_{G_k H_k}(\tilde{v}, \tilde{w})$ *can be expressed as*

$$C_{G_k H_k}(\tilde{v}, \tilde{w}) = \sum_{i=1}^{k} D(\tilde{g}_i)C(\tilde{v}, \tilde{w})D(\tilde{h}_i) = \left(\frac{\tilde{g}_i^T \tilde{h}_j}{v_i - w_j} \right)_{i,j=0}^{n-1}, \tag{2.4}$$

where $C(\tilde{v}, \tilde{w})$ *is a Cauchy matrix and* $k << n$.

By Combining lemma 1 and proposition 1, we obtain the following results:

Corollary 1 *Let* $\tilde{v} \in \mathbf{F}^{n \times 1}$ *be a vector and* $C_{G_r \underset{\sim}{R}_r}(\tilde{d}, \tilde{p}) \in \mathbf{F}^{n \times n}$ *be a Cauchy-like matrix. Then the product* $\tilde{y} = C_{G_r H_r}(\tilde{d}, \tilde{p})\tilde{v}$ *can be computed in*

$$A_{C_{like}\tilde{v}}(n) = O(rA_{C\tilde{v}}(n)) = O(rn \log^2 n) \tag{2.5}$$

ao. *Furthermore, if the elements of the vectors* $\tilde{d}$ *and* $\tilde{p}$ *are roots of unity, then the complexity can be accelerated to*

$$A_{C_{like}\tilde{v}}(n) = O(rA_{C_s\tilde{v}}(n)) = O(rn \log n) \tag{2.6}$$

ao.

Theorem 1 *Let* $\Delta^{[E,F]}(\cdot) : \mathbf{F}^{n\times n} \to \mathbf{F}^{n\times n}$ *be the Sylvester operator. Let* $C_{G_aH_a}(\tilde{z},\tilde{x})$, $C_{G_bH_b}(\tilde{x},\tilde{y})$ *and* $C_{G_dH_d}(\tilde{y},\tilde{w})$ *denote the Cauchy-like matrices according to Definition 3 with* $\Delta^{[D(\tilde{z}),D(\tilde{x})]}(C_{G_aH_a}(\tilde{z},\tilde{x})) = G_aH_a^T$, $\Delta^{[D(\tilde{x}),D(\tilde{y})]}(C_{G_bH_b}(\tilde{x},\tilde{y})) = G_bH_b^T$ *and* $\Delta^{[D(\tilde{y}),D(\tilde{w})]}(C_{G_dH_d}(\tilde{y},\tilde{w})) = G_dH_d^T$ *respectively, where the matrices* $G_a, H_a \in \mathbf{F}^{n\times a}$, G_b, $H_b \in \mathbf{F}^{n\times b}$, G_d, $H_d \in \mathbf{F}^{n\times d}$ *are generators repectively, for* $a, b, d << n$, *and all values of* x_i, y_j, z_k, w_m *are all distinct in a field* $\mathbf{F}$. *Then the product matrix*

$$C_{G_sR_s}(\tilde{z},\tilde{w}) = C_{G_aH_a}(\tilde{z},\tilde{x})C_{G_bH_b}(\tilde{x},\tilde{y})C_{G_dH_d}(\tilde{y},\tilde{w}), \tag{2.7}$$

is a Cauchy-like matrix with the $[D(\tilde{z}), D(\tilde{w})]$*-displacement rank* s ($s = a + b + d$) *such that*

$$\Delta^{[D(\tilde{z}),D(\tilde{w})]}(C_{G_sH_s}(\tilde{z},\tilde{w})) = G_sH_s^T, \tag{2.8}$$

where the pair of matrices G_s, H_s *are* $[D(\tilde{z}), D(\tilde{w})]$*-generators of the size* $n \times s$,

$$G_s = [G_a, C_{G_aH_a}(\tilde{z},\tilde{x})G_b, C_{G_aH_a}(\tilde{z},\tilde{x})C_{G_bH_b}(\tilde{x},\tilde{y})G_d],$$

$$H_s = [C_{G_aH_a}(\tilde{y},\tilde{w})^T C_{G_bH_b}(\tilde{x},\tilde{y})^T H_a, C_{G_dH_d}(\tilde{y},\tilde{w})^T H_b, H_d].$$

Proof. We have $\Delta^{[D(\tilde{z}),D(\tilde{w})]}(C_{G_sH_s}(\tilde{z},\tilde{w})) = G_aH_a^TC_{G_bH_b}(\tilde{x},\tilde{y})\, C_{G_aH_a}(\tilde{y},\tilde{w}) + C_{G_aH_a}(\tilde{z},\tilde{x})\, G_bH_b^TC_{G_dH_d}(\tilde{y},\tilde{w}) + C_{G_aH_a}(\tilde{z},\tilde{x})\, C_{G_bH_b}(\tilde{x},\tilde{y})\, G_dH_d^T = G_sH_s^T$, where $s = a+b+d$ and $G_s = [G_a, C_{G_aH_a}(\tilde{z},\tilde{x})G_b, C_{G_aH_a}(\tilde{z},\tilde{x})\, C_{G_bH_b}(\tilde{x},\tilde{y})G_d]$, $H_s = [C_{G_aH_a}(\tilde{y},\tilde{w})^TC_{G_bH_b}(\tilde{x},\tilde{y})^TH_a, C_{G_dH_d}(\tilde{y},\tilde{w})^TH_b, H_d]$. Q.E.D.

Theorem 2 *Let* $C_{G_rH_r}(\tilde{e},\tilde{f})$ *be an* $n \times n$ *generalized Cauchy-like matrix associated with the symmetric Sylvester operator* $\Delta^{[D(\tilde{e}),D(\tilde{f})]}(\cdot) : \mathbf{F}^{n\times n} \to \mathbf{F}^{n\times n}$ *of (2.3) such that* $\Delta^{[D(\tilde{e}),D(\tilde{f})]}(C_{G_rH_r}(\tilde{e},\tilde{f})) = G_rH_r^T$. *Let* $C(\tilde{p},\tilde{e}), C(\tilde{f},\tilde{q})$ *be a pair of Cauchy matrices, where* e_i, f_j, p_k, q_l *are all distinct. Then we have the following matrix equations*

$$C_{G_rH_r}(\tilde{e},\tilde{f})^{-1} = C(\tilde{f},\tilde{q})C_{G_{r+2}H_{r+2}}(\tilde{p},\tilde{q})^{-1}C(\tilde{p},\tilde{e}), \tag{2.9}$$

where $C_{G_{r+2}H_{r+2}}(\tilde{p},\tilde{q})$ *is a Cauchy-like matrix associated with* $[D(\tilde{p}), D(\tilde{q})]$*-displacement generators,*

$$G_{r+2} = [\tilde{1}, C(\tilde{p},\tilde{e})G_r, C(\tilde{p},\tilde{e})C_{G_rH_r}(\tilde{e},\tilde{f})\tilde{1}],$$

$$H_{r+2} = [C(\tilde{f},\tilde{q})^TC_{G_rH_r}(\tilde{e},\tilde{f})^T\tilde{1}, C(\tilde{f},\tilde{q})^TH_r, \tilde{1}],$$

for $\tilde{1}^T = (1, \ldots, 1)$. *Furthermore, the complexity of our computations is* $O(nr \log^2 n)$ *arithmetic operations.*

Proof: Choose the $[D(\tilde{p}), D(\tilde{e})]$, $[D(\tilde{f}), D(\tilde{q})]$-displacement rank a, d of the Cauchy-like matrices $C_{G_a H_a}(\tilde{p}, \tilde{e})$, $C_{G_d H_d}(\tilde{f}, \tilde{q})$ respectively equal to 1; i.e., $a = d = 1$. We now have $C(\tilde{p}, \tilde{e}) = C_{G_a H_a}(\tilde{p}, \tilde{e})$, $C(\tilde{f}, \tilde{q}) = C_{G_d H_d}(\tilde{f}, \tilde{q})$ from the lemma 2. From the theorem 1, it follows that $C(\tilde{p}, \tilde{e}) C_{G_r H_r}(\tilde{e}, \tilde{f}) C(\tilde{f}, \tilde{q}) = C_{G_{r+2} H_{r+2}}(\tilde{p}, \tilde{q})$. Substitute this matrix equation into the trivial matrix identity $C_{G_r H_r}(\tilde{e}, \tilde{f}) = C(\tilde{p}, \tilde{e})^{-1} C(\tilde{p}, \tilde{e}) C_{G_r H_r}(\tilde{e}, \tilde{f})\ C(\tilde{f}, \tilde{q}) \quad C(\tilde{f}, \tilde{q})^{-1}$ and obtain equations $C_{G_r H_r}(\tilde{e}, \tilde{f}) = C(\tilde{p}, \tilde{e})^{-1} C_{G_{r+2} H_{r+2}}(\tilde{p}, \tilde{q}) C(\tilde{f}, \tilde{q})^{-1}$. By inverting both sides of this equation, we immediately obtain the equation (2.9). The complexity is immediately followed by the corollary 1. *Q.E.D.*

Theorem 3 *(cf. [1]) Given the* $[D(\tilde{m}), D(\tilde{n})]$*-displacement generators* G_u, H_u *with displacement rank* u *of a Cauchy-like matrix* $C_{G_u H_u}(\tilde{m}, \tilde{n})$, $x < u < n$, *it requires* $O(u^2 n)$ *operations to compute the* $D[(\tilde{m}), (\tilde{n})]$*-displacement generators* G_x, H_x *of the Cauchy matrix* $C_{G_x H_x}(\tilde{m}, \tilde{n})$.

Corollary 2 *Let the matrix* $C_{G_{r+2} H_{r+2}}(\tilde{p}, \tilde{q})$ *be the Cauchy-like matrix as in Theorem 3. Then we use* $O(r^2 n)$ *to compute the new generator* $G_{r'}$ *and* $H_{r'}$ *for the matrix* $C_{G_{r'} H_{r'}}(\tilde{p}, \tilde{q})$, *where* $r' \leq r < r + 2 < n$.

Proof. The running time of computing the displacement generators $G_{r'}$, $H_{r'}$ is immediately followed by Theorem 3. *Q.E.D.*

3. Recursive Triangular Factorization of a matrix.

In this section, we will recall the known method of the Recursive Triangular Factorization to a strongly nonsingular matrix [20], [21].

Definition 4 *A matrix* M *is strongly nonsingular if all its leading principal submatrices are nonsingular.*

Let us denote I identity matrix, O denotes a null matrix. Given an $n \times n$ strongly nonsingular matrix M, we partition the matrix M into four $n/2 \times n/2$ blocks, e.g.

$$M = \begin{pmatrix} A & B \\ D & F \end{pmatrix}. \tag{3.1}$$

The matrix S is called the *Schur complement* of A in M and can be computed as

$$S = F - DA^{-1}B. \tag{3.2}$$

we may write its inverse as

$$M^{-1} = \begin{pmatrix} A^{-1} + A^{-1}BS^{-1}DA^{-1} & -A^{-1}BS^{-1} \\ -S^{-1}DA^{-1} & S^{-1} \end{pmatrix}. \tag{3.3}$$

Let matrices A and S be investable. We can continue this process recursively for the submatrices A and S. Based on the following propositions, we may do the factorization to all leading principal submatrices, $1\times 1, 2\times 2, 4\times 4, \ldots, n\times n$. The total numbers of the steps for the Recursive Triangular Factorization is $\lceil \log_2 n \rceil$. Each step of the computations cost matrices multiplications and subtractions.

Proposition 2 *([20], [21]) If $M \in \mathbf{F}^{n\times n}$ is matrix with leading principal submatrices being nonsingular, so are A and S.*

Proposition 3 *([20], [21]) Suppose M is a matrix with leading principal submatrices being nonsingular and write S as (3.2). Let M_1 be a leading principal submatrix of S and let S_1 denote the Schur complement of M_1 in S. Then S^{-1} and S_1^{-1} are located in the southeastern blocks of M^{-1}.*

Here is the algorithms of the Recursive Triangular Factorization for inversion.

Algorithms 3.1. *Recursive Triangular Factorization for inversion*
Input: an matrix $M \in \mathbf{F}^{n\times n}$.
Output: the inversive matrix M^{-1}.
Computations:

1. Use Algorithms 3.1 Compute A^{-1} as (3.1).
2. Compute the Schur complement $S = F - DA^{-1}B$.
3. Use Algorithms 3.1 to the inverse matrix S^{-1}
4. Compute M^{-1} from (3.3).

4. Modified Recursive Triangular Factorization to a Cauchy-like matrix

In this section, we will describe the techniques of the modified Recursive Triangular Factorization to a Cauchy-like matrix. As we can see from the equation (1.2), the complexity of the Recursive Factorization depends on the complexity of a Cauchy matrix by a vector. It needs only $O(n \log n)$ *ao* to compute a special Cauchy matrix times a vector. It motivates us to modify the process of the Recursive Factorization so that the complexity bound of [20] can be improved by a factor of $\log n$.

Let us recall some well known results first.

Lemma 3 *([20], [21]) Let $\tilde{b}_j \in \mathbf{F}^{n\times 1}$, $j = 1,2,3$, be tree distinct vectors and $M_i \in \mathbf{F}^{n\times n}$ be cauchy-like matrices such that $\Delta^{[D(\tilde{b}_i),D(\tilde{b}_{i+1})]}(M_i) = G_i H_i^T$, where G_i, $H_i \in \mathbf{F}^{n\times r_i}$, $r_i < n$, for $i = 1,2$. The matrix $M = M_1 M_2$ is a Cauchy-like matrix with $\Delta^{[D(\tilde{b}_1),D(\tilde{b}_3)]}(M) = G_r H_r^T$, $G_r = [G_1, M_1 G_2]$, $H_r = [M_2^T H_1, H_2]$, G_r, $H_r \in \mathbf{F}^{n\times r}$, $r = r_1 + r_2$. Furthermore, $O(r_1 r_2 A_{C\tilde{v}}(n))$* ao *suffice to compute G_r and H_r.*

By multiplying $C_{M_t W_t}(\tilde{x},\tilde{b})^{-1}$ on both sides of the symmetric Sylvester equation for $C_{M_t W_t}(\tilde{x},\tilde{b})$, it follows:

Lemma 4 *Given an $n\times n$ Cauchy-like matrix $C_{M_t W_t}(\tilde{x},\tilde{b})$ as defined in definition 3 such that (2.3) holds for $t << n$, the matrix $C_{M_t W_t}(\tilde{x},\tilde{b})^{-1}$ satisfies $\Delta^{[D(\tilde{b}),D(\tilde{x})]}\ (C_{M_t W_t}(\tilde{x},\tilde{b})^{-1}) = G_r H_r^T$, where $G_r = [C_{M_t W_t}(\tilde{x},\tilde{b})^{-1} M_t]$, $H_r = [(C_{M_t W_t}(\tilde{x},\tilde{b})^{-1})^T W_t]$.*

Fact 1 *By Lemma 4, the* rank *$\Delta^{[D(\tilde{b}),D(\tilde{x})]}\ (C_{M_t W_t}(\tilde{x},\tilde{b})^{-1}) << t$.*

Lemma 5 *([20], [21]) Let M and N be $n\times n$ matrices which satisfy $\Delta^{[D(\tilde{p}),D(\tilde{q})]}\ (M) = G_{r_1} H_{r_1}^T$ and $\Delta^{[D(\tilde{p}),D(\tilde{q})]}(N) = G_{r_2} H_{r_2}^T$ respectively. Then the matrices $M+N$ and $M-N$ are Cauchy-like matrices associated with a $[D(\tilde{p}), D(\tilde{q})]$ -generator of length at most $r_2 + r_1$.*

Lemma 6 *([20], [21]) Let $C_{G_{r'} H_{r'}}(\tilde{p},\tilde{q})$ be an $n\times n$ generalized Cauchy-like matrix associated with the symmetric Sylvester operator $\Delta^{[D(\tilde{p}),D(\tilde{q})]}$ $(\cdot) : \mathbf{F}^{n\times n} \to \mathbf{F}^{n\times n}$ such that $\Delta^{[D(\tilde{p}),D(\tilde{q})]}(C_{G_{r'} H_{r'}}(\tilde{p},\tilde{q})) = G_{r'} H_{r'}^T$. Let*

$$C_{G_{r'} H_{r'}}(\tilde{p},\tilde{q}) = \begin{pmatrix} A & B \\ D & F \end{pmatrix}, \tag{4.1}$$

where A, B, D, F and S as (3.1) and (3.2). Then the displacement ranks of A, B, D, F and S are all less than r'.

Fact 2 *([3]) Every Cauchy matrix is nonsingular and every square submatrix of Cauchy matrix is also nonsingular.*

Fact 3 *([20]) The matrix $C_{G_{r+2} H_{r+2}}(\tilde{p},\tilde{q})$ as in Theorem 2 is a strongly nonsingular matrix.*

Now let us describe the complexity of the modefied Recursive Triangular Factorization to Cauchy-like matrices.

Theorem 4 *The arithmetic complexity of the solution of a strongly nonsingular generalized Cauchy-like equations is bounded by $O(nr^2 \log^2 n)$ operations.*

Proof. Let $\tilde{v}$ be a given vector and $C_{G_rH_r}(\tilde{e},\tilde{f}) \in \mathbf{F}^{n\times n}$ be a given generalized Cauchy-like matrix satifies $\Delta^{[D(\tilde{e}),D(\tilde{f})]}(C_{G_rH_r}(\tilde{e},\tilde{f})) = G_rH_r^T$. By the theorem 2, the computations of $\tilde{x} = C_{G_rH_r}(\tilde{e},\tilde{f})^{-1}\tilde{v}$ are equivalent to the computations of $\tilde{x} = C(\tilde{f},\tilde{q})C_{G_{r+2}H_{r+2}}(\tilde{p},\tilde{q})^{-1}C(\tilde{p},\tilde{e})\tilde{v}$. We choose the new vectors $\tilde{p}$, $\tilde{q}$, where the elements of the vectors are n-th roots of 1 and -1, e.g. $\tilde{p}$=$\tilde{\alpha}$, $\tilde{q}$=$\tilde{\beta}$, where α_i and β_j of definition 1. Compute the generators of the matrix $G_{r'}$ and $H_{r'}$ by corollary 2. Then, we apply the Recursive Triangular Factorization Algorithms 3.1 to the matrix $C_{G_{r'}H_{r'}}(\tilde{\alpha},\tilde{\beta})$. Partition the matrix into $\frac{n}{2}\times\frac{n}{2}$ blocks,

$$C_{G_{r'}H_{r'}}(\tilde{\alpha},\tilde{\beta}) = \begin{pmatrix} A & B \\ D & F \end{pmatrix}, \tag{4.2}$$

where A, B, D, F and S as (3.1) and (3.2). We compute the generators of A, B, D, F, A^{-1}, S as (3.2), S^{-1}. This process continues until we have the generators of the inverse $C_{G_{r'}H_{r'}}(\tilde{\alpha},\tilde{\beta})^{-1}$. Finally, we compute $\tilde{x} = C(\tilde{f},\tilde{q})C_{G_{r'}H_{r'}}(\tilde{\alpha},\tilde{\beta})^{-1}C(\tilde{p},\tilde{e})\tilde{v}$. The computations of the generators of $C_{G_{r'}H_{r'}}(\tilde{\alpha},\tilde{\beta})^{-1}$ in each step involve with the multiplications of a special Cauchy matrix times a vector. The complexity of $O(nr^2\log^2 n)$ operations is followed by Corollary 1, 2, Lemma 1 and Theorem 2. *Q.E.D.*

Acknowledgments

I wish to thank Professor Victor Pan. I would like to thank Professor Henry Africk, Peter Deraney, Joel Greenstein, Sandie Han, Earl Hill, Sonja Jackson, Janet Liou-Mark, Anshel Michael, Jonathan Natov, Melvyn Nathanson, Rhona Noll, Richard Patterson, Joann Perla, Estela Rojas, Annette Schaefer, Ely Stern, Arnavaz Taraporevala and all other faculty members.

Notes

1. Established in 1946, New York City Technical College is the senior college of technology at the City University of New York. It is a recognized national model for urban technological education and a pioneer in integrating technology into the teaching/learning experience. More than 11,000 students currently are enrolled in 50 career-specific baccalaureate, associate and specialized certificate programs in the technologies of art and design, business, computer systems, engineering, health care, hospitality, human services, the law-related professions, occupational and technology teacher education, and the liberal arts and sciences. Lo-

cated at 300 Jay Street in Downtown Brooklyn, City Tech is part of the MetroTech Center academic and commercial complex.

References

[1] D. Bini, V. Y. Pan, (1994), Polynomial and Matrix Computations, v. 1: Fundamental Algorithms, *Birkhaeuser* , Boston.

[2] R. R. Bitmead, B. D. O. Anderson, (1980), Asymptotically Fast Solution of Toeplitz and Related Systems of Linear Equations, *Linear Algebra Appl.*, **34**, 103-116.

[3] A. L. Cauchy, (1841), Mémorie sur les Fonctions Alternées et sur les Somme Alternées, *Exercises d' Analyse et de Phys. Math.*, **II**, 151-159.

[4] W.F. Donoghue, (1974), Monotone Matrix Functions and Analytic Continuation, *Springer, Berlin.*

[5] W. Gautschi, (1975), Norm Estimates for the Inverses of Vandermonde Matrices, *Numer. Math.*, **23**, 337-347.

[6] I. Gohberg, V. Olshevsky, (1994), Fast state space algorithms for matrix Nehari and Nehari-Takagi interpolation problems, *Integral Equations and Operator Theory*, **20**, 1, 44-83.

[7] I. Gohberg, V. Olshevsky, (1994), Complexity of Multiplication with Vectors for Structured Matrices, *Linear Algebra Appl.*, **202**, 163-192.

[8] I. Gohberg, V. Olshevsky, (1994), Fast Algorithms with Preprocessing for Matrix-Vector Multiplication Problems, *J. of Complexity*, **10, 4**, 411-427.

[9] I. Gohberg, T. Kailath, V. Olshevsky, (1995), Fast Gausian Elimination with Partial Pivoting for Matrices with Displacement Structure, *Math. of Computation*, **64**, 1557-1576.

[10] G. Heinig, (1995), Inversion of Generalized Cauchy Matrices and the Other Classes of Structured Matrices, *Linear Algebra for Signal Processing, IMA Volume in Math. and Its Applications*, **69**, 95-114, Springer.

[11] T. Kailath, A. Viera, and M. Morf, (1978), Inverses of Toeplitz Operators, Innovations, and Orthogonal Polynomials, *SIAM Review*, 20, 1, pp. 106-119.

[12] T. Kailath, A. Sayed (editors), (1999), Fast Reliable Algorithms for Matrices with Structure, *SIAM Publications*, Philadelphia.

[13] M. Morf, (1980), Doubling Algorithms for Toeplitz and Related Equations, *Proc. IEEE Internat. Conf. on ASSP*, 954-959, IEEE Computer Society Press.

[14] V. Olshevsky, V. Y. Pan, (1998), A Unified Superfast Algorithm for Boundary Rational Tangential Interpolation Problem, *Proc. 39th Ann. IEEE Symp. Foundations of Comp. Sci.*,192-201, IEEE Comp. Soc. Press.

[15] V. Olshevsky, V. Y. Pan, (1999), Polynomial and Rational Evaluation and Interpolation (with Structured Matrices), *Proc. 26th Ann. Intern. Colloq. on Automata, Languages, Programming (ICALP'99), Lecture Notes in Computer Science*, Springer.

[16] V. Olshevsky, M. A. Shokrollahi, (1999), A Displacement Approach to Efficient Decoding of Algebraic-Geometric Codes, *Proc. 31st Ann. Symp. on Theory of Computing*, 235-244, ACM Press, New York.

[17] V. Y. Pan, M. AbuTabanjeh, Z. Chen, S. Providence, A. Sadikou, (1998), Transformations of Cauchy Matrices for Trummer's Problem and a Cauchy-like Linear Solver, *Proc. of 5th Annual International Symposium on Solving Irregularly Structured Problems in Parallel, (Irregular98)*, (A. Ferreira, J. Rolim, H. Simon, S.-H. Teng Editors), *Lecture Notes in Computer Science*, **1457**, 274-284, Springer.

[18] V. Y. Pan, M. AbuTabanjeh, Z. Chen, E. Landowne, A. Sadikou, (1998), New Transformations of Cauchy Matrices and Trummer's Problem, *Computers and Math. (with Applics.)*, **35**, **12**, 1-5.

[19] V. Y. Pan, A. Sadikou, E. Landowne, O. Tiga, (1993), A New Approach to Fast Polynomial Interpolation and Multipoint Evaluation, *Computers & Math. (with Applications)*, **25**, **9**, 25-30.

[20] V. Y. Pan, A. Zheng, (2000), Superfast algorithms for Cauchy-like matrix computations and extensions. *Linear Algebra Appl.* 310, no. 1-3, 83-108.

[21] V. Y. Pan, A. Zheng, M. Abutabanjeh, Z. Chen, S. Providence, (1999), Superfast Computations with Singular Structured Matrices over Abstract Fields, *Proc. 2rd Workshop on Computer Algebra in Scientific Computing (CASC'99)*, (V. G. Ganzha, E. W. Mayr, and E. V. Vorontsov, Editors), 323-338, Springer, Berlin.

[22] V. Y. Pan, A. Zheng, X. Huang, Y. Yu, (1997), Fast Multipoint Polynomial Evaluation and Interpolation via Computations with Structured Matrices, *Annals of Numerical Math.*, **4**, 483–510.

[23] L. Reichel, (1990), A Matrix Problem with Application to Rapid Solution of Integral Equations, *SIAM J. on Scientific and Stat. Computing*, **11**, 263-280.

[24] F. Rokhlin, (1985), Rapid Solution of Integral Equations of Classical Potential Theory, *J. Comput. Physics*, **60**, 187-207.

[25] M. Trummer, (1986), An Efficient Implementation of a Conformal Mapping Method Using the Szegö Kernel, *SIAM J. on Numerical Analysis*, **23**, 853-872.

ALGORITHMIC COMPLEXITY OF PROTEIN IDENTIFICATION: SEARCHING IN WEIGHTED STRINGS

Mark Cieliebak, Zsuzsanna Lipták, Emo Welzl
ETH Zurich, Institute of Theoretical Computer Science
cielieba,zsuzsa,emo@inf.ethz.ch

Thomas Erlebach
ETH Zurich, Computer Engineering and Networks Laboratory
erlebach@tik.ee.ethz.ch

Jens Stoye*
Max Planck Institute of Molecular Genetics, and Konrad-Zuse-Zentrum (ZIB), Berlin
stoye@techfak.uni-bielefeld.de

Abstract We investigate a problem which arises in computational biology: Given a constant–size alphabet $\mathcal{A}$ with a weight function $\mu : \mathcal{A} \to \mathbb{N}$, find an efficient data structure and query algorithm solving the following problem: For a string σ over $\mathcal{A}$ and a weight $M \in \mathbb{N}$, decide whether σ contains a substring with weight M (ONE–STRING MASS FINDING PROBLEM). If the answer is **yes**, then we may in addition require a witness, i.e., indices $i \leq j$ such that the substring beginning at position i and ending at position j has weight M. We allow preprocessing of the string, and measure efficiency in two parameters: storage space required for the preprocessed data, and running time of the query algorithm for given M. We are interested in data structures and algorithms requiring subquadratic storage space and sublinear query time, where we measure the input size as the length of the input string. Among others, we present two non–trivial efficient algorithms: LOOKUP solves the problem with $O(n)$ space and $O(\frac{n}{\log n} \cdot \log\log n)$ time; INTERVAL solves the problem for binary alphabets with $O(n)$ storage space in $O(\log n)$ query time. Finally, we introduce other variants of the problem and sketch how our algorithms may be extended for these variants.

Keywords: Weighted Strings, Protein Identification, Database Searching

---

*Present address: Bielefeld University, Faculty of Technology

1. Introduction

In the present paper, we introduce a combinatorial problem which originates from computational biology: Given a string σ over a weighted alphabet $\mathcal{A}$, find a data structure and a query algorithm which, for a given weight $M \in \mathbb{N}$, decides whether σ has a (contiguous) substring of weight M. If the answer is yes, we may in addition ask for a *witness*, i.e., two positions within σ where a substring with weight M begins and ends. The actual problem in computational biology is to find several masses $M_1, \ldots, M_m$ in a database of strings. We concentrate on the one–string problem because algorithms can be easily extended to the multiple–string problem. We formally define the other problem variants at the end of the paper and sketch how extensions may be designed. There are two simple algorithms which solve the one–string problem: One uses linear time for a query and no additional storage space; the other has logarithmic query time, but requires a preprocessing step and additional storage space for the resulting data structure which may be quadratic. Hereby, space and time complexities are measured in the length of the string. We are interested in algorithms that are better than these two: i.e., we allow preprocessing and look for an algorithm where the data structure needs subquadratic space *and* the running time for a query is sublinear.

Formulated in this way, the problem is a purely combinatorial and algorithmic problem: Are there algorithms which allow searching in weighted strings of size n with $o(n^2)$ additional storage space and $o(n)$ query time? If so, can we find a tradeoff between space and time?

The problem differs from traditional string searching problems in one important aspect: While those look for substructures of strings (substrings, non–contiguous subsequences, particular types of substrings such as repeats, palindromes etc.), we are interested only in *weights* of substrings. This means that, on the one hand, we lose a lot of the structure of strings: e.g. the weight of a string is invariant under permutation of letters; on the other, we gain the additional structure of the weight function, such as its additivity. For instance, the problem of searching in $X+Y$, where X and Y are two sets of numbers, turns out to be closely related to our problem (see [Fre75] and [HPSS75]). However, we have been able to extend negative results which have been reached for that problem ([CDF90]): We can show that this approach (using the naïve solution without preprocessing) cannot lead to an efficient algorithm for our problem. We have been unable to find any treatment of our problem in the vast amount of literature on strings (e.g. [AG95, Gus97, Lot97, RS97, CR94]).

Our problem is positioned between the areas of string algorithms, search algorithms, and algebra. We believe that it is not only relevant for computational biology, but that it is also of theoretical interest to the field of combinatorial searching. As far as we are aware, no efficient algorithms have so far been presented for this problem.

We would like to stress at this point that, even though the source of our problem is a biological question, the results we present here are primarily of theoretical interest. The reason is twofold: First, none of the algorithms we present are efficiently applicable in their current form. LOOKUP requires sublinear query time, but this is a mainly asymptotic result, since the query time only improves for very long input strings. Algorithm INTERVAL, on the other hand, is very efficient both in query time and storage space, but it only works for alphabets of size 2, a case which does not occur in the usual biological setting. The second reason is that all biological data are prone to errors; in fact, there is no such thing as error–free data. Thus, all applications in computational biology need to be highly fault tolerant. Our algorithms can be straightforwardly adapted to become tolerant to measurement errors. However, this aspect is not included in this paper.

Thus, the present paper demonstrates that efficient algorithms for the problem presented are possible; it remains a great challenge to actually find algorithms that are also of practical value.

1.1. Biological Motivation

Proteomics is the field that investigates the nature and function of proteins. As in molecular biology in general, large amounts of data are being accumulated at present, which presents particular computational and mathematical challenges.

Proteins are large molecules that play a fundamental role in all living organisms. They are made up of smaller molecules (amino acids) that are linked together in a certain order. The sequence of amino acids constitutes the so–called *primary structure* of a protein. Protein size ranges from below 100 to several thousand amino acids, where a typical protein has length 300 to 600. Most proteins are made up of the 20 most common amino acids. For the purposes of this paper, we will view a protein as a finite string over an alphabet of size 20.

The information about known proteins is stored in large databases, such as SWISS-PROT (nearly 100,000 proteins) or PIR (more than 200,000 proteins). When a protein is isolated, one would like to know whether it is already known and if so, to identify it. An obvious way is to establish its primary structure: This is called *de novo protein sequencing*. However, protein sequencing, unlike DNA sequencing, is very expensive (both in time and money!). E.g. identifying *one* amino acid with Edman Degradation, one standard method for protein sequencing, takes about 45 minutes, which makes this approach unfeasible in a high–throughput context.

Therefore, methods are required that test the protein against a database without having to sequence it first. One such method—which we will investigate here—makes use of the differences in molecular weights of amino acids: The protein is broken up into smaller pieces and these pieces are then weighed[1], using a mass spectrometer. This will yield a "fingerprint" of the protein that

[1] Biologists will excuse some rough simplifications.

can then be tested against the database: The goal is to find a protein in the database that has substrings matching each of the input masses.

The method used for breaking up the protein into smaller pieces is referred to by biologists as *digestion*: a so-called cleavage agent such as an enzyme, e.g. trypsin, is used which literally cuts the protein in certain well-defined places. The whole process is called *mass spectrometry*. Using digestion is algorithmically rather simple, at least with error-free data, since the breakup points are known in advance; it is thus possible to preprocess the database in an appropriate way. The complications arise due to measurement errors and post-translational modifications. There is a large amount of literature on mass spectrometry [YISGH93, HBS+93, JQCG93, PHB93, MHR93, EMYI94]; some papers dealing with different aspects and modifications of the problem, e.g. the minimum number of masses needed to identify a protein [PHB93], combinatorial [PDT00] or probabilistic [BE01] models for scoring the difference of two mass spectra, or approaches for a correct identification even in the presence of post-translational modifications of the protein [MW94, YEM95, PMDT01]. The review [YI98] as well as chapter 11 of the book [Pev00] contain more detailed introductions to this topic. For an introduction to computational biology in general, see [SM97]; for more on molecular biology, [Str88]; while [GW91] is an easy-going introduction to genetics for non-biologists.

In this paper, we deal with algorithmic questions that arise if nothing is known about the breaking points, i.e., we assume random fragmentation. Testing for random weights is algorithmically far more complex than the digestion method, because the cutting places are not known in advance. This approach makes sense because it allows combination of several cleavage agents and it eliminates problems such as incomplete digestion. In addition, since we never make any assumptions about the probability distribution of breaking points, any algorithm for the random fragmentation method can be used for digested inputs, too. In the long run, however, for the biological application, algorithms are needed that are not only efficient, but also fault tolerant: They need to be tolerant both to measurement errors ($M \pm \epsilon$; missing or additional masses in the spectrum) and to sequencing errors.

1.2. Overview

The paper is organized as follows. We first introduce the problem and all necessary definitions in Section 2, where we also present some simple ideas that motivate our efficiency requirements. In Section 3, we design an algorithm (LOOKUP) that is asymptotically efficient, with linear storage space and a query running time that is only just sublinear. LOOKUP thus serves to demonstrate that the requirements we defined earlier can be met. Section 4 contains an algorithm (INTERVAL) that solves the problem for alphabets of size 2 and has a very good performance. However, we do not think that it can be generalized to larger alphabets. In Section 5, we present two other problem variants and discuss how algorithms for the original problem can be extended to these. In addition, we sketch improvements of our algorithms for special cases. Finally, in Section 6, we sketch ongoing work.

2. Problem and Simple Solutions

Fix an *alphabet* $\mathcal{A}$ of size $|\mathcal{A}| = s$ and a *mass function* $\mu : \mathcal{A} \to \mathbb{N}$. The *mass* (or the *weight*) of a string (or a *sequence*) σ over $\mathcal{A}$ is defined as the sum of the individual masses $\mu(\sigma) := \sum_{i=1}^{n} \mu(\sigma(i))$, where $\sigma(i)$ denotes the i'th letter of σ, and $n = |\sigma|$ is the length of σ. For a mass $M \in \mathbb{N}$ and a string σ of length n, we say that M is a *submass* of σ if σ has a substring of mass M, i.e., if there are indices $1 \leq i \leq j \leq n$ s.t. $\mu(\sigma(i,j)) = M$, where $\sigma(i,j)$ is the substring of σ starting in position i and ending in position j. For $a \in \mathcal{A}$, let us denote the *multiplicity* of a in σ by $|\sigma|_a := |\{i \mid \sigma(i) = a\}|$.

The ONE–STRING MASS FINDING PROBLEM is defined as follows:

Given a string σ of length $|\sigma| = n$ and a mass M, is M a submass of σ?

A simple algorithm to solve the problem is LINSEARCH, which performs a linear search through the string: For given σ, start at position $\sigma(1)$ and add up masses until reaching the first position j s.t. $\mu(\sigma(1,j)) \geq M$. If the mass of the substring $\sigma(1,j)$ equals M, then output **yes** and stop; else start subtracting masses from the beginning of the string until the smallest index i s.t. $\mu(\sigma(i,j)) \leq M$ is reached. Repeat until finding a pair of indices (i,j) s.t. $\mu(\sigma(i,j)) = M$, or until reaching the end of the string (i.e., until the current substring is $\sigma(i,n)$ for some i and $\mu(\sigma(i,n)) < M$). The algorithm can be visualized as shifting two pointers ℓ and r through the string, where ℓ points to the beginning of the current substring and r to its end. LINSEARCH takes $O(n)$ time, since it looks at each letter at most twice. If we do not allow any preprocessing, this is asymptotically optimal, since it may be necessary to look at each letter at least once.

On the other hand, if preprocessing of σ is allowed, then there is another simple algorithm for the ONE–STRING MASS FINDING PROBLEM which uses binary search: in a preprocessing step, it calculates the set of all possible submasses of σ (i.e., $\mu(\sigma(i,j))$ for all $1 \leq i \leq j \leq n$) and stores them in a sorted array. Given a query mass M, it performs binary search for M in this array. We will refer to this algorithm as BINSEARCH. The space required to store the sorted array is proportional to the number of different submasses in σ, which is bounded by $O(n^2)$. The time for answering a query is thus $O(\log n)$.

From now on, an algorithm for the ONE–STRING MASS FINDING PROBLEM will consist of three components: a preprocessing phase, a data structure in which the result of the preprocessing is stored, and a query method. For a string σ, the preprocessing will be done only once, while the query step will typically be repeated many times. For this reason, we are interested in algorithms with fast query methods, whereas we ignore time and space required for the preprocessing step (as long as they are within reasonable bounds). Space efficiency is measured in storage space required by the data structure.

We are looking for algorithms that are better than LINSEARCH and BINSEARCH, i.e., require storage space $o(n^2)$ for the data structure, and query time $o(n)$. We will call an algorithm *skinny* if the associated data structure requires $o(n^2)$ space, and *speedy* if the query method runs in time $o(n)$.

In this context, the question naturally arises whether a given mass M can be the weight of a string. If the size of the alphabet is variable, then this question is a variant of the INTEGER KNAPSACK PROBLEM, and is NP–complete (cf. [GJ79]). If the alphabet size is constant, the question can be solved with a simple Integer Linear Program.

A third simple algorithm for the ONE–STRING MASS FINDING PROBLEM, which we will call BOOLEANARRAY, works as follows: In the preprocessing phase, define $W := \max\{\mu(a) \mid a \in \mathcal{A}\}$ and let B be a Boolean array of length $n \cdot W$. Set $B[k]$ to `true` if and only if k is a submass of σ. Given a query mass M, we output $B[M]$. This algorithm has query time $O(1)$, while the data structure B requires space $O(n \cdot W)$ bits. Thus, the algorithm is speedy and, if $W = o(n)$, it is skinny, too. However, this does not solve the ONE–STRING MASS FINDING PROBLEM in general, since we do not want to restrict the size of W.

In the following, we assume that the alphabet $\mathcal{A}$ is of constant size and we do not restrict the maximum weight W of a letter. We assume a machine model with word size $\Omega(\log n + \log W)$ in which arithmetic operations on numbers with $O(\log n + \log W)$ bits can be executed in constant time; storage space is measured in terms of the number of words used. Without this assumption, we would get an extra factor $O(\log W)$ in the query time and in the storage space.

3. An Algorithm that is Both Skinny and Speedy

In this section, we present algorithm LOOKUP that solves the ONE–STRING MASS FINDING PROBLEM with storage space $O(n)$ and query time $O(\frac{n}{\log n} \cdot \log\log n)$. The idea is as follows. Similar to the simple linear search algorithm LINSEARCH introduced in Section 2, LOOKUP shifts two pointers along the sequence which point to the potential beginning and end of a substring with mass M. However, $c(n)$ steps of the simple algorithm are bundled into one step here. If $c(n)$ is chosen appropriately, i.e. approximately $\log n$, then this will reduce the number of steps from $O(n)$ to $O(\frac{n}{\log n})$, while each step will now require $O(\log\log n)$ time instead of constant time. We will hereby heavily exploit the fact that the alphabet has constant size.

3.1. An Example

Example 1. Let $\mathcal{A} = \{a, b, c\}, \mu(a) = 1, \mu(b) = 2, \mu(c) = 5$. Let us assume that we are looking for $M = 14$ in $\sigma = abbcabccaabb$. LINSEARCH would shift two pointers ℓ and r through the sequence until reaching positions 5 and 9 respectively. Here, it would stop because the substring $\sigma(5, 9) = abcca$ has weight 14. Let us assume that $c(n) = 3$. We divide the sequence σ into blocks of size $c(n)$. Now, rather than shifting the two pointers letter by letter, we will shift them by a complete block at a time. In order to do this, for each block we store a pointer to an index I which corresponds to the substring which starts with the first letter of the block and ends with the last. Let us assume now that ℓ is at the beginning of the first block, and r is at the end of the second block, as indicated in Figure 1. We are interested in the possible *changes* to the current submass if we shift the two pointers at most $c(n)$ to the right. Given a list of these, we could search for $M - \mu(\sigma(\ell, r))$. For example, the current submass in Figure 1 is $\mu(\sigma(1,6)) = 13$, and we want to know whether, by moving ℓ and r at most 3 positions to the right, we can achieve a change of $14 - 13 = 1$.

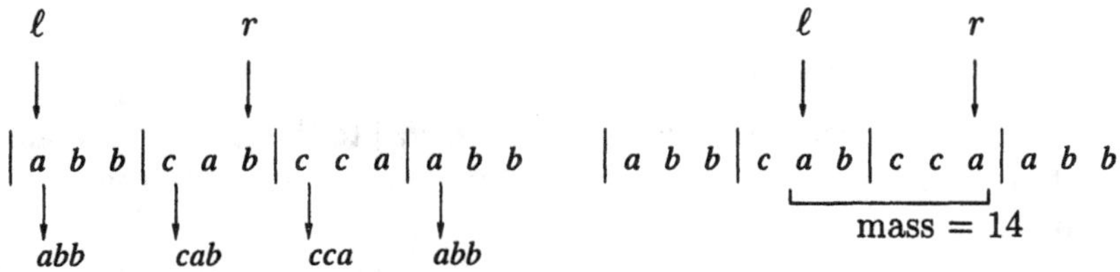

Figure 1. Example 1 – LOOKUP searching for $M = 14$

We can calculate these possible changes and store them in a $(c(n)+1) \times (c(n)+1)$ matrix T whose (i,j)–entry holds the submass change when ℓ is moved $i-1$ positions to the right, and r is moved $j-1$ positions to the right:

$$T[abb, cca] \quad = \quad \begin{pmatrix} 0 & 5 & 10 & 11 \\ -1 & 4 & 9 & 10 \\ -3 & 2 & 7 & 8 \\ -5 & 0 & 5 & 6 \end{pmatrix}$$

In order to be able to do fast search, we store the entries of the matrix in a sorted array: $S[abb, cca] = [-5, -3, -1, 0, 2, 4, 5, 6, 7, 8, 9, 10, 11]$. Now we can find out in time $O(\log(\text{size of array}))$ whether the difference we are looking for is there. In the present case, 1 is not in the array, which tells us that we have to move one of the two pointers to the next block.

To determine which pointer to move, we consider what the linear search algorithm LINSEARCH would do when searching for M and starting in the current positions of the left resp. right pointer. Since M is not present within these two blocks, at least one of the two pointers would reach the end of its current block. Here, we want to move the pointer which would first reach the end of its block. We can determine which pointer this is if we compare the difference $M - \mu(\sigma(\ell, r))$ with the matrix entry $T(c(n), c(n))$ corresponding to $c(n) - 1$ moves of both the left and the right pointer (in this case 7). If the difference is smaller, we move the left pointer to the next block, otherwise we move the right one. In our example, we have a difference of 1, thus we move the left pointer to the next block.

This will change the current submass by -5 (the minimum of the array), yielding $\mu(\sigma(4,6)) = 13 - 5 = 8$. Thus, we now look for $M - \mu(\sigma(4,6)) =$

$14 - 8 = 6$. The sorted array for this pair of positions is $S[cab, cca] = [-8, -6, -5, -3, -1, 0, 2, 3, 4, 5, 6, 10, 11]$, and the matrix is as follows:

$$T[cab, cca] = \begin{pmatrix} 0 & 5 & 10 & 11 \\ -5 & 0 & 5 & 6 \\ -6 & -1 & 4 & 5 \\ -8 & -3 & 2 & 3 \end{pmatrix}$$

Value 6 is in the array: By looking in the matrix, we can see that a difference of 6 can be achieved by moving the left pointer by 1 position and the right pointer by 3 positions. The algorithm outputs positions 5 and 9 and then terminates.

3.2. Algorithm LOOKUP

We postpone the exact choice of the function $c(n)$ to the analysis, but assume for now that it is approximately $\log n$. For simplicity, we assume that $c(n)$ is a divisor of n.

Preprocessing: Given σ of length n, first compute $c(n)$. Next, build a table T of size $|\mathcal{A}|^{c(n)} \times |\mathcal{A}|^{c(n)}$. Each row resp. column of T will be indexed by a string from $\mathcal{A}^{c(n)}$. For $I, J \in \mathcal{A}^{c(n)}$, the table entry $T[I, J]$ contains the matrix and the sorted array as described above. The matrix contains all differences $\mu(\text{prefix}(J)) - \mu(\text{prefix}(I))$. Note that the table T depends only on n and $\mathcal{A}$, and not on the sequence σ itself. Next, divide σ into blocks of length $c(n)$. For each block, store a pointer to an index I that will be used to look up table T. Each such index I represents one string from $\mathcal{A}^{c(n)}$.

Query Algorithm: Given M, set $\ell := 1$ and $r := 0$. Repeat the following steps until M has been found or $r > n$:

Step 1: Say ℓ is set to the beginning of the i'th block and r to the end of the $(j-1)$'th block. Then look in the sorted array in $T(I, J)$ where the pointer of block i resp. j points to index I resp. J. Find whether $M - \mu(\sigma(\ell, r))$ is in the array with binary search.

Step 2: If $M - \mu(\sigma(\ell, r))$ is in the array, search for an entry (u, v) in the matrix $T(I, J)$ which equals $M - \mu(\sigma(\ell, r))$ by exhaustive search[2], and return **yes**, along with the witness $i' := (i-1) \cdot c(n) + u, j' := (j-1) \cdot c(n) + (v-1)$, since $\mu(\sigma(i', j'))$ has mass M.

Step 3: Otherwise, $M - \mu(\sigma(\ell, r))$ is not in the array. If $M - \mu(\sigma(\ell, r))$ is less than the matrix entry at position $(c(n), c(n))$, then increment ℓ by $c(n)$ and set $\mu(\sigma(\ell, r)) := \mu(\sigma(\ell, r)) + \min(\text{array})$; otherwise, increment r by $c(n)$ and set $\mu(\sigma(\ell, r)) := \mu(\sigma(\ell, r)) + \max(\text{array})$.

Analysis: First we derive formulas for space and time, and then we show how to choose $c(n)$. The space needed for storing table T is $|\mathcal{A}|^{2c(n)} \cdot ((c(n) + 1)^2 + (c(n) + 1)^2) = O(|\mathcal{A}|^{2c(n)} \cdot c(n)^2)$. Space needed for storing the pointer at each block is $\frac{n}{c(n)} \cdot \log(|\mathcal{A}|^{c(n)}) = O(n)$. For the last equality, recall that $\mathcal{A}$ is of constant size. For the query time, observe that after each iteration

[2] Alternatively, we could have stored (u, v) during the preprocessing in the sorted array, too.

(consisting of Steps 1 to 3), either ℓ or r is advanced to the next block. As each of the pointers can advance at most $\frac{n}{c(n)}$ times, there can be at most $2\frac{n}{c(n)}$ iterations. Each iteration except the last one takes time $O(\log c(n)^2) + O(1)$. The last iteration may take time $O(c(n)^2)$.

In total, the algorithm requires storage space $O(n+|\mathcal{A}|^{2c(n)} \cdot c(n)^2)$ and time $O(\frac{n}{c(n)}+\frac{n}{c(n)} \log c(n)+c(n)^2)$. Now, if we choose $c(n) := \frac{\log_{|\mathcal{A}|} n}{4}$, then we obtain $|\mathcal{A}|^{c(n)} = n^{\frac{1}{4}}$. This yields a storage space of $O(n + n^{\frac{1}{2}} \cdot \log^2 n) = O(n)$ and query time $O(\frac{n}{\log n} \log\log n)$, which is both skinny and speedy. Other choices of $c(n)$ do not asymptotically improve time and space at the same time.

Theorem 1. *Algorithm* LOOKUP *solves the* ONE–STRING MASS FINDING PROBLEM *with storage space* $O(n)$ *and query time* $O(\frac{n}{\log n} \log\log n)$.

Thus, LOOKUP beats both the query time of LINSEARCH and the storage space of BINSEARCH. However, its practical use is limited to very long sequences: In order to obtain a block size of, say, $c(n) = 10$, the input string would have to have length $n = |\mathcal{A}|^{40}$. In the next section we present a practical algorithm for binary alphabets.

4. A Speedier Algorithm for Binary Alphabets

In this section, we present algorithm INTERVAL which solves the ONE–STRING MASS FINDING PROBLEM for an alphabet of size 2. It uses storage space $O(n)$ and has query time $O(\log n)$. The algorithm *decides* whether a given mass is a submass of σ, but does not return a witness.

Let σ be a string over $\mathcal{A} := \{a, b\}$ of length n and fix $k \leq n$. Observe that, when sliding a window of size k over σ, then in one step, the multiplicities of a and b within the window change at most by one. We represent substrings of σ by points in the $\mathbb{Z} \times \mathbb{Z}$ lattice, where the two coordinates signify the multiplicities of a and b:

$$S_k := \{(i,j) \in \mathbb{Z} \times \mathbb{Z} \mid i+j = k, \text{ there is a substring } \tau \text{ of } \sigma : |\tau|_a = i, |\tau|_b = j\}.$$

All points in S_k will lie on a line (a diagonal), and moreover, they will be neighbours. We will refer to such a set of neighbours on a line as an *interval*. Each such interval has two extremal points.

Example 2. $\sigma = aaaaabaabb$. The figure shows the representation of all substrings of length $k = 8$. Extremal points of this interval are $(5,3)$ and $(7,1)$.

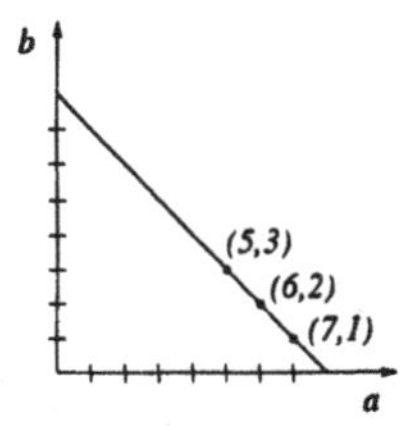

Assume for a moment that we know the multiplicities of a and b in M, e.g. $M = i \cdot \mu(a) + j \cdot \mu(b)$. Then we can easily find out whether M is a submass of σ: We store the S_k's, for $1 \leq k \leq n$ by their extremal points during the preprocessing phase. Now we only have to check whether $(i,j) \in S_{i+j}$, which takes $O(1)$ time. This requires storage space linear in n. If, in addition, i and j were known to be the only feasible multiplicities of a and b (i.e., the unique solution of the equation $x \cdot \mu(a) + y \cdot \mu(b) = M$), then this algorithm would even decide whether M is a submass of σ, and we would be done.

Unfortunately, we do not know the multiplicities of a and b in M. We define $d := \mu(b) - \mu(a)$ (w.l.o.g., assume $\mu(a) < \mu(b)$) and use the residue of $M \bmod d$ to look up a table. The table, generated during the preprocessing phase, contains representations of all submasses of σ.

Let $M_k := \{\mu(\tau) \mid \tau \text{ is a } k\text{-length substring of } \sigma\}$. Observe that consecutive elements of M_k (when sorted) differ by exactly d. Therefore, we can write $M_k = \{c_k + \ell \cdot d \mid \ell = 0, \ldots, n_k - 1\}$, where $c_k = \min M_k$ and $n_k = |M_k|$. Furthermore, $M_k = \{r_k + \ell \cdot d \mid \ell = a_k, \ldots, b_k\}$, where $r_k := (c_k \bmod d)$, $a_k := \lfloor \frac{c_k}{d} \rfloor$ and $b_k := a_k + n_k - 1$. This says that all submasses of the same length have the same residue modulo d.

Observe that $r_k = (k \cdot \mu(a) \bmod d)$. Thus, we may have the same residue modulo d for different values of k. Instead of storing a_k and b_k for each r_k individually (which could result in linear query time), we will store the union of all intervals which belong to the same residue r, sorted by their endpoints. For an example, see [CEL+01].

4.1. Algorithm INTERVAL

In the preprocessing phase, we calculate the r_k's, a_k's, and b_k's as above. We then sort the r_k's, thus obtaining a sorted array $q_1, \ldots, q_m$, where $m \leq n$ (since different S_k's may have the same residue). For each q_l, we compute a list of interval endpoints which represents the union of all intervals $[a_k, b_k]$ with $r_k = q_l$. This list consists of one or more disjoint intervals, which we store in sorted order in an array A_l.

Now, when querying whether a given mass M is contained in σ, first decompose $M = g \cdot d + r$, where $r = (M \bmod d)$ and $g \in \mathbb{N}$; then find index $l \in \{1, \ldots, m\}$ such that $r = q_l$, using binary search; if no such index can be found, then M is not a submass of σ, and the algorithm outputs **no**; otherwise, find whether there is an interval $[a,b]$ in array A_l such that $g \in [a,b]$, using binary search on (the left endpoints of) the intervals; M is a submass of σ if and only if such an interval exists.

Since the total number of intervals to be stored is n, the storage space needed is $O(n)$. The first step of the query algorithm takes time $O(1)$. The second step takes time $O(\log n)$, since the number of different residues is at most n. The third step takes time $O(\log n)$, since the maximum number of intervals stored in one array A_l is n. We obtain a total query time $O(\log n)$.

Theorem 2. *Algorithm* INTERVAL *solves the* ONE–STRING MASS FINDING PROBLEM *for binary alphabets with storage space $O(n)$ and query time $O(\log n)$.*

The problem in generalizing this approach to larger alphabets is that the algorithm relies on the crucial fact that points representing substrings of the same length lie on a line and form an interval. This does not generalize to higher dimensions, since there we only know that the points representing substrings of the same length are connected.

5. Problem Variants

The MULTIPLE–STRING MASS FINDING PROBLEM is defined as follows:

> Given k strings $\sigma_1, \ldots, \sigma_k$ and a mass $M \in \mathbb{N}$, return a list $i_1, \ldots, i_r$ of those strings σ_{i_j} which have M as a submass.

An algorithm Ψ for the ONE–STRING MASS FINDING PROBLEM can be extended to an algorithm for the MULTIPLE–STRING MASS FINDING PROBLEM by running Ψ on each string σ_i one by one. Required storage space and query time simply sum up.

Alternatively, we can adapt an approach from Group Testing (cf. [DH00]): We define a new string $\sigma := \sigma_1 \omega \sigma_2 \omega \ldots \omega \sigma_k$, where ω is a new letter with mass $\mu(\omega) := \max\{\mu(\sigma_i) \mid 1 \leq i \leq k\} + 1$. Before applying Ψ to σ, we check whether $M \geq \mu(\omega)$. If so, then M cannot be a submass of any of the strings, and we are done. Otherwise, we know that whenever Ψ finds mass M in σ, then it is a submass of σ_i for some index i. If algorithm Ψ can output *all* positions of M in σ, this solves the MULTIPLE–STRING MASS FINDING PROBLEM. If Ψ only *decides* whether M is a submass of σ (i.e. it outputs only **yes** or **no**), we use a kind of "binary tree search" BINTREESEARCH to find all σ_i with submass M as follows. First, we run Ψ on σ as described above. If it outputs **no**, then no string σ_i has submass M, and we are done. Otherwise, we divide σ into two new strings $\sigma_l := \sigma_1 \omega \ldots \omega \sigma_{\lfloor \frac{k}{2} \rfloor}$ and $\sigma_r := \sigma_{\lfloor \frac{k}{2} \rfloor + 1} \omega \ldots \omega \sigma_k$ and run Ψ on both strings separately. We repeat the division step until the new strings cover exactly one σ_i, in which case the answer of Ψ determines whether σ_i has a submass M. Analysis of BINTREESEARCH depends heavily on storage space and query time required by Ψ. For instance, if algorithm Ψ requires storage space linear in the length of the string, then the storage space of BINTREESEARCH is $O((\log k) \cdot \sum_{i=1}^{k} |\sigma_i|)$. Query time of BINTREESEARCH depends on the number of strings with submass M, in contrast to the simple idea of applying Ψ to each string separately.

Given a specific algorithm for the ONE–STRING MASS FINDING PROBLEM, there might be even better ways to extend it to the MULTIPLE–STRING MASS FINDING PROBLEM: E.g. for BINSEARCH, we can use *one* sorted array to store all submasses of all strings. For each submass x we store the set of indices I_x of all those strings which have a submass x. Given mass M, we perform binary search in the array and output all indices stored in I_M. Required storage space remains unchanged, but the running time becomes $O(\log(\sum_{i=1}^{k} |\sigma_i|) + |I_M|)$, where $|I_M| \leq k$ is the size of the output. A similar idea applies to LOOKUP, where we could store only one table T of size $|\mathcal{A}|^{c_{max}} \times |\mathcal{A}|^{c_{max}}$ where $c_{max} = \max_{i=1}^{k} c(|\sigma_i|)$, and use it for all runs of the algorithm. However, this does not decrease the asymptotic space required, which still remains linear.

We define a third problem variant, the MULTIPLE–STRING MULTIPLE–MASS FINDING PROBLEM:

> Given k strings $\sigma_1, \ldots, \sigma_k$, m masses $M_1, \ldots, M_m \in \mathbb{N}$, and a threshold $1 \le t \le m$, return a list $i_1, \ldots, i_r$ of those strings σ_{i_j} which have at least t of the masses as submasses.

In the setting of our application in computational biology, this will be a more realistic formulation, since typically, one breaks a given protein in several pieces and wants to find the protein in the database which contains all (or at least many) of these pieces. Obviously, the MULTIPLE–STRING MULTIPLE–MASS FINDING PROBLEM can be solved by applying algorithms for the MULTIPLE–STRING MASS FINDING PROBLEM m times. We are investigating the question whether concurrently searching for m masses can be performed more efficiently.

Finally, we present an improvement of all our algorithms for "short masses": Let the *length* of a mass M be defined as $\lambda(M) := \max(\{|\tau| \mid \tau \in \mathcal{A}^*, \mu(\tau) = M\} \cup \{-1\})$. Here, $\lambda(M) = -1$ means that there is no string with mass M. Suppose that we know in advance that all query masses are short in comparison to n, i.e., that there is a function $f(n)$ such that $\lambda(M) \le f(n) = o(n)$ for all queries M. Then there is a simple algorithm to solve the ONE–STRING MASS FINDING PROBLEM, which is a variant of BINSEARCH: In the preprocessing, we store all submasses of σ of length $\ell \le f(n)$ in a sorted array. This requires storage space $O(n \cdot f(n))$, since for each position i in σ, at most $f(n)$ substrings of length $\ell \le f(n)$ start in i. For a query, we do binary search in this array. This takes time $O(\log n)$, which is speedy. Since $f(n) = o(n)$, the algorithm is skinny, too. We can use this approach to improve our algorithms in the sense that they will run faster on short masses.

6. Conclusion

For some algorithms (e.g. BINSEARCH), storage space and query time for a string σ depend on the number of different submasses of σ. This number is bounded from below by $\Omega(n)$ and from above by $O(n^2)$, and there are examples which meet these boundaries: For instance, let $\mathcal{A} = \{a, b\}, \mu(a) = 1$ and $\mu(b) = n + 1$. The string a^n has n different submasses, while the string $a^n b^n$ has $(n+1)^2 - 1$ different submasses. It may be interesting to explore properties of the number of different submasses, e.g. its expected value or a characterization of all strings with $\Theta(n^2)$ submasses. Furthermore, we are looking for efficient algorithms to compute this number.

With LOOKUP, we presented an algorithm for the ONE–STRING MASS FINDING PROBLEM which is both skinny and speedy. This proves that it is asymptotically possible to beat both LINSEARCH and BINSEARCH at the same time. This raises the question whether there are more practical algorithms that are skinny and speedy. In the long run, we are interested in the tradeoff between query time and storage space for the ONE–STRING MASS FINDING PROBLEM. Do algorithms exist that can be parametrized to allow for adjustment of this tradeoff?

Acknowledgments

We would like to thank Juraj Hromkovic who read an earlier version of this paper and made many helpful suggestions, and Sacha Baginsky for comments on the biological introduction.

References

[AG95] A. Apostolico and Z. Galil, editors. *Combinatorial Algorithms on Words.* Springer, 1995.

[BE01] V. Bafna and N. Edwards. SCOPE: A probabilistic model for scoring tandem mass spectra against a peptide database. *Bioinformatics,* 17(Supplement 1):S13–S21, 2001.

[CDF90] M. Cosnard, J. Duprat, and A. G. Ferreira. The complexity of searching in $X + Y$ and other multisets. *Information Processing Letters,* 34:103–109, 1990.

[CEL+01] M. Cieliebak, T. Erlebach, Zs. Lipták, J. Stoye, and E. Welzl. Algorithmic complexity of protein identification: Combinatorics of weighted strings. Technical Report 361, ETH Zurich, Department of Computer Science, 2001.

[CR94] M. Crochemore and W. Rytter. *Text Algorithms.* Oxford University Press, New York, NY, 1994.

[DH00] D. Du and F. K. Hwang, editors. *Combinatorial Group Testing and its Applications.* World Scientific, second edition, 2000.

[EMYI94] J. Eng, A. McCormack, and J. R. Yates III. An approach to correlate tandem mass spectral data of peptides with amino acid sequences in a protein database. *J. Amer. Soc. Mass Spect.,* 5:976–989, 1994.

[Fre75] M. L. Fredman. Two applications of a probabilistic search technique: Sorting $X+Y$ and building balanced search trees. In *Conference Record of Seventh Annual ACM Symposium on Theory of Computing (STOC),* pages 240–244, 1975.

[GJ79] M. R. Garey and D. S. Johnson. *Computers and Intractability: A Guide to the Theory of NP-Completeness.* Freeman, 1979.

[Gus97] D. Gusfield. *Algorithms on Strings, Trees, and Sequences.* Cambridge University Press, 1997.

[GW91] L. Gonick and M. Wheelis. *The Cartoon Guide to Genetics.* Harper-Perennial, updated edition, 1991.

[HBS+93] W. J. Henzel, T. M. Billeci, J. T. Stults, S. C. Wong, C. Grimley, and C. Watanabe. Identifying proteins from two-dimensional gels by molecular mass searching of peptide fragments in protein sequence databases. *Proc. Natl. Acad. Sci. USA,* 90(11):5011–5015, 1993.

[HPSS75] L.H. Harper, T.H. Payne, J.E. Savage, and E. Straus. Sorting $X + Y$. *Communications of the ACM,* 18(6):347–349, 1975.

[JQCG93] P. James, M. Quadroni, E. Carafoli, and G. Gonnet. Protein identification by mass profile fingerprinting. *Biochem. Biophys. Res. Commun.,* 195(1):58–64, 1993.

[Lot97] M. Lothaire. *Combinatorics on Words.* Cambridge University Press, second edition, 1997.

[MHR93] M. Mann, P. Højrup, and P. Roepstorff. Use of mass spectrometric molecular weight information to identify proteins in sequence databases. *Biol. Mass Spectrom.*, 22(6):338–345, 1993.

[MW94] M. Mann and M. Wilm. Error-tolerant identification of peptides in sequence databases by peptide sequence tags. *Anal. Chem.*, 66(24):4390–4399, 1994.

[PDT00] P. A. Pevzner, V. Dančík, and C. L. Tang. Mutation-tolerant protein identification by mass spectrometry. *J. Comp. Biol.*, 7(6):777–787, 2000.

[Pev00] P. A. Pevzner. *Computational Molecular Biology: An Algorithmic Approach.* MIT Press, 2000.

[PHB93] D. J. C. Pappin, P. Højrup, and A. J. Bleasby. Rapid identification of proteins by peptide-mass fingerprinting. *Curr. Biol.*, 3(6):327–332, 1993.

[PMDT01] P. A. Pevzner, Z. Mulyukov, V. Dančík, and C. L. Tang. Efficiency of database search for identification of mutated and modified proteins via mass spectrometry. *Genome Res.*, 11(2):290–299, 2001.

[RS97] G. Rozenberg and A. Salomaa, editors. *Handbook of Formal Languages*, volume 1-3. Springer, 1997.

[SM97] J. Setubal and J. Meidanis. *Introduction to Computational Molecular Biology.* PWS Boston, 1997.

[Str88] L. Stryer. *Biochemistry.* Freeman, 1988.

[YEM95] J. R. Yates, J. K. Eng, and A. L. McCormack. Mining genomes: Correlating tandem mass-spectra of modified and unmodified peptides to sequences in nucleotide databases. *Anal. Chem.*, 67(18):3202–3210, 1995.

[YI98] J. R. Yates III. Database searching using mass spectrometry data. *Electrophoresis*, 19(6):893–900, 1998.

[YISGH93] J. R. Yates III, S. Speicher, P. R. Griffin, and T. Hunkapillar. Peptide mass maps: A highly informative approach to protein identification. *Anal. Biochem.*, 214:397–408, 1993.

AN EFFICIENT PARALLEL POINTER MACHINE ALGORITHM FOR THE NCA PROBLEM

A. Dal Palú, E. Pontelli, D. Ranjan
Department of Computer Science
New Mexico State University
Las Cruces, NM 88003
{apalu,epontell,dranjan}@cs.nmsu.edu

1. Introduction

The Nearest Common Ancestor (NCA) Problem can be broadly defined as follows: Given a rooted tree T and two nodes $x, y \in T$, find the common ancestor of x and y in T that is furthest from the root. In the *static* version of the problem, T is known in advance. In the *dynamic* version T is modified via some pre-defined operations. In the *offline* version, T as well all the NCA queries are known in advance. NCA problem has been studied extensively [16, 21, 15, 3, 25, 1, 6, 5, 8, 4] .

We present an efficient parallel pointer machine algorithm for the NCA problem for trees in the static case. The algorithm assumes that the tree T is known in advance. It requires $O(\lg n)$ parallel time and $O(n)$ processors for pre-processing the tree, where n is the number of nodes. Thereafter, the algorithm can answer any nca query in $O(\lg \lg n)$ time using a single processor. To our knowledge, this is the best known parallel pointer machine algorithm for the NCA problem. Our NCA algorithm requires an efficient parallel solution of the temporal precedence (TP) problem [20]. We provide an efficient parallel pointer machine algorithm to solve this problem as well.

The paper is organized as follows. In the next section, we give a brief description of the pointer machine and the parallel pointer machine models. In Section 3, we present an arithmetic-free compression scheme that was first introduced and used in [8] to obtain an optimal sequential solution for the NCA Problem on Pure Pointer Machines. We then discuss why straightforward parallelizations of this scheme fail (Section 4). In Section 5 we present our parallel algorithm. We show that our algorithm works correctly and that it requires $O(\lg n)$ parallel time. In Section 6, we compare and constrast our algorithm with other parallel NCA algorithms. Our NCA algorithm requires an efficient parallel solution of the TP problem [20]. An efficient parallel pointer machine algorithm for this problem is presented in Section 7.

2. Pointer Machines

Pointer Machines have been defined in various different ways [2]. All models of pointer machines share the common characteristic of disallowing indexing into an array (i.e., pointer arithmetic), as opposed to RAM models. The *Pure Pointer Machine (PPM)* model also disallows constant-time arithmetic operations. The PPM model is essentially the Linking Automaton model proposed by Knuth [18]. Further details on PPMs can be found in [2, 18, 23, 22].

As with sequential pointer machine model, various versions of parallel pointer machines have been proposed. They all share the common characteristic that no pointer arithmetic is allowed; these models commonly differ in the way interprocessor communication is realized (see [7] for an extensive discussion). All models rely on the presence of a number of processors; each processor is essentially a sequential pointer machine, and all processors execute the same program in a synchronous fashion. At one end of the spectrum we have the *CRCW Parallel Pointer Machine* [14], where arbitrary (concurrent) read and write operations on a shared memory are allowed (although the shared memory cannot be accessed as an array). At the other end of the spectrum, we have the *Parallel PPM* model [7]. The Parallel PPM is defined by a collection of finite state synchronous machines (thus ruling out the use of constant time arithmetic), each of which can rearrange its communication links by a bounded amount in one step. Each finite state machine has an ordered set of input lines (also called links), that can be thought as taps on other processors' outputs. The usual parallel PPM model allows for unbounded fan-out but only constant fan-in. Each finite state machine has the ability to change its links in a restricted way: a finite state machine may redirect one of its links to point to another unit at a "pointer distance" no more than two from it. It has been shown that Parallel PPMs are surprisingly powerful. The details of what exactly constitutes a parallel PPM can be found in [7].

There is a number of models whose computational power lies between that of the two models defined above, e.g., the CREW/EREW Parallel pointer machines, the CROW (Concurrent-Read Owner-Write model), and the SIMDAG model with its variants [13]. Several interesting results regarding their computational power have been established. In particular, an $n-$processor CROW PRAM running in time $O(\lg n)$ can be simulated by a Parallel PPM in time $O(\lg n \lg\lg n)$ using polynomially many processors. In addition any step-by-step simulation of an n processors CROW PRAM by a Parallel PPM requires time $\Omega(\lg\lg n)$ per step [10].

3. A Sequential Compression Scheme for Trees

The starting point of our parallel algorithm is an optimal sequential method for solving the NCA problem on PPMs that was first presented in [8]. This method is based on a *compression scheme* aimed at creating a new tree with logarithmic depth that preserves the ancestor structure of the original tree. It starts from the initial tree $T = T_0$ and repeatedly performs two types of compressions, thus generating a sequence of trees: $T_0, T_1, T_2, \ldots$ until a tree T_k containing a single node is obtained. The trees in this sequence are used to build a second tree structure (called H-tree), that summarizes the nearest common ancestor information of T. The key property of the H-tree is that its depth is at most logarithmic in the number of nodes of T. This allows a fast solution of nca queries.

Given T_i, T^L_{i+1} the result of *leaf-compression* of T_i: it is obtained by merging each leaf of T_i with its parent. If a leaf ℓ is merged with its parent $parent(\ell)$, then $parent(\ell)$ is said to be the *direct representative* of ℓ. A *path-compression* of a tree T^L_{i+1} returns a tree T_{i+1}, where each path containing only nodes with a single child and ending in a leaf of T^L_{i+1} is replaced by the head of such path. If a path containing nodes $v_0, v_1, \ldots, v_k$ is compressed to the node v_0, then v_0 is said to be the *direct representative* of $v_0, \ldots, v_k$. A *compression* of a tree T_i is the tree T_{i+1}, where T_{i+1} is the path-compression of T^L_{i+1}, and T^L_{i+1} is the result of a leaf-compression on T_i. Fig. 1 shows an example of repeated compression of T.

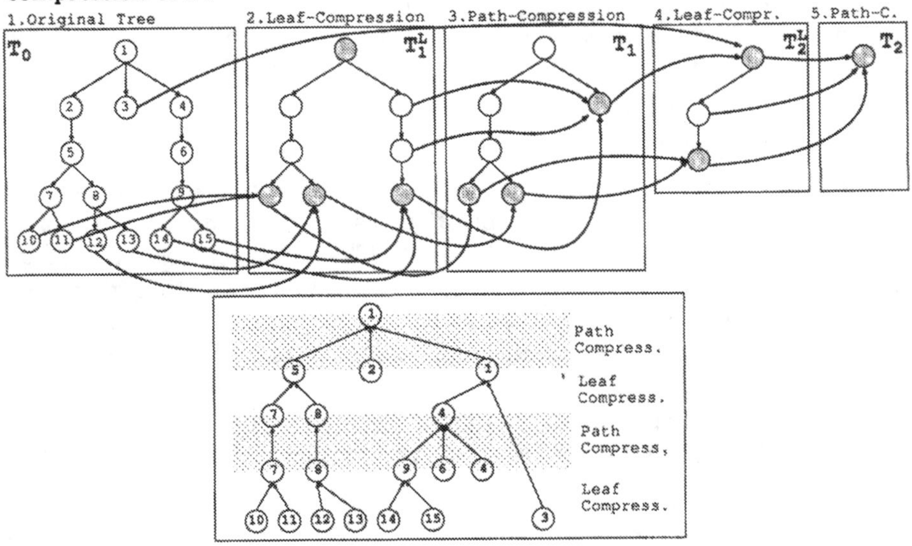

Figure 1. Building the H-tree

The H-Tree

In order to compute *nca* queries in optimal time, it is useful to collect the information about representatives in a separate tree, called *Horizontal Tree (H-tree)*. The H-tree, H, can be constructed from the sequence of trees obtained during the compression process (e.g., the sequence of trees shown in Fig. 1).

If a leaf-compression is applied to node v in tree T_i and ℓ is the direct representative of v in such compression, then node v is connected to the last occurrence of ℓ in a tree T_j ($i < j$), where ℓ appears in T_j as a direct representative of a leaf-compression. If all the children of a node w in T_i are leaf-compressed at the same time, then the representative of such children is node w in T^L_{i+1} (as for leaves $10, 11$ in Fig. 1). If the children of w are leaf-compressed at different points in time (e.g., the children of 1 in Fig. 1), then the representative of such leaf is the last occurrence of its direct representative in a tree as representative in a leaf-compression. If a path-compression is applied, then all nodes in the path are connected to the head of the list in the next tree, as shown in Fig. 1. Such node is said to be the representative of all nodes in the path.

H is obtained using the single node in the last compressed tree (e.g., the node in T_2 in Fig. 1) as the root and using the links between nodes and representatives as edges (e.g., the dark edges in Fig. 1). In [8] the following result was established:

Theorem 1 *Let n be the number of nodes in T and let k be the minimum integer such that T_k has a single node. Then $k \leq \lg n$. In other words, T gets compressed to a single node within* $\lg n$ *compressions.*

The H-tree preserves enough *nca* information from T, so that it is possible to answer the query $nca(x, y)$ in T by answering a suitable NCA query in H. In fact, since the height of H is $O(\lg n)$, it is possible to compute the *nca* of any two nodes in T with worst-case time complexity $O(\lg \lg n)$[8].

4. From Sequential to Parallel

The direct simulation of the sequential algorithm requires $O(\lg^2 n)$ parallel time. Unfortunately, this direct simulation may also require $\Omega(\lg^2 n)$ time. Consider, for example, the situation in Fig. 2. The tree is composed of a main path, with a number of complete trees (of depth $k, k-1, \ldots, 1$) hanging from it. In this situation, at every leaf compression, a path of length l is created in the main branch, allowing for the the next path compression to take place; this path compression will require $\lg l$ time. The process is repeated k times, hence the total parallel time is $k \lg l$. If l is chosen equal to 2^k, then the total number of nodes $n = \Theta(2^{2k})$, thus $k = \Theta(\lg n)$ and the parallel time is $k^2 = \Theta(\lg^2 n)$.

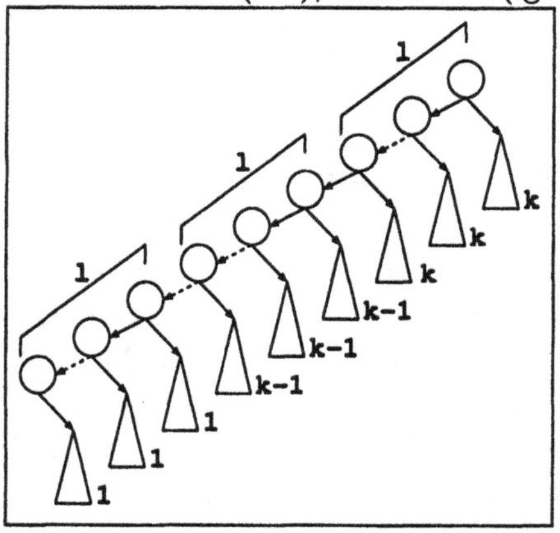

Figure 2. An example of slow computation

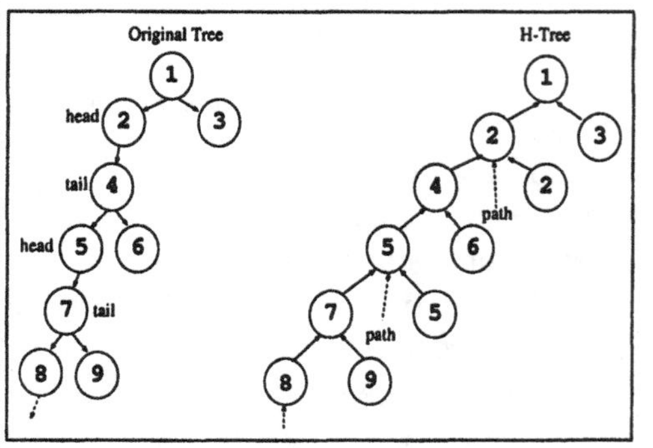

Figure 3. An example of bad H-tree

We could attempt to improve this running time by allowing path compressions to occur also in the internal paths (i.e., paths that do not end in a leaf); similarly we could allow leaf compression to be performed at all the leaves and heads of paths detected at each parallel step. Unfortunately this will not help our case either. As illustrated by the example in Fig. 3, the H-tree resulting from these compressions can have linear depth, thus preventing us from using the H-tree to perform fast computation of nca queries.

However, these considerations do suggest a possible way to improve parallel running time without loosing the efficient computation of nca queries. The idea is that the scheme should compress all paths present in the tree (even the internal ones), but leaf compressions should not be performed on nodes that are currently not leaves. This idea is translated into a concrete parallel algorithm in the next section.

5. A Parallel Compression Scheme for Trees

The compression algorithm is a sequential iteration of parallel phases. Each parallel phase is composed of two parallel steps. The first step is compression of leaves (*leaf compression*) in the current tree and the second step contributes to the compression of paths (*path compression*) in the current tree using a step of pointer doubling [11].

Additionally, our efficient parallel solution for the NCA problem requires the ability to efficiently solve the TP problem [20] in parallel. A parallel version of the problem and an efficient parallel solution to it is presented in Sect. 7.

5.1. The Algorithm

We start by introducing some notation. For a node v in T, T_v denotes the subtree of T rooted at node v. A parallel *phase* i of the algorithm is the sequence of two parallel *steps* called a and b, which are executed at parallel time $i(a)$ and $i(b)$ respectively. For an integer i, T_i denotes the tree after the ith parallel phase. Given a tree T_i, the result of step a applied to T_i is the tree T^a_{i+1} and the result of step b applied to T^a_i is T_i.

During the processing, nodes in the tree may get marked with the symbol L; if node v in T is marked L at parallel time $i(a)$ $(i(b))$, then we denote this with $m^a_i(v) = L$ $(m_i(v) = L)$. We will often refer to this marking as $m(v)$ when the time is clear from the context. If v is not marked then $m(v) =?$. Every node v in T has a pointer π to an ancestor of v at parallel time $i(a)$ $(i(b))$ and we denote it with $\pi^a_i(v)$ $(\pi_i(v))$.

A *leaf compression* of a tree T_i is executed in step $i(a)$ and returns a tree T^a_{i+1} such that for each node v in T_i (see Fig. 4):

(i) if ($m_i(v) = L$ and v currently has no sibling) then
$$\pi^a_{i+1}(v) \leftarrow \pi_i(p(v)) \text{ and } m^a_{i+1}(\pi^a_{i+1}(v)) \leftarrow L;$$

(ii) if ($m_i(v) = L$ and v has a sibling z and $m_i(z) =?$) then
$$v \text{ is merged with its parent } \pi^a_{i+1}(v) \leftarrow NULL;$$

(*iii*) if ($m_i(v) =?$ and ((v has a sibling z and $m_i(z) = L$) or (v currently has no sibling))) then

$$\pi^a_{i+1}(v) \leftarrow \pi_i(p(v));$$

(*iv*) if ($m_i(v) = L$ and v has a left sibling z and $m_i(z) = L$) then
v is merged with its parent and $\pi^a_{i+1}(v) \leftarrow NULL$;

(*v*) if ($m_i(v) = L$, v has a right sibling z and $m_i(z) = L$) then
$\pi^a_{i+1}(v) \leftarrow \pi_i(p(v))$ and $m^a_{i+1}(\pi^a_{i+1}(v)) \leftarrow L$.

A *path compression* of a tree T^a_i is executed in step $i(b)$ and returns a tree T_i, such that for each v in T^a_i, $\pi_i(v) \leftarrow \pi^a_i(\pi^a_i(v))$ and if $m^a_i(v) = L$ then $m_i(\pi_i(v)) \leftarrow L$ (see Fig. 5).

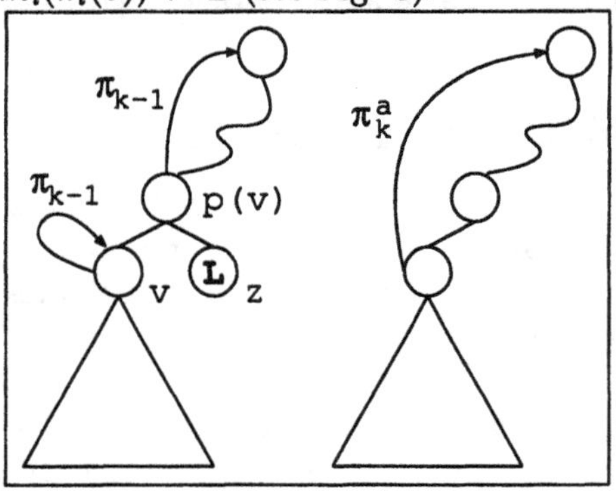

Figure 4. Example of Leaf compression for node x

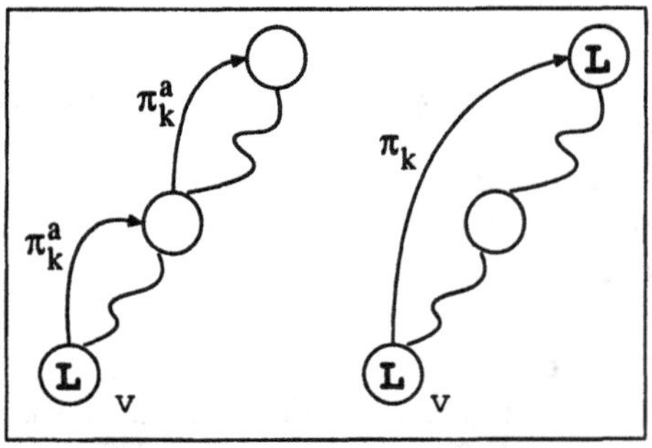

Figure 5. Example of Path compression for node x

If T is the initial tree, then the tree T_0 is a copy of T, such that for each v in T_0, $\pi_0(v) = v$ if v has a sibling, else $\pi_0(v) = p(v)$; in addition, for each leaf l of T_0, $m_0(l) = L$. The root is the only exception: $\pi_0(root) = root$.

Fig. 6 provides an example of a compression. The nodes marked represent the nodes labeled L and the dashed pointers are the π pointers. The pointers π pointing to $NULL$ are not shown.

Definition 2 *A node x is* finished *after step k if one of the following holds:*

1. *x is root and $m_k(x) = L$;*
2. *$\exists y$ y is a proper ancestor of x and $m_k(y) = L$;*
3. *$\pi_k(x) = NULL$.*

The theorem below provides a result that is critical for establishing the the efficiency of the compression scheme.

Theorem 3 *For each parallel time step k and for each node x in T one of the following holds:*

1. *x is finished before or at the end of parallel step k;*
2. *x is marked L during parallel step k, it is unfinished after parallel step k, $|T_x| \geq 2^{k-1}$ and $|T_{p(x)}| \geq 2^k + 1$;*

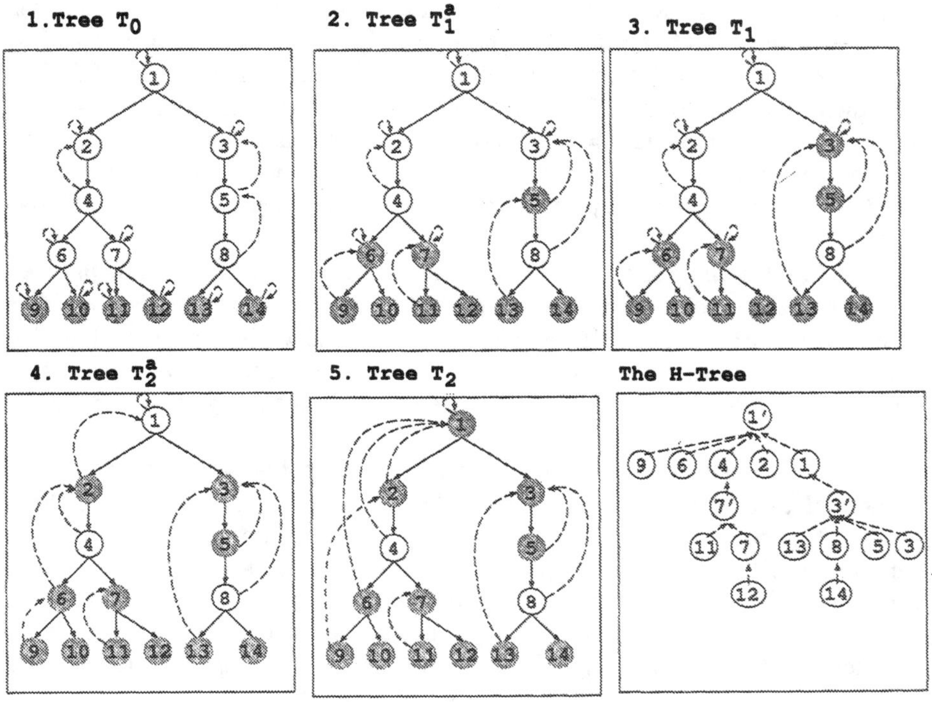

Figure 6. An example of parallel compression

3. *x is unmarked and unfinished after parallel step k, and either $\pi_k(x) =$ root or ($|T_{p(\pi_k(x))}| - |T_x| \geq 2^{k-1}$ and $|T_x| \geq 2^k + 1$).*

The complete proof can be found in [9].

Corollary 1 *Let n be the number of nodes in T and let k be the smallest integer such that the root is finished after phase k. Then $n \geq 2^{k-1}+1$. In other words, the algorithm requires at most* $\lg(n-1)+1$ *phases.*

If we have n processors that have been assigned to the n different nodes of T, then both leaf and path compressions will be performed in constant parallel time. From Corollary 1, the total number of compressions is at most $\lg n$, thus the total parallel time required by the algorithm is $O(\lg n)$.

5.2. The *H*-tree

The H-Tree built in the parallel scheme is the same described in Section 3. It is possible to reuse the π pointers to build H in constant parallel time. Once a node v is leaf compressed into its parent, $\pi(v)$ is set to $NULL$. At that time for every node w in the path having v as head, we have $\pi(w) = v$. The goal is to maintain this information in the successive phases, avoiding avoiding pointer doubling if a node is finished (line 26 in Fig. A.1). Once the compression is completed, every head of a path x has $\pi(x) = NULL$ and the π pointer for every other node y points to the head of the path containing y.

Let us introduce another pointer p_H, that will be used as the parent pointer in H. During a leaf compression the pointer $p_H(v)$ is set to point to $p(v)$. Since the root is not leaf compressed, $p_H(root)$ is set to point to $NULL$. For each head x of a path l in T, a new node x' is created in H, with $\pi(x') = x'$, $p_H(x') = p_H(x)$ and $\pi(x) = x'$. After each step of pointer doubling (applied to the π pointers), every node in a path points to a newly created copy of the head. For every node x in a path $p_H(x) = \pi(x)$, and this completes the building of H.

Finally, for each path the auxiliary data structure for the $\mathcal{TP}$ Problem is set up. It is possible to identify the tails of path in $O(1)$ time as a node v is a tail iff for all children w of v $\pi(v) \neq \pi(w)$. Given a tail t, the corresponding list is processed as described in Section 7. The complete algorithm is presented in Appendix A. The lines marked with * are the ones necessary to set up the H tree.

The H-tree can be used to answer nca queries in the same way as in the sequential case. In [19] it was shown that there is a PPM algorithm that, given a tree with height h, preprocesses the tree in time $O(n \lg h)$ and then can compute the nca of any two given nodes in the tree in worst case time complexity $O(\lg h)$ per query. The sequential scheme presented in [19] can be easily translated into a parallel scheme that uses n processors and $O(\lg h)$ parallel time for preprocessing. Using this result, we can preprocess the H-tree in parallel time $O(\lg \lg n)$ using n processors. Then, the nca of in H, and hence in T, can be computed in time $O(\lg \lg n)$ using a single processor.

6. Discussion

The algorithm described above clearly can be directly implemented on a CRCW Parallel PPM. A problem arises if we were not allowed concurrent writes, because too many processors may attempt to update the L mark of the same node in the tree at the same time (e.g., line 10 in Fig. A.1). This will not be allowed in the CREW/EREW/CROW parallel pointer machines. This is also not allowed in the Parallel PPM (as described in Sect. 2) because it would correspond to an unbound fan-in. However, it is possible to modify the algorithm to overcome this problem. This is essentially obtained by concurrently performing a pointer doubling in the reverse direction along the branches of the tree. More precisely, each node u of the tree maintains a pointer $\pi_{down}(u)$ which is updated to point to $\pi_{down}(v)$ whenever the node has only one child v. In addition, if $\pi_{down}(v)$ is marked L, then u will mark itself L as well. With this addition, the fan-in of each unit is restricted to be finite. Moreover, the algorithm still requires only $O(\lg n)$ parallel phases. Hence, the algorithm can be modified to correctly work on Parallel PPMs.

The algorithm requires n processors to perform the $O(\lg n)$ parallel time preprocessing. After preprocessing, a single processor can answer an nca query in time $O(\lg \lg n)$. It is interesting to compare this result with the other parallel algorithms proposed for the NCA problem. The best known PRAM algorithms require $O(n/\lg n)$ processors and work in $O(\lg n)$ parallel time for preprocessing. After preprocessing, a single processor can answer an nca query in time $O(1)$. Hence, going from a PRAM to a Parallel PPM we incur a penalty of $O(\lg n)$ in number of processors and total time taken, and a penalty of $O(\lg\lg n)$ time to answer a query. Observe that we do not incur *any* penalty in parallel time for preprocessing.

It is also important to observe that if we have any CROW PRAM NCA algorithm which solves the problem in parallel time $O(\lg n)$ with $f(n)$ processors and answers a query in $O(1)$ time, then for a generic translation of this algorithm to a Parallel PPM algorithm (as illustrated in [10]) one can only claim that it requires parallel time $O(\lg n \lg\lg n)$ with *polynomially* many processors, and answers an nca query in time $O(\lg\lg n)$. Hence, the algorithm presented here is substantially better than a generic translation of any PRAM NCA algorithm presented in the literature to date to a Parallel PPM algorithm.

It is interesting to note that if we have simple arithmetic capabilities (actually only constaint-time addition is needed), then we can compute the centroid path and the H-tree based on it in $O(\lg n)$ parallel time. This is obtained by keeping a count of the number of nodes in the subtree rooted in each node during the algorithm execution. Each time we have a leaf compression phase where both children of a node are marked L, instead of leaf compressing the right child, we leaf compress always the child with a smaller count. It is easy to show that this will build the centroid path tree.

Note that if we are allowed only one processor to answer an nca query, then the time required must be at least $\Omega(\lg\lg n)$ [20]. Hence our algorithm is optimal in that regard. Observe also that the parallel time $O(\lg n)$ used to perform preprocessing is the best known for any parallel NCA algorithm (including PRAM algorithms). If one were allowed arbitrary (e.g., n^3) number of processors, then it is possible to devise a Parallel PPM algorithm that requires $O(\lg n)$ parallel time for preprocessing and answers nca queries in time $O(1)$ [10]. This can be simply accomplished by precomputing all the answers in parallel (in time $O(\lg\lg n)$) and making a different processor responsible for each different possible query.

7. A Parallel Algorithm For The TP Problem

The TP Problem, first defined in [20], can be reformulated in the context of parallel computations as follows: given a list L with l nodes representing an ordered sequence of objects, we want to answer the query precedes(x, y), where x, y are pointers to nodes in that list. We present a solution to this problem on Parallel PPMs that requires l processors, $O(\lg l)$ parallel preprocessing time, and $O(\lg\lg l)$ time to answer each query using a single processor thereafter.

The basic idea is to create an auxiliary complete binary tree BT, such that each leaf is assigned to an element of L. If BT maintains a left to right ordering in each level, then the precedes(x, y) query can be answered comparing the children of $nca(x, y)$ in BT. We maintain this order in each level of BT using *sibl* pointers. BT is constructed via a parallel level-by-level construction. During the construction, each node of BT has one processor associated to it. The root of BT is created in the first step. Then, in parallel (for $\lg l$ steps), each new processor p associated to a node v in BT executes the following operations: create two new nodes (v_l and v_r) with new processors associated to them, set *sibl* pointer of v_l to v_r and set $sibl(v_r)$ to left child of $sibl(v)$.

The last level of BT contains a list of nodes S. Since L is the input, from the way inputs are presented in the Parallel PPM model, we can assume that active processors can be assigned to each element of L in time $O(\lg l)$. We can also assume that these processors have pointers that points to the *previous* element in the list (see Fig. 7). The elements of the original list L are mapped to the elements of S in $O(\lg l)$ parallel time with $O(l)$ processors using a pointer doubling scheme, which modifies the sibling list of S and *previous* pointers of L. Each node of L contains a pointer *map* that is used to point to the corresponding node in S. Initially none of the *map* pointers is set. In the first step processor assigned to the head of L sets its *map* pointer to the head of S. At the same time, the second element of L sets its *map* pointer to the second element of S (see Fig. 8). This is followed by a step of pointer doubling in both S and L. In S pointer doubling is accomplished using the *sibl* pointers, while in L it is performed using the *previous* pointers. After a step of pointer doubling, if the *previous* pointer of a node v in L whose *map* pointer is not set points to a node u whose *map* pointer is set, then $map(v)$ is set to $sibl(map(u))$. Note that the *map* pointer of a node in L is set only once. The process continues until all the nodes in L have their *map* pointers set. The whole process requires $O(\lg l)$ parallel time.

The last step of the preprocessing constructs in $O(\lg l)$ parallel time the auxiliary data structures (called *p-lists*) using a straightforward parallelization of the algorithm presented in [19]. A precedes(x, y) query is then answered by a single processor in $O(\lg\lg l)$ time using the algorithm presented in [19].

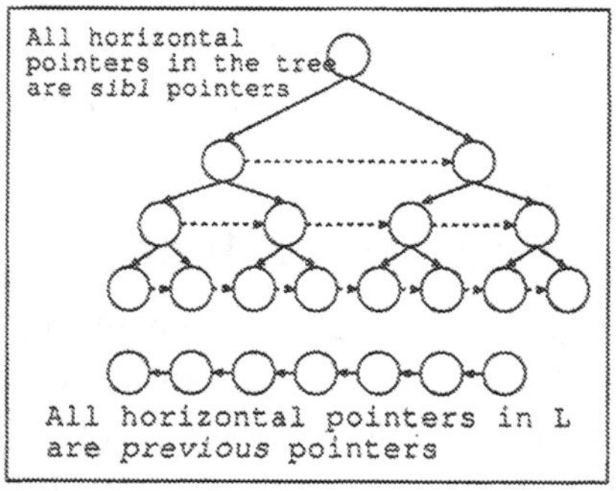

Figure 7. BT tree and list L

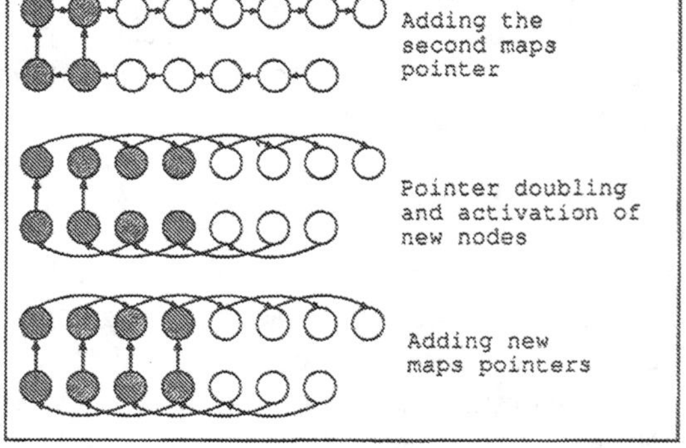

Figure 8. Example of mapping

Acknowledgments

The research was supported by NSF grants CCR-9900320, EIA-0130887, CCR-9875279, CCR-9820852, and EIA-9810732.

Appendix: Appendix A: Compression Algorithm

```
   pardo
     if ($v$=root or $v$ has a sibling) $\pi_0(v) = v$ else $\pi_0(v) = p(v)$;
     if ($v$ is leaf) $m_0(v) = L$ else $m_0(v) =?$;
   i:=0;
   iterate
     % leaf compression
     pardo
        if ($m_i(v) = L$ & $v$ has currently no sibling)
            $\pi^a_{i+1}(v) = \pi_i(p(v))$;
            $m^a_{i+1}(\pi^a_{i+1}(v)) = L$;
        elseif ($m_i(v) = L$ & $v$ has a sibling $z$ & $m_i(z) =?$)
            $\pi^a_{i+1}(v) = NULL$;
*           $p_H(v) = p(v)$;
        elseif ($m_i(v) =?$ & (($v$ has a sibling $z$ & $m_i(z) = L$)
                 or ($v$ currently has no sibling)))
            $\pi^a_{i+1}(v) = \pi_i(p(v))$;
        elseif ($m_i(v) = L$ & $v$ has a left sibling $z$ & $m_i(z) = L$)
            $\pi^a_{i+1}(v) = NULL$;
*           $p_H(v) = p(v)$;
        elseif ($m_i(v) = L$ & $v$ has a right sibling $z$ & $m_i(z) = L$)
            $\pi^a_{i+1}(v) = \pi_i(p(v))$;
            $m^a_{i+1}(\pi^a_{i+1}(v)) = L$;
        else $\pi^a_{i+1}(v) = \pi_i(v)$;
     % path compression
     pardo
        $\pi_{i+1}(v) = \pi^a_{i+1}(\pi^a_{i+1}(v))$;
*       if ($\pi_{i+1}(v) = NULL$) $\pi_{i+1}(v) = \pi^a_{i+1}(v)$;
        if ($m^a_{i+1}(v) = L$) $m_{i+1}(\pi_{i+1}(v)) = L$;
     i:=i+1;
   loop until $m_i(root) = L$;
*  $p_H(root) = NULL$;
*  pardo $x$ head of a path
*    create $x'$ & $\pi(x') = x'$;          % $x'$ is a copy of $x$
*    $p_H(x') = p_H(x)$;
*    $\pi(x) = x'$;
*  pardo $p_H(x) = \pi(\pi(x))$; % each node in the path of $x$ points to $x'$
*  pardo if ($v$ is tail)
     TP preprocess on the list starting at $v$ and ending at $\pi_i(v)$
```

Figure A.1. Preprocessing Algorithm

References

[1] S. Alstrup and M. Thorup. Optimal Pointer Algorithms for Finding Nearest Common Ancestors in Dynamic Trees. *Journal of Algorithms*, 35:169–188, 2000.

[2] A.M. Ben-Amram. What is a Pointer Machine? In *SIGACT News*, 26(2), 1995.

[3] M.A. Bender and M. Farach-Colton. The LCA Problem Revisited. In *Proceedings of LATIN 2000*, Springer Verlag, 2000.

[4] O. Berkman & U. Vishkin. Recursive *-Tree Parallel Data Structure. *FOCS*, 1989.

[5] A.L. Buchsbaum et al. Linear-Time Pointer-Machine Algorithms for Least Common Ancestors, MST Verification, and Dominators. In *STOC*, ACM Press, 1998.

[6] R. Cole and R. Hariharan. Dynamic LCA Queries on Trees. In *Proceedings of the Symposium on Discrete Algorithms (SODA)*, pages 235–244. ACM/SIAM, 1999.

[7] S.A. Cook and P.W. Dymond. Parallel Pointer Machines. Computational Complexity, 3:19–30, 1993.

[8] A. Dal Palú, E. Pontelli, D. Ranjan. An Optimal Algorithm for Finding NCA on Pure Pointer Machines. Tech. Rep., TR-CS-007/2001, NMSU, 2001.

[9] A. Dal Palú, E. Pontelli, D. Ranjan. An Efficient Parallel Pointer Machine Algorithm for Nearest-Common Ancestor Problem. Tech. Rep., TR-CS-009/2001, NMSU, 2001.

[10] P.W. Dymond, F.E. Fich, N. Nishimura, P. Ragde, W.L. Ruzzo. Pointers versus Arithmetic in PRAMs. In *Structures in Complexity Theory Conf.*, IEEE, 1993.

[11] S. Fortune and J. Wyllie. Parallelism in RAMs. In *STOC*, ACM Press, 1978.

[12] H.N. Gabow and R. E. Tarjan A linear-time algorithm for a special case of disjoint set union *J. Comput. System Sci* 30 (1985), 209-221.

[13] L.M. Goldschlager. A Universal Interconnection Pattern for Parallel Computers. In Journal of the ACM, 29, 1982.

[14] M.T. Goodrich and S.R. Kosaraju. Sorting on a Parallel Pointer Machine with Applications to Set Expression Evaluation. In FOCS, IEEE, 1989.

[15] D. Gusfield. *Algorithms on Strings, Trees, and Sequences.* Cambridge University Press, 1999.

[16] D. Harel and R.E. Tarjan. Fast Algorithms for Finding Nearest Common Ancestor. *SIAM Journal of Computing*, 13(2):338–355, 1984.

[17] J.W. Hong. On Similarity and Duality of Computation. In FOCS, 1980.

[18] D.E. Knuth. *The Art of Computer Programming.* Addison-Wesley, 1968.

[19] E. Pontelli and D. Ranjan. Ancestor Problems on Pure Pointer Machines. In *LATIN*, Springer Verlag, 2002 (to appear).

[20] D. Ranjan, E. Pontelli, L. Longpre, and G. Gupta. The Temporal Precedence Problem. *Algorithmica*, 28:288–306, 2000.

[21] B. Schieber and U. Vishkin. On Finding Lowest Common Ancestors. *SIAM J. Comp.*, 17:1253–1262, 1988.

[22] A. Schönhage. Storage Modification Machines. *SIAM Journal of Computing*, 9(3):490–508, August 1980.

[23] R.E. Tarjan. A Class of Algorithms which Require Nonlinear Time to Maintain Disjoint Sets. *Journal of Computer and System Sciences*, 2(18):110–127, 1979.

[24] A. Tsakalidis. Maintaining Order in a Generalized Linked List. *ACTA Informatica*, (21):101–112, 1984.

[25] A.K. Tsakalidis. The Nearest Common Ancestor in a Dynamic Tree. *ACTA Informatica*, 25:37–54, 1988.

RANDOMIZED DINING PHILOSOPHERS WITHOUT FAIRNESS ASSUMPTION

Marie Duflot, Laurent Fribourg and Claudine Picaronny
LSV, CNRS & ENS de Cachan,
61 av. du Prés. Wilson, 94235 Cachan cedex, France
{duflot,fribourg,picaro}@lsv.ens-cachan.fr

Abstract We consider Lehmann-Rabin's randomized solution to the well-known problem of the dining philosophers. Up to now, such an analysis has always required a "fairness" assumption on the scheduler: if a philosopher is continuously hungry then he must eventually be scheduled. In contrast here, we modify the algorithm in order to get rid of the fairness assumption. We claim that the spirit of the original algorithm is preserved. We prove that, for *any* (possibly unfair) scheduler, the modified algorithm converges: every computation reaches with probability 1 a configuration where some philosopher eats. Furthermore, we are now able to evaluate the expected time of convergence as a number of transitions. We show that, for some "malicious" scheduler, this expected time is at least exponential in the number N of philosophers.

1. Introduction

Recently, due to the rising risk of traffic congestion, there has been an increasing interest in providing differentiated Internet services, departing from the traditional notion of fairness for bandwidth allocations [1, 6]. This motivates reconsidering the need for the fairness assumption, which is classical in resource-allocation algorithms (see [9] chap 11). Here we consider Lehmann-Rabin's randomized solution to a special case of resource-allocation problem: the dining philosophers. N philosophers, $P_1, \cdots, P_N$, are seated around a table, and variously think or try to eat by using some shared forks. The problem is to find a distributed protocol guaranteeing that some philosopher will eventually eat. A philosopher is only able to execute a step provided he is selected by a general mechanism called *scheduler*. When he is selected by the scheduler, he executes exactly one action (and nothing is done by the others). Let $\mathcal{L}$ be the set of configurations, called here "legitimate", where some philosopher eats. We show here that the algorithm reaches $\mathcal{L}$ within a finite time with probability 1. In the following, we will call this property *convergence*. (This is sometimes called *progress* in the literature, see e.g. [9].) Up to now, such a proof has always required a "fairness" assumption on the scheduler: if a philosopher is continuously hungry (i.e., trying to eat) then he must eventually be scheduled.

Fairness guarantees that there exist *rounds*, intervals in which each philosopher has been scheduled at least once. It is shown in [10, 9, 13, 11, 14] that within a constant number of rounds, the probability of reaching $\mathcal{L}$ is greater than 0. It follows that the algorithm converges towards a configuration of $\mathcal{L}$ with probability 1.

In contrast here, we consider *arbitrary* schedulers, without any fairness assumption (so we do not use the notion of rounds). We modify the original Lehmann-Rabin's algorithm by removing self-looping actions. We show that the new algorithm still converges towards $\mathcal{L}$ with probability 1. This is done by constructing a measure Δ over configurations that decreases with positive probability at each computation step (that does not reach $\mathcal{L}$). We thus propose a solution to the resource allocation problem for dining philosophers under *arbitrary* scheduler. We also show that the expected time of convergence, in terms of individual actions, is at least exponential in N, for some "malicious" scheduler.

The plan of the paper is as follows. After some preliminaries on Lehmann-Rabin's original algorithm (Section 2), we explain how we modify the algorithm (Section 3). We then prove the convergence of the modified algorithm for an arbitrary scheduler without fairness assumption (Section 4). In Section 5, we show that, for some malicious scheduler, the expected time of convergence is at least exponential in N. We conclude in Section 6.

2. Randomized Distributed Algorithms

2.1. Randomized Uniform Ring Systems

A *randomized uniform ring system* is a triple $(N, \rightarrow, Q)$ where N is the number of processes in the system, $\rightarrow$ is a state transition algorithm, and Q is the *alphabet*, i.e. a finite set of process states. The N processes $P_1, ..., P_N$ form a ring: there is an edge between two consecutive processes, which means that P_i can observe the states q_{i-1} and q_{i+1} of P_{i-1} and P_{i+1} respectively. Let Q be the state set of P_i. The system is *uniform* in the sense that $\rightarrow$ and Q are common to all processes. A *configuration* is an N-tuple of process states (or letters); if the current state of process P_i is $q_i \in Q$, then the configuration of the system is $x = q_1 q_2 \cdots q_N$. The state transition algorithm $\rightarrow$ is given as a set $\mathcal{R}$ of rules, consisting of *deterministic* or *probabilistic* rewrite rules. A deterministic rule is here of one the following forms:

- $q \rightarrow q'$
- $qr \rightarrow q'r$,
- $rq \rightarrow rq'$

where q, q', r denote states of Q. A probabilistic rule is here simply of the form:

$q \rightarrow q'_1$ with probability $1/2$
or q'_2 with probability $1/2$.

where q, q'_1, q'_2 denote states of Q.The letter q of the left-hand side is the *old* letter of the rule. The letter q' (with possible subscripts) of the right-hand side is the *new* letter of the rule. For readability, the old and new letters will often be written in bold font within rules. A rewrite rule R of left-hand side $\mathbf{q}$ is *applicable* at position i of a configuration x if the i-th letter of x is q. Likewise, a rewrite rule R of left-hand side $\mathbf{q}r$ (resp. $r\mathbf{q}$) is *applicable* at position i if the i-th letter of x is q and the $(i+1)$-th (resp. $(i-1)$-th) letter is r.

We say that process P_i is *enabled* if at least one rewrite rule is applicable to the i-th letter of x. Let $\mathcal{E}(x)$ be the set of indices of the enabled processes of x.

Given x and an enabled position i of x, a *transition* leads from x to the configuration x' obtained from x by changing the i-th letter of x equal to the old letter of some applicable rule, say R, into the new letter. Such a transition is written $x \xrightarrow[\mathtt{R}]{i} x'$. Notation $x \xrightarrow[\mathcal{R}]{i} x'$ means $x \xrightarrow[\mathtt{R}]{i} x'$ for some rule $\mathtt{R} \in \mathcal{R}$.

A *(central) scheduler* is a mechanism that selects one enabled process at each step. The distributed system corresponds to repeated application of transition rules according to the philosopher chosen by the scheduler at each step. Given a scheduler $\mathcal{A}$, we are interested in proving the following *convergence* property (see [7, 2]): No matter which initial configuration x_0 one starts from, the probability that $\rightarrow$ under $\mathcal{A}$ reaches a legitimate configuration in a finite number of transitions is 1. This will be written: $Pr(x_0 \xrightarrow[\mathcal{R}]{\mathcal{A}} {}^*\mathcal{L})$. (See [4] for a formal definition.)

2.2. Lehmann-Rabin's algorithm

We present Lehmann-Rabin's algorithm [8] along the lines of [11]. Since we are only interested in the time before some philosopher eats, we disregard what follows this event (after state E has been reached).

The state set of each philosopher is $Q = \{T, H, \overleftarrow{W}, \overrightarrow{W}, \overleftarrow{S}, \overrightarrow{S}, \overleftarrow{D}, \overrightarrow{D}, E\}$. The letter T represents thinking, H that a philosopher is hungry, $\overleftarrow{W}$ (resp. $\overrightarrow{W}$) that a philosopher waits in order to attempt to pick up the left (resp. right) fork next time he is scheduled, $\overleftarrow{S}$ (resp. $\overrightarrow{S}$) that he is holding only the left (resp. right) fork, $\overleftarrow{D}$ (resp. $\overrightarrow{D}$) that he will put down the left (resp. right) fork next time he is scheduled and E that he eats. The details relating to the shared forks are omitted. Thus, for example, if P_i is in state $\overleftarrow{S}$ or P_{i-1} is in state $\overrightarrow{S}$, it means the variable representing the shared fork (between P_i and P_{i-1}) has been set to a value 'taken'. In this modeling, not all configurations are possible. This is because a fork can be taken by at most one process. More generally, we say that a configuration is *admissible* iff it does not contain any substring of the form $\overrightarrow{S}\overleftarrow{S}$, $\overrightarrow{S}\overleftarrow{D}$, $\overrightarrow{D}\overleftarrow{S}$ or $\overrightarrow{D}\overleftarrow{D}$. Henceforth, we will always implicitly focus on admissible configurations. The set $\mathcal{R}$ of transition rules is:

Q0: $\mathbf{T} \rightarrow \mathbf{T}$
Q1: $\mathbf{T} \rightarrow \mathbf{H}$
R0: $\mathbf{H} \rightarrow \overleftarrow{\mathbf{W}}$ with probability 1/2
or $\overrightarrow{\mathbf{W}}$ with probability 1/2.
R1: $\neg\overrightarrow{SD}\ \overleftarrow{\mathbf{W}} \rightarrow \neg\overrightarrow{SD}\ \overleftarrow{\mathbf{S}}$
R2: $\overrightarrow{SD}\ \overleftarrow{\mathbf{W}} \rightarrow \overrightarrow{SD}\ \overleftarrow{\mathbf{W}}$
R3: $\overrightarrow{\mathbf{W}}\ \neg\overleftarrow{SD} \rightarrow \overrightarrow{\mathbf{S}}\ \neg\overleftarrow{SD}$
R4: $\overrightarrow{\mathbf{W}}\ \overleftarrow{SD} \rightarrow \overrightarrow{\mathbf{W}}\ \overleftarrow{SD}$
R5: $\overleftarrow{\mathbf{S}}\ \neg\overleftarrow{SD} \rightarrow \mathbf{E}\ \neg\overleftarrow{SD}$
R6: $\overleftarrow{\mathbf{S}}\ \overleftarrow{SD} \rightarrow \overleftarrow{\mathbf{D}}\ \overleftarrow{SD}$
R7: $\neg\overrightarrow{SD}\ \overrightarrow{\mathbf{S}} \rightarrow \neg\overrightarrow{SD}\ \mathbf{E}$
R8: $\overrightarrow{SD}\ \overrightarrow{\mathbf{S}} \rightarrow \overrightarrow{SD}\ \overrightarrow{\mathbf{D}}$
R9: $\overleftarrow{\mathbf{D}} \rightarrow \mathbf{H}$
R10: $\overrightarrow{\mathbf{D}} \rightarrow \mathbf{H}$

where $\neg\overrightarrow{SD}$ (resp. $\neg\overleftarrow{SD}$) denotes any letter of Q distinct from $\overrightarrow{S}$ and $\overrightarrow{D}$ (resp. $\overleftarrow{S}$ and $\overleftarrow{D}$), and $\overrightarrow{SD}$ (resp. $\overleftarrow{SD}$) denotes $\overrightarrow{S}$ or $\overrightarrow{D}$ (resp. $\overleftarrow{S}$ or $\overleftarrow{D}$).

The rules describe the behaviour of a selected philosopher as follows: initially he thinks "repeatedly" (Q0); he becomes hungry (Q1); he decides randomly which fork to pick up first (R0); next he persists with his decision (R2 or R4) until he finally picks it up when available (R1 or R3), only putting it down later if he finds that his other fork is already held by his neighbour (R6 followed by R9, or R8 followed by R10); if he finds that his other fork is not held, he takes it and eats (R5 or R7).

This behaviour is depicted on figure 1 (drawn from [13]).

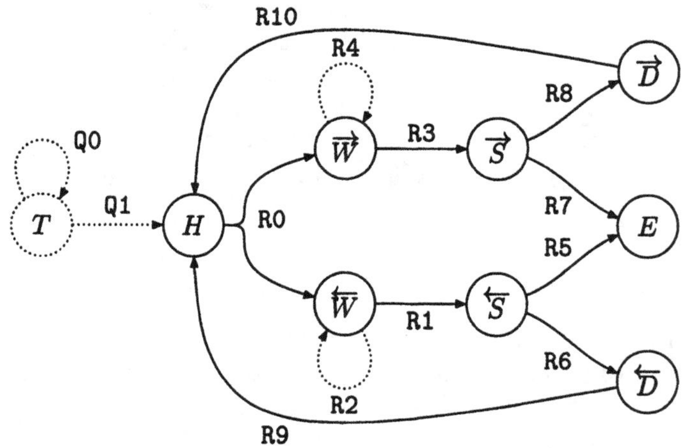

Figure 1. Illustration of the original algorithm. The dashed state and transitions are the ones removed in our variant.

3. Removal of stuttering rules

Let us observe that rule Q0 (resp. R2, R4) is "stuttering" in the sense that the old and new letters of the rule coincide. When a selected philosopher is thinking (resp. waiting for picking up a first fork held by a neighbour), a transition that does not change the configuration, may occur. This is depicted by a self-loop on state T (resp. $\overleftarrow{W}$, $\overrightarrow{W}$) in figure 1. We modify Lehmann-Rabin's algorithm by removing stuttering rules Q0, R2 and R4:

- without Q0, when a philosopher in state T is selected by the scheduler, his state always becomes H via Q1. States T and H then play the same role and will be merged together in the following;
- without R2 (resp. R4), when a philosopher waits for a first fork that is held by a neighbour, i.e., is in state $\overleftarrow{W}$ (resp. $\overrightarrow{W}$) and his left (resp. right) neighbour in state $\overrightarrow{SD}$ (resp. $\overleftarrow{SD}$), he is no longer enabled: no rule applies to him. In such a situation, the philosopher cannot be selected by the scheduler. Note that this differs from Lehmann-Rabin's original algorithm where every process can always be selected.

The rewrite system $\mathcal{R}$ is transformed into $\mathcal{R}' = \mathcal{R} - \{\texttt{Q0}, \texttt{Q1}, \texttt{R2}, \texttt{R4}\}$. The behaviour of $\mathcal{R}'$ is depicted on figure 1 without the dashed node and transitions. The new state set Q' is $Q - \{T\}$. Accordingly, the new legitimate set $\mathcal{L}'$ is $Q'^*EQ'^*$.

Discussion.
In Lehmann-Rabin's algorithm, a non-eating philosopher either thinks (state T) or tries to eat (states $\{H, W, S, D\}$). In our version of the algorithm, as the state T has been dropped, this philosopher can only try to eat. This feature may be seen as a limitation. Actually, since we have no fairness assumption, a philosopher can be indefinitely ignored by the scheduler, thus behaving in state H as he used to do originally in state T (i.e., not trying to pick up a fork). We thus claim that our modified algorithm is similar in spirit to the original one.

The original convergence property of Lehmann-Rabin's algorithm can be stated:
for any *fair* scheduler $\mathcal{A}$ and every $x \in Q^*\{H, W, S, D\}Q^*$, $Pr(x \xrightarrow[\mathcal{R}]{\mathcal{A}}{}^* \mathcal{L}) = 1$.
Surprisingly, as shown in section 4, for our modified version $\mathcal{R}'$ of $\mathcal{R}$, the convergence property holds with *no* fairness assumption on the scheduler, i.e.:

Theorem 1 *For any* arbitrary *scheduler $\mathcal{A}$ and every $x \in \{H, W, S, D\}^*$,*
$$Pr(x \xrightarrow[\mathcal{R}']{\mathcal{A}}{}^* \mathcal{L}') = 1.$$

4. Convergence of $\mathcal{R}'$

4.1. Scheme of the proof

We are going to prove Theorem 1 by using the following property proved in [4] (cf Theorem 1 of [2] and Theorem 5 of [3]):

Theorem 2 *Suppose that there exists a measure Δ and an ordering $\ll$ such that:*
$$\forall x \notin \mathcal{L}' \ \forall i \in \mathcal{E}(x) \quad \exists x' \ (x \xrightarrow[\mathcal{R}']{i} x' \wedge (\Delta(x') \ll \Delta(x) \vee x' \in \mathcal{L}'))$$
Then, for any (central) scheduler $\mathcal{A}$: $\forall x \ \ Pr(x \xrightarrow[\mathcal{R}']{\mathcal{A}}{}^* \mathcal{L}') = 1$.

More precisely, we will find an appropriate "rewriting strategy" for probabilistic rule R0 (i.e., a fixed choice of the new letter $\overleftarrow{W}$ or $\overrightarrow{W}$, depending on the context of the old letter H in x, when rewriting x via R0), and a measure Δ such that

- the application of R0 under this strategy makes Δ decrease,
- the application of any other rule makes Δ decrease or leads to $\mathcal{L}'$.

Actually, a configuration x will be given with two lists π and ψ such that for every choice of the scheduler, there exists a configuration x' and two lists of indices π', ψ' constructed from π and ψ such that $x \rightarrow x'$ and $\Delta(x', \pi', \psi') \ll \Delta(x, \pi, \psi)$ (see Sections 4.2 and 4.3 for definitions of π and ψ).

In the following, the configurations are implicitly non-legitimate (i.e, belong to $(Q' - \{E\})^*$). Symbol W denotes a letter of $\{\overleftarrow{W}, \overrightarrow{W}\}$. Likewise, S (resp.

D) denotes a letter of $\{\overleftarrow{S}, \overrightarrow{S}\}$ (resp. $\{\overleftarrow{D}, \overrightarrow{D}\}$). Let $\overleftarrow{Q} = \{H, \overleftarrow{W}, \overleftarrow{S}, \overleftarrow{D}\}$ and $\overrightarrow{Q} = \{H, \overrightarrow{W}, \overrightarrow{S}, \overrightarrow{D}\}$. (Note that $\overleftarrow{Q} \cap \overrightarrow{Q} = \{H\}$.)

In order to define Δ, we exploit the fact that any non-legitimate configuration can be decomposed into:

- "bonds", i.e. strings of two letters in $\overleftarrow{Q}\,\overrightarrow{Q}$.
- "anti-bonds", that are, roughly speaking, strings of two letters in $\overrightarrow{Q}\,\overleftarrow{Q}$.
- letters belonging to neither a bond nor an anti-bond. (These letters are in $\{W, S, D\}^*$, since every H belongs to a bond or an anti-bond; see Prop. 4 below.)

4.2. Bonds

A *bond* in a configuration x is a substring of x made of two consecutive letters in $\overleftarrow{Q}\,\overrightarrow{Q}$. The *index* of a bond $\overleftarrow{\alpha}\,\overrightarrow{\beta}$ is the position of its first letter $\overleftarrow{\alpha}$. Note that, due to letter H, two bonds may overlap: for example, in expression $\overleftarrow{W}H\overrightarrow{S}$ there are two overlapping bonds $\overleftarrow{W}H$ and $H\overrightarrow{S}$. In the following, given a configuration x, we focus on a sequence π of indices of *disjoint* bonds of x, i.e., such that $i + 2 \leq j$, for all consecutive indices i and j of π. We suppose also that π is *maximal*, i.e., such that between two consecutive indices $i, j \in \pi$ there is no bond of index k with $i + 2 \leq k \leq j - 2$. A maximal sequence π of indices of disjoint bonds of x is called a ***bond list*** of x. Note that such a list is not unique. *Bond*(π) is defined as the set of letters of x at position ℓ such that $\ell = i$ or $\ell = i + 1$ for some $i \in \pi$.

Example: For the configuration $\overleftarrow{W}\overrightarrow{W}\overleftarrow{S}\overleftarrow{W}H\overrightarrow{D}$, there are two possible bond lists $\pi_1 = \{1, 4\}$ and $\pi_2 = \{1, 5\}$. Bonds are $\overleftarrow{W}\overrightarrow{W}$ and $\overleftarrow{W}H$ for π_1, and $\overleftarrow{W}\overrightarrow{W}$ and $H\overrightarrow{D}$ for π_2.

Henceforth, every non-legitimate configuration x will be provided with a bond list π. The bond list π_0 of the initial configuration x_0 is arbitrary. Given a configuration x, a bond list π of x, and a rewriting of x into x' via some rule of $\mathcal{R}'$, the bond list π' associated with x' is constructed from π as follows:

- If the rewriting changes a $H \in$ *Bond*(π) via probabilistic rule R0, we apply the following strategy:
  - if H is the first letter of a bond of π, then H is changed into $\overleftarrow{W}$,
  - if H is the second letter of a bond of π, then H is changed into $\overrightarrow{W}$.

Bonds of π are thus preserved, and we let: $\pi' = \pi$.

- If the rewriting creates a new bond of index k disjoint from every bond of π, then π' is $\pi \cup \{k\}$.
- In all other cases, we let $\pi' = \pi$.

Let $\Delta_1(x, \pi)$ be N minus the number of elements of π. The *bond coefficient* is 3 for D, 2 for H, 1 for W and 0 for S. The *weight of a bond* $\overleftarrow{\alpha}\,\overrightarrow{\beta}$ is the sum of the bond coefficients of $\overleftarrow{\alpha}$ and $\overrightarrow{\beta}$: for example, the weight of bond HH is 4. Then $\Delta_2(x, \pi)$ is the sum of the weights of all the bonds of x indexed by π.

Example: In the configuration $\overleftarrow{W}\overrightarrow{W}\overleftarrow{S}\overleftarrow{W}H\overrightarrow{D}$ of the previous example, we have $\Delta_1 = 4$, $\Delta_2 = 5$ for $\pi_1 = \{1, 4\}$, and $\Delta_1 = 4$, $\Delta_2 = 7$ for $\pi_2 = \{1, 5\}$.

$\Delta_3(x)$ is defined as the number of two-letter strings of the form $\overleftarrow{D}H$, $\overleftarrow{D}\overleftarrow{W}$, $H\overrightarrow{D}$ or $\overrightarrow{W}\overrightarrow{D}$ of x.

4.3. Anti-bonds

Given a configuration x and a bond list π of x, an ***anti-bond*** is a substring of two letters of x of one of the three forms:

1. $\overrightarrow{\gamma}\overleftarrow{\delta}$ with $\overrightarrow{\gamma} \notin Bond(\pi)$, $\overleftarrow{\delta} \notin Bond(\pi)$ and $\overrightarrow{\gamma} \in \overrightarrow{Q}$, $\overleftarrow{\delta} \in \overleftarrow{Q}$.
2. $\overrightarrow{\gamma}\overleftarrow{\alpha}$ with $\overrightarrow{\gamma} \notin Bond(\pi)$, $\overrightarrow{\gamma} \in \overrightarrow{Q}$ and $\overleftarrow{\alpha}$ 1st letter of a bond of π,
3. $\overrightarrow{\beta}\overleftarrow{\delta}$ with $\overleftarrow{\delta} \notin Bond(\pi)$, $\overleftarrow{\delta} \in \overleftarrow{Q}$ and $\overrightarrow{\beta}$ 2nd letter of a bond of π.

The *index of an anti-bond* of x is the position of its first letter ($\overrightarrow{\gamma}$ in cases 1-2, $\overrightarrow{\beta}$ in case 3).

Consider two bonds $\overleftarrow{\alpha}\overrightarrow{\beta}$ and $\overleftarrow{\alpha}'\overrightarrow{\beta}'$ indexed by consecutive index i and i' of π. Then either:

- $\overleftarrow{\alpha}\overrightarrow{\beta}$ and $\overleftarrow{\alpha}'\overrightarrow{\beta}'$ are contiguous (i.e: $i' = i + 2$) and there is no anti-bond between them (i.e: no anti-bond indexed by j with $i + 1 \le j \le i' - 1$), or
- $\overleftarrow{\alpha}\overrightarrow{\beta}$ and $\overleftarrow{\alpha}'\overrightarrow{\beta}'$ are not contiguous (i.e: $i' \ge i + 3$).

In the latter case, it is easy to see that, between $\overleftarrow{\alpha}\overrightarrow{\beta}$ and $\overleftarrow{\alpha}'\overrightarrow{\beta}'$, there is no substring of the form $\cdots\overleftarrow{\gamma}\cdots\overrightarrow{\delta}\cdots$ with $\overleftarrow{\gamma} \in \overleftarrow{Q}$ and $\overrightarrow{\delta} \in \overrightarrow{Q}$. (Otherwise, there would be a disjoint bond between $\overleftarrow{\alpha}\overrightarrow{\beta}$ and $\overleftarrow{\alpha}'\overrightarrow{\beta}'$, and π would not be maximal.) Hence the substring between $\overleftarrow{\alpha}\overrightarrow{\beta}$ and $\overleftarrow{\alpha}'\overrightarrow{\beta}'$ is of the form $\overrightarrow{Q}^*\overleftarrow{Q}^*$ with either no H (case H0) or just one H (case H1). More precisely, the substring delimited by the two bonds is of the form:

- H0: $\overleftarrow{\alpha}\overrightarrow{\beta}\overrightarrow{I}\overleftarrow{I}\overleftarrow{\alpha}'\overrightarrow{\beta}'$, or
- H1: $\overleftarrow{\alpha}\overrightarrow{\beta}\overrightarrow{I}H\overleftarrow{I}\overleftarrow{\alpha}'\overrightarrow{\beta}'$,

with $\overrightarrow{I} \in \{\overrightarrow{W}, \overrightarrow{S}, \overrightarrow{D}\}^*$ and $\overleftarrow{I} \in \{\overleftarrow{W}, \overleftarrow{S}, \overleftarrow{D}\}^*$. Let $\overrightarrow{\lambda}$ and $\overleftarrow{\mu}$ be the last letter of $\overrightarrow{\beta}\overrightarrow{I}$ and the first letter of $\overleftarrow{I}\overleftarrow{\alpha}'$ respectively. Between these bonds, there is:

- H0: a single anti-bond, viz: $\overrightarrow{\lambda}\overleftarrow{\mu}$, or
- H1: two overlapping anti-bonds, viz: $\overrightarrow{\lambda}H$ and $H\overleftarrow{\mu}$.

Given π, we construct an ***anti-bond list*** ψ of x as a set of indices obtained by putting, for every couple of non-contiguous consecutive bonds indexed by π:

- the index of $\overrightarrow{\lambda}\overleftarrow{\mu}$ in case H0,
- the index of either $\overrightarrow{\lambda}H$ or $H\overleftarrow{\mu}$ in case H1.

Given a configuration x and a bond list π of x, an anti-bond list ψ of x, is thus a maximal set of indices j of anti-bonds of (x, π). More precisely:

Proposition 3 *Given a configuration x and a bond list π of x, an anti-bond list ψ of x, is such that, for any couple of consecutive indices $i, i' \in \pi$, either:*

- the bonds indexed by i and i' are contiguous ($i' = i + 2$), in which case no anti-bond of ψ lies between them (i.e., no $j \in \psi$ such that $i + 1 \le j \le i' - 1$), or

- they are not contiguous ($i' \ge i + 3$), in which case exactly one anti-bond indexed by ψ lies between them (i.e., $\exists!\ j \in \psi :\ i + 1 \le j \le i' - 1$).

Note that, in any case, the occurrence of H (if any) between $\overleftarrow{\alpha}\overrightarrow{\beta}$ and $\overleftarrow{\alpha}'\overrightarrow{\beta}'$ always belongs to an anti-bond indexed by ψ. Formally:

Proposition 4 *For a given configuration $x \notin \mathcal{L}'$, a bond list π of x and an anti-bond list ψ, every H of x belongs to a bond of π or an anti-bond of ψ.*

Example: For the configuration $\overrightarrow{W}H\overleftarrow{S}\,\overrightarrow{D}\overrightarrow{W}\overleftarrow{W}\overleftarrow{D}$, we have one bond list $\pi = \{3,7\}$ and two possible anti-bond lists $\psi_1 = \{1,5\}$ and $\psi_2 = \{2,5\}$. Anti-bonds are $\overrightarrow{W}H$ and $\overrightarrow{W}\overleftarrow{W}$ for ψ_1, $H\overleftarrow{S}$ and $\overrightarrow{W}\overleftarrow{W}$ for ψ_2.

Henceforth, every configuration x coupled with a bond list π, will be provided with an anti-bond list ψ. The anti-bond list ψ_0 associated with the initial couple (x_0, π_0) is arbitrary. Given a couple (x, π) and an associated anti-bond list ψ, the rewriting of x into x' via probabilistic rule R0 preserves π when a bond is rewritten, using the strategy described in section 4.2. Rewriting via R0 also preserves ψ using the following strategy:

- if H is the 1st letter of an anti-bond of ψ, then H is changed into $\overrightarrow{W}$;
- if H is the 2nd letter of an anti-bond of ψ, then H is changed into $\overleftarrow{W}$.

It is easy to see that this strategy is compatible with the one for bonds: if H is shared by a bond of π and an anti-bond of ψ, both strategies agree for rewriting H either into $\overleftarrow{W}$ or $\overrightarrow{W}$. For example, if H is in $\overleftarrow{S}\mathbf{H}\overleftarrow{S}$, the expression rewrites to $\overleftarrow{S}\overrightarrow{\mathbf{W}}\overleftarrow{S}$. The rewriting of x into x' via the other rules transforms π into π' as explained in section 4.2, and ψ into ψ' where $\psi' = \psi$ except in some cases where D is replaced by H via R9 or R10 (see [5], for details).

We say that an anti-bond is *oriented leftwards* (resp. *oriented rightwards*) if it is of the form $\{\overrightarrow{S},\overrightarrow{D}\}\{\overleftarrow{W},H\}$ (resp. $\{\overrightarrow{W},H\}\{\overleftarrow{S},\overleftarrow{D}\}$).

Given a bond list π and an anti-bond A of index k oriented leftwards (resp. rightwards), the *π-distance* of A is $k - i$ (resp. $i - k$) where i is the index of the closest bond of π to the left (resp. right) of A.

Let $\Delta_4(x, \psi)$ be N minus the number of oriented anti-bonds of x indexed by ψ and $\Delta_5(x, \pi, \psi)$ be the sum of π-distances of the oriented anti-bonds of ψ.

The *anti-bond coefficient* is 3 for H, 2 for W, 1 for S and 0 for D. The *weight of an anti-bond* $\overrightarrow{\alpha}\overleftarrow{\beta}$ is the sum of the anti-bond coefficients of $\overrightarrow{\alpha}$ and $\overleftarrow{\beta}$: for example, the weight of $H\overleftarrow{W}$ is 5.

Let $\Delta_6(x, \psi)$ be the sum of the weights of the anti-bonds of x indexed by ψ.

Example: In the configuration $\overrightarrow{W}H\overleftarrow{S}\,\overrightarrow{D}\overrightarrow{W}\overleftarrow{W}\overleftarrow{D}$ of the previous example, we have $\Delta_4 = 7 (= N)$, $\Delta_5 = 0$, $\Delta_6 = 9$ for $\psi_1 = \{1,5\}$, and $\Delta_4 = 6$, $\Delta_5 = 1$, $\Delta_6 = 8$ for $\psi_2 = \{2,5\}$. In ψ_1 no anti-bond is oriented. In ψ_2, the anti-bond $H\overleftarrow{S}$ is oriented rightwards. Its π-distance with token $\overleftarrow{S}\overrightarrow{D}$ is 1.

4.4. Measure Δ

The *WSD-coefficient* is 2 for W, 1 for S and 0 for D. Let $\Delta_7(x)$ be the sum of the WSD-coefficients of all the letters of x distinct from H.

Given a configuration $x \notin \mathcal{L}'$, a bond list π and an anti-bond list ψ, measure Δ is defined as 7-tuple $(\Delta_1, \Delta_2, \Delta_3, \Delta_4, \Delta_5, \Delta_6, \Delta_7)$. The evolution of Δ along a computation is illustrated on an example in the first appendix.

By Proposition 4, any configuration can be decomposed into bonds, anti-bonds and W, S, D-letters. Using this fact, it can be shown by case analysis that, under the strategies defined above for RO, Δ decreases whenever x rewrites to x' for lists π' and ψ' constructed from π and ψ as described in sections 4.2 and 4.3. (The full proof is given in [5].) Formally, let $\ll$ be the lexicographic extension of $<$; we have:

Proposition 5 *For every $x \notin \mathcal{L}'$, every bond list π and anti-bond list ψ of x, and every position i in $\mathcal{E}(x)$, there exists a configuration x', a bond list π' and an anti-bond list ψ' of x' such that:*

$$x \xrightarrow[\mathcal{R}']{i} x' \wedge (x' \in \mathcal{L}' \ \vee \ \Delta(x', \pi', \psi') \ll \Delta(x, \pi, \psi)).$$

Theorem 1 then follows from proposition 5 and Theorem 2.

5. Expected Time of Convergence

In traditional approaches (see e.g. [9]) the time is measured in terms of rounds (intervals in which each process has been scheduled at least once). The time is never evaluated as a number of transitions.

With our approach, we do not make any assumption on this round time. We evaluate the expected time of convergence as a number of transitions. It turns out that, for some "malicious" scheduler such a time can be "very" long. In 2nd appendix, we show indeed that, for some configuration x_0 (of the form $\overrightarrow{S}\,\overrightarrow{S} \cdots \overrightarrow{S}$) and some scheduler $\mathcal{A}_0$, one can stay out of $\mathcal{L}'$ during an expected time at least exponential in N. This gives us an exponential lower bound for the expected time of convergence. In other words, the expected number of transitions in a round can be exponential.

6. Conclusion

We have shown in this paper that a modified version of Lehmann-Rabin's algorithm always converges, whatever the (possibly unfair) scheduler is. We claim that our modified algorithm preserves the spirit of the original one.

With our approach, we are also able to evaluate the expected time of convergence as a number of transitions. We have shown that this time may be exponential for some malicious scheduler.

An interesting further work would be to remove the fairness assumption with more complex connection topologies of dining philosophers as, e.g., in [12].

References

[1] T. Bonald and L. Massouli. Impact of fairness on Internet performance. In *Proc. of Joint International Conference on Measurements and Modeling of Computer Systems (SIGMETRICS/Performance 2001)*, Cambridge, USA, 2001, pp. 82–91.

[2] J. Beauquier, J. Durand-Lose, M. Gradinariu, and C. Johnen. Token based self-stabilizing uniform algorithms. *J. of Parallel and Distributed Systems*, To appear.

[3] S. Dolev, A. Israeli, and S. Moran. Analyzing expected time by scheduler-luck games. *IEEE Transactions on Software Engineering*, 21(5), 1995, pp. 429–439.

[4] M. Duflot, L. Fribourg, and C. Picaronny. Randomized distributed algorithms as Markov chains. In *Proc. 15th Int. Conf. on Distributed Computing (DISC'01)*, Lisbon, 2001, pp. 240–254. (Extended version available on `http://www.lsv.ens-cachan.fr/Publis/`).

[5] M. Duflot, L. Fribourg, and C. Picaronny. Randomized dining philosophers without fairness assumption. Research Report LSV-01-11, Cachan, France, December 2001. Available on `http://www.lsv.ens-cachan.fr/Publis/`

[6] P. Gevros, F. Risso, and P. Kirstein. Analysis of a Method for Differential TCP Service. In *Proc. of the IEEE GLOBECOM'99*, Rio de Janeiro, 1999.

[7] H. Kakugawa and M. Yamashita. Uniform and Self-Stabilizing Token Rings Allowing Unfair Daemon. *IEEE Trans. Parallel and Distributed Systems*, 8(2), 1997, pp. 154–163.

[8] D. Lehmann and M.O. Rabin. The advantages of free choice: a symmetric and fully-distributed solution to the dining philosophers problem. In *Proc. 8th Annual ACM Symposium on Principles of Programming Languages(POPL'81)*, Williamsburg, U.S.A., 1981, pp. 133–138.

[9] N.A. Lynch. *Distributed Algorithms.* Morgan Kaufmann Publishers, Inc., 1996.

[10] N.A. Lynch, I. Saias, and R. Segala. Proving time bounds for randomized distributed algorithms. In *Proc. 13th Annual ACM Symposium on Principles of Distributed Computing (PODC'94)*, Los Angeles, USA, 1994, pp. 314–323.

[11] A.K. McIver. Quantitative program logic and efficiency in probabilistic distributed algorithms. Tech. report, Computing Lab., Oxford U., UK, 1998. (Extended version of "Quantitative program logic and performance", Proc. of 5th Int AMAST Workshop, ARTS '99).

[12] O.M. Herescu and C. Palamidessi. On the Generalized Dining Philosophers Problem. In Proc. 20th Annual ACM Symposium on Principles of Distributed Computing (PODC'01), Newport, USA, 2001, pp. 81–89.

[13] A. Pnueli and L. Zuck. Verification of multiprocess probabilistic protocols. *Distributed Computing*, 1(1), 1986, pp. 53–72.

[14] A. Pogosyants and R. Segala. Formal verification of timed properties for randomized distributed algorithms. In *Proc. 14th Annual ACM Symposium on Principles of Distributed Computing (PODC'95)*, Ontario, Canada, 1995, pp. 174–183.

Appendix: An example of computation

The general behaviour of bonds should be clear: with our strategy, once a bond is created, it never disappears (Δ_1 never increases) and stays in fixed position. On the example below, the two leftmost and two rightmost letters of each configuration are the bonds of π (π is here the only possible bond list).

Let us explain the general evolution of an anti-bond A located between two consecutive (non contiguous) bonds B_1 and B_2. Once oriented, say rightwards, A moves towards B_2 until it overlaps with it. It may then lose its orientation and become oriented later in the other direction. Such a U-turn is possible only if the left letter of B_2 is rewritten. A thus oscillates between B_1 and B_2. The total number of U-turns is finite as each bond coefficient can decrease at most 3 times. For example the first letter of a bond can change from $\overleftarrow{D}$ to H then to $\overleftarrow{W}$ and to $\overleftarrow{S}$. Once it is $\overleftarrow{S}$, if this process is selected by the scheduler and as the right fork is not taken (the right neighbour is in state $\overrightarrow{D}$, H, $\overrightarrow{W}$ or $\overrightarrow{S}$) the philosopher enters state E and $\mathcal{L}$ is reached. A computation with Δ_1 and Δ_3 constant illustrated on Figure A.1.

$$\overleftarrow{W}\underset{\Delta_5=3}{\underline{\overrightarrow{W}\overleftarrow{\mathbf{D}}}}\overleftarrow{D}\overleftarrow{D}\overrightarrow{S} \xrightarrow{\Delta_5} \overleftarrow{W}\overrightarrow{W}\underset{\Delta_5=2}{\underline{\mathbf{H}\overleftarrow{D}}}\overleftarrow{D}\overrightarrow{S} \xrightarrow{\Delta_6} \overleftarrow{W}\overrightarrow{W}\underset{\Delta_5=2}{\underline{\overrightarrow{W}\overleftarrow{\mathbf{D}}}}\overleftarrow{D}\overrightarrow{S} \xrightarrow{\Delta_5} \overleftarrow{W}\overrightarrow{W}\overrightarrow{W}\underset{\Delta_5=1}{\underline{H\overleftarrow{\mathbf{D}}}}\overrightarrow{S}$$

$$\xrightarrow{\Delta_2} \overleftarrow{W}\overrightarrow{W}\overrightarrow{W}\underset{\Delta_5=0}{\underline{H\mathbf{H}}}\overrightarrow{S} \xrightarrow{\Delta_2} \overleftarrow{W}\overrightarrow{W}\overrightarrow{W}\underset{\Delta_5=0}{\underline{\mathbf{H}\overleftarrow{W}}}\overrightarrow{S} \xrightarrow{\Delta_6} \overleftarrow{W}\overrightarrow{W}\overrightarrow{W}\underset{\Delta_5=0}{\underline{\overrightarrow{\mathbf{W}}\overleftarrow{W}}}\overrightarrow{S} \xrightarrow{\Delta_4} \overleftarrow{W}\overrightarrow{W}\overrightarrow{\mathbf{W}}\underset{\Delta_5=3}{\underline{\overrightarrow{S}\overleftarrow{W}}}\overrightarrow{S}$$

$$\xrightarrow{\Delta_7} \overleftarrow{W}\overrightarrow{W}\overrightarrow{S}\underset{\Delta_5=3}{\underline{\overrightarrow{\mathbf{S}}\overleftarrow{W}}}\overrightarrow{S} \xrightarrow{\Delta_6} \overleftarrow{W}\overrightarrow{W}\overrightarrow{S}\underset{\Delta_5=3}{\underline{\overrightarrow{\mathbf{D}}\overleftarrow{W}}}\overrightarrow{S} \xrightarrow{\Delta_5} \overleftarrow{W}\overrightarrow{\mathbf{W}}\underset{\Delta_5=2}{\underline{\overrightarrow{S}H}}\overleftarrow{W}\overrightarrow{S} \xrightarrow{\Delta_2} \overleftarrow{W}\overrightarrow{S}\underset{\Delta_5=2}{\underline{\overrightarrow{\mathbf{S}}H}}\overleftarrow{W}\overrightarrow{S}$$

$$\xrightarrow{\Delta_6} \overleftarrow{W}\overrightarrow{S}\underset{\Delta_5=2}{\underline{\overrightarrow{D}\mathbf{H}}}\overleftarrow{W}\overrightarrow{S} \xrightarrow{\Delta_6} \overleftarrow{W}\overrightarrow{S}\underset{\Delta_5=2}{\underline{\overrightarrow{\mathbf{D}}\overleftarrow{W}}}\overleftarrow{W}\overrightarrow{S} \xrightarrow{\Delta_5} \overleftarrow{W}\underset{\Delta_5=1}{\underline{\overrightarrow{S}\mathbf{H}}}\overleftarrow{W}\overleftarrow{W}\overrightarrow{S} \xrightarrow{\Delta_6} \overleftarrow{W}\underset{\Delta_5=1}{\underline{\overrightarrow{S}\overleftarrow{W}}}\overleftarrow{W}\overleftarrow{W}\overrightarrow{S}$$

Figure A.1. A computation: each configuration has a single anti-bond, which is underlined and labelled with the value of the Δ_5-distance; the bold letter is the letter to be changed; the arrow depicting each transition is labelled with the first decreasing Δ-component.

Each transition is represented by an arrow labelled with the first decreasing component of Δ. For each configuration, we also give the corresponding measure Δ_5. As there is just one anti-bond A, $\Delta_5(x, \pi, \psi)$ is either the π-distance of A when it is oriented, or 0 otherwise. Note that, in the last configuration, the anti-bond $\overrightarrow{S}\overleftarrow{W}$ cannot be rewritten without reaching $\mathcal{L}'$.

Appendix: Expected Time of Convergence

Let us show on a example that the expected time of convergence is at least exponential in the number N of processes for some "malicious" scheduler. We exhibit a scheduler which, starting from the uniform configuration $x_0 = \overrightarrow{S}^N$, goes to $x_1 = \overrightarrow{S}^{N-2}\overleftarrow{W}\overleftarrow{S}$ within a constant expected time. Then, from x_1, it can stay in the set of configurations $\{\overrightarrow{S}^i\overleftarrow{W}\overleftarrow{S}^j | i+j+1 = N\} \cup \{\overrightarrow{S}^i\overrightarrow{W}\overleftarrow{S}^j | i+j+1 = N\}$ during an expected time exponential in N.

Consider $x_0 = \overrightarrow{S}^N$ as a starting configuration. Let us first show that the scheduler may reach the configuration $x_1 = \overrightarrow{S}^{N-2}\overleftarrow{W}\overleftarrow{S}$ in a finite amount of time: It selects a $\overrightarrow{S}$ and applies R8.R10.R0 to obtain $\overrightarrow{S}^{N-1}\overrightarrow{W}$ or $\overrightarrow{S}^{N-1}\overleftarrow{W}$. In the first case, it applies R3 and goes on. In the second case, it selects the last $\overrightarrow{S}$ and applies R8.R10.R0 to obtain $\overrightarrow{S}^{N-2}\overrightarrow{W}\overleftarrow{W}$ or $\overrightarrow{S}^{N-2}\overleftarrow{W}\overleftarrow{W}$. In the first case, it applies R3 to $\overrightarrow{W}$ and begins again. In the second case, it applies R2 to the right $\overleftarrow{W}$. The expected time E of going from x_0 to x_1 may be easily computed: $E = 15$.

Now, we describe two possible choices of the scheduler on a configuration of the form $\overrightarrow{S}^i\overleftarrow{W}\overleftarrow{S}^j$ with $i \geq 2$ and $i+j+1 = N$. These two choices are the application of four consecutive rules, one of which is a probabilistic one. The three first rules are the same: select the rightmost $\overrightarrow{S}$ and applies R8.R10.R0. This yields: $\overrightarrow{S}^{i-1}\overleftarrow{W}\overleftarrow{W}\overleftarrow{S}^j$ or $\overrightarrow{S}^{i-1}\overrightarrow{W}\overleftarrow{W}\overleftarrow{S}^j$.
From $\overrightarrow{S}^{i-1}\overleftarrow{W}\overleftarrow{W}\overleftarrow{S}^j$, apply R1 to the rightmost $\overleftarrow{W}$ to get $\overrightarrow{S}^{i-1}\overleftarrow{W}\overleftarrow{S}^{j+1}$.
From $\overrightarrow{S}^{i-1}\overrightarrow{W}\overleftarrow{W}\overleftarrow{S}^j$, the scheduler can choose $\overrightarrow{W}$ or $\overleftarrow{W}$, which leads to the two cases:
(A) The scheduler applies R1 to $\overleftarrow{W}$ and yields $\overrightarrow{S}^{i-1}\overrightarrow{W}\overleftarrow{S}^{j+1}$,
(B) The scheduler applies R3 to $\overrightarrow{W}$ and yields $\overrightarrow{S}^i\overleftarrow{W}\overleftarrow{S}^j$.

Symmetrically, using R6.R9.R0, on a configuration $\overrightarrow{S}^i\overrightarrow{W}\overleftarrow{S}^j$ (with $j \geq 2$ and $i+j+1=N$), one goes to $\overrightarrow{S}^i\overrightarrow{W}\overleftarrow{W}\overleftarrow{S}^{j-1}$ or $\overrightarrow{S}^i\overrightarrow{W}\overrightarrow{W}\overrightarrow{S}^{j-1}$.
From $\overrightarrow{S}^i\overrightarrow{W}\overrightarrow{W}\overrightarrow{S}^{j-1}$, apply R3 to the leftmost $\overrightarrow{W}$, to get $\overrightarrow{S}^{i+1}\overrightarrow{W}\overleftarrow{S}^{j-1}$.
From $\overrightarrow{S}^i\overrightarrow{W}\overleftarrow{W}\overleftarrow{S}^{j-1}$, the scheduler can choose $\overrightarrow{W}$ or $\overleftarrow{W}$, which leads to the two cases:
(A') The scheduler applies R3 to $\overrightarrow{W}$ and yields $\overrightarrow{S}^{i+1}\overleftarrow{W}\overleftarrow{S}^{j-1}$.
(B') The scheduler applies R1 to $\overleftarrow{W}$ and yields $\overrightarrow{S}^i\overrightarrow{W}\overleftarrow{S}^j$.

The scheduler can iterate such application of four consecutive rules until the system reaches an "end" configuration, i.e, a configuration of the form:
$x_{end} = \overrightarrow{S}^{N-2}\overrightarrow{W}\overleftarrow{S}$ or $y_{end} = \overrightarrow{S}\overleftarrow{W}\overleftarrow{S}^{N-2}$.

Let us consider the following "malicious" scheduler: From $\overrightarrow{S}^i\overleftarrow{W}\overleftarrow{S}^j$ (with $i \geq 2$ and $i+j+1=N$), it chooses (A) or (B) according to the compared values of i and j. Precisely: it chooses (A) if $j \geq i$, and (B) if $j < i$.
Symmetrically, from $\overrightarrow{S}^i\overrightarrow{W}\overleftarrow{S}^j$ (with $j \geq 2$ and $i+j+1=N$): it chooses (A') if $i \geq j$, and (B') if $i < j$.
For this scheduler, let us now compute the expected time for reaching x_{end} or y_{end}, starting from a configuration $\overrightarrow{S}^i\overrightarrow{W}\overleftarrow{S}^j$ (resp. $\overrightarrow{S}^i\overleftarrow{W}\overleftarrow{S}^j$), with $j \geq 2$ (resp. $i \geq 2$) and $i+j+1=N$. Let us abbreviate this expected time as $E[\overrightarrow{i}]$ (resp. $E[\overleftarrow{i}]$). We have:

$E[\overrightarrow{i}] = 4 + 1/2\ E[\overrightarrow{i+1}] + 1/2\ E[\overleftarrow{i+1}]$, for $1 \leq i$ and $2i \geq N-2$.
$E[\overrightarrow{i}] = 4 + 1/2\ E[\overrightarrow{i+1}] + 1/2\ E[\overrightarrow{i}]$, for $1 \leq i$ and $2i < N-2$.
$E[\overleftarrow{i}] = 4 + 1/2\ E[\overrightarrow{i-1}] + 1/2\ E[\overleftarrow{i-1}]$, for $2 \leq i$ and $2i \leq N-2$.
$E[\overleftarrow{i}] = 4 + 1/2\ E[\overleftarrow{i-1}] + 1/2\ E[\overleftarrow{i}]$, for $2 \leq i$ and $2i > N-2$.
$E[\overrightarrow{N-2}] = 0$.
$E[\overleftarrow{1}] = 0$.

We then solve this linear system. The symmetry shows that $E[\overrightarrow{i}] = E[\overrightarrow{N-1-i}]$, for all $1 \leq i < N-1$. Let m be the integral part of $(N-1)/2$. The result is:
$E[\overrightarrow{i}] = 8(3 \times 2^{m-1} + 5m - 6 - i)$, for $1 \leq i < m$.
$E[\overrightarrow{i}] = 8(3(2^{m-1} - 2^{i-m}) - 2m + i + 1))$ for $m \leq i < N-1$.

In particular, the time to go from x_1 to x_{end} (hence x_0 to x_{end}) is $E[\overleftarrow{N-2}] = E[\overrightarrow{1}] = 8(3 \times 2^{m-1} + 5m - 7) \simeq 3 \times 2^{m+2}$.
This shows that we can stay out of $\mathcal{L}'$ an exponential expected number of steps, hence the upper bound for the expected time of convergence for a general scheduler is at least exponential.

GUARDING GALLERIES AND TERRAINS

Alon Efrat
Department of Computer Science, University of Arizona
WWW: `http://www.cs.arizona.edu/people/alon`

Sariel Har-Peled
Department of Computer Science, University of Illinois
WWW: `http://www.cs.uiuc.edu/contacts/faculty/harpeled.html`.

Abstract Let P be a simple polygon with n vertices. We say that two points of P see each other if the line segment connecting them lies inside (the closure of) P. In this paper we present efficient approximation algorithms for finding the smallest set S of points of P so that each point of P is seen by at least one point of S. We also present similar algorithms for terrains and polygons with holes.

1. Introduction

The *art gallery problem* [O'R87] is stated as follows: Given a polygon P (the *gallery*), find a smallest set G of points (*guards*) inside P, such that each point in P is seen by at least one of the guards. This problem has been studied extensively in recent years, see, e.g., [O'R83, Agg84, Gho87, Hof90, HKK91, JL93, BS93, BG95], and the recent survey paper by Urrutia [Urr00].

This problem is known to be NP-hard even when P is simple [OS83], and even finding an $(1+\varepsilon)$-approximation (that is, finding a set of guards whose cardinality is at most $1+\varepsilon$ times the optimum) is NP-hard [Eid00]. Ghosh [Gho87] presented a (multiplicative) $O(\log n)$-approximation algorithm that runs in $O(n^5 \log n)$ time, for the case where guards located on vertices (as well as of other types of visibility). Recently, Gonzalez-Banos and Latombe [GBL01] presented an algorithm for a rather restricted version of the art-galley problem, and with much larger set of guards.

Our contribution We present an algorithm for finding in time $O(nc_{opt}^2 \log^4 n)$ a set of vertices that sees P, and its cardinality is within a factor of $O(\log c_{opt})$ from the optimum.

If one allows guards to be placed arbitrarily (not only on vertices), the problem seems to be considerably harder. We present in Section 4 an exact algorithm for this problem, that runs in $O((nc_{opt})^{3(2c_{opt}+1)})$ time. To the best of our knowledge, this is the first exact solution to the problem. The proof follows from recent results in algorithmic real algebraic geometry. Thus if the optimum number of guards is a constant then we obtain a polynomial algorithm.

In Section 5 we present an efficient implementation of our approximating algorithm, for the case where the guards are restricted to lie on an arbitrarily dense grid. For this case, we get an $O(\log c_{opt})$-approximation in $O(nc_{opt}^2 \log n \log(nc_{opt}) \log^2 \Delta)$ time, where Δ is the ratio between the diameter of the polygon and the grid size, and c_{opt} is the cardinality of the smallest set of grid-points that sees P. Note that the running time depends on Δ only logarithmically, which implies that we can choose a rather fine grid without paying too large a penalty, so the resulting set of grid-points is likely to cover all of P.

The new algorithms can be extended to handle polygons with h holes, as their VC-dimension is $O(\log h)$ [Val98], yielding an approximation factor of $O(\log h \log(c_{opt} \log h))$. We also show how to solve related problems on terrains: Given a terrain T, find a small set of vertices that see every point of T. This problem has numerous applications in Geographic Information Science (GIS). Our approximation algorithm can be modified for this setting, yields an $O(\log n \log(c_{opt} \log n)) = O(\log n \log\log n)$ approximation factor. Analogous to the case of a simple polygon, these extensions can be modified to find a set of guards that see the whole polygon or terrain, respectively, where the guards are taken from the set of vertices of an arbitrary dense grid. These extensions are described in Section 6.

Our efficient algorithms are the result of obtaining data structures for carefully counting and maintaining the weights of sets of grid points, as described below.

2. Preliminaries

For a point $q \in P$, the *visibility polygon* of q in P, denoted by $Vis(q)$, is the region of all the points of P that q sees. The following observation appeared in [GMMN90].

Observation 1. Let q be any point in P, and let $s \subseteq P$ be a segment. If P is a simple polygon, then the intersection between s and $Vis(q)$ is a (possibly empty) segment.

Lemma 2. Let $G = \{g_1 \dots g_k\}$ be a set of k points in P, and let Vis_i denote the visibility polygon of g_i, for $i = 1, \dots, k$. Let $Vis(G) = \cup_{q \in G} Vis(q)$. Then the complexity of the arrangement $\mathcal{A} = \mathcal{A}(G)$ formed by $\partial Vis_1, \dots, \partial Vis_k$ is $O(nk^2)$. Furthermore, the complexity of the zone of ∂Vis_i in $\mathcal{A}$ is $O(nk\alpha(k))$, for $i = 1, \dots, k$. Here $\alpha(n)$ is the inverse Ackermann function, and is an extremely slowly growing function.

Theorem 1 ([GMMN90]). *$Vis(G)$ is bounded by $O(nk + k^2)$ edges, and this bound is tight in the worst case.*

Efficient construction of *Vis*(G)**.** The bound of Theorem 1 yields the following simple but efficient divide-and-conquer algorithm for constructing *Vis*(G).

If $|G| = 1$, one can construct *Vis*(G), the visibility polygon from a single point, in $O(n)$ time [EA81]. Otherwise, divide G into two subsets G_1, G_2 of roughly $k/2$ guards each. Compute recursively the visibility polygons *Vis*(G_1) and *Vis*(G_2), and merge them, using a standard line-sweeping procedure [dB-vKOS00] to obtain *Vis*(G). It is easy to see that the running time of this procedure is $O(nk \log k \log n)$.

3. Finding a Small Set of Vertices that Sees P

Let V be the set of n vertices of P. For a point $q \in P$, let $V_q = V \cap Vis(q)$ denote the set of vertices of P that q sees. Let $\mathcal{X} = (V, \mathcal{V})$ be the range space defined by the visibility inside P, where $\mathcal{V} = \left\{ \mathcal{V}_q \,\middle|\, q \in P \right\}$. Valtr [Val98] showed that 23 is a upper bound on the VC-dimension of the more general space $\mathcal{Y} = \left(P, \left\{ Vis(q) \,\middle|\, q \in P \right\} \right)$.

Finding a set of guards on the vertices of P that sees all of P, is equivalent to finding a subset U of the vertices of P that hit all the ranges of $\mathcal{V}$. That is $\forall X \in \mathcal{V}, X \cap U \neq \emptyset$. However, since the VC-dimension of $\mathcal{X}$ is bounded, we can use the property that this space has a small ε-net to get an efficient approximation algorithm (see [Cla93, BG95]). We describe next an efficient implementation of this general method for the case of computing a guarding set, i.e. a set of points that sees P. Assume that we have a guess k of the value of c_{opt}. We initialize the value of k to one. We now call the procedure `ComputeGuards`(P, k), depicted in Figure 1, repeatedly. The procedure `ComputeGuards`(P, k) tries to compute a guarding set of P with $O(k \log k)$ guards. If such a call fails, we know that with high probability, our guess of the number of guards needed to guard P (i.e., k) is too small. Thus, we double its value and iterate. Overall, we would perform $O(\log c_{opt})$ calls to `ComputeGuards`. The correctness of this algorithm, and the values of the constants in the big-O — notations follow from the analysis of Clarkson [Cla93] (see also [BG95], and a slightly different presentation in [EHKKRW02]).

We implement `ComputeGuards` using the algorithm of Section 2 to compute the union, and to pick a point outside it, in $O(nk \log n \log k)$ time. Computing the *Vis*(q) can be done in linear time, using the algorithm of [EA81].

In each call to `ComputeGuards`, the algorithm performs $O(k \log(n/k))$ iterations. Overall, the running time of the algorithm is thus

$$O\left(\sum_{i=1}^{\log c_{opt}} n(2^i)^2 \log n \log \frac{n}{2^i} \log^2 2^i \right) = O\left(n c_{opt}^2 \log n \log \frac{n}{c_{opt}} \log^2 c_{opt} \right).$$

We conclude:

Theorem 2. *Given a simple polygon P with n vertices, one can compute, in $O(n c_{opt}^2 \log n \log(n/c_{opt}) \log^2 c_{opt})$ expected time, a set S of $O(c_{opt} \log c_{opt})$ ver-*

Procedure **ComputeGuards**(P - simple polygon, k - number of guards)

1 Assign weight 1 to all the elements of V, the set of vertices of P.

2 For $i := 1$ to $O(k \log(n/k))$ do:

(a) Pick randomly a set S of $O(k \log k)$ vertices, by choosing each guard randomly and independently from V, according to the weights of the vertices.

(b) Check if the points of S see all of P, if so, terminate and return S as the set of guards.

(c) Else, find a point $q \in P$ which is not visible from S, and compute Vis_q.

(d) Compute Ω, the sum of weights of vertices in $V \cap Vis(q)$. If $2k\Omega \leq$ the sum of weights of all vertices of P, double the weight of every vertex of $V \cap Vis(q)$.

3 Failure — no solution found.

Figure 1. **ComputeGuards**(P, k) computes with high probability a guarding set of P of $O(k \log k)$ guards, if $k \geq c_{opt}$

tices of V that seems P, where c_{opt} is the cardinality of the minimal set. The quality of approximation is correct with high probability.

Remark: Theorem 2 provides an $O(\log c_{opt})$-approximation to the optimal solution. Previously, only $O(\log n)$-approximation was known [Gho87] in $O(n^5 \log n)$ time. This is especially striking, when one observes that the dependency of the running time of the new algorithm on n is near linear.

4. Exact Algorithm for Fixed Number of Guards

Theorem 3. *A smallest set of guards that can see a given simple polygon P with n edges can be computed in time $O((nk)^{3(2k+1)})$, where k is the size of such an optimal set.*

Proof. This is an easy consequence of known techniques in algorithmic real algebraic geometry. Suppose first that we wish to determine whether there exists a set of k guards that can see the whole of P. This is equivalent to deciding the truth of the following predicate in the first-order theory of the reals:

$$\exists x_1, y_1, x_2, y_2, \ldots, x_k, y_k \forall u, v \mid$$

$$\Big[\mathsf{InP}(u,v) \Longrightarrow (\mathsf{Visib}(x_1,y_1;u,v) \vee \mathsf{Visib}(x_2,y_2;u,v) \vee \cdots \vee \mathsf{Visib}(x_k,y_k;u,v))\Big],$$

where $\mathsf{InP}(u,v)$ is a predicate that is true iff $(u,v) \in P$, and $\mathsf{Visib}(x,y;u,v)$ is a predicate that is true iff (x,y) and (u,v) are visible to each other within

P. Clearly, InP is a Boolean combination of $O(n)$ linear inequalities, whereas Visib$(x, y; u, v)$ is a Boolean combination of $O(n)$ quadratic inequalities. Hence the whole predicate involves $O(nk)$ polynomials of maximum degree 2, and has only one alternation of quantifiers. Applying the result of [BPR96], deciding the truth of this predicate can be done in time $O((nk)^{3(2k+1)})$. Finding the optimal value of k can then be done by a straightforward unbounded linear search, within asymptotically the same complexity bound. □

5. Unconstrained Locations of Guards

We consider in this section the art gallery problem where the location of the guards inside the polygon is not restricted to vertices. Instead, their location is restricted to lie on a dense grid inside the polygon. Intuitively, if the polygon P is "well-behaved", such a minimum set of guards would be a good approximation (in its cardinality) to the optimal guarding set.

The main idea of our algorithm is that, instead of maintaining the weight of the relevant grid points explicitly, as done in the algorithm for the case of vertices, we exploit the special properties of the grid, and of the weight function defined over the grid points, to maintain those weights implicitly.

Suppose that we are given a simple n-gon P with diameter ≤ 1, a parameter $\varepsilon > 0$, and Γ a grid of square-length ε inside P; that is $\Gamma = P \cap \{(i\varepsilon, j\varepsilon) \mid i, j \in \mathbb{Z}\}$. We present an algorithm that finds a set $G \subseteq \Gamma$ of guards that see all the points of Γ, and its cardinality is $O(c_{opt} \log c_{opt})$ where c_{opt} is the cardinality of a smallest set of vertices of Γ that sees P.

We apply the algorithm of the previous section, with a different scheme for maintaining the weights over the points of Γ, and picking a set of guards in each stage of `ComputeGuards`.

The range space for this problem is defined as follows: Let $\tilde{V}$ denote the set of vertices of P, and the set of vertices of Γ. Let L be the set of lines passing through pairs of vertices of $\tilde{V}$. Let $\mathcal{X}$ be the set of all intersection points of lines of L. Let the range space $\Sigma = (\tilde{V}, \{\tilde{V} \cap \mathit{Vis}(p) | p \in \Gamma\})$. We do not construct Σ explicitly, as it is not necessary. It is not hard to see that $S \subseteq \Gamma$ sees P if and only if S sees $\mathcal{X}$. Assume that $S \subseteq \Gamma$ does not see P. Let K be a connected component of $P \setminus \mathit{Vis}(S)$. Observe that since P is a close set, the edges of the closure of $\mathit{Vis}(S)$ which are not edges of P, are not edges of $\mathit{Vis}(S)$. Thus each vertex of K (which is also a vertex of $\mathcal{X}$) is not seen by any guard of S.

The weights of the points of Γ are maintained by a subdivision $\mathcal{A}_i$ of P, so that the weight $w(f)$ assigned to all the points of Γ inside a face f of $\mathcal{A}_i$ is the same, where i is the current iteration of `ComputeGuards`. We associate with a face f of $\mathcal{A}_i$ the quantities $n(f) = \Gamma \cap f$, namely, the number of grid points of Γ inside f, $w(f)$, which is the weight assigned to each point of $\Gamma \cap f$, and $W(f) = w(f) \cdot n(f)$, which is the overall weight of f. Initially $\mathcal{A}_0$ consists of a single cell, namely all of P. In the i-th iteration of `ComputeGuards`, we pick at random as set S_i of vertices of Γ, according to their weights. This is done by first picking the face f of $\mathcal{A}_{i-1}$ from which a point $g \in S_i$ is to be picked, and then picking g uniformly from $f \cap \Gamma$. Next we compute the polygon $\mathit{Vis}(S_i)$

and check as in Section 4 if it covers P. If $P \neq Vis(S_i)$ we find a vertex q_i of $P \setminus Vis(S_i)$. As mentioned above, $q_i \in \tilde{V}$. We compute the visibility polygon $Vis(q_i)$, computes the total weight Ω of points of $\Gamma \cap Vis(q_i)$ (details described below) and if $2k\Omega \leq W(\Gamma)$, we insert $\partial Vis(q_i)$ into $\mathcal{A}_{i-1}$, splitting some faces of $\mathcal{A}_{i-1}$ and forming a new arrangement $\mathcal{A}_i$. We double (in an implicit fashion) the weight of $\Gamma \cap Vis(q_i)$. In Section 5.1 we explain how to find the number of grid-point of Γ inside a face f, how to split f and and how to pick a grid point at random from $\Gamma \cap f$ uniformly (note that all grid-points of $\Gamma \cap f$ have the same weight). We next explain how to insert $Vis(q_i)$ and maintain the weights of the faces of $\mathcal{A}_i$.

We assume for simplicity of exposition that each face f is a triangle (if not, when we compute f we also compute a triangulation of it, and pick a triangle from this triangulation. This does not effect the overall complexity of the algorithm, and we omit the tedious but straightforward details).

To explain how to efficiently maintain the weights, we need the following lemma, whose proof is postponeded to the end of this section.

Lemma 3. Let $S = \{x_i\}_1^n$ be a set of n points on a line, where each point x_i is associated with a weight w_i. There is an augmented search tree $\mathcal{T}$ that supports the following operations in time $O(\log n)$:

insert(x_i, w_i) — insert a new point x_i into S, with an assisted weight w_i.

modify(x_i, w_i, w_i') — change the weight of x_i from w_i to w_i'.

pick — pick a point x_i at random from S, with probability $w_i / \sum_{j=1}^n w_j$.

interval_sum(x, y) — report the sum of weights of the points in $S \cap [x, y]$.

interval_double(x, y) — double the weight of each point of $S \cap [x, y]$.

Let the weight of a cell of $\mathcal{A}_{i-1}$ be the sum of weights of grid points inside this cell. By Theorem 1 the arrangement $\mathcal{A}$ consists of $O(nk^2 + k^2)$ edges, which we call *arrangement-edges.* These edges lie on one of the $O(nk)$ edges of the original polygons $Vis(q_j)$, $(1 \leq j < i)$ which we call *long edges.* We replace long edge e by two copies of e, so that each copy bounds faces of $\mathcal{A}_{i-1}$ only on one of its sides (analogously to half-edges in the description of the DCEL data structure [dBvKOS00]). We denote these edges *polygon-edges.* We construct the tree $\mathcal{T}_i$ of Lemma 3 for each polygon edge e_i, where the keys stored in that tree are the vertices of $\mathcal{A}_{i-1}$ along e_i. Each vertex v of $\mathcal{A}$ appears on four polygon-edges adjacent to v. In each of them, v is stored twice (with the same coordinate), corresponding to two of the four cells of $\mathcal{A}_{i-1}$ adjacent to v. The weight of the copy of v corresponding to a cell c is $\omega_c/(2m_c)$, where ω_c is the total weight of c, and m_c is the number of vertices of c. As easily checked, the sum of weights of vertices corresponding to c, summed over all data structure $\mathcal{T}_i$ for all edges e_i in $\mathcal{A}_{i-1}$ is ω_c. Let the *total weight* of a polygon-edge e_i denote the sum of weights of vertices on e_i. To pick a face of $\mathcal{A}_{i-1}$ at random, we first pick a polygon edge polygon edge bounding the face.

Picking an polygon edge e. This is accomplished by maintaining a tree $\tilde{\mathcal{T}}$ storing a representative point x_i for each polygon-edge e_i, where the weight of x_i is the total-weight of e_i. Similarly to the data structure of Lemma 3, $\tilde{\mathcal{T}}$ stores for each node μ the variable W_μ maintaining the sum of total-weights of the polygon-edges stored at the subtree rooted by μ. Maintaining W_μ upon changing the total weight of one of the polygon-edges in μ's subset is done in a routine bottom up fashion. Picking an edge e_i is done similar to Lemma 3. Both operation are doable in time $O(\log n)$.

Inserting a new polygon $Vis(q_i)$. We find a cell c of $\mathcal{A}$ containing a point of ∂Vis_i, which is also a vertex of P. This is easy to accomplish by maintaining which cell of $\mathcal{A}$ contains every vertex of P, so all is left to do is finding a vertex of P that sees the point corresponding (the "center" of) $Vis(q_i)$. It is known that the complexity of the zone of $\partial Vis(q_i)$ in $\mathcal{A}$ is only $O(ni\alpha(n))$. We follow $\partial Vis(q_i)$ through these cells that it intersects, splitting each cell we pass through. We compute, using the operations on the discrete hull described in Lemma 4 the number of points in the new cells, and update the weights accordingly, and the number of vertices along the boundaries of these cells.

Next we apply, for each tree $\mathcal{T}_i$ associated with e_i, the operation $sum(x_1, x_2)$ in order to compute the value of Ω defined above. If $2c_{opt}\Omega \leq$ the sum of weights of all points of Γ we double the weight of all the vertices of cells encapsulated in Vis_i, but applying *Interval_double*(x_1, x_2) operations described above to each of the trees associated with polygon edges. After a triangle is split, we need to compute the number of grid points inside each of the new triangles. This is required for calculating the weights of the new triangles. This is accomplished by the data structure of Lemma 4, and add a factor of $O(\log^2 \Delta)$, where Δ is the ratio between the diameter of the polygon and the grid size. Thus the time needed for the ith iteration is $O(ni\alpha(n) \log n(\log i + \log^2 \Delta))$.

We perform exponential search for the value of c_{opt} by performing $O(\log c_{opt})$ calls to `ComputeGuards`(P, k), where k is always $O(c_{opt})$, we conclude after omitting details due to lack of space

Theorem 4. *Given a simple polygon P with n vertices, one can spread a grid Γ inside P, and compute an $O(\log c_{opt})$-approximation to the smallest subset of Γ that sees P. The expected running time of the algorithm is $O\left(nc_{opt}^2 \log c_{opt} \log(nc_{opt}) \log^2 \Delta\right)$, where Δ is the ratio between the diameter of the polygon and the grid size.*

5.1. Range Searching on a Grid

Lemma 4. Let T be a triangle in the plane, and let Γ be a grid inside T. The boundary of DiscreteHull$(T) = \mathcal{CH}(\Gamma \cap T)$ and the number of points of Γ inside T can be computed in $O(\log \Delta)$ time, where Δ is the ratio between the diameter of the T and the grid size.

Proof. The boundary of the discrete hull $C_T = \mathcal{CH}(T \cap \Gamma)$ can be computed in $O(\log \Delta)$ time [KS96, HP98]. One can compute (in the same time complexity), the number M_T of points of Γ on the boundary of C_T, and Area(C_T). Now, using Pick's Theorem one can now derive a precise closed formula on the number of grid points in $\Gamma \cap T$. Thus, the number of points of Γ inside T can be computed in $O(\log \Delta)$ time. □

Lemma 5. Let T, Γ and Δ be as in Lemma 4. One can pick randomly and uniformly a point from $T \cap \Gamma$ in $O(\log^2 \Delta)$ time.

Proof. Let B_0 be a bounding box of T. In the i-th stage, we split B_{i-1} vertically in the middle by a line ℓ_i that does not pass through points of the grid Γ, let B_i^R, B_i^L be the resulting two boxes to the right and left of ℓ_i, respectively. We can compute in $O(\log \Delta)$ time, by Lemma 4, the number of grid points in $w_i^R = B_i^R \cap T \cap \Gamma$, and in $w_i^L = B_i^L \cap T \cap \Gamma$. We now decide to B_i to be either B_i^R or B_i^L randomly according to their weights w_i^L, w_i^R. We can stop as soon as a single vertical grid line crosses our box B_i, as we can uniformly pick a grid point along this vertical grid line that lies inside T. Overall, this process clearly takes $O(\log^2 \Delta)$ time. □

5.2. Proof of Lemma 3

Proof. We maintain a sorted balanced tree $\mathcal{T}$, whose leaves are associate with the values x_i. Let $\pi(v, \mu)$ denote the path connecting node v to node μ, where v is an ancestor of μ. Each internal node μ maintains its *multiplicative factor* M_μ, initially 1.

The weight of a point x_i associated with the leaf μ equals $\prod_{\xi \in \pi(root(\mathcal{T}),\mu)} M_\xi$. We assign for each internal node μ the variable σ_μ, which equals

$$\sigma_\mu = \sum_{v \text{ descendent leaf of } \mu} \prod_{\xi \in \pi(\mu, v)} M_\xi$$

As easily observed, the sum of weights of the leaves in the subtree rooted at a node μ equals $\sigma_\mu \cdot \prod_{\xi \in \pi(root(\mathcal{T}),\mu)} M_\xi$. We next explain how to perform a "*pick*" operation: Assume that we already decided that point x_i to be picked belongs to the subtree $\mathcal{T}_\mu$ of a node μ, and we next decide whether x_i belongs to the left subtree of $\mathcal{T}_{left(\mu)}$, where $left(\mu)$ is the left child of μ. Observe that the probability of picking a point from the left subtree of μ equals

$$\frac{\sum \text{weights of leaves in } \mathcal{T}_{left(\mu)}}{\sum \text{weights of leaves in } \mathcal{T}_\mu} = \frac{\sigma_{left(\mu)} \prod_{\xi \in \pi(root(\mathcal{T}), left(\mu))} M_\xi}{\sigma_\mu \prod_{\xi \in \pi(root(\mathcal{T}),\mu)} M_\xi} = \frac{\sigma_{left(\mu)} M_{left(\mu)}}{\sigma_\mu}$$

This suggests the following approach to find a leaf x_i. We branch from the root to one of its children μ with the probabilities given above. Thus we perform a pick operation in time $O(\log n)$.

To support *Interval_double*(x_1, x_2), and *Interval_sum*(x_1, x_2) we first locate the set X of *canonical* nodes μ with the property that all descendent leaves of μ lie in the range $[x_1, x_2]$, but the parent of μ does not have this property. It is well known (see e.g. [dBvKOS00]) that we can visit all nodes in X in time $O(\log n)$. In the case of *Interval_double*(x_1, x_2) we just double M_μ for each $\mu \in X$.

In the case of *Interval_sum*(x_1, x_2) we use the equation above for computing the sum of weights of the points of each subtree μ for $\mu \in X$. Since we can visit all of them in $O(\log n)$ time, this is also the time required for this operation. □

6. Polygons with Holes and Terrains

The algorithms introduced in the previous sections, can easily be modified to solve visibility problems in more complicated "galleries', such as polygons with holes, or terrains. The modifications needed are only in the bounds on the complexity of the arrangements of visibility regions, and in way we compute them, and in the approximation factor.

Visibility in a Polygon with holes Let P be a polygon with n vertices and h holes. Let $\{q_1 \dots q_k\}$ be a set of points and let Vis_i denote the visibility polygon of q_i,

We claim that the complexity of the arrangement forms by the visibility polygons $\{Vis_1 \dots Vis_k\}$ is $O(nk^2h)$. This follows from the following argument. The boundary of Vis_i consists of $n+h$ edges which are not on ∂P. Every such edge can intersect ∂Vis_j in $\leq 2h$ points. Thus the total number of intersection points on ∂Vis_i is $\leq nhk$, and summing this bounds for all i yields the asserted bound.

Next we consider vertex visibility in a polygon with holes. Since the VC-dimension of the problem is $\Theta(1+\log h)$, as shown in [Val98] the approximation factor increases to $O(\log h \log(c_{opt} \log h))$. See [BG95] for details. Regarding the case where guards can be located inside P, Analogous version to Section 5 yields, after the obvious modification

Theorem 5. *Let P be a given polygon with n vertices in total and h holes.*

- *We can find a set G of $O(c_{opt} \log n \log(c_{opt} \log n))$ vertices of P that sees P, where c_{opt} is the cardinality of the optimal solution. The running time is $O(nh\, c_{opt}^3 polylog\, n)$.*
- *Let Γ be a grid inside P. Then we can find a set G of $O(c_{opt} \log h \log(c_{opt} \log n))$ vertices of Γ that sees P. The running time is $O(nhc_{opt}^3 polylog\, n \log^2 \Delta))$, where Δ is the ratio between the diameter of the polygon and the grid size.*

Visibility in Terrains Let T be a (triangulated) terrain of n triangles. We can also modify our algorithm in order to find a set S of vertices of T that sees T. Clearly, $O(\log n)$ bounds the VC-dimension of the set system obtained by assigning to each point $q \in T$ the points of T that q sees. This follows from the following observation: Assume d is the VC-dimension of the problem, and let S be a set of points d of T which is shatterable under visibility. That is, for every $S' \subseteq S$ there is a point $g_{S'}$ on T such that $S' = S \cap Vis(g_{S'})$.

The visibility region $Vis(g_{S'})$ can be described as the union of $\Theta(n^2)$ triangles in T, each fully contained inside a face of T, where the boundary of each such triangle Δ is either the boundary of a triangle of T, or of the intersection of T with the the plane h_r passing through p and through an edge r of T. Since there are $O(n)$ edges r in T, and each plane h_r intersects a triangle of T along a straight segment, the $\Theta(n^2)$ bound on the complexity of $Vis(p)$ follows. The total number of edges of $Vis(q)$ for all $q \in S$ is $O(dn^2)$, and overlaying the boundaries of $Vis(p)$ for each $q \in S$ imposes a subdivision $\mathcal{S}$ of T into $O(d^2n^4)$ regions, where if two points x_1, x_2 of T lie in the same region of $\mathcal{S}$, then they see the same subset of S. Since S is shattered under visibility, the number of regions in $\mathcal{S}$ is at least 2^d implying $d = O(\log n)$.

In [dB93] de Berg showed that the complexity of the arrangement $\mathcal{A}$ forms by the visibility polygons of a set G of k guards is $O(n^2k^2)$. Plugging the upper bound into our algorithm, and skipping obvious details, we obtain a running time of $O(n^2 c_{opt}^3 polylog\, n \log^2 \Delta)$ where Δ is the ratio between the diameter of the terrain and the grid size. This improves the recent $O(n^8)$-algorithm of Eidenbenz [Eid02], who obtained a slightly better approximation factor of $O(\log n)$.

Note that if guards are allowed to be located only on vertices of the terrain, then the use of the grid is not needed, and the running time is improved to $O(n^2 c_{opt}^3 polylog\ n)$. To summerize

Theorem 6.
- *Given a terrain T of n triangles, we can find in time $O(n^2 c_{opt}^3 polylog\ n)$ a set S of vertices of T that see T, where $|S|$ is within a factor of $O(\log n \log\log n)$ of the minimum.*
- *Given a terrain T of n triangles, and a grid Γ placed on each triangle of T, we can find in time $O(n^2 c_{opt}^3 polylog\ n \log^2 \Delta)$ a set S of vertices of Γ that see T, where $|S|$ is within a factor of $O(\log n \log\log n)$ of the minimum, and Δ is the ratio between the diameter of the terrain and the grid size.*

Remark: Currently we are working on methods which allow us to find small sets of guards inside polygons and terrains, *without* using the grid. This would improve the running time, and simplify the algorithm significantly. Preliminary results obtained so far look very promising, and will be reported separately.

Acknowledgments We would like to thank Will Evans, Stephen Kobourov, T.M. Murali and Micha Sharir for very helpful discussions. We also thank Micha Sharir for the result of Section 4.

References

[Agg84] A. Aggarwal. *The art gallery problem: Its variations, applications, and algorithmic aspects.* PhD thesis, Dept. of Comput. Sci., Johns Hopkins University, Baltimore, MD, 1984.

[BG95] H. Brönnimann and M. T. Goodrich. Almost optimal set covers in finite VC-dimension. *Discrete Comput. Geom.*, 14:263–279, 1995.

[BPR96] S. Basu, R. Pollack, and M.-F. Roy. On the combinatorial and algebraic complexity of quantifier elimination. *J. ACM*, 43:1002–1045, 1996.

[BS93] I. Bjorling-Sachs, *Variations on the Art Gallery Theorem.* PhD thesis, Rutgers University, 1993.

[Cha91] B. Chazelle, Triangulating a simple polygon in linear time. *Discrete Comput. Geom.* , 6 (1991) 485–524.

[Cla93] K.L. Clarkson. Algorithms for polytope covering and approximation. In *Proc. 3rd Workshop Algorithms Data Struct.*, LNCS 709, 246–252, 1993.

[dB93] M. de Berg. Generalized hidden surface removal. In *Proc. 9th Annu. ACM Sympos. Comput. Geom.*, pages 1–10, 1993.

[dBvKOS00] M. de Berg, M. van Kreveld, M. H. Overmars, and O. Schwarzkopf. *Computational Geometry: Algorithms and Applications.* Springer-Verlag, 2nd edition, 2000.

[EHKKRW02] A. Efrat, F. Hoffmann, K. Kriegel, C. Knauer, G. Rote and C. Wenk, Covering Shapes by Ellipses *ACM-SIAM Symposium on Discrete Algorithms*, 2002, 453–454.

[EA81] H. ElGindy and D. Avis. A linear algorithm for computing the visibility polygon from a point. *J. Algorithms*, 2:186–197, 1981.

[Eid00] S. Eidenbenz. *(In-)Approximability of Visibility Problems on Polygons and Terrains.* Phd thesis, diss. ETH no. 13683, 2000.

[Eid02] S. Eidenbenz. Approximation algorithms for Terrain Guarding *Information Processing Letters (IPL)* 82 (2002) 99-105.

[GBL01] Hector Gonzalez-Banos and Jean-Claude Latombe. A randomized art-gallery algorithm for sensor placement. In *Proc. 17th Annu. ACM Sympos. Comput. Geom.*, pages 232–240, 2001.

[Gho87] S. K. Ghosh. Approximation algorithms for art gallery problems. In *Proc. Canadian Inform. Process. Soc. Congress*, 1987.

[GMMN90] L. Gewali, A. Meng, Joseph S. B. Mitchell, and S. Ntafos. Path planning in $0/1/\infty$ weighted regions with applications. *ORSA J. Comput.*, 2(3):253–272, 1990.

[HKK91] F. Hoffmann, M. Kaufmann, and K. Kriegel. The art gallery theorem for polygons with holes. In *Proc. 32nd Annu. IEEE Sympos. Found. Comput. Sci.*, (1991) 39–48.

[Hof90] F. Hoffmann. On the rectilinear art gallery problem. In *Proc. 17th Internat. Colloq. Automata Lang. Program.*, LBNCS 443, 717–728, 1990.

[HP98] S. Har-Peled. An output sensitive algorithm for discrete convex hulls. *Comput. Geom. Theory Appl.*, 10:125–138, 1998.

[JL93] G. F. Jennings and W. J. Lenhart. An art gallery theorem for line segments in the plane. In G. T. Toussaint, editor, *Pattern Recognition Letters Special Issue on Computational Geometry*, 1993.

[KS96] S. Kahan and J. Snoeyink. On the bit complexity of minimum link paths: Superquadratic algorithms for problems solvable in linear time. In *Proc. 12th Annu. ACM Sympos. Comput. Geom.*, 151–158, 1996.

[O'R83] J. O'Rourke. An alternative proof of the rectilinear art gallery theorem. *J. Geom.*, 21:118–130, 1983.

[O'R87] J. O'Rourke. *Art Gallery Theorems and Algorithms.* The International Series of Monographs on Computer Science. Oxford University Press, New York, NY, 1987.

[OS83] J. O'Rourke and K. J. Supowit. Some NP-hard polygon decomposition problems. *IEEE Trans. Inform. Theory*, IT-30:181–190, 1983.

[Urr00] J. Urrutia. Art gallery and illumination problems. In Jörg-Rüdiger Sack and Jorge Urrutia, editors, *Handbook of Computational Geometry*, pages 973–1027. North-Holland, 2000.

[Val98] P. Valtr. Guarding galleries where no point sees a small area. *Israel J. Math*, 104:1–16, 1998.

GOSSIPING WITH UNIT MESSAGES IN KNOWN RADIO NETWORKS

Leszek Gąsieniec* and Igor Potapov
Department of Computer Science, University of Liverpool
Liverpool L69 7ZF, UK.
leszek@csc.liv.ac.uk, igor@csc.liv.ac.uk

Abstract A *gossiping* is a communication primitive in which each node of the network possesses a unique message that is to be communicated to all other nodes in the network. We study the gossiping problem in known *ad hoc* radio networks, where during each transmission only *unit* messages originated at any node of the network can be transmitted successfully. We survey a number of radio network topologies. Assuming that the size (a number of nodes) of the network is n we show that the exact complexity of radio gossiping in stars is $2n-1$, in rings is $2n \pm O(1)$, and on a line of processors is $3n \pm O(1)$. We later prove that radio gossiping in free trees is harder and it requires at least $3\frac{1}{6}n - 16$ time steps to be completed. For free trees we also show a gossiping algorithm with time complexity $5n + 8$. In conclusion we prove that in general graphs radio gossiping requires $\Omega(n \log n)$ time, and we propose radio gossiping algorithm that works in time $O(n \log^2 n)$.

1. Introduction

The importance of communication networks and their use on a daily basis has been steadily growing over the past few decades. One of the most striking examples of modern networking technology is the Internet with its diverse applications in research, business, education, and entertainment. Mobile *radio networks* [19] are expected to play an important role in future commercial and military applications. These networks are suitable in situations where instant infrastructure is needed and no central system administration (such as base stations in a cellular system) is available.

There are two important communication primitives used in the process of dissemination of information in networks: *broadcasting* and *gossiping*. In the *broadcasting problem*, a distinguished *source* node has a message that needs to

*Supported in part by EPSRC grants GR/N09855 and GR/R85921.

be sent to all other nodes. The *gossiping* is a communication primitive in which each node of the network possesses a unique message that is to be communicated to all other nodes in the network. The gossiping problem raises naturally both in the theoretical as well as more applied setting. It is a part of several multiprocessor computation tasks, such as global processor synchronization, linear system solving, Discrete Fourier Transform, and parallel sorting, e.g.,see [4, 13].

Most of the work in the field has been done under the assumption that processors can transmit messages of an arbitrary size in a single time step. The gossiping problem with bounded (size) messages was previously studied in the matching model, e.g., see [3]. In this model during every time step nodes organize themselves and exchange information in independent pairs. The results presented in [3] include the study of the exact complexity of the gossiping problem in Hamiltonian graphs and k-ary trees, and optimal asymptotic bounds for general graphs, in the matching model with unit messages. Their paper contains also a number of asymptotically optimal results in the matching model with messages of arbitrarily bounded size. Another interesting study of the gossiping problem with limited size messages in graphs with bounded degree can be also find in [12].

In this paper we study the gossiping problem with unit messages in known *ad hoc* radio networks. We adopt here a communication model used previously, e.g., in [1, 14, 6]. A radio network is modeled as an undirected graph $G = (V, E)$. The nodes in set $V = \{v_0, .., v_{n-1}\}$ are interpreted as processors (transmitter/receiver devices) while undirected edges in set E indicate that every neighboring node in the graph is in the transmission range. The processors work synchronously. In each time step, any processor can either transmit or receive a message. A message transmitted by processor v reaches all its neighbors in the same time step. However, any neighbor w can receive it only if no message from another processor reaches it at this time step. Otherwise a *collision* occurs and none of the messages is delivered to w. The *size* of the network corresponds to the number of nodes in the underlying graph of connections. In what follows we assume that the size of the network is n.

It is only recently that studies on radio gossiping have been intensified, see [7, 9, 10, 15, 11, 16, 17, 18]. However this work is devoted to the case when the messages used in the gossiping process can be of an arbitrary size. Under this strong assumption Chrobak *et al.* [9] showed that deterministic gossiping can be performed in unknown directed *ad-hoc* radio networks in time $O(n^{3/2} \log^2 n)$. A constructive version of their algorithm was recently proposed by Indyk, see [16]. This result was recently improved by Gąsieniec and Lingas [15] for networks with diameter $D = n^\alpha$, for $\alpha < 1$. They presented an alternative gossiping algorithm working in time $O(n\sqrt{D} \log^2 n)$. These results show that radio networks with a long diameter constitute a bottleneck in deterministic radio gossiping with messages of an arbitrary size. An alternative approach to the radio gossiping problem was presented by Clementi *et al.* in [11]. They proposed deterministic gossiping algorithm with running time $O(Dd^2 \log^3 n)$, where d stands for the maximum in-degree of the underlying graph of connections. Chrobak *et al.* in [10] proposed also a randomized radio gossiping algorithm with expected running time $O(n \log^4 n)$. A study on oblivious gossiping in *ad hoc* radio networks can be found in [7]. An alternative radio model was studied by Ravishankar and Singh. They presented distributed gossiping algorithms for networks with nodes placed randomly on a line [17] and a ring [18].

In this paper we initiate a discussion on gossiping in *known* radio networks with messages of a limited size. In what follows we assume that the messages originated in all nodes of the network are unique and they are of the same size. Moreover each transmission performed by any node of the network can contain *only one*, a *unit*, message originated in some node of the network. A similar concept of communication in *unknown* radio networks with messages of a limited size has been recently adopted by Christersson *et al.* in [8]. Another interesting study of randomized multiple communication in unknown radio networks with messages limited to $O(\log n)$ bits can be found in [2].

2. Radio gossiping in stars and rings

A *star* of size n is a free tree $S = \{V, E\}$, where $V = \{v_0, .., v_{n-1}\}$ and $E = \{(v_0, v_1), .., (v_0, v_{n-1})\}$. Node v_0 is called a *central node* (*center*) and all other nodes form *arms* of the star.

Theorem 1 *The exact complexity of radio gossiping in stars of size n is $2n-1$.*

A *ring* of size n is a graph $R = (V, E)$, s.t., $V = \{v_0, .., v_{n-1}\}$ and $E = \{(v_i, v_{(i+1) \bmod n}) : i = 0, ..n-1\}$.

Theorem 2 *The exact complexity of radio gossiping in rings of size n is $2n \pm O(1)$.*

Lower bound As in other instances of the gossiping problem also in this case every unit message has to be transmitted to all other nodes of the network. We say that a *delivery* occurs at node v at time step t if node v receives a message at time step t. We show that during a single time step any communication algorithm performs at most $\lfloor n/2 \rfloor$ deliveries. Assume opposite, i.e., at least $\lfloor n/2 \rfloor + 1$ deliveries have occurred. It means that there exist three nodes with consecutive labels v_{i-1}, v_i, v_{i+1} to which some messages have been delivered. However this leads to a contradiction since node v_i is not able to receive any messages when its both (and only) neighbors are in the receiving mode. Finally since every node is expecting $n-1$ deliveries the total number of steps required in this case is $\geq n(n-1)/\lfloor n/2 \rfloor \geq 2n-2$.

Upper bound We show that there exists an algorithm performing radio gossiping on a ring of size n in time $2n+9$. The algorithm consists of a number of steps. During each step consecutive processors along the ring are grouped into alternating pairs, i.e., *active* (transmitting) pairs alternated with *dormant* (expecting messages) pairs. We place as many as possible alternating pairs on the ring assuring that all active pairs are at least at distance 2 apart. Note that in the worst case at one point on the ring there may be two active pairs at distance 5 apart (if the distance was six we could introduce another active pair). We call this phenomenon as a *gap*, see Figure 1b. The pattern of alternating pairs is fixed however it is rotated along the ring by one position, e.g., in clockwise order at the end of each step. Note that two elements of any active pair are responsible for transmissions to opposite (clockwise and anti-clockwise) directions. A *round* for a node is a number of steps between two consecutive clockwise transmissions. If the gap was of size 2 (n is a multiple of 4) all rounds would consist of 4 steps, see Figure 1a. However some *long* rounds can have as many as 7 steps when the gap is of size 5. Since every original message has to be transmitted through at most $\lceil (n-1)/2 \rceil \leq n/2$ nodes in clockwise (and anti-clockwise) order and each long round occurs with periodicity at least $\lceil (n-3)/4 \rceil \geq n/4 - 1$, each traversing message can experience at most 3 long rounds. Thus the time complexity of radio gossiping in rings of size n is bounded by $4 \cdot n/2 + 3 \cdot 3 = 2n + 9$.

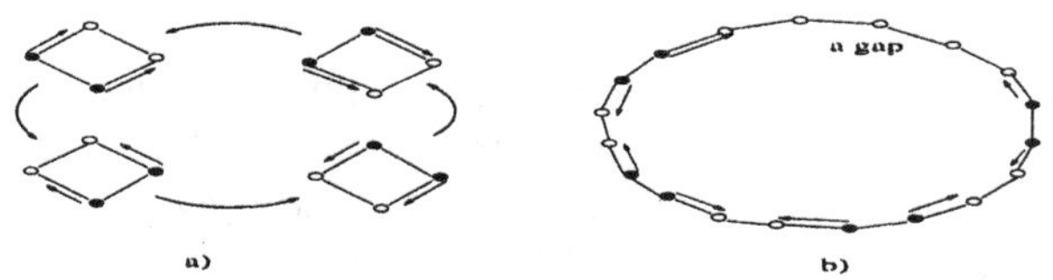

Figure 1. a) 4 consecutive rounds, b) maximum gap

3. Radio gossiping on a line

A *line* of size n is a free tree $L = \{V, E\}$, where $V = \{v_0, .., v_{n-1}\}$ and $E = \{(v_i, v_{i+1}) : i = 0, ..n-2\}$. Nodes v_0 and v_{n-1} are called the *left end* and the *right end* respectively.

Theorem 3 *The exact complexity of radio gossiping on a line of size n is* $3n \pm O(1)$.

Lower bound To prove the lower bound $3n - O(1)$ we will consider a similar but easier problem called here a *monotonic gossiping*. The task is to send all original messages only to the nodes being on their right hand side. We will need the following definitions.

Stack-and-knot problem A *stack and knot* $SK(n)$ is an object that consists of a short line of length 4, a *knot* K, and a stack S of n messages available at node v_0, see Figure 2a. A *stack-and-knot problem* is to move all messages from the stack at node v_0 to node v_3 using radio transmissions with unit messages.

Lemma 1 *The exact complexity of stack-and-knot problem in* $SK(n)$ *is* $3n$.

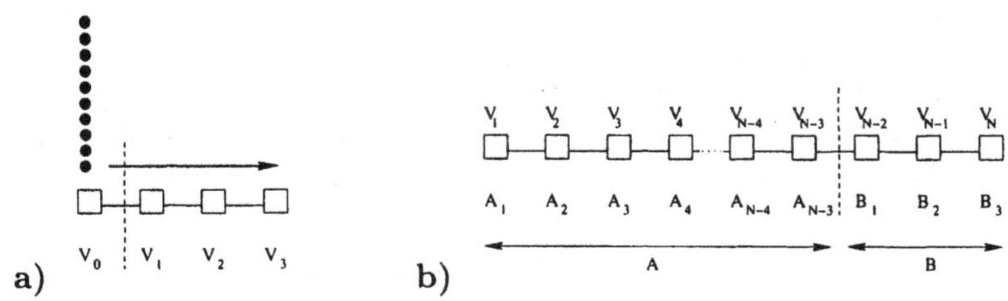

Figure 2. a) Gossiping in part B ; b) Line L of n nodes

Proof: Due to a collision problem during each step of any algorithm we can perform at most one successful transmission towards the right end of the knot. Thus the process of transmission of n messages from stack S at node v_0 to node v_4 requires at least $3n$ independent steps.

Note also that the algorithm with transmission pattern, s.t., node v_i transmits in time step t if $(i = t) \mod 3$, where initial time step $t = 0$, solves stack-and-knot problem in $SK(n)$ in time $3n$. □

Lemma 2 *Monotonic gossiping on a line of size n requires* $3n - 6$ *steps.*

Proof: Let L be a line of size n. Lets split line L into two parts A and B, where part A is formed by the left end of L of size $n-3$ and part B corresponds to the right end of size 3, see Figure 2b. According to Lemma 1 a transmission of all $(n-3)$ messages from part A to the node B_3 takes at least $3(n-3)$ steps. Note that to complete monotonic gossiping the additional three steps are required to transmit to B_3 messages originated in nodes B_1 and B_2. It means that to deliver all messages from the nodes of line L to the node B_3 (which is equivalent to monotonic gossiping) we need at least $3(n-3)+3 = 3n-6$ rounds. □

Corollary 1 *Radio gossiping on a line of size n requires* $3n - 6$ *steps.*

Lemma 3 *Monotonic gossiping on a line of size n can be completed in* $3n - 3$ *steps.*

Proof: The algorithm runs in rounds. Each round consists of 3 steps. During step i, for $i = 0, 1, 2$, all nodes with index $(j = i) \mod 3$, for $j = 0, .., n-1$, transmit and all other nodes remain silent (expecting messages). Following this pattern during each round every node sends one message towards the right end of the line. And after $n-1$ rounds the monotonic gossiping is completed. □

Upper bound We present here a gossiping algorithm with running time $3n + O(1)$ that is a combination of algorithms on a ring and quick pipelining on a line. However before we start the presentation we give an outline of two simpler algorithms working respectively in $4(n-1)$ and $3\frac{1}{2}n - 3$ steps. Note that applying the idea of alternating pairs (see section 2, the gap problem does not exist here) on a line, each round is of size 4 and the gossiping problem can be solved in $4(n-1)$ steps. The efficiency of this algorithm can be improved when we notice that gossiping into two directions is required only during initial $\lfloor n/2 \rfloor$ rounds. Afterwards sequences of messages traversing to the left and to the right become disjoint. This allows to use monotonic gossiping on both of them. Thus the total time of improved gossiping algorithm is bounded $\lfloor n/2 \rfloor \cdot 4 + \lfloor n/2 \rfloor \cdot 3 - 3 \leq 3\frac{1}{2}n - 3$.

In what follows we present further improvement showing that monotonic gossiping is actually performed on a sparser sequence of messages allowing us to achieve a gossiping algorithm with running time $3n + O(1)$. The gossiping algorithm works in two phases: *Phase 1* and *Phase 2*. The main goal of *Phase 1* is to move all messages originated in the left half of L to its right half (as far as possible) and vice versa. During *Phase 1* nodes run a code of two different processes. *Process 1* is responsible for transmission of original messages in two opposite directions. Initially all nodes and then gradually decreasing number of central nodes are involved in that process. *Process 2* is responsible for efficient monotonic gossiping (on a sparse sequence of messages) on both ends of the line. During *Phase 2* all nodes take part in efficient monotonic gossiping and they run the code of *Process 3*.

More formally we define sets of pairs (i, t), where i stands for a label of a node and t is a number of a time step, as follows:

- $G_1 = \{(i,t) | (0 < t \leq 2n) \text{ and } (r_1 \leq i \leq n - r_1)\}$,
- $G_{2L} = \{(i,t) | (12 < t \leq 2n + 12) \text{ and } (i \leq n/2 - (r_1 - 3))\}$,
- $G_{2R} = \{(i,t) | (12 < t \leq 2n + 12) \text{ and } (i \geq n/2 + (r_1 - 3))\}$,
- $G_{3L} = \{(i,t) | (2n + 12 < t \leq 3n + 12) \text{ and } (i \leq n/2 - r_2)\}$,
- $G_{3R} = \{(i,t) | (2n + 12 < t \leq 3n + 12) \text{ and } (i \geq n/2 + r_2)\}$.

Phase 1 corresponds to sets G_1, G_{2L}, and G_{2R}, where pairs in set G_1 represent nodes running the code of *Process 1* and pairs in sets G_{2L} and G_{2R} represent nodes running the code of *Process 2* on the left end and on the right end of a line respectively. Similarly *Phase 2* corresponds to sets G_{3L} and G_{3R} whose elements represent nodes running the code of *Process 3* on the left end and on the right end of a line respectively, see Figure 3a.

Phase 1 and *Phase 2* run in rounds. Each round of *Phase 1* consists of 4 steps and each round of *Phase 2* consists of 2 steps. During *Phase 1* a number of a round r_1 is defined as $\lceil t/4 \rceil$, where t is a current time step. We show later that *Phase 1* takes at most $2n + 12$ steps, i.e., $\lfloor n/2 \rfloor + 3$ rounds. Similarly, during *Phase 2* a number of a round r_2 corresponds to the value $\lceil (t - (2n + 12))/2 \rceil$, where t is a current time step.

The nodes governed by *Process 1* execute a fixed pattern of transmissions during each round. The pattern is based on values of ($t \mod 4$) and ($i \mod 4$), where t is a current time step and i is an index of a node. In contrast *Process 2* and *Process 3* have more complex the transmission selection mechanism that is based on offset values b, c, d and e, within sets G_*, compare with Figure 3b. The transmission selection in *Process 2* is based on values ($b \mod 3$) in G_{2L} and ($c \mod 3$) in G_{2R} as well as the number of an internal step $t \mod 4$ of

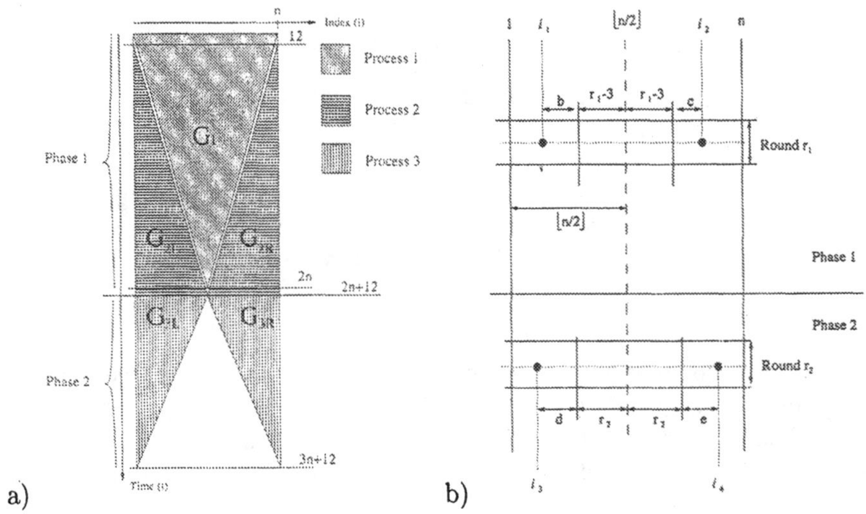

Figure 3. Gossiping on a line: a) Sets $G_1, G_{2L}, G_{2R}, G_{3L}$, and G_{3R}; b) Offset calculation.

round $\lceil t/4 \rceil$. Similarly the transmission selection in *Process 3* is based on values $(d \mod 3)$ in G_{3L} and $(e \mod 3)$ in G_{3R} as well as the number of an internal step of round $\lceil (t-(2n+12))/2 \rceil$.

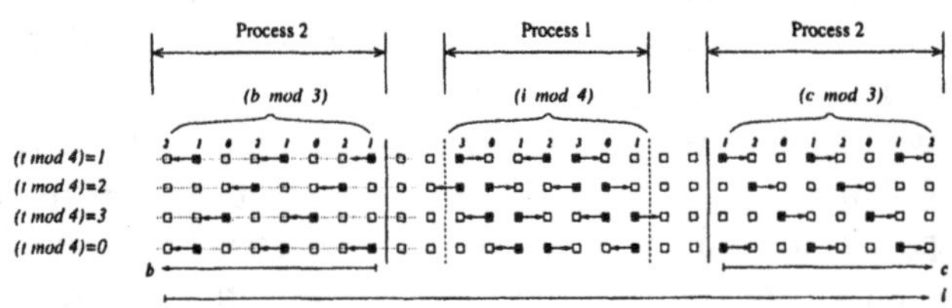

Figure 4. One round of Phase 1

Analysis of Phase 1: Initially each node contains a unit message. This message is kept in two copies. One will be send to the left and the other one to the right neighbour. The following invariant holds. During an execution of one *round* (four consecutive steps) of *Process 1* each node receives two new messages: one from its left and one from its right neighbour. It also transmits two messages: one message to its right neighbour (received during the last round from its left neighbour) and one message to its left neighbour (received during the last round from its right neighbour), see middle part in Figure 4. Note also that shrinking middle part of the line executing code of Process 1 after each round leaves always 2 messages on its left and 2 messages on its right border. These two messages are later dispatched and transmitted to the end of the line by *Process 2*.

Both ends of the line are managed by *Process 2* as follows. An *invariant* at the beginning of each round states that every third node is empty (starting with

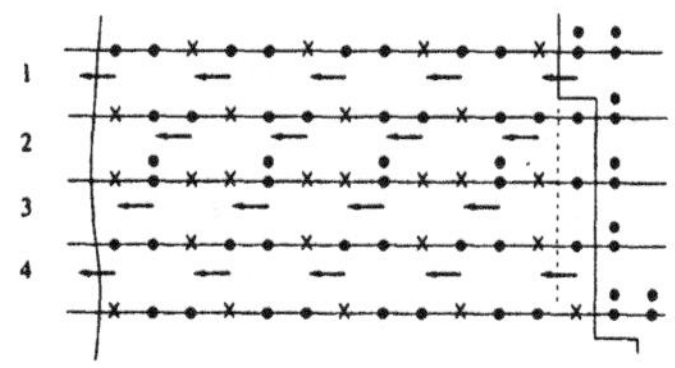

Figure 5. One round of Process 2

the most central one) and all other nodes contain a unit message, see Figure 5. During initial 3 steps of each round every node $\in G_{2L} \cup G_{2R}$ receives one message from its right neighbour and sends one message to its left neighbour. The last step of each round is used to move to the left (right) additional one half of messages available at nodes in G_{2L} (G_{2R}). This is to move a pattern of empty nodes towards the center of a line in order to maintain a validity of the invariant at the beginning of the next round. Note also that at any node the execution of *Process 2* starts three rounds after the execution of *Process 1* is completed in order to avoid collisions caused by nodes executing codes of different processes. In total, *Phase 1* consists of $\lfloor n/2 \rfloor + 3$ rounds which is $(\lfloor n/2 \rfloor + 3) \cdot 4 \leq 2n + 12$ time steps.

Analysis of Phase 2: The order in which messages are positioned on the line L at the end of Phase 1 allows to move them to the end of the line (in each half independently) in time $2 \cdot \lfloor n/2 \rfloor$. This is done by the execution of the algorithm used in the proof of Lemma 1 as follows. In each half of L all messages are grouped into two disjoint sequences S_1 and S_2 in which consecutive messages are at distance 3 apart. In odd steps of *Phase 2* the monotonic gossiping algorithm is applied on sequence S_1 and in even steps on sequence S_2. The monotonic gossiping algorithm applied on each sequence requires (and it can be completed in) $\lfloor n/2 \rfloor$ steps since this is the distance that has to be traversed by the most centrally positioned message and all messages traverse with the speed one position per one time step. Thus the total time of monotone gossiping performed in *Phase 2* is bounded by $2 \cdot \lfloor n/2 \rfloor \leq n$.

Lemma 4 *Radio gossiping on a line of size* n *can be performed in time* $3n+12$.

Proof: *Phase 1* requires $\leq 2n+12$ time steps. The partial monotonic gossiping in *Phase 2* takes time $\leq n$. Thus the total time complexity is bounded by $3n + 12$. □

Theorem 3 follows from Lemma 4 and Corollary 1.

4. Gossiping in trees

In this section we discuss the time complexity of the radio gossiping problem in general free trees.

Theorem 4 *The exact complexity of radio gossiping in trees of size n is* $\geq 3\frac{1}{6}n - 16$ *and* $\leq 5n + 8$.

Lower bound We use again the stack-and-knot argument, see section 2. Consider a line L of length $2m+1$ with a *knot* of length 4 in the middle of the line, see Figure 6, where $n = 2m+5$. In order to solve the gossiping problem on line L all messages from the left half of L have to be moved to the right half of L and vice versa. All these messages have to traverse via node a. This requires at least $4m$ independent time steps: $2m$ steps in which node a receives unit messages from its left and from its right neighbors, and $2m$ steps devoted to appropriate transmissions to the left and to the right hand side. Note that line L has knot K, and we have to guarantee that all messages are transmitted to the nodes on the knot too. Since node b (in K) may interfere in the process of moving elements along line L, the last message m_x transmitted through a can be send in time $4m + k$, for some $k \geq 0$. In order to complete the gossiping on knot K it is still required another m steps to send m_x to the end of the line. Thus the cost of gossiping on line L takes at least time $4m + k + m = 5m + k$. On the other hand note also that during $2m$ rounds (when $2m$ messages originated in L have been delivered to a) out of initial $4m+k$ rounds both node a and node b didn't transmit. The other $2m+k$ rounds could have been used either for transmissions from a to b or from b towards the end of the knot K. Note also that the number of messages transmitted beyond b in K cannot exceed the number of messages delivered from a to b. Which means that the number of messages transmitted beyond b is bounded by $(2m + k)/2$. It means that there are at least $2m - (2m + k)/2 = m - k/2$ other messages that have to be transmitted beyond b in knot K. Due to Lemma 1, this process requires at least $3(m - k/2)$ additional steps. Thus to complete the gossiping task in the knot we need at least $4m + k + 3(m - k/2) = 7m - k/2$ time steps in total. We need to find a balance between the two bounds. They meet each other when $2m = 3k/2$. And this happens when $k = 4m/3$. Thus the minimal number of steps required by gossiping on a line with one knot is $5m + 4m/3 = 6\frac{1}{3}m \geq 3\frac{1}{6}n - 16$. Since a line with one knot is a simple form of a free tree we have proved that the gossiping problem in trees is harder than its counterpart on a line.

Upper bound We show here that the gossiping in any tree of size n can be performed in time $O(n)$. We start the presentation with a simple gossiping

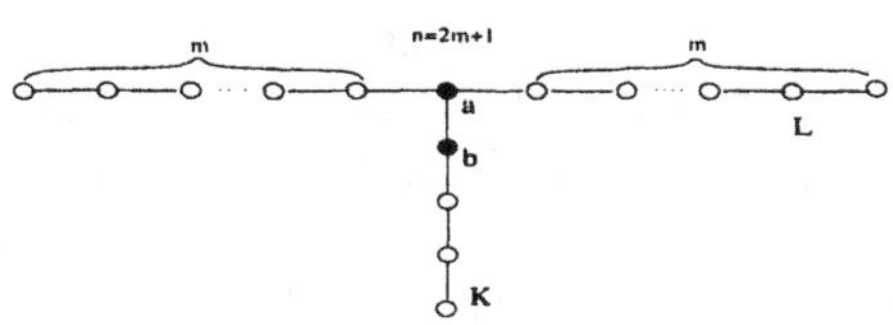

Figure 6. Line with a knot

algorithm with running time $6\frac{1}{2}n$, and then we present its tuned version with running time $5n + 8$.

We need the following definitions. Let $G = (V, E)$ be an undirected graph and let $r_i = \max_{v_j \in V} dist(v_i, v_j)$, for each $v_i \in V$. A *radius* of G is defined as $r_c = \min_{i=1,..,|V|} r_i$ where node v_c is called a *central node*. Recall also that $r_c \leq \lfloor n/2 \rfloor$.

Gossiping in time 5n+8 The algorithm behaves similarly to our best gossiping algorithm on a line. It collates messages in the center of the tree and simultaneously broadcasts them along its all branches. Similarly, as it happens on the line, when two sequences of messages (collated and broadcasted) work against each other, all messages are traversing with a speed 1/4, i.e. one position ahead in every 4 consecutive time steps (see page 198, Process 1). However in the final stage when only the broadcast sequence is available the speed of it is much higher, and it is 1/2 (see page 198, Process 2). The algorithm collecting all messages in the central node can be completed in time $4(n-1)+12 = 4n+8$. At this time only one message has not been broadcasted from the center (all other messages are already traversing with a speed 1/2 along all branches of the tree). The time remaining to complete the gossiping process is bounded by $2 \cdot \lfloor n/2 \rfloor \leq n$, where $\lfloor n/2 \rfloor$ is the length of the longest possible branch (a radius of a tree). In total the complexity of the algorithm is bounded by $4n + 8 + n = 5n + 8$.

5. General undirected graphs

In this section we study the time complexity of gossiping in general undirected graphs.

Theorem 5 *The asymptotic complexity of radio gossiping in general undirected graphs of size n is $\Omega(n \log n)$ and $O(n \log^2 n)$.*

Upper bound The algorithm is based on a broadcasting algorithm due to Chlamtac and Weinstein [5] that works in time $O(D \log^2 n)$, where D is a diameter of the graph. Their broadcasting algorithm goes along consecutive BFS levels in the graph spending at most $O(\log^2 n)$ time at each level. Our gossiping algorithm runs (similarly as in trees) in two stages. In *Stage 1* we collect all messages in some distinguished node z transporting them (in the form of a pipeline) along branches of any BFS spanning tree rooted in z. This can be done in time $O(n)$. The second stage (broadcasting) is performed via pipelined distribution of n unit messages from node z. The broadcasts are scheduled such that when the ith broadcasted message is at BFS level j (formed by nodes at distance j from z), the wave with $(i+1)$th broadcasted message is at level $j-3$. This allows to avoid collisions caused by messages at different BFS levels. All collisions caused by nodes on the same BFS level are handled in time $O(\log^2 n)$

using techniques from [5]. In total, the pipelined gossiping process can be accomplished in time $O(n \log^2 n)$.

Lower bound We need the following definitions.

Definition The undirected graph $G = \{V = L \cup R, E\}$ is *proper* if:

1 G is bipartite, with partition L and R,

2 $|L| \leq |R|$, and

3 for each $w \in R$ there exists $v \in L$, s.t., $(v, w) \in E$.

We say that the size of a proper graph $G = \{V = L \cup R, E\}$ is $|R|$. Let $n = |R|$. A *one-way* broadcasting problem in proper graphs is to transmit one message originally stored in all nodes in L to all nodes in R. A *one-way* gossiping problem in proper graphs is defined as follows. Assume that initially each node in L has the same set S of n messages. The task is to deliver one copy of each of n messages from S to every node in R. (Note that it doesn't matter which node in L transmits this copy)

Consider one (atomic) step of any gossiping algorithm. Two cases apply:

Case 1 Assume that there exists a proper graph G with $|R| = n$, s.t., one can deliver at most $O(n/\log n)$ messages to nodes in R during a single step of any communication procedure. We show that in this graph the one-way gossiping problem cannot be solved in time $o(n \log n)$. And indeed, the property of the graph imposes that the number of messages received in R at any time cannot be greater than $O(n/\log n)$. Note that to accomplish one-way gossiping we need to deliver from L to R $n \cdot n = n^2$ messages in total. However this process requires at least $n^2/O(n/\log n) = \Omega(n \log n)$ steps.

Update now G to G', s.t., nodes in L form a complete graph (to guarantee that G' is connected and gossiping in G' is feasible). We claim that gossiping in G' is harder than one-way gossiping in G, where set S is formed by n messages originated in R in gossiping problem. And indeed, any algorithm that solves gossiping problem in G can mimic one-way gossiping in G'. This means that gossiping in G' requires $\Omega(n \log n)$ steps, and the lower bound follows in this case.

Case 2 This is a complementary case, i.e., we assume that for any proper graph there exists a way to inform in a single step $\omega(n/\log n) = n \cdot f(n)/\log n$ nodes in R, for some asymptotically growing positive integer function $f(n)$, s.t., $f(n) = \omega(1)$ and $f(n) = o(\log n)$ (e.g. $f(n) = \sqrt{\log n}$). We show that under this assumption we can solve one-way broadcasting problem in any proper graph of size n in time $o(\log^2 n)$.

According to our assumption, we can inform in one round at least $n \cdot f(n)/\log n$ nodes in R. When this is done we remove all informed nodes from R and select appropriate number of nodes in L to much the property of proper graphs, i.e., if $|L| > |R|$ we pick at most $|R|$ nodes in L to match property 3 in proper graphs. When this is done we perform another step of one-way broadcasting, remove informed nodes in R and again select appropriate nodes in L. We repeat this process as long as the size of L is grater than, e.g., $\log n$. Afterwards all uninformed nodes in R receive broadcasting message sequentially in separate steps. Let $T(n)$ be the maximum number of time steps required by the one-way broadcasting algorithm in proper graph with $|R| = n$. The recurrence on $T(n)$ is defined as follows

1 $T(i) = T(i - i \cdot f(i)/\log i) + 1$, for $i > \log n$,

2 $T(i) = i$, for $i \leq \log n$.

In order to solve the recurrence we can find, e.g., a good estimation on a number of steps after which a value of argument n will be decreased by half. Note that after each step the value of an argument is decreased by some value. Initially by value $\frac{n \cdot f(n)}{\log n}$, and after some number of steps when the value of the argument will drop to $\frac{n}{2}$, by value $\frac{\frac{n}{2} \cdot f(\frac{n}{2})}{\log(\frac{n}{2})}$. Since all consecutive decrements had values $\geq \frac{\frac{n}{2} \cdot f(\frac{n}{2})}{\log(\frac{n}{2})}$ the number of decrements can be bounded by $\frac{\frac{n}{2}}{\frac{\frac{n}{2} \cdot f(\frac{n}{2})}{\log(\frac{n}{2})}} \leq \frac{\log n}{2f(\frac{n}{2})}$. Hence after at most $\frac{\log n}{2f(\frac{n}{2})}$ recursive steps the value of argument n will be decreased by half. Thus

$$T(n) \leq T(\tfrac{n}{2}) + \frac{\log n}{2f(\frac{n}{2})},$$

and

$$T(n) \leq \sum_{i=1}^{\log n} \frac{i}{2f(2^{i-1})} \leq \log n \cdot o(\log n) = o(\log^2 n)$$

since $f(n) = o(\log n)$ and $f(n) = \omega(1)$.

This means that we can perform one-way broadcasting in any proper graph G of size n in time $o(\log^2 n)$. For any proper graph $G = \{V = L \cup R, E\}$ of size n we define its extension $G'' = \{V'', E''\}$, where $V'' = V \cup \{\bar{v}\}$ and $E'' = E \cup \{(\bar{v}, v) : v \in L\}$. Note that due to our result for any G'' there should exist a broadcasting procedure (with source node $\bar{v}$) in time $o(\log^2 n)$. However we also know that this is not possible due to the construction of a family of graphs (extension of proper graphs) requiring $\Omega(\log^2 n)$ broadcasting time, by Alon *et al.* in [1]. This means that case 2 is not feasible and that the minimum gossiping time in undirected graphs is bounded from below by $\Omega(n \log n)$.

References

[1] N. Alon, A. Bar-Noy, N. Linial and D. Peleg, A Lower Bound for Radio Broadcast, *Journal of Computer and System Sciences* 43, (1991), pp. 290–298.

[2] R. Bar-Yehuda, A. Israeli, and A. Itai, Multiple communication in multi-hop radio networks, *SIAM Journal on Computing* 22 (1993), pp. 875–887.

[3] J.C. Bermond, L. Gargano, A.A. Rescigno, and U. Vacarro, Fast Gossiping by Short Messages, In Proc *22th Int. Colloq. on Automata, Languages and Programming*, (ICALP'1995), LNCS 944, pp. 135–146.

[4] D.P. Bertsekas and J.N.Tsitsiklis, Parallel and Distributed Computation: Numerical Methods, Prentice-Hall, Englewood Cliffs, NJ, 1988.

[5] I. Chlamtac and O. Weinstein, The Wave Expansion Approach to Broadcasting in Multihop Radio Networks, In Proc. *INFOCOM*, 1987.

[6] B.S. Chlebus, L. Gąsieniec, A.M. Gibbons, A. Pelc, and W. Rytter, Deterministic broadcasting in unknown radio networks, In Proc. 11*th ACM-SIAM Symp. on Discrete Algorithms*, (SODA'2000), pp. 861–870.

[7] B.S. Chlebus, L. Gąsieniec, A. Lingas, and A. Pagourtzis, Oblivious gossiping in ad-hoc radio networks, In Proc *5th Int. Workshop on Discrete Algorithms and Methods for Mobile Computing and Communications*, (DIALM'2001), pp. 44-51.

[8] M. Christersson, L. Gąsieniec, and A. Lingas, Gossiping with bounded size messages in *ad hoc* radio networks, In Proc 29*th Int. Colloq. on Automata, Languages and Programming*, (ICALP'2002), to appear.

[9] M. Chrobak, L. Gąsieniec, and W. Rytter, Fast broadcasting and gossiping in radio networks, In Proc. 41*st IEEE Symp. on Found. of Comp. Sci.*, (FOCS'2000), pp. 575–581.

[10] M. Chrobak, L. Gąsieniec, and W. Rytter, A randomized algorithm for gossiping in radio networks, In Proc. *7th Annual Int. Computing and Combinatorics Conference*, (COCOON'2001), pp. 483-492.

[11] A.E.F. Clementi, A. Monti, and R. Silvestri, Selective families, superimposed codes, and broadcasting in unknown radio networks, In Proc. 12*th ACM-SIAM Symp. on Discrete Algorithms*, (SODA'2001), pp. 709–718.

[12] M. Flamini and S. Perennes, Lower Bounds on the Broadcasting and Gossiping Time of Restricted Protocols, *Tech Report*, INRIA, January 1999.

[13] G. Fox, M. Johnson, G. Lyzenga, S. Otto, J. Salmon, and D. Walker, Solving Problems on Concurrent Processors, Volume 1, Prentice-Hall, 1988.

[14] I. Gaber and Y. Mansour, Broadcast in radio networks, in *Proc.* 6*th Ann. ACM-SIAM Symp. on Discrete Algorithms*, (SODA'1995), pp. 577–585.

[15] L. Gąsieniec and A. Lingas, On adaptive deterministic gossiping in *ad hoc* radio networks, In Proc. *13th ACM-SIAM Symp. on Discrete Algorithms* (SODA'2002), pp. 689–690.

[16] P. Indyk, Explicit constructions of selectors and related combinatorial structures, with applications, In Proc. 13*th ACM-SIAM Symp. on Disc. Alg.*, (SODA'2002), pp. 697–704.

[17] K. Ravishankar and S. Singh, Asymptotically optimal gossiping in radio networks, *Discrete Applied Mathematics* 61 (1995), pp 61-82.

[18] K. Ravishankar and S. Singh, Gossiping on a ring with radios, *Par. Proc. Let.* 6, (1996), pp 115-126.

[19] S. Tabbane, *Handbook of Mobile Radio Networks*, Artech House Publishers, 2000.

MEMORYLESS SEARCH ALGORITHMS IN A NETWORK WITH FAULTY ADVICE

Nicolas Hanusse*

LaBRI - CNRS - Université Bordeaux I, 351 Cours de la Liberation, 33405 Talence, France.

hanusse@labri.fr

Dimitris Kavvadias

Department of Mathematics, University of Patras, Rio, Greece.

kavadias@ceid.upatras.gr

Evangelos Kranakis†

School of Computer Science, Carleton University, Ottawa, ON, K1S 5B6, Canada.

kranakis@scs.carleton.ca

Danny Krizanc‡

Department of Mathematics, Wesleyan University, Middletown CT 06459, USA.

dkrizanc@caucus.cs.wesleyan.edu

Abstract In this paper, we present a randomized algorithm for a *mobile agent* to search for an item t stored at a node of a network, without prior knowledge of its exact location. Each node of the network has a database that will answer queries of the form "how do I get to t?" by responding with the first edge on a shortest path to t. It may happen that some nodes, called *liars*, give bad advice. We investigate a simple memoryless algorithm which follows the advice with some fixed probability $q > 1/2$ and otherwise chooses a random edge. If the degree of each node and number of liars k are bounded, we show that the expected number of edges to follow in order to reach t is bounded from above by $O(d + r^k)$,

*Research supported in part by CNRS-Jemstic "Mobicoop" grant.

†Research supported in part by NSERC (Natural Sciences and Engineering Research Council of Canada) and MITACS (Mathematics of Information Technology and Complex Systems) grants.

‡Research supported in part by NSERC and MITACS grants.

where d is the distance between the initial and target node and $r = \frac{q}{1-q}$. We also show that this expected number of steps can be significantly improved for particular topologies such as the complete graph, the torus, and the spider graph.

Keywords: Data Structure and Algorithms, Distributed Computing, Randomized Algorithms, Graph , Random Walks, Faulty Networks

1. Introduction

Searching for a piece of information is one of the most common tasks in a distributed environment. The evolution of the World Wide Web, for example, has led to the development and the wide use of specialized sites called search engines, that are able to provide the user with information on almost any possible query. Therefore, seeking a piece of information in a distributed environment is an algorithmic problem of major significance.

The usual mathematical model of such an environment is a graph (directed or undirected) whose nodes are computers that perform routing of message packets and/or store information. This information can be retrieved by an interrogating program that can be seen as an agent (*mobile agent*) that visits the nodes of the graph in a certain way moving from node to node using the edges. The purpose of the mobile agent is to locate a specific piece of information or *token* that resides in a certain node [KK99, KKKS00].

Graph searching is a central procedure in many computational problems. There are numerous techniques and algorithms for this problem that vary from purely deterministic (e.g., depth or breadth-first search) to purely randomized, (e.g. random walks, [AKL+79, KR95]). A variety of versions of the problem have been proposed that capture the requirements of different problem settings. For example, the agent may have limited knowledge of the topology of the network (e.g., searching in an unknown graph [MHG+88]) or it may exploit some geometric properties of the network (e.g., searching in the plane [BYCR93]). The efficiency of the search algorithm that the agent executes is measured, as usual, by the time it takes to find the token and/or its memory requirements.

In this paper we study the efficiency of searching algorithms where the advice is based on *shortest path* information that an agent receives when it arrives at a node. That is, the computer that resides at the node advises the agent by revealing to it an edge that is the beginning of a shortest path from the current node to the token. This kind of advice is common in the IP protocol where routing is done usually through the shortest available path. However, as is common in practice, we also assume that certain nodes–called *liars*–may give faulty information. This may model a situation where a computer is malfunctioning or its knowledge about the topology of the graph and/or the position of the token, is outdated. The searching problem in the presence of liars has been proposed and studied in [HKK00]. In that paper, several deterministic algorithms were proposed for various network topologies. In the present paper we propose a randomized algorithm for the same problem focusing mainly on the memory requirements. The reason for this is simple: the size of a network such as the Web is such that any agent that accumulates information in its traversal of the network can end up having devastating memory requirements. We are therefore interested in *memoryless* algorithms. It is easily seen however, that any memoryless algorithm that only receives shortest path information has to use randomization in order to avoid deadlocks. For specific topologies, our randomized algorithm is more competitive than any deterministic algorithm. For example, for the worst-case distribution of k liars, to find deterministically a token in the complete graph of n nodes, we need at least $O(k)$ steps and $O(\log k)$ bits of memory. Using our memoryless randomized algorithm, on average $O(1)$ steps are sufficient.

1.1. Definitions

A *network* is for our purposes an undirected graph $G = (V, E)$ where V is the set of nodes and E the set of edges. There are two specific nodes in the graph, the *start node* which we denote by s and the *token* which we denote by t. All nodes point to an edge incident to them. Each node is either a *truth–teller* or a *liar.* A node of the former kind points to an edge which is the start of a shortest path from that node to the token. A node of the latter kind may point to any incident edge except the start of a shortest path.

In our model, once a node is characterized it always gives the same response throughout the execution of the algorithm. Before the execution of the algorithm, all nodes are truth-tellers. An adversary then selects k nodes out of the n and substitutes their advice with bad advice. We will be working mainly in the *strong adversary* model in which the adversary can also modify the advice of each truthteller as long as they continue to provide an edge on a shortest path. We show in Section 4.2 a spectacular difference between a weak and a strong adversary model whenever there are many paths to go to the destination. Intuitively, a strong adversary can break any coordination of truth-tellers that might be of help to the agent. The agent is unaware of the kind of node in which it arrives.

Parameters that will be used are the total number of liars k and the initial distance d between s and t. We shall also be interested in the topology of the graph and in its diameter which we denote by D.

1.2. Markov Chains and zero memory algorithms

We use the terminology of [Nor97] of Markov Chains and random walks. Let $P = (0 \leq p_{x,y} \leq 1 | x, y \in V)$ be a *stochastic* matrix, i.e. the sum of every row is equal to 1. A *discrete-time Markov Chain* on a finite set of *states* V is a sequence of random variables $V_0, V_1, \ldots$ where $V_i \in V$ and such that V_{i+1} depends only on V_i and $\Pr(V_{i+1} = x | V_i = y) = p_{x,y}$. The matrix P is called the *transition probability matrix.* In the context of a biased random walk in a graph, $p_{x,y}$ denotes the probability to go from node x to node y in one step.

A node x *leads* to a node y if $\Pr(V_j = y \text{ for some } j \geq i | V_i = x) > 0$. A state y is *absorbing* if y leads to no other state. The *expected hitting time* or *hitting time*[1] $\mathbb{E}_x^y$ is the expected or mean number of steps starting from node x to reach node y. In our paper, the token t is the only absorbing state and the expected number of steps to reach t from s is noted $\mathbb{E}_s^t$.

We will make use of the following well-known theorem for Markov chains:

Theorem 1 *The vector of hitting times $\mathbb{E}^t = (\mathbb{E}_x^t : x \in V)$ is the minimal non-negative solution to the system of linear equations:*

$$\begin{cases} \mathbb{E}_t^t = 0 \\ \mathbb{E}_x^t = 1 + \sum_{y \neq t} p_{xy} \mathbb{E}_y^t \ \ \textit{for } x \in V \end{cases}$$

1.3. Description of the Algorithm

In the model above the mobile agent executes a simple, memoryless algorithm whose goal is to take advantage of the advice that it gets in a node but at the same time avoid deadlocks that may arise in some cases. The algorithm SEARCH is as follows:

1 The agent arrives at a node of degree, say, Δ and if it discovers the token it halts. Otherwise, it asks the node for advice.

2 The node responds by pointing to one of the edges incident to it.

3 The agent then flips a biased coin and with probability q it follows the advice. That is, it moves to the adjacent node which is the other endpoint of the edge. If it decides not to follow the advice (an event of probability $1 - q$), it selects uniformly another edge among the remaining $\Delta - 1$ incident edges.

4 It then moves to the other endpoint of the selected edge.

5 The above steps are repeated at the new node.

The idea behind this algorithm is that the agent expects, as a general rule, the advice to be correct. That is, the agent assumes that the majority of the nodes are truth tellers and that following their advice will bring the agent faster to the token. Therefore the probability q is taken to be more than $1/2$. The agent however cannot trust completely the advice that it receives as this may lead to a deadlock. Consider for example a case where the endpoints of an edge are nodes pointing to each other. It is clear that at least one of the nodes is a liar and that if the agent chooses to always follow the advice it will move back-and-forth between these two nodes forever. By allowing the agent not to trust the advice with some positive probability, we expect that, eventually, it will be able to get out of situations like this.

[1]In Norris' book [Nor97], the hitting time is the minimum amount of time needed to reach a node.

We note that the actions of the algorithm SEARCH resembles that of a *biased random walk* as studied by Azar et al. [ABK+96]. In that case however, on each step a coin is flipped which determines whether the current node a liar or not. Thus the number of liars is a random variable which is a function of the bias, and a node can in one instance be a liar on the next visit of the agent, a truthteller. In our case, the number and positions of the liars remains fixed throughout the execution of the algorithm. The analysis of both cases relies on Markov chains.

1.4. Results and structure of the paper

Our main task is to analyze the probabilistic algorithm SEARCH presented in Section 1.3. Our algorithm is memoryless and we are interested in the time complexity to reach the destination for the worst distribution of liars. We count each edge used by the mobile agent as a single step. We assume the majority of the nodes are truth tellers. As a consequence, we consider that $q > 1/2$.

In Section 2, we consider the chain graph where we show the expected number of steps of our algorithm is $O(d + r^k)$, for $r = \frac{q}{1-q}$. In Section 3 we show that we can generalize this result to an arbitrary graph. We use this to show that if the mobile agent has approximate knowledge of the distance and the number of liars, the token can be reached on average in $d + O(dk/\ln(dk^{-6}))$ steps.

In Section 4, we consider specific topologies. In Section 4.1, we deal with the complete graph for which we prove that, even if the number of liars is large, the expected number of steps to reach the destination is a constant. However, in the complete graph, the distance is the constant 1. In the torus, studied in Section 4.2, we prove a lower bound of $\Omega(d + r^k)$ steps in the strong adversary model. For the same distribution of liars, the expected number of steps is $O(d)$ in the weak adversary model. In Section 4.3 we study the case of having different shortest paths to reach the token. For spider graphs of N nodes per layer, the algorithm leads to an expected number of steps of $O(d + N^2 + r^{O(k/N)})$ using the fact that the mobile agent has the possibility to avoid obstacles, i.e. liars, to reach an area with only truth-tellers.

Due to space limitations, the proofs are not given.

2. Chain

It turns out that a very simple graph, namely the chain – or line – behaves as bad as any graph with the same number of liars. This is of no surprise as in a line all nodes –and therefore all liars– have to be visited as there is only one way to reach the token. We begin our study therefore with the chain graph.

Consider the n-vertex chain graph $0, 1, \ldots, n-1$. Each node i except the first and the last, is only connected to its predecessor $i-1$ and its successor $i+1$. We suppose that $0 \leq s \leq t$. If i is therefore a truth teller, $i < t$, the probability that the agent will move to node $i+1$ is $p_{i,i+1} = q$ (trust the advice) and to move to node $i-1$ is $p_{i,i-1} = 1-q$ (not trust the advice). The corresponding probabilities for i being a liar are of course $p_{i,i+1} = 1 - q$ and $p_{i,i-1} = q$. We assume that $p_{0,1} = p_{n-1,n-2} = 1$.

Set $r = \frac{q}{1-q}$. Since $q > 1/2$, $r > 1$. To each edge i between node i and $i+1$, we assign a weight w_{i+1} with $w_1 = 1$ and $w_{i+1} = \frac{p_{i,i+1}}{p_{i,i-1}} w_i$. So $w_{i+1} = rw_i$ if i is a truth teller and $w_{i+1} = \frac{w_i}{r}$ otherwise. For convenience, let $W_j = \sum_{i=0}^{j} w_i$.

The following result is given in [AF][Chapter 5], and constitutes a version of theorem 1 mentioned above:

Lemma 1 (Essential edge lemma) *Let $s < t$ be two nodes of the weighted chain $C = (w_1, w_2, \ldots, w_{n-1})$. For a random walk in C,*

$$\mathbb{E}_s^t = t - s + 2 \sum_{j=s+1}^{t} \sum_{i=1}^{j-1} w_i w_j^{-1}$$

Our main result for the chain follows from the following two lemmas.

Lemma 2 *The worst case distribution of liars before s that maximizes $\mathbb{E}_s^{s+1}$ is to place them consecutively and close to s.*

Lemma 3 *Let l_i be the number of liars between nodes 0 and i. Let β_i be the number of consecutive truth-tellers between the last liar and i. Then, for $l_i \geq \beta_i$, we have $\mathbb{E}_i^{i+1} < 1 + (\frac{6r^{l_i - \beta_i + 1}}{r-1})$ and for $l_i < \beta_i$, $\mathbb{E}_i^{i+1} < 1 + \frac{6}{r-1}$. For r constant, it turns out $\mathbb{E}_i^{i+1} = O(1 + r^{l_i - \beta_i})$.*

Theorem 2 *Let $s < t$ be two nodes of the chain with k liars and q be the probability of trusting the advice. Let $r = \frac{q}{1-q}$ and let d be the distance between s and t. Then the expected number of steps to reach t from s using algorithm* SEARCH *is less than $d\left(1 + \frac{6}{r-1}\right) + \frac{6r^{k+3}}{(r-1)^3}$. If q is constant this is $O(d + r^k)$.*

Theorem 3 *For r constant and for the worst distribution of k liars in the line, $\mathbb{E}_s^t = \Omega(d + r^k)$.*

3. Arbitrary Graph

In this section we study the expected number of steps to reach the token in an arbitrary graph. Let G be any graph. Starting from t, arrange all nodes of G in layers according to their distance from t. Denote by $L_i, i = 0, \ldots, m$ the set of nodes that are at distance i from t. Here, m is the maximum distance between t and any other node. Construct the following Markov chain $(Q_j), j = 0, \ldots, m$. Q has $m+1$ states each corresponding to a layer of the graph. As for the transition probabilities let the probability of moving from state Q_i to state Q_j, $p_{i,j}$, be:

$$p_{i,j} = \begin{cases} q & : \text{if } i = j+1 \text{ and all nodes in layer } L_i \text{ are truth-tellers;} \\ \frac{1-q}{\Delta_i - 1} & : \text{if } i = j+1 \text{ and there exists at least one liar in layer } L_i; \\ 1 - p_{i,i-1} & : \text{if } i = j-1; \\ 0 & : \text{if } |i-j| \neq 1. \end{cases}$$

In the above, we denote by Δ_i the maximum degree among all nodes of layer L_i. In effect, the Markov chain that we defined has one state for each layer of the graph and its transition probability from state (layer) Q_i to state Q_{i-1} is the minimum among the probabilities to move from any node in layer L_i to a node in layer L_{i-1}, that is, one step closer to the token: When the agent is on a layer with truth-tellers only, the probability of moving closer is of course q and moving further is $1-q$. When the agent is on a layer with at least one liar then it may happen that it resides on a node i which is actually a liar, in which case the probability of moving closer is at least $\frac{1-q}{\Delta_i-1}$. The initial state of (Q) is of course the state that corresponds to the layer which includes the starting node s. This is state Q_d since we have assumed that the token is at distance d from our initial position.

By our choice of transition probabilities the event of moving closer to state Q_0 from any state of Q, is less probable than moving from any node of G to another node closer to t. It is therefore clear that the expected number of steps to reach node t when starting from s, is less or equal than the expected number of steps to reach state Q_0 when starting from state Q_d.

The so constructed Markov chain (Q) is a line in which the distance between the initial and target nodes is d and with at most k liars. Notice that while the truth-tellers in (Q) preserve their usual probabilities (i.e. q to go closer and $1-q$ to go furher), the liars are "stronger" in the sense that the probability of going closer is now only $p_{i,i-1} = \frac{1-q}{\Delta_i-1}$. (Notice the change in the numbering of nodes in the current chain (Q).) Let $\Delta = \max_i \Delta_i$. Going back to the proof of Theorem 2 of the previous section, we see that the effect of a liar in the application of Lemma 1 is to divide the weight of the next edge by r. In our case $r = \frac{p_{i,i+1}}{p_{i,i-1}} = \frac{\Delta_i-2+q}{1-q}$. In the case of the current chain (Q) the parameter r is always less than $\frac{\Delta-1}{1-q}$. The above observation along with Theorem 2 therefore gives:

Theorem 4 *Let G be any network of maximal degree Δ with k liars in which the distance between the initial node s and the token t is d. Then the expected number of steps of a mobile agent to reach t is less than $d\left(1+\frac{6}{r-1}\right)+\frac{6r^{k+3}}{(r-1)^3}$ where $r = \frac{\Delta-1}{1-q}$.*

3.1. Tuning the mobile agent

As algorithm SEARCH is memoryless, the only parameter that can be adjusted to improve performance is the amount of "belief" of the agent to the advice it receives, that is, probability q. Even so, there can be many different versions of this problem depending on what we might consider known to the agent. Interesting choices include the distance, the number of liars, the distribution of liars, the location of the initial node in the network, the topology etc. As an easy example, we consider the worst topology, i.e the line, the worst distribution of liars and we assume that the order of the distance and the number of liars are known.

Theorem 5 *Let G be any network with k liars in which the distance between the initial node s and the token t is d. For $d > \frac{(k+3)^6}{2}$, taking $r = 1 + \frac{\ln(2d/(k+3)^6)}{2(k+3)}$, the expected number of steps of a mobile agent to reach t is $d + O(dk/\ln(dk^{-6}))$.*

4. Special Graphs

The above disapointing bound of $O(d + r^k)$ comes from the fact that there may exist a bottleneck between the initial node and the token where all k liars may reside. If however the topology of the graph allows multiple paths, then things can become much better as for example in the complete graph. Moreover multiple paths give rise to new interesting problems as a truth teller may now have a choice of shortest paths to point to. We show by giving a specific example in the torus that different advice from the truth tellers do make a difference in the expected number of steps.

4.1. Complete Graph

For the complete graph, we prove that, even when the number of liars is large, the expected number of steps to reach the destination is a constant.

Theorem 6 *Assume that in K_n (the complete graph with n nodes) the number of liars is $k = cn$, where c is a constant, $0 < c < 1$. Then the expected number of steps to reach the token is $\frac{1}{(q-qc)(1-q)} + O(\frac{1}{n})$.*

Corollary 1 *In the complete network K_n containing $\Theta(n)$ truth-tellers, the expected number of steps to reach the destination is $O(1)$.*

4.2. Multiple paths. The case of the torus

The effect of liars on the required time to reach the token may sometimes be affected by the advice of the truth tellers. This can happen in cases where multiple shortest paths exist from some nodes to the token and consequently multiple possibilities exist for the advice that the agent receives from a truth teller. The purpose of this section is to demonstrate this by studying the running time of the algorithm for a specific setup on the torus. A torus of $n = n_1 n_2$ nodes is a the graph obtained by the cartesian product of two cycles of length n_1 and n_2. For convenience, we label each node $u = (i, j)$ with $i = [-\lceil n_1/2 \rceil, \lfloor n_1/2 \rfloor]$ and $j = [-\lceil n_2/2 \rceil, \lfloor n_2/2 \rfloor]$. A node $v = (i', j')$ is a neighbor of u if $|i - i'| = 1(n_1)$ and $|j - j'| = 1(n_2)$. The torus is an appropriate example as it is a common network architecture with symmetry that allows relatively easy calculations.

Consider the following situation: Graph G is a torus with diameter D much larger than the distance d between the starting node and the token. The token is placed at the origin $(0, 0)$ while the k liars occupy the nodes from $(0, 1)$ to $(0, k)$. The starting node s is the node $(0, d)$. We distinguish two cases:

Case 1 All truth tellers with coordinates (x, y) point left if $x > 0$ and right if $x < 0$. They point to the appropriate (unique) direction if they lie on the y axis.

Comments: This case is a lower bound for this kind of adversary.

Case 2 All truth tellers with $y \neq 0$ point either up or down depending on their position. That is, all truth tellers with $y > 0$ point down and all truth tellers with $y < 0$ point up. They point to the appropriate direction if they lie on the x axis.

We next study the expected number of steps to reach the token in each of the above cases. We have the following result:

Theorem 7 *The expected number of steps to reach the token in Case 1 is* $\Omega(d + r'^k)$ *where* r' *is a constant depending on* q*. In contrast, in the second case the expected number of steps is proportional to* d*.*

4.3. Spider Graph

We now turn our attention to a particular graph which reveals some additional merits of our searching algorithm. We call this graph *spider graph* because it resembles a spider's web. A node u of the spider graph has the polar coordinates $(x, y) \in \{1, .., n/N\} \times \{0, \ldots, N-1\} \cup \{(0,0)\}$ where x represents the *radius*, i.e. the distance to the token and y the angle. Node $u \neq (0,0)$ has four neighbors $(x, y+1)$, $(x, y-1)$, $(x-1, y)$ and $(x+1, y)$. The node $(0,0)$ is called the *center* of the spider graph. We assume that the token lies on the center node and is connected to N nodes $(1,0), \ldots, (1, N-1)$. N is called the *density* which is the number of nodes on a cycle of the graph. Less important parameter is the number of cycles, or equivalently the number of nodes on any radius of the graph. This is because we assume, as usual, that the majority of the nodes are truth tellers and consequently far enough from the center all nodes are truth tellers. We are therefore interested in the part of the spider graph around the center and at a radius approximately the maximum distance between d (the distance of the initial node) and the distance of the furthest liar.

All truth tellers point of course at the center of the graph while we assume for simplicity that all liars point at the opposite direction. Hence with probability $p_r = q + (1-q)/3$ the agent will move along its current radius and with probability $p_c = 1 - p_r = 2(1-q)/3$ it will change radius and move to a neighboring node on the current cycle. Since the probabilities to move on a cycle clockwise and counterclockwise are equal, if we ignore the moves that the agent does along a radius, its resulting motion is an unbiased random walk on a ring with N nodes. This last observation explains how the agent is able to avoid areas with high concentration of liars: It "probes" the area in front (toward the token) in an attempt to get closer to the token. We know from Section 2 that if there are k liars in front, then it will take the agent $O(r^k)$ steps to pass them. But since the agent also moves around on the current ring continuously trying to get closer to the token, it will eventually discover areas with no liars or with only a few liars and it will pass from there.

We label the set of radii from 0 to $N-1$ in the clockwise direction. We denote by (X_i, Y_i) the polar coordinates of the mobile agent at time i where X_i corresponds to the distance from the center and Y_i is the label of the current radius. By convention, $(X_0, Y_0) = (d, 0)$. The following result comes from the *mixing time* in a ring of N nodes [AF]:

Lemma 4 *If q is constant then for $i = \Omega(N^2)$ and for all $y \in [0, N-1]$, $|\Pr(Y_i = y) - 1/N| = \epsilon$.*

Let t_x be the number of truth-tellers of layer x (at distance x from the destination). Let p_x^i be the probability, knowing that $X_i = x$, to go from layer x to layer $x-1$ at time i.

Lemma 5 *After $i = \Omega(N^2)$ steps, $p_x^i > 1/2$ if $t_x > cN$ with $c = \frac{1+2q}{2(4q-1)}$.*

The next result shows that if the density of the spider graph is significant, the time to reach the token can become much better.

Theorem 8 *In a spider graph of density N, if q is constant, the expected number of steps to reach the token is upper bounded by $O(d + N^2 + r^{\lfloor \frac{2k(4q-1)}{(6q-3)N} \rfloor})$.*

5. Conclusion

We have presented "memoryless" randomized algorithms to search for an item t contained in a node of a network, without prior knowledge of its exact location and under the assumption that some nodes, called *liars*, may give bad advice. We have provided an algorithm and studied its performance in an arbitrary network and also considered different topologies like the complete graph, the torus, and the spider graph. It would be interesting to consider techniques similar to [KR95, BKRU89] and [MR95][Chapter 6] in order to study time-memory tradeoffs, as well as search algorithms for multiple mobile agent systems in our model.

References

[ABK+96] Y. Azar, A. Broder, A. Karlin, N. Linial, and S. Phillips. Biased random walks. *Combinatorica*, 16:1–18, 1996.

[AF] D.J. Aldous and J.A. Fill. Reversible markov chains and random walks on graphs. (book in preparation, available on the web from http://www.stat.berkeley.edu/users/aldous/book.html).

[AKL+79] R. Aleliunas, R. M. Karp, R. J. Lipton, L. Lovász, and C. Rackoff. Random walks, universal traversal sequences and the complexity of maze problems. In *Proc. 20th FOCS*, pages 218–223, 1979.

[BKRU89] A. Broder, A. K. Karlin, P. Raghavan, and E. Upfal. Trading space for time in undirected $s-t$ connedtivity. In *Proc. STOC*, pages 543–549, 1989.

[BYCR93] R. Baeza-Yates, J. Culberson, and G. Rawlins. Searching in the plane. *Information and Computation*, 1993.

[HKK00] N. Hanusse, E. Kranakis, and K. Krizanc. Searching with mobile agents in networks with liars. In *Proc. EUROPAR'2000, LNCS 1900*, pages 583–590, Munich, 2000.

[KK99] E. Kranakis and D. Krizanc. Searching with uncertainty. In *Proc. SIROCCO'99, Carleton Scientific*, pages 194–203, 1999.

[KKKS00] L. M. Kirousis, E. Kranakis, D. Krizanc, and Y. Stamatiou. Locating information with uncertainty in fully interconnected networks. In *Proc. DISC'2000, LNCS 1914*, pages 283–296, Toledo, Spain, 2000.

[KR95] A. R. Karlin and P. Raghavan. Random walks and undirected graph connectivity: A survey. In *Discrete Probability and Algorithms, Institute of Mathematics and its Applications, Vol., 72, Springer Verlag, 1995*, pages 95–101, 1995.

[MHG+88] N. Megiddo, S. Hakimi, M. Garey, D. Johnson, and C. Papadimitriou. The complexity of searching a graph. *Journal of the ACM*, 1988.

[MR95] R. Motwani and P. Raghavan. *Randomized Algorithms.* Cambridge University Press, 1995.

[Nor97] J.R. Norris. *Markov Chains.* Cambridge University Press, 1997.

LOWER BOUNDS AND THE HARDNESS OF COUNTING PROPERTIES

Lane A. Hemaspaandra *and* Mayur Thakur
Department of Computer Science
University of Rochester, Rochester NY 14627, USA
{lane,thakur}@cs.rochester.edu

Abstract Rice's Theorem states that all nontrivial language properties of recursively enumerable sets are undecidable. Borchert and Stephan [BS00] started the search for *complexity-theoretic* analogs of Rice's Theorem, and proved that every nontrivial counting property of boolean circuits is UP-hard. Hemaspaandra and Rothe [HR00] improved the UP-hardness lower bound to $\mathrm{UP}_{\mathcal{O}(1)}$-hardness. The present paper raises the lower bound for nontrivial counting properties from $\mathrm{UP}_{\mathcal{O}(1)}$-hardness to FewP-hardness, i.e., from constant-ambiguity nondeterminism to polynomial-ambiguity nondeterminism. Furthermore, we prove that this lower bound is rather tight with respect to relativizable techniques, i.e., no relativizable technique can raise this lower bound to FewP-$\leq^p_{1\text{-}tt}$-hardness. We also prove a Rice-style theorem for NP, namely that every nontrivial language property of NP sets is NP-hard.

Keywords: ambiguity-bounded computation, boolean circuits, computational complexity, counting properties, lower bounds, Rice's Theorem.

1. Introduction

The relationship between languages and the machines used to recognize them plays an important role in both computability theory and complexity theory. Languages are semantic objects with which computability and complexity theories deal. Machines are syntactic objects used to describe the languages.

Rice's Theorem ([Ric53], see also [Ric56]) links, in a rather thrilling and broad way, these semantic and syntactic objects. Rice's Theorem says that for any language class $\mathcal{C}$, $\emptyset \subsetneq \mathcal{C} \subsetneq$ RE, the set of all machines whose languages belong to $\mathcal{C}$ is highly noncomputable, in particular is RE-$\leq_m$-hard or coRE-$\leq_m$-hard. Note that Rice's Theorem not only bridges between semantic and syntactic aspects, but also in its statement displays a theme that is central in both computability and complexity theory: the study of which languages hard for which classes with respect to which types of reductions. This theme will also be important in the present paper.

"Rice's Theorem" is commonly used to refer both to the strong form just mentioned and the weaker form that speaks just of undecidability.

Theorem 1 (Rice's Theorem) *Let A be a nonempty, proper subset of the recursively enumerable sets. Then the language $\{M \mid L(M) \in A\}$ is* RE-$\leq_m$-*hard or is* coRE-$\leq_m$-*hard.*

Corollary 2 (Rice's Theorem, second version) *Let A be a nonempty, proper subset of the recursively enumerable sets. Then the language $\{M \mid L(M) \in A\}$ is undecidable.*

Rice's Theorem may be viewed as a statement about the remarkable nontransparency of programs. Rice's Theorem says that no total Turing machine can test any nontrivial language property of programs.

Borchert and Stephan [BS00] raise the question of whether complexity-theoretic analogs of Rice's Theorem hold. Rice's Theorem deals with recursively enumerable languages, and any such language is accepted by some Turing machine. Borchert and Stephan show that a related result holds for the case of boolean formulas and boolean circuits. Their result deals with counting properties of boolean formulas—those properties that depend solely on the number of satisfying assignments of a boolean formula. In particular, Borchert and Stephan prove that any nontrivial *counting* property of circuits is UP-hard. (Throughout this paper $\mathcal{C}$-hard always means $\mathcal{C}$-$\leq_T^p$-hard, unless some other reduction is explicitly inserted as in, for example, $\mathcal{C}$-$\leq_{1\text{-}tt}^p$-hardness.)

Hemaspaandra and Rothe [HR00] improve the UP-hardness lower bound of Borchert and Stephan [BS00] to $\mathrm{UP}_{O(1)}$-hardness. That is, they prove that every nontrivial counting property of circuits is $\mathrm{UP}_{O(1)}$-hard. In the same vein, they ask if it is possible to improve the lower bound beyond $\mathrm{UP}_{O(1)}$, and they show that relativizable techniques cannot raise the $\mathrm{UP}_{O(1)}$-hardness lower bound to SPP-hardness, where SPP [OH93,FFK94] is the gap analog of UP. In particular, they note that if every nontrivial counting property of circuits is SPP-hard, then $\mathrm{SPP} \subseteq \Delta_2^p$.

The class FewP, of Allender and Rubinstein [AR88], is the collection of all NP sets acceptable via polynomial-ambiguity nondeterminism; $\mathrm{UP} \subseteq \mathrm{UP}_{O(1)} \subseteq \mathrm{FewP} \subseteq \mathrm{NP}$. We prove that every nontrivial counting property of circuits is FewP-hard (equivalently, is Few-hard), and indeed even is FewP-$\leq_{tt}^p$-hard and Few-$\leq_{tt}^p$-hard. We thus raise Hemaspaandra and Rothe's constant-ambiguity nondeterminism lower bound for nontrivial counting properties to polynomial-ambiguity nondeterminism. We prove that no relativizable technique can improve that lower bound to FewP-$\leq_{1\text{-}tt}^p$-hardness. We also prove an analog of Rice's Theorem for NP, namely, that all language properties of NP are NP-$\leq_m^p$-hard or coNP-$\leq_m^p$-hard.

Due to space limitations, the proofs are omitted; they can be found in [HT02].

2. Preliminaries

This section presents the notation and definitions used in the paper. All sets, unless otherwise stated, are considered subsets of Σ^*, where Σ is the standard alphabet $\{0,1\}$. The length of a string x is denoted by $|x|$. Σ^n denotes the set of strings in Σ^* of length exactly n. For any C and n, $C^{=n} = \{a \mid a \in C \wedge |a| = n\}$. We say that a set A is a *nontrivial subset* of B if $\emptyset \subsetneq A \subsetneq B$. $\langle \cdot, \ldots, \cdot \rangle$ usually denotes a standard, fixed, easily computable and invertible multi-arity pairing function (see [HHT97]) or a standard, fixed, easily computable and invertible 2-ary pairing function (which holds will be clear from context).

For any set A, χ_A denotes the characteristic function of A. That is, for any $x \notin A$, $\chi_A(x) = 0$, and for any $x \in A$, $\chi_A(x) = 1$. A boolean predicate Q is a total function from Σ^* to $\{0,1\}$. SAT denotes the set of all satisfiable boolean formulas. FP denotes the class of all (total) polynomial-time computable functions.

For any Turing machine N and any $x \in \Sigma^*$, we will use $N(x)$ as an abbreviation for "the computation of N on x." We will use DPTM as an abbreviation for "deterministic polynomial-time Turing machine." We will use NPTM as an abbreviation for "nondeterministic polynomial-time Turing machine." For any NPTM N, $\#\mathrm{acc}_N$ is the function such that, for any $x \in \Sigma^*$, $\#\mathrm{acc}_N(x)$ is equal to the number of accepting computation paths of $N(x)$. We will use UPTM as an abbreviation for "unambiguous, nondeterministic polynomial-time Turing machine." That is, N is a UPTM if and only if N is an NPTM and, for all $x \in \Sigma^*$, the number of accepting paths of N on input x is at most 1.

We now define some standard counting-based limited-ambiguity classes. We make use of the ambiguity-limited counting operator $\#_g$ [HR00] in defining these classes. In the definitions below, for simplicity of notation, we use $\#_k$ when we actually mean $\#_{\lambda x.k}$.

Definition 3

1. [Val79] #P *is the class of all functions* $f : \Sigma^* \to \mathbb{N}$ *such that there exists an* NPTM N *such that, for all* $x \in \Sigma^*$, *the number of accepting paths of* N *on input* x *is exactly* $f(x)$.
2. [HR00] *For any total function* $f : \mathbb{N} \to \mathbb{N}$, *and for any complexity class* $\mathcal{C}$, $\#_{\mathrm{f}} \cdot \mathcal{C}$ *is the set of all functions* $g : \Sigma^* \to \mathbb{N}$ *such that there exist a language* $L \in \mathcal{C}$ *and a polynomial* p *such that the following hold for each* $x \in \Sigma^*$:

   (a) $g(x) \leq f(|x|)$, *and*

   (b) $||\{y \mid |y| = p(|x|) \wedge \langle x, y \rangle \in L\}|| = g(x)$.
3. [HR00] *For each class* $\mathcal{C}$, *let* $\#_{\mathrm{const}} \cdot \mathcal{C} = \{g : \Sigma^* \to \mathbb{N} \mid (\exists k)[g \in \#_k \cdot \mathcal{C}]\}$.
4. [HV95] *For each class* $\mathcal{C}$, *let*

$$\#_{\mathrm{few}} \cdot \mathcal{C} = \{g : \Sigma^* \to \mathbb{N} \mid (\exists \text{ polynomial } q)[g \in \#_{\mathrm{q}} \cdot \mathcal{C}]\}.$$

5 [Val76] UP $= \{L \mid (\exists g \in \#_1 \cdot \mathrm{P})(\forall x \in \Sigma^*)[x \in L \iff g(x) > 0]\}$.

6 [AR88] FewP $= \{L \mid (\exists g \in \#_{\mathrm{few}} \cdot \mathrm{P})(\forall x \in \Sigma^*)[x \in L \iff g(x) > 0]\}$.

7 [CH90] Few $= \mathrm{P}^{\#_{\mathrm{few}} \cdot \mathrm{P}[1]}$, *i.e., the class of languages accepted by* P *machines that on each input are allowed at most one query to a function from* $\#_{\mathrm{few}} \cdot \mathrm{P}$.

8 [OH93,FFK94] SPP *is the class of all languages such that there exist a function* $f \in \#\mathrm{P}$ *and a polynomial-time computable function* $g : \Sigma^* \to \mathbb{N}$ *such that, for all* x*, the following hold:*

(a) $x \notin L \implies f(x) = g(x)$*, and*

(b) $x \in L \implies f(x) = g(x) + 1$.

FewP-hardness and Few-hardness are known to coincide (e.g., by using prefix search to pull down certificates one at a time, bit by bit, but note that doing so is truly using the adaptive nature of Turing reductions). FewP-$\leq_{tt}^p$-hardness and Few-$\leq_{tt}^p$-hardness are not known to coincide (and the "obvious" proof that they coincide, namely guessing all census values in parallel, does not seem to work—informally speaking, due to the fact that $\binom{q(n)}{q(n)/2}$, where q is a nonconstant polynomial, may be exponentially large), though certainly all Few-$\leq_{tt}^p$-hard sets are FewP-$\leq_{tt}^p$-hard.

We now define the standard reductions used in the paper.

Definition 4 *Let* A *and* B *be arbitrary sets.*

1 *We say that* $A \leq_m B$ *(*A recursively many-one reduces to B*) if there exists a recursive function* σ *such that, for all* x, $x \in A$ *if and only if* $\sigma(x) \in B$.

2 *We say that* $A \leq_T B$ *(*A recursively Turing reduces to B*) if there exists an oracle Turing machine* M*, such that* $L(M^B) = A$ *and, for each* x, $M^B(x)$ *halts.*

3 *We say that* $A \leq_m^p B$ *(*A polynomial-time many-one reduces to B*) if there exists a total, polynomial-time computable, function* σ *such that, for all* x, $x \in A$ *if and only if* $\sigma(x) \in B$.

4 *We say that* $A \leq_T^p B$ *(*A polynomial-time Turing reduces to B*) if there exists an oracle* DPTM M*, such that* $L(M^B) = A$.

5 *We say that* $A \leq_{tt}^p B$ *(*A polynomial-time truth-table reduces to B*) if there exists a* DPTM M *and a polynomial-time computable function* f *such that, for any* x*, there exists an integer* m *such that*

(a) $f(x) = \langle q_1, q_2, \ldots q_m \rangle$*, and*

(b) $M(\langle x, \chi_B(q_1), \chi_B(q_2), \ldots, \chi_B(q_m) \rangle)$ *accepts if and only if* $x \in A$.

6 *For any* $h : \mathbb{N} \to \mathbb{N}$, *we say that* $A \leq^p_{h(n)\text{-}tt} B$ *(A* polynomial-time $h(n)$-truth-table reduces to B*) if there exists a* DPTM M *and a polynomial-time computable function* f *such that, for any* x, *there exists an integer* $m \leq h(|x|)$ *such that*

(a) $f(x) = \langle q_1, q_2, \ldots q_m \rangle$, *and*

(b) $M(\langle x, \chi_B(q_1), \chi_B(q_2), \ldots, \chi_B(q_m)\rangle)$ *accepts if and only if* $x \in A$.

A set B is $\mathcal{C}$-hard exactly if $(\forall C \in \mathcal{C})[C \leq^p_T B]$.

Next we present some notations about circuits and boolean formulas that will be used in the paper.

Definition 5 (see [BS00]) *For any boolean formula (respectively, boolean circuits)* x, $\#_b(x)$ *(respectively,* $\#_c(x)$*) denotes the number of satisfying assignments of* x *(respectively, the number of appropriate-length input bit vectors that make the output of the circuit* 1*).*

In light of the existence of parsimonious versions of Cook's reduction (see [Gal74,Sim75]) and of efficient, parsimonious transformations between formulas and circuits, it holds that for each #P function f there exist functions $\hat{c}_f \in \mathrm{FP}$ and $\hat{b}_f \in \mathrm{FP}$ such that, for each x, $\hat{c}_f(x)$ is a boolean circuit satisfying $f(x) = \#_c(\hat{c}_f(x))$ and $\hat{b}_f(x)$ is a boolean formula satisfying $f(x) = \#_b(\hat{b}_f(x))$. For each $f \in \#\mathrm{P}$, arbitrarily choose one such $\hat{c}_f$ and one such $\hat{b}_f$ and denote these henceforward by $\hat{c}_f$ and $\hat{b}_f$.

We now present definitions related to the counting properties of circuits.

Definition 6 *Let* $A \subseteq \mathbb{N}$.

1 [BS00] Counting(A) *is the set of all boolean circuits such that the number of satisfying assignments of the circuit is a member of* A. *That is,*

$$\mathrm{Counting}(A) = \{\hat{c} \mid \#_c(\hat{c}) \in A\}.$$

2 *We say that* $T_0(A)$ *holds if and only if there exists an* n *such that* $n \in A \iff n+1 \notin A$, *and the least such* n *belongs to* $\overline{A}$.

3 *We say that* $T_1(A)$ *holds if and only if there exists an* n *such that* $n \in A \iff n+1 \notin A$, *and the least such* n *belongs to* A.

4 *For each* $A \subseteq \mathbb{N}$, *we say that* Counting(A) *is a* counting property *of* circuits.

5 *For each* $\emptyset \subsetneq A \subsetneq \mathbb{N}$, *we say that* Counting(A) *is a* nontrivial counting property of circuits.

Let $M_1, M_2, \ldots$ be any acceptable enumeration of Turing machines. The halting problem, which is RE-$\leq_m$-complete, is $\mathrm{HP} = \{x \mid M_{\mathrm{rank}(x)}(x) \text{ halts}\}$, where rank($x$) denotes the lexicographic rank of x, i.e., $\mathrm{rank}(\epsilon) = 1, \mathrm{rank}(0) = 2, \mathrm{rank}(1) = 3, \mathrm{rank}(00) = 4$, etc.

3. USAT_Q and Hardness for Polynomial Ambiguity

Hemaspaandra and Rothe [HR00] prove that every nontrivial counting property of circuits is $\text{UP}_{O(1)}$-hard. They also prove that it is unlikely that the $\text{UP}_{O(1)}$ lower bound can be raised much higher: If every nontrivial counting property of circuits is SPP-hard, then $\text{SPP} \subseteq \text{P}^{\text{NP}}$. (Fortnow [For97] provides a relativization in which SPP is not contained in P^{NP}.) In the light of these two results, it is natural to examine the complexity classes that fall between $\text{UP}_{O(1)}$ and SPP, and to ask whether it is possible to raise the $\text{UP}_{O(1)}$-hardness lower bound that holds for nontrivial counting properties.

Two natural complexity classes that lie between $\text{UP}_{O(1)}$ and SPP are FewP and Few. FewP is the polynomial-ambiguity version of UP, and Few is the class of languages accepted by polynomial-ambiguity nondeterministic Turing machines operating under any polynomial-time computable counting acceptance mechanism (see [CH90] for full details, or see Definition 3 for a simple alternate definition/characterization of the class). It is known that $\text{UP}_{O(1)} \subseteq \text{FewP} \subseteq \text{Few} \subseteq \text{SPP}$ [KSTT92,FFK94].

In this section we prove that every nontrivial counting property of circuits is Few-hard, thus raising the lower bound. We first prove that for any nontrivial property A, there exists a predicate Q such that at least one of Counting(A) and $\overline{\text{Counting}(A)}$ is $\leq_m^p$-hard for USAT_Q, where, for any boolean predicate Q, USAT_Q is defined (see [VV86]) as follows.

$$\chi_{\text{USAT}_Q}(x) = \begin{cases} \chi_{\text{SAT}}(x) & \text{if } \#_b(x) \in \{0,1\}, \\ Q(x) & \text{otherwise.} \end{cases}$$

The flavor of the following lemma, which we state here for completeness, is implicit in the comments at the end of Section 5.1 of [BS00].

Lemma 7 *Let* $A \subseteq \mathbb{N}$.

1 $(\exists n, m : n < m)[n \notin A \wedge m \in A] \implies (\exists Q)[\text{USAT}_Q \leq_m^p \text{Counting}(A)]$.

2 $(\exists n, m : n < m)[n \in A \wedge m \notin A] \implies (\exists Q)[\overline{\text{USAT}_Q} \leq_m^p \text{Counting}(A)]$.

Glaßer and Hemaspaandra [GH00] prove that for every Q, USAT_Q is Few-$\leq_{tt}^p$-hard.

Theorem 8 ([GH00]) *If* $L \in \text{Few}$, *then* $(\forall Q)[L \leq_{tt}^p \text{USAT}_Q]$.

We now can state the strengthening of the lower bound on the hardness of nontrivial counting properties of circuits from constant-ambiguity nondeterminism to polynomial-ambiguity nondeterminism.

Theorem 9 *For any nontrivial* ($\emptyset \subsetneq A \subsetneq \mathbb{N}$) *A*, Counting($A$) *is* Few-$\leq_{tt}^p$*-hard (and thus certainly* Few*-hard,* FewP*-hard, and* FewP-$\leq_{tt}^p$*-hard).*

We can prove Theorem 9 by noting that using Theorem 8 and Lemma 7 it follows; alternatively we can (and in the full version do) give a direct proof that gives more intuition about what is going on.

One might wonder whether it is possible to prove a strengthened version of Theorem 9 in which the "Few- $\leq_{tt}^p$" in the statement of Theorem 9 is changed to "NP-$\leq_T^p$" (or even to "NP- $\leq_{tt}^p$"). In fact, as is essentially noted by Borchert and Stephan [BS00, p. 492], if such a strengthened claim were true then it would follow that NP $\leq_T^p \oplus$P, where $\oplus$P [PZ83,GP86] is the class of languages L such that there exists a #P function f_L such that, for each $x \in \Sigma^*$, x is in L if and only if $f_L(x)$ is odd; and so since $\oplus$P is closed downward under Turing reductions [PZ83], if such a claim were true then NP $\subseteq \oplus$P. However, Torán [Tor88,Tor91] constructed a relativized world in which NP is not contained in $\oplus$P. Thus, Theorem 9's Few- $\leq_{tt}^p$ -hardness lower bound, relativized in the natural way we will discuss in the next section, cannot be strengthened to NP- $\leq_{tt}^p$ -hardness (or even to NP-$\leq_T^p$-hardness) using any relativizable technique.

We mention in passing that Theorem 9 seems neither to imply nor to be implied by a result of Borchert, Hemaspaandra, and Rothe that shows that certain "restricted counting classes" contain FewP [BHR00, Theorem 3.4]. On the one hand, the result of Borchert et al. applies only to promise classes; but on the other hand, the result of Borchert et al. (conditionally) concludes containment results rather than hardness results.

Valiant and Vazirani [VV86] prove that, for every Q, USAT$_Q$ is $\leq_{randomized}^p$-hard for NP, where we are using $\leq_{randomized}^p$ to denote the Valiant–Vazirani ([VV86], see also [BS00]) notion of randomized reduction. So the following result (which is a more refined, detailed statement of the flavor of of [BS00], Theorem 5.2) follows from Lemma 7.

Proposition 10 ([BS00]) *Let* $A \subseteq \mathbb{N}$.

1 $(\exists n, m \in \mathbb{N})[n < m \wedge n \notin A \wedge m \in A] \implies \text{NP} \leq_{randomized}^p$ Counting(A).

2 $(\exists n, m \in \mathbb{N})[n < m \wedge n \in A \wedge m \notin A] \implies \text{coNP} \leq_{randomized}^p$ Counting(A).

Proposition 10 gives a lower bound on the hardness of Counting(A). It is thus natural to seek an interesting upper bound. Theorem 11 states that under certain assumptions, Counting(A) is as easy as detecting unique solutions. In particular, for each nontrivial A, at least one of Counting(A) and $\overline{\text{Counting}(A)}$ $\leq_m^p$ -reduces to USAT$_Q$, for some Q.

Theorem 11 *Let* $A \subseteq \mathbb{N}$.

1 $T_0(A) \implies (\exists Q)[\text{Counting}(A) \leq_m^p \text{USAT}_Q]$.

2 $T_1(A) \implies (\exists Q)[\overline{\text{Counting}(A)} \leq_m^p \text{USAT}_Q]$.

Corollary 12 *For each* $\emptyset \subsetneq A \subsetneq \mathbb{N}$, *there exists a* Q *such that at least one of* Counting(A) *or* $\overline{\text{Counting}(A)}$ $\leq_m^p$ *-reduces to* USAT$_Q$.

Lemma 7 proves that under suitable conditions Counting(A) is $\leq_m^p$-hard for USAT_Q, for some predicate Q. On the other hand, Theorem 11 proves that under suitable conditions, Counting(A) is $\leq_m^p$-easy for $\mathrm{USAT}_{Q'}$, for some predicate Q'.

4. A Relativized Upper Bound on the Complexity of Counting Properties

Theorem 9 proves that all nontrivial counting properties of circuits are Few-$\leq_{tt}^p$-hard (and thus, FewP-$\leq_{tt}^p$-hard). Can the FewP-$\leq_{tt}^p$-hardness lower bound of nontrivial counting properties of circuits be improved? Hemaspaandra and Rothe [HR00] proved that raising the lower bound to SPP-hardness (a) would imply an unexpected complexity class containment, and (b) cannot be proven via relativizable proof techniques. However, in light of the fact that the previous UP-hardness and $\mathrm{UP}_{O(1)}$-hardness results in fact achieve in each case not just hardness (i.e., $\leq_T^p$-hardness) but even $\leq_{1\text{-}tt}^p$-hardness, it would be natural to hope that the FewP-$\leq_{tt}^p$-hardness result of Theorem 9 can at least be improved to FewP-$\leq_{1\text{-}tt}^p$-hardness. Nonetheless, we prove (in Theorem 17) that relativizable proof techniques cannot improve the FewP-$\leq_{tt}^p$-hardness lower bound of nontrivial counting properties to FewP-$\leq_{1\text{-}tt}^p$-hardness. In particular, we prove that there is a relativized world in which the following statement is false: "All nontrivial counting properties are FewP-$\leq_{1\text{-}tt}^p$-hard."

Before we state Theorem 17, we need to state what we mean by "counting property relative to an oracle." Counting, as defined and used in earlier sections, is based on the number of appropriate-length bit vectors that make the output of the circuit 1. For the purpose of relativizing counting properties we will define and use another equivalent, easily relativizable version of counting based on the number of accepting paths of NPTMs. For any $A \subseteq \mathbb{N}$, we call this version of counting PathCounting(A) and define it as follows. In what follows, let $N_1, N_2, \ldots$ be a fixed, nice enumeration of NPTMs such that, for each $x \in \Sigma^*$, N_i on input x robustly (i.e., for all oracles) runs within time $|x|^i + i$.

Definition 13 *Let* $A \subseteq \mathbb{N}$. *Then* PathCounting(A) *is defined as follows.*

$$\mathrm{PathCounting}(A) = \{\langle i, x, 1^{|x|^i+i}\rangle \mid \#\mathrm{acc}_{N_i}(x) \in A\}.$$

It follows from the existence of parsimonious versions of Cook's reduction that, for any A, PathCounting(A) $\leq_m^p$ Counting(A). It is also easy to see that, for any A, Counting(A) $\leq_m^p$ PathCounting(A). It easily follows that, for each $A \subseteq \mathbb{N}$ and $B \subseteq \Sigma^*$, (a) $B \leq_m^p$ Counting(A) $\iff$ $B \leq_m^p$ PathCounting(A), and (b) $B \leq_T^p$ Counting(A) $\iff$ $B \leq_T^p$ PathCounting(A). In fact, the reductions can be chosen so as to be independent of A, as the following proposition notes.

Proposition 14 *There exist polynomial-time computable functions f and g such that, for every A, the following hold.*

1 $(\forall x \in \Sigma^*)[x \in \mathrm{Counting}(A) \iff f(x) \in \mathrm{PathCounting}(A)]$, *and*

2 $(\forall x \in \Sigma^*)[x \in \text{PathCounting}(A) \iff g(x) \in \text{Counting}(A)]$.

Next, we define the relativized version of PathCounting.

Definition 15 *For each* $B \subseteq \Sigma^*$ *and each* $A \subseteq \mathbb{N}$, *we define* $\text{PathCounting}^B(A)$ *(*$\text{PathCounting}(A)$ *relative to oracle* B*) as follows.*

$$\text{PathCounting}^B(A) = \{\langle i, x, 1^{|x|^i+i}\rangle \mid \#\text{acc}_{N_i^B}(x) \in A\}.$$

Since we will need it in the statement of Theorem 17, we explicitly state the definition of relativized truth-table reductions.

Definition 16 *For any* $h : \mathbb{N} \to \mathbb{N}$, *and any* $A, B, C \subseteq \Sigma^*$, $A \leq_{h(n)\text{-}tt}^{p,C} B$ *(A* polynomial-time $h(n)$-truth-table reduces to B relative to oracle C*) if there exists an oracle* DPTM M *and a deterministic polynomial-time transducer* M_1 *such that, for all* x, *there exist an integer* $m \in \mathbb{N}$ *and strings* $q_1, q_2, \ldots, q_m$ *such that the following hold:*

1 $m \leq h(|x|)$,

2 $M_1^C(x)$ *produces* $\langle q_1, q_2, \ldots q_m\rangle$ *as its output, and*

3 $M^C(\langle x, \chi_B(q_1), \chi_B(q_2), \ldots, \chi_B(q_m)\rangle)$ *accepts if and only if* $x \in A$.

Next we turn to the following result, which shows that relativizable proof techniques cannot improve the FewP-$\leq_{tt}^p$-hardness lower bound of nontrivial counting properties to FewP-$\leq_{1\text{-}tt}^p$-hardness.

Theorem 17 *There is an oracle* $B \subseteq \Sigma^*$ *and a set* A, $\emptyset \subsetneq A \subsetneq \mathbb{N}$, *such that* $\text{PathCounting}^B(A)$ *is not* FewP^B-$\leq_{1\text{-}tt}^{p,B}$*-hard.*

The proof is by a diagonalization-based oracle construction involving counting and invoking the Party Lemma of Cai et al. [CGH+89].

5. The Natural NP Analog of Rice's Theorem

Rice's Theorem deals with language properties of RE sets. Borchert and Stephan [BS00] started the search for complexity-theoretic analogs of Rice's Theorem. They proved an analog of Rice's Theorem in circuit complexity that deals with the *counting* properties of circuits. In this section, we state an analog of Rice's theorem that deals with the *language* properties of NP. To be clear, let us specify more clearly our terminology. Let $N_1, N_2, \ldots$ be a fixed, nice enumeration of NPTMs. For specificity, let the enumeration be that of Du and Ko [DK00, Section 1.5] (though any *effective enumeration of languages in* NP in the formal time-sensitive sense of [DK00, Section 1.5] would work equally well). A *property of* NP is any subset of NP. A set $A \subseteq \mathbb{N}$ is said to be a *language property of* NP if there exists a property ρ of NP such that $A = \{i \in \mathbb{N} \mid L(N_i) \in \rho\}$.

We prove that any nontrivial language property of NP sets is NP-hard. Note that this is, in some sense, the exact analog of Rice's Theorem for NP:

Any nontrivial language property of NP is NP- $\leq_m^p$ -hard or coNP- $\leq_m^p$ -hard (compare this with Theorem 1).

Theorem 18 *Let A be any nonempty, proper subset of the* NP *sets. Then* $\{i \mid L(N_i) \in A\}$ *is* NP- $\leq_m^p$ *-hard or is* coNP- $\leq_m^p$ *-hard.*

As an immediate corollary, we have the following result.

Corollary 19 *Let A be a nonempty, proper subset of the* NP *sets. Then* $\{i \in \mathrm{N} \mid L(N_i) \in A\}$ *is* NP*-hard.*

Note that Theorem 18 is a natural complexity-theoretic analog of Theorem 1 and Corollary 19 is a natural complexity-theoretic analog of Theorem 2.

However are these two results trivial in light of the following fact which states that every nontrivial language property of NP is undecidable?

Fact 20 *Let $N_0, N_1, \ldots$ be an enumeration of* NPTMs*. Let A be a nontrivial subset of* NP*. Then* $\{i \mid L(N_i) \in A\}$ *is undecidable (in fact, is either* RE- $\leq_m$ *-hard or* coRE- $\leq_m$ *-hard).*

It might seem that Theorem 18 follows from Fact 20. However, as we will show (as Theorem 23), under reasonable complexity-theoretic assumptions, RE- $\leq_m$ -hardness does not imply NP- $\leq_m^p$ -hardness (though, in fact, under other complexity-theoretic assumption we will see that RE- $\leq_m^p$ -hardness does imply NP- $\leq_m^p$ -hardness). We first state a useful definition and result due to Karp and Lipton [KL80].

Definition 21 ([KL80])

1 *For each language class $\mathcal{C}$ and each function $f : \mathrm{N} \to \mathrm{N}$, $\mathcal{C}/f$ is defined as follows.* $\mathcal{C}/f = \{L \mid (\exists g)(\exists L_1 \in \mathcal{C})(\forall x)[|g(1^{|x|})| \leq f(|x|) \wedge (x \in L \iff \langle x, g(1^{|x|})\rangle \in L_1)]\}$.

2 *For each language class $\mathcal{C}$ and each function class $\mathcal{F}$, $\mathcal{C}/\mathcal{F}$ is defined as follows.* $\mathcal{C}/\mathcal{F} = \{L \mid (\exists f \in \mathcal{F})[L \in \mathcal{C}/f]\}$.

Theorem 22 [KL80] *If* SAT $\in$ P$/O(\log n)$*, then* P $=$ NP.

We now have the following result, which says that the issue of whether RE- $\leq_m^p$ -hardness implies NP-hardness is completely controlled by the P $=$ NP question.

Theorem 23

1 *If* P $=$ NP*, then every* RE- $\leq_m$ *-hard set is* NP- $\leq_m^p$ *-hard (and thus certainly* NP*-hard).*

2 *If* P $\neq$ NP*, then there is some* RE- $\leq_m$ *-hard set that is not* NP*-hard (and thus certainly not* NP- $\leq_m^p$ *-hard).*

In this paper we have been discussing analogs, of a nonprobabilistic nature, of Rice's Theorem. We mention that one can investigate analogs of Rice's Theorem that use probabilistic notions in their attempts to frame analogs of Rice's Theorem [BGI+01] or that are aimed at handling probabilistic complexity classes [HT01].

6. Conclusions and Open Issues

This paper improved the lower bound for nontrivial counting properties of circuits from $\mathrm{UP}_{O(1)}$-hardness to Few-hardness. It showed that relativizable techniques cannot improve the Few-hardness lower bound of nontrivial counting properties of circuits to Few-$\leq^{p}_{1\text{-}tt}$-hardness. The paper also proved a Rice-style theorem for language properties of NP sets.

Can the Few-$\leq^{p}_{tt}$-hardness result of the present paper be improved to Few-$\leq^{p}_{n^k\text{-}tt}$-hardness, for some fixed k? Or conversely, given an arbitrary k, does there exist a relativization—even stronger than the one mentioned earlier in the paper—such that there exists a nontrivial subset, A, of $\mathbb{N}$, such that A is not Few-$\leq^{p}_{n^k\text{-}tt}$-hard, or better yet, can one show some unexpected complexity class collapse that would follow were every nontrivial counting property of circuits Few-$\leq^{p}_{n^k\text{-}tt}$-hard? We conjecture that the Few-$\leq^{p}_{tt}$-hardness result cannot be improved to Few-$\leq^{p}_{n^k\text{-}tt}$-hardness.

Acknowledgments: We thank H. Hunt for commenting that an alternate proof of Corollary 19 might be obtained using techniques similar to those of [Hun82], and we thank an anonymous IFIP TCS 2002 conference referee for helpful comments. This work was supported in part by grants NSF-CCR-9322513 and NSF-INT-9815095/DAAD-315-PPP-gü-ab.

References

[AR88] E. Allender and R. Rubinstein. P-printable sets. *SIAM Journal on Computing*, 17(6):1193–1202, 1988.

[BGI+01] B. Barak, O. Goldreich, R. Impagliazzo, S. Rudich, A. Sahai, S. Vadhan, and K. Yang. On the (im)possibility of obfuscating programs. Report 2001/069, Cryptology ePrint Archive, August 2001. Preliminary version appears in *Advances in Cryptology - CRYPTO '2001*.

[BHR00] B. Borchert, L. Hemaspaandra, and J. Rothe. Restrictive acceptance suffices for equivalence problems. *London Mathematical Society Journal of Computation and Mathematics*, 3:86–95, 2000.

[BS00] B. Borchert and F. Stephan. Looking for an analogue of Rice's Theorem in circuit complexity theory. *Mathematical Logic Quarterly*, 46(4):489–504, 2000.

[CGH+89] J. Cai, T. Gundermann, J. Hartmanis, L. Hemachandra, V. Sewelson, K. Wagner, and G. Wechsung. The boolean hierarchy II: Applications. *SIAM Journal on Computing*, 18(1):95–111, 1989.

[CH90] J. Cai and L. Hemachandra. On the power of parity polynomial time. *Mathematical Systems Theory*, 23(2):95–106, 1990.

[DK00] D. Du and K. Ko. *Theory of Computational Complexity*. John Wiley and Sons, 2000.

[FFK94] S. Fenner, L. Fortnow, and S. Kurtz. Gap-definable counting classes. *Journal of Computer and System Sciences*, 48(1):116–148, 1994.

[For97] L. Fortnow. Counting complexity. In L. Hemaspaandra and A. Selman, editors, *Complexity Theory Retrospective II*, pages 81–107. Springer-Verlag, 1997.

[Gal74] Z. Galil. On some direct encodings of nondeterministic Turing machines operating in polynomial time into P-complete problems. *SIGACT News*, 6(1):19–24, 1974.

[GH00] C. Glaßer and L. Hemaspaandra. A moment of perfect clarity I: The parallel census technique. *SIGACT News*, 31(3):37–42, 2000.

[GP86] L. Goldschlager and I. Parberry. On the construction of parallel computers from various bases of boolean functions. *Theoretical Computer Science*, 43(1):43–58, 1986.

[HHT97] Y. Han, L. Hemaspaandra, and T. Thierauf. Threshold computation and cryptographic security. *SIAM Journal on Computing*, 26(1):59–78, 1997.

[HR00] L. Hemaspaandra and J. Rothe. A second step towards complexity-theoretic analogs of Rice's Theorem. *Theoretical Computer Science*, 244(1–2):205–217, 2000.

[HT01] L. Hemaspaandra and M. Thakur. Rice-style theorems for complexity theory. Technical Report TR-757, Department of Computer Science, University of Rochester, Rochester, NY, September 2001.

[HT02] L. Hemaspaandra and M. Thakur. Lower bounds and the hardness of counting properties. Technical Report TR-768, Department of Computer Science, University of Rochester, Rochester, NY, January 2002.

[Hun82] H. Hunt. On the complexity of flowchart and loop program schemes and programming languages. *Journal of the ACM*, 29(1):228–249, 1982.

[HV95] L. Hemaspaandra and H. Vollmer. The Satanic notations: Counting classes beyond #P and other definitional adventures. *SIGACT News*, 26(1):2–13, 1995.

[KL80] R. Karp and R. Lipton. Some connections between nonuniform and uniform complexity classes. In *Proceedings of the 12th ACM Symposium on Theory of Computing*, pages 302–309. ACM Press, April 1980. An extended version has also appeared as: Turing machines that take advice, *L'Enseignement Mathématique*, 2nd series, 28, 1982, pages 191–209.

[KSTT92] J. Köbler, U. Schöning, S. Toda, and J. Torán. Turing machines with few accepting computations and low sets for PP. *Journal of Computer and System Sciences*, 44(2):272–286, 1992.

[OH93] M. Ogiwara and L. Hemachandra. A complexity theory for feasible closure properties. *Journal of Computer and System Sciences*, 46(3):295–325, 1993.

[PZ83] C. Papadimitriou and S. Zachos. Two remarks on the power of counting. In *Proceedings 6th GI Conference on Theoretical Computer Science*, pages 269–276. Springer-Verlag *Lecture Notes in Computer Science #145*, January 1983.

[Ric53] H. Rice. Classes of recursively enumerable sets and their decision problems. *Transactions of the AMS*, 74:358–366, 1953.

[Ric56] H. Rice. On completely recursively enumerable classes and their key arrays. *Journal of Symbolic Logic*, 21:304–341, 1956.

[Sim75] J. Simon. *On Some Central Problems in Computational Complexity.* PhD thesis, Cornell University, Ithaca, N.Y., January 1975. Available as Cornell Department of Computer Science Technical Report TR75-224.

[Tor88] J. Torán. *Structural Properties of the Counting Hierarchies.* PhD thesis, Universitat Politècnica de Catalunya, Barcelona, Spain, 1988.

[Tor91] J. Torán. Complexity classes defined by counting quantifiers. *Journal of the ACM*, 38(3):753–774, 1991.

[Val76] L. Valiant. The relative complexity of checking and evaluating. *Information Processing Letters*, 5(1):20–23, 1976.

[Val79] L. Valiant. The complexity of enumeration and reliability problems. *SIAM Journal on Computing*, 8(3):410–421, 1979.

[VV86] L. Valiant and V. Vazirani. NP is as easy as detecting unique solutions. *Theoretical Computer Science*, 47(3):85–93, 1986.

FRAMEWORK FOR ANALYZING GARBAGE COLLECTION *

Matthew Hertz Neil Immerman J Eliot B Moss
Dept. of Computer Science University of Massachusetts Amherst, MA 01003
{hertz, immerman, moss}@cs.umass.edu

Abstract While the design of garbage collection algorithms has come of age, the analysis of these algorithms is still in its infancy. Current analyses are limited to merely documenting costs of individual collector executions; conclusive results, measuring across entire programs, require a theoretical foundation from which proofs can be offered. A theoretical foundation also allows abstract examination of garbage collection, enabling new designs without worrying about implementation details. We propose a theoretical framework for analyzing garbage collection algorithms and show how our framework could compute the efficiency (time cost) of garbage collectors. The central novelty of our proposed framework is its capacity to analyze costs of garbage collection over an entire program execution.

In work on garbage collection, one frequently uses heap traces, which require determining the exact point in program execution at which each heap allocated object "dies" (becomes unreachable). The framework inspired a new trace generation algorithm, Merlin, which runs more than 800 times faster than previous methods for generating accurate traces [7]. The central new result of this paper is using the framework to prove that Merlin's asymptotic running time is optimal for trace generation.

1. Introduction

Most modern computer languages use a heap to hold objects allocated dynamically during the running of a program and a garbage collector to re-

*A version of this paper including the proofs, related works, and proposed refinements to the framework can be found as UMass CMPSCI Tech Report TR-02-016 at ftp://ftp.cs.umass.edu/pub/techrept/techreport/2002/UM-CS-2002-016.ps

This work is supported by NSF ITR grant CCR-0085792, NSF CCR-9877078, and IBM. Any opinions, findings, conclusions, or recommendations expressed in this material are the authors' and do not necessarily reflect those of the sponsors.

move no longer needed objects from the heap. As use of these modern computer languages is increasing dramatically, it is very important that garbage collection run quickly. Towards this goal, many different garbage collection algorithms, optimizations, and techniques have been proposed and studied, e.g., [1, 2, 3, 6, 8, 9]. The experimental results document running times, relative volume of objects examined and copied, and other dynamically generated metrics from a small set of benchmarks. While this information illustrates and convinces, it is unable to *prove* the arguments being made. For conclusive results, arguments need a theoretical foundation from which proofs can be offered. A strong theoretical foundation also supports abstract analyses of garbage collection, enabling a more thoughtful look at where opportunities for improvement exist.

This paper presents a new theoretical framework for examining garbage collection. Our framework is robust enough to capture the behavior of a garbage collector over the execution of a program, but abstract enough not to require any specific fashion of implementation. This allows the framework to be used with most garbage collection algorithms, to prove that collectors have several important properties, and to be expanded easily to include other analyses. While this is an important feature of our framework, its key feature is computing a garbage collector's asymptotic running time (or other costs) over an entire program execution, and not just a single invocation. With this, our framework can be used to prove optimality for garbage collection and related algorithms.

As an example of our framework's usefulness, we analyze the Merlin trace generation algorithm. Prior research has described Merlin and discussed its running time [7]; using our framework, we formally prove its asymptotic running time and that this time is optimal for trace generation.

Sections 2 and 3 of this paper discuss the structures and graphs that make up our framework. Section 4 uses the framework in a series of proofs of algorithmic requirements, asymptotic running times, and optimal running times. Finally, Section 6 summarizes these results.

2. Structures Used

Many modern computer languages allow objects to be allocated dynamically into a heap during program execution. This makes writing the program easier, but requires that objects be freed during program execution to avoid running out of memory. Many languages (such as Lisp, Smalltalk, and Java) use garbage collection (GC) to reclaim memory automatically, because GC increases program safety and makes the programmer's job easier. Difficulty arises in limiting the amount of time used for automated memory reclamation.

It has long been understood that a program's heap memory can be envisioned as a graph. Thus, our framework presents the analyses as graph theoretic problems. We first explain how a program allocates objects dynamically and how garbage collectors determine which objects may be freed. We then propose a series of graphs that model the heap and capture the execution of the program and garbage collector.

2.1. Program Structures

Before explaining our framework, it is important to understand what "objects" are and how programs use objects allocated dynamically into the heap.

Objects and References— We model each object as a finite mapping of keys to reference values, i.e., locations in the heap; each key names a unique field of the object. The fields are typed; they may contain values of primitive data types (such as integers) or may reference other objects. A *referring field* may refer to an object within the heap or may be null (i.e., does not refer to any object). Because garbage collection is concerned only with how objects refer to one another, our framework considers the content only of fields that may contain references to other objects. When primitive data types and referring types cannot be disambiguated (e.g., C++ integer fields may contain references), the framework must consider the contents of all fields.

The Heap and Program Roots— A program's *heap* exists within an address space and holds the objects dynamically allocated during the program's execution. When allocating an object into the heap, the memory manager reserves space in the heap for each field and sets the map from the object's keys. Typically, the memory for an object is contiguous and a field's mapping key is the offset from the start of the object to that field. Objects reside in the heap because the program needs them; unneeded objects may be removed from the heap. A *garbage collector* reclaims these obsolete objects.

Determining which objects are no longer needed (will not be used again) requires knowledge of the future. As this is not always possible (true liveness is an undecidable property), *reachability-based* garbage collectors reclaim only objects they can demonstrate that the program will not use.

Objects within the heap are allocated during program execution; the program, not knowing where in memory the objects reside, cannot access them directly. To access the heap the program relies on its *root set*, locations the program accesses directly that may hold references into the heap. Like an object, the root set is a mapping of keys (each representing a unique *root* with a referring type) to reference values. Unlike an object, this mapping is neither bounded nor fixed. Keys may be added and removed, because *root references* are found in, for example, the program stack and the static and global variable table, which change size during execution. The program uses dynamically allocated objects only through the root set; objects not reachable from the root set cannot be used by the program. Garbage collectors determine which objects the program may use and which may be safely removed from the heap by analyzing reachability from the root set. For convenience we use "reachable" and "live" interchangeably, and likewise "unreachable" and "dead".

2.2. Heap State

The current state of the heap is defined by the objects and references within the heap and the program's root set. To allow for further analyses, we also include the set of objects that have been identified by the collector as unreachable. We represent the *heap state* (the current state of the heap) as a rooted, directed multi-graph. We call this multi-graph the *heap state multi-graph*, and express it as $H = (L, D, r, E)$.[1]

The set of vertices of H, $V = L \cup D$, includes a vertex for each object allocated in the heap, and a special vertex, called the root vertex and designated as r, representing the root set. The set D contains the vertices identified as dead, while L includes the remaining vertices ($L = V - D$, so $L \cap D = \emptyset$). Since roots are always reachable, r must be in L. Vertices represent only the existence of an object. This representation makes modeling garbage collectors easier by abstracting away implementation details of objects' actual locations in the address space.

Edges in the multi-graph represent references to objects in the heap. The edge multiset, E, contains an edge $(\langle v, n \rangle, o)$ if and only if object v at the field mapped to by key n contains a reference to object o. Notice that v may be r, in which case n is the key to a root that refers to o.

This structure is not a graph, but a multi-graph, because an object may have multiple fields that refer to the same object. Likewise, there may be multiple root locations that refer to the same object. Just as a garbage collector analyzes the heap using the root set and objects, this multi-graph can be analyzed for the relationships between the root vertex and other vertices.

2.3. Reachability

Given $H = (L, D, r, E)$, we say v refers to o (in H), written $refers_H(v, o)$, if and only if v has at least one field that refers to object o: $refers_H(v, o) = (\exists n)((\langle v, n \rangle, o) \in E)$.

Extending this definition, we say v *reaches* o (in H), written $reaches_H(v, o)$, if and only if v and o are equal or o is in the transitive closure of objects to which v refers: $reaches_H(v, o) = (v = o \vee (\exists p)(refers_H(v, p) \wedge reaches_H(p, o))$.

Given the importance of objects that the program can access from the root set, an object o is *reachable* (in H) if and only if r reaches o ($reachable_H(o) = reaches_H(r, o)$). Reachable objects may be used in the future by the program. Objects not reachable in the heap state (e.g., objects not in the transitive closure of the root set) cannot be accessed by the program, so garbage collectors remove only *unreachable* objects.

A heap state multi-graph $H = (L, D, r, E)$ is *well-formed* if and only if all reachable objects are in L: $L \supseteq \{o | reachable_H(o)\}$. From here on we are concerned only with well-formed heap state multi-graphs.

2.4. Program Actions

The heap state and its corresponding graph are useful for analyzing snapshots of a program. But programs and their heaps are dynamic entities: the

[1]Other elements could exist within the heap state multi-graph; we include only those elements needed for this paper.

Action Name	*Effect*	*Precondition*
Object Allocation	$L_{t+1} = L_t \cup \{o\}$; $E_{t+1} = E_t \cup \{(\langle r, n\rangle, o)\}$	$o \notin L_t \cup D_t \wedge$ $\neg(\exists o')((\langle r, n\rangle, o') \in E_t$
Root Creation	$E_{t+1} = E_t \cup \{(\langle r, n\rangle, o)\}$	$reachable_{H_t}(o) \wedge$ $\neg(\exists o')((\langle r, n\rangle, o') \in E_t)$
Root Deletion	$E_{t+1} = E_t - \{(\langle r, n\rangle, o)\}$	$(\langle r, n\rangle, o) \in E_t$
Heap Creation	$E_{t+1} = E_t \cup \{(\langle v, n\rangle, o)\}$	$reachable_{H_t}(v) \wedge reachable_{H_t}(o)$ $\wedge\neg(\exists o')((\langle v, n\rangle, o') \in E_t)$
Heap Deletion	$E_{t+1} = E_t - \{(\langle v, n\rangle, o)\}$	$reachable_{H_t}(v) \wedge (\langle v, n\rangle, o) \in E_t$
Program Termination	$E_{t+1} = E_t -$ $\{(\langle r, n\rangle, o) \mid (\langle r, n\rangle, o) \in E_t\}$	None

Table 1. Definition of action a_t. Only changes from H_t to H_{t+1} are listed.

program mutates its heap as it runs. These changes include objects being dynamically allocated, fields of objects being updated, and objects being passed to and from functions. These changes occur throughout program execution and a dynamic memory manager must be capable of handling all of them.

Interesting Program Actions— While programs perform many operations, our framework is interested only in those actions that affect the heap. Though the causes of these mutations are language-specific, we group actions by their effect on the heap: object allocation, root creation, root deletion, field reference creation, field reference deletion, and program termination.[2]

We describe the effect each action has on the heap state with respect to the heap state multi-graph at time t, $H_t = (L_t, D_t, r, E_t)$, and the multi-graph following the action, $H_{t+1} = (L_{t+1}, D_{t+1}, r, E_{t+1})$.[3] We additionally describe the preconditions necessary within H_t for the possible actions a_t. A precise mathematical definition of these effects and limitations can be found in Table 1. The following paragraphs provides a simple description of each of the actions:

Object Allocation actions occur when an object is allocated in the heap. An object allocation action defines a new vertex to be added to the set of live vertices and the key value of the root that references the newly allocated vertex.

Root Creation actions occur when a root reference to an object is created. The root creation action defines an new edge to be added to the edge set from the root vertex to a reachable vertex.

[2]Other actions could be included; these primitives are quite general and can be combined. We distinguish root and heap reference actions because many algorithms treat them differently. We specifically exclude GC behavior from program actions.

[3]Actions may also be considered functions producing H_{t+1} from H_t; we do this when it is convenient.

$$\begin{array}{ccccccc} H_0 & & H_1 & & & & H_T \\ (L_0, D_0, r, E_0) & \xrightarrow{a_1} & (L_1, D_1, r, E_1) & \xrightarrow{a_2} & \ldots & \xrightarrow{a_T} & (L_T, D_T, r, E_T) \end{array}$$

Figure 1. Example Program History

Root Deletion actions remove an existing edge from the root vertex to a vertex in the set of live vertices. This action is equivalent to deleting a root reference or making a root null.

Heap Reference Creation actions occur when the program updates a heap object's unused field updated to reference another object. These actions add an edge to the edge set from the source vertex to the target vertex.

Heap Reference Deletion actions specify an edge in the edge set between two vertices that is removed. Heap reference deletion actions occur whenever an object in the heap has a non-null field made null.

Program Termination actions occur when the program execution ends. Program termination may occur at any time and deletes the root set (removes any edge whose source is the root vertex).

In our framework, as in program execution, only existing references may be removed, each object field contains at most one reference, and only reachable objects may be involved in actions. Since the heap state multi-graph reflects the heap, the program actions mirror the changes in the heap. *By construction* programs cannot attempt actions whose preconditions are not met, so we need not consider the possibility. Further, it is easy to see that program actions preserve well-formedness of heap state multi-graphs (since they cannot remove vertices nor cause unreachable objects to become reachable).

2.5. Program History

Every program begins with the same heap. This *initial heap state* is represented by the multi-graph $H_0 = (L_0, D_0, r, E_0)$. This graph has only one vertex (the root vertex) and an empty edge multiset ($L_0 = \{r\}, D_O = \emptyset, E_0 = \emptyset$).

We also consider a program's heap state following program termination. *Final heap state* multi-graphs, designated $H_T = (L_T, D_T, r, E_T)$, have a vertex (in $L_T \cup D_T$) for each object allocated into the heap, plus the root vertex. Final heap state multi-graphs cannot contain edges from the root vertex, so only the root vertex is reachable in these multi-graphs.

The initial and final heap states do not contain much information about the program execution, but the entire run is necessary to analyze GC algorithms and optimizations. To record this information, our framework uses a *program history*. The program history begins with the initial heap state multi-graph, H_0, and the first program action, a_1. From this multi-graph and action, we build the successor heap state multi-graph, H_1, then add action, a_2, and so on. The program history continues up to the program termination action (a_T) and final heap state multi-graph, H_T. Figure 1 illustrates a program history.

Since all programs start from the initial heap state, and each program action is deterministic, the actions alone are sufficient to recreate the program history.

By replaying the actions, GC can be simulated or, via profile feedback [4], tuned. Files called *heap traces* store the actions (and an additional piece of information, as we explain in Section 3) for these purposes. Since it works in a manner similar to heap traces, our program history is intuitive to use.

2.6. Null and Reachable Multi-Graphs

Using the program history, our framework can compute the *null heap state multi-graph* for each time step. The null heap state multi-graph at time t, $H_t^\emptyset = (L_t^\emptyset, D_t^\emptyset, r, E_t)$, is the heap state multi-graph where no objects have been determined to be dead (e.g., $D_t^\emptyset = \emptyset$).

Using the program history, the time when each object becomes unreachable can be determined. Given $H_t = (L_t, D_t, r, E_t)$, the *reachable heap state multi-graph* is $H_t^* = Live(H_t) = (L_t^*, D_t^*, r, E_t)$. The reachable multi-graph specially defines L and D: $L_t^* = \{v \in L_t | reachable_{H_t}(v)\}$; $D_t^* = V_t - L_t^*$, that is, L_t^* is exactly the set of reachable vertices, and D_t^* the remainder (the $*$ superscript is intended to suggest the optimal, i.e., smallest possible, L_t set.)

Given a heap state multi-graph H_t, we define the *reduced heap state multi-graph*, $H_t^R = Reduce(H_t)$, as the heap state multi-graph (L_t^R, D_t^R, r, E_t^R), where: $L_t^R = L_t$, $D_t^R = \emptyset$, $E_t^R = \{(\langle v, n\rangle, o) \in E_t | v \in L_t\}$. This reduction removes those vertices identified as dead, and any edges from these vertices, from the multi-graph. By removing edges and vertices that are known to be unnecessary, the reduced heap state multi-graph resembles the physical heap following GC.

Finally, given a heap state multi-graph H_t, we define the *reduced reachable heap state multi-graph*, H_t^-, as the heap state multi-graph $Reduce(Live(H_t))$.

The null, reachable, and reduced reachable heap state multi-graphs are similar: their reductions via $Reduce \circ Live$ are the same. Further, since program actions can manipulate only reachable objects, if we apply a_{t+1} to $H_t^\emptyset$, H_t^*, and H_t^-, we get analogous results, a fact we state precisely in a moment. First we argue that if we have a well-formed heap state $H_t^\emptyset$ and corresponding action a_{t+1}, then a_{t+1} is legal for H_t^* and H_t^-.

Lemma 1 *If $H_t^\emptyset$ fulfills the preconditions of a_{t+1}, then so do H_t^* and H_t^-.*

Now we state a stronger relationship for program histories and their corresponding reachable and reduced reachable heap states:

Theorem 2 *If a_{t+1} takes H_t to H_{t+1}, then $a_{t+1} \circ Live$ takes H_t^* to H_{t+1}^* and $a_{t+1} \circ Live \circ Reduce$ takes H_t^- to H_{t+1}^-.*

2.7. Modeling Collector Behavior

We model garbage collector behavior by following each program action a_t with a garbage collector action g_t. Thus, we form H_t by first applying program action a_t to H_{t-1}, and then applying collector action g_t. A collector action potentially identifies some unreachable objects as dead. In fact, we will equate g_t with the set of objects it identifies as dead, and define its effect on the heap

$$\begin{array}{ccccccc}
H_0^{\emptyset} & & H_1^{\emptyset} & & & & H_T^{\emptyset} \\
(L_0^{\emptyset}, D_0^{\emptyset}, r, E_0^{\emptyset}) & \xrightarrow{a_1, g_1^{\emptyset}} & (L_1^{\emptyset}, D_1^{\emptyset}, r, E_1^{\emptyset}) & \xrightarrow{a_2, g_2^{\emptyset}} & \ldots & \xrightarrow{a_T, g_T^{\emptyset}} & (L_T^{\emptyset}, D_T^{\emptyset}, r, E_T^{\emptyset}) \\
\downarrow Live() & & \downarrow Live() & & & & \downarrow Live() \\
H_0^{*} & & H_1^{*} & & & & H_T^{*} \\
(L_0^{*}, D_0^{*}, r, E_0^{*}) & \xrightarrow{a_1, g_1^{*}} & (L_1^{*}, D_1^{*}, r, E_1^{*}) & \xrightarrow{a_2, g_2^{*}} & \ldots & \xrightarrow{a_T, g_T^{*}} & (L_T^{*}, D_T^{*}, r, E_T^{*})
\end{array}$$

Figure 2. Expanded Program History, including the null ($H^{\emptyset}$) and reachable (H^*) heap states.

state as mapping heap state $H = (L, D, r, E)$ to $(L - g_t, D \cup g_t, r, E)$, with the precondition that $g_t \subseteq L - \{o \in L | reachable_H(o)\}$. When convenient, we will also use the notation g_t for the function that the collector action induces on heap states.

The simplest collector, which we call the *null collector*, never identifies any objects as unreachable. We write its actions as $g_t^{\emptyset}$; it induces the identity function on heap states.

The most "aggressive" collector, which we call the *comprehensive collector*, always identifies all unreachable objects. We write it as g_t^* and the function it induces is complementary to *Live* (i.e., it identifies the unreachable objects).

Figure 2 shows how the null ($H^{\emptyset}$) and reachable (H^*) heap state multigraphs relate to null and comprehensive collector actions.

Real collectors are bounded (in what they reclaim) by the null and comprehensive collectors. Further, many collectors identify unreachable objects only occasionally. For example, they may allow a portion of the heap to fill, and identify unreachable objects in a batch only after the space is full. While we model only the end result (e.g., the set of objects identified as dead), in applying the framework it is easy to associate costs with collector action g_t and derive the effort taken by the collector at each time step.

3. Heap Traces

When exploring the performance of a new garbage collector, one can often work faster by using a simulator. For this, one runs a program in a system instrumented to produce a collector-neutral *heap trace*. The simulator accepts the trace as an input and, given the GC algorithm, system parameters, and algorithm tuning parameters (such as the maximum heap size allowed), estimates the work needed by the algorithm for the traced instance of the program.

A heap trace is a time-ordered sequence of records. The records are of these kinds: *object allocation*, giving the new object's size and a unique identifier; *object death*, giving the dying object's unique identifier; *heap reference update*, giving the source object, field key, and target object or null value; and *root reference update*, giving a location of the root, and the target object or null value. (The update records also implicitly define a reference deletion if the field/root previously contained a reference). With a perfectly accurate trace the simulator *could* determine when each object dies, but it is easier to write the simulator, and the simulator runs much faster, if the death times are provided in the trace. This is since a single trace file is used in many different simulations, it is cheaper to compute the death times once in advance.

One way to obtain death times for traces is to perform a comprehensive collection whenever a collection could occur in practice. Since most collectors attempt collection only in response to an allocation (i.e., when they require additional space in the heap), this requires doing a collection just prior to each allocation. This is the *brute force* approach to trace generation. As these constant collections take a substantial amount of time, researchers often use traces generated with less frequent comprehensive collections, resulting in traces that may distort simulator results significantly [7]. We now extend our framework to model object death times, in preparation for presenting and analyzing the Merlin trace generation algorithm.

3.1. Object Death Time Multi-Graph

We add to our framework the *object death time multi-graph.* This multi-graph differs from the others because it concerns only the efficiency of a collector. It is not related to the heap at any moment of the program history; rather it exists to prove the minimum information and work needed to determine the earliest time each object could be reclaimed. This multi-graph can also compare the efficiency of comprehensive collectors by analyzing the relative work needed to populate and analyze this multi-graph. Before describing the new graph, we discuss a concept upon which it relies: *final reference deletion time.*

An object's final reference deletion time is the last time at which the object has an incoming reference deleted (by a root or heap reference deletion action or at program termination). Because each object is allocated with a reference from the root set, and the program termination action removes any root references that exist, each object has a final reference deletion time. This time occurs between the object's allocation and program termination. We define the function f to map each vertex to its final reference deletion time. Given a vertex v, f is defined as: $f(v) = \max_{i<T}((\exists o, n)((\langle o, n\rangle, v) \in E_i \wedge (\langle o, n\rangle, v) \notin E_{i+1}))$. An object may have incoming references at its final reference deletion time, provided that the remaining incoming references are not deleted by a program action; in the name, "final" modifies "deletion", not "reference".

With this definition, we present the last graph structure of the paper. The object death time multi-graph is also a directed, rooted multi-graph, $F = (V, E_T, f)$, where f is the final reference deletion time function from above. The multi-graph's set of vertices V contains a vertex for each object allocated in the heap, i.e., $V = V_T - \{r\}$. The multi-graph's edge multiset, E_T, is the edge multiset of the final heap state.

As the name implies, the multi-graph determines the *death time* for each object. An object's death time is the time at which it becomes unreachable—the time the corresponding vertex would be included in a g^* action. As the following theorem shows, this time can be computed in the object death time multi-graph as each vertex's latest *reaching final reference deletion time*, the latest final reference deletion time among the vertices that reach each vertex.

Theorem 3 *The latest reaching final reference deletion time to a vertex in F is the time the corresponding object in the heap became unreachable.*

We note that this multi-graph is similar to H_T^* ($V = D_T^*$ and $E_T = E_T^*$) as one would expect, given its function. Using this multi-graph, we can compute object death times in asymptotically optimal time, as we show in the next section.

4. Merlin Algorithm Analysis

The Merlin trace generation algorithm, proposed by Hertz *et al.*, generates traces over 800 times faster than the previous method of trace generation [7]. The Merlin algorithm, shown in Figure 3, achieves its speedup by performing a small amount of work with (some) program actions and can thus delay performing more costly analyses until necessary. Designed using insights gained from this research, the Merlin algorithm computes each object's final reference deletion time in conjunction with program actions and analyzes the object death time multi-graph that it constructs with this information. From this analysis, the Merlin algorithm can easily and quickly determine each object's death time for trace generation.

This section shows that our framework can help one to discover new insights about garbage collection and also prove asymptotic running times for garbage collection algorithms. We begin by proving the asymptotic running time for the Merlin algorithm. We then prove that this running time is optimal for heap trace generation.

4.1. Merlin's Running Time

Brute force trace generation, discussed in Section 3, computes object death times by performing a reachability analysis whenever it wants object death times. Computing which objects are reachable in the heap state is equivalent to solving the single-source/multi-sink reachability problem from the root vertex, which requires $O(L_t^R + E_t^R)$ time. Since this reachability analysis must be repeated throughout the running of the program, brute force trace generation can require up to $\sum_{t=1}^{T} O(L_t^R + E_t^R)$ time!

To generate traces, Merlin must compute when each object becomes unreachable. For this, Merlin builds and analyzes the object death time multi-graph as the program runs. Using the object death time multi-graph, finding when objects become unreachable requires comparing the reaching final reference deletion times for each vertex; this processing seems analogous to computing all of the transitive closures sets, requiring $O(V_T \cdot E_T)$ time [5, p. 766].

Solving the transitive closures determines when each object can be reclaimed, but requires more work than is needed. Object death times are the *latest* reaching final reference deletion times; to find death times, Merlin requires

```
if (action = create root reference to o) then
  rootReferences[o]++
else if (action = create heap reference) then
  No action needed
else if (action = delete root reference to o) then
  rootReferences[o]--
  finalReferenceDeletionTime[o] ← current time
else if (action = delete heap reference to o) then
  finalReferenceDeletionTime[o] ← current time
else if (action = allocate object o) then
  rootReferences[o] ← 1
  finalReferenceDeletionTime[o] ← 0
else if (action = program termination) then
  for each vertex v do
    if rootReferences[v] ≠ 0 then
      finalReferenceDeletionTime[v] ← current time
  for each vertex v by decreasing final reference deletion time do
    push v onto the stack
  while (the stack is not empty) do
    pop v from the stack
    workingTime ← finalReferenceDeletionTime[v]
    for each non-null w to which v refers do
      if (finalReferenceDeletionTime[w] < workingTime) then
        finalReferenceDeletionTime[w] ← workingTime
        push w onto the stack
```

Figure 3. The Merlin Trace Generation Algorithm

a single depth-first search from each vertex in reverse order of vertex final reference deletion times.

Lemma 4 *When computing object death times, each vertex and each edge need be processed only once.*

With the framework, we can now prove Merlin's asymptotic running time.

Theorem 5 *Merlin requires $O(T)$ time to create the Object Death Time Multi-Graph and then $O(V_T + E_T)$ time to compute the object death times.*

This leads to an additional theorem.

Theorem 6 *The Merlin algorithm computes object death times in asymptotically optimal time for trace generation.*

5. Future Work

Our framework has been designed to be abstract and yet robust. We demonstrate how it can be used to easily prove asymptotic running times, optimal running times, and space bounds for garbage collection. It is our hope that we, and other researchers, use and extend this framework to fully develop a theoretical underpinning for garbage collection. There are, in particular, several areas in which we hope to expand upon the work in this paper.

Collector Efficiency Analysis
In this research we analyzed properties of comprehensive garbage collection. Using this framework, we hope to perform a similar analysis for non-comprehensive collectors and help focus research seeking faster algorithms. One approach for this future work is to find and recast the important features for the collector state multi-graph much as the object death time multi-graph distills the important features of the reachable multi-graph.

Collection Optimization Analysis
Another direction in which we wish to expand this work is determining the value of different optimizations. By using our framework to measure both the asymptotic cost and potential savings for a variety of optimizations, we could better understand what limits garbage collector speed and how best to overcome it.

6. Summary

This paper introduces several structures and analyses for garbage collectors. With the new structures, we prove the running time of the Merlin trace generation algorithm and show that this time is optimal for computing object death times.

Acknowledgments

We would like to thank Steve Blackburn and Darko Stefanovic for providing valuable insights. We would also like to thank J. M. Robson for his initial encouragement into this research.

References

[1] Barrett, D. A., and Zorn, B. G. Using lifetime predictors to improve memory allocation performance. In *Proceedings of SIGPLAN 1993 Conference on Programming Language Design and Implementation* (Albuquerque, NM, June 1993), vol. 28(6) of *ACM SIGPLAN Notices*, ACM Press, pp. 187–196.

[2] Blanchet, B. Escape analysis for object oriented languages. applications to Java. In *Proceedings of SIGPLAN 1999 Conference on Object-Oriented Programming, Languages, & Applications* (Denver, CO, Oct. 1999), vol. 34(10) of *ACM SIGPLAN Notices*, ACM Press, pp. 20–34.

[3] Cannarozzi, D. J., Plezbert, M. P., and Cytron, R. K. Contaminated garbage collection. In *Proceedings of SIGPLAN 2000 Conference on Programming Language Design and Implementation* (Vancouver, Canada, June 2000), vol. 35(5) of *ACM SIGPLAN Notices*, ACM Press, pp. 264–273.

[4] Chilimbi, T., Jones, R. E., and Zorn, B. Designing a trace format for heap allocation events. In *ISMM 2000 Proceedings of the Second International Sympo-*

sium on Memory Management (Minneapolis, MN, Oct. 2000), vol. 36(1) of *ACM SIGPLAN Notices*, ACM Press, pp. 35–49.

[5] CORMEN, T. H., LEISERSON, C. E., AND RIVEST, R. L. *Introduction to algorithms.* MIT Press, Cambridge, MA, 1990.

[6] HAYES, B. Using key object opportunism to collect old objects. In *Proceedings of SIGPLAN 1991 Conference on Object-Oriented Programming, Languages, & Applications* (Phoenix, AZ, Oct. 1991), vol. 26(11) of *ACM SIGPLAN Notices*, ACM Press, pp. 33–40.

[7] HERTZ, M., BLACKBURN, S. M., MOSS, J. E. B., MCKINLEY, K. S., AND STEFANOVIĆ, D. Error-free garbage collection traces: How to cheat and not get caught. To Appear at Sigmetrics 2002. Available at ftp://ftp.cs.umass.edu/pub/osl/papers/sigmetrics-2002-merlin.ps.gz.

[8] HUDSON, R. L., AND MOSS, J. E. B. Incremental garbage collection for mature objects. In *Proceedings of the International Workshop on Memory Management* (St. Malo, France, Sept. 1992), vol. 637 of *Lecture Notes in Computer Science*, Springer-Verlag, pp. 388–403.

[9] SHAHAM, R., KOLODNER, E. K., AND SAGIV, M. On the effectiveness of GC in Java. In *ISMM 2000 Proceedings of the Second International Symposium on Memory Management* (Minneapolis, MN, Oct. 2000), vol. 36(1) of *ACM SIGPLAN Notices*, ACM Press, pp. 12–17.

ONE-WAY PERMUTATIONS AND SELF-WITNESSING LANGUAGES

Christopher M. Homan*
Department of Computer Science
University of Rochester, Rochester NY 14627
choman@cs.rochester.edu

Mayur Thakur
Department of Computer Science
University of Rochester, Rochester NY 14627
thakur@cs.rochester.edu

Abstract A desirable property of one-way functions is that they be total, one-to-one, and onto—in other words, that they be permutations. We prove that one-way permutations exist exactly if $P \neq UP \cap coUP$. This provides the first characterization of the existence of one-way permutations based on a complexity-class separation and shows that their existence is equivalent to a number of previously studied complexity-theoretic hypotheses.

We also study permutations in the context of witness functions of nondeterministic Turing machines. A language is in PermUP if, relative to some unambiguous, nondeterministic, polynomial-time Turing machine accepting the language, the function mapping each string to its unique witness is a permutation of the members of the language. We show that, under standard complexity-theoretic assumptions, PermUP is a nontrivial subset of UP.

We study SelfNP, the set of all languages such that, relative to some nondeterministic, polynomial-time Turing machine that accepts the language, the set of all witnesses of strings in the language is identical to the language itself. We show that SAT $\in$ SelfNP, and, under standard complexity-theoretic assumptions, SelfNP $\neq$ NP.

*Supported in part by Dept. of Education (GAANN program) grant EIA-0080124, and by grant NSF-INT-9815095/DAAD-315-PPP-gü-ab.

Keywords: One-way functions, permutations, one-to-one functions, complexity-theoretic cryptography, self-witnessing languages.

1. Introduction

Until the results in this paper were known the question, "What complexity class separations, if any, characterize the existence of one-way permutations (i.e., total, one-to-one, onto, one-way functions)?" has remained open. We prove that one-way permutations exist exactly if $\mathrm{P} \neq \mathrm{UP} \cap \mathrm{coUP}$. UP [Val76] is the class of all languages accepted by a nondeterministic Turing machine that runs in polynomial time and has on any input at most one accepting path. Such Turing machines are called "UPTMs," or *unambiguous, polynomial-time Turing machines.*

Grollmann and Selman [GS88] and, independently, Ko [Ko85] (see also work by Berman [Ber77]) show that $\mathrm{P} \neq \mathrm{UP}$ exactly if total, one-to-one, (but not necessarily onto,) one-way functions exist and that $\mathrm{P} \neq \mathrm{UP} \cap \mathrm{coUP}$ exactly if *partial*, one-to-one, onto, one-way functions exist. In this paper, we extend their results to *total*, one-to-one, *onto*, one-way functions. The existence of one-way permutations is thus equivalent to a number of hypotheses [Ko85,GS88, HH88,Grä94,FFNR96,HR00,RH02], including the following.

1 The weak definability principle does not hold for some logic that is closed under first order operations [Grä94].

2 $\mathrm{EASY}_{\forall}^{\forall}(\mathrm{UP}) \neq \mathrm{UP}$ [RH02].

3 There exist UPTMs M and N such that $L(M) \subseteq L(N)$ and $f \notin \mathrm{FP}$, where f is any function having the property that, for all $x \in L(M)$, $f(\langle x, \mathrm{wit}_{\mathrm{M}}(x)\rangle) = \mathrm{wit}_{\mathrm{N}}(x)$.

$\mathrm{EASY}_{\forall}^{\forall}(\mathrm{UP})$ [RH02] is the class of all languages in P for which there is an unambiguous Turing machine U accepting the language such that the mapping between a member of the language and its unique witness (i.e., the bits guessed by a nondeterministic accepting path of U) is not polynomial-time computable. Item 3 above follows by analogy to a result by Fenner et al. [FFNR96], who show that onto, one-way functions do not exist exactly if all nondeterministic polynomial-time Turing machines accepting a language have roughly (i.e., modulo a polynomial-time transformation) the same witnesses. Theorem 7, stated in Section 3 of this paper, lists all hypotheses known to be equivalent to the existence of one-way permutations.

We also study permutations in the context of witness functions of nondeterministic Turing machines. In this context, a function f is a *permutation* of a language L if f is a (partial) one-to-one function defined over exactly the members of L such that the image of f is exactly L. We say a function *permutes* a language if the function is a permutation of the language. Let PermUP, or *self-permuting* UP, be the set of all languages such that there is a UPTM accepting the language, where the mapping between each string in the language and the unique witness used by the UPTM to accept the string permutes the language. That is, PermUP = $\{L \mid$ there exists a UPTM U such that $L(U) = L$ and wit_U permutes $L\}$, where wit_U denotes a function that maps each $x \in L$ to its unique witness in U.

Clearly PermUP $\subseteq$ UP. Furthermore, it is easy to see that any language $L \in$ P is in PermUP via the following "simple UPTM" (i.e., a UPTM whose witness function wit_U is computable in polynomial-time).

> On input x nondeterministically guess a string y of length exactly $|x|$ and accept if and only if $x \in L$, and $y = x$.

Based on the the above example, one might be tempted to regard PermUP as a characterization of simplicity. We show, however, that the closure of PermUP under polynomial-time one-to-one reductions is UP. Thus, PermUP captures the most complex languages in UP. In other words, assuming P $\neq$ UP some languages in PermUP are accepted via "complex UPTMs" (i.e., UPTMs whose witness functions are not polynomial-time computable).

On the other hand, we show that it is unlikely that all languages in UP are in PermUP (i.e., that PermUP is closed under polynomial-time, many-one reductions), for if this is the case then E = UE. Thus, under standard complexity-theoretic assumptions, PermUP is a nontrivial subset of UP that captures the hardest languages in UP.

It appears then that PermUP contains languages that are either simple (i.e., accepted by some simple UPTM) or complex (i.e., accepted by some complex UPTM). But is there a language in PermUP that is accepted by both a simple and a complex UPTM? The answer to this question is linked to the existence of one-way permutations. Consider the class HiPermUP = $\{L \mid$ there exists a UPTM U such that $L = L(U)$, wit_U permutes L, and wit_U is not polynomial-time computable$\}$, which is a (possibly empty) subset of PermUP. Intuitively, languages in HiPermUP are in PermUP via a complex UPTM. We show that one-way permutations exist exactly if there is a language $L \in$ P $\cap$ HiPermUP. Such an L is thus in PermUP via both a simple and a complex UPTM.

We also study the class SelfNP, or *self-witnessing* NP, which we regard as the natural NP analog of PermUP. SelfNP is defined as $\{L \mid$ there exists a nondeterministic Turing machine N that runs in polynomial time such that $L(N) = L$ and $\bigcup_{x \in L} \text{wit}_N(x) = L\}$, where wit_N is a function that maps each $x \in L$ to its set of witnesses relative to N. We show that the relationship between SelfNP and NP is basically analogous to the relationship between PermUP and UP (i.e., the closure of SelfNP under polynomial-time many-one reductions is NP, and if SelfNP = NP, then E = NE). Furthermore, SAT $\in$ SelfNP.

The remainder of the paper is organized as follows. Section 2 presents definitions and notations. Section 3 proves results on the existence of one-way permutations. Section 4 proves results related to PermUP and SelfNP. Section 5 is the conclusion. *(Note: Due to space limitations, some of the proofs have been omitted. All omitted proofs can be found in [HT01].)*

2. Definitions and Notations

All sets, unless otherwise stated, are subsets of Σ^*, where Σ is the standard alphabet $\{0,1\}$. The length of a string x is denoted by $|x|$. For any set S, $||S||$ denotes the cardinality of S. For any string $w \in \Sigma^*$ and any $S \subseteq \Sigma^*$, $wS = \{ws \mid s \in S\}$. For any two sets $S, T \subseteq \Sigma^*$, $ST = \{uv \mid u \in S \wedge v \in T\}$.

For each Turing machine N and each $x \in \Sigma^*$, $N(x)$ means "the computation of N on input x." NPTM (respectively, NETM) means "nondeterministic polynomial-time (respectively, exponential-time) Turing machine." NP (respectively, NE) denotes the class of all languages L such that L is accepted by an NPTM (respectively, NETM). UPTM (respectively, UETM [RRW94, HJ95]) means "unambiguous, polynomial-time (respectively, exponential-time) Turing machine," that is, N is a UPTM if and only if N is an NPTM and, on any input $x \in \Sigma^*$, N has at most one accepting path. UP [Val76] (respectively, UE [RRW94,HJ95]) is the class of all languages L such that L is accepted by some UPTM (respectively, UETM). TALLY is the class of all tally languages, that is, TALLY $= \{L \mid L \subseteq 1^*\}$. The set of total, deterministic, polynomial-time computable functions is called FP.

For each complexity class $\mathcal{C}$ and all a, b such that $\leq_b^a$ is defined, the *reduction closure* of $\mathcal{C}$ relative to $\leq_b^a$, denoted $\mathrm{R}_b^a(\mathcal{C})$, is the set $\{L \subseteq \Sigma^* \mid (\exists L' \in \mathcal{C})[L \leq_b^a L']\}$.

For each NPTM N and each $x \in L(N)$, a string y is said to be a *witness* for x relative to N if there is an accepting path of $N(x)$ on which the sequence of nondeterministic guess bits is exactly y. Note that y (unless by coincidence) does not necessarily encode x, i.e., it is *naked*. Denote the mapping from $x \in L(N)$ to the set of witnesses for x relative to N as $\mathrm{wit}_N : \Sigma^* \to 2^{\Sigma^*}$ (where 2^{Σ^*} is the set of all subsets of Σ^*). When N is a UPTM, since each $x \in \Sigma^*$ has at most one witness relative to such an N, for convenience we regard wit_N as a mapping from $L(N)$ to Σ^*.

For each function f let $\mathrm{im}(f)$ denote the image of f. For each function $f : D \to R$ and each $S \subseteq D$ such that f is defined on each element in S let $f(S)$ denote the set $\{r \in R \mid (\exists s \in S)[f(s) \text{ is defined, and } r = f(s)]\}$.

For any set $S \subseteq \Sigma^*$ a function $f : \Sigma^* \to \Sigma^*$ is a *permutation* of S if the set of all strings in Σ^* on which f is defined is exactly S, $\mathrm{im}(f) = S$, and f is one-to-one. We say f *permutes* S if f is a permutation of S.

A function $f : \Sigma^* \to \Sigma^*$ is *polynomial-time-invertible* (alternatively, *FP-time invertible*) if there is a function $g \in \mathrm{FP}$ such that for every $y \in \mathrm{im}(f)$, $f(g(y)) = y$. Let $\langle \cdot, \cdot \rangle$ denote a standard, fixed, polynomial-time computable, polynomial-time invertible, total, one-to-one, onto function from $\Sigma^* \times \Sigma^*$ to Σ^* such that the output of the function is strictly length increasing in the lengths of either of its arguments when the other argument is fixed. Such a function is called a *pairing function*.

Definition 1 [Ber77,Ko85,GS88,Sel92] *A function $f : \Sigma^* \to \Sigma^*$ is* honest *if there exists a polynomial p (called the* honesty polynomial*) such that for all $y \in \mathrm{im}(f)$ there exists an $x \in \Sigma^*$ such that $f(x)$ is defined, $y = f(x)$, and*

$|x| \leq p(|y|)$.

Intuitively, if an honest function is hard to invert, then it is so "honestly", i.e., not merely because the function shrinks the input too much. Grollmann and Selman [GS88] provided the first independent study of complexity-theoretic, (one-to-one,) one-way functions (see also [Ber77,Ko85]). The definition below is for complexity-theoretic one-way functions of arbitrary ambiguity [Wat88].

Definition 2 [Wat88] (see [Ber77,Ko85]) *A function $f : \Sigma^* \to \Sigma^*$ is* one-way *if f is honest, polynomial-time computable, and not polynomial-time invertible.*

Definition 3 (see [Wat88,Hom00]) *For $g : \mathbb{N} \to \mathbb{N}$, we say a function $\sigma : \Sigma^* \to \Sigma^*$ is* g-to-1 *if*

$$(\forall y \in \mathrm{im}(\sigma))[\ \| \{x \in \Sigma^* \ \mid\ \sigma(x) = y\} \| \leq g(|y|)].$$

3. One-Way Permutations

In this section we prove the main result of this paper.

Theorem 4 P $\neq$ UP $\cap$ coUP *if and only if one-way permutations exist.*

As noted in the introduction, it is known (see [Ko85,GS88,Ber77]) that P $\neq$ UP $\cap$ coUP exactly if partial, one-to-one, onto, one-way functions exist. The independent existence of both partial, one-to-one, onto, one-way functions and one-way permutations has been studied in a variety of settings (see [Ko85, GS88,HH88,Grä94,FFNR96,HR00,RH02]). Theorem 5 below collects results either previously known to be equivalent to the existence of partial, one-to-one, onto, one-way functions or that can be obtained by arguments analogous to previously known proofs. In particular, the equivalence of items 1 and 2 is due to Ko [Ko85], of items 1–4 to Grollmann and Selman [GS88], of items 2 and 5 to Hartmanis and Hemachandra [HH88], of items 3 and 5–8 to Rothe and Hemaspaandra [RH02], and of 3 and 9 to Grädel [Grä94]. The equivalence of 1, 2, 5, 10, and 11 is analogous to results for the existence of onto, one-way functions by Fenner et al. [FFNR96].

Theorem 5 *The following are equivalent.*

1 *There exists a partial, one-to-one, onto, one-way function.*

2 P $\neq$ UP $\cap$ coUP.

3 *There exists a partial, one-to-one, one-way function f such that* $\mathrm{im}(f) \in$ P.

4 *There exists a total, one-to-one, one-way function f such that* $\mathrm{im}(f) \in$ P.

5 $\mathrm{EASY}^{\forall}_{\forall}(\mathrm{UP}) \neq$ P.

6 $1\text{–}\mathrm{EASY}^{\forall}_{\forall}(\mathrm{UP}) \neq$ P.

7 $\Sigma^* \notin \mathrm{EASY}_{\forall}^{\forall}(\mathrm{UP})$.

8 $\mathrm{EASY}_{\forall}^{\forall}(\mathrm{UP})$ *is not closed under complementation.*

9 The weak definability principle does not hold for some logic that is closed under first order operations.

10 $\mathrm{UPSV_t} \not\subseteq \mathrm{FP}$.

11 There exist UPTMs *M and N such that $L(M) \subseteq L(N)$ and $f \notin \mathrm{FP}$, where f is any function having the property that for all $x \in L(M)$, $f(\langle x, \mathrm{wit_M}(x)\rangle) = \mathrm{wit_N}(x)$.*

Rothe and Hemaspaandra [RH02] define $1{-}\mathrm{EASY}_{\forall}^{\forall}(\mathrm{UP})$ as the set of all languages L in UP for which there exists a UPTM U that accepts L and a polynomial-time computable function f_U such that, for all $x \in L$, $f_U(x)$ outputs the first bit of the accepting path of $U(x)$. Fenner et al. [FFNR96] consider the "NP version" of $1{-}\mathrm{EASY}_{\forall}^{\forall}(\mathrm{UP})$, however their results regarding this "NP version" [FFNR96] are not known to be analogous to the above result for $1{-}\mathrm{EASY}_{\forall}^{\forall}(\mathrm{UP})$.

Regarding the existence of one-way permutations, the following equivalence, due to Hemaspaandra and Rothe, is known.

Theorem 6 [HR00] *The following are equivalent.*

1 There exists a one-way permutation.

2 There exists a total, one-to-one, one-way function whose range is P*-rankable.*

As a result of Theorem 4, we have the following.

Theorem 7 *The following are equivalent.*

1 Any item from Theorem 5.

2 Any item from Theorem 6.

3 $\Sigma^* \in \mathrm{HiPermUP}$.

4 There exists $L \in \mathrm{P}$ such that $L \in \mathrm{HiPermUP}$.

We now prove Theorem 4.

Proof of Theorem 4 The proof follows immediately from the equivalence of items 1 and 2 of Theorem 5 and the application of Lemma 8 for $g(n) = 1$.

Lemma 8 *Let g be a nondecreasing function from $\mathbb{N}$ to $\mathbb{N}^+$. There exists a partial, g-to-1, onto, one-way function if and only if there exists a total, g-to-1, onto, one-way function.*

Proof The ($\Leftarrow$) direction is easy, since all total functions are partial functions. For the ($\Rightarrow$) direction, let h be a partial, g-to-1, onto, one-way function for some nondecreasing $g : \mathbb{N} \to \mathbb{N}^+$. We claim that f, defined on input x as

$$f(x) = \begin{cases} 1^n 0y & \text{if } (x = 1^{n+1}0y) \wedge (h(y) \text{ is defined}) \wedge (n > 0), \\ 01h(y) & \text{if } (x = 10y) \wedge (h(y) \text{ is defined}), \\ 0^{n+1}1y & \text{if } (x = 0^n 1y) \wedge (n > 0), \\ x & \text{otherwise}, \end{cases}$$

is total, g-to-1, onto, and one-way.

The intuition behind the proof is captured by Figures 1 and 2. Figure 1 is

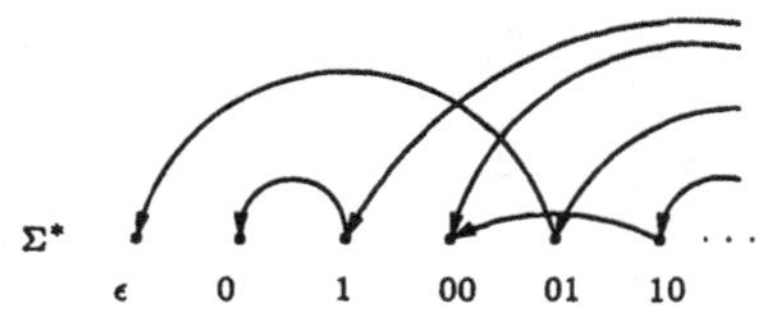

Figure 1. A graph where the vertices are the elements of Σ^*, and the edges are the mapping defined by h.

a graphical representation of an example of h, where the vertices of the graph are the strings in Σ^* and the edges are the relations defined by h, i.e., there is an edge from x to y if and only if $f(x) = y$. It is easy to see that a function is total precisely when the outdegree of every vertex in the graph representing the function is exactly one and is onto whenever the indegree of every vertex is at least one. Since h is not necessarily total, each vertex of the graph in Figure 1 has an outdegree of either zero or one, however since h is onto, every vertex has an indegree of at least one.

The construction of f essentially imposes a two-dimensional structure on Σ^*, as shown in Figure 2, and embeds h into this structure by identifying the domain of h with the elements in row $10\Sigma^*$ and the image of h with the elements in row $01\Sigma^*$. The construction then fills the graph in with additional edges in a way that guarantees that "one-way"-ness is preserved and that every vertex x has an outdegree of exactly one and an indegree of at least one and at most $g(|x|)$ (since the indegree represents the "many-to-one"-ness of the function). By checking that the graph in Figure 2 has these properties, it is easy to see that f is a total, g-to-one, onto, one-way function.

Continuing with the formal proof, it is easy to see that f is total and polynomial-time computable. To see that f is onto, note that

$$f(11^*0\{y \mid h(y) \text{ is defined}\}) \quad = \quad 11^*0\{y \mid h(y) \text{ is defined}\} \cup 01\Sigma^*,$$

by the first two conditions in the definition of f, and because h is onto. Furthermore,

$$f(00^*1\Sigma^*) \quad = \quad 000^*1\Sigma^*,$$

by the third condition in the definition of f, and

$$\begin{aligned} & f(\Sigma^* - 11^*0\{y \mid h(y) \text{ is defined}\} - 00^*1\Sigma^*) \\ & \quad = \quad \Sigma^* - 11^*0\{y \mid h(y) \text{ is defined}\} - 00^*1\Sigma^*, \end{aligned}$$

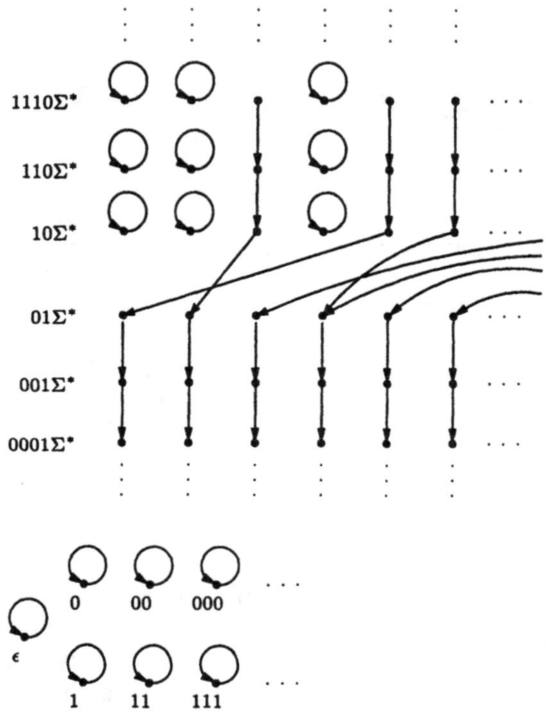

Figure 2. A graph where the vertices are the elements of Σ^*, and the edges are the mapping defined by f.

by the last condition in the definition of f, so clearly $\mathrm{im}(f) = \Sigma^*$.

To see that $f(x)$ is g-to-1, choose an arbitrary $z \in \Sigma^*$. If for some $z' \in \Sigma^*$, $z = 01z'$, then $||f^{-1}(z)|| = ||h^{-1}(z')||$. Since g is nondecreasing, $||h^{-1}(z')|| \leq g(|z'|) \leq g(|z|)$. If $z \notin 01\Sigma^*$ then by the definition of f, $||f^{-1}(z)|| = 1$, thus f is g-to-1.

To see that f is one-way, first note that f is honest. Next, suppose that there exists a function $\gamma \in \mathrm{FP}$ that inverts f. We could then invert h in polynomial time as follows.

On input y, compute $\gamma(01y) = 10w$ and output w.

We thus conclude that f is total, g-to-1, onto, and one-way. ∎

4. Self-Witnessing Languages

In this section, we study properties of PermUP and SelfNP. First, we show that the closure of PermUP under polynomial-time one-to-one reductions is UP.

Theorem 9 $\mathrm{UP} = \mathrm{R}^{p}_{1\text{-}1}(\mathrm{PermUP})$.

As an immediate corollary to Theorem 9, we get the following.

Corollary 10 $\mathrm{P} \neq \mathrm{UP} \iff \mathrm{P} \neq \mathrm{PermUP}$.

The following theorem shows that $\mathrm{P} \neq \mathrm{UP} \cap \mathrm{coUP}$ if and only if $\mathrm{P} \neq \mathrm{PermUP} \cap \mathrm{coPermUP}$.

Theorem 11 $\mathrm{P} \neq \mathrm{UP} \cap \mathrm{coUP} \iff \mathrm{P} \neq \mathrm{PermUP} \cap \mathrm{coPermUP}$.

We wish to define a natural relaxation of PermUP that would include languages in NP − UP (if indeed UP ≠ NP). One important distinction between UPTMs and NPTMs is that the witness functions of UPTMs are single-valued (because there is at most one accepting path), but those of NPTMs may be multivalued (since there can be more than one accepting path). SelfNP, as defined in the introduction, is a natural NP analog of PermUP, where instead of requiring the witness function to be a permutation, we only require that the set of witnesses is the same as the language. Note that the "self-witnessing property," i.e., the property that witnesses themselves be part of the language, also holds for languages in PermUP, since for these languages the witness function is a permutation of the language.

Theorem 12 shows that SAT is a member of SelfNP.

Theorem 12 $\mathrm{SAT} \in \mathrm{SelfNP}$.

From this result, the corollary below easily follows.

Corollary 13 *1* $\mathrm{P} \neq \mathrm{NP} \iff \mathrm{P} \neq \mathrm{SelfNP}$.

2 $\mathrm{NP} = \mathrm{R}^p_m(\mathrm{SelfNP})$.

3 $\mathrm{P} \neq \mathrm{NP} \cap \mathrm{coNP} \iff \mathrm{P} \neq \mathrm{SelfNP} \cap \mathrm{coSelfNP}$.

4 *For all $L \subseteq \Sigma^*$, if there is a polynomial-time computable, honest, onto reduction from L to* SAT *then $L \in$* SelfNP.

Corollaries 10 and 13, part 2 show that PermUP and SelfNP capture the hardest problems in UP and NP, respectively. Theorem 14, which follows from Lemma 15 below (both are due to Hemaspaandra [Hem00]), show that it is unlikely that either SelfNP = NP or PermUP = UP.

Theorem 14 *1* $\mathrm{PermUP} = \mathrm{UP} \Rightarrow \mathrm{E} = \mathrm{UE}$.

2 $\mathrm{SelfNP} = \mathrm{NP} \Rightarrow \mathrm{E} = \mathrm{NE}$.

Lemma 15 $\mathrm{TALLY} \cap \mathrm{SelfNP} \subseteq \mathrm{P}$.

We conclude with a relevant oracle result.

Theorem 16 *There exists an oracle B such that*

1 $\mathrm{SelfNP}^B \neq \mathrm{UP}^B$,

2 $\mathrm{PermUP}^B \neq \mathrm{UP}^B$,

3 $\mathrm{SelfNP}^B \neq \mathrm{NP}^B$, *and*

4 $\mathrm{P}^B \neq \mathrm{PermUP}^B$.

5. Conclusions and Open Questions

We showed that one-way permutations exist if and only if $P \neq UP \cap coUP$. Thus, the existence of one-way permutations is equivalent to a number of previously studied hypotheses [Ko85,GS88,HH88,FFNR96,RH02,HR00].

We studied the self-witnessing language classes PermUP and SelfNP. We showed that the closure of PermUP under polynomial-time one-to-one reductions is UP and that if PermUP = UP, then E = UE. We showed that SAT ∈ SelfNP (thus NP is the closure of SelfNP under polynomial-time many-one reductions) and that if SelfNP = NP, then E = NE. SelfNP can thus be viewed as a natural NP analog of PermUP.

Figure 3 shows the known containment relations between the main classes studied in this paper.

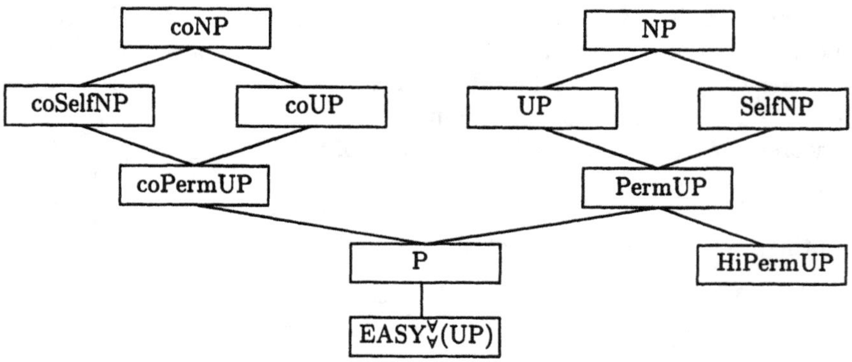

Figure 3. The known containment relationships between the classes studied in this paper.

Having developed a theory of self-witnessing languages, we hope it will be useful in studying additional open problems in complexity theory. For instance, Corollary 13, part 4 shows that all languages reducible to SAT via a polynomial-time computable, honest, onto reduction are in SelfNP. Berman and Hartmanis famously conjectured [BH77] that all NP-complete languages are pairwise reducible to each other via a polynomial-time computable, polynomial-time invertible, onto, one-to-one reduction. This is known as the *Isomorphism Conjecture.* It could be the case that all NP-complete languages are self-witnessing, even if the Isomorphism Conjecture fails. This leads to the following conjecture.

Conjecture 17 *All* NP-*complete languages are in* SelfNP.

Note that if Conjecture 17 does not hold then, by Corollary 13, part 4, the Isomorphism Conjecture does not hold. Conversely, we ask, "If Conjecture 17 holds, does the Isomorphism Conjecture necessarily hold?" As noted by Berman and Hartmanis [BH77], if the Isomorphism Conjecture holds, then $P \neq NP$. It follows that if the answer to our question is "yes" and Conjecture

17 holds, then P $\neq$ NP.

Another idea is to explore generalizations of SelfNP and PermUP. For instance, let SelfUP = $\{L \mid$ there exists a UPTM U such that $L(U) = L$ and $\mathrm{wit}_{\mathrm{U}}(L) = L\}$ and Self$_{\subseteq}$NP = $\{L \mid$ there exists an NPTM N such that $L(N) = L$ and $\bigcup_{x \in L} \mathrm{wit}_{\mathrm{N}}(x) \subseteq L\}$. Does PermUP = SelfUP? Does SelfNP = Self$_{\subseteq}$NP? What are the complexity-theoretic consequences of either equality holding?

Acknowledgments

We thank Lane Hemaspaandra for helpful comments and for allowing us to include here Theorem 14 and Lemma 15. We also thank Jörg Rothe for key insights, Alina Beygelzimer and William Scherer for helpful discussions, and an anonymous referee for helpful suggestions and for pointing that in Corollary 13, number 4 we could change "polynomial-time invertible" to "honest" and still have a valid claim.

References

[Ber77] L. Berman. *Polynomial Reducibilities and Complete Sets.* PhD thesis, Cornell University, Ithaca, NY, 1977.

[BGS75] T. Baker, J. Gill, and R. Solovay. Relativizations of the P=?NP question. *SIAM Journal on Computing*, 4(4):431–442, 1975.

[BH77] L. Berman and J. Hartmanis. On isomorphisms and density of NP and other complete sets. *SIAM Journal on Computing*, 6(2):305–322, 1977.

[Boo74] R. Book. Tally languages and complexity classes. *Information and Control*, 26(2):186–193, 1974.

[FFNR96] S. Fenner, L. Fortnow, A. Naik, and J. Rogers. On inverting onto functions. In *Proceedings of the 11th Annual IEEE Conference on Computational Complexity*, pages 213–222. IEEE Computer Society Press, May 1996.

[Grä94] E. Grädel. Definability on finite structures and the existence of one-way functions. *Methods of Logic in Computer Science*, 1(3):299–314, 1994.

[GS88] J. Grollmann and A. Selman. Complexity measures for public-key cryptosystems. *SIAM Journal on Computing*, 17(2):309–335, 1988.

[Hem00] L. Hemaspaandra. Personal Communication, October 2000.

[HH88] J. Hartmanis and L. Hemachandra. Complexity classes without machines: On complete languages for UP. *Theoretical Computer Science*, 58(1-3):129–142, 1988.

[HJ95] L. Hemaspaandra and S. Jha. Defying upward and downward separation. *Information and Computation*, 121(1):1–13, 1995.

[Hom00] C Homan. Low ambiguity in strong, total, associative, one-way functions. Technical Report TR734, University of Rochester, Computer Science Department, August 2000. Thu, 10 Aug 00 13:36:44 GMT.

[HR00] L. Hemaspaandra and J. Rothe. Characterizing the existence of one-way permutations. *Theoretical Computer Science*, 244(1-2):257–261, 2000.

[HT01] C. Homan and M. Thakur. One-way permutations and self-witnessing languages. Technical Report 760, University of Rochester, 2001.

[Ko85] K. Ko. On some natural complete operators. *Theoretical Computer Science*, 37(1):1–30, 1985.

[RH02] J. Rothe and L. Hemaspaandra. On characterizing the existence of partial one-way permutations. *Information Processing Letters*, 82(3):165–171, 2002.

[RRW94] R. Rao, J. Rothe, and O. Watanabe. Upward separation for FewP and related classes. *Information Processing Letters*, 52(4):175–180, 1994. Corrigendum appears in same journal, Volume 74, number 1-2, page 89, 2000.

[Sel92] A. Selman. A survey of one-way functions in complexity theory. *Mathematical Systems Theory*, 25(3):203–221, 1992.

[Val76] L. Valiant. The relative complexity of checking and evaluating. *Information Processing Letters*, 5(1):20–23, 1976.

[Wat88] O. Watanabe. On hardness of one-way functions. *Information Processing Letters*, 27(3):151–157, 1988.

APPROXIMATION ALGORITHMS FOR GENERAL PACKING PROBLEMS WITH MODIFIED LOGARITHMIC POTENTIAL FUNCTION *

Klaus Jansen and Hu Zhang

Institute of Computer Science and Applied Mathematics,
University of Kiel, Germany
kj,hzh@informatik.uni-kiel.de

Abstract In this paper we present an approximation algorithm based on a Lagrangian decomposition via a logarithmic potential reduction to solve a general packing or min-max resource sharing problem with M nonnegative convex constraints on a convex set B. We generalize a method by Grigoriadis et al to the case with weak approximate block solvers (i.e. with only constant, logarithmic or even worse approximation ratios). We show that the algorithm needs at most $O(M(\varepsilon^{-2}\ln\varepsilon^{-1} + \ln M))$ calls to the block solver, a bound independent of the data and the approximation ratio of the block solver. For small approximation ratios the algorithm needs at most $O(M(\varepsilon^{-2}+\ln M))$ calls to the block solver.

1. Introduction.

We consider the following general *packing problem* or *convex min-max resource-sharing problem* of the form:

$$(P) \qquad \lambda^* = \min\{\lambda | f(x) \le \lambda, i.e, x \in B\},$$

where $f : B \rightarrow \mathbb{R}^M$ is a vector of M nonnegative continuous convex functions defined on a nonempty convex compact set B, and e is the vector of all ones. Without loss of generality we assume $\lambda^* > 0$. The functions f_m,

*This research was supported in part by the DFG - Graduiertenkolleg, Effiziente Algorithmen und Mehrskalenmethoden, by the EU Thematic Network APPOL, Approximation and Online Algorithms, IST-1999-14084 and by the EU Research Training Network ARACNE, Approximation and Randomized Algorithms in Communication Networks, HPRN-CT-1999-00112.

$1 \le m \le M$, are the packing or coupling constraints. In addition we denote $\lambda(x) = \max_{1\le m\le M} f_m(x)$ for any given $x \in B$. There are many applications of the general packing problem. Typical examples are scheduling on unrelated machines, job shop scheduling, network embeddings, Held-Karp bound for TSP, multicommodity flows, maximum concurrent flow, spreading metrics, approximation of metric space, and graph partitioning [1, 2, 3, 7, 9, 10].

Grigoriadis and Khachiyan [4, 5] proposed an elegant method to compute an ε-approximate solution for this problem; *i.e.* for a given accuracy $\varepsilon > 0$,

$$(P_\varepsilon) \qquad \text{compute } x \in B \text{ s.t. } f(x) \le (1+\varepsilon)\lambda^* e.$$

The approach is based on the Lagrangian duality relation:

$$\lambda^* = \min_{x\in B} \max_{p\in P} p^T f(x) = \max_{p\in P} \min_{x\in B} p^T f(x),$$

where $P = \{p \in \mathbb{R}^M | \sum_{m=1}^M p_m = 1, p_m \ge 0\}$. Denoting $\Lambda(p) = \min_{x\in B} p^T f(x)$, a pair $x \in B$ and $p \in P$ is optimal, if and only if $\lambda(x) = \Lambda(p)$. The corresponding ε-approximate dual problem has the form:

$$(D_\varepsilon) \qquad \text{compute } p \in P \text{ s.t. } \Lambda(p) \ge (1-\varepsilon)\lambda^*.$$

The Lagrangian or price-directive decomposition method is an iterative strategy that solves the primal problem (P_ε) and its dual problem (D_ε) by computing a sequence of pairs x, p to approximate the exact solution from above and below, respectively.

Grigoriadis and Khachiyan [5] proved that the primal problem (P_ε) and the dual problem (D_ε) can be solved with a t-approximate block solver that solves the block problem for a given tolerance t:

$$ABS(p,t) \qquad \text{compute } \hat{x} = \hat{x}(p) \in B \text{ s.t. } p^T f(\hat{x}) \le (1+t) \min_{y\in B} p^T f(y).$$

Usually the number of the calls to the approximate block solver to obtain the solution required is applied as a criterion of the complexity of the approximation algorithm. The reason is that we consider the general case of these problems where the approximate block solver is unknown to us (used as an oracle in the approximation algorithm). Therefore, the running time of the approximation algorithm is just the product of the number of calls and the running time of the block solver. The number of iterations (or calls to the block solver) in [5] is $O(M(\varepsilon^{-2} \ln \varepsilon^{-1} + \ln M))$.

The method uses either the exponential or standard logarithmic potential function [4, 5] and is based on a ternary search procedure that repeatedly scales the functions f. Villavicencio and Grigoriadis [11] proposed a modified logarithmic potential function to avoid the scaling phases and to simplify the analysis. The number of iterations used in [11] is also $O(M(\varepsilon^{-2} \ln \varepsilon^{-1} + \ln M))$. Furthermore, Grigoriadis et al [6] studied a general covering or concave max-min resource sharing problem (that is orthogonal to the packing problem studied here). They showed that the number of iterations or block optimization steps is only $O(M(\varepsilon^{-2}+\ln M))$ for the general covering problem. Therefore it is natural to ask whether one can improve the number of iterations for the packing problem. In fact we reduce the number of iterations for the general packing problem (P_ε) and dual problem (D_ε) to $O(M(\varepsilon^{-2}+\ln M))$.

On the other hand, in general the block problem is hard to approximate [1, 2]. This means that the assumption to have a block solver with accuracy $t = O(\varepsilon)$ is too strict. Therefore we consider in this paper the case where we have only a weak approximate block solver. A (t, c)-approximate block solver is defined as follows:

$$ABS(p,t,c) \qquad \text{compute } \hat{x} = \hat{x}(p) \in B \text{ s.t. } p^T f(\hat{x}) \leq c(1+t)\min_{y\in B} p^T f(y),$$

where $c \geq 1$ is the approximation ratio. The main goal is now to solve the following primal problem (using the weak block solver)

$$(P_{\varepsilon,c}) \qquad \text{compute } x \in B \text{ s.t. } f(x) \leq c(1+\varepsilon)\lambda^* e.$$

The corresponding dual problem has the form:

$$(D_{\varepsilon,c}) \qquad \text{compute } p \in P \text{ s.t. } \Lambda(p) \geq \frac{1}{c}(1-\varepsilon)\lambda^*.$$

New results. Our main result is an approximation algorithm that for any accuracy $\varepsilon \in (0,1)$ solves the problem $(P_{\varepsilon,c})$ in

$$N = O(M(\varepsilon^{-2}\ln\varepsilon^{-1}+\ln M))$$

iterations or coordination steps. Each step requires a call to the weak block solver $ABS(p, O(\varepsilon), c)$ and an overhead of $O(M\ln\ln(M/\varepsilon))$ arithmetic operations. Furthermore for small ratios c with $\ln c = O(\varepsilon)$ we improve the number of iterations to $O(M(\varepsilon^{-2}+\ln M))$.

Related results. Plotkin et al [10] considered the linear feasibility variants of both problems: either to find a point $x \in B$ such that $f(x) = Ax \geq (1-\varepsilon)b$ or to find a point $x \in B$ such that $f(x) = Ax \leq (1+\varepsilon)b$ where A is the coefficient matrix with M rows and b is an M-dimensional vector. The problems are solved in [10] by Lagrangian decomposition using exponential potential reductions. The number of iterations (calls to the corresponding block solver) in these algorithms are $O(\varepsilon^{-2}\rho\ln(M\varepsilon^{-1}))$ and $O(M+\rho\log^2 M+\varepsilon^{-2}\rho\ln(M\varepsilon^{-1}))$, where $\rho = \max_{1\leq m\leq M}\max_{x\in B} a_m^T x/b_m$ is the width of B relative to $Ax \geq b$. Garg and Könemann [3] proposed a $(1+\varepsilon)$ approximation algorithm to solve the linear packing problem within $O(M\varepsilon^{-2}\ln M)$ iterations which is independent of the width. Recently Young [13] has proposed an approximation algorithm for a mixed linear packing and covering problem (with convex set $B = \mathbb{R}_+^N$) with running time $O(Md\ln M/\varepsilon^2)$ where d is the maximum number of constraints any variable appears in. Young [12] studied also the linear case of the packing problem but with weak block solvers. He proposed an algorithm that uses $O(\rho' \ln M/(\lambda^* \varepsilon^2))$ calls to the block solver, where $\rho' = \max_{1 \le m \le M} \max_{x \in B} a_m^T x / b_m - \min_{1 \le m \le M} \min_{x \in B} a_m^T x / b_m$ and λ^* is the optimal value of the packing problem. Furthermore, Charikar et al [1] noticed that the result in [10] for the packing problem can be extended also to the case with weak block solvers and the same number $O(\varepsilon^{-2} \rho \ln(M \varepsilon^{-1}))$ of iterations. Recently Jansen and Porkolab [8] studied the general covering problem and showed that at most $O(M(\ln M + \varepsilon^{-2} + \varepsilon^{-3} \ln c)$ coordination steps are necessary.

Main ideas. Our paper is strongly based on ideas in [4, 5, 6, 11]. We use the modified logarithmic potential function proposed in [11]. Our algorithm is based on the scaling phase strategy such that the relative error tolerances σ in all phases achieve the given relative tolerance ε gradually. The oracle $ABS(p,t,c)$ is called once in each iteration. We found that the stopping rules proposed in [4, 5, 11] are too strict, and that the number of iterations would be at least $O(Mc^2(\ln Mc + \varepsilon^{-3} \ln c))$. Therefore we analyzed a combination of two stopping rules in order to obtain a running time independent of c. In fact our result is the first one independent of the width ρ, the optimal value λ^* and the approximation ratio c. For certain c small enough, we use an upper bound for the difference of potential function values $\phi_t(x) - \phi_t(x)$ for two points $x, x' \in B$ similar to [6]. This enables us to show that the original method in [11] (with a slightly modified stopping rule) uses only $O(M(\varepsilon^{-2} + \ln M))$ coordination steps for c with $\ln c = O(\varepsilon)$.

The paper is organized as follows. In Section 2 the modified logarithmic potential function and price vector are introduced and their properties are presented. The algorithm is described in details in Section 3. Finally, in Section 4 the correctness and the complexity of the algorithm are analyzed.

2. Modified logarithmic potential function.

In order to solve the min-max resource sharing problem (P), we use the Lagrangian decomposition method that is based on a special potential function. Villavicencio and Grigoriadis [11] proposed the following modification of Karmarkar's potential function to relax the coupling constraints of (P):

$$\Phi_t(\theta, x) = \ln \theta - \frac{t}{M} \sum_{m=1}^{M} \ln(\theta - f_m(x)), \tag{1}$$

where $\theta \in \mathbb{R}_+$ and $x \in B$ are variables and $t \in \mathbb{R}_+$ is a fixed tolerance parameter (that is used in the approximate block solver $ABS(p,t,c)$). In our algorithm, we shall set values of t from $1/6$ initially down to $O(\varepsilon)$, where ε is the desired relative accuracy for the solution. The function Φ_t is well-defined for $\lambda(f(x)) < \theta < \infty$ where $\lambda(x) = \max\{f_1(x), \ldots, f_M(x)\}$ and has the *barrier property*: $\Phi_t(\theta, x) \to \infty$ for $\theta \to \lambda(x)$ and $\theta \to \infty$.

We define the reduced potential function as the minimal value $\Phi_t(\theta, x)$ over $\theta \in (\lambda(x), \infty)$ for a given fixed $x \in B$, *i.e.*

$$\phi_t(x) = \min_{\lambda(x)<\theta<\infty} \Phi_t(\theta, x). \tag{2}$$

The minimum $\theta(x)$ of $\Phi_t(\theta, x)$ (using the first-order optimality condition) is the solution of the equation:

$$\frac{t}{M}\sum_{m=1}^{M} \frac{\theta}{\theta - f_m(x)} = 1. \tag{3}$$

The function $g(\theta) = \frac{t}{M}\sum_{m=1}^{M} \frac{\theta}{\theta - f_m}$ is a strictly decreasing function of θ in $(\lambda(x), \infty)$. Therefore the implicit function $\theta(x)$ is the unique root of (3) in the interval $(\lambda(x), \infty)$. The following two lemmas by the definition of $\theta(x)$ show the bounds of $\theta(x)$ and $\phi_t(x)$, similar to those in [11] and [6].

Lemma 1 $\lambda(x)/(1 - t/M) \leq \theta(x) \leq \lambda(x)/(1 - t)$ *for any* $x \in B$.

Lemma 2 $(1 - t)\ln \lambda(x) \leq \phi_t(x) \leq (1 - t)\ln \lambda(x) + t\ln(e/t)$ *for any* $x \in B$.

Remark. Lemmas 1 and 2 show (for certain sufficiently small values of t) that the minimum value $\theta(x)$ approximates $\lambda(x)$ and that the potential function $\phi_t(x)$ approximates $\ln \lambda(x)$ closely. This gives us the possibility to solve the approximation problem $(P_{\varepsilon,c})$ by minimizing the smooth function $\phi_t(x)$ over $x \in B$ based on these two Lemmas.

2.1. Price vector function.

The price vector $p(x) \in \mathbb{R}^M$ is defined as follows [11]:

$$p_m(x) = \frac{t}{M}\frac{\theta(x)}{\theta(x) - f_m(x)}, \qquad m = 1, \ldots, M. \tag{4}$$

According to equation (3), each price value $p_m(x)$ is nonnegative and for any $x \in B$, $\sum_{m=1}^{M} p_m(x) = 1$, which are the properties desired. Thus, we can just simply compute the price vector p from (4), which is easier to calculate compared with other methods (e.g. using the exponential potential function). In view of the definition above, the following lemma holds.

Lemma 3 $p(x)^T f(x) = \theta(x)(1 - t)$ *for any* $x \in B$.

Remark. By Lemma 3, for t small enough, the dual value $p^T f(x)$ is only slightly less than $\theta(x)$, the minimum of the potential function. This shows that the dual value $p^T f(x)$ is also an approximation of $\lambda(x)$. Therefore the primal problem can also be solved by obtaining the solution of the dual value and in fact that is what we need to construct our algorithm.

3. The approximation algorithm.

The exact dual value $\Lambda(p)$ can be approximated by the dual value $p^T f(\hat{x})$, where $\hat{x}$ is the block solution computed by the (t,c)-approximate block solver for the current price vector p. Furthermore, to establish the stopping rules of the scaling phases in the approximation algorithm, the value of duality gap should be estimated in each iteration. For our first stopping rule we use the following parameter ν:

$$\nu = \nu(x,\hat{x}) = \frac{p^T f(x) - p^T f(\hat{x})}{p^T f(x) + p^T f(\hat{x})}. \tag{5}$$

If ν is $O(\varepsilon)$, then the duality gap is also quite small. But for larger ν close to 1, the gap can be extremly large (see also Section 4). Therefore, we have to define a second parameter. Let σ be the relative error of a scaling phase. Then the parameter w is given by:

$$w = \begin{cases} \frac{1+\sigma}{(1+\sigma/6)M} & \text{for the first } \sigma\text{-scaling phase,} \\ \frac{1+\sigma}{1+2\sigma} & \text{otherwise.} \end{cases} \tag{6}$$

Let x_s be the solution of sth scaling phase. Then the two stopping rules used in the sth scaling phase are:

$$\begin{array}{ll} \textit{Rule } 1: & \nu \le \sigma/6, \\ \textit{Rule } 2: & \lambda(x) \le w\,\lambda(x_{s-1}). \end{array}$$

Remark. Grigoriadis et al [6, 11] used either only the first stopping rule or the rule $p^T f(x) - p^T f(\hat{x}) \le \sigma\theta(x)/2$ that is similar to the former. In the case of a weak block solver, such stopping rules are not sufficient to obtain a running time independent of the ratio c. It may happen that the block solver is called more times than what necessary. Therefore we have introduced the second stopping rule to make sure that the scaling phase stops as soon as the solution meets the requirement of the phase. On the other hand the first stopping rule is needed to have always a constant decrement in the potential function (see Section 4).

The algorithm works now as follows. First we apply the scaling phase strategy. In each scaling phase the relative error tolerance σ is set. Then based on the known pair of x and p, a solution $\hat{x}$ is generated by the approximate block solver. Afterwards an appropriate linear combination of the old solution x and $\hat{x}$ is computed as the new solution. The iteration stops when the solution satisfies one of the stopping rules. After one scaling phase, the error tolerance σ is halved and the next scaling phase is started until the error tolerance $\sigma \le \varepsilon$. The solution generated in the last scaling phase solves the primal problem $(P_{\varepsilon,c})$ (see also Section 4).

In our algorithm we set $t = \sigma/6$ for the error tolerance in the block solver $ABS(p,t,c)$. To run the algorithm, we need an initial solution $x_0 \in B$ in advance. Here we use as x_0 the solution of the block solver $ABS(e/M, \sigma_0/6, c)$

where the price vector e/M is the vector of all $1/M$'s and the initial error tolerance $\sigma_0 = 1$.

Algorithm $\mathcal{L}(f, B, \varepsilon, c)$:

```
initialize: s := 0, σ := σ_0 := 1, t := σ/6, p := e/M, x := ABS(p, t, c)
        and x_s := x,
  while σ > ε/2  do                                    /* σ-scaling phase */
   s := s + 1, x := x_{s-1}, finished := false,
    while not(finished) do                              /* coordination step */
     computer θ(x) from (3) and p = p(x) ∈ P from (4),
     x̂ := ABS(p, t, c),
     compute ν = ν(x, x̂) from (5) and w from (6),
      if either Stopping Rule 1 or 2 is satisfied then
        x_s := x, finished := true,
      else
        x := (1 − τ)x + τx̂ for an appropriate step length τ ∈ (0, 1],
      end
    end
   σ := σ/2, t := σ/6,
  end
```

We set the step length τ as:

$$\tau = \frac{t\theta\nu}{2M(p^T f + p^T \hat{f})}. \tag{7}$$

We note that the step length τ can be computed also by a line search to minimize the potential value ϕ_t. The algorithm and the analysis remains valid if we use such a line search to compute τ (see also Section 4).

4. Analysis of the algorithm $\mathcal{L}$.

In this section we verify the convergence of algorithm $\mathcal{L}$ by proving that the algorithm stops in each scaling phase after a finite number of iterations. Furthermore we show that the vector x computed in the final phase solves the primal problem $(P_{\varepsilon,c})$. From now on we denote $\theta = \theta(x)$, $\theta' = \theta(x')$, $f = f(x)$, $f' = f(x')$ and $\hat{f} = f(\hat{x})$. Before proving the correctness of the approximation algorithm we have the following bound for the initial solution x_0.

Lemma 4 *If x_0 is the solution of $ABS(e/M, t, c)$ with $t = 1/6$, then $\lambda(x_0) \leq (7/6)cM\lambda^*$.*

Lemma 5 *If algorithm $\mathcal{L}$ stops, then the computed $x \in B$ solves $(P_{\varepsilon,c})$.*

Proof: We shall consider both stopping rules. If the first stopping rule is satisfied, then using the definition of ν and $\nu \leq t$ we have

$$(1-t)p^T f \leq (1+t)p^T \hat{f}.$$

According to Lemma 3, the inequality above, and the value $p^T f(\hat{x})$ of the solution $\hat{x}$ computed by the block solver $ABS(p,t,c)$, we have

$$\theta = \frac{p^T f}{1-t} \leq \frac{1+t}{(1-t)^2} p^T \hat{f} \leq c\left(\frac{1+t}{1-t}\right)^2 \Lambda(p).$$

Denote by σ_s the value of σ in the sth scaling phase, using the fact that $\Lambda(p) \leq \lambda^*$ and $t = \sigma_s/6$, we obtain $\theta \leq c(1+\sigma_s)\lambda^*$. Then by Lemma 1, $f_m(x_s) \leq \lambda(x_s) \leq \theta(x_s) \leq c(1+\sigma_s)\lambda^*$.

If the second stopping rule is satisfied, then we consider two further cases. If the rule is satisfied in the first phase $s = 1$, then according to the value of w in formula (6) and $\sigma = \sigma_0$,

$$\lambda(x_1) \leq \frac{1+\sigma_0}{(1+\sigma_0/6)M} \lambda(x_0).$$

Since x_0 is the solution of $ABS(e/M,t,c)$ we have $\lambda(x_0) \leq c(1+\sigma_0/6)M\lambda^*$. This implies that $\lambda(x_1) \leq c(1+\sigma_0)\lambda^*$.

Now let us assume (by induction) that $\lambda(x_{s-1}) \leq c(1+\sigma_{s-1})\lambda^*$ after scaling phase $s-1$. Thus if the second stopping rule is satisfied in phase s, then according to the definition of w and $\sigma_s = \sigma_{s-1}/2$, we have

$$\lambda(x_s) \leq \frac{(1+\sigma_s)}{(1+2\sigma_s)} \lambda(x_{s-1}) \leq (1+\sigma_s)c\lambda^*.$$

Therefore in both cases we have $\lambda(x_s) \leq (1+\sigma_s)c\lambda^*$ for any $s > 0$. Since $\sigma \leq \varepsilon$ when the algorithm $\mathcal{L}$ halts, the solution $x = x_s$ computed in the last phase solves $(P_{\varepsilon,c})$. □

In the next Lemma we show that the potential function ϕ_t decreases by a constant factor in each coordination step. This enables us later to prove an upper bound for the number of iterations.

Lemma 6 *For any two consecutive iterates x, $x' \in B$ within a scaling phase of Algorithm $\mathcal{L}$,*

$$\phi_t(x') \leq \phi_t(x) - \frac{t\nu^2}{4M}.$$

The proof is similar to that in [6] and we do not give the details in this paper. Now we are ready to estimate the complexity of Algorithm $\mathcal{L}$.

Theorem 1 *For a given relative accuracy $\varepsilon \in (0,1]$, Algorithm $\mathcal{L}$ finishes with a solution x that satisfies $\lambda(x) \leq c(1+\varepsilon)\lambda^*$ and performs a total of*

$$N = O(M(\ln M + \varepsilon^{-2}\ln \varepsilon^{-1}))$$

coordination steps.

Proof: If algorithm $\mathcal{L}$ finishes after a finite number of iterations with a solution x, then using Lemma 5 $\lambda(x) \leq c(1+\varepsilon)\lambda^*$. First we calculate a bound on the number N_σ of coordination steps performed in a single scaling phase. Afterwards we obtain a bound for all scaling phases. This shows also that algorithm $\mathcal{L}$ halts and proves the theorem.

Let x, x' denote the initial and final iterates of a σ-scaling phase. In addition let $\bar{x}$ be the solution after $\overline{N}_\sigma = N_\sigma - 1$ iterations in the same scaling phase. During the phase from x to x' we have $\nu > t$; otherwise the phase would finish earlier. Furthermore (by the same reason) the objective value $\lambda(\bar{x}) > w\lambda(x)$. Then Lemma 6 provides

$$\phi_t(x) - \phi_t(\bar{x}) \geq \overline{N}_\sigma \frac{t\nu^2}{4M} \geq \overline{N}_\sigma \frac{t^3}{4M}.$$

By the left and right inequality of Lemma 2 we have $(1-t)\ln\lambda(\bar{x}) \leq \phi_t(\bar{x})$ and $\phi_t(x) \leq (1-t)\ln\lambda(x) + t\ln(e/t)$, respectively. Hence,

$$\overline{N}_\sigma \frac{t^3}{4M} \leq (1-t)\ln\frac{\lambda(x)}{\lambda(\bar{x})} + t\ln\frac{e}{t}.$$

This gives directly a bound for N_σ,

$$N_\sigma = \overline{N}_\sigma + 1 \leq 4Mt^{-3}\left((1-t)\ln\frac{\lambda(x)}{\lambda(\bar{x})} + t\ln\frac{e}{t}\right) + 1.$$

Next we shall bound the term $\lambda(x)/\lambda(\bar{x})$. For the first scaling phase we have $\lambda(\bar{x}) > \frac{1+\sigma}{(1+\sigma/6)M}\lambda(x)$. In this case

$$\frac{\lambda(x)}{\lambda(\bar{x})} \leq \frac{(1+\sigma/6)M}{1+\sigma} < \frac{2M}{1+\sigma}.$$

Since $t = 1/6$ and $\sigma = 1$ during the first scaling phase, there are only $O(M\ln M)$ coordination steps in the first phase. For the other scaling phases, we have $\lambda(\bar{x}) > \frac{1+\sigma}{1+2\sigma}\lambda(x)$. Therefore,

$$\frac{\lambda(x)}{\lambda(\bar{x})} \leq \frac{1+2\sigma}{1+\sigma} = 1 + \frac{\sigma}{1+\sigma}.$$

Then according to the elementary inequality $\ln(1+u) \leq u$ for $u \geq 0$,

$$\ln\frac{\lambda(x)}{\lambda(\bar{x})} \leq \ln(1+\frac{\sigma}{1+\sigma}) \leq \frac{\sigma}{1+\sigma} < \sigma.$$

Substituting this in the bound for N_σ above and using $t = \sigma/6$, we have the expression

$$N_\sigma = O(M\sigma^{-2}\ln\sigma^{-1}).$$

Finally, the total number of coordination steps N is obtained by summing the N_σ over all scaling phases:

$$N = O(M(\ln M + \sum_{\sigma} \sigma^{-2} \ln \sigma^{-1})).$$

Since σ is halved in each scaling phase, the sum above is further bounded by

$$\begin{aligned}\sum_{\sigma} \sigma^{-2} \ln \sigma^{-1} &= O(\textstyle\sum_{q=0}^{\lceil \log \varepsilon^{-1} \rceil} 2^{2q} \ln(2^q)) \\ &= O(\log \varepsilon^{-1} \textstyle\sum_{q=0}^{\lceil \log \varepsilon^{-1} \rceil} 2^{2q}) \\ &= O(\varepsilon^{-2} \log \varepsilon^{-1}),\end{aligned}$$

which provides the claimed bound. □

4.1. Analysis for small approximation ratio c.

In this subsection we analyze the case that the approximation ratio c of the block solver is small enough, *i.e.*, $\ln c = O(\varepsilon)$. In this case we modify the algorithm $\mathcal{L}$ as follows: we use only the first stopping rule $\nu \leq t$. This rule is similar to the one in [11]. Let $\mathcal{L}'$ be the modified algorithm. First it can be proved that the pair x and p computed in the last scaling phase is a solution of both primal problem $(P_{\varepsilon,c})$ and dual problem $(D_{\varepsilon,c})$, respectively. Furthermore we prove an upper bound on the difference of the potential function for any two arbitrary points x and x' in B. The proof is similar to one in [6] for the general covering problem.

Lemma 7 *For any two iterates x, $x' \in B$ within a scaling phase with $\lambda(x) > 0$ and $\lambda(x') > 0$,*

$$\phi_t(x) - \phi_t(x') \leq (1-t) \ln \frac{p^T f}{p^T f'} \leq (1-t) \ln \frac{p^T f}{\Lambda(p)},$$

where $p = p(x)$ is defined by (4).

For the complexity of the algorithm $\mathcal{L}'$ we have the following theorem.

Theorem 2 *In the case of $\ln c = O(\varepsilon)$, algorithm $\mathcal{L}'$ performs a total of*

$$N = O(M(\ln M + \varepsilon^{-2}))$$

coordination steps.

The proof will be given in the full version of the paper.

4.2. Analysis of accuracy.

In the analysis above we assumed that the price vector $p \in P$ can be computed exactly from (4). However, in practice this is not true since $\theta(x)$ is needed in (4). $\theta(x)$ is the root of (3) that in general can only be computed approximately. We can prove that if p has a relative accuracy bounded by $O(\varepsilon)$, the algorithm can still generate the desired solution. To obtain that requirement, the absolute error of θ must be bounded by $O(\varepsilon^2/M)$. Then $O(M\ln(M/\varepsilon))$ arithmetic operations are sufficient to compute the price vector p with binary search. If the Newton's method is applied, only $O(M\ln\ln(M/\varepsilon))$ operations are enough. The detailed analysis will be presented in the full version of this paper. A similar argument how to resolve this problem can be found in [4, 5, 6, 11].

5. Conclusion.

In this paper we have proposed an approximation algorithm for the general packing problem that needs only $O(M(\ln M + \varepsilon^{-2}\ln\varepsilon^{-1}))$ coordination steps or calls to a weak block solver with approximation ratio c. This bound is independent of the approximation ratio c. However, if the original methods in [4, 5, 11] are used, then the number of iterations can be bounded only by $O(Mc^2(\ln Mc + \varepsilon^{-3}\ln c))$, which is larger than our bound here significantly. The reason is that the stopping rules in [4, 5, 11] are too strict and that the duality gap is too large.

But we note that the original method could be faster, since we don't have a lower bound on the number of iterations. Another interesting point is that our algorithm does not automatically generate a dual solution for general c. We can prove that only $O(M\varepsilon^{-3}\ln(M/\varepsilon))$ coordination steps are necessary to get a dual solution of $(D_{\varepsilon,c})$. It would be interesting whether this bound can be improved. We will focus on these questions in our future research.

References

[1] M. Charikar, C. Chekuri, A. Goel, S. Guha and S. Plotkin, Approximating a finite metric by a small number of tree metrics, *Proceedings of the 39th Annual IEEE Symposium on Foundations of Computer Science*, FOCS 1998, 379-388.

[2] G. Even, J. S. Naor, S. Rao and B. Schieber, Fast approximate graph partitioning algorithms, *SIAM Journal on Computing*, 6 (1999), 2187-2214.

[3] N. Garg and J. Könemann, Fast and simpler algorithms for multicommodity flow and other fractional packing problems, *Proceedings of the 39th IEEE Annual Symposium on Foundations of Computer Science*, FOCS 1998, 300-309.

[4] M. D. Grigoriadis and L. G. Khachiyan, Fast approximation schemes for convex programs with many blocks and coupling constraints, *SIAM Journal on Optimization*, 4 (1994), 86-107.

[5] M. D. Grigoriadis and L. G. Khachiyan, Coordination complexity of parallel price-directive decomposition, *Mathematics of Operations Research*, 2 (1996), 321-340.

[6] M. D. Grigoriadis, L. G. Khachiyan, L. Porkolab and J. Villavicencio, Approximate max-min resource sharing for structured concave optimization, *SIAM Journal on Optimization*, 11 (2001), 1081-1091.

[7] K. Jansen, Approximation algorithms for fractional covering and packing problems, and applications, Manuscript, (2001).

[8] K. Jansen and L. Porkolab, On preemptive resource constrained scheduling: polynomial-time approximation schemes, *Proceedings of the 9th Conference on Integer Programming and Combinatorial Optimization*, IPCO 2002.

[9] J. K. Lenstra, D. B. Shmoys and E. Tardos, Approximation algorithms for scheduling unrelated parallel machines, *Mathematical Programming*, 24 (1990), 259-272.

[10] S. A. Plotkin, D. B. Shmoys and E. Tardos, Fast approximation algorithms for fractional packing and covering problems, *Mathematics of Operations Research*, 2 (1995), 257-301.

[11] J. Villavicencio and M. D. Grigoriadis, Approximate Lagrangian decomposition with a modified Karmarkar logarithmic potential, *Network Optimization, P. Pardalos, D. W. Hearn and W. W. Hager, Eds, Lecture Notes in Economics and Mathematical Systems 450, SpringerVerlag, Berlin*, (1997), 471-485.

[12] N. E. Young, Randomized rounding without solving the linear program, *Proceedings of the 6th ACM-SIAM Symposium on Discrete Algorithms*, SODA 1995, 170–178.

[13] N. E. Young, Sequential and parallel algorithms for mixed packing and covering, *Proceedings of the 42nd Annual Symposium on Foundations of Computer Science*, FOCS 2001, 538-546.

ON RANDOMNESS AND INFINITY

Grégory Lafitte
Ecole Normale Supérieure de Lyon, Laboratoire d'Informatique et Parallélisme, 46 allée d'Italie, 69364 Lyon Cedex 07, France
glafitte@ens-lyon.fr

Abstract In this paper, we investigate refined definitions of random sequences. Classical definitions have always the shortcome of making use of the notion of an algorithm. We discuss the nature of randomness and different ways of obtaining satisfactory definitions of randomness after reviewing previous attempts at producing a non-algorithmical definition. We present alternative definitions based on infinite time machines and set theory and explain how and why randomness is strongly linked to *strong axioms of infinity*.

Keywords: Randomness, infinite time machines, large cardinals.

1. Introduction

Various attempts at outlining, understanding and formalizing randomness have been carried out. One major approach stems from probability theory and statistics. It is based essentially on statistical properties such as stability of relative frequencies. Sequences produced by fairly tossing a coin is the core idea of random sequences. This approach merely describes the properties that should have a random sequence; it does not provide a definition or notion of randomness.

It should be mentioned that many people in statistics and probability object to thinking of points in a probability space as being random and prefer to talk of random processes for pickings points instead. (This viewpoint is the one of H. Rubin as expressed to A.H. Kruse in [Kruse, 1967].) We tend to agree totally with this. It encourages us in thinking that randomness has not much to do with the theory of probabilities apart from the trivial statistical facts concerning "random objects". Nevertheless, it is certainly worthwhile to investigate where random objects appear (*e.g.*, Rado's graph, randomness in complexity theory, ...) and find coherent randomness definitions verified by those objects.

The other major approach is of an algorithmic nature. It is sometimes mixed with the previous approach. This approach is based on *unpredictability*. It does provide some way to define randomness but then it is rather surprising to have algorithms involved since probability theory does not use the notion of an algorithm. Is it then possible to find a mathematical definition for random sequences not based on algorithmic unpredictability?

Durand *et al.* [Durand et al., 2001] propose such a definition by opting for a randomness whose strong properties are *relatively consistent* after showing that it is impossible to get a notion of randomness having *provably* those properties. In their definition, they have to take a basis for randomness and use "arithmetical randomness" to be just that. The algorithmic nature has then not completely disappeared from the definition. Other set-theoretical approaches are those of A.H. Kruse [Kruse, 1967], with the use of an appropriate class theory instead of ZFC, and M. van Lambalgen [van Lambalgen, 1992], adding to ZFC an extra atomic predicate of randomness and some axioms which govern its use.

What is randomness after all? We could define randomness as the absence of any law. The problem is again in the fact that we will want (to be able to do something from the definition and not to get an inconsistency) the avoided laws to be definable in some way and somehow we will come back to some algorithmically-based definition.

Another way of presenting randomness is the complete *independence* between any two terms, or even only between any one term and its predecessors, of a random sequence. Usually some independence is sought by using some algorithmic method. We propose a method based completely on set-theoretic independence and its strong link with the theory of large cardinals. This has prompted more and more the author to believe in a strong connection between large cardinality concepts and randomness occurrences. This idea is somehow also at the basis of M. van Lambalgen's study in [van Lambalgen, 1992] apart from the fact that van Lambalgen searches for new axioms, giving predicates for randomness, to add to ZFC while we think the axioms are somewhat already there in the set theory literature.

Let us now proceed to a brief description of the contents of this paper. In the first section, we discuss classical definitions of randomness through a definition, using games, of Muchnik *et al.* and obtain a simple characterization of this classic randomness using infinite time Turing machines. This is what we call the *unfeasibility approach.*

We then continue in section 2 by giving several methods making it possible to introduce some "non provable" randomness. We present Durand *et al.*'s method, generalize it and also introduce in the same direction some randomness notions, based on *independence* from ZFC, using original infinite time machines tailored to be able to capture all the properties of sets of reals. We call this the *unprovability approach.*

Using those notions, we construct in section 3 randomness notions hopefully meeting our goal. We call this the *unknowability approach.* Our ultimate randomness notion seems to be *unknowably randomness* (Randomness notion 5) along Randomness hierarchies 1 or 2. Surely, this is the strongest form one could wish for a randomness notion since then there is no way (in our base theory ZFC or even in ZFC + $\exists$ some large cardinal) to *connect* any one value to the other values of such a random sequence.

2. Notations

In this paper, a sequence (as in *random sequence*) is an infinite binary sequence, *i.e.*, belonging to $\{0,1\}^{\omega} = \{0,1\}^{\mathbb{N}}$, which can also be seen as a real, *i.e.*, belonging to $\mathbb{R}$. $\{0,1\}^{<\omega}$ is the set of finite sequences, which can also be seen as $\mathbb{N}$. For any $s \in \{0,1\}^{<\omega}$, $\{0,1\}^{\omega}_s$ denotes the set of infinite binary sequences that extend s. For a sequence a, a_k denotes the $k+1$-th term of the sequence.

3. Unfeasibility-based randomness

3.1. Algorithmic randomness ...

Muchnik *et al.* [Muchnik et al., 1998] have given various temptative definitions of randomness and compared them making all the while sure that they verify several properties that everyone believes a random sequence should verify. It turns out that the two most restrictive definitions of randomness are *chaotic* and *unpredictable*. The former is based on those sequences whose initial segments' entropies grow sufficiently fast. All chaotic sequences turn out to be unpredictable, the truth of the converse is an open question. Because every other known definition of randomness reduces to the notion of unpredictableness, we will use it as our base definition and we call it *Muchnik randomness*.

Definition 1 *Let a be a sequence, $\mu : \{0,1\}^{<\omega} \to \mathbb{R}^+$ a computable quasi-measure*[1] *that we extend to intervals $\{0,1\}^{\omega}_s$ by having $\mu(\{0,1\}^{\omega}_s) = \mu(s)$ and $C \in \mathbb{Q}^{+\star}$ called the capital*[2].

*A one-player gambling game is played against the sequence a using the quasi-measure μ. We call it a μ-*game. *At the start of the game, the player has his wallet W_0 equal to C. At the k-th move, the player plays by giving $n = n(k)$ and a guessed value $i = i(k)$ for $a_{n(k)}$. As this is a gambling game, he also makes a bet $w = w(k) \in \mathbb{Q}^{+\star}$ such that $w(k) \leq W_{k-1}$.*

If the player was incorrect about the guessed value, he loses his bet : $W_k = W_{k-1} - w(k)$. Otherwise

$$W_k = W_{k-1} + w(k)\frac{\mu(\mathfrak{A}_{1-i(k)})}{\mu(\mathfrak{A}_{i(k)})}$$

where for $j = 0, 1$,

$$\mathfrak{A}_j = \{a' \in \{0,1\}^{\omega} \mid a'_{n(k)} = j \text{ and } a'_{n(l)} = a_{n(l)} \text{ for } l = 1, 2, \ldots, k-1\}$$

*The sequence a is called μ-*predictable *if there is computable winning strategy for winning μ-games against a. Otherwise, it is called μ-*unpredictable.

The sequence a is Muchnik random *if it is μ-unpredictable for some computable μ.*

Theorem 1 *A Muchnik random sequence is Martin-Löf random*[3].

PROOF. See theorem 7.4 in [Muchnik et al., 1998]. ∎

3.2. ... and infinite time machines

Finite automata on infinite sequences have been introduced by Büchi in [Büchi, 1962] to prove the decidability of the monadic second order theory of $\langle \omega, < \rangle$. Büchi automata differ from finite automata on finite sequences only by its condition of acceptance of a word. A word is *accepted* by a Büchi automata if and only if the set of states, through which the automata goes an infinite number of times during an *execution* (there may be several executions if the automata is nondeterministic), contains at least a final state. Then Büchi introduced in [Büchi, 1965] finite automata that are able to describe transfinite sets of sequences. Using those automata, he proved the decidability of the monadic second order theory of $\langle \alpha, < \rangle$, where α is a countable ordinal. It featured special transitions for limit ordinal stages such that the state reached at that limit stage ξ depends only on previously reached states $\{s \in S \mid \forall \beta < \xi \, \exists \gamma > \beta \, \varphi(\gamma) = s\}$. He modified again the definition, using still other special transitions for particular limit ordinal stages, to use those automata to prove the decidability of the monadic second order theory of $\langle \omega_1, < \rangle$. For a survey of Büchi automata, see [Büchi, 1973].

In [Lafitte, 2001], we defined a variant of Büchi automata making it as powerful as infinite time Turing machines as defined by Hamkins and Lewis in [Hamkins and Lewis, 2000]. We will now propose a definition of machines working with infinite time on reals but with access to transfinite tapes. It borrows ideas from Büchi automata deciding the monadic second order theory of $\langle \omega_1, < \rangle$ and $\mathcal{W}^2$-automata as studied in [Lafitte, 2001]. Much of the idea about the use of *stationary*[4] *sets* is due to Menachem Magidor.

Definition 2 *Fix a $n \in \mathbb{N}$. We will work with time and tapes of cardinality*[5] $\aleph_n$.

An enhanced tape *is a function from ω_n to $\{0, 1\}$.*

A continuum machine, *or* $\mathfrak{c}$-machine[6], *is a Turing machine with $k \geq 3$ separate enhanced tapes, one for input, $k-2$'s for scratch work, and one for output. The scratch and output tapes are filled with zeros at the beginning of any computation. At non-limit stages, it behaves like a normal Turing machine according to its transition relation. At limit stages, if the transition says so, the head is plucked from wherever it might have been racing towards, and placed on top of the first cell. Moreover, it enters* a limit state. *For a given cell of the tape, at a limit stage it takes the value of the lim sup of the cell values before the limit.*

A $\mathfrak{c}$-machine distinguishes different kinds of limit states. At a limit stage, it is in a composition of limit states

$$\mathfrak{Q}_0 \times \mathfrak{Q}_1 \times \cdots \times \mathfrak{Q}_n$$

where each $\mathfrak{Q}$ is defined as

$$\mathfrak{Q}_i = \{q \in Q \mid \exists \alpha \text{ of cofinality } \omega_i \text{ with } \{\beta < \alpha \mid q_\beta = q\} \text{ stationary in } \alpha\}$$

A $\mathfrak{c}$-automaton is a $\mathfrak{c}$-machine that only reads and never writes.

The output of a $\mathfrak{c}$-machine can be considered as a real when considering only the "first" ω terms of the tape. Assuming the very reasonable "$2^{\aleph_0} < \aleph_\omega$", with this definition of continuum machines, we can effectively work on $\mathbb{R}$ using and comprehending completely its *power*, *i.e.*, properties concerning sets of reals.

The notion of a $\mathfrak{c}$-machine is clearly a generalization of infinite time Turing machines[7] and by the simple techniques used in [Lafitte, 2001], has at least the same power of computation. Hence the following theorem also applies to $\mathfrak{c}$-machines.

Theorem 2 *If $r \in \mathbb{R}$ is not writable by an infinite time Turing machine, then it is Muchnik random.*

PROOF. Take $r \in \{0,1\}^\omega$ such that it is not Muchnik random. For every computable measure μ, there is thus a strategy to win the μ-game. The strategy is necessarily computable by an infinite time Turing machine.

We translate the strategy in an infinite time Turing machine, that will be able to write r since the strategy generates winning games. ∎

Theorem 2 is our cornerstone theorem for characterizing simple randomness in terms of machine computability of reals.

The following theorem is the Lost Melody Theorem of [Hamkins and Lewis, 2000]. It shows for our purpose that Muchnik random reals are not so much *random* as some can be recognized as such.

Theorem 3 *There are random reals that are still singleton recognizable by infinite time Turing machines.*

PROOF. We sketch the proof.

Consider the ordinal stages of repeat-points, that is by which it either halts or repeats, of computations with a null input. Let δ be the supremum of those repeat-points. By the nature of infinite time Turing machines, δ is a countable ordinal in L, Gödel's class of constructible sets. Let $\beta_0 = \delta$, β_n be a countable ordinal appearing first in $L_{\beta_{n+1}}$ (not in L_{β_n}) and $\beta = \sup_n \beta_n$. β is then the smallest ordinal $\geq \delta$ such that $L_{\beta+1} \models \beta$ is countable.

Since $L_{\beta+1}$ has a canonical well-ordering, there is some real $r \in L_{\beta+1}$, which is least with respect to the canonical L order, such that r codes β. The real r is the one we are looking for.

The real r is not writable because if it were, then we could solve the halting problem by searching for a real, appearing at some moment on a tape, that codes an ordinal large enough to see the repeat-point of the computation in question. Since r codes β, which is as large as δ, r is big enough and so our algorithm succeeds. And this contradicts the undecidability of the halting problem.

By usual techniques for coding the L_α's, the singleton $\{r\}$ is decidable : given a real, one must verify that it codes an ordinal, that this ordinal is larger than δ and then by the coding techniques, check whether our real really is the least code in $L_{\alpha+1}$ for some α and whether α really is the least ordinal above δ such that α is countable in $L_{\alpha+1}$. ∎

We now have a way of obtaining randomness through the use of infinite time machines. How can we get stronger notions of randomness, while not excluding reals that are actually "random"?[8]

One way is by fixing some strong requirements for our randomness notion; so strong that there is no such notion. We then loosen up the conditions by requiring that "the randomness notion satisfies the requirements" is merely relatively consistent. This is the Durand *et al.* approach. We can actually try to go on like this and obtain stronger and stronger notions. We will go through this method in the following section and apply it to other randomness notions than the Durand *et al.* one. The main problem of this way of obtaining stronger randomness is that it has to be based on another randomness notion and this, of course, doesn't help in obtaining a non-algorithmical and *independence*-based randomness notion.

Another way is by having more powerful infinite time machines. We must however make sure that the gain in power is really a gain in excluding non-random reals. This will be the second part of the following section.

4. Unprovability-based randomness

4.1. Durand et al.

Durand *et al.* in [Durand et al., 2001] were looking for a non-algorithmically-based randomness definition. They proposed the following definition.

Definition 3 *Let x be an infinite binary sequence.*

The sequence x is Solovay random *over L if it avoids any null G_δ (countable intersections of open sets) set with a code*[9] *in L. We note ρ_L the predicate for this randomness, and $R_L = \{x \in \{0,1\}^\omega \mid \rho_L(x)\}$.*

The sequence x is arithmetically random *if it avoids any null arithmetically coded G_δ set. We have also the similar notations ρ_A and R_A.*

The sequence x is consistently random *if $x \in R_A$ and if R_L is of full measure, then $x \in R_L$. We have also the similar notations ρ_C and R_C.*

A randomness predicate ρ is said to be consistent *if it verifies the following conditions :*

(1) *ZFC proves that $\{x \in \{0,1\}^\omega \mid \rho(x)\}$ is a full set;*

(2) *$\forall\Psi(x)$, if ZFC proves that $\{x \in \{0,1\}^\omega \mid \Psi(x)\}$ is null, then ZFC does not prove that there is an $x \in \{0,1\}^\omega$ satisfying $\rho(x) \wedge \Psi(x)$;*

(3) *ZFC proves that $\forall x \in \{0,1\}^\omega$, if $\rho(x)$, then x is Martin-Löf random.*

Theorem 4 ([Durand et al., 2001]) *In the Solovay model*[10]*, R_L is a full G_δ set and ρ_L verifies* **(2)**[11].

Corollary 5 *The randomness ρ_C is a consistent randomness.*

PROOF. The randomness ρ_C satisfies obviously **(1)** and **(3)** because of Theorem 4 and of the definition of arithmetical randomness.

In the Solovay model, R_L is of full measure, so $\rho_C(x) \leftrightarrow \rho_L(x)$. Hence ρ_C satisfies $\mathbf{(2)}_{\mathrm{plain}}$. ■

This study prompts a way of obtaining always finer randomness notions.

Take a randomness predicate ρ and the corresponding set of random sequences R. Set some requirements (predicates $\{P_1, P_2, \ldots, P_k\}_{k\geq 2}$) for the quality of randomness desired. To be able to operate our method, R has to verify the requirements *only*[12] in a certain model. Fix an $l \leq k$, the new randomness predicate ρ' (R') is the *consistent realisation* of ρ on top of some randomness notion ρ_{base}, noted ${\rho_{\mathrm{base}}}^\rho$ and it is defined by :

$$\rho'(x) \text{ if and only if } x \in R_{\mathrm{base}} \text{ and if } P_l(R), \text{ then } x \in R.$$

This is what Durand *et al.* did. We can go even further[13] by then defining the following randomness notion ρ^+, noted $\rho_{\text{base}_+^\rho}$:

$$\rho^+(x) \text{ if and only if } x \in R_{\text{base}} \text{ and if } \text{cons}(P_l(R)), \text{ then } x \in R'.$$

And we can continue like that and obtain[14] ρ^{++} ($\rho_{\text{base}_{++}^\rho}$), ρ^{+3} ($\rho_{\text{base}_{+3}^\rho}$), ... The drawback is that we don't know much of the obtained gain in strongness for our randomness notion. We will complement this idea in the second part of this section.

The Durand *et al.* randomness is based on unprovability on top of common randomness notion. Do we really still need this arithmetical/computational basis for randomness? We aim at answering this question in section 3. But first, we look for still stronger randomness notions with a precise measure of the gain.

4.2. More unprovability-based randomness

We saw that $\mathfrak{c}$-machines have very peculiar properties and we will base some new randomness notions on them. To reach those notions, we need to introduce some set-theoretical material.

An uncountable cardinal number κ is *inaccessible* if it is regular and a strong limit cardinal. An immediate consequence of inaccessibility is that V_κ, the collection of all sets of rank less than κ, is a model of ZFC; another immediate consequence is that $\kappa = \aleph_\kappa$ is a fixed point of the aleph sequence. By Gödel's second incompleteness theorem, it follows that the existence of inaccessible cardinals is unprovable in ZFC. In fact, a slightly more involved argument shows that the relative consistency of inaccessible cardinals is unprovable. Thus the existence of inaccessibles is to ZFC as the existence of an infinite set is to Peano arithmetic. For that reason, large cardinal axioms are sometimes referred to as *strong axioms of infinity.*

Modern set theory recognizes a substantial number of large cardinal axioms. Interestingly enough, these axioms form a linearly ordered scale, on which the relation of a stronger axiom to the weaker theories is just as described above in the case of inaccessible cardinals and ZFC. This scale of large cardinals serves as a measure of consistency strength of various set theoretic assumptions.

One of those large cardinals is a *weakly compact* cardinal and we introduce them now. Every weakly compact is not only inaccessible, but on the scale of large cardinals is also above Mahlo cardinals. Among large cardinals stronger than weakly compact cardinals and even Woodin and measurable cardinals, *supercompact* cardinals are most prominent. Above supercompact and huge cardinals, the scale approaches its end with the existence of a non-trivial elementary embedding $j : V_\lambda \to V_\lambda$, as by a theorem of Kunen, $j : V \to V$ is inconsistent. For more on large cardinals and set theory at large, see [Jech, 1978], [Kanamori, 1994] and [Kanamori and Magidor, 1978].

Definition 4 *Let κ be a regular uncountable cardinal. We call a set $C \subseteq \kappa$* closed unbounded *in κ if*

1 for every sequence $\alpha_0 < \alpha_1 < \cdots < \alpha_\xi < \cdots$ ($\xi < \gamma$) of elements of C, of length $\gamma < \kappa$, we have $\lim_{\xi \to \gamma} \alpha_\xi \in C$ (closed);

2 for every $\alpha < \kappa$, there is $\beta > \alpha$ such that $\beta \in C$ (unbounded).

We say that $S \subseteq \kappa$ is stationary *in κ if $S \cap C \neq \emptyset$ for every closed unbounded subset C of κ.*

A cardinal κ is weakly compact *if it is uncountable and satisfies the partition property $\kappa \to (\kappa)^2$.*

Jensen [Jensen, 1972] proved the following :

Theorem 6 *Assuming $V = L$, a regular cardinal κ is weakly compact if and only if for every stationary $A \subseteq \kappa$, such that every $\alpha \in A$ is of cofinality ω, $\{\alpha \mid cf(\alpha) > \omega, \alpha < \sup A$, and $A \cap \alpha$ is a stationary subset of $\alpha\} \neq \emptyset$.*

Baumgartner [Baumgartner, 1976] studied the question for $\kappa = \aleph_2$ and obtained a relative consistency result with ZFC+"$\exists$ weakly compact cardinal". Magidor in [Magidor, 1982] obtained an equiconsistency result that we use to prove the following theorem.

Theorem 7 *There is a $\mathfrak{c}_2$-machine $\mathfrak{M}$ such that the ouput real $\mathfrak{r}$ of $\mathfrak{M}$ (on a blank input) is such that "$\mathfrak{r} \neq 0$" is equiconsistent with the existence of a weakly compact cardinal.*

PROOF. From the study in [Gurevich et al., 1983], we can easily construct a $\mathfrak{c}_2$-automaton such that the language recognized by this automaton is nonempty if and only if $\{\alpha < \omega_2 \mid \mathrm{cf}(\alpha) = \omega_1$ and $\alpha \cap X$ is stationary in $\alpha\}$ is nonempty for every $X \subseteq \{\alpha < \omega_2 \mid \mathrm{cf}(\alpha) = \omega_0\}$. We can code this language (or a countable part of it) in a real $\mathfrak{r}$ such that $\mathfrak{r} \neq 0$ if and only if the language is not empty. Using Baumgartner's and Jensen's results (Theorem 6), it is clear that "$\mathfrak{r} \neq 0$" is independent of ZFC.

Using Magidor's result in [Magidor, 1982], in the same manner, we construct a $\mathfrak{c}_2$-automaton such that "$\mathfrak{r} \neq 0$" is equiconsistent with the existence of a weakly compact cardinal. ■

The first part of Theorem 3 can be extended[15] to general $\mathfrak{c}$-machines to give finer randomness definitions.

REMARK. We can also get the other half of Theorem 3 by using core model theory but we won't enter into such troubled waters. It is important to notice that this second part of the theorem tells us that somehow there will always be some reals non writable by such machines that will not be really random (because they are singleton recognizable) and that we always need to seek a stronger randomness. This, of course, prompts the importance of the randomness notions of the last section. □

Randomness notion 1 *The sequence $\mathfrak{x} \in \{0,1\}^\omega$ is $\mathfrak{c}_n$-random if it is not writable by a $\mathfrak{c}_n$-machine. We use the notation $\rho_{\mathfrak{c}_n}$ for this randomness' predicate.*

Theorem 7 implies :

Corollary 8 *$\mathfrak{c}_3$-randomness is strictly stronger than $\mathfrak{c}_2$-randomness.*

PROOF. $\mathfrak{c}_3$ can decide and thus write $\mathfrak{r}$ from Theorem 7. The nice thing is that the gain in randomness is quantified by a "$\exists$ weakly-compact cardinal". ■

Jech and Shelah [Jech and Shelah, 1990], using supercompact cardinals, generalized Magidor's result to $\aleph_n$ and that enables us to prove the following.

Theorem 9 *For any $n \in \mathbb{N}$, there is a $\mathfrak{c}_n$-machine $\mathfrak{M}$ such that the ouput real $\mathfrak{r}$ of $\mathfrak{M}$ (on a blank input) is such that "$\mathfrak{r} \neq 0$" is implied by the existence of n supercompact cardinals.*

PROOF. As in Theorem 7, using the generalization of Magidor's result by Jech and Shelah, we can construct for any $n \in \mathbb{N}$ a suitable $\mathfrak{c}_n$-automaton and obtain the required $\mathfrak{r}$. ■

Theorem 10 *The hierarchy of randomness given by the hierarchy of $\mathfrak{c}$-machines is at least as strict as some part of the large cardinal hierarchy.*

PROOF. By Theorem 9, this hierarchy of randomness is as strict as : "$\exists\, n+1$ supercompact cardinals" is stronger consistency-wise than "$\exists\, n$ supercompact cardinals". ■

We can consider taking those mysterious reals $\mathfrak{r}$ (of Theorems 7 and 9) as oracles for our $\mathfrak{c}$-machines. We are not sure if there is a gain in randomness doing this.

But one can still do as in the first part of this section and define for any $m \in \mathbb{N}$:

Randomness notion 2 *The randomness $\rho_A{}^{\rho_{\mathfrak{c}_n}}_{+m}$ using as the P_l requirement: "$\exists\, n$ supercompact cardinals",*

or perhaps more interestingly, $\rho_{\mathfrak{c}_n}{}^{\rho_A}_{+m}$ using as the P_l requirement: "$\nexists\, n$ supercompact cardinals".

The advantage of the latter notion is that we are guaranteed, with the randomness base ρ_A, not to put aside any real that should be considered as *random*.

REMARK. Note that $\rho_A{}^{\rho_{\mathfrak{c}_{n+1}}}$ is a stronger notion than $\rho_A{}^{\rho_{\mathfrak{c}_n}}_{+}$. □

5. Unknowability-based randomness

The previous randomness notions still lack the *unknowability* (using *independence* from ZFC) that we are looking for.

We propose a hierarchy of randomness definitions based on the results of the previous section using the large cardinal *empirical* hierarchy.

Randomness notion 3 *A real $\mathfrak{x} \in \{0,1\}^\omega$ is a large cardinal random real if there is a $\mathfrak{c}$-machine $\mathfrak{M}$ with metamathematical[16] ouput $\mathfrak{x}$ such that in ZFC, "the ouput real of $\mathfrak{M}$ is non zero" is equiconsistent with the existence of a large cardinal.*

It is clearly quite, and perhaps too, restrictive but at least the notion is really of the "unknowable" nature and is not based on algorithmic notions. It has also the advantage of relying on the only well-understood notion of "objects beyond ZFC", *i.e.*, large cardinals.

Using the "unknowable" method, the most natural definition seems to be

Randomness notion 4 *Each bit is* unknowable *from the previous ones:* $\mathfrak{x} \in \{0,1\}^\omega$ *is* increasingly unknowably random *if there is a countable family of propositions* $\{Q_i\}_{i\in\mathbb{N}}$ *such that*

$$\forall n \in \mathbb{N},\quad Q_n \textit{ is independent of ZFC} + \bigwedge_{i<n} Q_i$$

and

$$\mathfrak{x}_i = \begin{cases} 1 & \textit{if } Q_i \textit{ is true,} \\ 0 & \textit{otherwise.} \end{cases}$$

It is not an effective definition but it can be in part realized using $\mathfrak{c}$-machines : let Q_i be "$\mathfrak{r}_i$ is null", where $\mathfrak{r}_i$ is the problematic real for $\mathfrak{c}_i$-machines in Theorem 9.

We can propose a variant of this definition by requiring that

$$\forall n \in \mathbb{N},\quad Q_n \text{ is independent of ZFC} + \bigwedge_{i\neq n} Q_i$$

but we don't know of any realization of such a strong definition.

From our different hierarchies of randomness of the previous section, we can define an *unknowable* randomness by using *generic sets.*

Definition 5 *Let* $\langle \mathbb{P}, \leq, 1\rangle$ *be a partial order.*

1 $D \subset \mathbb{P}$ *is* dense *in* $\mathbb{P}$ *iff* $\forall p \in \mathbb{P}\, \exists q \leq p\;\; q \in D$.

2 $G \subset \mathbb{P}$ *is a* filter *in* $\mathbb{P}$ *iff*

(a) $\forall p, q \in G\, \exists r \in G\;\; r \leq p \wedge r \leq q,$

(b) $\forall p \in G\, \forall q \in \mathbb{P}\;\; q \leq p \rightarrow q \in G.$

3 G *is* $\mathbb{P}$-generic *on* $\mathbb{D}$ *iff* G *is a filter on* $\mathbb{P}$ *and for all* $\mathbb{P}$*-dense* $D \in \mathbb{D}$, $D \cap G \neq \emptyset$.

The existence of a generic set for a particular partial order $\mathbb{P}$ is not necessarily trivial. Nevertheless there is a much studied proposition that guarantees us (if, of course, we suppose it to hold) that most of the generic sets that we consider exists. It is called Martin's Axiom and it is independent of ZFC. See [Jech, 1978] for more on this.

Fix a strict randomness hierarchy $\{\rho_i^H\}_{i\in\mathbb{N}}$ ($\{R_i^H\}_{i\in\mathbb{N}}$) from the previous section. Take for partial order $\mathbb{P}$ the set $\subset \{0,1\}^\omega$ of infinite binary random sequences (somewhere in our fixed randomness hierarchy) with the order $\prec$:

$$p \prec q \text{ iff } \hat{p} > \hat{q} \text{ and there is } i > 0 \text{ such that } p\restriction_{\mathbb{N}\setminus\{0,\ldots,i-1\}} = q$$

where $\hat{p}$ is the smallest n such that $p \in R_n^H$.

Randomness notion 5 *Fix a hierarchy of randomness notions whose strictness is according to the hierarchy of some large cardinals. The sequence* $\mathfrak{x} \in \{0,1\}^\omega$ *is* unknowably random along *this hierarchy if* $\mathfrak{x}$ *is the union*[17] *of a* $\langle \mathbb{P}, \prec \rangle_{\{\rho_i^H\}_{i \in \mathbb{N}}}$*-generic set.*

If we take for example the Randomness hierarchy definition 1, by the previous randomness definition, we obtain random sequences where each bit is *unknowable* from the other ones in the strong sense that the hierarchy is strict because of the supposed strictness of the hierarchy of the large cardinals used. It thus has the advantage of being in compliance with classical notions of randomness while making sure it verifies our "unknowability" requirement.

Acknowledgments

We wish to express our gratitude to Menachem Magidor for all his thoughtful ideas about infinite time machines and for pointing out the article [Gurevich et al., 1983] of Yuri Gurevich, Saharon Shelah and himself. We are also greatly indebted to Jacques Mazoyer for his advice and his never-failing enthusiasm.

Notes

1. For any $s \in \{0,1\}^{<\omega}$, $\mu(s) = \mu(s \frown 0) + \mu(s \frown 1)$ and $\mu(\epsilon) = 1$ where ϵ is the empty sequence.

2. Without loss of generality, we can assume $C = 1$.

3. A sequence $x \in \{0,1\}^\omega$ is *Martin-Löf random* if it avoids all effectively null sets. It is one of the classical definitions of randomness. Algorithms appear in the word 'effectively'. For more on Martin-Löf randomness, see [Martin-Löf, 1966].

4. See Definition 4.

5. For any ordinal α : $\aleph_\alpha$ denotes the α-th cardinal and ω_α is the smallest ordinal of cardinality $\aleph_\alpha$. Following von Neumann, we identify an ordinal β with the set of ordinals $\alpha < \beta$.

6. We use the notation $\mathfrak{c}_n$-machine to indicate that we work on $\aleph_n$.

7. An *infinite time Turing machine* is a $\mathfrak{c}$-machine (with $k = 3$) working with countable tapes and (of course) countable time. The difference with $\mathfrak{c}$-machines is in the limit stages' behaviour : It is placed in a special unique *limit state*. For a given cell of the tape, at a limit stage it takes the value of the lim sup of the cell values before the limit.

8. Nevertheless, stronger notions of randomness still are better notions if such strong random reals exist.

9. $C \subseteq \omega \times 2^{<\omega}$ is a code for a G_δ set $U \subseteq \{0,1\}^\omega$ if $U = \bigcap_n \bigcup_{\langle n,u \rangle \in C} \{0,1\}^\omega_u$.

10. Let $\mathcal{M}$ be a transitive model of ZFC and let κ be an inaccessible cardinal in $\mathcal{M}$. The Solovay model is $\mathcal{M}[G]$ where G is an $\mathcal{M}$-generic ultrafilter on P, the notion of forcing that collapses each $\lambda < \kappa$ onto $\aleph_0$.

11. Actually, it verifies $\mathbf{(2)}_{\text{plain}}$: $\forall \Psi(x)$, if $\{x \in \{0,1\}^\omega \mid \Psi(x)\}$ is null, then there is no $x \in \{0,1\}^\omega$ satisfying $\rho(x) \wedge \Psi(x)$.

12. As much as we know or can prove.

13. If $P_l(R)$ is relatively consistent (cons($P_l(R)$)), there is a model in which $P_l(R)$ is true. Living in this model, we can now consider taking R' instead of R in our "then $x \in R$". And so on ...

14. $\rho^{++}(x)$ if and only if $x \in R_{\text{base}}$ and if cons(cons($P_l(R)$)), then $x \in R^+$.

15. by replacing in the proof each occurrence of the L_α's by V_α's.

16. the *truth* about the ouput of $\mathfrak{M}$.

17. By definition of $\prec$, all the infinite sequences are mutually compatible. There is a sequence that *contains* all of them, called the union.

References

Baumgartner, J. E. (1976). A new class of order types. *Annals of Mathematical Logic*, 9:187–222.

Büchi, J. R. (1962). On a decision method in the restricted second-order arithmetic. In *Logic, Methodology, and Philosophy of Science: Proc.* **1960** *Intern. Congr.*, pages 1–11. Stanford University Press.

Büchi, J. R. (1965). Decision methods in the theory of ordinals. *Bulletin of the American Mathematical Society*, 71:767–770.

Büchi, J. R. (1973). The monadic second-order theory of ω_1. In Büchi, S., editor, *Decidable theories. II*, volume 328 of *Lecture Notes in Mathematics*, pages 1–127. Springer-Verlag, Berlin and New York.

Durand, B., Kanovei, V., Uspensky, V. A., and Vereshagin, N. (2001). Do stronger definitions of randomness exist? *Theoretical Computer Science*. to appear.

Gurevich, Y., Magidor, M., and Shelah, S. (1983). The monadic theory of ω_2. *Journal of Symbolic Logic*, 48(2):387–398.

Hamkins, J. D. and Lewis, A. (2000). Infinite time Turing machines. *Journal of Symbolic Logic*, 65(2):567–604.

Jech, T. (1978). *Set Theory*. Academic Press, New York.

Jech, T. and Shelah, S. (1990). Full reflection of stationary sets below $\aleph_\omega$. *Journal of Symbolic Logic*, 55:822–830.

Jensen, R. B. (1972). The fine structure of the constructible hierarchy. *Annals of Mathematical Logic*, 4:229–308.

Kanamori, A. (1994). *The Higher Infinite*. Springer Verlag.

Kanamori, A. and Magidor, M. (1978). The evolution of large cardinal axioms in set theory. In Muller, G. H. and Scott, D. S., editors, *Higher Set Theory*, volume 669 of *Lecture Notes in Mathematics*, pages 99–275. Springer Verlag, Berlin.

Kruse, A. H. (1967). Some notions of random sequence and their set-theoretic foundations. *Zeitschift mathematische Logik und Grundlagen der Mathematik*, 13:299–322.

Lafitte, G. (2001). How powerful are infinite time machines? In Freivalds, R., editor, *Thirteenth International Symposium on Fundamentals of Computation Theory*, volume 2138 of *Lecture Notes in Computer Science*, pages 252–263. Springer-Verlag.

Magidor, M. (1982). Reflecting stationary sets. *Journal of Symbolic Logic*, 47(4):755–771.

Martin-Löf, P. (1966). The definition of random sequences. *Information and Control*, 9:602–619.

Muchnik, A. A., Semenov, A. L., and Uspensky, V. A. (1998). Mathematical metaphysics of randomness. *Theoretical Computer Science*, 207:263–317.

van Lambalgen, M. (1992). Independence, randomness, and the axiom of choice. *Journal of Symbolic Logic*, 57:1274–1304.

SERVER PLACEMENTS, ROMAN DOMINATION AND OTHER DOMINATING SET VARIANTS*

Aris Pagourtzis[§¶] Paolo Penna[§] Konrad Schlude[§]
Kathleen Steinhöfel[||] David Scot Taylor[§] Peter Widmayer[§]

Abstract Dominating sets in their many variations model a wealth of optimization problems like facility location or distributed file sharing. For instance, when a request can occur at any node in a graph and requires a server at that node, a minimum dominating set represents a minimum set of servers that serve an arbitrary single request by moving a server along at most one edge. This paper studies domination problems for two requests. For the problem of placing a minimum number of servers such that two requests at different nodes can be served with two different servers (called win-win), we present a logarithmic approximation, and we prove that nothing better is possible. We show that the same is true for Roman domination, the well studied problem variant that asks for each vertex to either possess its own server or to have a neighbor with two servers. Still the same is true if each idle server can move along one edge while the first of both requests is being served. For planar graphs, we propose a PTAS for Roman domination (and show that nothing better exists), and we get a constant approximation for win-win.

1. Introduction

In this paper, we study a generalization of the *dominating set problem* [GJ79]. We are given a graph, and at every node of this graph a request can appear. We want to service such requests. To do so, we place servers at nodes. The request at a node v is serviced, if there is a server on v, or if a server in its neighborhood is moved to v. Clearly, if we want to be able to service one

*This work was partially supported by the Swiss Federal Office for Education and Science under the Human Potential Programme of the European Union under contract no. HPRN-CT-1999-00104 (AMORE).

[§]Institute of Theoretical Computer Science, ETH Zentrum, 8092 Zürich, Switzerland. {lastname}@inf.ethz.ch.

[¶]Part of this work was done while this author was with the University of Liverpool, supported by EPSRC grant GR/N09855, and also with the National Technical University of Athens.

[||]GMD - FIRST, Berlin, Germany. Kathleen.Steinhoefel@gmd.de.

request, then the multiset of server locations must contain a dominating set of nodes.

However, there are applications in which we want to ensure that more than one request can be serviced. In this paper, we study the case of two requests. Imagine, e.g., that two requests occur simultaneously and a server can satisfy only one at a time. We view our problem as a member of the large family of dominating set problems, of which [HHS98] already cite more than 75 different variants. These may depend on conditions on the dominating set DS (e.g. connectivity) or on the other nodes (e.g. a node is dominated if there is a node in DS at distance at most k, or each vertex is dominated at least k times, etc.). The study of such dominating set problems is motivated by their applications to facility location (minimizing the number of facilities, subject to every demand being close enough to some facility), file sharing in distributed systems [NR95], game theory [dJ62], etc. Interestingly, some very old questions have also triggered new research on the topic [AF95, RR00, Ste99]:

> ROMAN DOMINATION : Where should the armies of the Roman Empire be placed so that a smallest number of armies can protect the whole Empire (see Figure 1)?

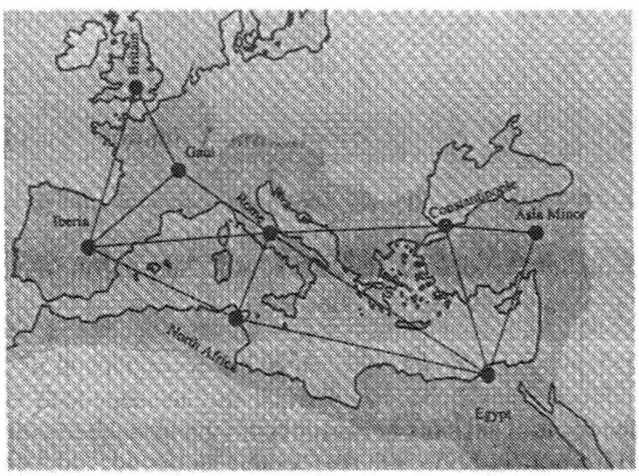

Figure 1. The Roman Empire around 300 A.C.

The assumption is that an area can be protected either by one army located inside the area, or by an army in a neighbor area that comes over for the defense. In the latter case it is required that a second army remains in the neighbor area, so that it can quickly confront a second attack. A reason for the historical 1-2 requirement (one army here or two at a neighbor) is that we want to be able to service two requests in one time unit (provided that no two requests can come from the same point at the same time).

In this paper we deal with variants of the ROMAN DOMINATING SET [Dre00, Ste99]. In particular, we consider the case in which there are two requests we want to service and no two requests appear at the same node. Moreover, a server that is used to service the first request cannot be used to service another request. A solution to our problem for a given graph is a set of servers at nodes;

since all servers are identical, a multiset of nodes (where the multiplicity of a node is the number of servers at that node) represents a server placement.

Two factors we will consider are: (i) whether the two requests are known before the first one must be serviced (OFFLINE), or the first one must be serviced before the second one is known (ONLINE), and (ii) whether servers must stay in place unless they service a request (STATIC), or we allow for a rearrangement (DYNAMIC): as one server services the first request, all other servers are allowed to move to a neighbor node. The goal of the move is to guarantee that any second request can be handled, too, in the ONLINE case (that is, the resulting server placement is a dominating set if we ignore the first requesting node and its server). The ONLINE STATIC WIN-WIN version has been discussed earlier [Och96] and called *Win-Win* there. (Unlike in Roman Domination, in this case we only require to be able to win against any two consecutive attacks.) Since our problems also deal with two consecutive requests, we adopt the name terminology and we denote the four problem variants as ONLINE STATIC WIN-WIN, ONLINE DYNAMIC WIN-WIN, OFFLINE STATIC WIN-WIN, and OFFLINE DYNAMIC WIN-WIN.

1.1. Our (and Previous) Results

In this paper we investigate the relationships between the above problems (including ROMAN DOMINATION), as well as the complexity of computing exact and approximate solutions. In particular, we consider the following questions:

1 Given a multiset S, is S a feasible solution to (one of) the above problem variants? Is there a combinatorial characterization for those S?

2 Let VARA WIN-WIN and VARB WIN-WIN denote any two problem variants. If S is a solution for VARB WIN-WIN, does this imply that S is also a solution to VARA WIN-WIN?

3 A positive answer to the above question implies that $\mathsf{opt}_\mathrm{A}(G) \leq \mathsf{opt}_\mathrm{B}(G)$, where opt_A and opt_B denote the minimum size multiset solving the two variants, resp. Is there a graph for which the inequality is strict?

Let VARA WIN-WIN$\preceq$ VARB WIN-WIN denote the fact that Question 2 has a positive answer, and let VARA WIN-WIN$\prec$ VARB WIN-WIN denote the fact that Question 3 does too. It turns out that the problems we look at form the partial order in Figure 2 (Figure 2 contains a new problem (DOMINATING 2-SET) which we introduce to prove some of our results). Noticeably, this relationship also holds when we restrict ourselves to planar graphs.

As for Question 1, for two out of the four win-win problems we provide a characterization of those multisets corresponding to each problem. For the DYNAMIC WIN-WIN, we prove the NP-hardness of the rearrangement step after the first request. This result seems to denote that such a characterization for this problem version does not exist, or at least is different from those given for the other two problems (those can be checked in polynomial time).

This leads us to complexity and (non-) approximability issues. Intuitively, the $\preceq$ relationship may have some consequences on the (non-) approximability

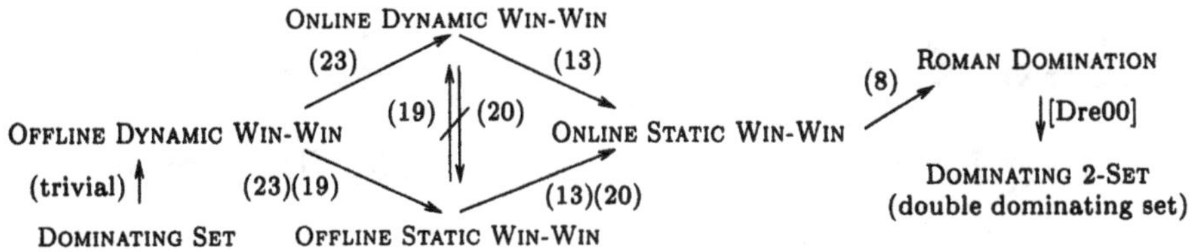

Figure 2. Relationships between the problems: arrows represent '$\prec$' and they are numbered according to the corresponding theorem.

Problem Version	General Graphs	Planar Graphs
ROMAN DOMINATION	$(2+2\ln n)$-APX, not $c \log n$-APX (NP-hard [Dre00])	PTAS, in P for r-outerplanar (NP-hard [Hed00], in P for trees & $(r \times n)$-grids [Dre00])
ONLINE STAT. WIN-WIN, ONLINE DYN. WIN-WIN, OFFLINE STAT. WIN-WIN	$(2+2\ln n)$-APX, not $c \log n$-APX	$(2+\epsilon)$-APX, for any $\epsilon > 0$

Table 1. Hardness and approximability: Our and previous results. (All NP-hardness results are in strong sense, thus implying the non-existence of a FPTAS. Previous results are displayed inside parentheses.)

of those problems. Indeed, the order in Figure 2, combined with the fact that "doubling" a dominating set (the DOMINATING 2-SET problem in Figure 2) yields a feasible solution for *all* of the problems, implies an approximation preserving reduction ($\leq_{\mathsf{AP}}$, see [ACG+99]) between all these problems. Let $f(n)$-APX denote the class of problems that admit a polynomial-time $f(n)$-approximation algorithm [ACG+99]. In Table 1 we summarize the complexity and (non-) approximability results of this work.

As for the results on planar graphs, our technical contribution is a Polynomial-Time Approximation Scheme (PTAS) for ROMAN DOMINATION. This result is based on an exact polynomial-time algorithm for r-outerplanar graphs. The latter improves over the previous results in [Dre00]: in this work only trees and $r \times n$-grids (for any *fixed* r) are shown to be exactly solvable. Our result subsumes both of them (an $r \times n$-grid is clearly an r-outerplanar graph).

2. Online Static Win–Win

In the sequel, given a multiset S, uniq(S) denotes the set resulting by removing multiplicities.

Definition 1 (online static) *Given a graph $G = (V, E)$, a* server placement *for G is a multiset S of nodes. A server placement S is a* win–win *for G, if for all $v \in V$ there is an $u_v \in S$ with the properties:*

1 $v = u_v$ or $(u_v, v) \in E$,

2 for all $v' \in V \setminus \{v\}$ there is an $u_{v'} \in S \setminus \{u_v\}$ with

$$v' = u_{v'} \text{ or } (u_{v'}, v') \in E.$$

Figure 3. A win–win.

Figure 4. Not a win–win.

Lemma 2 (sandwich) *Any graph G has the following properties:*

1 *For every dominating set DS, the server placement $SP := DS \uplus DS$ is a* win–win *for G, where $\uplus$ denotes the multi-union.*

2 *For every* win–win *WW, the set* uniq(*WW*) *is a dominating set of G.*

3 *For every minimum dominating set MDS and for every minimum* win–win *MWW, $|MDS| \le |MWW| \le 2|MDS|$ hold.*

Proof. For Property 1, let $v_1, v_2 \in V$ be a pair of nodes with $v_1 \ne v_2$. Since DS is a dominating set, there are $u_{v_1}, u_{v_2} \in DS$, such that

$$\begin{aligned} v_1 &= u_{v_1} \text{ or } (v_1, u_{v_1}) \in E, \text{ and} \\ v_2 &= u_{v_2} \text{ or } (v_2, u_{v_2}) \in E \end{aligned}$$

hold. Due to the definition of SP, $\{u_{v_1}, u_{v_2}\} \subset SP$ holds. This implies that requests at v_1, v_2 can be serviced.

For Property 2, let $v \in V$ be a node. There is a $u_v \in WW$ with $v = u_v$ or $(v, u_v) \in E$. Since $u_v \in \text{uniq}(WW)$, uniq(WW) is a dominating set.

For Property 3, it suffices to consider the win–win $WW := MDS \uplus MDS$ and the dominating set $DS := \text{uniq}(MWW)$. Clearly, $|MDS| \le |DS| \le |MWW| \le |WW| = 2|MDS|$. □

2.1. Characterization of win–win Multisets

The property of being a win–win does not depend only on a node and its neighbors. Furthermore, it is not enough that for every pair of nodes there are two different adjacent servers. This is illustrated by the example in Figure 4. The server placement $S = \{v_2, v_3\}$ is not a win–win. If the first request is at v_2, then there are two cases. Case 1, the request is serviced by v_2, then a second request at v_1 cannot be serviced. Case 2, the request is serviced by v_3, then a second request at v_4 cannot be serviced.

This observation lead us to the following characterization of the server placements that are win–win.

Definition 3 *Given a graph $G(V,E)$ and a multiset D for it, a vertex $v \in V$ is* weak *if D dominates v only once. A vertex $u \in D$ is* safe *if every $v \in N(u)^+$ is not weak, where $N(u)^+ = N(u) \cup \{u\}$.*

Lemma 4 *A multiset D for $G(V,E)$ is a* win–win *if and only if the following two properties hold:*

at-most-1-weak *Every $u \in D$ does not dominate more than one weak node;*

at-least-1-safe *Every non weak node $v \in V$ is dominated by at least one safe node $u \in D$.*

Proof. (⇒) By contradiction, assume that some $u \in D$ does not satisfy Property at-most-1-weak. Then, there exist two weak nodes w_1 and w_2 dominated *only* by u. After a first request at w_1, w_2 is no longer dominated (we must have used u for the first request). This contradicts the hypothesis that D is a win–win. Now suppose (again by contradiction) that a non weak node v is not adjacent to any safe node (thus contradicting Property at-least-1-safe). Let $u_1, \ldots, u_k$ be the nodes of D adjacent to v, for some $k \geq 2$ (this follows from the fact that v is not weak). By hypothesis, none of $u_1, \ldots, u_k$ is safe. So, there exist $w_1, \ldots, w_k$ distinct weak nodes, with w_i adjacent to u_i, for $1 \leq i \leq k$. Now consider a first request at node v. For this request we must use one among $u_1, \ldots, u_k$, let us say u_j. Then, if the second request is at the weak node w_j we do not have any server to react. Again, this contradicts the hypothesis.
(⇐) Let v_1 be the position of the first request. We have two cases: v_1 is weak, or v_1 is not weak. In the first case, we must use the only node $u \in D$ that is adjacent to v_1; Property at-most-1-weak guarantees that every node in $N(u)^+ \setminus \{v_1\}$ will still be dominated. So, any second request can be handled. Otherwise, that is, v_1 is not weak, Property at-least-1-safe implies that there exists a $u \in D$ which is safe; we use such a u for this request. At this point all the nodes in $N(u)^+ \setminus \{v_1\}$ will still be dominated by some $u' \in D$. Also in this case any second request can be handled. □

2.2. Complexity

We are interested in the complexity of the ONLINE STATIC WIN-WIN problem. We discuss hardness and approximation of this problem. Both NP-hardness and approximation hardness can be proved using the following lemma.

Lemma 5 *Any $f(n)$-approximation algorithm A for* MIN DOMINATING SET *implies a $2f(n)$-approximation algorithm for* MIN ONLINE STATIC WIN-WIN. *Conversely, any $g(n)$-approximation algorithm B for* MIN ONLINE STATIC WIN-WIN *implies a $2g(n)$-approximation algorithm for* MIN DOMINATING SET.

Proof. Applying A to any graph G we can find a dominating set DS of size $|DS| \leq f(n)|MDS_G|$. By Lemma 2 the server placement $SP = DS \uplus DS$ is a win–win for G of size $|SP| = 2|DS| \leq 2f(n)|MDS_G| \leq 2f(n)|MWW_G|$.

Conversely, applying B to any graph G we obtain a win–win SP of size $|SP| \leq g(n)|MWW_G|$. Then, according to Lemma 2 the set $DS = \text{uniq}(SP)$ is a dominating set of size $|DS| \leq |SP| \leq g(n)|MWW_G| \leq 2g(n)|MDS_G|$. □

We know that MIN DOMINATING SET is not approximable within $c \log n$ for some $c > 0$ [RS97] (unless P=NP) and that it is approximable within $1 + \ln n$ [Joh74]. From these facts and the above lemma one can easily prove the following.

Theorem 6 *The* MIN ONLINE STATIC WIN-WIN *problem in general graphs can be approximated within $2 + 2\ln n$, but (unless* P=NP*) cannot be approximated within $c \log n$ for some $c > 0$.*

For MIN DOMINATING SET in planar graphs a Polynomial Time Approximation Scheme (PTAS) is known [Bak94]. Therefore, Lemma 5 implies an approximation algorithm for MIN ONLINE STATIC WIN-WIN in planar graphs, called MIN PLANAR ONLINE STATIC WIN-WIN, with ratio $2+\epsilon$ for every $\epsilon > 0$.

Moreover, this approximation ratio is tight for the approach of "doubling" a dominating set to construct the solution. We illustrate this by the example in Figure 5. For this graph, the set $M := \{v_1, \ldots, v_8\}$ is a minimum dominating set. Doubling it gives a solution WW with $|WW| = 16$. On the other hand, the server placement $MWW = \{w, v_1, v_2, \ldots, v_8\}$ is a minimum win–win with $|MWW| = 9$. In this case, the approximation ratio is 16/9. If we increase the number of rays from 8 to k, then we get $|WW|/|MSP| = 2k/(k+1)$. This shows that there exist graphs for which the simple doubling algorithm has approximation ratio greater than $2-\epsilon$, for any $\epsilon > 0$.

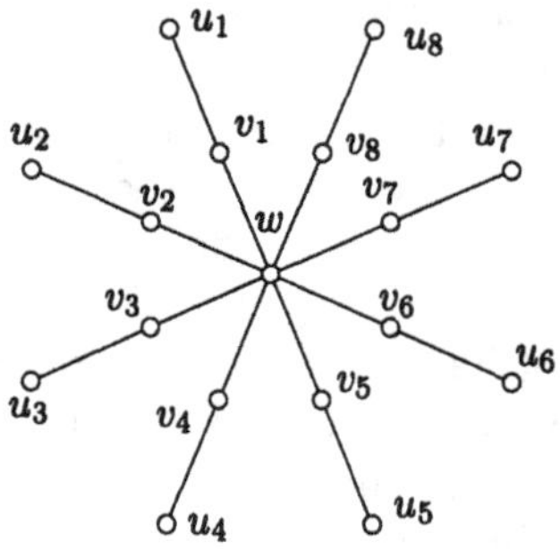

Figure 5. Doubling a dominating set gives a win–win of cost roughly twice the optimum.

3. Roman Domination

We come back to the original problem of the so called ROMAN DOMINATION. On every node, we can place none, one, or two servers.

Definition 7 (roman domination) *Given a graph $G = (V, E)$, a* roman *for G is a server placement S such that every node v in V either belongs to S or has a neighbor u in S whose multiplicity in S is at least 2. Formally, $\forall v \in V, v \notin S \rightarrow \exists u : (v, u) \in E \wedge \{u, u\} \subset S$.*

Clearly, every roman S is a win–win: If the first request is at a node $v \in S$, then v is serviced by its own server; if $v \notin S$, then v is serviced by a neighbor u with $\{u, u\} \in S$. This implies that a minimum win–win does not have cardinality larger than a minimum roman. The next result shows that the '$\preceq$' relationship between those two problems is actually strict; since in all cases '$\preceq$' is trivial, in the sequel we will only show that '=' does not hold.

Strict Inclusion **8** ONLINE STATIC WIN-WIN$\prec$ ROMAN DOMINATION*:*
For the graph in Figure 3, the server placement $S' = \{v_2, v_2, v_4, v_4\}$ is a minimum roman. On the other hand, $S = \{v_2, v_3, v_4\}$ is a minimum win–win: if the first request is at v_2, then this request is serviced by v_3; if after that the second request is at v_3, then it is serviced by v_2 or by v_4. $\diamondsuit$

It is known that MIN ROMAN DOMINATION is NP-hard for arbitrary graphs [Dre00]. We strengthen this result and show that the problem is also hard to approximate. As a by–product, we get a new proof for the NP-hardness. In particular, Lemma 2 remains true if we replace the notion of win–win by roman (see also [Dre00, Proposition 2.1]). Hence, we get the following theorem:

Theorem 9 *The* MIN ROMAN DOMINATION *problem in general graphs can be approximated within* $2 + 2\ln n$, *but (unless* $\mathsf{P} = \mathsf{NP}$*) cannot be approximated within* $c \log n$ *for some* $c > 0$.

3.1. Planar Graphs

Often, our problem instances are not arbitrary graphs; planarity is quite a natural condition (see Figure 1). It is therefore interesting to study the problem complexity for planar graphs, since we know that minimum dominating set can be approximated well for planar graphs. It turns out that MIN ROMAN DOMINATION is NP-hard for planar graphs. A simple reduction from PLANAR VERTEX COVER (shown NP-hard in [GJ79]) is: for each edge of the given graph we add two nodes and connect them with the endpoints of the edge; see [PPS+01] for more details. (In [Dre00, page 68] it is mentioned that NP-hardness of the planar case has been also stated in [Hed00]; however, the latter reference is not published yet.)

Theorem 10 MIN ROMAN DOMINATION *is strongly* NP*-hard even if the input graph* G *is planar.*

The results from the previous section show that the planar MIN ROMAN DOMINATION can be approximated within $2 + \epsilon$. The next theorem shows that we can find a better approximation. Its proof follows the ideas from [Bak94, ABFN00] which have become a well known standard method to get PTASs for many problems on planar graphs. Those approximations schemes look very similar; the only specific part is that the problem has to be solved optimally on r–outerplanar graphs. We use dynamic programming and the notion of *bounded treewidth* [ABFN00] to show how this can be done for the MIN ROMAN DOMINATION problem.

Theorem 11 (PTAS) MIN PLANAR ROMAN DOMINATION *has a Polynomial Time Approximation Scheme* (PTAS), *but (unless* $\mathsf{P} = \mathsf{NP}$*) it does not have a Fully Polynomial Time Approximation Scheme* (FPTAS).

Proof. Let G be a r–outerplanar graph. This implies that G has a treewidth l of at most $3r - 1$ [ABFN00]. A tree decomposition $\langle\{X_i | i \in I\}, T\rangle$, with width at most $3r - 1$ and with $|I| = O(|V|)$ of G, can be found in $O(r|V|)$ time [ABFN00].

Let $\langle\{X_i | i \in I\}, T\rangle$ be a tree decomposition for the graph $G = (V, E)$. Let $X_i = \{x_1^{(i)}, \ldots, x_{n_i}^{(i)}\}$ be a bag [ABFN00] with $n_i := |X_i|$. A number $j \in \{0, \ldots, 3^{n_i} - 1\}$ can be identified with a server placement $S_j^{(i)}$ in the following way. We write j in ternary arithmetic, i.e., $j = \sum_{\nu=1}^{n_i} 3^{\nu-1} j_\nu$, where $j_\nu \in \{0, 1, 2\}$. Every node $x_\nu \in X_i$ occurs with multiplicity j_ν in $S_j^{(i)}$.

The algorithm we will describe visits the vertices of T from the leaves to the root. For every server placement $S_j^{(i)}$ of a bag X_i, the algorithm computes a server placement $\overline{S}_j^{(i)}$ for the bags in the subtree rooted at i as a partial solution.

The dynamic programming algorithm proceeds in three steps.

Step 1: For every leaf X_i, for every $j \in \{0, \ldots, 3^{n_i} - 1\}$, we define $\overline{S}_j^{(i)} := S_j^{(i)}$.

Step 2: After this initialization, we visit the vertices of our tree decomposition from the leaves to the root. Suppose node i has a child k in the tree T. In the case that i has several children $k_1, \ldots, k_s$ in the tree T, this step has to be repeated for each child.

1 Determine the intersection $Y := X_i \cap X_k$.

2 For every server placement $S_j^{(i)}$ of X_i, we choose a server placement $S_{j'}^{(k)}$ of X_k such that the following properties hold:

   (a) $S_j^{(i)}{}_{|Y} = S_{j'}^{(k)}{}_{|Y}$.

   (b) For every $v \in X_k \setminus Y$ with $v \notin S_{j'}^{(k)}$, there is a u_v with $\{u_v, u_v\} \subset \overline{S}_{j'}^{(k)}$ and $(v, u_v) \in E$.

   (c) The number $|(S_j^{(i)} \uplus \overline{S}_{j'}^{(k)}) \setminus (S_j^{(i)}{}_{|Y})|$ is minimized.

   Then, we define $\overline{S}_j^{(i)} := (S_j^{(i)} \uplus \overline{S}_{j'}^{(k)}) \setminus S_j^{(i)}{}_{|Y}$. For different $j_1, j_2 \in \{0, \ldots, 3^{n_i}\}$ with $S_{j_1}^{(i)}{}_{|Y} = S_{j_2}^{(i)}{}_{|Y}$, the same $j_1' = j_2'$ can be chosen.

   Note that, by properties of tree decomposition, we know that none of the nodes $v \in X_k \setminus Y$ will appear in a bag that has not been visited up to this point. Otherwise, such a node would also appear in X_i.

Step 3: Let X_R be the root of T, let $n := |X_R|$. Choose a $j \in \{0, \ldots, 3^n - 1\}$, such that

1 $\overline{S}_j^{(R)}$ is a roman for G, and

2 $|\overline{S}_j^{(R)}|$ is minimum.

The algorithm described above runs in time polynomial in the size of G and in 3^{3r}. Due to construction, for every vertex $i \in T$ and for every $j \in \{0, \ldots, 3^{n_i} - 1\}$, $\overline{S}_j^{(i)}$ is a smallest server placement such that property 2 (b) of step 2 is fulfilled. This implies that $\overline{S}_j^{(R)}$ is a minimum roman for G.

Finally, the strong NP-hardness proof of Theorem 10 implies that ROMAN DOMINATION is not in FPTAS (see [GJ79] for the definition of strong NP-hardness and its implications). □

4. Online Dynamic Win–Win

In this section, we assume that after the first request, there is enough time to move the servers from one node to a neighbor before the second request occurs. This leads to the following definition.

Definition 12 (online dynamic) *Given a graph $G = (V, E)$ and a server placement S. A function $r : S \to V$ is called* rearrangement *for G, S, if for*

every server $v \in S$

$$r(v) = v \text{ or } (v, r(v)) \in E$$

holds. We say that S is a dynamic win–win *for G, if for every $u \in V$ there is a rearrangement r_u with the properties:*

- *There is $v \in S$ with $r_u(v) = u$, i.e., the first request at u can be serviced.*
- *For all $u' \in V \setminus \{u\}$, there is a $v' \in S \setminus \{v\}$ with $r_u(v') = u'$ or $(r_u(v'), u') \in E$.*

***Strict Inclusion* 13** ONLINE DYNAMIC WIN-WIN$\prec$ ONLINE STATIC WIN-WIN*:*
Consider the cycle of length 4, $(v_1, \ldots, v_4, v_1)$. By one hand, the server placement $S = \{v_1, v_3\}$ is a dynamic win–win. For instance, if the first request is at v_2, then this request is serviced by v_1 and v_3 moves to v_4. On the other hand, there is no server placement S' which is a win–win with $|S'| = 2$. To see this, we consider two cases. Case 1, $S' = S$. A first request at v_2 must be serviced by v_1 or v_3, let us say v_1. Then a second request occurring at v_1 cannot be serviced. Case 2, $S' = \{v_1, v_4\}$. Consider a request at v_1. If we use the server at v_1, then v_2 is no longer dominated. Similarly, using the server at v_4 leaves v_3 undominated. $\diamond$

Again, the methods from Section 2 can be used to show the complexity of MIN ONLINE DYNAMIC WIN-WIN.

Theorem 14 *The* MIN ONLINE DYNAMIC WIN-WIN *problem is* NP*-hard. It can be approximated within $2 + 2\ln n$, but (unless $\mathsf{P} = \mathsf{NP}$) cannot be approximated within $c \log n$ for some $c > 0$.*

We know that finding a minimum dominating set is hard to do. But what happens if we are given a server placement, and are asked if the arrangement is 'close to' a dominating set – that is, if each server is allowed to move at most 1 step, can a dominating set be obtained?

Definition 15 *Let r be a rearrangement for $\langle G, S\rangle$; r is called* dominating rearrangement *for $\langle G, S\rangle$, if the server placement $\{r(v) | v \in S\}$ contains a dominating set for G.*

Given a graph G and a server placement S, the DOMINATING REARRANGEMENT problem asks whether there is a dominating rearrangement for $\langle G, S\rangle$. We give two theorems related to this problem; proofs are omitted due to space limitations and the reader is referred to [PPS+01].

Theorem 16 DOMINATING REARRANGEMENT *is* NP*-complete. This remains true, even if the input graph is planar.*

Theorem 17 *Given a graph G and a server placement S. The problem to decide whether S is a* dynamic win–win *for G is* NP*-complete.*

5. Offline Static/Dynamic Win–Win

In this section, we consider the situation in which both requests occur at the same time (equivalently, as the first request must be serviced, it is already known where the second one will be).

Definition 18 (offline static) *Let $G = (V, E)$ be a graph. A server placement S is an* offline win–win *if for every pair of nodes $v_1, v_2 \in V$, $v_1 \neq v_2$, there is a pair $\{u_{v_1}, u_{v_2}\} \subset S$ with*

- $v_1 = u_{v_1}$ *or* $(v_1, u_{v_1}) \in E$, *and*
- $v_2 = u_{v_2}$ *or* $(v_2, u_{v_2}) \in E$.

***Non-Inclusion* 19** ONLINE DYNAMIC WIN-WIN$\not\prec$ OFFLINE STATIC WIN-WIN*:*
For the graph in Figure 4 the set $\{v_2, v_3\}$ is an offline win–win. For the same graph, no dynamic win–win can have size 2. Indeed, consider a first request at v_2. No matter what server we use to service this request, the remaining one cannot cover the nodes $\{v_1, v_3, v_4\}$, where a second request can occur.

***Non-Inclusion* 20** OFFLINE STATIC WIN-WIN$\not\prec$ ONLINE DYNAMIC WIN-WIN*:*
It is easy to verify that $\{u, v_1, v_2\}$ is a dynamic win–win for the graph in Figure 6. On the other hand, there is no offline win–win multiset of size less than 4: each of the subtrees rooted at v_1 or v_2 must contain at least two servers. ◊

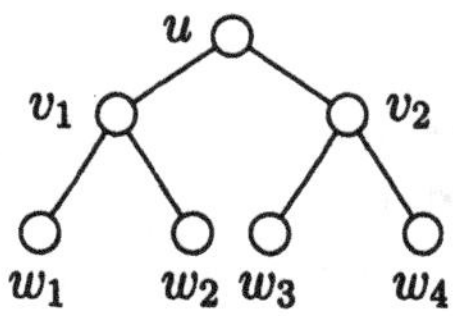

Figure 6. Proof of ***Non-Inclusion 20***

Again, MIN OFFLINE STATIC WIN-WIN is an NP-hard problem, illustrated by the techniques of Section 2. Moreover, we can give the following characterization of the offline win–win multisets:

Lemma 21 *A server placement S is an* offline win–win, *iff for every pair of two different nodes there is one server in the neighborhood of one node and a different server in the neighborhood of the other node.*

We conclude this section with OFFLINE DYNAMIC WIN-WIN. Here we combine the fact that servers can be rearranged before serving the second request (DYNAMIC) with the fact that the second request is known by the time we have to serve the first one (OFFLINE). Therefore, we have the following definition for the corresponding server placement:

Definition 22 (offline dynamic) *Let $G = (V, E)$ be a graph. A server placement S, is an* offline dynamic win–win *for G, if for every pair of nodes $v_1, v_2 \in V$, with $v_1 \neq v_2$, there is a pair of distinct nodes $u_{v_1}, u_{v_2} \in V$ such that v_i is at distance at most i from u_{v_i}, for $i = 1, 2$.*

***Strict Inclusion* 23** OFFLINE DYNAMIC WIN-WIN$\prec$ ONLINE DYNAMIC WIN-WIN*:*
Consider the cycle of length 5, $(v_1, v_2, \ldots, v_5, v_1)$. It is easy to verify that the set $S = \{v_1, v_3\}$ is an offline dynamic win–win (S is a dominating set and both servers are at distance at most 2 from any other non-server node). To prove that no multiset of size 2 can be a dynamic win–win we use the following argument.

After the first request has been serviced, the set of nodes to be considered as possible positions for the second request induce a path of length 4; therefore, no matter where we place the remaining server, there is no way to dominate all such nodes. ◇

References

[ABFN00] J. Alber, H.L. Bodlaender, H. Fernau, and R. Niedermeier. Fixed Parameter Algorithms for Planar Dominating Set and Related Problems. In *Proc. 7th Scandinavian Workshop on Algorithm Theory (SWAT)*, volume 1851 of *LNCS*, pages 97–110, 2000.

[ACG+99] G. Ausiello, P. Crescenzi, G. Gambosi, V. Kann, A. Marchetti-Spaccamela, and M. Protasi. *Complexity and Approximation – Combinatorial optimization problems and their approximability properties.* Springer Verlag, 1999.

[AF95] J. Arquilla and H. Fredricksen. Graphing an Optimal Grand Strategy. *Military Operations Research*, pages 3–17, Winter 1995.

[Bak94] B. Baker. Approximation Algorithms for NP–Complete Problems on Planar Graphs. *J. ACM*, 41(1):153–180, 1994.

[dJ62] C.F. de Jaenish. Trait des Applications de l'Analyse Mathematique au Jeau des Echecs. Petrograd, 1862.

[Dre00] P.A. Dreyer. *Applications and Variations of Domination in Graphs.* PhD thesis, Rutgers University, New Jersey, 2000.

[GJ79] M.R. Garey and D.S. Johnson. *Computers and Intractability / A Guide to the Theory of NP–Completeness.* Freeman and Company, 1979.

[Hed00] S.T. Hedetniemi. Roman domination in graphs ii. Slides and notes from presentation at 9th Quadrienn. Aint. Conf. on Graph Theor., Combinatorics, Algorithms, and Applications, June 2000.

[HHS98] T.W. Haynes, S.T. Hedetniemi, and P.J. Slator. *Fundamentals of Domination in Graphs.* Marcel Dekker, New York, 1998.

[Joh74] D.S. Johnson. Approximation algorithms for combinatorial problems. *J. Comput. System Sci*, 9:256–278, 1974.

[NR95] M. Naor and R.M. Roth. Optimal file sharing in distributed networks. *SIAM J. Comput.*, 24(1):158–183, 1995.

[Och96] D. Ochmanek. Time to Restructure U.S. Defense Forces. *ISSUES in Science and Technology*, Winter 1996.

[PPS+01] A. Pagourtzis, P. Penna, K. Schlude, K. Steinhöfel, D.S. Taylor, and P. Widmayer. Server placements, roman domination and other dominating set variants. Technical Report 365, Department of Computer Science, ETH Zürich, 2001.

[RR00] C.S. ReVelle and K.E. Rosing. Defendens Imperium Romanum: A Classical Problem in Military Strategy. *American Mathematical Monthly*, pages 585–594, 2000.

[RS97] R. Raz and S. Safra. A Sub–Constant Error–Probability Low–Degree Test, and a Sub–Constant Error–Probability PCP Characterization of NP. In *Proc. 29th ACM STOC*, pages 475–484, 1997.

[Ste99] I. Stewart. Defend the roman empire! *Scientific American*, pages 94–95, December 1999.

A LINEAR TIME ALGORITHM FOR FINDING TREE 3-SPANNER ON 2-TREES *

B. S. Panda
Department of Mathematics
Indian Institute of Technology, Delhi
Hauz Khas, New Delhi 110 016, INDIA

Sajal K. Das
Department of Computer Science and Engineering
The University of Texas at Arlington, Arlington TX 76019, USA
das@cse.uta.edu

Abstract A spanning tree T of a graph G is said to be a **tree t-spanner** if the distance between any two vertices in T is at most t times their distance in G. A graph that has a tree t-spanner is called a **tree t-spanner admissible graph**. The complexity of recognizing tree 3-spanner admissible graphs is still unknown. In this paper, a characterization of tree 3-spanner admissible 2-trees is presented. Linear time algorithms for recognizing tree 3-spanner admissible 2-trees and for constructing tree 3-spanners in such 2-trees are also proposed.

Keywords: Tree t-spanner, k-trees, Computational Complexity, Algorithms.

1. Introduction

A spanning subgraph H of a graph G is called a **t-spanner** if the distance between every pair of vertices in H is at most t times their distance in G. For a t-spanner H of G, t is called the **stretch factor** and $|E(H)|$, the number of edges in H, is called the **size** of the spanner. A t-spanner H of G is called a tree t-spanner if H is a tree. The notion of t-spanner was intro-

*Part of the work was done when the first author was at the Department of Computer and Information Sciences, University of Hyderabad and was visiting the Department of Computer Science and Engineering, The University of Texas at Arlington. This work was supported by NASA Ames Research Center under Cooperative Agreement Number NCC 2-5395.

cation networks, message routing, data analysis, motion planning, computational geometry, image processing, network design, and phylogenetic analysis (see [1,2,3,9,10,13,17-21]). The study of graph spanners has attracted many researchers and is currently an active area of research (see [4-9,12,15,19,22]).

The goal behind the notion of spanners is to find a sparse spanner H of a given graph G such that the distance between every pair of vertices in H is relatively close to the corresponding distance in the original graph G. Therefore, one of the fundamental problems in the study of spanners is to find a minimum t-spanner, i.e., a t-spanner having minimum number of edges, for every fixed i integer $t \geq 1$. Unfortunately, the problem of finding a minimum t-spanner is NP-Hard for $t = 2$ [18] and for $t \geq 3$ [8]. For a minimum t-spanner H of G, $|E(H)| \geq |V(G)| - 1$ with equality holding if and only if H is a tree t-spanner, where $|V(G)|$ is the number of vertices of G. The problem of determining whether an arbitrary graph admits a tree t-spanner has been studied in detail.

Cai and Corneil [9] have shown that for a given graph G, the problem of deciding whether G has tree t-spanner is NP-Complete for any fixed $t \geq 4$ and is linearly solvable for $t = 1, 2$. The status of the case $t = 3$ is still open for arbitrary graphs. They have also observed that split graphs, co-graphs, and complement of bipartite graphs always have tree 3-spanner. Madanlal *et al.* [14] have shown that interval graphs and permutation graphs admit tree 3-spanner which can be constructed in linear time. They have also characterized regular bipartite graphs which admit tree 3-spanner. Recently, Brandstadt *et al.* [4] have shown that strongly chordal graphs and dually chordal graphs admit tree 4-spanner which can be computed in linear time.

Let $G[S], S \subseteq V$, be the induced subgraph of $G = (V, E)$ on S. A subset $C \subseteq V$ is said to be a **clique** if $G[C]$ is a maximal complete subgraph of G. A clique C is called a k-clique if $|C| = k$. A 3-clique is called a **triangle**. A graph G is called a k**-tree** if it can be obtained by the following recursive rules.

- Start with any k-clique as the basis graph. A k-clique is a k-tree.
- To any k-tree H add a new vertex and make it adjacent to a k-clique of H, to form a $(k + 1)$-clique.

Note that a tree is nothing but a 1-tree. A 2-tree is a k-tree for $k = 2$.

In this paper, we, first, observe that 2-trees in general do not admit tree 3-spanner. We, then, characterize those 2-trees that admit tree 3-spanner. We also present a linear time algorithm for recongnizing a tree 3-spanner admissible 2−tree and for constructing a tree 3-spanner of a tree 3-spanner admissible 2-tree.

The main idea behind our characterization of tree 3-spanner admissible 2-tree is the identification of forced edges, i.e, edges that will appear in any tree 3-spanner. We show that a 2-tree G admits a tree 3-spanner if and only if it does not contain any triangle containing all three forced edges. We show that this characterization and the technique for identifying forced edges enable us to recognize tree 3-spanner admissible 2-trees in linear time. A 2-tree G, as seen from the definition, consists of $|V(G)| - 2$ triangles. We show that every triangle of a 2-tree G contributes at least one edge to every tree 3-spanner of G. Given a tree 3-spanner admissible 2-tree G, we employ a *D-search* (a search similar to the classical *BFS* but differs from *BFS* in that the next element to explore is the element most recently added to the list of unexplored elements) to search the triangles of G. Using this search, we explore all the triangles and keep on adding some edges of the triangles using certain rules to construct a spanning tree of G. We show that this tree is indeed a tree 3-spanner of G.

The rest of the paper is organized as follows. Section 2 presents some pertinent definitions and results. Section 3 presents the characterization of tree 3-spanner admissible 2-trees. Section 4 presents the recognition algorithms of tree 3-spanner admissible 2-trees and an algorithm for constructing a tree 3-spanner of a tree 3-spanner admissible 2-tree. The proof of correctness of these algorithms are presented in this section. Section 5 presents the complexity analysis of these algorithms. Finally, section 6 concludes the paper.

2. Preliminaries

For a graph $G = (V, E)$, let $N_G(v) = \{ w \in V \mid vw \in E \}$ be the set of neighbors of v. If $G[N_G(v)]$ is a complete subgraph of G, then v is called a **simplicial vertex** of G. An ordering $\alpha = (v_1, v_2, \ldots, v_n)$ is called a **perfect elimination ordering** (PEO) of G if v_i is a simplicial vertex of $G[\{v_i, v_{i+1}, \ldots, v_n\}]$ for all i, $1 \le i \le n$. Let $d_G(v)$ denote the **degree** of v in G. Let $d_G(u, v)$ denote the **distance** from u to v in G. Unless otherwise stated the graph G is assumed to be connected. A triangle $\{a, b, c\}$ is said to be **simplicial triangle** if one of its vertices is simplicial. An edge of a simplicial triangle is called **simplicial** if it is incident on a simplicial vertex of the triangle. A triangle is said to be **interior** if all of its edges are shared by at least two triangles. A triangle is said to be **double interior** if the triangle is interior and two of its adjacent triangles on different edges are interior. A triangle is said to be **triple interior** if the triangle is interior and its adjacent triangles on each of the three edges are also interior.

A graph is said to be **chordal** if every cycle of length at least four has a chord. k-trees are a subclass of chordal graphs. So, every k-tree has a PEO. A 2-tree is said to be a **minimal triple interior** (respectively, **double or single interior**) 2-tree if it contains a triple interior (respectively, double or single interior) triangle but none of its proper subgraph contains a triple interior (respectively, double interior or interior) triangle.

Let Tr be a triangle of a minimal triple interior 2-tree. Tr is called an outer triangle if it contains a simplicial vertex. Tr is called an innermost triangle if it is triple interior. Tr is called inner triangle if it is neither innermost nor outer triangle. The **multiplicity** $M(e)$ of an edge e is defined to be the number of triangle containing e. Let Tr be a triangle and e_1, e_2, e_3 be the three edges of Tr. Tr is said to be one sided developing with respect to e_1 if either $M(e_2) = 1$ and $M(e_3) > 1$ or $M(e_3) = 1$ and $M(e_2) > 1$. Suppose $M(e_2) = 1$ and $M(e_3) > 1$. In this case, e_3 is said to be a developing edge of Tr with respect to e_1. Tr is said to be double side developing with respect to e_1 if $M(e_2) > 1$

and $M(e_3) > 1$. In this case, e_2 and e_3 are said to be developing edges of Tr with respect to e_1.

If $G - C$ is disconnected for a clique C with components $H_i = (V_i, E_i)$, $1 \leq i \leq r$, $r \geq 2$, then C is said to be a separating clique and $G_i = G[(V_i \cup C)]$, is called a **separated graph** of G with respect to C, $1 \leq i \leq r$, and $r \geq 2$. Let $W(G_i) = \{v \in C |$ there is a $w \in V_i$ with $vw \in E(G)\}$. Cliques of G other than C which intersect C are called **relevant cliques** of G with respect to C. A relevant clique C_i of G_i for which $(C_i \cap C) = W(G_i)$ is called a **principal clique** of G_i.

The existence of a principal clique of every separated graph of a chordal graph is guaranteed by the following result due to Panda et al [16].

Lemma 2.1 [16]: Every separated graph

G_i of a chordal graph has a principal clique.

Let H be a spanning subgraph of G. Since the $d_H(x,y) \leq t \times d_G(x,y)$ for every $x, y \in V(G)$ if and only if $d_H(x,y) \leq t$ for every edge $xy \in E(G)$, we have the following useful lemma.

Lemma 2.2 A spanning subgraph H of G is a t-spanner if and only if $d_H(x,y) \leq t$ for every edge $xy \in E(G)$.

In view of Lemma 2.2, in the rest of the paper we assume that a spanning subgraph H (a spanning tree T) of G is a t-spanner (tree t-spanner) if $d_H(x,y) \leq t$ ($d_T(x,y) \leq t$) for every edge $xy \in E(G)$.

3. Characterization of Tree 3-spanner Admissible 2-trees

In this section, we present a characterization of tree 3-spanner admissible 2-tree. We do this by identifying the forced edges, i.e., edges which will appear in every tree 3-spanner.

Proposition 3.1: Let G be a tree 3-spanner admissible 2-tree having at least 3 vertices. Then $G - v$ is a tree 3-spanner admissible 2-tree for every simplicial vertex v of G.

Proof: Let T be a tree 3-spanner of G. Let v be any simplicial vertex of G. If $d_T(v) = 1$, then $T - v$ is a tree 3-spanner for $G - v$. Suppose $d_T(v) = 2$. Let $N_G(v) = \{x, y\}$. Then $vx, vy \in E(T)$ but $xy \notin E(T)$. Let $T_1 = T - \{xv\} \cup \{xy\}$. Then, T_1 is a tree 3-spanner of $G - v$. □

The following proposition, whose proof is omitted, follows from the above proposition.

Proposition 3.2: Let G be a tree 3-spanner admissible 2-tree with at least 3 vertices and let $\alpha = (v_1, v_2, \ldots, v_n)$ be a PEO of G. Then $G[\{v_i, v_{i+1}, \ldots, v_n\}]$ is a tree 3-spanner admissible 2-tree for all i, $1 \leq i \leq n - 2$.

Note that every induced sub 2-tree of a 2-tree can be obtained by successively deleting simplicial vertices. Therefore, we have the following.

Proposition 3.3: Every induced sub 2-tree of a tree 3-spanner admissible 2-tree is tree 3-spanner admissible.

The following lemma implies that if a triangle in G satisfies certain properties, then two out of its three edges are present in every tree 3-spanner of G.

The proof of the following lemma, which uses the method of contradiction, can be found in the appendix.

Lemma 3.4: Let G be a tree 3-spanner admissible 2-tree. If G has an induced interior triangle, say $\{a, b, c\}$, then exactly two out of the three edges ab, bc, and ca are present in every tree 3-spanner T of G.

Let G be a tree 3-spanner admissible 2-tree. Below, we show that every triangle Tr of G contributes at least one edge to every tree 3-spanner T of G.

Lemma 3.5: Let T be a tree 3-spanner of a 2-tree G. Then, T contains at least one edge of every triangle $\{a, b, c\}$ of G.

Proof: (By contradiction) If $\{a, b, c\}$ is an interior triangle, then by Lemma 3.4, T contains two edges of $\{a, b, c\}$. If $\{a, b, c\}$ is a simplicial triangle, then T must contain one of the simplicial edges of $\{a, b, c\}$. If $\{a, b, c\}$ is neither a simplicial triangle nor an interior triangle, then, wlg, $M(ab) > 1$ and $M(ac) > 1$. Since, $M(bc) = 1$, b, a, c is the only path of length 2 in G from b to c. Since, T does not contain bc, T contains a path of length at most 3 from b to c. Since, the only path of length 2 from b to c is b, a, c, T has no path of length 2 from b to c as T does not contain ab and ac. But, every path of length 3 from b to c contains either ab or ac. So, there is no path of length at most 3 from b to c in T. This is a contradiction to the fact that T is a tree 3-spanner of G. So, T must contain one of the edges ab, bc, and ac. □

Let G be a tree 3-spanner admissible 2-tree. An edge e of G is said to be a **forced edge** if it belongs to every tree 3-spanner of G. An edge which is common to two interior triangle is called a **strong edge**. Strong edges are forced edges as shown below.

Lemma 3.6: Every strong edge of a tree 3-spanner admissible 2-tree is a forced edge.

Proof: Let bc be a strong edge. So, bc is common to two interior triangles, say $\{a, b, c\}$, and $\{b, c, d\}$ of a tree 3-spanner admissible 2-tree G. Let T be a tree 3-spanner of G. If possible, T does not contain bc. By Lemma 3.4, two of the three edges ab, bc, and ca are present in T and two of the three edges bc, cd, and bd are also present in T. Since, T does not contain bc, T contains the edges ab, ac, cd, and bd. Now, a, b, d, c, a is a cycle in T. Hence a contradiction. So, T contains bc. Hence, bc is a forced edge. □

Since, all the three edges of a triple interior triangles are strong edges, and hence forced edges by Lemma 3.6, we have the following corollary.

Corollary 3.7: Let G be a 2-tree containing a triple interior triangle. Then G can not have a tree 3-spanner.

A triangle having two forced edges is called a **semi-forced triangle**. Let $\{a, b, c\}$ be a semi-forced triangle having forced edges ab and bc. A triangle $\{x, a, c\}$ is said to be **dependent** on the triangle $\{a, b, c\}$ if either $M(xa) > 1$ or $M(xc) > 1$. Suppose $\{x, a, c\}$ is dependent on the triangle $\{a, b, c\}$ and $M(xc) > 1$. Then, the edge xc of the dependent triangle is called a **semi-strong edge**. The motivation behind introducing the concept of semi-strong edge is that semi-strong edges are also forced edges.

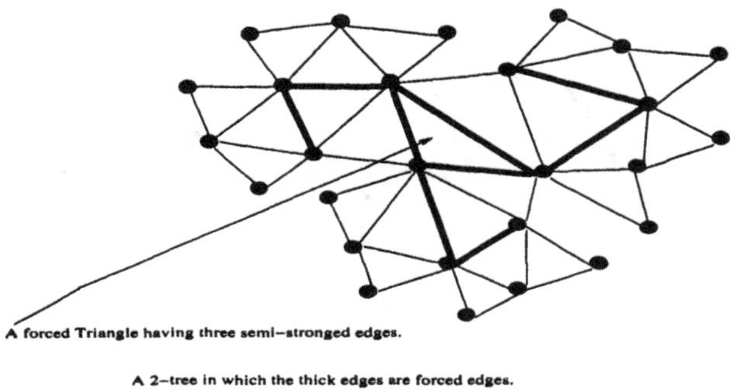

Figure 1. A 2-Tree without any triple interior triangle that has a triangle containing forced edges.

The following lemma whose proof can be found in the appendix, proves that semi-strong edges are forced edges.

Lemma 3.8: Every semi-strong edge of a tree 3-spanner admissible 2-tree G is a forced edge.

We have seen earlier that if a 2-tree G has a triple interior triangle, then it can not have a tree 3-spanner. A 2-tree may contain triangle consisting of forced edges that is not an interior triangle. Figure 1 contains a 2-tree which does not have any triple interior triangle but it has a triangle consisting of semi-strong edges. So, the graph does not have any tree 3-spanner.

The characterization theorem (Theorem 3.11) for tree 3-spanner admissible 2-trees is proved using induction principle and the following lemma is a key in achieving that. The proof of the following lemma can be found in the appendix.

Lemma 3.9: Let G be a tree 3-spanner admissible 2-tree. Let $\{a, b, c\}$ be an interior triangle of G and H_i, $1 \le i \le 3$, be the connected components of $G - \{a, b, c\}$. Let $V(H_2) = \{e\}$ and $V(H_3) = \{f\}$. Let $d \in V(H_1)$ be such that $\{d, a, b\}$ is a triangle in $G_1{=}G[V(H_1) \cup \{a, b, c\}]$.

(α) If $\{d, a, b\}$ is an interior triangle in G_1, then G_1 contains two tree 3-spanners T_1 and T_2 such that T_1 contains the edges ab and ac, and T_2 contains the edges ab and bc.

(β) If $\{d, a, b\}$ is not an interior triangle of G_1 and ab is not a semi-strong edge, then G_1 has a tree 3-spanner T_3 containing any two edges of the three edges ab, bc, and ac.

The following lemma shows that a 2-tree, that does not have any interior triangle, admits a tree 3-spanner. The proof of the lemma indeed constructs a tree 3-spanner of a 2-tree that does not have any interior triangle. The proof of the following lemma can be found in the appendix.

Lemma 3.10: If a 2-tree is free from induced interior triangle, then it admits a tree 3-spanner.

Figure 2 contains a 2-tree which does not contain any interior triangle. So, by Lemma 3.10, it has a tree 3-spanner. The method which is employed in the proof of Lemma 3.10 is illustrated for the 2-tree G of figure 2. The edges which

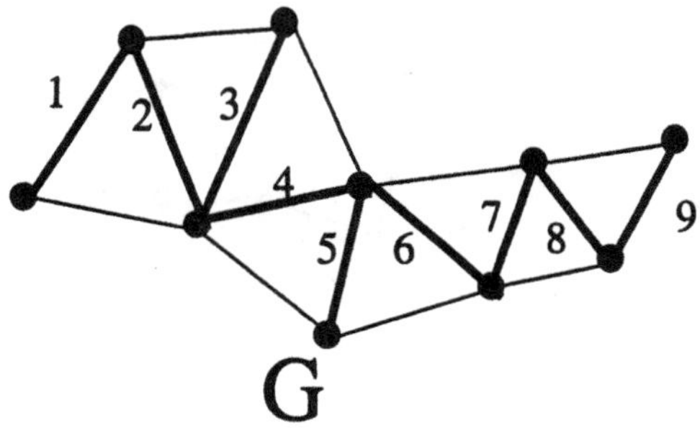

The thick edges form a tree 3–spanner.

Figure 2. Construction of a tree 3-spanner of a 2-tree with out having any interior triangle

are selected by the methods are the thick edges of G and are numbered in the order they are selected. We employ a similar kind of method, which is more involved, to construct a tree 3-spanner of a tree 3-spanner admissible 2-tree.

A triangle of a 2-tree G is said to be a **strong triangle** if each of its edges is either a strong edge or a semi-strong edge.

We are now in a position to characterize tree 3-spanner admissible 2-tree. The theorem is proved using induction principle and the proof is given in the appendix.

Theorem 3.11 (Characterization Theorem): A 2-tree G admits a tree 3-spanner if and only if it does not contain a strong triangle as an induced subgraph.

4. Algorithms and Proof of Correctness

In this section, we first present a recognition algorithm for tree 3-spanner admissible 2-trees. Given a tree 3-spanner admissible 2-tree G, we next present an algorithm to find a tree 3-spanner of G. We also present the proof of correctness of these algorithms.

Algorithm Tree 3-spanner Recognition

Input: A 2-tree G;

Output: If G admits a tree 3-spanner then output **"G admits a tree 3-spanner" and E, the set of forced edges** else output **"G has no tree 3-spanner"**.

{

1. Find all the triangles of G. $E = \phi$
2. For each strong edge e
   Mark e as forced edge.
   $E = E \cup \{e\}$;
3. Find all semi-strong edges, mark them as forced edges, and add them to E.
4. If G has a triangle containing all three edges in E,

then output **"G has no tree spanner"** else
output **"G is a tree 3-spanner admissible 2-tree", and** E.
}

Since, **Algorithm tree 3-spanner Recognition** checks for a strong triangle and declares that the 2-tree without having any strong triangle is tree 3-spanner admissible 2-tree, the proof of correctness follows from Theorem 3.11.

In view of the above, we have the following theorem.

Theorem 4.1: Algorithm tree 3-spanner recognition correctly recognize whether a 2-tree T is tree 3-spanner admissible.

Next, we present an algorithm to construct a tree 3-spanner of a tree 3-spanner admissible 2-tree T. We, then, prove the correctness of this algorithm.

The algorithm maintains a stack of edges, the current edge and a triangle containing the current edge as the current triangle. In every iteration, the stack is popped and the popped edge is made the current edge, and an unmarked triangle containing the current edge is made the current triangle. Based on the current edge and the current triangle, the algorithm pushes one or more edges to the stack. The algorithm also maintains two arrays, namely *CUR* and *NUM* to maintain the information of the triangles of G. $CUR[Tr] = 1$ if the triangle Tr is made current triangle at some iteration of the algorithm. Otherwise, $CUR[Tr] = 0$. $NUM[Tr]$ represents the number when the triangle Tr was marked. The information stored in these arrays will be used for the proof of correctness of the algorithm. The exact rules for pushing the edges into the stack are given in the following algorithm.

Algorithm Tree 3-Spanner Construction
Input: A tree 3-spanner admissible 2-tree G.
Output: A tree T which is a tree 3-spanner of G.

```
{
1. Find all the triangles of G.
2. Q = φ;
   T = φ;
   CUR[Tr] = NUM[Tr] = 0 for all triangles of G;
   count=1;
   Let ab be a simplicial edge of G;
   Push(Q,ab);
3. while (Q ≠ φ) /* start of while loop 1 */
   {
   CE=Pop(S); T = T ∪ {CE};
   while(there is any unmarked triangle, say Tr, containing CE)
    /* start of while loop 2 */
       {
        CT=Tr; CUR[Tr]=1; NUM[Tr]=count++; Mark CT;
        Let the edges of Tr be CE, e and f;
         Case I: if (CT is one side developing w.r.t. CE)
          Let CT be developing on e;
```

```
            if ( there is a unmarked triangle Tr' = {e, f, g} such that f
and g are
              forced edges )
                { T = T ∪ {g, h}; Push(Q,g); Push(Q,h);
                  Mark Tr'; CUR[Tr']=0;
                 NUM[Tr']=count++;
                  COVER(CT,e); }
            else
                { T = T ∪ {e}; Push(Q,e);}
            Case II: if ( CT is double side developing on e and f
w.r.t. CE)
             if ( either e or f is a forced edge )
               { Wlg, let e be a forced edge;
               T = T ∪ {e}; Push(Q,e);
                COVER(CT,f);}
             else
               { T = T ∪ {e};
                Push(Q,e);
                COVER(CT,f); }
             Case III: if CT is zero-side developing
             T = T ∪ {e};
             Push(Q,e);
             }   /* end of while loop 2 */
             }   /* end of while loop 1 */
       }
```

Procedure COVER(CT,e)

```
       {
        for each one-side developing triangle Tr ≠ CT containing e
        { Let the edges of the triangle be e, x, and y such that
      Tr is developing on x;
          T = T ∪ {x}; Push(Q,x); Mark Tr;
         CUR[Tr]=0; NUM[Tr]=count++;}
          for each zero-side developing triangle Tr ≠ CT containing e
         { Let the edges of the triangle be e, x, and y;
          T = T ∪ {x}; Push(Q,x); Mark Tr ;
         CUR[Tr]=0; NUM[Tr]=count++;}
       }
```

Next, we prove that **Algorithm Tree 3-spanner Construction** correctly constructs a tree 3-spanner of a tree 3-spanner admissible 2-tree. We do this by first showing that T, which is constructed by this algorithm, is a spanning tree of G, and then show that it is indeed a tree 3-spanner of G.

Theorem 4.2: Algorithm tree 3-spanner Construction produces a spanning tree T of G.

Proof: Note that **Algorithm tree 3-spanner Construction** consists of several iterations. In each iteration, a triangle is made current triangle, and based on whether it is one side developing, two side developing or zero side developing, certain triangles are marked. Let T^i be the graph formed by the set of edges selected on or before ith iteration, and S_i be the set of vertices spanned by the triangles which are marked on or before ith iteration.

Claim: T^i is a spanning tree of G$[S_i]$ for each i.

Proof of Claim: Assume that the set of edges included in T by the end of $(i-1)$th iteration is a spanning tree of the set of vertices spanned by the set of triangles which are marked on or before $(i-1)$th iteration. Let $CT = \{CE, e, f\}$, and CE be the current triangle, and current edge in ith iteration, respectively. Edges are added to T based on whether CT is zero side, one side or two side developing.

Case I: CT is zero-side developing.

In this case, only one edge e is added to T. So, it is easy to see that our claim is true in this case.

Case II: CT is developing one-side on e.

In this case, only one edge e is added if there is no triangle $\{e, g, h\}$ containing two strong edges g and h. In this case, our claim is easily seen to be true. Suppose, there is a triangle $\{e, g, h\}$, where g and h are strong edges. In such a situation, g and h are added to T and CT and $\{e, g, h\}$ are marked. Moreover, for all one side developing triangles $\{e, x, y\}$ on x, the edge x is added and $\{e, x, y\}$ is marked and for all zero-side developing triangles $\{e, x_1, y_1\}$, x_1 is added to T and $\{e, x_1, y_1\}$ is marked. So, T^i forms a spanning tree on $G[S_i]$ in this case.

Case III: CT is two-side developing.

Assume one of the edges e and f is a forced edge. Wlg, e is a forced edge. In this case, e is added to T and CT is marked. Also for all one-side developing triangles $\{f, g, h\}$ on g and for all zero-side developing triangles $\{f, g, h\}$ the edge g is added to T, and all such triangles are marked. So, T^i is a spanning tree of $G[S_i]$ in this case.

Now, assume that neither e nor f is a forced edge. In this case as well, e is added to T and for all one-side developing triangles $\{f, g, h\}$ on g, the edge g is added and $\{e, f, g\}$ is marked, and for all zero-side developing triangles $\{f, g_1, h_1\}$ the edge g_1 is added to T, and all such triangles are marked. So, T^i is a spanning tree of $G[S_i]$ in this case.

So, our claim is true. Hence, T is a spanning tree of G. □

Next, we show that T is a tree 3-spanner of G.

Theorem 4.3: Let G be a tree 3-spanner admissible 2-tree. Then algorithm tree 3-spanner construction successfully constructs a tree 3-spanner of G.

Proof: By Theorem 4.2, the graph T produced by the algorithm is a spanning tree of G.

Let $e \in E(G)$ and $e = xy$. If $e \in T$, then $d_T(x, y) = 1$. So, assume that $e \notin T$. Let Tr be the triangle containing e such that $NUM[Tr]$ is minimum.

Case I: $CUR[Tr] = 1$.

Let $Tr = \{CE, e, f\}$. Since, $e \notin T$, $e \neq CE$. If Tr is zero-sided developing, then f is added to T. So, $d_T(x, y) = 2$. If Tr is one-side developing, then either f is added or g and h are added to T, where either $\{e, g, h\}$ or $\{f, g, h\}$ is a triangle containing two forced edges g and h. In this case, $d_T(x, y) \leq 3$. If Tr is two-side developing, then $d_T(x, y) = 2$ as f is added to T in this case. So, $d_T(x, y) \leq 3$ if $CUR[Tr] = 1$.

Case II: $CUR[Tr] = 0$.

Let $Tr1$ be the current triangle when Tr was marked. So, $Tr1$ was either one-side developing or two-side developing. In either case, it can be seen using the arguments employed above that $d_T(x, y) \leq 3$.

So, T is tree 3-spanner of G. □

So, by Theorem 4.2 and Theorem 4.3, we have the following result.

Theorem 4.4: Algorithm tree 3-spanner construction correctly constructs a tree 3-spanner of a tree 3-spanner admissible 2-tree.

5. Complexity Analysis

In this section, we show that **Algorithm Tree 3-spanner Recognition** and **Algorithm Tree 3-spanner Construction** can be implemented in $O(n + m)$ time, where, n and m are the number of vertices, and number of edges of the input 2-tree.

Assume that the input graph G which is a 2-tree is given in adjacency list representation. First we do some preprocessing and compute certain information which will make the implementation of the above mentioned algorithms easy.

First, find all the triangles of G. Since, cliques in chordal graphs can be found in $O(n+m)$ time [11], and cliques in 2-tree (which is a chordal graph) are triangles, all the triangles can be found in $O(n + m)$ time. Scan the adjacency lists and find a numbering of the edges of G. Modify the adjacency list of G such that for each $v \in V$, $L(v)$, the adjacency list of v, contains an adjacent vertex, say, w, edge number of the edge vw as obtained in the previous step, and a pointer to the next cell. This step takes $O(n+m)$ time. Next, number all the triangles and construct an array TN of pointers such that $TN[i]$ contains the list of edges of the triangle having number i. This takes $O(n + m)$ time. Construct an array A of pointers such that $A[i]$ contains the list of triangle containing the edge having number i. This can be done as follows. From the list of all triangles, construct a list of order pair by replacing a triangle, say Tr, by (e,Tr), (f,Tr), and (g,Tr), where e, f, and g are the edges of Tr. Now, sort this list in non-decreasing order on the first component. All the triangles containing an edge appear consecutively on this list. Since each edge has a unique number, bucket sort can be used to sort the above list. So, this takes $O(n + m)$ time. Now, from this sorted list, the array A can be constructed in $O(n + m)$ time. Now, from the lists TN and A, we can construct an array N such that $N[i] = 1$ if the triangle having number i is interior, else $N[i] = 0$. This takes $O(n + m)$ time. Using the list N, it is easy to find all the strong edges of G in $O(n + m)$ time.

We, now, describe the implementation of **Algorithm tree 3-spanner recognition**. As discussed above, step 1 takes $O(n+m)$. Implementation of step 2 is straightforward once the list of interior triangles is given. Let S_i, $0 \leq i \leq 3$, be the set of triangles having i strong edges. Recall that, for each edge e, $M(e)$ is the number of triangles containing e. Then, step 3 can be implemented as follows:

```
while ( S2 ≠ φ)
        {
        Let Tr ∈ S2;
        S2 = S2 − Tr;
        Let e, f, and g be the edges of Tr and e and f be
        the forced edges;
        for each Tr' ≠ Tr containing g
      {
      Let g, g1, g2 be the edges of Tr';
      if (M(g1) > 1) then
      { mark g1 as forced edge; E' = E' ∪ {g1};
      delete Tr' from Si and insert Tr' into Si+1;}
    if (M(g2) > 1) then
      { mark g2 as forced edge;E' = E' ∪ {g2};
      delete Tr' from Si and insert Tr' into Si+1;}
      }
    }
```

It is easy to see that this takes $O(n+m)$ time. Since, $S_3 \neq \phi$ if and only if the 2-tree G has a triangle consisting of three forced edges, implementation of step 4 is easy.

Since, 2-trees can be recognized in linear time [11], in view of the above and by Theorem 4.1, we have the following theorem.

Theorem 5.1: Tree 3-spanner admissible 2-trees can be recognized in linear time.

Next, we show that **Algorithm tree 3-spanner Construction** can also be implemented in $O(n+m)$ time.

Algorithm tree 3-spanner starts by selecting a simplicial edge. This can be done by selecting an edge e with $A[e] = 1$. This takes $O(n+m)$ time. Next, the algorithm marks the triangle containing the edge e. Whether the current triangle develops in one direction of two direction can be tested by checking the array A for the edges of the triangle. The edge number of the edges can be found out by scanning the appropriate list of the modified adjacency list of G obtained above. Again, whether a triangle is two-side developing, one-side developing or zero-side developing can be tested in $O(n+m)$ time for all triangles. Now it is easy to see that other operations of Algorithm tree 3-spanner construction can be implemented in $O(n+m)$ time. So, algorithm tree 3-spanner construction takes $O(n+m)$ time. In view of the above and by Theorem 4.3, we have the following theorem.

Theorem 5.2: Tree 3-spanner in a tree 3-spanner admissible 2-tree can be constructed in linear time.

6. Conclusion

In this paper, we have observed that 2-trees in general do not admit tree 3-spanner. We characterize those 2-trees that admit tree 3-spanner. We have also presented linear time algorithms for recognizing tree 3-spanner admissible 2-tree and for constructing a tree 3-spanner in a tree 3-spanner admissible 2-tree. It would be interesting to study the tree 3-spanner problem on chordal graphs and in particular on k-trees for $k \geq 3$.

References

[1] I.Althöfer, G.Das, D.Dobkin, D. Joseph, and J. Soares, On sparse spanner of weighted graphs, Discrete Comput. Geom. 9 (1993), 81-100.

[2] J.P.Barthélemy and A. Guénoche, Trees and Proximity Representations, Wiley, New Yark, 1991.

[3] S.Bhatt, F.Chung, F.Leighton, and A. Rosenberg, Optimal simulation of tree machines, in "27th IEEE Foundations of Computer Science, Toronto, 1986, " pp. 274-282.

[4] A. Brandstadt, V. Chepoi, and F. Dragan, Distance approximating trees for chordal and dually chordal graphs, Journal of Algorithms, 30 (1999) 166-184.

[5] L.Cai and D.G. Corneil, Tree Spanners: an Overview, Congressus Numerantium 88 (1992), 65-76.

[6] L. Cai and J. M. Keil, Degree-Bounded Spanners, Parallel Processing Letters, 3(1993), 457-468.

[7] L. Cai and J. M. Keil, Spanners in Graphs of Bounded Degree, Networks, 24(1994),187-194.

[8] L.Cai, NP-completeness of minimum spanner problems, Disc. Appl. Math., 48 (1994), 187-194.

[9] L. Cai and D.G.Corneil, Tree Spanners, SIAM J. Discrete Ma th. 8 (1995) 359-387.

[10] L.P.Chew, There are planar graphs almost as good as the complete graph, J. Comput. Syst. Sci. 39 (1989), 205-219

[11] M. C. Golumbic, Algorithmic Graph Theory and Perfect Graphs. (Academic Press, New York, 1980).

[12] A.L.Liestman and T.C. Schermer, Grid Spanners, Networks, 23 (2) (1993) 123-133.

[13] A. L. Liestman and T. C. Shermer, Additive graph Spanner, Networks, 23 (1993), 343-364.

[14] M.S.Madanlal, G. Venkatesan, and C. Pandu Rangan, Tree 3-spanners on interval, permutation and regular bipartite graphs, Infor. Proc. Lett. 59 (1996) 97-102.

[15] G. Narasimhan, B. Chandra, D. Gautam, and J. Soares, New sparseness results on graph spanners, in 8th Annual ACM Symposium on Computational Geometry (1992) 192-201.

[16] B. S. Panda and S.P.Mohanty, Intersection graphs of vertex disjoint paths in a tree, Discrete Math. 146 (1995) 179-209.

[17] D. Peleg and J. D. Ullman, An optimal Synchronizer for the hypercube, proceedings of the 6th ACM Symposium on principles of Distributed computing, Vancouver (1987) 77-85.

[18] D. Peleg and E. Upfal, A trade off between space and efficiency for routing tables, Proceedings of the 20th ACM Symposium on Theory of Computing, Chicago (1988), 43-52.

[19] D. Peleg and A. A. Schäffer, Graph spanners. J. Graph Theory 13 (1989) 99-116.

[20] P.H.A.Sneath and R.R. Sokal, Numerical Taxonomy, San Francisco, 1973.

[21] D.L.Swofford and G.J.Olsen, Phylogeny reconstruction, in (D.M.Hills and C. Moritz, eds.), Molecular Systematics, pp. 411-501, Sinauer Associates, Sunderland, MA, 1990.

[22] G. Venkatesan, U.Rotics, M. Madanlal, J.A. Makowsky, and Pandu Rangan, Restrictions of minimum spanner problems, Information and computation, 136(2)(1997)143-164.

Appendix

Proof of Lemma 3.4:
Proof:(By contradiction) Let $\{a, b, c, d, e, f\}$ be the vertices of an interior triangle $\{a, b, c\}$ such that d is adjacent to a and b, e is adjacent to b and c, and f is adjacent to a and c.
Case 1: None of the edges ab, bc and ca is present in T.

So, $d_T(a, b) \geq 2$ and $d_T(b, c) \geq 2$. Let $P(a, b)$ and $P(b, c)$ be the paths in T from a to b and b to c, respectively. If $P(a, b)$ and $P(b, c)$ have no edges in common, then clearly $d_T(a, c) > 3$, a contradiction to the fact that T is a tree 3-spanner of G. So, assume that $P(a, b)$ has an edge in common with $P(b, c)$. If $d_T(a, b) = d_T(b, c) = 2$ and xb is the common edge, then $\{a, b, c, x\}$ is a K_4, a complete graph on four vertices, which is a contradiction to the fact that G is a 2-tree. So, either $d_T(a, b) > 2$ or $d_T(b, c) > 2$. If possible $d_T(a, b) = 3$ and $d_T(b, c) = 2$. Let $P(a, b) = a, x, y, b$ and $P(b, c) = b, y, c$. Now a, x, y, b, a is a cycle of length 4. So, either $ay \in E(G)$ or $bx \in E(G)$. If $ay \in E(G)$, then $\{a, y, b, c\}$ is a K_4, which is a contradiction to the fact that G is a 2-tree. If possible, $xb \in E(G)$. Now, a, x, y, c, a is a cycle of length 4. So, either $xc \in E(G)$ or $ay \in E(G)$. As we have seen $ay \in E(G)$ leads to a contradiction, so $xc \in E(G)$. Now $\{a, b, c, x\}$ is a K_4. Hence a contradiction.

So $d_T(a, b) = 3$ and $d_T(b, c) = 3$. Let $P(a, b) = a, x, y, b$. Suppose $P(b, c)$ has exactly one edge in common with $P(a, b)$. Since, ab and bx can not be present in T, $P(b, c)$ will be of the form b, y, z, c. Now, $d_T(a, c) > 3$, which is a contradiction to the fact that T is a tree 3-spanner of G. So, assume that $P(b, c)$ has two edges in common with $P(a, b)$. Let $P(b, c) = b, y, x, c$. Now, b, y, x, c, b is a cycle of length 4. So, either $xb \in E(G)$ or $cy \in E(G)$. If $bx \in E(G)$, then $\{a, b, c, x\}$ is a K_4, which is a contradiction to the fact that G is a 2-tree. So, $cy \in E(G)$. Again, a, x, y, b, a is a cycle of length 4. Since, bx is not an edge of G, $ay \in E(G)$. Now, $\{a, b, c, y\}$ is a K_4, which is a contradiction.
Case 2: Exactly one of the edges ab, bc and ca is present in T.

Wlg, $ab \in E(T)$. Now $d_T(b, c) \geq 2$. Suppose $d_T(b, c) = 2$. Let $P(b, c) = b, x, c$. Now, $d_T(f, a) \leq 3$. If $d_T(f, a) = 1$, then $d_T(f, c) > 3$, which is a contradiction. Suppose $d_T(f, a) = 2$. If $P(f, a)$, the path from f to a in T, contains the edge ab, then $P(f, a)$ will be of the form f, b, a. Now, $\{a, b, c, f\}$ is a K_4, which is a contradiction. If $P(f, a)$ does not contain the edge ab, then $d_T(f, c) \geq 4$, which is a contradiction. So, assume that $d_T(f, a) = 3$. If $P(f, a)$ has no edge in common with $P(b, c)$ and it does not contain the edge ab, then $d_T(f, c) > 3$, which is a contradiction. So, assume that $P(a, f) = f, x, b, a$. Now, f, x, b, a is a cycle of length 4. So, either $ax \in E(G)$ or$fb \in E(G)$. If $fb \in E(G)$, then $\{a, b, c, f\}$ is a K_4, which is a contradiction. If $ax \in E(G)$, then $\{a, b, c, x\}$ is a K_4, which is a contradiction. Next, assume that $d_T(b, c) = 3$. So, either $P(b, c) = b, x, y, c$ or $P(b, c) = b, a, x, c$. If $P(b, c) = b, x, y, c$, then $d_T(a, c) = 4$, which is a contradiction. So, $P(b, c) = b, a, x, c$. Now, $d_T(a, c) = 2$ and $ab \in E(T)$. So, we are in the same situation. (i.e, $d_T(b, c) = 2$, $ab \in E(T)$). So, using the similar arguments, it can be shown that this situation is not possible. Since, T can not contain all the three edges ab, bc and ca, exactly two out of these three edges are present in T. □

Proof of Lemma 3.8

Proof: Let xc be a semi-strong edge of a tree 3-spanner admissible 2-tree of G. So, there exists a triangle $\{x, a, c\}$ which is dependent on a semi-forced triangle $\{a, b, c\}$ having forced edges ab and bc. Let T be a tree 3-spanner of G. If possible, xc is not an edge of T. Since, by Lemma 3.5, T contains at least one edge of $\{a, c, x\}$, T contains ax as T can not contain ac. Since, $M(xc) > 1$, there is a triangle $\{x, c, d\}$. Now, T contains an edge of $\{x, c, d\}$. Since, xc is not in T, wlg, $cd \in T$. Now, x, a, b, c, d is the unique path in T from x to d. So, there is no path in T of length at most 3 from x to d. This is a contradiction to the fact that T is tree 3-spanner. So, T must contain xc. Hence, semi-strong edges are forced edges. □

Proof of Lemma 3.9

Proof: Since G is a tree 3-spanner admissible 2-tree, G_1 is also tree 3-spanner admissible 2-tree. Since $\{a, b, c\}$ is an interior triangle of G, by lemma 2.4, G has a tree 3-spanner T^* containing two of the three edges ab, bc, and ca. Let T=$T^* - \{e, f\}$. Then, T is a tree 3-spanner of H_1 containing two of the three edges ab, bc and ca.

Part (α) Assume that $\{a, b, d\}$ is an interior triangle in G_1.

Case 1: T contains ab and ac.

Let T_1=T and T_2=$(T-\{ac\})\cup\{bc\}$. Then clearly, T_1 and T_2 are the required trees.

Case II: T contains ab and bc.

Let T_2=T and T_1=$(T-\{bc\})\cup\{ac\}$. Then clearly T_1 and T_2 are the required trees.

Case III: T contains ac and bc.

Since, $\{a, b, d\}$ is an interior triangle, by lemma 2.4, T will contain two of the three edges ab, ad and bd. Since T contains ac and bc, T can not contain ab. So T has to contain ab and bd. Then a, c, b, d, a is a cycle in T. This is a contradiction to the fact that T is a tree. Hence, T can not contain ac and bc simultaneously.

So, Lemma 3.9 part (α) is proved. Next, we prove part (β).

Part (β) Now $\{a, b, d\}$ is not an interior triangle of G_1 and ab is not a semi-strong edge.

case I: T contains ab and ac.

Now, T_1=$(T - \{ac\}) \cup \{bc\}$ is a tree 3-spanner of G_1 containing ab and bc. So, G_1 has a tree 3-spanner containing ab and bc. Next, we show that T has a tree 3-spanner containing ac and bc. Since, $\{a, b, d\}$ is not an interior triangle, either ad or bd is not developing. Wlg, bd is not developing. If T contains bd, then it can not contain ad as it also contains ab. So, if T contains bd, consider the tree obtained from T by deleting bd and adding ad. Clearly, this tree is a tree 3-spanner of G. So, wlg, assume that T does not contain bd. If possible, let T do not contain da. Since, ab is not a semi-strong edge, none of the interior triangle containing ad contains two strong edges. So, if there is an interior triangle containing da, then there will be a tree 3-spanner of G_1 containing da. So, if possible, assume that there is no interior triangle containing da. Since, bd is not an edge in T, T contains two of the edges of some triangle containing da. Let T contain az and zd of the triangle $\{a, z, d\}$. Since, none of the triangle containing da is an interior triangle, either az or dz is not a developing edge. Wlg, az is not a developing edge. Now, $(T - \{az\}) \cup \{da\}$ is a tree 3-spanner containing the edge da. So, wlg, assume that T contains da. Consider T_3= i$(T - \{ab\}) \cup \{bc\}$. Now we claim that T_3 is tree 3-spanner of G_1. Let xy be an edge of G_1. Let $P(x, y)$ be the path from x to y in T. Since T is a tree 3-spanner of G_1, $length(P(x, y)) < 4$. If $P(x, y)$ does not contain ab, then $P(x, y)$ is also a path in T_3 of length at most 3. So assume that $P(x, y)$ contains ab. Let L be the length of $P(x, y)$. If $L = 1$, then wlg, $x = a$ and $y = b$. Now $P'(x, y) = x, c, y$ is a path of length 2 in T_3. If $L = 2$, then wlg, $y = b$. Let $P'(x, y) = x, a, c, y$. Then $P'(x, y)$ is a path of length 3 in T_3. Assume that $L = 3$. Since bz is not an edge in T for every $z \in V(G_1) - \{a, c, d\}$, ether $x = b$ or $y = b$ or$x, y \in \{e, c\}$. If either $x = b$ or $y = b$ then the other vertex has to be one of a, d, or c. Since, b, a, d is a path of length 2 in T_3, $\{x, y\} \neq \{b, d\}$. If $x = b$ and $y = c$, then it is easy to see that T_3 contains a path of length at most 3 from x to y. Hence, T_3 is a tree 3-spanner of G_1 containing ac and bc. So, if G_1 has a tree containing ab and ac, then it has also tree 3-spanners containing ab and bc, and it has a tree 3-spanner containing ac and bc.

case II: T contains ab and bc.

Now, T_1=(T-{bc}) ∪ {ac} is a tree 3-spanner of G_1 containing ab and ac. So, by case I of (b) above, G_1 will have a tree 3-spanner containing ac and bc.

case III: T contains ac and bc.

Let T_1=(T-{bc}) ∪ {ab}. It is easy to check that T_1 is a tree 3-spanner of G_1.

Hence, G_1 contains a tree 3-spanner containing any two of the three edges ab, bc, and ca. □

Proof of Lemma 3.10: We propose a method to construct a tree 3-spanner T. Let $T = \phi$. Start with some simplicial edge, say ab. Include ab to T. Let $\{a, b, c\}$ be the triangle containing ab. Since, G has no interior triangle, either ac or bc is not developing. If G has more than 3 vertices, then either ac or bc is a developing edge with respect to the current edge ab and current triangle $\{a, b, c\}$. Wlg, bc is developing. Then add bc to T. Now make bc as the current edge and consider all the triangles which are not considered. Let $\{b, c, d\}$ be any triangle containing bc. If this is a simplicial triangle, then either add bd or add cd to T. If $\{b, c, d\}$ is not a simplicial triangle, then add the developing edge of $\{b, c, d\}$ to T. Continue this process till all the triangles are considered. We claim that the tree T so obtained is a tree 3-spanner of G. Let xy be an edge of G. Let $\{x,y,z\}$ be the triangle which was considered earlier by the above method than any other triangle containing xy. If $\{x,y,z\}$ is simplicial, then two of the edges of the triangle $\{x,y,z\}$ are included in T. Then $d_T(x,y) < 3$. Suppose, $\{x,y,z\}$ is not a simplicial triangle. If xy is included in T, then $d_T(x,y) = 1$. If it is not included, wlg, yz was the current edge when $\{x,y,z\}$ was considered by the algorithm. Then, xz is a developing edges and so is included in T. So, $d_T(x,y) < 3$. So, T is a tree 3-spanner of G. □

Proof of Theorem 3.11:

Proof: Necessity: Necessity follows from Lemma 3.6 and Lemma 3.8.

Sufficiency: We will prove it by induction. Let n be the number of vertices of G. For $n = 4, 5$, G has at most three triangles. So, G has no interior triangles. Hence, by Lemma 3.10, G contains a tree 3-spanner. Assume that every 2-tree having n or fewer vertices without containing a strong triangle has a tree 3-spanner.

Let G be a 2-tree having $n+1$ vertices and does not have any induced strong triangle. If G has no interior triangle, then by Lemma 3.10, G has a tree 3-spanner. A Assume that G has an interior triangle, say $\{a,b,c\}$.

Let $G_1, G_2, \ldots, G_r$, $r \geq 3$ be the separated graphs of G with respect to $\{a,b,c\}$. Let $S_1 = \{G_i | W(G_i) = \{a,b\}\}$, S_2=$\{G_i | W(G_i) = \{a,c\}\}$, and S_3= $\{G_i | W(G_i) = \{b,c\}\}$. Note that none of the S_i is empty. Since, G does not contain any strong triangle, one of the edges ab, bc, and ca is neither a strong edge nor a semi-strong edge. Wlg, ab is neither a strong edge nor a semi-strong edge. So, for all j , $1 \leq j \leq r$, $G_j \in S_1$ implies the principal clique of G_j, which is a triangle, is not interior. Let $\{b,c,f\}$ and $\{a,c,e\}$ be the principal cliques of G_1 and G_2, where $G_1 \in S_2$ and $G_2 \in S_3$.

Case I: Either $|S_i| \geq 2$ for some i, $1 \leq i \leq 3$ or $|V(G_i) - \{a,b,c\}| \geq 2|$ for some i, $1 \leq i \leq r$, where G_i belongs to S_j, $j = 2, 3$.

Now, for each $G_i \in S_1, G' = G[\{e,f\} \cup V(G_i)]$ is a 2- tree without strong triangle with fewer vertices than G. So, by induction hypothesis, G' is tree 3-spanner admissible 2-tree. Now, the set of separated graphs of G' with respect to $\{a,b,c\}$ is $\{G_i, G_1', G_2'\}$, where $G_1' = G[\{a,b,c,e\}]$ and $G_2' = G[\{a,b,c,f\}]$. So, by Lemma 3.9, each $G_i \in S_1$ has a tree 3-spanner containing the edges ac and bc. By similar arguments and by Lemma 3.9, each $G_i \in S_2$ and each $G_j \in S_3$ has a tree 3-spanner containing the edges ac and bc. The union of these trees is a tree. This tree is the tree 3-spanner of G. Hence, by induction principle, every 2-tree which does not contain any strong triangle as an induced subgraph admits a tree 3-spanner.

Case II: Each $S_i, 1 \leq i \leq 3$ is singleton and $|V(G_j) - \{a,b,c\}| = 1$ for each G_j in S_i, $i = 2, 3$.

Consider $G' = G[\{e,f\} \cup V(G_1)]$, $G_1 \in S_1$. Note that, $G' = G$. Let $G'' = G - \{e\}$. Since, G'' is a 2-tree having n vertices without containing a strong triangle, G'' is a tree 3-spanner admissible by induction hypothesis. Let T be a tree 3-spanner of G''. Let $P(a,c)$ be the path from a to c in T. Now, length of $P(a,c) \leq 3$. If possible, length of $P(a,c) = 3$. Then, $P(a,c) = a,b,f,c$ or $P(a,c) = a,d,f,c$. If $P(a,c) = a,b,f,c$, then $T' = (T - \{bf\}) \cup \{bc\}$ is a tree

3-spanner of G'' and length of the path $P(a,c)$ in T' is 2. Suppose, $P(a,c) = a,d,b,c$. Since, $\{a,d,b\}$ is not an interior triangle, either bd or ad is not a developing edge. Wlg, bd is not a developing edge. Then, $T' = T - \{bd\} \cup \{ab\}$ is a tree 3-spanner of G'' such that there is a path of length 2 from a to c in T'. So, wlg, T has a path $P(a,c)$ from a to c of length one or two. Now $T \cup \{ae\}$ is a tree 3-spanner of G'. So, by induction principle, every 2-tree which does not contain a strong triangle as an induced subgraph admits a tree 3-spanner. □

EXACT COMPLEXITY OF EXACT-FOUR-COLORABILITY AND OF THE WINNER PROBLEM FOR YOUNG ELECTIONS*

Jörg Rothe and Holger Spakowski
Abteilung für Informatik
Heinrich-Heine-Universität Düsseldorf
40225 Düsseldorf
Germany
rothe@cs.uni-duesseldorf.de and spakowsk@cs.uni-duesseldorf.de

Jörg Vogel
Institut für Informatik
Friedrich-Schiller-Universität Jena
07740 Jena, Germany
Germany
vogel@minet.uni-jena.de

Abstract We classify two problems: `Exact-Four-Colorability` and the winner problem for Young elections. Regarding the former problem, Wagner raised the question of whether it is DP-complete to determine if the chromatic number of a given graph is exactly four. We prove a general result that in particular solves Wagner's question in the affirmative.

In 1977, Young proposed a voting scheme that extends the Condorcet Principle based on the fewest possible number of voters whose removal yields a Condorcet winner. We prove that both the winner and the ranking problem for Young elections is complete for $\mathrm{P}_{\|}^{\mathrm{NP}}$, the class of problems solvable in polynomial time by parallel access to NP. Analogous results for Lewis Carroll's 1876 voting scheme were recently established by Hemaspaandra et al. In contrast, we prove that the winner and ranking problems in Fishburn's homogeneous variant of Carroll's voting scheme can be solved efficiently by linear programming.

Keywords: Computational complexity; graph colorability; completeness; boolean hierarchy; voting schemes.

*Supported in part by grant NSF-INT-9815095/DAAD-315-PPP-gü-ab.

1. Introduction

In this paper, we classify two problems from different fields with respect to their computational complexity: `Exact-Four-Colorability` and the winner problem for Young elections.

Sections 2 and 3 are concerned with `Exact-Four-Colorability`. Let $M_k \subseteq \mathbb{N}$ be a given set that consists of k noncontiguous integers. Exact-M_k-Colorability is the problem of determining whether $\chi(G)$, the chromatic number of a given graph G, equals one of the k elements of the set M_k exactly. In 1987, Wagner [27] proved that Exact-M_k-Colorability is $\mathrm{BH}_{2k}(\mathrm{NP})$-complete, where $M_k = \{6k+1, 6k+3, \ldots, 8k-1\}$ and $\mathrm{BH}_{2k}(\mathrm{NP})$ is the $2k$th level of the boolean hierarchy over NP. In particular, for $k = 1$, it is DP-complete to determine whether $\chi(G) = 7$, where $\mathrm{DP} = \mathrm{BH}_2(\mathrm{NP})$. Wagner raised the question of how small the numbers in a k-element set M_k can be chosen such that Exact-M_k-Colorability still is $\mathrm{BH}_{2k}(\mathrm{NP})$-complete. In particular, for $k = 1$, he asked if it is DP-complete to determine whether $\chi(G) = 4$.

In Section 3, we solve this question of Wagner and determine the precise threshold $t \in \{4, 5, 6, 7\}$ for which the problem Exact-$\{t\}$-Colorability jumps from NP to DP-completeness: It is DP-complete to determine whether $\chi(G) = 4$, yet Exact-$\{3\}$-Colorability is in NP. More generally, for each $k \geq 1$, we show that Exact-M_k-Colorability is $\mathrm{BH}_{2k}(\mathrm{NP})$-complete for $M_k = \{3k+1, 3k+3, \ldots, 5k-1\}$.

Sections 4 and 5 are concerned with complexity issues related to voting schemes. More than a decade ago, Bartholdi, Tovey, and Trick initiated the study of electoral systems with respect to their computational properties. In particular, they proved NP hardness lower bounds [2] for determining the winner in the voting schemes proposed by Dodgson (more commonly known by his pen name, Lewis Carroll) and by Kemeny. Since then, a number of related results and improvements of their results have been obtained. Hemaspaandra, Hemaspaandra, and Rothe [15] classified both the winner and the ranking problem for Dodgson elections by proving them complete for $\mathrm{P}_{\|}^{\mathrm{NP}}$, the class of problems solvable in polynomial time by parallel access to an NP oracle. E. Hemaspaandra (as cited in [14]) and Spakowski and Vogel [26] obtained the analogous result for Kemeny elections; a joint paper by E. Hemaspaandra, Spakowski, and Vogel is in preparation. For many further interesting results and the state of the art regarding computational politics, we refer to the survey [14].

In this paper, we study complexity issues related to Young and Dodgson elections. In 1977, Young proposed a voting scheme that extends the Condorcet Principle based on the fewest possible number of voters whose removal makes a given candidate c the Condorcet winner, i.e., c defeats all other candidates by a strict majority of the votes. We prove that both the winner and the ranking problem for Young elections is complete for $\mathrm{P}_{\|}^{\mathrm{NP}}$. To this end, we give a reduction from the problem `Maximum Set Packing Compare`, which we also prove $\mathrm{P}_{\|}^{\mathrm{NP}}$-complete.

In Section 5, we study a homogeneous variant of Dodgson elections that was introduced by Fishburn [9]. In contrast to the above-mentioned result of Hemaspaandra et al. [15], we show that both the winner and the ranking problem for Fishburn's homogeneous Dodgson elections can be solved efficiently by a linear program that is based on an integer linear program of Bartholdi et al. [2].

2. Exact-M_k-Colorability and the Boolean Hierarchy over NP

To classify the complexity of problems known to be NP-hard or coNP-hard, but seemingly not contained in NP ∪ coNP, Papadimitriou and Yannakakis [22] introduced DP, the class of differences of two NP problems. They showed that DP contains various interesting types of problems, including *uniqueness problems*, *critical graph problems*, and *exact optimization problems*. For example, Cai and Meyer [5] proved the DP-completeness of `Minimal-3-Uncolorability`, a critical graph problem that asks whether a given graph is not 3-colorable, but deleting any of its vertices makes it 3-colorable. A graph is said to be *k-colorable* if its vertices can be colored using no more than k colors such that no two adjacent vertices receive the same color. The *chromatic number of a graph G*, denoted $\chi(G)$, is defined to be the smallest k such that G is k-colorable. Generalizing DP, Cai et al. [4] defined and studied the boolean hierarchy over NP. Their work initiated many further results on the boolean hierarchy; see e.g., [27, 20, 28, 17] to name just a few. To define the boolean hierarchy, we use the symbols $\wedge$ and $\vee$, respectively, to denote the *complex intersection* and the *complex union* of set classes.

Definition 1 [4] *The* boolean hierarchy over NP *is inductively defined as follows:*

$$\begin{aligned} \mathrm{BH}_1(\mathrm{NP}) &= \mathrm{NP}, \qquad \mathrm{BH}_2(\mathrm{NP}) = \mathrm{NP} \wedge \mathrm{coNP}, \\ \mathrm{BH}_k(\mathrm{NP}) &= \mathrm{BH}_{k-2}(\mathrm{NP}) \vee \mathrm{BH}_2(\mathrm{NP}) \quad \textit{for } k \geq 3, \textit{ and} \\ \mathrm{BH}(\mathrm{NP}) &= \bigcup_{k\geq 1} \mathrm{BH}_k(\mathrm{NP}). \end{aligned}$$

Equivalent definitions in terms of different boolean hierarchy normal forms can be found in the papers [4, 27, 20]; for the boolean hierarchy over arbitrary set rings, we refer to the early work by Hausdorff [13]. Note that $\mathrm{DP} = \mathrm{BH}_2(\mathrm{NP})$.

In his seminal paper [27], Wagner provided sufficient conditions to prove problems complete for the levels of the boolean hierarchy. In particular, he established the following lemma for $\mathrm{BH}_{2k}(\mathrm{NP})$.

Lemma 2 **[27, Thm. 5.1(3)]** *Let A be some* NP*-complete problem, let B be an arbitrary problem, and let $k \geq 1$ be fixed. If there exists a polynomial-time computable function f such that, for all strings $x_1, x_2, \ldots, x_{2k} \in \Sigma^*$ satisfying* *that $x_{j+1} \in A$ implies $x_j \in A$ for each j with $1 \leq j < 2k$, it holds that*

$$\|\{i \mid x_i \in A\}\| \text{ is odd} \iff f(x_1, x_2, \ldots, x_{2k}) \in B, \tag{1}$$

then B is $\mathrm{BH}_{2k}(\mathrm{NP})$*-hard.*

For fixed $k \geq 1$, let $M_k = \{6k+1, 6k+3, \ldots, 8k-1\}$, and define the problem `Exact-`M_k`-Colorability` $= \{G \mid \chi(G) \in M_k\}$. In particular, Wagner applied Lemma 2 to prove that, for each $k \geq 1$, `Exact-`M_k`-Colorability` is $\mathrm{BH}_{2k}(\mathrm{NP})$-complete. For the special case of $k = 1$, it follows that `Exact-`$\{7\}$`-Colorability` is DP-complete.

Wagner [27, p. 70] raised the question of how small the numbers in a k-element set M_k can be chosen such that `Exact-`M_k`-Colorability` still is $\mathrm{BH}_{2k}(\mathrm{NP})$-complete. Consider the special case of $k = 1$. It is easy to see that `Exact-`$\{3\}$`-Colorability` is in NP and, thus, cannot be DP-complete unless the boolean hierarchy collapses; see Proposition 3 below. Consequently, for $k = 1$, Wagner's result leaves a gap in determining the precise threshold $t \in \{4, 5, 6, 7\}$ for which `Exact-`$\{t\}$`-Colorability` jumps from NP to DP-completeness. Closing this gap, we show that it is DP-complete to determine whether $\chi(G) = 4$. More generally, answering Wagner's question for each $k \geq 1$, we show that `Exact-`M_k`-Colorability` is $\mathrm{BH}_{2k}(\mathrm{NP})$-complete for $M_k = \{3k+1, 3k+3, \ldots, 5k-1\}$.

3. Solving Wagner's Question

Proposition 3 *Fix any $k \geq 1$, and let M_k be any set that contains k noncontiguous positive integers including 3. Then,* `Exact-`M_k`-Colorability` *is in* $\mathrm{BH}_{2k-1}(\mathrm{NP})$*; in particular, for $k = 1$,* `Exact-`$\{3\}$`-Colorability` *is in* NP.

Hence, `Exact-`M_k`-Colorability` is not $\mathrm{BH}_{2k}(\mathrm{NP})$-complete unless the boolean hierarchy, and consequently the polynomial hierarchy, collapses. Proposition 3 easily follows from the fact that it can be tested in polynomial time whether a given graph is 2-colorable; for details of the proof, see [23].

Theorem 4 *For fixed $k \geq 1$, let $M_k = \{3k+1, 3k+3, \ldots, 5k-1\}$. Then,* `Exact-`$M_k$`-Colorability` *is* $\mathrm{BH}_{2k}(\mathrm{NP})$*-complete. In particular, for $k = 1$, it follows that* `Exact-`$\{4\}$`-Colorability` *is* DP*-complete.*

Proof. We apply Lemma 2 with A being the NP-complete problem `3-SAT` and B being `Exact-`M_k`-Colorability`, where $M_k = \{3k+1, 3k+3, \ldots, 5k-1\}$ for fixed k. The standard reduction σ from `3-SAT` to `3-Colorability` has the following property [10]:

$$\phi \in \texttt{3-SAT} \implies \chi(\sigma(\phi)) = 3 \quad \text{and} \quad \phi \notin \texttt{3-SAT} \implies \chi(\sigma(\phi)) = 4. \tag{2}$$

Using the PCP theorem, Khanna, Linial, and Safra [19] showed that it is NP-hard to color a 3-colorable graph with only four colors. Guruswami and Khanna [11] gave a novel proof of the same result that does not rely on the

PCP theorem. We use their direct transformation, call it ρ, that consists of two subsequent reductions—first from 3-SAT to the independent set problem, and then from the independent set problem to 3-Colorability—such that $\phi \in$ 3-SAT implies $\chi(\rho(\phi)) = 3$, and $\phi \notin$ 3-SAT implies $\chi(\rho(\phi)) \geq 5$. Guruswami and Khanna [11] note that the graph $H = \rho(\phi)$ they construct always is 6-colorable. In fact, their construction even gives that H always is 5-colorable; hence, we have:

$$\phi \in \text{3-SAT} \Longrightarrow \chi(\rho(\phi)) = 3 \quad \text{and} \quad \phi \notin \text{3-SAT} \Longrightarrow \chi(\rho(\phi)) = 5. \tag{3}$$

To see why, look at the reduction in [11]. The graph H consists of tree-like structures whose vertices are replaced by 3×3 grids, which always can be colored with three colors, say 1, 2, and 3. In addition, some leaves of the tree-like structures are connected by leaf-level gadgets of two types, the "same row kind" and the "different row kind." The latter gadgets consist of two vertices connected to some grids, and thus can always be colored with two additional colors. The leaf-level gadgets of the "same row kind" consist of a triangle whose vertices are adjacent to two grid vertices each. Hence, regardless of which 3-coloring is used for the grids, one can always color one triangle vertex, say t_1, with a color $c \in \{1,2,3\}$ such that c is different from the colors of the two grid vertices adjacent to t_1. Using two additional colors for the other two triangle vertices implies $\chi(H) \leq 5$, which proves Equation (3).

The join operation $\oplus$ on graphs is defined as follows: Given two disjoint graphs $A = (V_A, E_A)$ and $B = (V_B, E_B)$, their join $A \oplus B$ is the graph with vertex set $V_{A\oplus B} = V_A \cup V_B$ and edge set $E_{A\oplus B} = E_A \cup E_B \cup \{\{a,b\} \mid a \in V_A \text{ and } b \in V_B\}$. Note that $\oplus$ is an associative operation on graphs and $\chi(A \oplus B) = \chi(A) + \chi(B)$.

Let $\phi_1, \phi_2, \ldots, \phi_{2k}$ be $2k$ given boolean formulas satisfying $\phi_{j+1} \in$ 3-SAT $\Longrightarrow$ $\phi_j \in$ 3-SAT for each j with $1 \leq j < 2k$. Define $2k$ graphs $H_1, H_2, \ldots, H_{2k}$ as follows. For each i with $1 \leq i \leq k$, define $H_{2i-1} = \rho(\phi_{2i-1})$ and $H_{2i} = \sigma(\phi_{2i})$. By Equations (2) and (3),

$$\chi(H_j) = \begin{cases} 3 & \text{if } 1 \leq j \leq 2k \text{ and } \phi_j \in \text{3-SAT} \\ 4 & \text{if } j = 2i \text{ for some } i \in \{1,2,\ldots,k\} \text{ and } \phi_j \notin \text{3-SAT} \\ 5 & \text{if } j = 2i-1 \text{ for some } i \in \{1,2,\ldots,k\} \text{ and } \phi_j \notin \text{3-SAT}. \end{cases} \tag{4}$$

For each i with $1 \leq i \leq k$, define the graph G_i to be the disjoint union of the graphs H_{2i-1} and H_{2i}. Thus, $\chi(G_i) = \max\{\chi(H_{2i-1}), \chi(H_{2i})\}$, for each i with $1 \leq i \leq k$. The construction of our reduction f is completed by defining $f(\phi_1, \phi_2, \ldots, \phi_{2k}) = G$, where the graph $G = \bigoplus_{i=1}^{k} G_i$ is the join of the graphs $G_1, G_2, \ldots, G_k$. Thus,

$$\chi(G) = \sum_{i=1}^{k} \chi(G_i) = \sum_{i=1}^{k} \max\{\chi(H_{2i-1}), \chi(H_{2i})\}. \tag{5}$$

It follows from our construction that

$$
\begin{aligned}
&\|\{i \mid \phi_i \in \texttt{3-SAT}\}\| \text{ is odd} \\
&\iff (\exists i : 1 \leq i \leq k)\,[\phi_1, \ldots, \phi_{2i-1} \in \texttt{3-SAT} \text{ and } \phi_{2i}, \ldots, \phi_{2k} \notin \texttt{3-SAT}] \\
&\overset{(4),(5)}{\iff} (\exists i : 1 \leq i \leq k) \left[\sum_{j=1}^{k} \chi(G_j) = 3(i-1) + 4 + 5(k-i) = 5k - 2i + 1\right] \\
&\overset{(5)}{\iff} \chi(G) \in M_k = \{3k+1, 3k+3, \ldots, 5k-1\} \\
&\iff f(\phi_1, \phi_2, \ldots, \phi_{2k}) = G \in \texttt{Exact-}M_k\texttt{-Colorability}.
\end{aligned}
$$

Hence, Equation (1) is satisfied. Lemma 2 implies that Exact-M_k-Colorability is $\mathrm{BH}_{2k}(\mathrm{NP})$-complete. ∎

And now for something completely different [6]: Voting schemes.

4. Hardness of Determining Young Winners

We first give some background from social choice theory. Let C be the set of all candidates (or alternatives). We assume that each voter has strict preferences over the candidates. Formally, the preference order of each voter is strict (i.e., irreflexive and antisymmetric), transitive, and complete (i.e., all candidates are ranked by each voter). An election is given by a *preference profile*, a pair $\langle C, V\rangle$ such that C is a set of candidates and V is the multiset of the voters' preference orders on C. Note that distinct voters may have the same preferences over the candidates. A *voting scheme* (or *social choice function*, SCF for short) is a rule for how to determine the winner(s) of an election; i.e., an SCF maps any given preference profile to society's aggregate *choice set*, the set of candidates who have won the election. For any SCF f and any preference profile $\langle C, V\rangle$, $f(\langle C, V\rangle)$ denotes the set of winning candidates. For example, an election is won according to the *majority rule* by any candidate who is preferred over any other candidate by a strict majority of the voters. Such a candidate is called the Condorcet winner. In 1785, Marie-Jean-Antoine-Nicolas de Caritat, the Marquis de Condorcet, noted in his seminal essay [7] that whenever there are at least three candidates, say A, B, and C, the majority rule may yield cycles: A defeats B and B defeats C, and yet C defeats A. Thus, even though each individual voter has a rational (i.e., transitive or non-cyclic) preference order, society may behave irrationally and Condorcet winners do not always exist. This observation is known as the Condorcet Paradox. The *Condorcet Principle* says that for each preference profile, the winner of the election is to be determined by the majority rule. An SCF is said to be a *Condorcet SCF* if and only if it respects the Condorcet Principle in the sense that the Condorcet winner is elected whenever he or she exists. Note that Condorcet winners are uniquely determined if they exist. Many Condorcet SCFs have been proposed in the social choice literature; for an overview of the most central ones, we refer to the work of Fishburn [9]. They extend the Condorcet Principle in a way that avoids the troubling feature of the majority rule. In this paper, we will focus on only two such Condorcet SCFs, the Dodgson voting scheme [8] and the Young voting scheme [29].

In 1876, Charles L. Dodgson (better known by his pen name, Lewis Carroll) proposed a voting scheme [8] that suggests that we remain most faithful to the Condorcet Principle if the election is won by any candidate who is "closest" to being a Condorcet winner. To define "closeness," each candidate c in a given election $\langle C, V\rangle$ is assigned a score, denoted DodgsonScore(C, c, V), which is the smallest number of sequential interchanges of adjacent candidates in the voters' preferences that are needed to make c a Condorcet winner. Here, one interchange means that in (any) one of the voters two adjacent candidates are switched. A *Dodgson winner* is any candidate with minimum Dodgson score. Using Dodgson scores, one can also tell who of two given candidates is ranked better according to the Dodgson SCF.

Young's approach to extending the Condorcet Principle is reminiscent of Dodgson's approach in that it is also based on altered profiles. Unlike Dogson, however, Young [29] suggests that we remain most faithful to the Condorcet Principle if the election is won by any candidate who is made a Condorcet winner by *removing the fewest possible number of voters*, instead of doing the fewest possible number of switches in the voters' preferences. For each candidate c in a given preference profile $\langle C, V\rangle$, define YoungScore(C, c, V) to be the size of a largest subset of V for which c is a Condorcet winner. A *Young winner* is any candidate with a maximum Young score. Homogeneous variants of these voting schemes will be defined in Section 5.

To study computational complexity issues related to Dodgson's voting scheme, Bartholdi, Tovey, and Trick [2] defined the following decision problems.

Dodgson Winner

Instance: A preference profile $\langle C, V\rangle$ and a designated candidate $c \in C$.
Question: Is c a Dodgson winner of the election? That is, is it true that for all $d \in C$, DodgsonScore$(C, c, V) \leq$ DodgsonScore(C, d, V)?

Dodgson Ranking

Instance: A preference profile $\langle C, V\rangle$ and two designated candidates $c, d \in C$.
Question: Does c tie-or-defeat d in the election? That is, is it true that

$$\text{DodgsonScore}(C, c, V) \leq \text{DodgsonScore}(C, d, V)\ ?$$

Bartholdi et al. [2] established an NP-hardness lower bound for both these problems. Their result was optimally improved by Hemaspaandra, Hemaspaandra, and Rothe [15] who proved that **Dodgson Winner** and **Dodgson Ranking** are complete for $\mathrm{P}_{\|}^{\mathrm{NP}}$, the class of problems solvable in polynomial time with parallel (i.e., truth-table) access to an NP oracle. As above, we define the corresponding decision problems for Young elections as follows.

`Young Winner`

Instance: A preference profile $\langle C, V\rangle$ and a designated candidate $c \in C$.
Question: Is c a Young winner of the election? That is, is it true that for all $d \in C$, YoungScore$(C, c, V) \geq$ YoungScore(C, d, V)?

`Young Ranking`

Instance: A preference profile $\langle C, V\rangle$ and two designated candidates $c, d \in C$.
Question: Does c tie-or-defeat d in the election? That is, is it true that

$$\text{YoungScore}(C, c, V) \geq \text{YoungScore}(C, d, V) \text{ ?}$$

The main result in this section is that the problems `Young Winner` and `Young Ranking` are complete for $P_{||}^{NP}$. In Theorem 6 below, we give a reduction from the problem `Maximum Set Packing Compare` defined below. For a given familiy $\mathcal{S}$ of sets, let $\kappa(\mathcal{S})$ be the maximum number of pairwise disjoint sets in $\mathcal{S}$.

`Maximum Set Packing Compare`

Instance: Two families $\mathcal{S}_1$ and $\mathcal{S}_2$ of sets such that, for $i \in \{1, 2\}$, each set $S \in \mathcal{S}_i$ is a nonempty subset of a given set B_i.
Question: Does it hold that $\kappa(\mathcal{S}_1) \geq \kappa(\mathcal{S}_2)$?

Theorem 5 `Maximum Set Packing Compare` *is* $P_{||}^{NP}$*-complete.*

Theorem 5 is proven (see the full version [24] for details) via a reduction from `Independence Number Compare`, which in turn can be shown $P_{||}^{NP}$-complete by the techniques of Wagner [27]; see [25, Thm. 12] for an explicit proof of this result. `Independence Number Compare` has also been used in [16]. To define the problem, let G be an undirected, simple graph. An *independent set of* G is any subset I of the vertex set of G such that no two vertices in I are adjacent. For any graph G, let $\alpha(G)$ be the *independence number of* G, i.e., the size of a maximum independent set of G.

`Independence Number Compare`

Instance: Two graphs G_1 and G_2.
Question: Does it hold that $\alpha(G_1) \geq \alpha(G_2)$?

Now, we prove the main result of this section.

Theorem 6 `Young Ranking` *and* `Young Winner` *are* $P_{||}^{NP}$*-complete.*

Proof. It is easy to see that `Young Ranking` and `Young Winner` are in $P_{||}^{NP}$. To prove the $P_{||}^{NP}$ lower bound, we first give a polynomial-time many-one reduction from `Maximum Set Packing Compare` to `Young Ranking`.

Let $B_1 = \{x_1, x_2, \ldots, x_m\}$ and $B_2 = \{y_1, y_2, \ldots, y_n\}$ be two given sets, and let $\mathcal{S}_1$ and $\mathcal{S}_2$ be given families of subsets of B_1 and B_2, respectively. Recall that $\kappa(\mathcal{S}_i)$, for $i \in \{1, 2\}$, is the maximum number of pairwise disjoint sets in $\mathcal{S}_i$; w.l.o.g., we may assume that $\kappa(\mathcal{S}_i) > 2$. We define a preference profile $\langle C, V \rangle$ such that c and d are designated candidates in C, and it holds that:

$$\text{YoungScore}(C, c, V) = 2 \cdot \kappa(\mathcal{S}_1) + 1; \quad (6)$$
$$\text{YoungScore}(C, d, V) = 2 \cdot \kappa(\mathcal{S}_2) + 1. \quad (7)$$

Define the set C of candidates as follows: Create the two designated candidates c and d; for each element x_i of B_1, create a candidate x_i; for each element y_i of B_2, create a candidate y_i; finally, create two auxiliary candidates, a and b.

Define the set V of voters as follows:

- **Voters representing $\mathcal{S}_1$:** For each set $E \in \mathcal{S}_1$, create a single voter v_E as follows:
  - Enumerate E as $\{e_1, e_2, \ldots, e_{\|E\|}\}$ (renaming the candidates e_i from $\{x_1, x_2, \ldots, x_m\}$ for notational convenience), and enumerate its complement $\overline{E} = B_1 - E$ as $\{\overline{e}_1, \overline{e}_2, \ldots, \overline{e}_{m-\|E\|}\}$.
  - To make the preference orders easier to parse, we use

    "$\overrightarrow{E}$" to represent the text string "$e_1 > e_2 > \cdots > e_{\|E\|}$";
    "$\overrightarrow{\overline{E}}$" to represent the text string "$\overline{e}_1 > \overline{e}_2 > \cdots > \overline{e}_{m-\|E\|}$";
    "$\overrightarrow{B_1}$" to represent the text string "$x_1 > x_2 > \cdots > x_m$";
    "$\overrightarrow{B_2}$" to represent the text string "$y_1 > y_2 > \cdots > y_n$".

  - Create one voter v_E with preference order:

$$\overrightarrow{E} > a > c > \overrightarrow{\overline{E}} > \overrightarrow{B_2} > b > d. \quad (8)$$

- Additionally, create two voters with preference order:

$$c > \overrightarrow{B_1} > a > \overrightarrow{B_2} > b > d, \quad (9)$$

  and create $\|\mathcal{S}_1\| - 1$ voters with preference order:

$$\overrightarrow{B_1} > c > a > \overrightarrow{B_2} > b > d. \quad (10)$$

- **Voters representing $\mathcal{S}_2$:** For each set $F \in \mathcal{S}_2$, create a single voter v_F as follows:
  - Enumerate F as $\{f_1, f_2, \ldots, f_{\|F\|}\}$ (renaming the candidates f_j from $\{y_1, y_2, \ldots, y_n\}$ for notational convenience), and enumerate its complement $\overline{F} = B_1 - F$ as $\{\overline{f}_1, \overline{f}_2, \ldots, \overline{f}_{n-\|F\|}\}$.
  - To make the preference orders easier to parse, we use

    "$\overrightarrow{F}$" to represent the text string "$f_1 > f_2 > \cdots > f_{\|F\|}$";
    "$\overrightarrow{\overline{F}}$" to represent the text string "$\overline{f}_1 > \overline{f}_2 > \cdots > \overline{f}_{n-\|F\|}$".

– Create one voter v_F with preference order:

$$\overrightarrow{F} > b > d > \overleftarrow{F} > \overrightarrow{B_1} > a > c. \quad (11)$$

- Additionally, create two voters with preference order:

$$d > \overrightarrow{B_2} > b > \overrightarrow{B_1} > a > c, \quad (12)$$

and create $\|\mathcal{S}_2\| - 1$ voters with preference order:

$$\overrightarrow{B_2} > d > b > \overrightarrow{B_1} > a > c. \quad (13)$$

We now prove Equation (6): $\text{YoungScore}(C, c, V) = 2 \cdot \kappa(\mathcal{S}_1) + 1$.

Let $E_1, E_2, \ldots, E_{\kappa(\mathcal{S}_1)} \in \mathcal{S}_1$ be $\kappa(\mathcal{S}_1)$ pairwise disjoint subsets of B_1. Consider the following subset $\widehat{V} \subseteq V$ of the voters. $\widehat{V}$ consists of: (a) every voter v_{E_i} corresponding to the set E_i, where $1 \leq i \leq \kappa(\mathcal{S}_1)$; (b) the two voters given in Equation (9); and (c) $\kappa(\mathcal{S}_1) - 1$ voters of the form given in Equation (10).

Then, $\|\widehat{V}\| = 2 \cdot \kappa(\mathcal{S}_1) + 1$. Note that a strict majority of the voters in $\widehat{V}$ prefer c over any other candidate, and thus c is a Condorcet winner in $\langle C, \widehat{V} \rangle$. Hence,

$$\text{YoungScore}(C, c, V) \geq 2 \cdot \kappa(\mathcal{S}_1) + 1.$$

Conversely, to prove that $\text{YoungScore}(C, c, V) \leq 2 \cdot \kappa(\mathcal{S}_1) + 1$, we need the following lemma. The proof of Lemma 7 can be found in the full version [24].

Lemma 7 *For any λ with $3 < \lambda \leq \|\mathcal{S}_1\| + 1$, let V_λ be any subset of V such that V_λ contains exactly λ voters of the form (9) or (10) and c is the Condorcet winner in $\langle C, V_\lambda \rangle$. Then, V_λ contains exactly $\lambda - 1$ voters of the form (8) and no voters of the form (11), (12), or (13). Moreover, the $\lambda - 1$ voters of the form (8) in V_λ represent pairwise disjoint sets from $\mathcal{S}_1$.*

To continue the proof of Theorem 6, let $k = \text{YoungScore}(C, c, V)$. Let $\widehat{V} \subseteq V$ be a subset of size k such that c is the Condorcet winner in $\langle C, \widehat{V} \rangle$. Suppose that there are exactly $\lambda \leq \|\mathcal{S}_1\| + 1$ voters of the form (9) or (10) in $\widehat{V}$. Since c, the Condorcet winner of $\langle C, \widehat{V} \rangle$, must in particular outpoll a, we have $\lambda \geq \lceil \frac{k+1}{2} \rceil$. By our assumption that $\kappa(\mathcal{S}_1) > 2$, it follows from $k \geq 2 \cdot \kappa(\mathcal{S}_1) + 1$ that $\lambda > 3$. Lemma 7 then implies that there are exactly $\lambda - 1$ voters of the form (8) in $\widehat{V}$, which represent pairwise disjoint sets from $\mathcal{S}_1$, and $\widehat{V}$ contains no voters of the form (11), (12), or (13). Hence, $k = 2 \cdot \lambda - 1$ is odd, and $\frac{k-1}{2} = \lambda - 1 \leq \kappa(\mathcal{S}_1)$, which proves Equation (6). Equation (7) can be proven analogously. Thus, we have $\kappa(\mathcal{S}_1) \geq \kappa(\mathcal{S}_2)$ if and only if $\text{YoungScore}(C, c, V) \geq \text{YoungScore}(C, d, V)$. Hence, `Young Ranking` is $\mathrm{P}_{\|}^{\mathrm{NP}}$-complete. Modifying the above reduction, we can also prove `Young Winner` $\mathrm{P}_{\|}^{\mathrm{NP}}$-complete, which completes the proof of Theorem 6. For details of the modified reduction, we refer to the full version [24]. ∎

5. Homogeneous Young and Dodgson Voting Schemes

Social choice theorists have studied many "reasonable" properties that any "fair" election procedure arguably should satisfy, including very natural properties such as nondictatorship, monotonicity, the Pareto Principle, and independence of irrelevant alternatives. One of the most notable results in this regard is Arrow's famous Impossibility Theorem [1] stating that the just-mentioned four properties are logically inconsistent, and thus no "fair" voting scheme can exist. In this section, we are concerned with another quite natural property, the homogeneity of voting schemes (see [9, 29]).

Definition 8 *A voting scheme f is said to be* homogeneous *if and only if for each preference profile $\langle C, V\rangle$ and for all positive integers q, it holds that $f(\langle C, V\rangle) = f(\langle C, qV\rangle)$, where qV denotes V replicated q times.*

Homogeneity means that splitting each voter $v \in V$ into q voters, each of whom has the same preference order as v, yields exactly the same choice set of winning candidates. Fishburn [9] showed that neither the Dodgson nor the Young voting schemes are homogeneous. For the Dodgson SCF, he presented a counterexample with seven voters and eight candidates; for the Young SCF, he modified a preference profile constructed by Young with 37 voters and five candidates. Fishburn [9] provided the following limit devise in order to define homogeneous variants of the Dodgson and Young SCFs. For example, the Dodgson scheme can be made homogeneous by defining from the function DodgsonScore for each preference profile $\langle C, V\rangle$ and designated candidate $c \in C$ the function

$$\text{DodgsonScore}^*(C, c, V) = \lim_{q\to\infty} \frac{\text{DodgsonScore}(C, c, qV)}{q}.$$

The resulting SCF is denoted by Dodgson* SCF, and the corresponding winner and ranking problems are denoted by `Dodgson* Winner` and `Dodgson* Ranking`. Analogously, the Young voting scheme defined above can be made homogeneous by defining YoungScore*. Remarkably, Young [29] showed that the corresponding problem `Young* Winner` can be solved by a linear program. Hence, the problem `Young* Winner` is efficiently solvable, since the problem `Linear Programming` can be decided in polynomial time [12], see also [18]. We establish an analogous result for the problems `Dodgson* Winner` and `Dodgson* Ranking`. The proof of Theorem 9 can be found in the full version [24].

Theorem 9 `Dodgson* Winner` *and* `Dodgson* Ranking` *can be solved in polynomial time.*

Acknowledgments

The first author gratefully acknowledges interesting discussions with Klaus Wagner, Venkatesan Guruswami, Edith and Lane Hemaspaandra, Dieter

Kratsch, and Gerd Wechsung. We thank an anonymous referee for his or her nice comments.

References

[1] K. Arrow. *Social Choice and Individual Values.* John Wiley and Sons, 1951 (revised editon 1963).

[2] J. Bartholdi III, C. Tovey, and M. Trick. Voting schemes for which it can be difficult to tell who won the election. *Social Choice and Welfare*, 6:157–165, 1989.

[3] D. Black. *The Theory of Committees and Elections.* Cambridge University Press, 1958.

[4] J. Cai, T. Gundermann, J. Hartmanis, L. Hemachandra, V. Sewelson, K. Wagner, and G. Wechsung. The boolean hierarchy I: Structural properties. *SIAM Journal on Computing*, 17(6):1232–1252, 1988.

[5] J. Cai and G. Meyer. Graph minimal uncolorability is D^P-complete. *SIAM Journal on Computing*, 16(2):259–277, April 1987.

[6] G. Chapman. *The Complete Monty Python's Flying Circus: All the Words.* Pantheon Books, 1989.

[7] M. J. A. N. de Caritat, Marquis de Condorcet. Essai sur l'application de l'analyse à la probabilité des décisions rendues à la pluraliste des voix. 1785. Facsimile reprint of original published in Paris, 1972, by the Imprimerie Royale. English translation appears in I. McLean and A. Urken, *Classics of Social Choice*, University of Michigan Press, 1995, pages 91–112.

[8] C. Dodgson. A method of taking votes on more than two issues. Pamphlet printed by the Clarendon Press, Oxford, and headed "not yet published" (see the discussions in [21, 3], both of which reprint this paper), 1876.

[9] P. Fishburn. Condorcet social choice functions. *SIAM Journal on Applied Mathematics*, 33:469–489, 1977.

[10] M. Garey and D. Johnson. *Computers and Intractability: A Guide to the Theory of NP-Completeness.* W. H. Freeman and Company, New York, 1979.

[11] V. Guruswami and S. Khanna. On the hardness of 4-coloring a 3-colorable graph. In *Proceedings of the 15th Annual IEEE Conference on Computational Complexity*, pages 188–197. IEEE Computer Society Press, May 2000.

[12] L. Hačijan. A polynomial algorithm in linear programming. *Soviet Math. Dokl.*, 20:191–194, 1979.

[13] F. Hausdorff. *Grundzüge der Mengenlehre.* Walter de Gruyten and Co., 1914.

[14] E. Hemaspaandra and L. Hemaspaandra. Computational politics: Electoral systems. In *Proceedings of the 25th International Symposium on Mathematical Foundations of Computer Science*, pages 64–83. Springer-Verlag *Lecture Notes in Computer Science #1893*, 2000.

[15] E. Hemaspaandra, L. Hemaspaandra, and J. Rothe. Exact analysis of Dodgson elections: Lewis Carroll's 1876 voting system is complete for parallel access to NP. *Journal of the ACM*, 44(6):806–825, 1997.

[16] E. Hemaspaandra, J. Rothe, and H. Spakowski. Recognizing when heuristics can approximate minimum vertex covers is complete for parallel access to NP. In

Proceedings of the 28th International Workshop on Graph-Theoretical Concepts in Computer Science (WG 2002), June 2002. To appear.

[17] L. Hemaspaandra and J. Rothe. Unambiguous computation: Boolean hierarchies and sparse Turing-complete sets. *SIAM Journal on Computing*, 26(3):634–653, 1997.

[18] N. Karmarkar. A new polynomial-time algorithm for linear programming. *Combinatorica*, 4(4):373–395, 1984.

[19] S. Khanna, N. Linial, and S. Safra. On the hardness of approximating the chromatic number. *Combinatorica*, 20(3):393–415, 2000.

[20] J. Köbler, U. Schöning, and K. Wagner. The difference and truth-table hierarchies for NP. *R.A.I.R.O. Informatique théorique et Applications*, 21:419–435, 1987.

[21] I. McLean and A. Urken. *Classics of Social Choice.* University of Michigan Press, Ann Arbor, Michigan, 1995.

[22] C. Papadimitriou and M. Yannakakis. The complexity of facets (and some facets of complexity). *Journal of Computer and System Sciences*, 28(2):244–259, 1984.

[23] J. Rothe. Exact complexity of Exact-Four-Colorability. Technical Report cs.CC/0109018, Computing Research Repository (CoRR), September 2001. 5 pages. Available on-line at http://xxx.lanl.gov/abs/cs.CC/0109018.

[24] J. Rothe, H. Spakowski, and J. Vogel. Exact complexity of the winner problem for Young elections. Technical Report cs.CC/0112021, Computing Research Repository (CoRR), December 2001. 10 pages. Available on-line at http://xxx.lanl.gov/abs/cs.CC/0112021.

[25] H. Spakowski and J. Vogel. Θ_2^p-completeness: A classical approach for new results. In *Proceedings of the 20th Conference on Foundations of Software Technology and Theoretical Computer Science*, pages 348–360. Springer-Verlag *Lecture Notes in Computer Science #1974*, December 2000.

[26] H. Spakowski and J. Vogel. The complexity of Kemeny's voting system. In *Proceedings of the 5th Argentinian Workshop on Theoretical Computer Science*, pages 157–168, 2001.

[27] K. Wagner. More complicated questions about maxima and minima, and some closures of NP. *Theoretical Computer Science*, 51:53–80, 1987.

[28] K. Wagner. Bounded query classes. *SIAM Journal on Computing*, 19(5):833–846, 1990.

[29] H. Young. Extending Condorcet's rule. *Journal of Economic Theory*, 16:335–353, 1977.

QUANTUM NP AND A QUANTUM HIERARCHY

(Extended Abstract)

Tomoyuki Yamakami
School of Information Technology and Engineering
University of Ottawa, Ottawa, Ontario, Canada K1N 6N5
yamakami@site.uottawa.ca

Abstract The complexity class NP is quintessential and ubiquitous in theoretical computer science. Two different approaches have been made to define "Quantum NP," the quantum analogue of NP: NQP by Adleman, DeMarrais, and Huang, and QMA by Knill, Kitaev, and Watrous. From an operator point of view, NP can be viewed as the result of the $\exists$-operator applied to P. Recently, Green, Homer, Moore, and Pollett proposed its quantum version, called the N-operator, which is an abstraction of NQP. This paper introduces the $\exists^{Q}$-operator, which is an abstraction of QMA, and its complement, the $\forall^{Q}$-operator. These operators not only define Quantum NP but also build a quantum hierarchy, similar to the Meyer-Stockmeyer polynomial hierarchy, based on two-sided bounded-error quantum computation.

Keywords: quantum quantifier, quantum operator, quantum polynomial hierarchy

1. What is Quantum NP?

Computational complexity theory based on a Turing machine (TM, for short) was formulated in the 1960s. The complexity class NP was later introduced as the collection of sets that are recognized by nondeterministic TMs in polynomial time. By the earlier work of Cook, Levin, and Karp, NP was quickly identified as a central notion in complexity theory by means of NP-completeness. NP has since then exhibited its rich structure and is proven to be vital to many fields of theoretical computer science. Meyer and Stockmeyer [13] further extended NP into a hierarchy, known as the *polynomial (time) hierarchy*. This hierarchy has inspired many tools and techniques, e.g., circuit lower-bound proofs and micro hierarchies within NP. There is known to be a relativized world where the hierarchy forms an infinite hierarchy. It is thus natural to consider a quantum analogue of NP, dubbed as "Quantum NP," and its extension. Several approaches have been made over the years to define Quantum NP.

As is known, NP can be characterized in several different manners. As the first example, NP can be characterized by probabilistic TMs with positive acceptance probability. Adleman et al. [1] introduced the complexity class NQP as a quantum extension of this probabilistic characterization. Subsequently, NQP (even with arbitrary complex amplitudes) was shown to coincide with the classical counting class co-$\mathrm{C_{=}P}$ [7, 6, 20]. This shows the power of quantum computation.

NP can be also characterized by logical quantifiers over classical (binary) strings of polynomial length. This is also known as the "guess-and-check" process. Knill [10], Kitaev [9], and Watrous [16] studied the complexity class QMA (named by Watrous), which can be viewed as a quantum extension of the aforementioned quantifier characterization of NP. In their definition, a quantifier bounds a quantum state instead of a classical string. We call such a quantifier a *quantum quantifier* to emphasize the scope of the quantifier being quantum states. Using this terminology, any set in QMA is defined with the use of a single quantum quantifier over polynomial-size quantum states. It appears that a quantum quantifier behaves in quite a distinctive manner. For instance, Kobayashi et al. [12] recently pointed out that allowing multiple quantum quantifiers may increase the complexity of QMA due to quantum entanglement (in [12], QMA(k) is defined with k quantum quantifiers).

From a different aspect, we can view the process of defining NP as an application of an operator that transforms a class $\mathcal{C}$ to another class $\mathcal{D}$. For example, we write co-$\mathcal{C}$ to denote the class $\{A \mid \overline{A} \in \mathcal{C}\}$, where $\overline{A}$ is the complement of A. This prefix "co" in co-$\mathcal{C}$ can be considered as the *complementation operator* that builds co-$\mathcal{C}$ from $\mathcal{C}$. Other examples are Schöning's BP-operator [14] and Wagner's C-operator [15]. The classical existential quantifier naturally induces the so-called $\exists$-*operator*. With this $\exists$-operator, NP is defined as $\exists \cdot \mathrm{P}$. Similarly, we can consider a quantum analogue of the $\exists$-operator. One possible analogue was recently proposed by Green et al. [8]. They introduced the N-operator, which is an abstraction of NQP.

To make the most of quantum nature, we define in this paper a quantum operator that expands the quantum existential quantifier used for QMA and QMA(k). This quantum operator is called the $\exists^{\mathrm{Q}}$-*operator* (whose complement is the $\forall^{\mathrm{Q}}$-*operators*). These quantum operators give a new definition for Quantum NP and its expansion, a quantum analogue of the polynomial hierarchy. Our quantum operators, however, require a more general framework than the existing one. In the subsequent section, we discuss a general framework for the quantum operators.

2. Toward a General Framework for Quantum Operators

Let our alphabet Σ be $\{0,1\}$ throughout this paper. Let $\mathbb{N}$ be the set of all nonnegative integers and set $\mathbb{N}^+ = \mathbb{N} - \{0\}$. To describe a quantum state, we use Dirac's ket notation $|\phi\rangle$. Write $\mathcal{H}_n$ to denote a Hilbert space of dimension n. In comparison with a classical (binary) string, we use the terminology, a *quantum string* (*qustring*, for short) *of size* n, to mean a unit-norm vector in $\mathcal{H}_{2^n}$. For such a qustring $|\phi\rangle$, $\ell(|\phi\rangle)$ denotes the *size* of $|\phi\rangle$. We use the notation Φ_n for each $n \in \mathbb{N}$ to denote the collection of all qustrings of size n and thus, $\Phi_n \subseteq \mathcal{H}_{2^n}$. Let $\Phi_\infty = \bigcup_{n\geq 0} \Phi_n$, the set of all finite-size qustrings.

We use a multi-tape quantum Turing machine (QTM), defined in [4, 17], as a mathematical model of quantum computations. A multi-tape QTM is equipped with two-way infinite tapes, tape heads, and a finite-control unit. We assume in this paper the following technical restriction on each QTM: a QTM is always designed so that all computation paths on each input terminate at the same time by entering its unique halting state after writing 0 (rejection) or 1 (acceptance) in the start cell of the designated output tape (see [17] for the discussion on the *timing problem*). Thus, the length of a computation path on input x is regarded as the *running time* of the QTM on x. The transition function δ of a QTM can be seen as an operator (called a *time-evolution operator*) that transforms a superposition of configurations at time t to another superposition of configurations at time $t+1$. A QTM is called *well-formed* if its time-evolution operator is unitary. Moreover, a QTM is said to have $\tilde{\mathbb{C}}$-*amplitudes* if all amplitudes in δ are drawn from set $\tilde{\mathbb{C}}$, where $\tilde{\mathbb{C}}$ is the set of all complex numbers whose real and imaginary parts are approximated deterministically to within 2^{-n} in time polynomial in n. For a well-formed QTM M and an input $|\phi\rangle$, the notation $\mathrm{Prob}_M[M(|\phi\rangle) = 1]$ denotes the acceptance probability of M on input $|\phi\rangle$. Similarly, $\mathrm{Prob}_M[M(|\phi\rangle) = 0]$ denotes the rejection probability of M on $|\phi\rangle$.

2.1. From Classical Inputs to Quantum Inputs

We have used classical (binary) strings as standard inputs given into quantum computations. As a result, any quantum complexity class, such as NQP or BQP [4], is defined to be a collection of subsets of Σ^*. Since a QTM acts as a unitary operator, it is legitimate to feed the QTM with a quantum state as an input. We call such an input a *quantum input* for clarity. As in the definition of QMA(k), for instance, such quantum inputs play an essential role. We thus need to expand a set of strings to a set of qustrings by considering a qustring as an input given to an underlying QTM. We use the following notation. For each $m, n \in \mathbb{N}^+$, let Φ_n^m denote the collection of all m-tuples $(|\phi_1\rangle, |\phi_2\rangle, \ldots, |\phi_m\rangle)$ such that each $|\phi_i\rangle$ is a qustring of size n. Such an m-tuple is expressed as $|\vec{\phi}\rangle$ and also seen as a tensor product $|\phi_1\rangle|\phi_2\rangle\cdots|\phi_m\rangle$ when the size of each $|\phi_i\rangle$ is known. For brevity, the notation $\ell(|\vec{\phi}\rangle)$ means the sum $\sum_{i=1}^m \ell(|\phi_i\rangle)$. We also set $\Phi_\infty^m = \bigcup_{n\geq 1} \Phi_n^m$ and $\Phi_\infty^* = \bigcup_{m\geq 1} \Phi_\infty^m$.

The introduction of quantum inputs gives rise to an important issue, which is not present in the classical framework: the duplication of an input. The repetition of a quantum computation on a classical input is seen in, e.g., the proof of $\mathrm{BQP}^{\mathrm{BQP}} = \mathrm{BQP}$ [3]. Nevertheless, the situation may change when we deal with a quantum input. Since a fundamental principle of quantum computation, the so-called *no-cloning theorem*, interdicts the duplication of an arbitrary quantum input, we cannot redo even the same quantum computation on a single quantum input unless the copies of the quantum input are given *a priori*. To establish a coherent but concise theory of quantum computation over Φ_∞^*, we need to allow the underlying quantum computation to access the quantum input repeatedly without disturbing other quantum states. Schematically, we supply a sufficient number of its copies as "auxiliary inputs." This guarantees the quantum extension of many existing complexity classes, such as BQP, to enjoy the same structural properties.

For later convenience, we first expand the function class #QP [18], which originally consists of certain quantum functions mapping from Σ^* to the unit real interval $[0,1]$. The notation ${}^*\#\mathrm{QP}$ is given in this paper to denote the corresponding extension—the collection of quantum functions mapping from Φ_∞^* to $[0,1]$. Since Φ_∞^* is a continuous space, these quantum functions are inherently continuous. For simplicity, write $|\vec{\phi}\rangle^{\otimes k}$ for k copies of $|\vec{\phi}\rangle$, which can be viewed as a tensor product of k identical $|\vec{\phi}\rangle$'s (as long as the size of $|\vec{\phi}\rangle$ is known).

Definition 1 *A function f from Φ_∞^* to $[0,1]$ is in ${}^*\#\mathrm{QP}$ if there exist a polynomial q and a polynomial-time, $\tilde{\mathbb{C}}$-amplitude, well-formed QTM M such that, for every $m \in \mathbb{N}^+$ and every $|\vec{\phi}\rangle \in \Phi_\infty^m$, $f(|\vec{\phi}\rangle) = \mathrm{Prob}_M[M(|\vec{\phi}\rangle^{\otimes q(\ell(|\vec{\phi}\rangle))}) = 1]$.*

We reserve the standard notation #QP to denote the class of quantum functions whose domains are Σ^* (i.e., those functions are obtained from Definition 1 by replacing $|\vec{\phi}\rangle$ with x from Σ^*).

To distinguish a set of qustrings from a set of classical strings, we use the terminology, a *quantum set*, for a set $A \subseteq \Phi_\infty^*$. A collection of quantum sets is called a *quantum complexity class* (which conventionally refers to any classical class related to quantum computations). From a different perspective, a classical set can be viewed as a "projection" of its corresponding quantum set. For a quantum set $A \subseteq \Phi_\infty^*$, its *classical part* $\breve{A}$ is given as follows:

$$\breve{A} = \{\langle s_1, s_2, \ldots, s_m\rangle \mid m \in \mathbb{N}^+, s_1, \ldots, s_m \in \Sigma^*, (|s_1\rangle, |s_2\rangle, \ldots, |s_m\rangle) \in A\},$$

where $\langle\ \rangle$ is an appropriate *pairing function* from $\bigcup_{m\geq 1}(\Sigma^*)^m$ to Σ^*. Thus, any quantum class $\mathcal{C}$ naturally induces its *classical part* $\{\breve{A} \mid A \in \mathcal{C}\}$. In a similar way, #QP is also viewed as the "projection" of ${}^*\#\mathrm{QP}$.

A relativized version of ${}^*\#\mathrm{QP}$ is defined by substituting oracle QTMs for non-oracle QTMs in Definition 1, where an *oracle QTM* can make a query of the form $|x\rangle|b\rangle$ ($x \in \Sigma^*$ and $b \in \{0,1\}$) by which oracle A transforms $|x\rangle|b\rangle$ into $(-1)^{b\cdot A(x)}|x\rangle|b\rangle$ in a single step.

2.2. From Decision Problems to Partial Decision Problems

We described in the previous subsection how to expand classical sets to quantum sets. The next step might be to expand well-known complexity classes, such as NQP and BQP, to classes of quantum sets. Unfortunately, since Φ_∞ is a continuous space, we cannot expand all classical classes in this way (for example, BQP). One of the resolutions is to consider "partial" decision problems. (See, e.g., [5] for classical partial decision problems.) In this paper, we define a *partial decision problem* to be a pair (A, B) such that $A, B \subseteq \Phi_\infty^*$ and $A \cap B = \emptyset$, where A indicates a set of accepted qustrings and B indicates a set of rejected qustrings. The *legal region* of (A, B) is $A \cup B$. For consistency with classical decision problems, we should refer A to as $(A, \overline{A})$, where $\overline{A} = \Phi_\infty^* - A$, and call it a *total decision problem.* The notions of *inclusion*, *union*, and *complement* are introduced in the following manner: let (A, B) and (C, D) be any partial decision problems and let E be the intersection of their legal regions; that is, $(A \cup B) \cap (C \cup D)$.

1. Inclusion: $(A, B) \subseteq (C, D)$ iff $A \subseteq C$ and $A \cup B = C \cup D$.
2. Intersection: $(A, B) \cap (C, D) \stackrel{def}{=} (A \cap C, (B \cup D) \cap E)$.
3. Union: $(A, B) \cup (C, D) \stackrel{def}{=} ((A \cup C) \cap E, B \cap D)$.
4. Complementation: $\overline{(A, B)} \stackrel{def}{=} (B, A)$.

Now, we focus on classes of partial decision problems. To denote such a class, we use the special notation $^*\mathcal{C}$, whose asterisk signifies the deviation from total decision problems. The *partial classical part* of a partial decision problem (A, B) is $(\breve{A}, \breve{B})$. When $\breve{A} \cup \breve{B} = \Sigma^*$, we call $(\breve{A}, \breve{B})$ the *total classical part* of (A, B) and simply write $\breve{A}$ instead of $(\breve{A}, \breve{B})$ as before. Notationally, let $\mathcal{C}$ denote the collection of *total* classical parts in $^*\mathcal{C}$. We call $\mathcal{C}$ the *total classical part* of $^*\mathcal{C}$.

For later use, we expand BQP to the class of partial decision problems. In a similar fashion, we can expand other classes, such as NQP and PQP [18].

Definition 2 *Let a, b be any two functions from $\mathbb{N}$ to $[0, 1]$ such that $a(n) + b(n) = 1$ for all $n \in \mathbb{N}$. A partial decision problem (A, B) is in* *BQP(a, b) *if there exists a quantum function $f \in {}^*\#$QP such that, for every $|\vec{\phi}\rangle \in \Phi_\infty^*$, (i) if $|\vec{\phi}\rangle \in A$ then $f(|\vec{\phi}\rangle) \geq a(\ell(|\vec{\phi}\rangle))$ and (ii) if $|\vec{\phi}\rangle \in B$ then $f(|\vec{\phi}\rangle) \leq b(\ell(|\vec{\phi}\rangle))$. For simplicity, write* *BQP *for* *BQP$(3/4, 1/4)$.

It is important to note that the total classical part of *BQP coincides with the standard definition of BQP, e.g., given in [4]. Since the duplication of a quantum input is available for free of charge, we can perform a standard majority-vote algorithm for a set in *BQP to amplify its success probability. Therefore, we obtain $^*\mathrm{BQP} = {}^*\mathrm{BQP}(1 - 2^{-p(n)}, 2^{-p(n)})$ for any polynomial p.

3. The $\exists^{\mathrm{Q}}$-Operator and the $\forall^{\mathrm{Q}}$-Operator

The process of defining a new complexity class $\mathcal{D}$ from a basis class $\mathcal{C}$ can be naturally viewed as an application of an operator, which maps $\mathcal{C}$ to $\mathcal{D}$. As seen in Section 1, the $\exists$-operator over classical sets is an abstraction of nondeterministic computation (as in NP $= \exists \cdot$ P) and its complement is called the *$\forall$-operator.* First, we generalize these operators to the ones whose scopes are classes of partial decision problems.

Definition 3 *Let $^*\mathcal{C}$ be any quantum complexity class of partial decision problems. A partial decision problem (A,B) is in $^*\exists\cdot{}^*\mathcal{C}$ if there exist a polynomial p and a partial decision problem (C,D) in $^*\mathcal{C}$ such that, for all vectors $|\vec{\phi}\rangle \in \Phi_\infty^*$,*

i) if $|\vec{\phi}\rangle \in A$ then $\exists x \in \Sigma^{p(\ell(|\vec{\phi}\rangle))}[(|x\rangle, |\vec{\phi}\rangle) \in C]$ and

ii) if $|\vec{\phi}\rangle \in B$ then $\forall x \in \Sigma^{p(\ell(|\vec{\phi}\rangle))}[(|x\rangle, |\vec{\phi}\rangle) \in D]$.

The class $^*\forall\cdot{}^*\mathcal{C}$ is defined similarly by exchanging the roles of the quantifiers in conditions i) and ii). In accordance to the standard notation, $\exists\cdot{}^*\mathcal{C}$ and $\forall\cdot{}^*\mathcal{C}$ denote the total classical parts of $^*\exists\cdot{}^*\mathcal{C}$ and $^*\forall\cdot{}^*\mathcal{C}$, respectively.

The class QMA uses a quantum quantifier, whose scope is qustrings of polynomial size instead of classical strings of polynomial length. Generalizing such a quantum quantifier, we introduce a quantum analogue of the $\exists$- and $\forall$-operators as follows. Our approach is quite different from that of Green et al. [8], who defined the N-operator as an abstraction of NQP.

Definition 4 *Let $^*\mathcal{C}$ be a quantum complexity class of partial decision problems. A partial decision problem (A,B) is in $^*\exists^{\mathrm{Q}}\cdot{}^*\mathcal{C}$ if there exist a polynomial p and a partial decision problem $(C,D) \in {}^*\mathcal{C}$ such that, for every $|\vec{\phi}\rangle \in \Phi_\infty^*$,*

i) if $|\vec{\phi}\rangle \in A$ then $\exists|\psi\rangle \in \Phi_{p(\ell(|\vec{\phi}\rangle))}[(|\psi\rangle, |\vec{\phi}\rangle) \in C]$ and

ii) if $|\vec{\phi}\rangle \in B$ then $\forall|\psi\rangle \in \Phi_{p(\ell(|\vec{\phi}\rangle))}[(|\psi\rangle, |\vec{\phi}\rangle) \in D]$.

Similarly, the class $^*\forall^{\mathrm{Q}}\cdot{}^*\mathcal{C}$ is defined by exchanging the roles of quantifiers in conditions i) and ii) above. The notations $\exists^{\mathrm{Q}}\cdot{}^*\mathcal{C}$ and $\forall^{\mathrm{Q}}\cdot{}^*\mathcal{C}$ denote the total classical parts of $^*\exists^{\mathrm{Q}}\cdot{}^*\mathcal{C}$ and $^*\forall^{\mathrm{Q}}\cdot{}^*\mathcal{C}$, respectively. More generally, write $^*\exists_1^{\mathrm{Q}}\cdot{}^*\mathcal{C}$ for $^*\exists^{\mathrm{Q}}\cdot{}^*\mathcal{C}$ and recursively define $^*\exists_{m+1}^{\mathrm{Q}}\cdot{}^*\mathcal{C}$ as $^*\exists^{\mathrm{Q}}\cdot({}^*\exists_m^{\mathrm{Q}}\cdot{}^*\mathcal{C})$. Similarly, $^*\forall_{m+1}^{\mathrm{Q}}\cdot{}^*\mathcal{C}$ is defined.

Obviously, if $^*\mathcal{C} \subseteq {}^*\mathcal{D}$ then $\exists^{\mathrm{Q}}\cdot{}^*\mathcal{C} \subseteq \exists^{\mathrm{Q}}\cdot{}^*\mathcal{D}$ and $^*\exists^{\mathrm{Q}}\cdot{}^*\mathcal{C} \subseteq {}^*\exists^{\mathrm{Q}}\cdot{}^*\mathcal{D}$.

We next show that the $\exists^{\mathrm{Q}}$- and $\forall^{\mathrm{Q}}$-operators indeed expand the classical $\exists$- and $\forall$-operators, respectively. Proving this claim, however, requires underlying class $^*\mathcal{C}$ to satisfy a certain condition, which is given in the following definition.

Definition 5 *1. A quantum set $B \subseteq \Phi_\infty^*$ is called* classically separable *if the following condition holds: for every $m,n \in \mathrm{N}^+$ and every $|\vec{\phi}\rangle \in \Phi_n^m$, if either $\langle\vec{x}|\vec{\phi}\rangle = 0$ or $(|\vec{x}\rangle, |\vec{\psi}\rangle) \in B$ for all $\vec{x} \in (\Sigma^n)^m$, then $(|\vec{\phi}\rangle, |\vec{\psi}\rangle) \in B$.*

2. A quantum complexity class $^*\mathcal{C}$ of partial decision problems is said to be classically simulatable *if, for every partial decision problem $(A, B) \in {}^*\mathcal{C}$, there exist a partial decision problem $(C, D) \in {}^*\mathcal{C}$ such that (i) C and D are classically separable and (ii) for all $m, n \in \mathbb{N}^+$ and all $\vec{x} \in (\Sigma^n)^m$, $(|\vec{x}\rangle, |\vec{\psi}\rangle) \in A \iff (|\vec{x}\rangle, |\vec{\psi}\rangle) \in C$ and $(|\vec{x}\rangle, |\vec{\psi}\rangle) \in B \iff (|\vec{x}\rangle, |\vec{\psi}\rangle) \in D$.*

The above notion stems from the proof of QMA containing NP. The classes of partial decision problems dealt with in this paper are indeed classically simulatable.

Lemma 6 *If a quantum complexity class $^*\mathcal{C}$ of partial decision problems is classically simulatable, then $^*\exists \cdot {}^*\mathcal{C} \subseteq {}^*\exists^{\mathrm{Q}} \cdot {}^*\mathcal{C}$ and $^*\forall \cdot {}^*\mathcal{C} \subseteq {}^*\forall^{\mathrm{Q}} \cdot {}^*\mathcal{C}$.*

The proof of the first claim of Lemma 6 easily follows from the definition of classical-simulatability. The second claim comes from the fact that if $^*\mathcal{C}$ is classically simulatable then so is co-$^*\mathcal{C}$, where co-$^*\mathcal{C}$ denotes the collection of partial decision problems whose complements belong to $^*\mathcal{C}$.

4. The Quantum Polynomial Hierarchy

The classical $\exists$- and $\forall$-operators are useful tools to expand complexity classes. In the early 1970s, Meyer and Stockmeyer [13] introduced the *polynomial hierarchy*, whose components are obtained from P with alternating applications of these operators; namely, $\Sigma_{k+1}^{\mathrm{P}} = \exists \cdot \Pi_k^{\mathrm{P}}$ and $\Pi_{k+1}^{\mathrm{P}} = \forall \cdot \Sigma_k^{\mathrm{P}}$ for each level $k \in \mathbb{N}^+$. The polynomial hierarchy has continued to be a center of research in complexity theory. The introduction of the $\exists^{\mathrm{Q}}$- and $\forall^{\mathrm{Q}}$-operators enables us to consider a quantum analogue of the polynomial hierarchy and explore its structural properties in light of the strength of quantum computability. We call this new hierarchy the *quantum polynomial hierarchy* (QP-hierarchy, for short). The basis of the QP hierarchy is *BQP opposed to P since two-sided bounded-error computations are more realistic in the quantum setting. Each level of the QP hierarchy is obtained from its lower level by a *finite number* of applications of the same quantum operator (either $\exists^{\mathrm{Q}}$- or $\forall^{\mathrm{Q}}$-operators). Although any repetition of the same classical operator has no significance in the polynomial hierarchy, as Kobayashi et al. [12] pointed out, there might be a potentially essential difference between a single quantum quantifier and multiple quantum quantifiers of the same type. The precise definition of the QP hierarchy is given as follows.

Definition 7 *Let a, b be two functions from $\mathbb{N}$ to $[0,1]$ such that $a(n)+b(n) = 1$ for all $n \in \mathbb{N}$. Let $k \in \mathbb{N}$ and $m \in \mathbb{N}^+$. The* quantum polynomial hierarchy *(QP hierarchy, for short) constitutes the following complexity classes of partial decision problems.*

i) $^*\Sigma_{0,m}^{\mathrm{QP}}(a, b) = {}^*\Pi_{0,m}^{\mathrm{QP}}(a, b) = {}^*\mathrm{BQP}(a, b)$.

ii) $^*\Sigma_{k+1,m}^{\mathrm{QP}}(a, b) = {}^*\exists_m^{\mathrm{Q}} \cdot {}^*\Pi_{k,m}^{\mathrm{QP}}(a, b)$.

iii) $^*\Pi_{k+1,m}^{\mathrm{QP}}(a, b) = {}^*\forall_m^{\mathrm{Q}} \cdot {}^*\Sigma_{k,m}^{\mathrm{QP}}(a, b)$.

Let ${}^*\Sigma_k^{QP}(a,b) = \bigcup_{m\geq 1} {}^*\Sigma_{k,m}^{QP}(a,b)$ *and* ${}^*\Pi_k^{QP}(a,b) = \bigcup_{m\geq 1} {}^*\Pi_{k,m}^{QP}(a,b)$. *Furthermore, let* ${}^*\mathrm{QPH}_m(a,b) = \bigcup_{k\geq 0}({}^*\Sigma_{k,m}^{QP}(a,b) \cup {}^*\Pi_{k,m}^{QP}(a,b))$ *and* ${}^*\mathrm{QPH}(a,b) = \bigcup_{m\geq 1} {}^*\mathrm{QPH}_m(a,b)$. *Their total classical parts are denoted (without asterisks)* $\Sigma_k^{QP}(a,b)$, $\Pi_k^{QP}(a,b)$, *and* $\mathrm{QPH}(a,b)$.

For brevity, we write ${}^*\Sigma_k^{QP}$ for ${}^*\Sigma_k^{QP}(3/4,1/4)$, ${}^*\Pi_k^{QP}$ for ${}^*\Pi_k^{QP}(3/4,1/4)$, and ${}^*\mathrm{QPH}$ for ${}^*\mathrm{QPH}(3/4,1/4)$. Likewise, we can define their total classical parts Σ_k^{QP}, Π_k^{QP}, and QPH. The choice of the value $(3/4,1/3)$ is artificial; however, a standard majority-vote algorithm can amplify $(3/4,1/4)$ to $(1 - 2^{-p(n)}, 2^{-p(n)})$ for an arbitrary polynomial p. Due to our general framework, it is likely that Σ_1^{QP} is strictly larger than $\bigcup_{k\geq 1} \mathrm{QMA}(k)$. From this reason, Σ_1^{QP} can be regarded as Quantum NP, as discussed in Section 1.

Several alternative definitions of the QP hierarchy are possible. Here, we present three alternatives. The first one uses the function class ${}^*\mathrm{Qopt}\#\Sigma_{k,m}^{QP}$—a generalization of Qopt#QP in [19]—introduced as follows: a quantum function f from Φ_∞^* to $[0,1]$ is in ${}^*\mathrm{Qopt}\#\Sigma_{k,m}^{QP}$ if there exist a polynomial p and a quantum function $g \in {}^*\#\mathrm{QP}$ such that, for every $|\vec{\phi}\rangle \in \Phi_\infty^*$,

$$f(|\vec{\phi}\rangle) = \sup_{|\vec{\psi}_1\rangle} \inf_{|\vec{\psi}_2\rangle} \cdots \mathrm{opr}^{(k)}_{|\vec{\psi}_k\rangle} \{g(|\vec{\phi}\rangle, |\vec{\psi}_1\rangle, |\vec{\psi}_2\rangle, \ldots, |\vec{\psi}_k\rangle)\},$$

where $\mathrm{opr}^{(k)} = \sup$ if k is odd and $\mathrm{opr}^{(k)} = \inf$ otherwise, and each $|\vec{\psi}_i\rangle$ is an m-tuple $(|\psi_{i,1}\rangle, |\psi_{i,2}\rangle, \ldots, |\psi_{i,m}\rangle)$ with each $|\psi_{i,j}\rangle$ running over all qustrings of size $p(\ell(|\vec{\phi}\rangle))$. The class ${}^*\mathrm{Qopt}\#\Sigma_{k,m}^{QP}$ gives a succinct way to define the kth level of the QP hierarchy.

Lemma 8 *Let* $k,m \geq 1$ *and let* a,b *be any two functions from* $\mathbb{N}$ *to* $[0,1]$ *satisfying* $a(n)+b(n) = 1$ *for all* $n \in \mathbb{N}$. *A partial decision problem* (A,B) *is in* ${}^*\Sigma_{k,m}^{QP}(a,b)$ *iff there exists a quantum function* f *in* ${}^*\mathrm{Qopt}\#\Sigma_{k,m}^{QP}$ *such that, for every* $|\vec{\phi}\rangle \in \Phi_\infty^*$, *(i) if* $|\vec{\phi}\rangle \in A$ *then* $f(|\vec{\phi}\rangle) \geq a(\ell(|\vec{\phi}\rangle))$ *and (ii) if* $|\vec{\phi}\rangle \in B$ *then* $f(|\vec{\phi}\rangle) \leq b(\ell(|\vec{\phi}\rangle))$.

In the second alternative definition, we use vectors whose components are described by classical strings. For each $n, r \in \mathbb{N}^+$, denote by $\tilde{\Phi}_n(r)$ the collection of all vectors $|\phi\rangle$ (not necessarily elements in a Hilbert space) such that $|\phi\rangle$ has the form $\sum_{s:|s|=n} \alpha_s |s\rangle$, where each complex number α_s is expressed as a pair of two binary fractions of r bits. If $r \geq \log(1/\epsilon) + n + 2$ for $\epsilon > 0$, such $|\phi\rangle$ satisfies $|\sum_{s:|s|=n} |\alpha_s|^2 - 1| \leq \epsilon$. Note that any element in $\tilde{\Phi}_n$ can be expressed as a binary string of length $r2^{n+1}$ (since each α_s needs $2r$ bits and we have exactly 2^n such α_s's). Thus, the cardinality of $\tilde{\Phi}_n$ is $2^{r2^{n+1}}$. For our purpose, we allow each ${}^*\#\mathrm{QP}$-function to take any vector in $\tilde{\Phi}_n(r)$ as its input.

Lemma 9 *Let* $k,m \in \mathbb{N}^+$. *A partial decision problem* (A,B) *is in* ${}^*\Sigma_{k,m}^{QP}$ *iff there exist a polynomial* p *and a quantum function* $f \in {}^*\#\mathrm{QP}$ *such that, for every series of qustrings* $|\vec{\phi}\rangle$ *in* Φ_∞^*,

i) if $|\vec{\phi}\rangle \in A$ then $\exists|\vec{\xi_1}\rangle\forall|\vec{\xi_2}\rangle\cdots Q_k|\vec{\xi_k}\rangle[f(|\vec{\phi}\rangle,|\vec{\xi_1}\rangle,|\vec{\xi_2}\rangle,\ldots,|\vec{\xi_k}\rangle)\geq 3/4]$ and

ii) if $|\vec{\phi}\rangle \in B$ then $\forall|\vec{\xi_1}\rangle\exists|\vec{\xi_2}\rangle\cdots \overline{Q}_k|\vec{\xi_k}\rangle[f(|\vec{\phi}\rangle,|\vec{\xi_1}\rangle,|\vec{\xi_2}\rangle,\ldots,|\vec{\xi_k}\rangle)\leq 1/4]$,

where each variable $|\vec{\xi_i}\rangle$ runs over all series of m vectors in $\tilde{\Phi}_{p(\ell(|\vec{\phi}\rangle))}(3p(\ell(|\vec{\phi}\rangle)))$, $Q_k = \forall$ if k is even and $Q_k = \exists$ otherwise, and $\overline{Q}_k$ is the opposite quantifier of Q_k.

The last alternative definition is much more involved and we need extra notions and notation. Firstly, we give a method of translating a qustring $|\phi\rangle$ into a series of unitary matrices that generate $|\phi\rangle$. Let $\mathbb{C}$ be the set of all complex numbers, I the 2×2 identity matrix, and λ the empty string. Fix $n \in \mathbb{N}$ and $|\phi\rangle \in \Phi_{n+1}$ and assume that $|\phi\rangle = \sum_{s:|s|=n+1}\gamma_s|s\rangle$, where each γ_s is in $\mathbb{C}$. For each $s \in \Sigma^{\leq n}$, set $g_s = \sqrt{\sum_t|\gamma_{s0t}|^2+\sum_t|\gamma_{s1t}|^2}$ and define a 2×2 matrix $U^{(s)}$ as follows: let $U^{(s)}|b\rangle = (\sqrt{\sum_t|\gamma_{s0t}|^2}/g_s)|0\rangle + (-1)^b(\sqrt{\sum_t|\gamma_{s1t}|^2}/g_s)|1\rangle$ if $|s|<n$; otherwise, let $U^{(s)}|b\rangle = (\gamma_{s0}/g_s)|0\rangle + (-1)^b(\gamma_{s1}/g_s)|1\rangle$. The series $\mathcal{U} = \langle U^{(s)} \mid s \in \Sigma^{\leq n}\rangle$ is called the *generator* of $|\phi\rangle$ since $|\phi\rangle = U_nU_{n-1}\cdots U_0|0^{n+1}\rangle$, where $U_0 = U^{(\lambda)}\otimes I^{\otimes n}$ and $U_k = \sum_{s:|s|=k}|s\rangle\langle s|\otimes U^{(s)}\otimes I^{\otimes n-k}$ for each k, $1\leq k\leq n$.

Secondly, we consider a good approximation of a given generator. For any 2×2 matrix $U = (u_{ij})_{1\leq i,j\leq 2}$ on $\mathbb{C}$ and any $\epsilon > 0$, $\tilde{U} = (\tilde{u}_{ij})_{1\leq i,j\leq 2}$ is called the *ϵ-fragment* of U if each $\tilde{u}_{ij}$ represents the first $\lceil\log(1/\epsilon)\rceil$ bits of the infinite binary fractions of the real and imaginary parts of u_{ij} (so that $|u_{ij}-\tilde{u}_{ij}|\leq 2\epsilon$). In this case, $\tilde{U}$ satisfies $\|U-\tilde{U}\|\leq 4\epsilon$. If $\tilde{U}^{(s)}$ is the ϵ-fragment of $U^{(s)}$ for all $s\in\Sigma^{\leq n}$, the series $\tilde{\mathcal{U}} \stackrel{def}{=} \langle\tilde{U}^{(s)} \mid s\in\Sigma^{\leq n}\rangle$ is also called the *ϵ-fragment* of $\mathcal{U}$. We assume a natural encoding scheme of $\tilde{\mathcal{U}}$ into oracle $\langle\tilde{\mathcal{U}}\rangle$ so that $\tilde{\mathcal{U}}$ can be retrieved by $O(2^n\log(1/\epsilon))$ queries to oracle $\langle\tilde{\mathcal{U}}\rangle$.

Lemma 10 *There exists a well-formed QTM M_0 that satisfies the following condition: for every $\epsilon > 0$, every $n\in\mathbb{N}$, and every generator $\mathcal{U}$ of a qustring $|\phi\rangle\in\Phi_{n+1}$, if $\tilde{\mathcal{U}}$ is the $\epsilon 2^{-n-4}$-fragment of $\mathcal{U}$, then M_0 with oracle $\langle\tilde{\mathcal{U}}\rangle$ halts on input $|0^{n+1}\rangle$ in time polynomial in $1/\epsilon$ and n and satisfies $\||\phi\rangle\langle\phi|-\rho\|_{tr}\leq\epsilon$, where ρ is the density matrix obtained from the final superposition of M_0 by tracing out all but the output-tape content and $\|A\|_{tr}$ denotes the trace of $\sqrt{A^\dagger A}$.*

Proof Sketch. The desired M_0 works as follows: at step 0, write $|0^{n+1}\rangle$ in the work tape. Let $s0^{n-k+1}$ be the string written in the work tape after step $k-1$. At step k, make appropriate queries to oracle $\langle\tilde{\mathcal{U}}\rangle$ to realize quantum gate $G^{(s)}$ that simulates $\tilde{U}^{(s)}$ with accuracy at most δ (i.e., $\|G^{(s)}-\tilde{U}^{(s)}\|\leq\delta$). Then, apply $|s\rangle\langle s|\otimes G^{(s)}\otimes I^{\otimes n-k}$ (or $G^{(\lambda)}\otimes I^{\otimes n}$ if $s=\lambda$) to $|s0^{n-k+1}\rangle$. ■

Finally, the third alternative definition of the QP hierarchy is given in Lemma 11. A merit of Lemma 11 is no need of the duplication of quantum inputs given to underlying QTMs. Note that Lemma 11 can be further generalized to non-generators.

Lemma 11 *Let $k \geq 1$. For any classical set $A \subseteq \Sigma^*$, A is in $\Sigma_{k,1}^{\mathrm{QP}}$ iff there exist two polynomials p, q and a polynomial-time well-formed oracle QTM M such that, for all $x \in \Sigma^*$,*

i) if $x \in A$ then $\exists \mathcal{U}_1 \forall \mathcal{U}_2 \cdots Q_k \mathcal{U}_k [\mathrm{Prob}_M[M^{\langle \tilde{\mathcal{U}}_1, \tilde{\mathcal{U}}_2, \ldots, \tilde{\mathcal{U}}_k \rangle}(x) = 1] \geq 3/4]$ and

ii) if $x \notin A$ then $\forall \mathcal{U}_1 \exists \mathcal{U}_2 \cdots \overline{Q}_k \mathcal{U}_k [\mathrm{Prob}_M[M^{\langle \tilde{\mathcal{U}}_1, \tilde{\mathcal{U}}_2, \ldots, \tilde{\mathcal{U}}_k \rangle}(x) = 1] \leq 1/4]$,

where $Q_k = \forall$ if k is even and $Q_k = \exists$ otherwise, $\overline{Q}_k$ is the opposite quantifier of Q_k, each variable $\mathcal{U}_i$ runs over all generators of qustrings of size $p(|x|)$, each $\tilde{\mathcal{U}}_i$ is the $2^{-q(|x|)}$-fragment of $\mathcal{U}_i$, and M on input x behaves as follows: whenever it makes a query, it writes $|1^i\rangle$ in an query tape and runs M_0 (defined in Lemma 10) on input $|0^{p(|x|)}\rangle$ with oracle $\langle \tilde{\mathcal{U}}_i \rangle$.

5. Fundamental Properties of the QP Hierarchy

Although the QP hierarchy looks more complex than its classical counterpart, the QP hierarchy shares many fundamental properties with the polynomial hierarchy. In the next proposition, we present without proofs a short list of fundamental properties of the QP hierarchy.

Proposition 12

1. *For every $k \in \mathbb{N}$, co-${}^*\Sigma_k^{\mathrm{QP}} = {}^*\Pi_k^{\mathrm{QP}}$ and co-${}^*\Pi_k^{\mathrm{QP}} = {}^*\Sigma_k^{\mathrm{QP}}$.*
2. *For each $k \in \mathbb{N}$, ${}^*\Sigma_k^{\mathrm{QP}}$ and ${}^*\Pi_k^{\mathrm{QP}}$ are closed under intersection and union.*
3. *For each $k \in \mathbb{N}$, ${}^*\Sigma_k^{\mathrm{QP}} \cup {}^*\Pi_k^{\mathrm{QP}} \subseteq {}^*\Sigma_{k+1}^{\mathrm{QP}} \cap {}^*\Pi_{k+1}^{\mathrm{QP}}$.*
4. *For every $k \in \mathbb{N}^+$, $\bigcup_{m>0} {}^*\exists_m^{\mathrm{Q}} \cdot ({}^*\Sigma_k^{\mathrm{QP}} \cap {}^*\Pi_k^{\mathrm{QP}}) = {}^*\Sigma_k^{\mathrm{QP}}$.*
5. *Let $k \in \mathbb{N}^+$. If ${}^*\Sigma_k^{\mathrm{QP}} = {}^*\Pi_k^{\mathrm{QP}}$ then ${}^*\Sigma_k^{\mathrm{QP}} = {}^*\mathrm{QPH}$.*
6. *For each $k \in \mathbb{N}^+$, ${}^*\exists \cdot {}^*\Sigma_k^{\mathrm{QP}} = {}^*\exists^{\mathrm{Q}} \cdot {}^*\Sigma_k^{\mathrm{QP}} = {}^*\Sigma_k^{\mathrm{QP}}$.*

Of the above items, item 6 is specifically meant for the QP hierarchy and requires the classical-simulatability of the QP hierarchy, which is shown below.

Lemma 13 *For each $k \in \mathbb{N}$, ${}^*\Sigma_k^{\mathrm{QP}}$ and ${}^*\Pi_k^{\mathrm{QP}}$ are classically simulatable.*

Proof. Since the base case $k = 0$ is easy, we skip this case and prove the general case $k > 0$. Let $m \in \mathbb{N}^+$ and (A, B) be any partial decision problem in ${}^*\Sigma_{k,m}^{\mathrm{QP}}$. There exists a function $f \in {}^*\mathrm{Qopt}\#\Sigma_{k,m}^{\mathrm{QP}}$ that satisfies Lemma 8 for (A, B). Assume that f has the form

$$f(|\vec{\phi}\rangle, |\vec{\psi}\rangle) = \sup_{|\vec{\xi}_1\rangle} \inf_{|\vec{\xi}_2\rangle} \cdots \mathrm{opr}^{(k)}_{|\vec{\xi}_k\rangle} \{h(|\vec{\xi}_1\rangle, |\vec{\xi}_2\rangle, \ldots, |\vec{\xi}_k\rangle, |\vec{\phi}\rangle, |\vec{\psi}\rangle)\}$$

for a certain quantum function h in ${}^*\#\mathrm{QP}$. For brevity, write $|\Xi\rangle$ for $(|\vec{\xi}_1\rangle, \ldots, |\vec{\xi}_k\rangle)$. Letting $h'(|\Xi\rangle, |\vec{\phi}\rangle, |\vec{\psi}\rangle) = \sum_{\vec{x}} |\langle \vec{x} | \vec{\phi} \rangle|^2 h(|\Xi\rangle, |\vec{x}\rangle, |\vec{\psi}\rangle)$, we define g as $g(|\vec{\phi}\rangle, |\vec{\psi}\rangle) = \sup_{|\vec{\xi}_1\rangle} \inf_{|\vec{\xi}_2\rangle} \cdots \mathrm{opr}^{(k)}_{|\vec{\xi}_k\rangle} \{h'(|\vec{\xi}_1\rangle, |\vec{\xi}_2\rangle, \ldots, |\vec{\xi}_k\rangle, |\vec{\phi}\rangle, |\vec{\psi}\rangle)\}$. Note that $g(|\vec{x}\rangle, |\vec{\psi}\rangle) \leq g(|\vec{\phi}\rangle, |\vec{\psi}\rangle)$ if $\langle \vec{x}|\vec{\phi}\rangle \neq 0$, since h' satisfies $h'(|\Xi\rangle, |\vec{x}\rangle, |\vec{\psi}\rangle) \leq h'(|\Xi\rangle, |\vec{\phi}\rangle, |\vec{\psi}\rangle)$ for any $|\Xi\rangle$. It follows that, for every $\vec{x}$, $g(|\vec{x}\rangle, |\vec{\psi}\rangle) = f(|\vec{x}\rangle, |\vec{\psi}\rangle)$.

To complete the proof, defining $C = \{(|\vec{\phi}\rangle, |\vec{\psi}\rangle) \mid g(|\vec{\phi}\rangle, |\vec{\psi}\rangle) \geq 3/4\}$ and $D = \{(|\vec{\phi}\rangle, |\vec{\psi}\rangle) \mid g(|\vec{\phi}\rangle, |\vec{\psi}\rangle) \leq 1/4\}$, we show that C and D are classically separable. Let $|\vec{\phi}\rangle$ be fixed arbitrarily. Assume that $\forall \vec{x}[\langle \vec{x}|\vec{\phi}\rangle = 0 \vee (|\vec{x}\rangle, |\vec{\psi}\rangle) \in C]$. We want to show that $(|\vec{\phi}\rangle, |\vec{\psi}\rangle) \in C$. Assume otherwise. We then have $g(|\vec{\phi}\rangle, |\vec{\psi}\rangle) < 3/4$. Take any element $\vec{x}$ such that $\langle \vec{x}|\vec{\phi}\rangle \neq 0$. It follows that $g(|\vec{x}\rangle, |\vec{\psi}\rangle) \leq g(|\vec{\phi}\rangle, |\vec{\psi}\rangle) < 3/4$, which implies $(|\vec{x}\rangle, |\vec{\psi}\rangle) \notin C$, a contradiction. The case for D is similar. Therefore, ${}^*\Sigma_k^{\mathrm{QP}}$ is classically simulatable. □

Next, we give rudimentary but meritorious upper and lower bounds of the QP hierarchy. The *exponential hierarchy* consists of the following classes: $\Delta_0^{\mathrm{EXP}} = \Sigma_0^{\mathrm{EXP}} = \Pi_0^{\mathrm{EXP}} = \mathrm{EXP}$ $(= \mathrm{DTIME}(2^{n^{O(1)}}))$, $\Delta_k^{\mathrm{EXP}} = \mathrm{EXP}^{\Sigma_k^{\mathrm{P}}}$, $\Sigma_k^{\mathrm{EXP}} = \mathrm{NEXP}^{\Sigma_{k-1}^{\mathrm{P}}}$, and $\Pi_k^{\mathrm{EXP}} = \text{co-}\Sigma_k^{\mathrm{EXP}}$ for every $k \in \mathbb{N}^+$. Let EXPH denote the union of Σ_k^{EXP} for all $k \in \mathbb{N}$. We show that $\mathrm{PH} \subseteq \mathrm{QPH} \subseteq \mathrm{EXPH}$.

Theorem 14 *For each* $k > 0$, $\Sigma_k^{\mathrm{P}} \subseteq \Sigma_k^{\mathrm{QP}} \subseteq \Sigma_k^{\mathrm{EXP}}$. *Thus,* $\mathrm{PH} \subseteq \mathrm{QPH} \subseteq \mathrm{EXPH}$.

Theorem 14 yields the following collapse: if $\Pi_k^{\mathrm{P}} \subseteq \Sigma_k^{\mathrm{QP}}$ then $\mathrm{PH} \subseteq \Sigma_k^{\mathrm{QP}}$.

Proof of Theorem 14. We show the first inclusion that $\Sigma_k^{\mathrm{P}} \subseteq \Sigma_k^{\mathrm{QP}}$. The proof is done by induction on $k \geq 0$. The base case $k = 0$ follows from $\mathrm{P} \subseteq \mathrm{BQP}$. Let $k > 0$. By the induction hypothesis, it follows that $\Sigma_{k-1}^{\mathrm{P}} \subseteq \Sigma_{k-1}^{\mathrm{QP}}$, which further implies $\Pi_{k-1}^{\mathrm{P}} \subseteq \Pi_{k-1}^{\mathrm{QP}}$. Since ${}^*\Pi_{k-1}^{\mathrm{QP}}$ is classically simulatable (Lemma 13), Lemma 6 yields $\exists \cdot {}^*\Pi_{k-1}^{\mathrm{QP}} \subseteq \exists^{\mathrm{Q}} \cdot {}^*\Pi_{k-1}^{\mathrm{QP}}$. Thus,

$$\Sigma_k^{\mathrm{P}} = \exists \cdot \Pi_{k-1}^{\mathrm{P}} \subseteq \exists \cdot {}^*\Pi_{k-1}^{\mathrm{QP}} \subseteq \bigcup_{m>0} \exists_m^{\mathrm{Q}} \cdot {}^*\Pi_{k-1}^{\mathrm{QP}} = \Sigma_k^{\mathrm{QP}}.$$

The second inclusion $\Sigma_k^{\mathrm{QP}} \subseteq \Sigma_k^{\mathrm{EXP}}$ follows from Lemma 9. Let $m \in \mathbb{N}^+$ and let A be any set in $\Sigma_{k,m}^{\mathrm{QP}}$. Take a polynomial p and a quantum function $f \in {}^*\#\mathrm{QP}$ guaranteed by Lemma 9 for $(A, \overline{A})$. We construct an alternating TM N as follows: on input x, start with an $\exists$-state, generate k vectors $|\vec{\xi_1}\rangle, |\vec{\xi_2}\rangle, \ldots, |\vec{\xi_k}\rangle$ in $(\tilde{\Phi}_{p(|x|)}(3p(|x|)))^m$ by alternately entering $\forall$- and $\exists$-states, and check if $f(|x\rangle, |\vec{\xi_1}\rangle, |\vec{\xi_2}\rangle, \ldots, |\vec{\xi_k}\rangle) \geq 3/4$. This last check is done in exponential time since f runs in time polynomial in $|x|$. Since any exponential-time alternating TM with k-alternation starting with the $\exists$-state is known to characterize Σ_k^{EXP}, A belongs to Σ_k^{EXP}. □

In the end of this section, we discuss the issue of complete problems. Each level of the polynomial hierarchy is known to have complete problems. Unfortunately, it is believed that classes like BQP and QMA lack such complete problems because of their acceptance criteria. Dealing with partial decision problems, however, allows us to go around this difficulty. (See, e.g., [5] for the NP-completeness for classical partial decision problems.) With an appropriate modification of classical completeness proofs, we can show that each level of the QP hierarchy indeed has a "complete" partial decision problem (under a deterministic reduction). An important open problem is to find natural complete partial decision problems for each level of the QP hierarchy.

6. Relativized QP Hierarchies

We have introduced a quantum analogue of the polynomial hierarchy and explored its basic properties and its relationship to the polynomial hierarchy. In this last section, we give simple relativized results related to the QP hierarchy. The *relativized QP hierarchy relative to* oracle A, $\{\Sigma_k^{\mathrm{QP}}(A), \Pi_k^{\mathrm{QP}}(A) \mid k \in \mathbb{N}\}$, is obtained simply by changing the basis class $^*\#\mathrm{QP}$ to its relativized version $^*\#\mathrm{QP}^A$.

Proposition 15 *1. There exists a recursive oracle A such that* $\mathrm{P}^A = \mathrm{PH}^A = \mathrm{QPH}^A$.

2. There exists a recursive oracle B such that $\Sigma_0^{\mathrm{QP}}(B) \neq \Sigma_1^{\mathrm{QP}}(B) \neq \Sigma_2^{\mathrm{QP}}(B)$.

3. There exists a recursive oracle C such that $\Sigma_k^{\mathrm{P}}(C) \neq \Sigma_k^{\mathrm{QP}}(C)$ for all $k \in \mathbb{N}^+$.

For the first claim of Proposition 15, it suffices to construct A such that $\mathrm{P}^A = \Sigma_1^{\mathrm{QP}}(A)$ since this yields $\Sigma_1^{\mathrm{QP}}(A) = \Pi_1^{\mathrm{QP}}(A)$, which further implies $\Sigma_1^{\mathrm{QP}}(A) = \mathrm{QPH}^A$. The desired set A is built by stages: at each stage, pick one relativized #QP-function f and encode its outcome into one string that cannot be queried by f. This is possible because an oracle QTM that witnesses f runs in polynomial time.

The second claim of Proposition 15 follows from a generalization of the result $\text{co-UP}^B \not\subseteq \bigcup_{k\geq 1} \mathrm{QMA}(k)^B$ [12]. In fact, we can show by modifying Ko's argument [11] a slightly stronger result: there exists an oracle B satisfying that $\text{co-UP}^B \not\subseteq \Sigma_1^{\mathrm{QP}}(B)$ and $\mathrm{BP}\cdot\Sigma_1^{\mathrm{QP}}(B) = \Sigma_1^{\mathrm{QP}}(B)$ with Schöning's BP-operator. It is easy to see that this yields the desired claim. Of particular interest is to show that $\text{co-UP}^B \not\subseteq \Sigma_1^{\mathrm{QP}}(B)$. This is done by cultivating a lower bound technique of a certain type of a real-valued circuit. Since the acceptance probability of an oracle QTM computation can be expressed by a multilinear polynomial of small degree [2], we can convert a $\Sigma_{1,m}^{\mathrm{QP}}(A)$-computation into a family of circuits C of depth-2 (that work on real numbers) such that (i) the top gate of C is a MAX-gate of fanin at most $m2^{n2^{n+1}}$ and (ii) all bottom gates of C are polynomial-gates of degree at most n with fanin 2^n, where a MAX-gate is a gate that takes real numbers as its inputs and outputs their maximal value and a *polynomial-gate* of degree k refers to a gate that computes a multilinear polynomial of degree exactly k. The *fanin* of such a polynomial gate is the number of variables actually appearing in its underlying polynomial.

The existence of B comes from the fact that such a family of circuits cannot approximate to within $1/3$ any Boolean function of large block sensitivity (e.g., a co-UP^B-computation). The oracle separation at higher levels of the QP hierarchy is one of the remaining open problems.

The third claim of Proposition 15 follows from the result $\mathrm{QMA}^C \not\subseteq \mathrm{MA}^C$ [16] and its generalization. The detail will appear in the complete version of this extended abstract.

References

[1] L. M. Adleman, J. DeMarrais, and M. A. Huang. Quantum computability, *SIAM J. Comput.* **26** (1997), 1524–1540.

[2] R. Beals, H. Buhrman, R. Cleve, M. Mosca, and R. de Wolf. Quantum lower bounds by polynomials, in *Proceedings of the 39th Annual Symposium on Foundations of Computer Science*, pp.352–361, 1998.

[3] C. H. Bennett, E. Bernstein, G. Brassard, and U. Vazirani. Strengths and weaknesses of quantum computing, *SIAM J. Comput.* **26** (1997), 1510–1523.

[4] E. Bernstein and U. Vazirani. Quantum complexity theory, *SIAM J. Comput.* **26** (1997), 1411–1473.

[5] D. Du and K. Ko. *Theory of Computational Complexity*, John Wiley & Sons, Inc., 2000.

[6] S. Fenner, F. Green, S. Homer, and R. Pruim. Determining acceptance probability for a quantum computation is hard for the polynomial hierarchy, *Proceedings of the Royal Society of London*, Ser.A, **455** (1999), 3953–3966.

[7] L. Fortnow and J. Rogers. Complexity limitations on quantum computation, *J. Comput. System Sci.* **59** (1999), 240–252.

[8] F. Green, S. Homer, C. Moore, and C. Pollett. Counting, fanout, and the complexity of quantum ACC, *Quantum Information and Computation*, **2** (2002), 35–65.

[9] A. Kitaev. "Quantum NP", Public Talk at AQIP'99: the 2nd Workshop on Algorithms in Quantum Information Processing, DePaul University, 1999.

[10] E. Knill. Quantum randomness and nondeterminism, Technical Report LAUR-96-2186, 1996. See also LANL quant-ph/9610012.

[11] K. Ko. Separating and collapsing results on the relativized probabilistic polynomial-time hierarchy, *J. ACM* **37** (1990), 415–438.

[12] H. Kobayashi, K. Matsumoto, and T. Yamakami. Quantum Merlin Arthur proof systems, manuscript, 2001. See also LANL quant-ph/0110006.

[13] A. R. Meyer and L. J. Stockmeyer. The equivalence problem for regular expressions with squaring requires exponential time, in *Proceedings of the 13th Annual Symposium on Switching and Automata Theory*, pp.125–129, 1972.

[14] U. Schöning. Probabilistic complexity classes and lowness, *J. Comput. System and Sci.* **39** (1989), 84–100.

[15] K. Wagner. The complexity of combinatorial problems with succinct input representation, *Acta Inf.* **23** (1986), 325–356.

[16] J. Watrous. Succinct quantum proofs for properties of finite groups, in *Proceedings of the 41st Annual Symposium on Foundations of Computer Science*, pp.537–546, 2000.

[17] T. Yamakami. A foundation of programming a multi-tape quantum Turing machine, in Proceedings of the 24th International Symposium on Mathematical Foundation of Computer Science, Lecture Notes in Computer Science, Vol.1672, pp.430–441, 1999.

[18] T. Yamakami. Analysis of quantum functions, in Proceedings of the 19th International Conference on Foundations of Software Technology and Theoretical Computer Science, Lecture Notes in Computer Science, Vol.1738, pp.407–419, 1999.

[19] T. Yamakami. Quantum optimization problems, manuscript, 2002. See LANL quant-ph/0204010.

[20] T. Yamakami and A. C. Yao. $NQP_{C} = co\text{-}C_{=}P$, *Inf. Process. Let.* **71** (1999), 63–69.

PROBABILISTICALLY CHECKABLE PROOFS THE EASY WAY

Marius Zimand

Department of Computer and Information Sciences, Towson University, Baltimore, MD, and Department of Computer Science, University of Bucharest, Bucharest, Romania.

mzimand@towson.edu

Abstract We present a weaker variant of the PCP Theorem that admits a significantly easier proof. In this variant the prover only has n^t time to compute each bit of his answer, for an arbitrary but fixed constant t, in contrast to being all powerful. We show that 3SAT is accepted by a polynomial-time probabilistic verifier that queries a constant number of bits from a polynomially long proof string. If a boolean formula ϕ of length n is satisfiable, then the verifier accepts with probability 1. If ϕ is not satisfiable, then the probability that a n^t-bounded prover can fool the verifier is at most 1/2. The main technical tools used in the proof are the "easy" part of the PCP Theorem in which the verifier reads a constant number of bits from an exponentially long proof string, and the construction of a pseudo-random generator from a one-way permutation.

Keywords: PCP Theorem, sampling, communication complexity

1. Introduction

The PCP Theorem (AS92, ALM+92) is one of the most important results in computer science. It gives an astonishing interactive proof protocol for any language in NP and it has been used to obtain impressive lower bounds on the approximation ratio that can be achieved in polynomial time for many important optimization combinatorial problems. The theorem states that for any language L in NP, a polynomial-time probabilistic verifier needs to check only a constant number of bits of a polynomially long proof of membership; if an input x is in L, then the verifier accepts with probability one (completeness condition), and if x is not in L, the verifier can be fooled to accept a (false) proof of membership with probability at most, say, 1/4 (the soundness condition).

The proof of the PCP Theorem is long and difficult. Moreover, the underlying construction is extremely complex and basically impractical. Some researchers (Goldreich (Gol99, pp.71), Sudan (Sud00b), Trevisan (Tre00)) have formulated as an open problem the discovery of a simpler proof. Trevisan (Tre00) says that a respectable task is to "focus on statements which (i) can be derived from (but are weaker than) the PCP Theorem, (ii) are already surprising enough to be interesting, (iii) are not known to have simple proofs; and to try and find simple proofs for such statements." Why would a weaker version of the PCP Theorem having a simpler proof and be worthy of interest?

One reason may be strictly utilitarian. A direct application of the PCP Theorem would be in program verification, that is in building certificates for the output of a program that can be checked by reading only a constant number of bits from them. The complexity of the construction in the proof of the PCP Theorem has deterred such an application. A simpler construction, even with weaker guarantees, may be useful here. Also, a simpler and weaker version of the PCP Theorem would be the natural choice in case we need to integrate the PCP into a larger protocol which assumes the weaker conditions anyway. Such a concrete application is discussed in the final paragraph of this section.

Another reason is pedagogical. A book entitled "PCP for Dummies" would be an immediate best-seller. Joke aside, given its stunning content, as well as its importance, it would be desirable to present the PCP Theorem in any course on computability or computational complexity at the graduate level (or even below). A weaker variant of the PCP Theorem that retains some of its striking characteristics, and whose proof can be covered in, say, 2-3 lectures would be, we think, an interesting alternative for standard Theory courses.

In this paper we present a weaker variant of the PCP theorem (the Light PCP Theorem) that is of interest because of the above reasons. In this variant, the verifier is still a polynomial-time probabilistic machine. The verifier still reads a constant number of bits from a membership proof that is polynomially long. The variant still applies to languages L in NP. The completeness condition is the same. The weakening is in the soundness condition. We will show that the protocol is safe against (dishonest) provers that can spend at most polynomial-time computational power to calculate each bit of the proof string (in the PCP theorem the dishonest provers can have unlimited power). Such provers are called *scribes*. The soundness condition says that if the input string x is not in L, no scribe can fool the verifier to accept x but with constant probability. More precisely, a scribe is a polynomial-size circuit that has as input (1) the string x the verifier wants to check if it is in L or not, (2) some other string of polynomial size in $|x|$ that may be the witness for the membership of x in L, and (3) a natural number i. On this input the scribe produces the i-th bit of the proof string. Thus in case x is in L, a scribe will be given a witness to show this and will produce a proof string to convince the verifier that x is in

L. Note that even though polynomially bounded, the computational power of the scribes can be much higher than the computational power of the verifier.

The merit of this result is that it has a simpler proof whose structure we sketch here. The proof has two parts. The first part is taken from the proof of the full PCP Theorem. Namely, it is the "easy" step which states that 3SAT $\in \mathrm{PCP}_{1,1/4}(O(n^3), O(1))$. This means that for a boolean formula ϕ in 3CNF, having length $|\phi| = n$, the verifier is using $O(n^3)$ random bits and reads $O(1)$ bits from a string w provided by the prover. If ϕ is satisfiable, then there is a string w (the correct proof string) so that the verifier accepts with probability 1. If ϕ is not satisfiable, then for .ny string w, the verifier accepts with probability at most 1/4. The proof of this step uses the main tools of the full PCP proof (i.e., arithmetization, error-correcting codes, consistency tests) in an elegant and relatively easy to understand way. However, the number of random bits is $poly(n)$ and, therefore, the length of the proof string w is exponential.

The second part reduces the length of the proof string w to $poly(n)$. The idea is to use sampling. The verifier selects a random subset $A \subset B$ of polynomial size, where B is the set of all $2^{O(n^3)}$ strings that can be chosen as random strings in the $\mathrm{PCP}_{1,1/4}(O(n^3), O(1))$ protocol. The verifier will always choose a random string from A, identified by its rank in A, and thus the number of random bits is $O(\log |A|) = O(\log n)$. If the prover does not know A, this does not reduce the length of the proof string, because the proof string must still contain the responses to all the queries that can be calculated by the verifier with the random strings chosen from the entire B. If the prover knows A, then the proof string can be of polynomial length (because the prover can prepare the answers to only those queries produced from the random strings in the sample set A), but, in this case, normal sampling is no longer guaranteed to give an accurate estimate of the probability in the $\mathrm{PCP}_{1,1/4}(O(n^3), O(1))$ protocol. Indeed, a dishonest prover, knowing the sample points in A, could provide some answers that lead the verifier to inadvertently accept with a probability much larger than in the case the random string is chosen from B. We need to produce sample points (i.e., the elements of the set A) that are good for estimating the average value of a function, even if the function is chosen afterwards and can depend on the sample points. In some circumstances this is possible: we show, roughly speaking, that a modified sampling procedure continues to be accurate if the prover knows A, provided that $|A| = n^{(4+\epsilon)t}$ and that the function that is sampled is computable by a circuit with oracle access to A and of size n^t, for an arbitrary t. The modified sampling procedure, dubbed "sampling under adverse conditions," has been introduced by us in (Zim99) but we present here a simpler proof. "Sampling under adverse conditions" relies on the construction of a pseudo-random generator from a one-way function (HILL99). Fortunately, what we need here is the easy case of that construction, when the one-way function is a permutation, and thus the whole proof remains relatively simple.

Interactive proof systems in which the prover is computationally bounded have been considered before. Argument systems have been introduced by Brassard, Chaum, and Crépeau (BCC88), and they require that the prover has access to an auxiliary input (as our scribe does) and that it runs in probabilistic polynomial-time. Computationally-Sound (CS) Proofs have been introduced by Micali (Mic00) to handle problems beyond NP. CS-proofs require that the prover writes down a proof in time polynomial in the decision time, and that the verifier works in time polynomial in the input length and polylog in the decision procedure (for example, for a language in EXP, the verifier works in polynomial time). Argument systems have been used to reduce the communication complexity of the interaction between the prover and the verifier. Kilian (Kil92) has shown that under a complexity assumption (existence of strong collision-free functions), for any $L \in$ NP, there is an argument system with communication complexity $polylog(n)$ (actually his protocol is also zero-knowledge, an aspect that we do not consider here). CS-proofs have been used to reduce the work of the verifier (for languages above EXP). Micali (Mic00) has shown that CS-proofs exist for any recursive language if the prover and the verifier have access to a random oracle. The proofs in both (Kil92) and (Mic00) work in two steps: In the first one, the PCP Theorem is used to produce a "holographic" proof, and then, in the second step, the proof string is shrunk by using cryptographic techniques which need the assumption that the prover is computationally bounded. Since this assumption is needed anyway for reducing the communication complexity (see (GH98) and (GVW01)), it is natural to ask whether it cannot be used to simplify the first step. Indeed, this is the case. We show that the Light PCP Theorem can be used to obtain an interactive protocol for any NP language with communication complexity $O(\log^2 n)$ that is sound against any prover that is bounded by a fixed polynomial.

2. The model

Let us first recall the standard model of PCP$[r(n), q(n)]$. A verifier V executing a PCP$[r(n), q(n)]$ protocol is a polynomial-time probabilistic Turing machine that in addition to its working tapes has three special tapes:

- the input tape, containing the input string x having length n,
- the random tape, containing the random bits forming a string ρ of length $r(n)$ that the verifier will use in its computation, and
- the proof tape, that contains the proof string w.

The verifier based on the input x and on ρ, first determines $q(n)$ bit positions in the proof string that it wants to read, reads these bits, and then performs some additional polynomial-time calculation at the end of which it accepts or it rejects the input.

Let the output of the above computation of the verifier be denoted by $V(\rho, w, x)$. A language L is in PCP$_{c(n),s(n)}[r(n), q(n)]$ if there is a verifier V executing a PCP$[r(n), q(n)]$ protocol with the following properties:

(i) If $x \in L$, then there is a proof string w such that $Prob_\rho(V(\rho, w, x) =$ "*accept*"$) \geq c(n)$, (completeness condition)

(ii) If $x \notin L$, then for any proof string w it holds that $Prob_\rho(V(\rho, w, x) =$ "*accept*"$) \leq s(n)$ (soundness condition)

Let us introduce our model called *probabilistically checkable proof with scribes*, abbreviated PCPS. For brevity, we consider the language 3SAT, but the model can be easily extended to any language in NP. A *scribe* of complexity $t(n)$ is an oracle circuit of size $t(n)$. A scribe has as input a boolean formula ϕ, an assignment for it called a, and an integer i. The scribe produces the i-th bit of the proof string. A verifier V is the same as above except that it has an extra tape called the oracle tape. A PCPS$[r(n), q(n), t(n)]$ protocol on input a formula ϕ in 3CNF of length n runs as follows: *Round 1:* The verifier writes on the oracle tape a random string R of polynomial size. The string R is called the public random string because the scribe has access to it.

Round 2: The verifier produces a random string r of length $r(n)$ that it keeps private. Based on ϕ, R, and r, the verifier selects $q(n)$ addresses in the proof string (that will be provided by the scribe in Round 3). The bits of the proof strings at these addresses will be queried in Step 4.

Round 3: A scribe of complexity $t(n)$ is using an assignment for ϕ and the string R as the oracle. The scribe produces bit by bit a proof string denoted $w^R(\phi, a)$ which is passed to the verifier.

Round 4: The verifier V reads from $w^R(\phi, a)$ the bits at the addresses selected at Round 2. At the end it accepts or it rejects the input.

We denote acceptance by 1, and rejection by 0, and we denote the output of the entire protocol by $V^R(r, w^R(\phi, a), \phi)$.

Remarks. Rounds 2 and 3 can be permuted. The random string R from Round 1 is public and, therefore, does not count for the communication complexity. Moreover, it can be seen from the proof of the Light PCP Theorem that the same R can be reused for all inputs of a given length and for all scribes of a given complexity.

The differences between a PCP and a PCPS protocol are: (a) the introduction of Round 1 in the PCPS protocol, which basically is used by the verifier to announce to the prover the subset A of B as discussed in the Introduction, and (b) the fact that in the PCPS protocol, the prover (called a scribe), after being given an assignment, has limited resources to produce each bit of the proof string, while in the PCP protocol, the provers have unlimited computational power. We can consider that behind the scene there is an all-powerful prover that passes n bits of information (encoded as an assignment) to the scribe to help him make the verifier accept the formula ϕ.

Definition 1 *A language L is in* PCPS$[r(n), q(n), t(n)]$ *if there is a verifier running a* PCPS$[r(n), q(n), t(n)]$ *protocol with the following properties:*

(1) (Completeness) If $x \in L$, then there is a scribe of complexity $t(n)$ and a string a of length poly($|x|$) such that for any R

$$Prob_r(V^R(r, w^R(x, a), x) = accept) = 1.$$

(2) (Soundness) If $x \notin L$, then with high probability of R (i.e., with probability at least $1 - 2^{-\Omega(n)}$), for any string a and for any scribe of complexity $t(n)$, it holds that

$$Prob_r(V^R(r, w^R(x, a), x) = accept) \leq 1/2.$$

3. Main result

Theorem 2 *(Light PCP Theorem) For any $t \geq 3$,*
3SAT $\in$ PCPS$[O(\log n), O(1), n^t]$.

This fact is an immediate consequence of the PCP Theorem (and of its proof), but we will show it without using the PCP theorem. Let us first clarify the meaning of Theorem 2. It shows that there is a probabilistic polynomial-time machine that on input a formula ϕ of length n, does the following. It uses two random strings, R of length $poly(n)$ and r of size $O(\log n)$. It sends R to the scribe. The scribe having access to R, to ϕ, and to an assignment a for ϕ, writes the proof string, spending time at most n^t in calculating each bit of the proof string. Next the verifier reads a constant number of bits from $w^R(\phi, a)$, and accepts or rejects. If ϕ is satisfiable, then there is a scribe of complexity n^t that for any R will determine V to accept with probability 1. On the other hand, if ϕ is not satisfiable, then with high probability of R, no scribe of complexity n^t can determine V to accept except with probability of r at most 1/2. It is noteworthy that the proof of Theorem 2 yields a stronger result in the sense that the verifier does not have to produce a random string R at Round 1 for each input. One public random string R is with high probability safe against all n^t scribes (with t fixed), on all strings of length n.

The proof of Theorem 2 relies on the following two theorems.

Theorem 3 $3SAT \in \mathrm{PCP}_{1,1/4}(O(n^3), O(1))$.

This is a well-known and relatively "easy" step in the proof of the PCP Theorem.

The second theorem that we use states that sampling is accurate even if the function that is sampled is chosen adversarially after the sample points have been selected, provided that the function is computable by a polynomial-size circuit.

Theorem 4 *For any $\alpha, \beta \geq 0$, for any τ sufficiently large, there is a function $f : \Sigma^* \times \Sigma^* \to \Sigma^*$, and a polynomial p, such that for any natural m*

(a) For R with $|R| = p(m)$ and for r with $|r| = (4 + \alpha) \cdot \tau \cdot \log m$, $f(R, r)$ has length m;

(b) With probability of R at least $1 - 2^{-poly(m)}$, for any oracle circuit C with inputs of length m, having size m^τ, and that outputs 1 or 0,

$$\left| \frac{1}{2^{|r|}} \sum_r C^R(f(R,r)) - \frac{1}{2^m} \sum_z C^R(z) \right| \leq m^{-\beta} \qquad (1)$$

where the first sum is taken over all the strings r of length $(4+\alpha)\cdot\tau\cdot\log m$, *and the second sum is over all the strings of length* m,

(c) *every bit of* $f(R,r)$ *can be computed in time polylog(m) independently of the other bits.*

The result has been established by us in (Zim99). In Section 4 we present a simpler proof.

Proof of Theorem 2. Let us fix a boolean formula ϕ in 3CNF, and let n be the length of ϕ. According to Theorem 3, there is a verifier V' running a $\text{PCP}_{1,1/4}(cn^3, q)$ protocol for some constants c and q. The computation of V' depends on the formula ϕ, the random string ρ, and the proof string w.

We now build a verifier V that simulates V', but runs a PCPS protocol with $O(\log n)$ private random bits. One obvious problem for the simulation is that V' is using cn^3 random bits that are not disclosed to the prover, while V can only use $O(\log n)$ private random bits. To solve this problem, we consider the function f promised by Theorem 4 with $m = cn^3, \alpha = 1, \beta = 1$ and a constant value of τ which will be specified later. In the first round, V writes a random string R of size $p(m)$ (p is the polynomial from Theorem 4), with the intention of using as the random string of length cn^3 needed in the simulation of V' only strings from the set $X_R = \{f(R,r) \mid |r| = 5\cdot\tau\cdot\log m\}$. Let ℓ denote $5\cdot\tau\cdot\log m$. In the second round, V selects uniformly at random a string r in $\{0,1\}^\ell$, calculates $f(R,r)$, simulates V' with the random string $f(R,r)$ having the desired length of $m = cn^3$ to determine the addresses in the proof string that V' is going to query later. In the third round, a scribe having ϕ and an assignment a for ϕ, and having access to R, calculates a proof string $w^R(\phi,a)$ spending no more than n^t time per bit of the proof string. It then passes $w^R(\phi,a)$ to the verifier. In the fourth round, V simulates V' with the queries established in round 2, and with the proof string $w^R(\phi,a)$. If some queried addresses are not in $w^R(\phi,a)$, then V rejects. Otherwise the simulation can be completed and V gives the verdict (1 for accept, 0 for reject) that the simulated V' gives.

Let us assume that ϕ is satisfiable and a is an assignment that makes ϕ to be true. According to Theorem 3, there is a proof string w such that $Prob_\rho(V'(\rho,w,\phi)=1)=1$. An inspection of the proof of Theorem 3 shows that given a, each bit of w can be calculated in $O(n^3)$ time. For any R written by V in the first round, a scribe, having the assignment a, produces the bits of w that V' queries when the random string ρ is taken from X_R. It follows that for every R, and for every r of length ℓ,

$$V^R(r, w^R(\phi,a),\phi) = V'(f(R,r),w,\phi) = 1.$$

Thus, if ϕ is satisfiable, for any string R,

$$Prob_{r\in\{0,1\}^\ell}(V^R(r,w^R(\phi,a),\phi)=1)=1.$$

This proves the completeness condition for the PCPS protocol.

Let us consider the case in which ϕ is not satisfiable. Then for any proof string w, $Prob_\rho(V'(\rho, w, \phi) = 1) \leq 1/4$. Let us fix an assignment a for ϕ and a scribe S producing each bit of the proof string in time n^t. The scribe S, on input ϕ and a, and with some R on the oracle tape, tries to convince V to accept. We build next an oracle circuit C that simulates the whole protocol run by the verifier V' and the scribe S. The circuit C has ϕ and a embedded into its circuitry and has as input a string ρ of size cn^3. C first simulates V' and determines the addresses $i_1, \ldots, i_q$ in the proof string that are queried by V' on input formula ϕ and random string ρ. Next it simulates the scribe S to determine what are the bits at addresses $i_1, \ldots, i_q$ of the proof string $w^R(\phi, a)$ that is produced by S. Finally, the circuit C simulates the last round of the computation of V' and accepts or rejects accordingly. Thus the circuit C simulates V' on the following input: The boolean formula ϕ, the random string ρ, and the bits of the proof string obtained as specified above. C also simulates the scribe S to determine the q bits of the proof string queried by V'. The simulation of V' takes $p_1(n)$ for some polynomial p_1, and S produces one bit in time n^t. Thus the size of the circuit C is bounded by $p_1(n) \log p_1(n) + qn^t \leq (cn^3)^\tau = m^\tau$ for some constant τ. This is the value of τ for which we use Theorem 4. We denote by $\mathcal{R}$ the set of strings R for which the equation (1) holds. Recall that the size of $\mathcal{R}$ represents a fraction of $1 - 2^{-poly(m)}$ from the set of strings of length $p(m)$.

Let $g^R(\rho)$ be the output of C on input ρ running with oracle R. From the simulation it holds that for every R,

$$g^R(\rho) = V'(\rho, w^R(\phi, a), \phi).$$

Also, g^R is calculated by the oracle circuit C of size n^τ. Thus, by Theorem 4, if $R \in \mathcal{R}$

$$\left|Prob_{r \in \{0,1\}^\ell}(g^R(f(R,r)) = 1) - Prob_{\rho \in \{0,1\}^m}(g^R(\rho) = 1)\right| \leq m^{-1} = \frac{1}{cn^3}.$$

Since

$$Prob_{\rho \in \{0,1\}^m}(g^R(\rho) = 1) = Prob_{\rho \in \{0,1\}^m}(V'(\rho, w^R(\phi, a), \phi) = 1) \leq \frac{1}{4},$$

it follows that for $R \in \mathcal{R}$,

$$Prob_{r \in \{0,1\}^\ell}(g^R(f(R,r)) = 1) \leq \frac{1}{4} + \frac{1}{cn^3} < \frac{1}{2}.$$

Note that $g^R(f(R,r))$ is exactly $V^R(r, w^R(\phi, a), \phi)$. Therefore, for any $R \in \mathcal{R}$, if $w^R(\phi, a)$ is produced by a scribe of complexity n^t, and a is any assignment for ϕ,

$$Prob_{r \in \{0,1\}^\ell}(V^R(r, w^R(\phi, a), \phi) < 1/2.$$

This proves the soundness condition for the PCPS protocol run by V. ∎

4. Sampling under adverse conditions

We now turn to Theorem 4. Note that Equation (1) means that, with high probability of R, the function $f(R,\cdot)$, which is constructed in the theorem, is a pseudo-random generator in the sense that no circuit oracle C of size m^τ can distinguish between the output of $f(R,\cdot)$ and a random string of length m with a bias larger than $m^{-\beta}$. Therefore, we need to build a function depending on a string R, that with high probability of R is a pseudo-random generator. There are basically two known approaches and both can be utilized in our context. The first one is to build a predicate that is hard on average and then to use the method of Nisan and Wigderson (NW94) and construct from it the pseudo-random generator. Using the random string R this is not hard to do, but it yields slightly weaker parameters (at the level of constants) than the second approach which we present next. This second method consists of a randomized construction of a one-way permutation and then of the standard transformation of a one-way permutation into a pseudo-random generator. The construction has four steps. In Step 1, we show that a random permutation from $\{0,1\}^n$ into $\{0,1\}^n$ is a one-way permutation. This result has been obtained by Gennaro and Trevisan (GT00), but we give here a different proof, which uses a technique of Impagliazzo (Imp96), and which allows more flexibility in the choice of the parameters. (We note however that the result from (GT00) would have been sufficient here.) In the second step, using the construction of Goldreich and Levin (GL89), we obtain a hidden bit, which, when appended to the one-way function from Step 1, yields a pseudo-random generator with expansion 1. In Step 3, using the hybrid technique (see for example (Gol93)), we make the pseudo-random generator to produce an output with length double the length of the input. Finally, in Step 4, we use the technique of Goldreich, Goldwasser and Micali (GGM86) to obtain a pseudo-random generator with exponential expansion. The constructions in Steps 2, 3, and 4, are well-known and therefore we will not present here the underlying proofs. For our claim for the relative simplicity and the pedagogical virtues of Theorem 2, it is however important to note that these proofs are reasonably short, self-contained, and important in their own right, and that the corresponding constructions are easy to implement.

Proof of Theorem 4 Let n be a natural number considered as a parameter. (This is not the n from the proof of Theorem 2; actually it is big-O of the log of that n.)

Step 1: Build a one-way permutation. We start by taking uniformly at random a permutation $h:\{0,1\}^n \rightarrow \{0,1\}^n$.

Proposition 5 *Let a and b be positive real numbers and let $s = 2^{an}, t = 2^{bn}$. Let C be an oracle circuit that on inputs of length n, makes at most s queries to the oracle. Let h be a random permutation, $h:\{0,1\}^n \rightarrow \{0,1\}^n$. Then with probability of h at least $1-2^{-t}$, $Prob_x(C^h(x) = h^{-1}(x)) < 2e \cdot 2^{-(1-a-b)n}$.*

Proof. Let $T = \{y_1,\ldots,y_t\} \subseteq \{0,1\}^n$ be a fixed set of size t. W.l.o.g., we can assume that the circuit C on an input y queries at some point its output x to check if $h(x) = y$. Let Q be the set of queries that C makes on inputs $y_1, \dots, y_t$. Clearly the size of Q, denoted $|Q|$, is at most st. The probability that C inverts $y_1, \dots, y_t$ is bounded from above by the probability that t queries from Q map via h respectively into $y_1, \dots, y_t$. The probability that t fixed queries from Q map in order into $y_1, \dots, y_t$ is $1/(N(N-1)\dots(N-t+1))$, where $N = 2^n$, and the number of ordered t-tuples chosen in Q is $|Q|(|Q|-1)\dots(|Q|-t+1) \leq st(st-1)\dots(st-t+1)$. Thus the probability that C inverts T is at most

$$\frac{st(st-1)\dots(st-t+1)}{N(N-1)\dots(N-t+1)} = \frac{\binom{st}{t}}{\binom{N}{t}} \leq \frac{(e \cdot s)^t}{\binom{N}{t}}.$$

There are $\binom{N}{t}$ ways to choose the set T in $\{0,1\}^n$, and, therefore, the expected number of sets of size t, which we denote by μ, that are inverted is at most

$$\binom{N}{t} \cdot \frac{(e \cdot s)^t}{\binom{N}{t}} = (e \cdot s)^t.$$

We take $u = 2e \cdot s \cdot t$. If there is a set of size u that is inverted, then all its subsets of size t are inverted, and there are $\binom{u}{t}$ such subsets. We have that

$$\binom{u}{t} \geq \left(\frac{u}{t}\right)^t = 2^t \cdot (e \cdot s)^t \overset{\text{def.}}{=} k.$$

By Markov Inequality, the probability that k sets of size t are inverted is at most $\mu/k \leq (e \cdot s)^t/k = 2^{-t}$. Thus, we have shown that the probability over h that C^h inverts $2e \cdot s \cdot t = 2e \cdot 2^{(a+b)n}$ strings of length n is at most 2^{-t}. ∎

Corollary 6 *Let $\gamma_1 > 0$. With probability of h at least $1 - 2^{-2^{\Omega(n)}}$, for any oracle circuit C of size at most $2^{(\frac{1}{2}-\gamma_1)n}$,*

$$Prob_{x \in \{0,1\}^n}(C^h(x) = h^{-1}(x)) \leq 2^{-\gamma_1 n}. \tag{2}$$

Proof. In Proposition 5, we take $t = 2^{\frac{1}{2}n}, s = 2^{(\frac{1}{2}-\gamma_1)n}$. Note that there are at most $(4s^2)^s$ circuits of size at most s and any such circuit can make at most s queries. It follows that the fraction of permutations h for which relation (2) does not hold is at most $(4s^2)^s \cdot 2^{-t} \leq 2^{-2^{\Omega(n)}}$. ∎

From now on we will consider only permutations h for which relation (2) holds. We call such permutations "good."

Step 2: Add one hidden bit. We take x and s two strings of length n which will be viewed as n-vectors over the field $\mathbf{Z_2}$. By a well-known result of Goldreich and Levin (GL89), the function $b(x, s) = x \cdot s$ (inner product in $(\mathbf{Z_2})^n$) provides a so-called hidden bit for a one-way permutation h. Formally, this means that for any "good" permutation h, for any oracle circuit C of size $2^{(\frac{1}{2}-\frac{5}{3}\gamma_1)n}$,

$$Prob_{x,s}(C^h(h(x), s) = b(x, s)) \leq \frac{1}{2} + \frac{1}{2^{\gamma_2 n}}.$$

By easy and well-known arguments, it follows that the function $g_h(x,s) = h(x) \odot s \odot b(x,s)$, where $\odot$ denotes concatenation, is a pseudo-random generator with expansion 1. That is, $g : \{0,1\}^{2n} \to \{0,1\}^{2n+1}$, and for any oracle circuit C of size $2^{(\frac{1}{2}-\frac{5}{3}\gamma_1)n}$, for any "good" permutation h,

$$\left|Prob_{x,s}(C^h(g_h(x,s)=1)) - Prob_{z\in\{0,1\}^{2n+1}}(C^h(z)=1)\right| < 2^{-\gamma_2 n}.$$

Step 3: Get double expansion. Based on the function g_h obtained at Step 2, we define $i_h : \{0,1\}^{2n} \to \{0,1\}^{4n}$ by $i_h(y) = (s_1, s_2, \ldots, s_{2|y|})$, where $s_1, \ldots, s_{2|y|}$ are bits defined inductively as follows: $y_0 = y$ and for $i = 1, \ldots, 2|y|$, $s_i =$ the first bit of $g_h(y_{i-1})$ and $y_i =$ the last $|y|$ bits of $g_h(y_{i-1})$. By an application of the hybrid method, there exists a positive constant γ_3 such that, for any "good" h, and for any oracle circuit C of size $2^{(\frac{1}{2}-2\gamma_1)n}$,

$$\left|Prob_{y\in\{0,1\}^{2n}}(C^h(i_h(y))=1) - Prob_{z\in\{0,1\}^{4n}}(C^h(z)=1)\right| < 2^{-\gamma_3 n}.$$

Step 4: Get exponential expansion. To simplify notation, we fix a "good" permutation h. Let $I_0(y)$ and $I_1(y)$ be the first and respectively the second half of the string $i_h(y)$ defined at Step 3. Let $j = cn$, where c is a constant such that $0 < c < \gamma_3$. Define $F_h : \{0,1\}^{2n} \to \{0,1\}^{2^{cn}}$ as follows. The $\alpha_1\alpha_2\ldots\alpha_j$ bit of $F_h(y)$ is the first bit of $I_{\alpha_1}(I_{\alpha_2}(\ldots(I_{\alpha_j}(y))\ldots)$. The techniques of Goldreich, Goldwasser, and Micali (GGM86) show that for $\gamma_4 = \gamma_3 - c$, for all good h, for any oracle circuit C of size at most $2^{(\frac{1}{2}-2\gamma_1-c)n}/poly(n)$ (for a fixed $poly$),

$$\left|Prob_{y\in\{0,1\}^{2n}}(C^h(F_h(y))=1) - Prob_{z\in\{0,1\}^{2^{cn}}}(C^h(z)=1)\right| < 2^{-\gamma_4 n}.$$

Observe that for all "good" h, F_h takes an input of size $2n$ and produces an output of size 2^{cn} that cannot be distinguished from a random string except with bias at most $2^{-\gamma_4 n}$ by any oracle circuit of size bounded by 2^{an}, for $a = 1/2 - \eta$ (for an arbitrarily small positive η), working with oracle h. To obtain Theorem 4, we only have to choose n such that $m \le 2^{cn}$, $2^{an} \ge m^{\tau}$ and $2^{-\gamma_4 n} \le m^{-\beta}$. If τ is sufficiently large, $n = (\tau/a)\log m$ satisfies all these conditions. Let R be the binary string that encodes in the natural way the permutation $h : \{0,1\}^n \to \{0,1\}^n$. We define $f(R,r)$ to be the first m bits of $F_h(r)$. Observe that $|R| = n2^n = poly(m)$ and $|r| = 2n = 2(\tau/a)\log m = 2/(1/2-\eta)\cdot\tau\log m = (4+\alpha)qm$, for an appropriately chosen value of η. It follows that with probability of R at least $1 - 2^{-2^{\Omega(n)}} = 1 - 2^{-poly(m)}$ for any oracle circuit C of size m^{τ}

$$\left|\frac{1}{2^{2n}}\sum_{r\in\{0,1\}^{2n}} C^R(f(R,r)) - \frac{1}{2^m}\sum_{z\in\{0,1\}^m} C^R(z)\right| < m^{-\beta}.$$

This concludes the proof of Theorem 4. ∎

5. Reducing the communication complexity

According to the standard definition, a language L is in NP if, for all inputs x, a prover can send to a polynomially time bounded verifier a proof string w, which the verifier accepts if and only if $x \in L$. Killian (Kil92) and Micali (Mic00), using the (full) PCP Theorem, have shown that the communication complexity in the above protocol (i.e., the length of the proof string transmitted by the prover to the verifier) can be reduced to $polylog(|x|)$, provided that the prover is poly-time bounded (but it has access to a witness string), the verifier is probabilistic, and a small error probability is admissible. Goldreich and Håstad (see also (GVW01)) have shown that bounding the power of the prover is necessary for reducing the communication complexity. Therefore, since provers need to be computationally bounded anyway, it seems an overkill to use the full PCP Theorem for reducing the communication complexity in the protocol. Indeed, the much simpler Light PCP Theorem can be used as well to implement the schema from (Kil92) and (Mic00). Moreover, the implementation can be done in a very natural way, since most of the needed algorithmical props are already in place.

We sketch next the modification of the protocol in Theorem 2 so that it incorporates the method from (Kil92) and (Mic00), achieves communication complexity $O((\log(n))^2)$, and is sound against any dishonest prover whose running time is bounded by n^t, for any fixed t.

The idea is that, since the verifier reads only $O(1)$ bits from the proof string, the prover does not need to send in Round 2 the entire proof string $w^R(\phi, a)$. It is enough if the prover *commits* to $w^R(\phi, a)$ and sends the verifier a certificate C of the commitment. Then, when the verifier requests the $O(1)$ bits, the prover delivers them together with some authentication information so that the verifier can check (with small error probability) that the bits really belong to the commited proof string. By using a Merkle tree (Mer90), the length of C is $O(\log n)$ and the authentication information for each revealed bit is $O(\log^2 n)$, and thus the entire communication complexity is $O(\log^2 n)$.

We present the technical details of the construction. Let ϕ be a boolean formula, $n = |\phi|$, let a be an assignment for ϕ, and let $k = (2t+1)\log n$ (recall that the prover is time bounded by n^t). A part of the public random string R (separate from the one used in the normal protocol) is used as a random function $f : \Sigma^{2k} \to \Sigma^k$. In the protocol given in Theorem 2, the prover prepares a proof string $w^R(\phi, a)$ having length $N = poly(n)$. The prover breaks the proof string $w^R(\phi, a)$ into consecutive, disjoint blocks of size k. Next, the prover builds a binary tree having each node labelled with a k-bit string as follows. The leaves of the tree are, in order, the N/k blocks resulted from the splitting. To keep notation simple, we assume that N/k is an integer power of two. The parent of two nodes labelled α and β is labelled $f(\alpha, \beta)$. Let C be the label of the root of the tree. Note that $|C| = k = O(\log n)$. The prover sends C to the verifier (as a certificate for his commitment to $w^R(\phi, a)$). Then, for each bit from $w^R(\phi, a)$ requested by the verifier, the prover will send the leaf block containing that bit together with all the siblings of the nodes located on the path from that leaf node to the root (this represents the authentication information). The verifier is now able to validate the entire path from the leaf block to the root, checking if it conforms to C. This works because in order for the prover to fool the verifier, he must find at least two strings $x, y \in \Sigma^{2k}$ such that $f(x) = f(y)$. Indeed, "fooling" means that the initial leaf in the authentication data is different from the genuine node (by modifying slightly the construction, we can easily ensure that the prover cannot substitute the initial block with another block of $w^R(\phi, a)$), and, at the other end, the root of the authentication data matches C. If $T = n^t$ is the running time of the prover, this can only happen with probability at most $T^2 \cdot 2^{-k}$, which is $1/n$. In this way, for any t, we have obtained an interactive protocol for 3SAT, in which the prover sends $O(\log^2 n)$ bits, and no dishonest prover that is time bounded by n^t can fool the verifier except with probability less than 1/2. Moreover, the protocol (for the verifier and the honest prover) is easy to implement avoiding the intricacies of the full PCP Theorem. We note that the honest prover, if given access to an assignment a, runs in time $n^{(4+\alpha)t}$, for any $\alpha > 0$, and this bound can be reduced to $n^{(2+\alpha)t}$, with some modifications in the protocol based on list decoding of error-correcting codes (as suggested by Sudan (Sud00a)). This may seem unsatisfactory because we require the honest prover to be stronger than the dishonest prover against whom the protocol is sound. However, if he is given access to an assignment a and to the commitment tree, the honest prover runs in $O(\log^2 n)$ steps and the protocol remains sound against dishonest provers that run n^t steps and have access to the same amount of information.

6. Final comments

Theorem 2 is obviously much weaker than the PCP Theorem and lacks the most important theoretical applications of the latter, namely proving inapproximability results. However, we think that it deserves attention for pedagogical reasons and for the possibility of being inserted in cryptographical applications in which the complexity of an adversary is assumed to be bounded anyway.

Acknowledgments

I thank William Gasarch and Lane Hemaspaandra for useful comments. I am grateful to Alina Beygelzimer, Richard Chang, Omer Horvitz, Bala Kalyanasundaram, and Joel Seiferas for helpful discussions.

References

[ALM+92] S. Arora, C. Lund, R. Motwani, M. Sudan, and M. Szegedy. Proof verification and intractability of approximation problems. In *Proceedings of the 32nd IEEE Symposium on Foundations of Computer Science*, pages 14–23, 1992.

[AS92] S. Arora and S. Safra. Probabilistic checkable proofs: A new characterization of NP. In *Proceedings of the 32nd IEEE Symposium on Foundations of Computer Science*, pages 1–13, 1992.

[BCC88] G. Brassard, D. Chaum, and C. Crépeau. Minumum disclosure proofs of knowledge. *Journal of Computer System Sciences*, 37:156–189, 1988.

[GGM86] O. Goldreich, S. Goldwasser, and S. Micali. How to construct a random functions. *Journal of the ACM*, 33(4):792–807, 1986.

[GH98] O. Goldreich and J. Håstad. On the complexity of interactive proofs with bounded communication. *Information Processing Letters*, 67(4):205–214, 1998.

[GL89] O. Goldreich and L. Levin. A hard-core predicate for all one-way functions. In *Proceedings of the 21st ACM Symposium on Theory of Computing*, pages 25–32, 1989.

[Gol93] O. Goldreich. Foundations of cryptography (fragments of a book), February 1993. ECCC Technical report, available at http://www.eccc.uni-trier.de/local/ECCC-Books/eccc-books.html.

[Gol99] O. Goldreich. *Modern Cryptography, Probabilistic Proofs and Pseudorandomness*. Springer Verlag, 1999.

[GT00] R. Gennaro and L. Trevisan. Lower bounds on the efficiency of generic cryptographic constructions. In *Proceedings of the 40th IEEE Symposium on Foundations of Computer Science*, 2000.

[GVW01] O. Goldreich, S. Vadhan, and A. Wigderson. On interactive proofs with a laconic prover, July 2001. ECCC Technical report TR01-046, available at http://www.eccc.uni-trier.de/eccc.

[HILL99] J. Håstad, R. Impagliazzo, L. Levin, and M. Luby. Construction of a pseudo-random generator from any one-way function. *SIAM Journal on Computing*, 28(4), 1999.

[Imp96] R. Impagliazzo. Very strong one-way functions and pseudo-random generators exist relative to a random oracle. (manuscript), January 1996.

[Kil92] J. Kilian. A note on efficient zero-knowledge proofs and arguments. In *Proceedings of the 24th ACM Symposium on Theory of Computing*, pages 723–732. ACM Press, 1992.

[Mer90] R. C. Merkle. A certified digital signature scheme. In Gilles Brassard, editor, *Advances in Cryptology — CRYPTO '89*, volume 435 of *Lecture Notes in Computer Science*, pages 218–238, Berlin, Germany / Heidelberg, Germany / London, UK / etc., 1990. Springer-Verlag.

[Mic00] S. Micali. Computationally sound proofs. *SIAM Journal on Computing*, 30(4):1253–1298, 2000.

[NW94] N. Nisan and A. Wigderson. Hardness vs. randomness. *Journal of Computer and System Sciences*, 49:149–167, 1994.

[Sud00a] M. Sudan. List decoding: Algorithms and applications (a survey). *Sigact News*, 31(1):16–27, 2000.

[Sud00b] M. Sudan. Probabilistically checkable proofs, July-August 2000. Lecture notes available at http://www.toc.lcs.mit.edu/ madhu/pcp/course.html.

[Tre00] L. Trevisan. Interactive and probabilistic proof-checking. *Annals of Pure and Applied Logic*, 2000. (to appear; available at http://www.cs.berkeley.edu/~luca).

[Zim99] M. Zimand. Sampling under adverse conditions with applications to distributed computing. In *Workshop on Parallel Algorithms, May 1999, Atlanta (satelite workshop of FCRC'99)*, 1999.

Track 2: Logic, Semantics, Specification and Verification

XML WEB SERVICES:

The Global Computer?

Andrew D. Gordon
Microsoft Research

Abstract The Web Services Description Language (WSDL) (Christensen et al., 2001) specifies how to implement remote procedure calls over the web via messages encoded in XML. WSDL promises a web-based programming model that works across multiple devices, multiple operating systems, and multiple organisations. Both commercial and open source implementations are available. There are critics, but momentum is building. We may say that WSDL is a detailed—though certainly partial—blueprint for the long awaited global computer. The first goal of my talk is simply to explain the basic ideas of WSDL and related specifications. My second goal is to explore some of the questions raised by WSDL, and some of the opportunities we have to apply ideas from theoretical computer science. In particular, I will report the results of an ongoing project to investigate security properties of XML web services.

References

Christensen, E., Curbera, F., Meredith, G., and Weerawarana, S. (2001). Web services description language (WSDL) 1.1. W3C Note, at *http://www.w3.org/TR/wsdl*, 15 March 2001.

MICRO MOBILE PROGRAMS

Carl A. Gunter
University of Pennsylvania
gunter@cis.upenn.edu

Abstract This paper describes a three-layer architecture for mobile code based on a distinction between code that is mobile versus code that is resident. We focus on code that is mobile but not resident and consider two contexts in which such code is constrained by its delivery mechanism to be small in size, resulting in *micro mobile programs.* The contexts we consider are programs carried in network communication packets and programs carried in two-dimensional barcodes.

Keywords: Mobile code, scripting language, active network, active barcode, PLAN, SwitchWare.

1. Introduction

The way computer programs are distributed is changing. Once programs came mainly from the vendor of the machine on which they ran. Open programming platforms allowed programs to be written by parties other than the vendor and possibly installed by a system administrator. Growing use of personal computers led to shrink-wrapped software, which could be bought at a store and installed by the user of a computer. Increasingly, however, programs are retrieved over the Internet by users and installed by users. This kind of installation generally takes two forms, implicit or explicit 'pull' retrieval. A user may visit a web page and implicitly download a Java applet, which is executed by the Java runtime system associated with his browser in order to provide customized functionality, like better graphics. Alternatively, a user may be told that a particular plugin is required in his browser in order to see a web page, so the user clicks on a button that instructs his browser to download and install the program that provides the desired functions.

Both of these new ways to obtain programs may be viewed as instances of *mobile* programs because their general distribution mode is to travel over the network. In the first case the program is *ephemeral,* because the program, an applet, is installed only temporarily while it is being used and disappears when the runtime completes running it. In the second case the program is *resident* in the form of a library like a Dynamic Link Library (DLL). There are still programs that are not really mobile, like, say, the operating system, which must generally be installed from media like a CD. We can call these *permanent* programs. Clearly there is a trend toward more minimal permanent programs in PCs. For instance, it is typical to download service packs for upgrading operating systems from web sites. Also, permanent programs are not really everlasting, since operating systems are reinstalled from CDs from time to time.

When we combine the concepts of mobile versus resident programs we obtain a three-layered architecture for mobile code as illustrated in Figure 1. The intersection of mobile and resident classes of programs yields a class of programs called *services,* consisting of programs like DLLs that are often or typically downloaded from the network. The class of permanent programs, which are resident but not mobile, forms an *infrastructure* layer. Finally, the class of ephemeral programs, which are mobile but not resident, forms an *invocation* layer. A simple example if this architecture arises with the postscript programming language [1]. A document is compiled into a postscript program. The program is sent to a printer at invocation layer to execute a printing using resident programs on the printer as libraries. After execution the ephemeral program (document) is discarded.

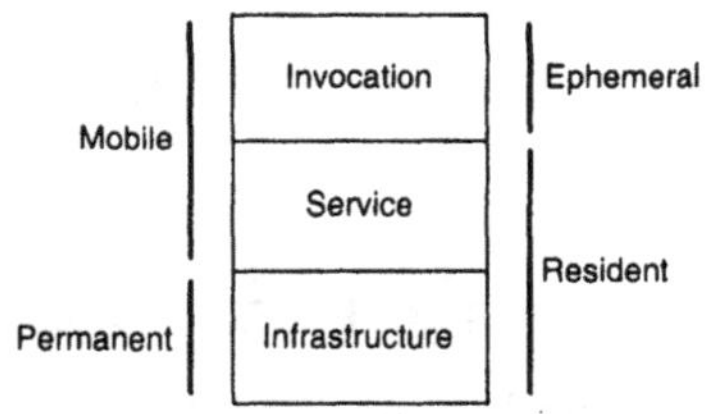

Figure 1. Three Layer Architecture

In this paper we consider the design issues associated with ephemeral mobile code within the three-layer architecture of Figure 1 in application contexts where the delivery mechanism for the mobile programs constrains the size of the programs in the invocation layer, resulting in what we will call *micro mobile programs.* We consider two such contexts, active packets and active barcodes. *Active networks,* as introduced in [17], allow users to program routers using active packets, which are packets that invoke custom processing functions on routers. If the programs that invoke such programs are to fit within packets, they are constrained by the packet sizes allowed by typical network path minimum transfer units, about 1300 bytes in the current Internet. *Active barcodes,* as introduced in [4], are 2D barcodes that contain computer programs. 2D barcodes provide for extremely cheap media used for delivering information in contexts like physical mail, where the barcode is printed on a letter or package. Such barcodes typically have a capacity of about 1-2 kilobytes.

The paper is divided into six sections. The second section discusses some application contexts that are used to motivate requirements and mechanisms described in the third section. The fourth and fifth sections discuss active packets and active barcodes respectively. The sixth section concludes.

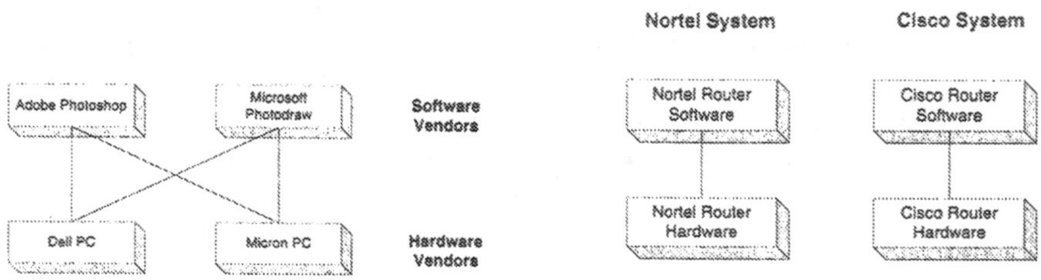

Figure 2. Horizontal Versus Vertical Programmability.

2. Application Contexts

A fundamental driver for software, especially for mobile code, is the issue of open Application Programming Interfaces (APIs). An open API enables third party vendors or users to write code for a platform. To see the issue in a networking context, consider IP, which assumes that communications are sent in packets that contain a small header used as data by routers. Users have little ability to program the way routers handle their packets, although facilities like ICMP and source routing provide some diagnostic and control capabilities. In the future internetworks are likely to offer more capable user customization functions such as RSVP, which enables Quality of Service (QoS) guarantees to be negotiated and allocated by routers. At a further level of programmability, routers may support the installation by owners of programs written by parties other than the router manufacturer. Consider Figure 2, which shows on the left the state of affairs for PCs, which typically provide an API usable by third party vendors. Even though this interface is generally Microsoft Windows, it is possible to run software from other vendors on top of it, resulting in a 'horizontal' software industry. By comparison, Figure 2 illustrates on the right the situation with routers, which generally provide only limited APIs for third parties, resulting in an essentially 'vertical' software industry. Developing a horizontal industry for programming routers could enable faster deployment of new functionality and more flexibility for users and owners.

Active networking concerns the idea of enabling users to run software on network elements the way one might be able to do on hosts with time-share operating systems. Research on active networking led to considerable exploration of the design options for the three-layer architecture for mobile code in Figure 1. Routers are viewed as supporting a layered model with the *NodeOS* at the infrastructure layer and a collection of *Execution Environments (EEs)* supporting various approaches to the service and invocation layers. Three examples illustrate some of the tradeoffs. The Active Network Transfer System (ANTS) [19] provides for packets that contain an identifier indicating which of a collection of resident Java program should handle a packet received by an active router supporting the ANTS EE. If the identifier is not recognized then the host or router sending the packet is asked to send the program for this identifier so it can be installed. Thus the ANTS EE has a minimal invocation layer consisting of identifiers and a rich service layer consisting of Java programs. By contrast, SmartPackets [15] sends programs in packets that are installed on routers only as long as they execute. These consist of very short programs written in a CISC specifically designed for collecting network diagnostic information. SmartPackets therefore provides an instance of the theme of this paper, micro mobile programs.

SwitchWare [2], a third active network architecture, is depicted in Figure 3. Its infrastructure is based on the Secure Active Network Environment [3], which features secure bootstrap and remote recovery capabilities that provide for a secure and minimal permanent infrastructure. Service layer programs are written in the OCaml, a General Purpose Language (GPL). Invocation layer programs can be written in a Domain Specific Language (DSL) called the Programming Language for Active Networks (PLAN). PLAN is a scripting language whose primary construct is a remote evaluation primitive. It can be viewed as a means of composing and invoking service layer functions (written in OCaml). PLAN is a micro mobile programming language; we discuss it it more detail in Section 4 below.

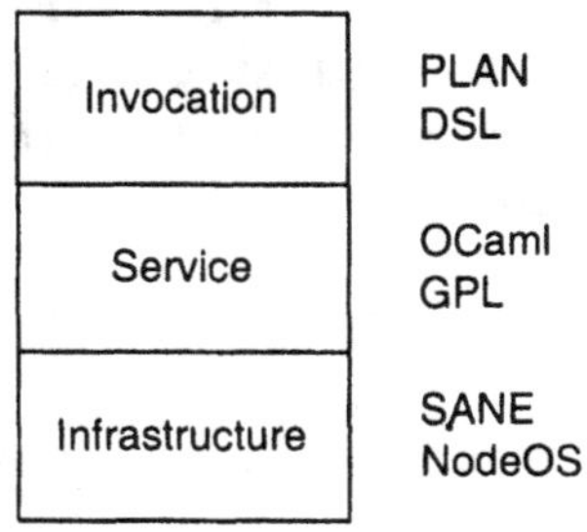

Figure 3. SwitchWare Architecture

Routers provide a good example of a programming application different from networked PCs, time-share servers, and web servers, but they are not the only one. Indeed, many computers are now present as *embedded* systems, that is, computers within other devices, often controlling processes these devices are involved in. A characteristic example is the software needed to control an aircraft. Embedded device programs are as diverse as the devices in which they are deployed so assumptions about networked PCs and routers are often inapplicable. In particular, embedded systems often have much different network connectivity than these other systems (especially routers). For example, airplane controller chips are likely to be connected across a networking system within the aircraft, but Internet connectivity is likely to be limited. Other programs, like those that control processes in automobiles may or may not be networked, whereas programs like the ones in a chip in a vacuum cleaner are not networked. The programs in a cell phone are an interesting example. Such programs may clearly control a communication link, but may or may not be downloadable through the network link. Another interesting dimension of embedded systems is the struggle for open APIs. Personal Digital Assistants PDAs, which are like small PCs, mostly provide an open API because PDA vendors would like to leverage an industry of independent PDA software vendors. Automobile software is typically not programmed with an open API, at least not for users, but there is an industry of chip replacements to help users circumvent tax and environmental regulations. Cell phone software is in a middle ground with most most cell phones being programmable only by their vendors, but with substantial development of open API platforms like the Java Mobile Information Device Protocol (MIDP) providing a path to cell phones that can download applet-like programs called 'midlets' using wireless web (typically cellular) networks.

```
1. Make 1 inch slit in plastic
2. 50% power for 5 minutes
3. Remove plastic overwrap
4. Rotate tray 1/2 turn
5. 100% for 1:45
```

Figure 4. Enchilada Recipe.

Because of complex connectivity issues and the challenge of open APIs, embedded systems provide a fertile ground for exploring new variations on the three-level mobile code architecture. This paper describes some of the progress that has been made in one specific context, that of programmable microwave ovens. Microwave ovens are familiar household and commercial cooking appliances. They often use very simple recipes, like 'cook at full power for 3-5 minutes'. Since the ovens vary from about 600 to 1000 watts in power, there is a compatibility problem that makes it necessary to provide such imprecise recipes. Moreover, recipes often involve programming the human operator as an additional actuator. As a running example consider the program in Figure 4. as studied in [4]. This is a recipe from a frozen food package. Note that the instructions 1, 3, and 4 are for the human operator and instructions 2 and 5 are for the microwave, as keyed in by the human. Recipes like this could be more sophisticated if the API of the microwave were known (eg. whether it cooks at 600 or 1000 watts) and the human did not need to key in the recipe.

Open APIs for microwaves have been attempted in various forms. For instance, a pair of patents (5,812,393 and 5,883,801) provides for recipes represented with 5 to 10 digits. The idea is to put these codes on packages and have the operator key them into the microwave. The five digit codes describe the time, power level, and pause period for the device. The ten digit codes provide for two phases of cooking, similar to what we have in the enchilada recipe in Figure 4. Another idea is to put the 10 digits into a bar code and put a scanner on the microwave so it read the recipe directly from the package. This approach has a variety of limitations, especially the need to get the food vendors to put the recipes on their packages. So, another idea is to put a database of recipes into the device and look up the proper recipe based on the Universal Product Code (UPC) generally found encoded in the linear barcodes that already appear on food packages. This has the limitation that the database may become out-of-date, but this problem can be addressed by putting the device on the Internet so recipes can be downloaded. This can be done on demand, or using occasional updates the way the Tivo television system downloads show schedules. Indeed, the Sharp Corporation demonstrated a programmable microwave at the 2000 International Housewares Show (`http://www.reviewsonline.com/IHS00.htm`) that interfaces with a PC. A descendent of this device is now being marketed in Japan. Another class of devices now approaching the market are called *multi-modal* ovens. These combine different oven technologies. A smart multi-modal device combining microwave and convection ovens was demonstrated at the 1999 International Housewares Show in Chicago (`http://www.foodtechsource.com/emag/004/gadgets.htm`). It was developed by Kit L. Yam in the Food Science Department of Rutgers University, with support from Samsung Electronics America. It reads bar codes and features a computer control with a touch screen and access to the Internet. A noteworthy aspect of multi-modal ovens is the fact that programming them is more complicated. Indeed, this complexity is a key impediment to selling them in the consumer market.

One fairly basic idea for programming a microwave to an open API is to allow recipes to be delivered using 2D barcodes. Current technology for 2D barcodes allows about 1-2 kilobytes of data to be transmitted in this way. This does not solve the problem of how to get food vendors to put recipes on packages and encounters the additional property that 2D barcodes require comparatively expensive Charged Coupled Devices (CCDs) as readers, but it is an interesting problem since it illustrates many of the issues that will arise with programming open APIs on a significant class of embedded systems. Microwaves are essentially required to have at least a rudimentary open API since food vendors are quite independent from microwave hardware vendors. Imagine the consequences of a vertical organization of this market, with Stouffer's frozen dinners that can only be cooked in a Stouffer's oven. The 2D barcode approach illustrates the idea of micro mobile programs. The programs are about the same size as those in active packets, but active barcodes are delivered by a human operator and read from a package. We discuss this in more detail in Section 5.

3. Requirements and Mechanisms

An interesting aspect of the postscript example above distinguishes it from Java applets and DLLs. Namely, the postscript program is 'pushed' to the printer. That is, the computer on which it runs is the server, not the client. In the case of Java, the client contacts the server and 'asks' for the Java applet, whereas the printer offers to the network the service of running the program provided to it by an authorized client. This distinction has various interesting security ramifications making it even more problematic in many ways than pull programs like Java and plugins already are. Mobile programming for network elements or embedded systems raises a collection of requirements of its own, and there are a variety of recurrent themes in the mechanisms that can be used to address these requirements.

3.1. Requirements

For network elements the primary challenges, as outlined in [9], are: flexibility, security, usability, and performance. Flexibility concerns just *how* programmable the network elements are. For instance, if the aim of the programmability is to allow flexible deployment of intrusion detection elements and firewalls, then there will be a need to allow the party that deploys these systems to authenticate themselves and gain access to routed packets for inspection and filtering. However, if the aim is only to provide limited diagnostic and customization features then it may be best to provide an interface that does not allow for inspection of the packets of other parties. Flexibility conflicts in general with the other requirements. Security is threatened by increased flexibility because attackers have more to work with. Even unintentional errors are more likely to cause significant harm to the network for this reason. Programmable routers are also challenged by usability in some of the same ways that microwave vendors are challenged by naive operators who must key in the recipes. Flexibility cannot appear to an endpoint as overwhelming complexity. An active network where getting a packet from its source to destination requires ingenious programming is not likely to be valuable. Performance is a key concern for active networks since custom processing times for packets must be proportionate to the benefit of custom processing. This is difficult for data path packets, so active network systems have often emphasized applications in the router control path, performing functions like configuration and diagnostics.

For embedded systems the primary challenges, as outlined in [5], are: flexibility, portability, extensibility, predictability, and deliverability. Flexibility covers quite a wide spectrum for embedded systems. In the case of microwaves it may be as simple as allowing two cooking phases, or as complex as code to control a family of sensors and actuators at a low level. Portability arises as a significant issue in these systems; for instance Java got its start as a language for portable programming of set-top boxes. Extensibility concerns the ability of the programming system to accommodate changes in the underlying device. For instance, the 10 digit recipes do not take account of whether the microwave has a turntable; indeed, one wonders if step 4 in the recipe in Figure 4 makes sense in this case. Ideally, programming APIs for embedded systems will be like those for PCs and assume that new peripherals will be introduced from time to time and access will be provided through some API. A rigid system may need to be completely redesigned to take advantage of a new sensor or actuator. Predictability is similar to the challenges with security and performance for programmable network elements. However, the concerns are often different. For example, a microwave may have little security risk but significant safety risk. Wireline routers are very performance sensitive, but typically do not care about power utilization in the way a cell phone might. Microwave ovens are not sensitive to power or performance, but typically are very sensitive to cost (in dollars) and convenience for unskilled operators. An interesting issue is whether domain-specific assumptions provide a new handle on predictability. For instance, it is undecidable whether a program in a GPL will cause a variable to exceed a given value, but it is trivial to tell how much cook time the recipe in Figure 4 will require. Finally, as argued earlier, the complex connectivity of embedded system raises deliverability mechanisms as an interesting question.

3.2. Mechanisms

Java provides a good example of several of the mechanisms that can address requirements for mobile code. A core question is whether a DSL is needed or whether an existing language can be used if it is provided with a suitable development environment or runtime analysis system. In the case of Java it was decided to produce a new general-purpose language that prioritized portability and security over performance and compatibility with existing C libraries. Subsequent efforts to use Java in diverse contexts have focused on using sandboxing and the JVM or something near to it, like the KVM, to address the needs of specialized contexts, like web browsers, mobile communication devices, or embedded systems. Much of the work on active networks has focused on the use of Java, particularly in the ANTS EE. ANTS provides a limited library API for Java and support for its code distribution system based on identifiers and on-demand installation. Many issues with security are addressed by the limitations imposed by sandboxing and limited APIs. For instance, a packet that wishes to be processed by a given program uses the cryptographic hash of that program to identify it; this prevents any confusion that might arise if an installation request attempted to spoof an often-requested network service. This approach exacts a price in configuration management, however, when a common service needs to be transparently upgraded: existing programs at endpoints will need to be modified to use the hash of the new program. The use of an existing scripting language or GPL can exploit existing support for development and runtime systems. A special-purpose approach like SmartPackets must build a new compiler and/or runtime system for its language, whereas ANTS can use off-the-shelf compilers and the JVM. A hybrid approach like SwitchWare requires an interpreter for PLAN, but can make use of the OCaml runtime system for its services. The hybrid strategy has the advantage of allowing functions to be 'pushed down' from the invocation layer to the service layer if they are more appropriate for implementation in a GPL.

The advantage of a DSL is the simplicity of the programs and the ability to exploit this simplicity to achieve other objectives like demonstrating security properties. Moreover, DSLs enable significant kinds of innovation in the way the system works without the baggage associated with a GPL. PLAN provides numerous examples of this, as we discuss further in Section 4. One specific area of challenge is in *resource control*, that is, the ability to determine and limit the use of a valuable resource by a mobile program. Java provides limited resource control by preventing programs like applets from carrying out potentially troublesome operations like accessing files on the disk of the browser. However, Java does little to prevent the use of space or cycles on the host machine. It is important to break this problem down into two architectural options to clarify the tradeoffs: *usage limitation* or *bound verification*. Under usage limitation the mobile program is given a collection of resources such as a given amount of space or cycles; if the program exceeds these limits it may be terminated or otherwise limited, for example, by having its priority reduced in cycle scheduling. In bound verification the program is checked in advance to determine whether it satisfies the necessary limitations. If it does, then it can be run with more limited runtime monitoring. Bound verification has significant advantages over usage limitation not only because it may impose less burden on the runtime system but also because usage limitation essentially begs the question of how it is known that a program will meet its usage limits and therefore perform properly. However, bound verification must be based on a technology for verifying the desired property. Static type checking is a major success of this approach, but verifying space and cycle usage are more stubborn problems. An additional problem that arises in programmable networks is the need to deal with decentralized replication of mobile programs that run on multiple routers. In this case usage limitation is problematic since no system has global knowledge of usage. This problem was recognized early in the TCP/IP system and addressed with the TTL value, which keeps track of how many more routers the packet should be allowed to visit.

In realizing bound verification there are essentially two options; these can be broadly classified as verification 'by others' versus verification 'by me' (that is, my local trusted computing base). The former is seen in systems like Microsoft Authenticode, which attaches a digital signature to code as proof of its conformance to requirements. That is, the code consumer trusts the code because it came from a trusted source. ActiveX controls aim for this kind of verification. By contrast Java applets aim more toward verification on the code consumer machine, eliminating the need for a trusted origin (at non-trivial cost to functionality). Efforts have been made to extend this approach by the use of proof-carrying code [13, 12], in which evidence of conformance is included with the mobile program. This evidence can be used to verify conformance by the code consumer.

Micro mobile programs provide another mechanism for addressing requirements. For delivery they offer the option of providing complex information through 'in-band' delivery. For instance, a small special-purpose diagnostic program that can be written in a few dozen lines can be sent in a packet to perform its task. In a system like ANTS, an 'out-of-band' delivery mechanism would install the program on each node and then send a new packet to invoke it by hash index. In a system like a microwave that reads a UPC barcode and looks up a program to match, the device requires connectivity to the Internet in order to maintain its selection of programs; a micro mobile program in a 2D barcode can provide all of the needed code without this connectivity. For predictability, micro mobile programs offer the prospect of carving out a class of programs for which analysis is feasible. This is the case for both micro DSL and GPL programs. The next two sections illustrate some of the ideas in each of these cases.

4. Active Packets

PLAN is a small scripting language with a syntax and semantics similar to Scheme and ML. Implementations have been carried out in several languages, but the reference implementation is written in OCaml and assumes a service layer of OCaml programs. The essential design goal was to balance ephemeral code in packets with resident service-layer code. Thus the nature of programming with PLAN is to decide what goes in the packet in PLAN versus what gets written in OCaml and installed on a node. For instance, a simple diagnostic or configuration packet that is meant to be executed once on each of a family of active nodes is written in PLAN, whereas a program that is complex, requires significant state or timers, or needs to be used many times is best coded in OCaml, installed as a library service on nodes and invoked from PLAN. PLAN programs therefore focus on simple invocations of service layer programs or provide discovery and set-up functions.

Perhaps the purest illustration of PLAN, and one of the purest illustrations of active networking generally, is the *PLANet* network testbed [7]. PLANet implements a range of internetworking functions, including both standard IP functions like distance vector routing and novel functions like Flow-Based Adaptive Routing (FBAR). FBAR allows PLAN agents to discover QoS properties and configure customized routes. The implementation of PLANet is in OCaml and essentially replaces the usual network layer with an active network functionality. Thus all packets are PLAN programs wrapped within link layer frames. The character of programs in PLANet reflects the tradeoffs between deployment of programs in the invocation versus service layers. For example, FBAR functions that perform diagnostic searches for good paths and set up labeled routing are written in PLAN, whereas distance vector routing, which involves state with tables and times, is written mainly in OCaml but uses PLAN functions to send routing table advertisements.

Another principal rationale for PLAN was the hope that a DSL would provide better support not only for convenient coding but also for reasoning about properties of programs. PLAN programs are comparatively simple and can be constrained to display desirable properties so specifying their semantics and reasoning about them is easier than doing so for OCaml or any similar GPL. Accomplishments along these lines include the specification of PLAN using a term rewriting model [9] and formal reasoning about FBAR [18]. Aside from this, there have been two main lines of investigation on reasoning about PLAN: resource control and the impact of programmability on communication privacy.

Resource control is a significant problem for programmable networks. The IP protocol provides for a TTL field to prevent packets from cycling indefinitely in the network. PLAN uses a similar concept and adds to this a guarantee that PLAN programs terminate if the services they invoke terminate. This is done simply by not including recursion or looping constructs in PLAN, a reasonable tradeoff given that network programs often do not need looping constructs and, when they do, these can be included in service layer functions. ANTS also provides for a TTL-like resource bound, but treats it more liberally than PLAN to support functions like multicast. This enables an exponential blow-up in program proliferation that would probably be as bad in practice as a completely unbounded program. As for PLAN, the resource bound on packet proliferation is more strict, but an individual packet with a nested collection of function definitions can display exponential use of time and space on a node [9]. A more recent direction is to use a special-purpose byte code called SNAP [10], which can be compiled from PLAN [8] or another source. SNAP shares with PLAN the property that resource bounds can be predicted from program lengths (so resource utilization is governed by bandwidth) but SNAP enables tighter estimation of the bounds [11].

Another line of study concerns the impact of active network functionality on guarantees of privacy. A PLAN packet can enter a network, gather diagnostic information, leave configuration state (if service layer functions support this), and return to its origin without needing to visit anything but active routers. This is clearly somewhat different from IP packets, which can invoke ICMP responses but otherwise have little ability otherwise to collect and return information from routers. There is work [6] exploring how to reason about the ability to collect information from active networks for various assumptions about available service layers, including reasoning about strategies for corrupting routing functions using active packets. More recent work investigates topics like anonymity and onion routing in active networks.

5. Active Barcodes

Figure 5. Enchilada Recipe as a Dataglyph Symbol

Barcodes provide an extremely cheap way to communicate bits. They can be printed on paper and therefore do not require any special material to be produced; a small, robust reader can be had for a modest cost. Linear barcodes are used very commonly for postal addressing, inventory management, and point-of-sale functions. 2D barcodes are a newer technology that has been making headway in various applications such as postage, where information in the barcode can be used to provide evidence of payment. The US Postal Service, for example, has explored the idea of digital signatures in 2D barcodes as part of its Information-Based Indicia Program (IBIP). Putting programs into barcodes is an idea explored in [4]. We discuss here some of the issues that arose in that study and how they compare to other applications for micro mobile programs.

The nature of active barcodes is likely to depend heavily on the application domain. In the case of microwave programs it is possible to put a program like the enchilada program of Figure 4 into a 2D barcode. Figure 5 shows the result of doing this, where the program has been coded in Java, reduced to bytecode, and compressed. The program in the barcode here is actually somewhat more sophisticated than the one in Figure 4. It includes a feature that modifies cooking times based on cooling that occurs after the user pauses the microwave: in particular, it keeps track of how long it takes the user to perform steps 3 and 4 and adjusts the cook time in 5 accordingly. It would be inconvenient for the user to do this himself. If the oven has a rotating platform then the program causes it to rotate the food and omits step 4, that is, does not ask the user to do the rotation. Thus the program exploits in interesting ways the flexibility offered by the opportunity to deliver a micro mobile program.

There are at least two interesting research problems that are suggested by this application. First, what can be done to compensate for the changed operator involvement? For example, if the recipe is miscoded to indicate cooking for 145 minutes rather than 1 minute and 45 seconds, a user is likely to notice this, but this sanity check may be missing if the program is not read by the operator. Second, how much of what kind of code can or needs to be used in the barcode? For example, it seems impractical to locate a Java JIT on the microwave, and it is not clear whether compression, for example, will be of any value for such small programs.

Interaction with human operators and the physical environment are two recurrent themes for embedded systems. The nature of these interactions is likely to be somewhat domain specific. In the microwave example, almost all of the non-determinism in the program is created by user actions. This particularly contrasts with assumptions in many other applications, where non-determinism arises from concurrency, and this has ramifications for predictability. As mentioned before, it is straight-forward to calculate maximum and minimum cooking times from the program in Figure 4. Doing this for the one in Figure 5 is harder, but not nearly as hard as a comparable task might seem for an arbitrary Java program. The problem is similar to array bounds checking: one needs to verify that a value is never more than a certain value in any run of the program. In this specific case this, depends on the operator interaction because the feature of the program that adds back cooking times would prevent the program from ever completing if the operator continued pausing it indefinitely. However, it is feasible to apply reasonable operator assumptions and prove, with off-the-shelf formal analysis tools, that the program does not cook the food for any more than a certain length or *any less* than a certain time (this is also a safety issue if the food is raw). These formal analysis techniques can be somewhat automated so it is possible to create an architecture in which the burden of verification is placed on the (sophisticated) development environment, and the code consumer can use mainly usage limitation to predict behavior. Thus the basic program is converted into one that keeps a cook time counter and rejects program runs that use less or more than a pre-specified range of times.

The question of what kind of code to put in the barcodes is interesting, but again domain-specific. For instance, even these simple recipe programs can be more clearly written in a reactive programming language like Esterel than in Java. However, portability is likely to be a key consideration in these applications (recall the leftist perspective in Figure 2) so the use of a highly portable GPL has advantages. This suggests that shipping JVM bytecode is a plausible approach, especially if the development and analysis environment and perhaps the target device can take advantage of the fact that the micro mobile programs can probably use only a modest fragment of Java. Shipping portable byte code rules out a number of options for how to compress the code since schemes based on source code or modified versions of the JVM can be ruled out. It is not obvious that a Java bytecode of only about a 1000 bytes will compress at all well given the overhead associated with compression. The situation is similar to that for IPSec-level compression [16], where each packet (of about 1300 bytes) must be individually compressed. Fortunately Java byte code recipes seem to display significant redundancy: for the example of Figure 5, a compression technique called Pack [14], specifically designed for Java bytecodes, compresses the 894 byte Java enchilada program to 60% of its original size.

6. Conclusions

Micro mobile programs are useful and feasible in a variety of contexts. Moreover, there are a number of recurrent themes that enable ideas in one context to be inspirational in others, even when there are significant differences in the kind of application involved. One central themes is the challenge of flexible open APIs that have predictable behavior. Micro mobile programs can help attain this objective, especially if domain-specific circumstances can be identified that aid the analysis of programs when they are being developed or at the time that a DSL for the micro mobile programs is designed.

Acknowledgments

The author is grateful for the insights and experimentation of participants in the SwitchWare and MiRL projects. The work was partially supported by DARPA (N66001-96-C-852), ONR (N00014-99-1-0403 and N00014-00-1-0641), and ARO (DAAG-98-1-0466 and DAAD-19-01-1-0473).

References

[1] Adobe. *PostScript Language Reference Manual.* Addison-Wesley, 1985.

[2] D. Scott Alexander, William A. Arbaugh, Michael Hicks, Pankaj Kakkar, Angelos Keromytis, Jonathan T. Moore, Carl A. Gunter, Scott M. Nettles, and Jonathan M. Smith. The switchware active network architecture. *IEEE Network Magazine*, 12(3):29–36, May/June 1998. Special issue on Active and Controllable Networks.

[3] D. Scott Alexander, William A. Arbaugh, Angelos D. Keromytis, and Jonathan M. Smith. A secure active network architecture: Realization in SwitchWare. *IEEE Network Special Issue on Active and Controllable Networks*, 12(3):37–45, 1998.

[4] Alwyn Goodloe, Michael McDougall, Rajeev Alur, and Carl A. Gunter. Predictable programs in barcodes. `http://www.cis.upenn.edu/sdrl/mirl/papers/predictable_barcodes.ps`, April 2002.

[5] Carl A. Gunter, Rajeev Alur, Alwyn Goodloe, and Michael McDougall. Third-party programmability for embedded processors, December 2001.

[6] Carl A. Gunter, Pankaj Kakkar, and Martín Abadi. Reasoning about secrecy for active networks. In Paul Syverson, editor, *13th IEEE Computer Security Foundations Workshop*, pages 118–131, Cambridge, England, July 2000. IEEE Computer Society.

[7] Michael Hicks, Jonathan T. Moore, D. Scott Alexander, Carl A. Gunter, and Scott Nettles. PLANet: An active internetwork. In *Proceedings of the Eighteenth IEEE Computer and Communication Society Infocom Conference*, pages 1124–1133, Boston, Massachusetts, March 1999. IEEE Communication Society Press.

[8] Michael Hicks, Jonathan T. Moore, and Scott Nettles. Compiling PLAN to SNAP. In *Proceedings of the IFIP-TC6 Third International Working Conference, IWAN 2001*, September/October 2001.

[9] Pankaj Kakkar, Michael Hicks, Jonathan T. Moore, and Carl A. Gunter. Specifying the PLAN networking programming language. In *Higher Order Operational Techniques in Semantics*, volume 26 of *Electronic Notes in Theoretical Computer Science*. Elsevier, September 1999. `http://www.elsevier.nl/locate/entcs/volume26.html`.

[10] Jonathan T. Moore. Safe and efficient active packets. Technical Report MS-CIS-99-24, Department of Computer and Information Science, University of Pennsylvania, October 1999.

[11] Jonathan T. Moore, Michael Hicks, and Scott Nettles. Practical programmable packets. In *Proceedings of the 20th Annual Joint Conference of the IEEE Computer and Communications Societies*, April 2001.

[12] George C. Necula. Proof-Carrying Code. In *Proceedings of the 24th Annual ACM SIGPLAN-SIGACT Symposium on Principles of Programming Languages (POPL '97)*. ACM Press, 1997.

[13] George C. Necula and Peter Lee. Safe Kernel Extensions Without Run-Time Checking. In *Second Symposium on Operating System Design and Implementation (OSDI '96)*, 1996.

[14] William Pugh. Compressing java clas files. In *ACM Sigplan Conference on Programming Language Design and Implementation*, pages 247–258. ACM Press, 1999.

[15] Beverly Schwartz, Wenyi Zhou, Alden W. Jackson, W. Timothy Strayer, Dennis Rockwell, , and Craig Partridge. Smart packets for active networks. In *Proceedings of the Second IEEE Conference on Open Architectures and Network Programming (OPENARCH)*, pages 90–97, March 1999.

[16] A. Shacham, R. Monsour, R. Pereira, and M. Thomas. IP payload compression protocol (IPComp). RFC 2923, IETF, December 1998.

[17] David L. Tennenhouse, Jonathan M. Smith, W. David Sincoskie, David J. Wetherall, and Gary J. Minden. A survey of active network research. *IEEE Communications Magazine*, 35(1):80–86, January 1997.

[18] Bow-Yaw Wang, José Meseguer, and Carl A. Gunter. Specification and formal verification of a PLAN algorithm in Maude. In Tenh Lai, editor, *Proceedings of the 2000 ICDCS Workshop on Distributed System Validation and Verification*, pages E:49–E:56. IEEE Computer Society, April 2000.

[19] David J. Wetherall, John Guttag, and David L. Tennenhouse. ANTS: A toolkit for building and dynamically deploying network protocols. In *Proceedings of the First IEEE Conference on Open Architectures for Signalling (OPENARCH)*, pages 117–129, April 1998.

CHECKING POLYNOMIAL TIME COMPLEXITY WITH TYPES

Patrick Baillot *

Laboratoire d'Informatique de Paris-Nord (UMR 7030 CNRS)

Institut Galile, 99 av. J.-B. Clment, 93430 Villetaneuse, France.

pb@lipn.univ-paris13.fr

Abstract Light Affine Logic (LAL) is a logical system due to Girard and Asperti offering a polynomial time cut-elimination procedure. It can be used as a type system for lambda-calculus, ensuring a well-typed program has a polynomial time bound on any input. Types use modalities meant to control duplication.

We consider parameterized types where parameters are on the number of modalities and the *type instantiation* problem: given a term and a parameterized type, does there exist a valuation of the parameters such that the term admits the corresponding type? We show that this type instantiation problem is decidable for normal terms.

1. Introduction:

Several authors have proposed these last years programming languages and calculi with intrinsic complexity property, for instance languages ensuring that all functions representable are polytime without refering to an explicit time measure (see for instance Hofmann, 2000, Leivant and Marion, 1993).

Light linear logic (Girard, 1998) is one of these systems: it is based on the proofs-as-programs approach to computation. Polynomial running time is ensured in this framework by control of duplication expressed by means of modalities in types. The system was later simplified by Asperti into Light *Affine* Logic (LAL) in Asperti, 1998.

LAL can be seen as a programming language using Curry-Howard isomorphism, and indeed Roversi developed a syntax and a type inference procedure (Roversi, 2000). However this approach requires the user to provide key information about the structure of the program (*boxes*) ensuring the time bound.

*This work was partly done while the author was at Laboratoire d'Informatique de Marseille, Universit Aix-Marseille II, Marseille.

Placing boxes makes programming difficult. We believe that it is worth trying to automatize this task and so we relegate the handling of boxes to the level of the type system. A motivation for that is the perspective of designing a language where control over time complexity would be done in a way transparent to the user: he would use a regular functional language and the compiler would in case of successful typing guarantee a bound for the running time on any input.

LAL types essentially differ from simple types by modalities. Recall that the idea of decorating intuitionistic proofs with linear logic modalities has been extensively studied by Schellinx *et al.* (Danos et al., 1994; Schellinx, 1994). Here we are reconsidering this idea in the more constraining setting of LAL. A related work has been done by Coppola and Martini (Coppola and Martini, 2001) who gave a type-inference procedure for lambda-terms in Elementary Affine Logic, a system which corresponds to elementary complexity.

In this paper we carry out a first step in this direction for LAL. We consider as source language lambda-calculus and introduce an intermediary typing syntax where boxes are not precised explicitly but can be recovered. Then we show how this typing can be extended with parameters on the number of modalities (parameterized graph terms). Finally we establish the following *type instantiation theorem*: given a term in normal form and a parameterized type does there exist an instance of the parameters making it a valid type for the term?

Due to space constraints proofs are omitted in this paper. The reader can refer to Baillot, 2001 where some examples are also given.

2. Light affine logic and lambda-term typing:

2.1. Typing with sequent calculus:

We use Light Affine Logic sequent calculus as a type derivation system for lambda-calculus. Lambda-calculus terms are given by: $t ::= x \mid \lambda x\, t \mid (t)t$.

The LAL types are obtained by the following grammar (we do not consider second-order quantifiers): $T ::= \alpha | T \multimap T | \,!T | \,\S T$.

Modalities ! and § are called *exponentials*. We write $!^n T$ for $! \cdots ! T$ with n repetitions of !. The typing rules are given on figure 1. In the paragraph promotion rule (§ prom.) we can have $n = 0$ and each $\Box_i$ is either § or !. As we deal only with terms in normal form we will not be using the *cut* rule.

For instance, types for tally integers and binary lists are given respectively by: $N_A^{LAL} = !(A \multimap A) \multimap \S(A \multimap A)$,

$Bin_A^{LAL} = !(A \multimap A) \multimap !(A \multimap A) \multimap \S(A \multimap A)$.

These are instances depending on A, as we are in a quantifier-free setting.

A variant of this system, Elementary Affine Logic (EAL), corresponds to elementary time. EAL types are obtained by: $T ::= \alpha | T \multimap T | \,!T$.

As to the typing rules, EAL typing is obtained by replacing (! prom. 1), (! prom. 2) and (§ prom) by a single rule:

$$\frac{x_1 : A_1, \ldots, x_n : A_n \vdash t : A}{x_1 : !A_1, \ldots, x_n : !A_n \vdash t : !A} \text{ (EAL prom.)}$$

$\dfrac{}{x : A \vdash x : A}$ (variable)	$\dfrac{\Gamma \vdash t : A \quad \Delta, x : A \vdash u : B}{\Gamma, \Delta \vdash u[t/x] : B}$ (Cut)
$\dfrac{\Gamma \vdash t : B}{\Gamma, x : A \vdash t : B}$ (weak.)	
$\dfrac{x_1 : !A, x_2 : !A, \Gamma \vdash t : B}{x : !A, \Gamma \vdash t[x/x_1, x/x_2] : B}$ (cont.)	$\dfrac{x_1 : A_1, \dots, x_n : A_n \vdash t : B}{x_1 : \Box_1 A_1, \dots, x_n : \Box_n A_n \vdash t : \S B}$ (§ prom.)
$\dfrac{x : A \vdash t : B}{x : !A \vdash t : !B}$ (! prom. 1)	$\dfrac{\vdash t : B}{\vdash t : !B}$ (! prom. 2)
$\dfrac{\Gamma, x : A \vdash t : B}{\Gamma \vdash \lambda x.t : A \multimap B}$ (right arrow)	$\dfrac{\Gamma, x : B \vdash t : C \quad \Delta \vdash u : A}{\Gamma, \Delta, y : A \multimap B \vdash t[(y)u/x] : C}$ (left arrow)

Figure 1. LAL sequent typing

Proposition 1 *If $\vdash_{LAL} t : A \multimap B$ then there exists a polynomial P such that: for any u such that $\vdash_{LAL} u : A$, the term $(t)u$ can be evaluated in $P(|u|)$ steps, where $|u|$ denotes the size of u.*

The evaluation mentioned in the result is performed by a graph-rewriting method (normalisation of proof-nets). Similarly, typing in EAL ensures an elementary recursive time bound.

With polymorphic types (LAL_2), a completeness result also holds (Roversi, 1999; Girard, 1998): for any polytime function f on binary lists there exists a term t representing f such that t is typable in LAL_2.

2.2. Translations:

One can see EAL and LAL types as refinements of simple types, adding some intensional information about which arguments can be duplicated and which ones are used linearly. To make this remark explicit, let us recall the natural forgetful functors from EAL to simple types (intuitionnistic logic IL) and from LAL to EAL: $[.]_0 : EAL \to IL$ and $[.]_1 : LAL \to EAL$. Their action on formulas is defined by:

$[\alpha]_0 = \alpha \quad [A \multimap B]_0 = [A]_0 \to [B]_0 \quad [!A]_0 = [A]_0$

$[\alpha]_1 = \alpha \quad [A \multimap B]_1 = [A]_1 \multimap [B]_1 \quad [!A]_1 = [\S A]_1 = ![A]_1$

Define $[.] = [.]_0 \circ [.]_1 : LAL \to IL$. These translations extend to sequent calculus proofs, hence to type derivations. So under each LAL/EAL type, a simple type is present... but while simple types only ensure termination, EAL/LAL types ensure termination with a time complexity bound.

It comes as no surprise that a LAL type can be weakened into an EAL type, as a polynomial bound surely yields an elementary bound, but in fact we can also give a translation $[.]_2 : EAL \to LAL$. For that we define by mutual induction two translations $(.)^p, (.)^n : EAL \to LAL$:

$(\alpha)^p = (\alpha)^n = \alpha \quad (A \multimap B)^p = (A)^n \multimap (B)^p \quad (!A)^p = \S(A)^p$

$(A \multimap B)^n = (A)^p \multimap (B)^n \quad (!A)^n = !(A)^n$

Then $[A]_2 = (A)^p$ and $[\Gamma \vdash A]_2 = (\Gamma)^n \vdash (A)^p$. So an EAL ! is transformed into a LAL ! if in negative position, and in § if in positive position. This gives a translation from EAL cut-free derivations into LAL (cut-free) derivations. Thus EAL typed normal terms are typable in LAL ... However, this translation is not compositional, and so this does not give nonsense on the complexity side.

In other words, we might in some cases be able to type a term t in LAL, and then be unable –because of its type – to apply it to any relevant argument ! For instance if $\vdash_{EAL} t : T$ with $T = N_\alpha^{EAL} \multimap N_\alpha^{EAL}$ we get in LAL a type $[T]_2 = (\S(\alpha \multimap \alpha) \multimap !(\alpha \multimap \alpha)) \multimap N_\alpha^{LAL}$. Thus with this typing t cannot be applied to a LAL integer ...So in general given a term t we will not be searching for *any* type, but for a type satisfying certain constraints allowing for a suitable use of t.

It should now be clear that the difficulty of typing a term in EAL/LAL lies in the problem of determining where to place modalities in the type, and *how many* modalities are needed. Given a term, there is no obvious direct way to bound the number of modalities needed for typing it.

2.3. Typing with dags:

Graph terms. Typing with sequent calculus is uneasy. However we cannot simply type lambda-terms using their syntactic tree, as explicit information about sharing of subterms as given by sequent calculus proofs is important. Therefore we will use directed acyclic graphs (*dags*) that we call *graph terms.* A graph term has one input (or root node) and a certain number of outputs (or variable nodes). On the figures, the implicit direction of edges is from top to bottom. There are four other nodes: λ-node (one premise and one conclusion), @-node (one premise and two conclusions), c-node (contraction: two premises and one conclusion), ax-node (one premise and one conclusion). The ax-node corresponds to sequent calculus axiom, and we will see its utility when we type the graphs.

A variable node can be *free* or *bound*, in which case we add a special edge (a *pointer*) from the variable node to the corresponding λ-node (there is no pointer to the λ-node if its variable is not used in the term).

Graph terms are defined inductively with these nodes according to the grammar on figure 2 (note that we do not represent root and variable nodes). Observe that in graph terms axiom nodes only appear (possibly preceded by a bunch of λ nodes) at the root or in the right conclusion of a @. The left conclusion of a @ can be followed by another @, a contraction or a variable.

A *contraction tree of the graph* is a maximal subgraph whose internal nodes are only c-nodes.

A *path* in a graph term is an oriented sequence of adjacent edges. We write $p = e_1 \dots e_n$ if p is obtained by concatenation of edges e_i. A *complete path* is a path going from the root node to a variable node. A path p is *prefix* of a path q (denoted $p \prec q$) if q is obtained by p followed by another path r (possibly empty, i.e. $p = q$).

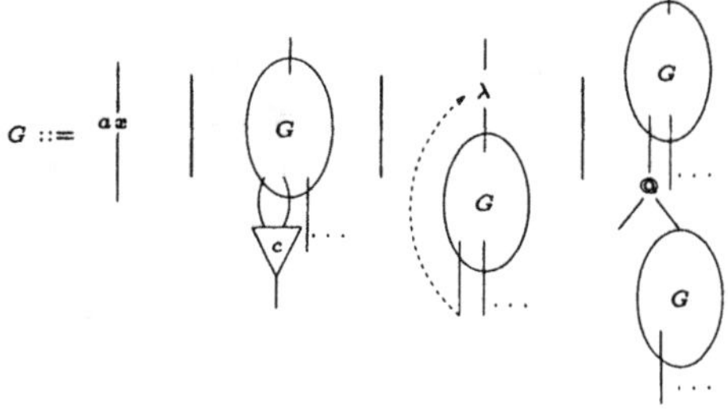

Figure 2. Graph terms inductive definition.

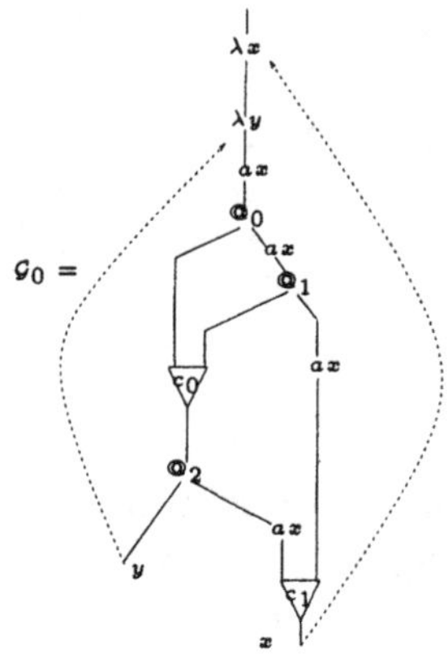

Figure 3. A graph term for $\lambda x \lambda y(y)x((y)x)x$

Given a graph term, we denote by $\leq$ the partial order given by the dag ($N \leq N'$ if N' is above N in the figure drawing of the graph term).

To each graph term we can associate a normal lambda-term in a natural inductive way (see figure 3). This mapping is not one-one though, as a lambda-term can allow various sharings of its subterms.

Now, given a graph term $\mathcal{G}$ and a contraction c of $\mathcal{G}$, we would like to be able to determine which is the minimal sub-graph term of $\mathcal{G}$ on which this contraction could be done. We define the *junction* of contraction c (denoted $jn(c)$) as the minimal node common to all complete paths containing c and strictly superior to c (for instance on fig.3, $jn(c_0) = @_0$ and $jn(c_1) = @_1$). This definition makes sense as: all these paths have at least one node in common (the root-node), and given two nodes common to all these paths they must be comparable. Note that a junction node is necessarily a @-node. Observe that we have:

Lemma 2 *if $\mathcal{G}$ is a graph term and x is a variable node bound by a λ node λ_0, then any complete path of $\mathcal{G}$ containing x also contains λ_0.*

Proposition 3

1. *If a variable node is inferior to a contraction c and this variable is bound, then the corresponding λ node is inferior to c or superior to $jn(c)$.*
2. *Given two contraction nodes c_1 and c_2, if $c_1 \leq c_2$ then either $jn(c_1) \leq c_2$ or $jn(c_1) \geq jn(c_2)$.*

Derivation graphs. We now want to define graph terms corresponding to LAL sequent calculus type derivations. For that we need to label edges with LAL types and introduce *boxes*. A box of a graph term $\mathcal{G}$ is a subgraph $\mathcal{G}'$ which is itself a graph term. The *input* of the box is the input edge of $\mathcal{G}'$ and its *outputs* are the premises of the free variable nodes of $\mathcal{G}'$. There are two kinds of boxes, corresponding respectively to the ! and § promotion rules of the sequent calculus: !-boxes and §-boxes.

LAL *derivation graphs* are defined inductively following the sequent calculus rules (figure 4); they are a subclass of *proof-nets* (Girard, 1998): nodes λ and @ correspond respectively to $\wp$ and $\otimes$.

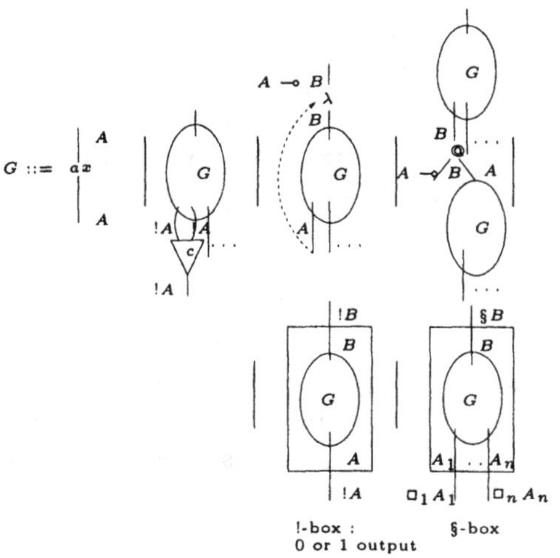

Figure 4. Derivation graphs inductive definition.

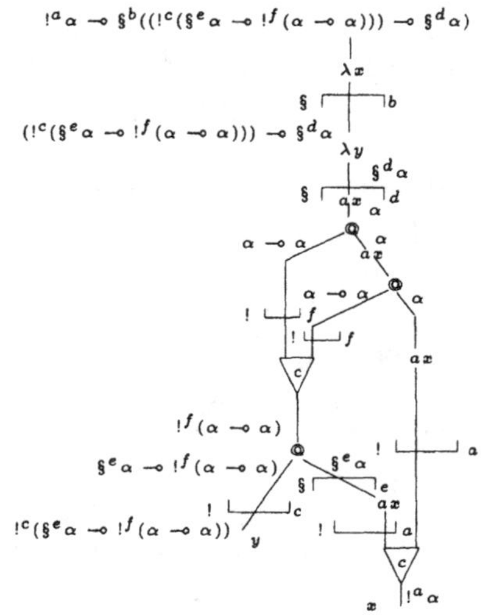

Figure 5. An example of a parameterized graph.

Note that in a derivation graph two distinct boxes either are disjoint or one is included in the other. Without loss of generality we restrict *ax* nodes to formulas A either atomic or of the form $B \multimap C$, as arbitrary axioms can be simulated with extra enclosing boxes. We keep this convention for the rest of the paper.

Let us call a *door* the crossing of a box by an edge; we say that the door is *opening* (resp. *closing*) if the edge enters (resp. exits) the box and we distinguish between ! doors and § doors.

Proposition 4 *In a LAL derivation graphs: root edge, right conclusion of @-node and conclusion of λ-node can have only opening doors, whereas conclusion of ax-node or c-node and left conclusion of @-node can have only closing doors. We say that these two categories of edges are respectively in* opening *and* closing mode.

Typed graphs. Our typing problem will be to turn a graph term into a derivation graph. For that we introduce an intermediary syntax, which forgets about the synchronization feature of boxes.

We consider doors as nodes with one premise and one conclusion and typing: a ! (resp. §) opening door has a premise $!A$ (resp. $\S A$) and a conclusion A; a ! (resp. §) closing door has a premise A and a conclusion $!A$ (resp. $\S A$).

Now, *LAL typed graph terms* are graphs built from the previous λ, @, c, *ax* and door nodes, with edges labeled by LAL formulas according to the typing conditions illustrated in figure 4 and such that: if we erase the doors and the labels we obtain a graph term, and only edges in opening mode (resp. closing mode) can have opening doors (resp. closing doors).

We introduce a function *el* from paths of a typed graph to $\mathbb{Z}$. It measures the *elevation* between the starting and the ending point of the path. Given a path p, $el(p)$ is defined in the following way: if p is an edge and n is the number of its doors, then if these are opening doors $el(p) = n$ and if they are closing doors $el(p) = -n$; otherwise if $p = e_1 \cdot e_2 \cdots e_k$, then $el(p) = \sum_{i=1}^{k} el(e_i)$.

One can obtain a LAL typed graph from a derivation graph by replacing each box by (disconnected) opening and closing doors: say a typed graph is *valid* if it can be obtained this way from a certain derivation graph.

Conversely, given a LAL typed graph we will study under which conditions we can associate opening and closing doors to define boxes so as to get a derivation graph. First let us examine the necessary conditions satisfied by valid typed graph terms.

Consider a path p in a typed graph term: we say it is *well-bracketed* if for any prefix q of p the number of closing doors in q is inferior or equal to the number of opening doors. Thus, each closing door can be matched in p with an opening door in the expected way (note that we do not require that each opening door is matched).

Now, the fact that boxes are disjoint or included one in the other ensures that valid typed graphs satisfy:

- **(C1) Bracketing condition**: any complete path p of $\mathcal{G}$ is well-bracketed.

  Second, if we consider two nodes N and N' and p, q two paths from N to N', they must cross the same boxes. Therefore we have:

- **(C2) Level condition**: given two nodes N, N' in $\mathcal{G}$ for any paths p, q from N to N' we have: $el(p) = el(q)$.

The level condition guarantees that the matching of closing doors with opening doors does not depend on the path chosen. Therefore, if a typed graph term satisfies the bracketing and level condition, then there is a unique way to associate doors in it to define boxes. Once boxes have been defined on a typed graph, we can examine further conditions:

- **(C3) Scope condition**: if x is a bound variable in $\mathcal{G}$, then the node corresponding to x and the binding λ-node belong to the same boxes.

- **(C4) Junction condition**: if a box contains a contraction node c, then it also contains the node $jn(c)$.

- **(C5) !-box conditions**: a box with opening !-door has no closing §-door and has at most one closing door.

We can check, using the inductive definition of derivations, that any valid typed graph satisfies these five conditions, but furthermore:

Theorem 5 (Synchronization) *A LAL typed graph term $\mathcal{G}$ is valid iff it satisfies conditions (C1) to (C5).*

The proof is given in Baillot, 2001.

3. Parameterized graphs

We want now to consider graph terms with a variable number of doors. This will be useful when we search for a valid type for a term. Therefore we will allow doors to be indexed by parameters, or even linear combination of parameters (on figures we write this index on the right-hand-side of the door).

We also have to change the types: intuitively, a LAL parameterized type is a type where exponentials can be indexed by variables, for instance:

$\S^{n_1}(\S^3 !^{n_2}\alpha \multimap \S^{n_2}\alpha)$.

Given a set of integer variables $n, m \ldots$, *parameterized LAL types* are given by the grammar:

$$T ::= \alpha | T \multimap T | !T | \S T | \S^n T | !^n T$$

The *parameters* of a parameterized type T are the integer variables appearing in it. We denote their set by $\mathcal{V}(T)$.

The former mapping from *LAL* types to simple types is naturally extended to parameterized types. An *instance* of a parameterized type T is an application $\phi : \mathcal{V}(T) \rightarrow \mathbb{N}$. By substituting in T each integer variable n by $\phi(n)$ we get a LAL type $\phi(T)$.

Parameterized graph terms are defined as typed graphs except that doors are parameterized: for instance an opening ! door with parameter n has a premise $!^n A$ and a conclusion A. See figure 5 for an example.

The set of parameters of the graph $\mathcal{G}$ is denoted by $\mathcal{V}(\mathcal{G})$. From an instance $\phi : \mathcal{V}(\mathcal{G}) \rightarrow \mathbb{N}$ we can define a typed graph $\phi(\mathcal{G})$ as expected.

The elevation function *el* is defined for a parameterized typed graph as before but its values are linear combinations over $\mathcal{V}(\mathcal{G})$ with integer coefficients.

We consider the following problem (synchronization of parameterized graph):

Problem 1 given a parameterized graph term $\mathcal{G}$, does there exist an instance $\phi : \mathcal{V}(\mathcal{G}) \rightarrow \mathbb{N}$ such that $\phi(\mathcal{G})$ is a valid typed graph?

We define *constraints* on a parameterized graph $\mathcal{G}$ as first-order arithmetic formulas over $\mathcal{V}(\mathcal{G})$ built from: (i) linear inequations, (ii) conjunction $\wedge$, disjonction $\vee$ and universal quantification $\forall$.

An instance ϕ of a parameterized graph $\mathcal{G}$ *satisfies* a constraint $\mathcal{C}$ if $\mathcal{C}$ evaluates to true when we replace each n by $\phi(n)$. We will express the conditions of theorem 5 by constraints.

(C1) Bracketing. Each complete path of the graph is well bracketed. This is expressed by: for any path p starting from the initial edge $el(p) \geq 0$. This gives a finite conjunction of inequalities $\mathcal{C}_1$.

(C2) Level. For any pair of paths p, q with same origin and target, we have $el(p) = el(q)$. As there is a finite number of paths in the dag, this gives a finite conjunction of inequalities $\mathcal{C}_2$.

(C3) Scope. We must express the scope condition for each bound variable. Take a bound variable x and the λ node N_0 in the graph corresponding to its abstraction. Each path p from N_0 to the variable edge of x corresponds to an occurrence of the variable. For each of these we should have: (1) each closing door in the path is associated to an opening door in the path, ie the path p is well-bracketed; this is expressed by: for all $q \prec p$, $el(q) \geq 0$, (2) all boxes opened along the path are closed along the path; this is expressed by: $el(p) = 0$.

These two conditions stated for all relevant path p for all bound variables of the graph yield a constraint C_3.

(C4) Junction. the problem of finding the junction of a given contraction in the graph is decidable, so we assume here that it has been done for all contractions of the graph. Now, given a contraction c and its junction j, we want to ensure that: each box containing c also contains j.

This is equivalent to the following condition: for any path p joining j to c, any opening door d_0 in p is associated to a closing door in p. We can express this condition by: for any suffix q of p, $el(q) \leq 0$.

Applied to each path from j to c, and so for all contractions of $\mathcal{G}$, this gives a constraint C_4.

(C5) Bang boxes conditions. We want to express the conditions: (1) a !-box does not have any § exit door, (2) a !-box has at most one ! exit door.

We have:

Proposition 6 *Let d_0 and d_1 be two (possibly) parameterized doors of the graph respectively opening and closing, such that d_0 is above d_1. Denote by n and m their respective parameters and by p a path from d_0 to d_1, including d_0 and d_1. Given k with $1 \leq k \leq n$, one of the doors of d_1 matches the k-th door of d_0 iff the following predicate evaluates to true:*

$$P(d_0, d_1, k) = (\textit{for all } q \prec p, q \neq p \Rightarrow el(q) - k + 1 > 0) \wedge (el(p) - k + 1 \leq 0).$$

$P(d_0, d_1, k)$ is a finite conjunction of inequations.

Condition (1) can then be expressed in the following way: for any pair of parameterized opening !-door d_0 with parameter n and closing §-door d_1 such that the first one is above the second one in the dag the following should hold:

$$\forall k \leq n, \ \neg P(d_0, d_1, k).$$

Condition (1) is thus expressed by a constraint C_{51}. Now, condition (2) is expressed in the following way:

for any d_0 parameterized opening !-door with parameter n and d_1, d_2 distinct closing !-doors such that d_0 is above d_1 and d_2 we have:

$$\forall k \leq n, \ \neg P(d_0, d_1) \vee \neg P(d_0, d_2).$$

Again, this yields a constraint C_{52}. The !-boxes conditions are thus expressed by $C_5 = C_{51} \wedge C_{52}$.

Solving. Let $\mathcal{C} = \wedge_{j=1}^{5}\mathcal{C}_j$. We have seen that $\mathcal{G}$ is a valid LAL derivation graph iff the constraint $\mathcal{C}$ is satisfied. Therefore an instance ϕ is a solution of the synchronization problem 1 for $\mathcal{G}$ iff it is a solution of $\mathcal{C}$. The problem of satisfiability for first-order formulas over linear inequations is decidable as these are part of Presburger arithmetic. Therefore we have:

Theorem 7 *Given a parameterized graph term $\mathcal{G}$, the problem of the existence of a valuation ϕ of $\mathcal{V}(\mathcal{G})$ such that $\phi(\mathcal{G})$ is a valid typed graph is decidable.*

Actually it is clear from the proof that we can even require the variables of $\mathcal{V}(\mathcal{G})$ to satisfy an initial constraint $\mathcal{C}_0$.

4. Type instantiation

Ideally, we would like starting from a lambda-term to decide whether it is typable in LAL. At the present we do not know whether this problem is decidable. It is the case if we consider a lambda-term in normal form. In general however we are not merely interested in obtaining a type for a normal lambda-term but we would like it to satisfy certain constraints. The most obvious requirement is that we want to be able to specify a data-type for arguments and result (recall the discussion in section 2.2). For instance:

does t admit a LAL type of the form $!^k N_A^{LAL} \multimap \S^l N_B^{LAL}$? where A, B are parameterized LAL types.

The compromise we adopt here is to ask the user to provide a parameterized type T, and to search whether there exists an instance of T which types t. In fact the user could also provide some conditions together with the parameterized type. For instance in the previous example he could require that $l \geq k$ holds.

Given a simple type T, the least constraining (and hence least informative) parameterized type over T is defined in the following way: decorate each positive (resp. negative) subformula of T with $\S^n$ (resp. $!^n$) where n is a fresh parameter. Observe that for checking typability in LAL for normal terms in a narrow-minded way such parameterized types suffice. Now we can state the problem we are considering:

Problem 2 (Type instantiation) *Given a (closed normal) lambda-term t, a LAL parameterized type T and a constraint $\mathcal{C}_0$ for T, does there exist an instance ϕ of T such that: $\phi(T)$ is a LAL type for t and ϕ satisfies $\mathcal{C}_0$?*

We now give the main lines of our decision procedure for problem 2. We start from a graph term $\mathcal{G}$ instead of a lambda-term, but since there is a finite number of graph terms corresponding to a lambda-term, this is not a problem for decidability. The dag specifies the information on sharing of subterms and order in which contractions are performed. We can also assume that $[T]$ is a valid simple type for T.

The algorithm will proceed in two phases: (i) *labeling phase*: from $\mathcal{G}$, T and $\mathcal{C}_0$ deduce all possible parameterized graphs with associated constraint $(\mathcal{G}', \mathcal{C}_1)$;

(ii) *solving phase*: apply theorem 7 to decide if one of these parameterized graphs admits an instance which makes it a valid typed graph.

Let us stress that the procedure we give has no pretention to efficiency, but aims at establishing decidability of problem 2 in a simple way.

Assume given a closed graph term $\mathcal{G}$ and a parameterized type T. We will give the corresponding parameterized graphs. We proceed in two steps: (a)(*labeling visit*) first we attribute (parameterized) types and doors to all edges of the graph but those belonging to contraction trees; this is done through one visit of the graph and there is only one possible labeling; (b)(*contraction trees decoration*) then we place (closing) doors with parameters in the contraction trees and add the corresponding types; there will be a finite number of ways to do this, and this step will determine the number of variables of our constraints system. At the end we obtain several possible parameterized graph terms with constraints.

(a) Labeling visit: For this first task, we perform a depth-first leftmost visit of the dag; more precisely we alternate downwards and upwards trips according to the following strategy:

- downwards trip: start from the root-edge and go down, choosing the left conclusion whenever you meet a @-node, until reaching an edge already labeled, then switch to upwards mode;
- upwards trip: go up until either meeting a @-node whose right conclusion has not been visited yet, in which case go down its right conclusion, or meeting the root edge, in which case the procedure is over. When meeting the root of a contraction tree go up its leaf edge who has been visited last (at least one leaf has already been visited);

It is easy to check that by this strategy we do visit the whole graph (but the edges in the contraction trees) and that we go through each edge two times, first down and then up.

Now we must say how we do the labeling during this visit. We denote by s an arbitrary sequence of indexed modalities.

Downwards trip. The root edge is labeled by T. For each λ node we cross: if the type before the node is not of the form $A = !^{n_1}\S^{n_2}\dots !^{n_{2k-1}}\S^{n_{2k}}(B \multimap C)$ for some k (some n_i's can be constants, in particular 0), then the procedure fails. Otherwise for each $1 \leq i \leq 2k$ such that $n_i \neq 0$ put an opening door (either ! or §) with parameter n_i. Then: the type before the doors is A; the type below the doors is $B \multimap C$; the conclusion of λ is typed by C and the variable edge bound by the λ, if there is one, is typed by B.

We necessarily meet an axiom node (by construction of the graphs). Denote by sC the current formula, where C is either atomic or of the form $A \multimap B$ and s is a sequence of modalities. Put the opening doors corresponding to s and type the premise of the axiom by C.

We then proceed with the trip but do not label anymore until switching to upwards mode.

Upwards trip.

- If we arrive to the root of a contraction tree with type sA where $A = \alpha$ or $B \multimap C$ we go up the leaf who has been visited last (it is a left conclusion of @ or a conclusion of ax) and we type it with A.

- If we arrive to a @ node from the left conclusion: if the type is not of the form $s(A \multimap B)$, the procedure fails; if it is, put the closing doors corresponding to s, type the premise of @ by B and the right conclusion by A; then go down the right conclusion of @.

- If we arrive to a @ node from the right conclusion: then this edge has already been typed; we continue upwards.

- If we arrive to an axiom node with type sC where C is either atomic or of the form $A \multimap B$, put the closing doors corresponding to s below the axiom, type their premise with C and proceed upwards.

Lemma 8 *After the labeling visit, all edges but those in contraction trees have been typed.*

(b) Contraction trees decoration: At this point, for each contraction tree we have a type for its root-edge and a type for each of its leaves, but not for the intermediary edges. We give a non-deterministic method for finding all possible contraction tree decorations.

Note that the leaves of the trees are conclusion of ax nodes or left conclusion of @ nodes and their types are of the form α or $B \multimap C$.

Say $\mathcal{C}_1$ is initially an empty constraints system. We will gradually need to add inequations to $\mathcal{C}_1$.

We will first deal with each branch of the tree separately. Consider one of these branches and denote by sB_1 (resp. B_2) the type of the leaf (resp. of the root-edge), where s is a sequence of indexed modalities and B_i is a type which does not start with a modality.

If $[B_1] = [B_2]$, identification of B_1 and B_2 can be expressed by a constraint which we add to $\mathcal{C}_1$, otherwise typing fails.

We then need to place along the branch enough doors to introduce the sequence s. As in s the number of alternances between ! and § is finite there is a finite number of ways to place parameterized closing doors along the branch to match this sequence. We choose one possibility, with a new parameter variable for each door, and add to $\mathcal{C}_1$ the equations expressing the fact that the sequence of doors on this branch introduces s. Finally we also add to $\mathcal{C}_1$ inequations imposing that the formula before each contraction starts with a ! (so that contraction is valid).

We proceed similarly for each branch of the contraction tree. Then for each contraction node, the types of both premises should be identified; again this is expressed by constraints which we include in $\mathcal{C}_1$.

Applying this method to each contraction tree and calling $\mathcal{C}_1$ again the resulting constraints system, we have completed the labeling phase. So we end up with a finite number of possible parameterized graph terms and associated constraints. The initial graph $\mathcal{G}$ is typable with the parameterized type T iff one of these systems admits a valuation making it valid. This can be decided according to theorem 7, hence we conclude:

Theorem 9 *The type instantiation problem for normal lambda-terms (problem 2) is decidable.*

Acknowledgments

The author wishes to thank Roberto Amadio and Laurent Rgnier for suggestions and encouragements, as well as Kazushige Terui for useful discussions. The results in this paper were first presented at the 'Linear' TMR network meeting held in Bertinoro in april 2001 and we wish to thank the organizers.

References

Asperti, A. (1998). Light affine logic. In *Proceedings LICS'98*. IEEE Computer Society Press.

Baillot, P. (2000). Stratified coherent spaces: a denotational semantics for light linear logic. LFCS Tech. report 0025, Univ. of Edinburgh. presented at ICC'00.

Baillot, P. (2001). Checking polynomial time complexity with types (extended version). Tech. report 2001-09, Laboratoire d'Informatique de Paris-Nord.

Bellantoni, S., Niggl, K.-H., and Schwichtenberg, H. (2000). Higher type recursion, ramification and polynomial time. *Annals of Pure and Applied Logic*, 104(1-3).

Coppola, P. and Martini, S. (2001). Typing lambda-terms in elementary logic with linear constraints. In *Proceedings TLCA'01*, volume 2044 of *LNCS*. Springer-Verlag.

Danos, V. and Joinet, J.-B. (1999). Linear logic and elementary time. First Workshop on Implicit Computational Complexity (ICC'99).

Danos, V., Joinet, J.-B., and Schellinx, H. (1994). On the linear decoration of intuitionistic derivations. *Archive for Mathematical Logic*, 33(6).

Girard, J.-Y. (1998). Light linear logic. *Information and Computation*, 143:175–204.

Hofmann, M. (2000). Safe recursion with higher types and BCK-algebra. *Annals of Pure and Applied Logic*, 104(1-3).

Leivant, D. and Marion, J.-Y. (1993). Lambda-calculus characterisations of polytime. *Fundamenta Informaticae*, 19:167–184.

Roversi, L. (1999). A P-time completeness proof for light logics. In *Proceedings CSL'99*, volume 1683 of *LNCS*. Springer-Verlag.

Roversi, L. (2000). Light affine logic as a programming language: a first contribution. *International Journal of Foundations of Computer Science*, 11(1).

Schellinx, H. (1994). *The Noble Art of Linear Decorating*. ILLC Dissertation Series 1994-1, Institute for Language, Logic and Computation, University of Amsterdam.

Terui, K. (2001). Light Affine Lambda-calculus and polytime strong normalization. In *Proceedings LICS'01*. IEEE Computer Society Press.

Wadler, P. (1991). Is there a use for linear logic? In *ACM Conference on Partial Evaluation and Semantics-Based Program Manipulation*, New Haven, Connecticut.

BOUNDARY INFERENCE FOR ENFORCING SECURITY POLICIES IN MOBILE AMBIENTS

Chiara Braghin, Agostino Cortesi, Riccardo Focardi*
Dipartimento di Informatica, Università Ca' Foscari di Venezia,
Via Torino 155, 30173 Venezia - Mestre (Italy)
{ braghin,cortesi,focardi } @dsi.unive.it

Steffen van Bakel
Department of Computing, Imperial College,
180 Queen s Gate, London SW7 2BZ, (UK)
svb@cs.ic.ac.uk

Abstract The notion of "boundary ambient" has been recently introduced to model multilevel security policies in the scenario of mobile systems, within pure Mobile Ambients calculus. Information flow is defined in terms of the possibility for a confidential ambient/data to move outside a security boundary, and boundary crossings can be captured through a suitable Control Flow Analysis. We show that this approach can be further enhanced to infer which ambients should be "protected" to guarantee the lack of information leakage for a given process.

Keywords: Mobile Ambients, Security, Static Analysis.

1. Introduction

A Trusted Computing Base is the set of protection mechanisms within a computer system the combination of which is responsible for enforcing a security policy [1]. One of the main challenges faced when building a TCB is deciding which parts of the system are security-critical. Our focus is on *Multilevel Security*, a particular *Mandatory Access Control* security policy: every entity is bound to a security level (for simplicity, we consider only two levels: *high* and *low*), and information may just flow from the low level to the high one. Typically, two access rules are imposed: (*i*) *No*

*Work partially supported by MURST Projects "Interpretazione Astratta, Type Systems e Analisi Control-Flow", and MEFISTO, and EU Contract IST-2001-32617.

Read Up, a low level entity cannot access information of a high level entity; (*ii*)*No Write Down*, a high level entity cannot leak information to a low level entity.

In order to detect information leakages, a typical approach (see, e.g., [2, 8, 9, 10, 12, 13]) consists in directly defining what is an information flow from one level to another one. Then it is sufficient to verify that, in any system execution, no flow of information is possible from level *high* to level *low*. This is the approach we follow also in this paper.

To model information flow security, we adopt the scenario of mobile systems. This particular setting, where code may migrate from one security level to another one, complicates even further the problem of capturing all the possible information leakages. As an example, confidential data may be read by an authorized agent which, moving around, could expose them to unexpected attacks. Moreover, the code itself could be confidential, and so not allowed to be read/executed by lower levels.

In order to study this problem in an as abstract manner as possible, we consider the "pure" Mobile Ambients calculus [5], in which no communication channels are present and the only possible actions are represented by the moves performed by mobile processes. This allows the study of a very general notion of information flow which should be applicable also to more "concrete" versions of the calculus.

The information flow property of interest is defined in terms of the possibility for a confidential ambient/data to move outside a *security boundary*. In [6], a very simple syntactic property is introduced that it is sufficient to imply the absence of unwanted information flow. In [3], a refinement of the control flow analysis defined in [11] is introduced that deals with the same property with improved accuracy.

As an example, consider two different sites *venice* and *montreal*, each with some set of confidential information that need to be protected. This can be modeled by just defining two boundary ambients, one for each site:

$$venice^b[\,P_1\,] \mid montreal^b[\,P_2\,] \mid Q^\ell,$$

where Q is an untrusted process. In order to make the model applicable, a mechanism for moving confidential data from one boundary to another one is certainly needed. This is achieved through another boundary ambient which moves out from the first protected area and into the second one. In the example, label b denotes a boundary, h a high-level ambient, ℓ a low-level ambient and c a capability. Consider the example depicted in Figure 1. Process

$$venice^b[\; send^b[\,\mathbf{out}^c\, venice.\mathbf{in}^c\, montreal\,] \mid hdata^h[\,\mathbf{in}^c\, send\,]\,] \mid \\ montreal^b[\,\mathbf{open}^c\, send\,] \mid Q^\ell$$

may evolve to (step (b))

$$venice^b[\;] \mid send^b[\,\mathbf{in}^c\, montreal \mid hdata^h[\;]\,] \mid montreal^b[\,\mathbf{open}^c\, send\,] \mid Q^\ell$$

then to (step (c))

$$venice^b[\;] \mid montreal^b[\,\mathbf{open}^c\, send \mid send^b[\; hdata^h[\;]\,]\,] \mid Q^\ell$$

and finally to

$$venice^b[\;] \mid montreal^b[\; hdata^h[\;]\,] \mid Q^\ell$$

Note that *send* is labeled as a boundary ambient. Thus, the high level data *hdata* is always protected by boundary ambients, during the whole execution.

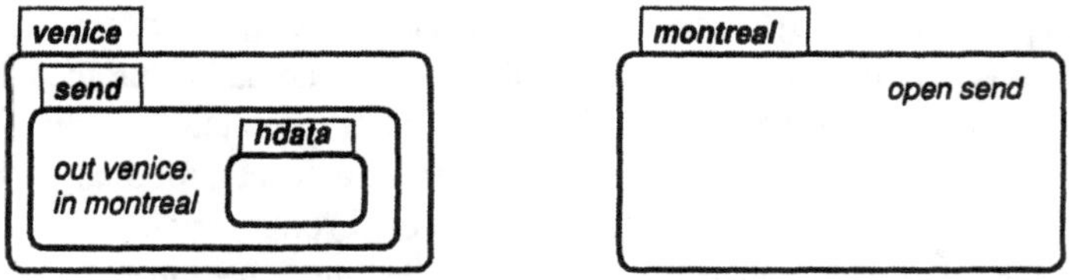

(a) *venice* needs to send confidential data *hdata* to *montreal.*

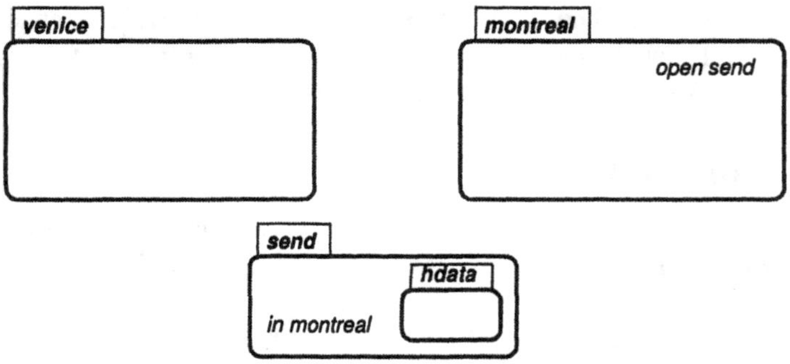

(b) The confidential data is sent inside the secure "envelope" *send.*

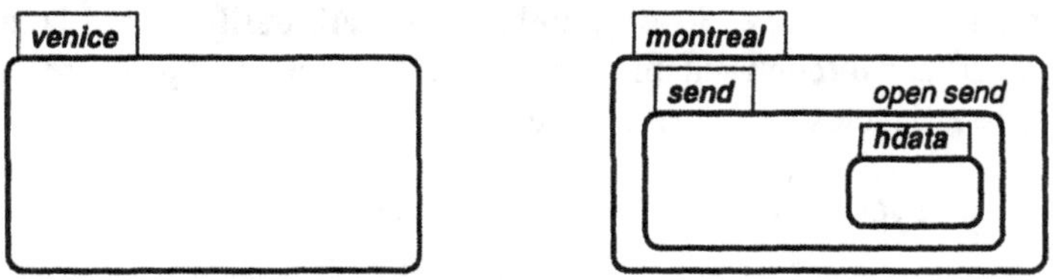

(c) The confidential data safely arrive in *montreal.*

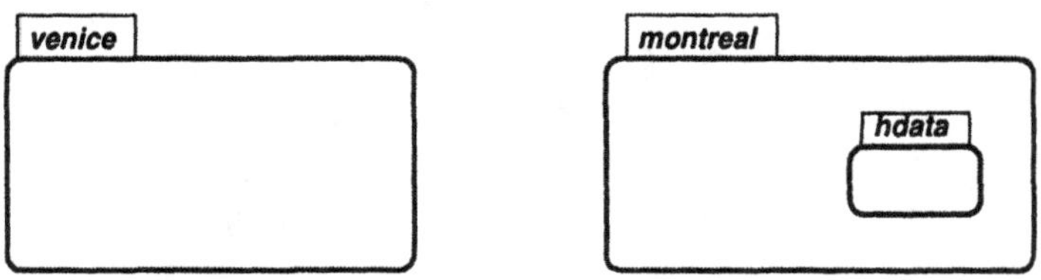

(d) The envelope is dissolved to allow confidential data to be accessed in *montreal.*

Figure 1. Venice and Montreal exchange confidential information.

The analysis developed in [3] allows to verify that no leakage of secret data/ambients outside the boundary ambients is possible. When applied to this example, it shows that h is always contained inside b, i.e., a boundary ambient. This basically proves that the system is secure and no leakage of h data may happen.

In this paper we are interested in merging these ideas towards the definition of a TCB, to a more ambitious perspective: which are the ambients that should be labeled “boundary”, to guarantee that the system is secure, i.e. that no h data may fall into an unprotected environment? Is there always a solution to this problem? Is there a minimal solution?

We show that these problems can be properly addressed by re-executing the Control Flow Analysis presented in [3]. A successful analysis infers boundary ambients until a fixed point is reached, returning the set of ambients that should be "protected".

In the example above, all we know is that *hdata* is information that must be protected during the whole execution of the process; thus, a successful analysis should infer *venice*, *montreal* and *send* as ambients to be labeled "boundary".

The rest of the paper is organized as follows. In Section 2 we introduce the basic terminology on ambient calculus, then we present the model of multilevel security for mobile agents and we show how to guarantee absence of unwanted information flows through the control flow analysis of [3]. In Section 3, we introduce the enhanced Control Flow Analysis. Section 4 concludes the paper.

2. Background

In this section we introduce the basic terminology on ambient calculus on multilevel security and we briefly recall the control flow analysis defined in [3].

2.1. Mobile Ambients

The Mobile Ambients calculus has been introduced in [5] with the main purpose of explicitly modeling mobility. Indeed, ambients are arbitrarily nested boundaries which can move around through suitable capabilities. The syntax of processes is given as follows, where n denotes an ambient name.

P, Q	$::=$	$(\nu n)P$	restriction
	$\mid$	$\mathbf{0}$	inactivity
	$\mid$	$P \mid Q$	composition
	$\mid$	$!P$	replication
	$\mid$	$n^{\ell^a}[P]$	ambient
	$\mid$	$\mathbf{in}^{\ell^t}\ n.P$	capability to enter n
	$\mid$	$\mathbf{out}^{\ell^t}\ n.P$	capability to exit n
	$\mid$	$\mathbf{open}^{\ell^t}\ n.P$	capability to open n

Labels $\ell^a \in \mathbf{Lab}^a$ on ambients and labels $\ell^t \in \mathbf{Lab}^t$ on transitions (capabilities), have been introduced in the control flow analysis proposed in [11]. This is just a way of indicating "program points" and will be useful in the next section when developing the analysis.

Intuitively, the restriction $(\nu n)P$ introduces the new name n and limits its scope to P; process $\mathbf{0}$ does nothing; $P \mid Q$ is P and Q running in parallel; replication provides recursion and iteration as $!P$ represents any number of copies of P in parallel. By $n^{\ell^a}[P]$ we denote the ambient named n with the process P running inside it. The capabilities $\mathbf{in}^{\ell^t}\ n$ and $\mathbf{out}^{\ell^t}\ n$ move their enclosing ambients in and out ambient n, respectively; the capability $\mathbf{open}^{\ell^t}\ n$ is used to dissolve the boundary of a sibling ambient n. The operational semantics [5] of a process P is given through a suitable reduction relation $\rightarrow$ and a structural congruence $\equiv$ between processes. Intuitively, $P \rightarrow Q$ represents the possibility for P of reducing to Q through some computation.

2.2. Modeling Multilevel Security

In order to define Multilevel security in Mobile Ambients we first need to classify information into different levels of confidentiality. We do that by exploiting the labeling of ambients. In particular, the set of ambient labels Lab^a will be partitioned into three mutually disjoint sets Lab^a_H, Lab^a_L and Lab^a_B, which stand for *high*, *low* and *boundary* labels. We denote by $\mathcal{L}$ the triplet $(\mathsf{Lab}^a_H, \mathsf{Lab}^a_L, \mathsf{Lab}^a_B)$.

Given a process, the multilevel security policy may be established by deciding which ambients are the ones responsible for confining confidential information. These will be labeled with boundary labels from set Lab^a_B and we will refer to them as *boundary ambients*. Thus, all the high level ambients must be contained in a boundary ambient, and labeled with labels from set Lab^a_H. On the other side, all the external ambients are considered low level ones and consequently labeled with labels from set Lab^a_L. This is how we will always label processes, and corresponds to defining the security policy (what is secret, what is not, what is a container of secrets). In all the examples, we will use the following notation for labels: $b \in \mathsf{Lab}^a_B, h \in \mathsf{Lab}^a_H$, $m, m' \in \mathsf{Lab}^a_L$ and $c, ch, cm, cm' \in \mathsf{Lab}^t$.

In [3] we introduced a refinement of the Control Flow Analysis of [11], in order to incorporate the ideas above, thus yielding to a more accurate tool for detecting unwanted boundary crossings. The main idea is to keep information about the nesting of boundaries, and about "unprotected" ambients.

Definition 1 The refined control flow analysis works on triplet $(\hat{I_B}, \hat{I_E}, \hat{H})$, where:

$(\hat{I_B})$: The first component is an element of $\wp(\mathsf{Lab}^a \times (\mathsf{Lab}^a \cup \mathsf{Lab}^t))$. If a process contains either a capability or an ambient labeled ℓ inside an ambient labeled ℓ^a which is a boundary or an ambient nested inside a boundary (referred as *protected ambient*) then (ℓ^a, ℓ) is expected to belong to $\hat{I_B}$. As long as high level data is contained inside a protected ambient there is no unwanted information flow.

$(\hat{I_E})$: The second component is also an element of $\wp(\mathsf{Lab}^a \times (\mathsf{Lab}^a \cup \mathsf{Lab}^t))$. If a process contains either a capability or an ambient labeled ℓ inside an ambient labeled ℓ^a which is not protected, then (ℓ^a, ℓ) is expected to belong to $\hat{I_E}$.

$(\hat{H})$: The third component keeps track of the correspondence between names and labels. If a process contains an ambient labeled ℓ^a with name n, then (ℓ^a, n) is expected to belong to $\hat{H}$.

The analysis is defined by a representation function and a specification, like in [11]. They are depicted, respectively, in Figure 2 and Figure 3, in which we consider a process P_* executing at the top-level environment labeled *env*.

Observe that within the specification of the analysis (depicted in Figure 3), some predicates are used to enhance readability, namely

$$
\begin{array}{llcl}
 & \beta^{\mathcal{L}}(P) & = & \beta^{\mathcal{L}}_{env,False}(P) \\
(res) & \beta^{\mathcal{L}}_{\ell,Proct}((\nu n)P) & = & \beta^{\mathcal{L}}_{\ell,Proct}(P) \\
(zero) & \beta^{\mathcal{L}}_{\ell,Proct}(\mathbf{0}) & = & (\emptyset,\emptyset,\emptyset) \\
(par) & \beta^{\mathcal{L}}_{\ell,Proct}(P \mid Q) & = & \beta^{\mathcal{L}}_{\ell,Proct}(P) \sqcup \beta^{\mathcal{L}}_{\ell,Proct}(Q) \\
(repl) & \beta^{\mathcal{L}}_{\ell,Proct}(!P) & = & \beta^{\mathcal{L}}_{\ell,Proct}(P) \\
(amb) & \beta^{\mathcal{L}}_{\ell,Proct}(n^{\ell^a}[\,P\,]) & = & \text{case } Proct \text{ of} \\
 & & & \quad \text{True}: \beta^{\mathcal{L}}_{\ell^a,Proct}(P) \sqcup (\{(\ell,\ell^a)\},\emptyset,\{(\ell^a,n)\}) \\
 & & & \quad \text{False: if } (\ell_a \in \mathbf{Lab}^a_B) \text{ then} \\
 & & & \qquad \text{let Proct}' = \text{True else Proct}' = \text{False in} \\
 & & & \qquad \beta^{\mathcal{L}}_{\ell^a,Proct'}(P) \sqcup (\emptyset,\{(\ell,\ell^a)\},\{(\ell^a,n)\}) \\
(\mathbf{in}) & \beta^{\mathcal{L}}_{\ell,Proct}(\mathbf{in}^{\ell^t}\ n.P) & = & \text{case } Proct \text{ of} \\
 & & & \quad \text{True}: \beta^{\mathcal{L}}_{\ell,Proct}(P) \sqcup (\{(\ell,\ell^t)\},\emptyset,\emptyset) \\
 & & & \quad \text{False}: \beta^{\mathcal{L}}_{\ell,Proct}(P) \sqcup (\emptyset,\{(\ell,\ell^t)\},\emptyset) \\
(\mathbf{out}) & \beta^{\mathcal{L}}_{\ell,Proct}(\mathbf{out}^{\ell^t}\ n.P) & = & \text{case } Proct \text{ of} \\
 & & & \quad \text{True}: \beta^{\mathcal{L}}_{\ell,Proct}(P) \sqcup (\{(\ell,\ell^t)\},\emptyset,\emptyset) \\
 & & & \quad \text{False}: \beta^{\mathcal{L}}_{\ell,Proct}(P) \sqcup (\emptyset,\{(\ell,\ell^t)\},\emptyset) \\
(\mathbf{open}) & \beta^{\mathcal{L}}_{\ell,Proct}(\mathbf{open}^{\ell^t}\ n.P) & = & \text{case } Proct \text{ of} \\
 & & & \quad \text{True}: \beta^{\mathcal{L}}_{\ell,Proct}(P) \sqcup (\{(\ell,\ell^t)\},\emptyset,\emptyset) \\
 & & & \quad \text{False}: \beta^{\mathcal{L}}_{\ell,Proct}(P) \sqcup (\emptyset,\{(\ell,\ell^t)\},\emptyset)
\end{array}
$$

Figure 2. Representation Function for the refined Control Flow Analysis

- $$path_B(\ell^a,\ell) = \begin{cases} \text{True} & \text{if } \ell^a = \ell \vee \exists \ell_1,\ell_2,\ldots,\ell_n \notin \mathbf{Lab}^a_B : n \geq 0 \wedge \\ & (\ell^a,\ell_1),(\ell_1,\ell_2),\ldots,(\ell_n,\ell) \in \hat{I}_B \ \wedge\ \ell^a,\ell \notin \mathbf{Lab}^a_B, \\ \text{False} & \text{otherwise.} \end{cases}$$

- $$path_E(\ell^a,\ell) = \begin{cases} \text{True} & \text{if } \ell^a = \ell \vee \exists \ell_1,\ell_2,\ldots,\ell_n \notin \mathbf{Lab}^a_B : n \geq 0 \wedge \\ & (\ell^a,\ell_1),(\ell_1,\ell_2),\ldots,(\ell_n,\ell) \in \hat{I}_E \ \wedge\ \ell^a,\ell \notin \mathbf{Lab}^a_B, \\ \text{False} & \text{otherwise.} \end{cases}$$

The representation function maps processes to their abstract representation, i.e. a triplet $(\hat{I}_B, \hat{I}_E, \hat{H})$ representing process P_*.

Example 2 Let P be a process of the form: $P = n^{\ell^a_1}[\ m^{\ell^a_2}[\mathbf{out}^{\ell^t}\ n\,]\ ]$, with $\ell^a_1 \in \mathbf{Lab}^a_B$ and $\ell^a_2 \in \mathbf{Lab}^a_L$, thus the representation function of P is the following: $\beta^{\mathcal{L}}(P) = (\{(\ell^a_1,\ell^a_2),(\ell^a_2,\ell^t)\},\{(env,\ell^a_1)\},\{(\ell^a_1,n),(\ell^a_2,m)\})$.

The specification of the analysis amounts to recursive checks of subprocesses, which provide constraints that the triplet $(\hat{I}_B, \hat{I}_E, \hat{H})$ should satisfy in order to be a correct solution for the analysis. It is possible to prove that a least solution of this analysis exists and it may be computed as follows: first apply the representation function

$$
\begin{array}{lll}
(res) & (\hat{I_B},\hat{I_E},\hat{H}) \models^{\mathcal{L}} (\nu n)P & \text{iff} \quad (\hat{I_B},\hat{I_E},\hat{H}) \models^{\mathcal{L}} P \\
(zero) & (\hat{I_B},\hat{I_E},\hat{H}) \models^{\mathcal{L}} 0 & \text{always} \\
(par) & (\hat{I_B},\hat{I_E},\hat{H}) \models^{\mathcal{L}} P \mid Q & \text{iff} \quad (\hat{I_B},\hat{I_E},\hat{H}) \models^{\mathcal{L}} P \wedge (\hat{I_B},\hat{I_E},\hat{H}) \models^{\mathcal{L}} Q \\
(repl) & (\hat{I_B},\hat{I_E},\hat{H}) \models^{\mathcal{L}} !P & \text{iff} \quad (\hat{I_B},\hat{I_E},\hat{H}) \models^{\mathcal{L}} P \\
(amb) & (\hat{I_B},\hat{I_E},\hat{H}) \models^{\mathcal{L}} n^{\ell^a}[P] & \text{iff} \quad (\hat{I_B},\hat{I_E},\hat{H}) \models^{\mathcal{L}} P \\
(\mathbf{in}) & (\hat{I_B},\hat{I_E},\hat{H}) \models^{\mathcal{L}} \mathbf{in}^{\ell^t}\, n.P & \text{iff} \quad (\hat{I_B},\hat{I_E},\hat{H}) \models^{\mathcal{L}} P \wedge
\end{array}
$$

$$
\begin{array}{l}
\forall \ell^a, \ell^{a'}, \ell^{a''} \in \mathbf{Lab}^a : \\
\text{case } ((\ell^a,\ell^t) \in \hat{I_B} \wedge (\ell^{a''},\ell^a) \in \hat{I_B} \wedge (\ell^{a''},\ell^{a'}) \in \hat{I_B} \wedge (\ell^{a'},n) \in \hat{H}) \\
\quad \Longrightarrow (\ell^{a'},\ell^a) \in \hat{I_B} \\
\text{case } ((\ell^a,\ell^t) \in \hat{I_B} \wedge (\ell^{a''},\ell^a) \in \hat{I_E} \wedge (\ell^{a''},\ell^{a'}) \in \hat{I_E} \wedge \ell^a \in \mathbf{Lab}^a_B \\
\qquad \wedge (\ell^{a'},n) \in \hat{H}) \Longrightarrow \\
\quad \text{if} (\ \ell^{a'} \in \mathbf{Lab}^a_B) \text{ then } (\ell^{a'},\ell^a) \in \hat{I_B} \\
\quad \text{else } (\ell^{a'},\ell^a) \in \hat{I_E} \\
\text{case } ((\ell^a,\ell^t) \in \hat{I_E} \wedge (\ell^{a''},\ell^a) \in \hat{I_E} \wedge (\ell^{a''},\ell^{a'}) \in \hat{I_E} \wedge (\ell^{a'},n) \in \hat{H}) \Longrightarrow \\
\quad \text{if} (\ \ell^{a'} \in \mathbf{Lab}^a_B) \\
\qquad \text{then } (\ell^{a'},\ell^a) \in \hat{I_B} \wedge \left\{(\ell,\ell') \in \hat{I_E} \mid path_E(\ell^a,\ell)\right\} \subseteq \hat{I_B} \\
\quad \text{else } (\ell^{a'},\ell^a) \in \hat{I_E}
\end{array}
$$

$$
\begin{array}{lll}
(\mathbf{out}) & (\hat{I_B},\hat{I_E},\hat{H}) \models^{\mathcal{L}} \mathbf{out}^{\ell^t}\, n.P & \text{iff} \quad (\hat{I_B},\hat{I_E},\hat{H}) \models^{\mathcal{L}} P \wedge
\end{array}
$$

$$
\begin{array}{l}
\forall \ell^a, \ell^{a'}, \ell^{a''} \in \mathbf{Lab}^a : \\
\text{case } ((\ell^a,\ell^t) \in \hat{I_B} \wedge (\ell^{a'},\ell^a) \in \hat{I_E} \cup \hat{I_B} \wedge (\ell^{a''},\ell^{a'}) \in \hat{I_E} \\
\qquad \wedge (\ell^{a'},n) \in \hat{H}) \Longrightarrow \\
\quad \text{if} (\ \ell^a \in \mathbf{Lab}^a_B) \text{ then } (\ell^{a''},\ell^a) \in \hat{I_E} \\
\quad \text{else } (\ell^{a''},\ell^a) \in \hat{I_E} \wedge \left\{(\ell,\ell') \in \hat{I_B} \mid path_B(\ell^a,\ell)\right\} \subseteq \hat{I_E} \\
\text{case } ((\ell^a,\ell^t) \in \hat{I_B} \wedge (\ell^{a'},\ell^a) \in \hat{I_B} \wedge (\ell^{a''},\ell^{a'}) \in \hat{I_B} \wedge (\ell^{a'},n) \in \hat{H}) \\
\quad \Longrightarrow (\ell^{a''},\ell^a) \in \hat{I_B} \\
\text{case } ((\ell^a,\ell^t) \in \hat{I_E} \wedge (\ell^{a'},\ell^a) \in \hat{I_E} \wedge (\ell^{a''},\ell^{a'}) \in \hat{I_E} \wedge (\ell^{a'},n) \in \hat{H}) \\
\quad \Longrightarrow (\ell^{a''},\ell^a) \in \hat{I_E}
\end{array}
$$

$$
\begin{array}{lll}
(\mathbf{open}) & (\hat{I_B},\hat{I_E},\hat{H}) \models^{\mathcal{L}} \mathbf{open}^{\ell^t}\, n.P & \text{iff} \quad (\hat{I_B},\hat{I_E},\hat{H}) \models^{\mathcal{L}} P \wedge
\end{array}
$$

$$
\begin{array}{l}
\forall \ell^a, \ell^{a'} \in \mathbf{Lab}^a : \\
\text{case } ((\ell^a,\ell^t) \in \hat{I_E} \wedge (\ell^a,\ell^{a'}) \in \hat{I_E} \wedge (\ell^{a'},n) \in \hat{H}) \Longrightarrow \\
\quad \text{if} (\ \ell^{a'} \in \mathbf{Lab}^a_B) \text{ then} \left\{(\ell^a,\ell^{a''}) \mid (\ell^{a'},\ell^{a''}) \in \hat{I_B}\right\} \subseteq \hat{I_E} \wedge \\
\qquad \left\{(\ell,\ell') \mid (\ell,\ell') \in \hat{I_B} \wedge (\ell^{a'},\ell'') \in \hat{I_B} \wedge path_B(\ell'',\ell)\right\} \subseteq \hat{I_E} \\
\quad \text{else } \left\{(\ell^a,\ell) \mid (\ell^{a'},\ell) \in \hat{I_E}\right\} \subseteq \hat{I_E} \\
\text{case } ((\ell^a,\ell^t) \in \hat{I_B} \wedge (\ell^a,\ell^{a'}) \in \hat{I_B} \wedge (\ell^{a'},n) \in \hat{H}) \\
\quad \Longrightarrow \left\{(\ell^a,\ell) \mid (\ell^{a'},\ell) \in \hat{I_B}\right\} \subseteq \hat{I_B}
\end{array}
$$

Figure 3. Specification of the Control Flow Analysis

to the process P_*, then apply the analysis to validate the correctness of the proposed solution, adding, if needed, new information to the triplet until a fixed point is reached.

Example 3 Let P be the process of Example 2. The least solution of P is the triplet $(\hat{I_B},\hat{I_E},\hat{H})$ where $\hat{I_B}=\{(\ell_1^a,\ell_2^a),(\ell_2^a,\ell^t)\}$, $\hat{I_E}=\{(env,\ell_1^a),(env,\ell_2^a),(\ell_2^a,\ell^t)\}$, and $\hat{H}=\{(\ell_1^a,n),(\ell_2^a,m)\}$. Observe that $(\hat{I_B},\hat{I_E},\hat{H})$ strictly contains $\beta^{\mathcal{L}}(P)$, as expected being $(\hat{I_B},\hat{I_E},\hat{H})$ a safe approximation.

More formally, the fixed point algorithm works as follows:

Algorithm 4 (Fixed Point Algorithm)
Input: a process P_* and a partition labeling $\mathcal{L}$.

(i) Apply the representation function $\beta^{\mathcal{L}}$ to process P_* to get a triplet $(\hat{I_B^o},\hat{I_E^o},\hat{H})$;

(ii) for all the constraints of the specification of the analysis, validate the triplet $(\hat{I_B^i},\hat{I_E^i},\hat{H})$ generated in (i):

1 if the constraint is satisfied, continue;

2 else, in case the constraint is not satisfied, this is due to the fact that either $\hat{I_B}$ or $\hat{I_E}$ do not consider nestings that may actually occur. In this case, modify $\hat{I_B}$ and $\hat{I_E}$ by adding the "missing" pairs, thus getting a new triplet $(\hat{I_B^{i+1}},\hat{I_E^{i+1}},\hat{H})$. Then, go back to (ii) with $i=i+1$.

The iterative procedure above computes the least solution independent of the iteration order.

The result of the analysis should be read, as expected, in terms of information flows.

Theorem 5 *No leakage of secret data/ambients outside the boundary ambients is possible if in the analysis no high level label appears in* $\hat{I_E}$.

Example 6 Consider, for instance, a process, which allows an *application* (say, an applet) to be downloaded from the *web* within *montreal*; then, the application may open the ambient *send* and disappear.

$$
\begin{gathered}
P_4 = venice^{b_1}[\,send^{b_3}[\,\mathbf{out}^c\, venice.\mathbf{in}^c\, montreal \mid hdata^h\,[\,\mathbf{in}^{ch}\, filter]\,]\mid \\
\mid download^{m'}\,[\,\mathbf{out}^{cm'}\, venice.\mathbf{in}^{cm'}\, web.\mathbf{in}^{cm'}\, montreal]\,]\mid \\
\mid montreal^{b_2}[\,\mathbf{open}^c\, web.\mathbf{open}^c\, application]\mid \\
web^m[\,application^m[\,\mathbf{open}^{cm}\, send.filter^m[\,]\,]\mid \mathbf{open}^{cm}\, download]
\end{gathered}
$$

In this case, there is no information flow, as the application is not exporting any data out of the *montreal* boundary. In this case, the refined CFA yields to positive information, namely:

$$
\begin{aligned}
\hat{I_B} &= \{(b_1,b_3),(b_1,m'),(b_3,h),(b_3,c),(h,ch),(m',cm'),(b_2,b_3),(b_2,h),\\
&\quad (b_2,m'),(b_2,b_2),(b_2,m),(b_2,c),(b_2,cm'),(b_2,cm),(m,h),(m,m'),\\
&\quad (m,b_2),(m,m),(m,cm'),(m,cm)\}\\
\hat{I_E} &= \{(env,b_1),(env,b_3),(env,m'),(env,b_2),(env,m),(m',cm'),(m,m'),\\
&\quad (m,b_2),(m,m),(m,cm'),(m,cm)\}\\
\hat{H} &= \{(b_1,\text{venice}),(b_3,\text{send}),(b_2,\text{montreal}),(h,\text{hdata}),(m',\text{download}),\\
&\quad (m,\text{web}),(m,\text{application}),(m,\text{filter})\}
\end{aligned}
$$

Observe that the result is also better than the Hansen-Jensen-Nielsons's CFA [11] as the latter does not capture the fact that h enters m only after it has crossed the boundary and can never return back.

3. Inferring Boundaries

Let us turn now to the boundary inference issue. By now, we consider a process P wherein high level data are known, i.e. $\mathbf{Lab}_H$ is fixed. We are interested to partition the set of ambient labels into $\mathbf{Lab}_L$ and $\mathbf{Lab}_B$ so that $\mathbf{Lab}_B$ is the minimal labeling that guarantees the absence of direct information flow concerning confidential data. In other words, the aim of the analysis is to detect which ambients among the "untrusted ones" should be protected (let's say by a firewall or by encryption) as they may carry sensitive data.

Since we want to infer a minimal set of boundary ambients it makes sense to discriminate all the ambients belonging to process P_*, thus we assume that initially all ambient occurrences have different labels. Note that this condition may not be verified during the execution of process P because of the replication operator. Given this initial labeling, a label has at most one parent, thus we can give the following definitions.

Definition 7 (Border of an ambient) *Given an ambient with label ℓ in a process P, we denote by $\mathcal{B}(\ell)$ the border of the ambient n labeled ℓ, i.e. the label of the ambient which n belongs to. Observe that $\mathcal{B}(\ell)$ is defined for all ambients but the environment* env.

For example, in process $P = p^t[\ m^k[\ n^\ell[\,0\,]\ ]\ \mid q^s[\,0\,]\ ]$, the border of the ambient labeled ℓ is $\mathcal{B}(\ell) = k$.

Definition 8 (Upward closure) *The upward closure of the border of an ambient labeled ℓ, $\overline{\mathcal{B}}(\ell)$, is the minimal set that contains $\mathcal{B}(\ell)$ and such that $m \in \overline{\mathcal{B}}(\ell) \Rightarrow \mathcal{B}(m) \in \overline{\mathcal{B}}(\ell)$.*

For instance, considering again process $P = p^t[\ m^k[\ n^\ell[\,0\,]\ ]\ \mid\ q^s[\,0\,]\ ]$, the upward closure of the border of ℓ is $\overline{\mathcal{B}}(\ell) = \{k, t, env\}$.

We have already observed that Algorithm 4 takes as input a labeling, where labels are partitioned into three distinct sets: high, low and boundary. Let us introduce this notion more formally in order to deal with a dynamic labeling, where only the high labels cannot change status.

Definition 9 (i-th Label Partitioning $\mathcal{L}_i$) *We denote by $\mathcal{L}_i$ and we call it the i-th Label Partitioning, the triplet $\mathcal{L}_i = (\mathbf{Lab}_H, \mathbf{Lab}^i_B, \mathbf{Lab}^i_L)$. We assume that $\mathbf{Lab}_H$, $\mathbf{Lab}_B$ and $\mathbf{Lab}_L$ are mutually disjoint, and that $\mathbf{Lab}_H \cup \mathbf{Lab}_B \cup \mathbf{Lab}_L = \mathbf{Lab}^a$.*

3.1. The Algorithm

The algorithm described below analyses process P starting from the initial labeling $\mathcal{L}_0$. It may either succeed (in this case a labeling $\mathcal{L}_k$ is reached that fulfills the security property we are interested in) or it may fail. The latter case simply means that the process P cannot be guaranteed to be secure by our analysis.

Initial Label Partitioning. Given a set of high level labels Lab_H, we initially partition the remaining ambients into the following sets:

$$\begin{aligned}\mathsf{Lab}_B^0 &= \{\ell \in \mathsf{Lab}^a \setminus \mathsf{Lab}_H \mid \exists h \in \mathsf{Lab}_H \wedge \ell = \mathcal{B}(h) \wedge \\ &\qquad \nexists h' \in \mathsf{Lab}_H : \mathcal{B}(h') \in \overline{\mathcal{B}}(h)\} \\ \mathsf{Lab}_L^0 &= \mathsf{Lab} \setminus (\mathsf{Lab}_H \cup \mathsf{Lab}_B^0)\end{aligned}$$

Through this step, the boundaries that guarantee the absence of information flow in the initial state of process P are defined. Observe that in this way we avoid initial boundary nesting. This is how the boundaries are inferred:
For all h ambients belonging to the process P:

1 compute the border $\mathcal{B}(h)$. If $\mathcal{B}(h) = env$, process P is insecure by construction, then stop with failure;

2 if $\mathcal{B}(h) \in \mathsf{Lab}^a \setminus \mathsf{Lab}_H$, compute the upward closure $\overline{\mathcal{B}}(h)$ and label $\mathcal{B}(h)$ as boundary iff $\nexists h' \in \mathsf{Lab}_H : \mathcal{B}(h') \in \overline{\mathcal{B}}(h)$.

Algorithm 10 (Boundary Inference Algorithm) The analysis is performed by the fixed point algorithm parameterized with respect to $\mathcal{L}_i$. At the beginning, $i = 0$.

(i) Compute Algorithm 4 with input $\mathcal{L}_i$ and P_*.

(ii) During the execution of Algorithm 4, whenever a high level ambient n labeled h gets into an unprotected environment, i.e. $\exists \ell\ :\ (\ell, h) \in \hat{I_E}$ do:

1 if $(env, h) \in \hat{I_E}$, the analysis terminates with failure, as it cannot infer a satisfactory labeling that guarantees absence of information leakage;

2 otherwise, if $(env, h) \notin \hat{I_E}$:

- a new labeling $\mathcal{L}_{i+1}$ should be considered, labeling every ℓ such that $(\ell, h) \in \hat{I_E}$ as a boundary. Let $\mathbf{L} = \{\ell \mid (\ell, h) \in \hat{I_E}\}$ then $\mathcal{L}_{i+1} = (\mathsf{Lab}_H, \mathsf{Lab}_B^i \cup \{\mathbf{L}\}, \mathsf{Lab}_L^i \setminus \{\mathbf{L}\})$.
- go to (i) with $i = i + 1$.

Refining the solution. Through this step, a more precise label partitioning $\mathcal{L}_*$ might be computed. From the set of boundaries inferred by the analysis, we take away, if possible, the set of boundaries $\mathbf{B}$ that are not needed to guarantee absence of information leakage (i.e. boundaries nested inside other boundaries). Observe that the set of boundaries nested inside other boundaries can be empty. This refinement procedure can be seen as a narrowing step in the sense of Abstract Interpretation. More formally:

$$\begin{aligned}\mathcal{L}_* &= (\mathsf{Lab}_H, \mathsf{Lab}_B \setminus \mathbf{B}, \mathsf{Lab}_L \cup \mathbf{B}) \\ \mathbf{B} &= \{\ell \in \mathsf{Lab}_B \mid \exists \ell' : (\ell', \ell) \in \hat{I_B} \wedge \nexists \ell'' : (\ell'', \ell) \in \hat{I_E})\}\end{aligned}$$

Before addressing termination, soundness and minimality issues, let us try to understand the behavior of this algorithm by looking at an example.

Example 11 *Let us consider again the example given in the Introduction:*

$$\text{venice}^x[\,\text{send}^y[\,\textbf{out}^c\ \text{venice}.\textbf{in}^c\ \text{montreal}\,]\ |\ \text{hdata}^h[\,\textbf{in}^c\ \text{send}\,]\,]\ |$$
$$|\ \text{montreal}^z[\,\textbf{open}^c\ \text{send}\,]\ |\ Q^\ell$$

- *Given the set of high level labels* Lab_H *in P, the initial label partitioning* $\mathcal{L}_0$ *is computed.* $\mathcal{L}_0 = (\mathsf{Lab}_H = \{h\}, \mathsf{Lab}_B^0 = \{x\}, \mathsf{Lab}_L^0 = \{y, z\})$
- *Applying the representation function* $\beta^{\mathcal{L}_0}$ *to* P_*, *it returns the triplet* $(\hat{I}_B^o, \hat{I}_E^o, \hat{H})$:

$$\begin{array}{rcl}\hat{I}_B^o & = & \{(x,y),(x,h),(y,c),(h,c)\}\\ \hat{I}_E^o & = & \{(env,x),(env,z),(z,c)\}\\ \hat{H} & = & \{(h,hdata),(x,venice),(y,send),(z,montreal)\}\end{array}$$

Executing Algorithm 4, the pair (y, h) *is introduced in* $\hat{I}_E$, *reflecting the fact that ambient send leaves ambient venice during the execution of process P.*

- *At this point, a new label partitioning should be considered:*
  $\mathcal{L}_1 = (\mathsf{Lab}_H = \{h\}, \mathsf{Lab}_B^1 = \{x, y\}, \mathsf{Lab}_L^1 = \{z\})$
  Algorithm 4 is computed again. During its execution, the pair $(z, h) \in \hat{I}_E$, *reflecting the fact that the boundary send, containing confidential data, is opened inside the low ambient montreal during the execution of process P.*
- *At this point, the following new label partitioning is considered:*
  $\mathcal{L}_2 = (\mathsf{Lab}_H = \{h\}, \mathsf{Lab}_B^2 = \{x, y, z\}, \mathsf{Lab}_L^2 = \emptyset)$
  Algorithm 4 is computed again, and a fixed point is finally reached. In this case, there is no need to refine the solution. Thus, the set of ambients that should be labeled as boundaries is {venice, send, montreal}.

3.2. Soundness and Minimality

In this final section we formally prove termination and correctness of the Boundary Inference Algorithm described in section 3.1. Moreover, we show a minimality result on the computed solution.

Theorem 12 (Termination) *The algorithm always terminates.*

Proof. Straightforward, as the number of labels is finite.

Theorem 13 (Soundness) *If there exists a label partitioning* $\mathcal{L}_k$ *such that the analysis of process P (with initial label partitioning* $\mathcal{L}_0$*) terminates with success and, in the resulting triplet* $(\hat{I}_B, \hat{I}_E, \hat{H})$, *no high level ambient does appear in the pairs of* $\hat{I}_E$, *then the labeling* $\mathcal{L}_k$ *is sufficient to guarantee the absence of direct leakage within the process P.*

We introduce a new predicate to formalize the notion of *protected ambient*. Given a label $\ell^a \in \mathsf{Lab}^a$, Protected($\ell^a$) is true iff $\nexists \ell^{a'} : [(\ell^{a'}, \ell^a) \in \hat{I}_E \wedge \ell^a \neq \text{env}] \vee \ell^a \in \mathsf{Lab}_B^a$.

The following result guarantees a minimality condition of the refined solution computed by Algorithm 10.

Lemma 14 In the solution $(\hat{I_B}, \hat{I_E}, \hat{H}) \models^{\mathcal{L}} P_*$:

(*i*) $\ell \in \mathbf{B} \Rightarrow$ Protected(ℓ) both in $\mathcal{L}$ and $\mathcal{L}_*$.
(*ii*) Protected(ℓ) in $\mathcal{L} \Leftrightarrow$ Protected(ℓ) in $\mathcal{L}_*$.
(*iii*) $\beta^{\mathcal{L}}(P_*) = \beta^{\mathcal{L}_*}(P_*)$ and Protected(ℓ) in $\beta^{\mathcal{L}}(P_*) \Leftrightarrow$ Protected(ℓ) in $\beta^{\mathcal{L}_*}(P_*)$.

Theorem 15 (Minimality) *Let $\mathcal{L}$ be the triplet $(\mathbf{Lab}^a_H, \mathbf{Lab}^a_L, \mathbf{Lab}^a_B)$ and $\mathcal{L}_*$ the label partitioning generated by the Boundary Inference Algorithm 3.1. Then, the Fixed Point Algorithm 4 parameterized with respect to $\mathcal{L}$ and to $\mathcal{L}_*$ compute the same solution $(\hat{I_B}, \hat{I_E}, \hat{H})$.*

Proof. Essentially the proof simply amounts to observing that, if computing both the algorithms step by step, the pairs added to $\hat{I_E}$ or $\hat{I_B}$ are the same. It is proven by induction on the steps computed by the algorithms. Only the cases for capabilities are non-trivial.

Base of the induction: performing one step from $\beta^{\mathcal{L}}(P_*) = \beta^{\mathcal{L}_*}(P_*)$ with label partitioning $\mathcal{L}$ and $\mathcal{L}_*$ has the same effect.

(**in**) : Protected($\ell^{a''}$) is the same both in $\beta^{\mathcal{L}}(P_*)$ and $\beta^{\mathcal{L}_*}(P_*)$ from point (iii) of Lemma 14, thus $\hat{I_B}$ is modified exactly in the same way with $\mathcal{L}$ and $\mathcal{L}_*$.
$\neg$ Protected($\ell^{a''}$) $\Rightarrow \ell^{a''}, \ell^{a'}, \ell^{a} \notin \mathbf{B}$, thus $\hat{I_E}$ is modified exactly in the same way with $\mathcal{L}$ and $\mathcal{L}_*$.

(**out**) : Protected($\ell^{a''}$) is the same both in $\beta^{\mathcal{L}}(P_*)$ and $\beta^{\mathcal{L}_*}(P_*)$ from point (iii) of Lemma 14, thus $\hat{I_B}$ is modified exactly in the same way with $\mathcal{L}$ and $\mathcal{L}_*$.

(**open**) : the case is analogous to the **out** one.

The inductive step is proved by exploiting the fact that the predicate Protected(ℓ) is the same for $\mathcal{L}$ and $\mathcal{L}_*$ in each step. Observe that minimality within each step is guaranteed by the fact that Algorithm 4 computes the least solution.

4. Conclusions

As far as we know, the idea of inferring a security policy that avoids direct information leakage when modeling mobility through Ambients, has not been investigated in the literature yet. Major emphasis, in fact, has been put on Access Control issues [7, 4] than in Information Flow properties. Most of the works in this area, in fact, focus more on enhancing the language to control how ambients may move in and out of other ambients, than on looking at how to "protect" high-data information from untrusted environments.

A few interesting open issues are under investigation to complete the picture we draw in this paper. In particular, it would be interesting to see if there is an ordering among labeling w.r.t. which the analysis behaves monotonically, and if optimizations can be applied to our algorithm to reduce the overall complexity.

References

[1] US Department of Defense. DoD Trusted Computer System Evaluation Criteria. *DOD 5200.28-STD*, 1985.

[2] C. Bodei, P. Degano, F. Nielson, and H.R.Nielson. Static Analysis of Processes for No Read-Up and No-Write-Down. In *Proc. FoSSaCS'99*, volume 1578 of *Lecture Notes in Computer Science*, pages 120–134, Springer-Verlag, 1999.

[3] Chiara Braghin, Agostino Cortesi, and Riccardo Focardi. Control Flow Analysis of Mobile Ambients with Security Boundaries. In Bart Jacobs and Arend Rensink, editors, *Proc. of Fifth IFIP International Conference on Formal Methods for Open Object-Based Distributed Systems (FMOODS'02)*, pages 197–212. Kluwer Academic Publisher, 2002.

[4] M. Bugliesi and G. Castagna. Secure Safe Ambients. In *Proc. 28th ACM Symposium on Principles of Programming Languages* (POPL'01), pp. 222-235, London. 2001.

[5] L. Cardelli and A. Gordon. Mobile Ambients. In *Proc. FoSSaCS'98*, volume 1378 of *Lecture Notes in Computer Science*, pages 140–155, Springer-Verlag, 1998.

[6] A. Cortesi, and R. Focardi. Information Flow Security in Mobile Ambients. In *Proc. of International Workshop on Cuncurrency and Coordination* CONCOORD'01, Lipari Island, July 2001, volume 54 of *Electronic Notes in Theoretical Computer Science*, Elsevier, 2001.

[7] P. Degano, F. Levi, C. Bodei. Safe Ambients: Control Flow Analysis and Security. In *Proceedings of ASIAN'00*, LNCS 1961, 2000, pages 199-214.

[8] R. Focardi and R. Gorrieri. A Classification of Security Properties for Process Algebras. *Journal of Computer Security*, 3(1): 5-33, 1995.

[9] R. Focardi and R. Gorrieri. The Compositional Security Checker: A Tool for the Verification of Information Flow Security Properties, *IEEE Transactions on Software Engineering*, Vol. 23, No. 9, September 1997.

[10] R. Focardi, R. Gorrieri, F. Martinelli. Information Flow Analysis in a Discrete Time Process Algebra, in *Proc. of 13th IEEE Computer Security Foundations Workshop* (CSFW13), (P.Syverson ed), IEEE CS Press, 170-184, 2000.

[11] R. R. Hansen, J. G. Jensen, F. Nielson, and H. R. Nielson. Abstract Interpretation of Mobile Ambients. In *Proc. Static Analysis Symposium* SAS'99, volume 1694 of *Lecture Notes in Computer Science*, pages 134–148, Springer-Verlag, 1999.

[12] M. Hennessy, J. Riely. Information Flow vs. Resource Access in the Asynchronous Pi-Calculus. *ICALP 2000*: 415-427.

[13] G. Smith, D.M. Volpano, Secure Information Flow in a Multi-Threaded Imperative Language. *In Proc. of POPL 1998: 355-364.*

DECIDABILITY AND TRACTABILITY OF PROBLEMS IN OBJECT-BASED GRAPH GRAMMARS *

Aline Brum Loreto

Leila Ribeiro

Laira Vieira Toscani

Instituto de Informática - PGCC

Universidade Federal do Rio Grande do Sul

{loreto,leila,laira}@inf.ufrgs.br

Abstract Object-based programming languages are being widely used to construct concurrent and distributed systems. In such systems, the complexity of performing some task is usually measured in terms of messages that are exchanged to perform the task, because communication is almost always the most time consuming operation in these environments. In this paper we analyze the problem of verifying if a message can be delivered. We model object-based systems as a graph grammars, and analyze this property as a graph grammar property. It turns out that this problem is NP-Hard. With suitable restrictions on the kind of graph grammar rules that are used, decidability can also be proven.

Keywords: object-based systems, graph grammars, complexity analysis

Introduction

Graph grammars have originated from the concept of formal grammars on strings by substituting strings by graphs [5]. Methods, techniques, and results for graph grammars have been studied since then, and applied in a variety of fields in computer science such as formal language theory, pattern recognition, software engineering, concurrent and distributed system modelling, database design and theory, etc. (see e.g., [4]). In particular, graph grammars are very well suited to the specification of concurrent and distributed systems: a (distributed) state of the system can naturally be represented by a graph and rules (where the left- and right-hand sides are graphs) describe possible state changes. The behavior of the system is then described via applications of these

*This work was partially supported by the projects PLATUS/ForMOS (CNPq and Fapergs), IQ-Mobile (CNPq and CNR) and GRAPHIT (CNPq and IB-DLR).

rules to graphs describing the actual states of a system. Rules operate locally on the state-graph, and therefore it is possible that many rules are applied at the same time. Graph grammars are appealing as a specification formalism because they are formal, they are based on simple but powerful concepts to describe behavior, and, at the same time, they have a nice graphical layout that helps non-theoreticians understand a graph grammar specification.

Practical applications involving concurrency and distribution aspects are hard to develop and reason about. Therefore, a specification formalism for this kind of application should offer verification methods (if possible, automated). For this, there must be a way to describe the desired properties of the system, and a way to prove them using the semantical models of the specification formalism. As graph grammars are an extension of string grammars (strings can be modeled by special graphs), we can try to generalize results from the theory of formal languages to graph grammars. However, as the main aim of a formal language is to describe a language and the aim of a graph grammar, as we are using them here, is to describe the behavior of a system, the kind of properties that are of interest are quite different. Graph grammars can also be seen as a generalization of Petri nets [11]. Therefore, we will also discuss in the conclusion the difficulties involved in the generalization of properties of Petri nets into properties of graph grammars, showing that read access to items, although very useful to let many actions occurr in parallel, provides a source of verification problems.

As formal grammars on strings and Petri nets are special cases of graph grammars, all undecidability results of these areas also hold for graph grammars. Here we consider some restrictions for graph grammars, still allowing read access to items. This restricted class of graph grammars is called object-based graph grammars [3]. The restrictions imposed in this model implement the usual requirements of an object-based model, like encapsulation and communication via message passing. Object-based specification/programming languages seem to be a natural choice to describe reactive systems: entities that compose the application are modeled as *objects* and the reactions are triggered by *messages* and implemented as *methods*.

Two measures are of great interest for concurrent object-based systems: the number of messages that must be exchanged to complete some task (this gives us a space bound to perform a task); and the lenght of a round, that is, the lenght of the biggest causally related chain of messages needed to complete a task (all messages in such chain can not be delivered in parallel, and therefore this gives us a kind of time bound to perform a task). For these two measures, one must define what *completion of a task* means. In this paper we investigate the possible answer to this question, based on a rule application problem (RAGG), and show results about decidability and complexity of this problem.

The paper is structured as follows: Sect. 1 brings an informal introduction to object-based graph grammars; in Sect. 2 we introduce the problem RAGG; Sect. 3 shows results about decidability and complexity of this problem; and Sect. 4 compares it to other related problems and concludes our work.

1. Object-Based Graph Grammars

In this paper we consider an object-based system as being a system consisting of autonomous entities called *objects* that communicate and cooperate with each other through *messages*. Objects may have an internal state and relate to other objects within the system. The behavior of an object is described through its *reactions* to the receipt of messages (triggers). This reaction may be to change the object's internal state and/or send messages to other objects. An object may perform many (re)actions in parallel. A way to model object-based systems using graph grammars have been presented [3] inspired in the actor model [1]. The basic idea was to use graph grammars as a (graphical) language to specify a kind of actor systems.

To define object-based graph grammars, we have to identify within a graph grammar what are the *objects*, *messages* and *attributes*, and then show how to specify *methods* within this formalism. The structural part will be modeled by distinguishing different kinds of vertices and edges within the graphs that model states of the system (see Figure 1 (a)). There are many ways to define typing mechanisms for graphs, here we will use the concept of a typed graph[2, 10]. The idea of a typed graph is to use a graph, called *type graph*, to define the possible kinds of vertices and edges of a system, and an actual graph is then a graph consisting of instances of elements of the type graph. A typed graph can be described by a graph homomorphism relating each instance with its type. Besides distinguishing types of vertices and edges, for practical applications we usually use values belonging to carrier sets of algebras. Graphs with such values associated to vertices and/or edges are called *attributed graphs* [13, 6].

In our specification formalism, objects and messages will be modeled as vertices. A message must have as destination an object and may have as parameters objects and/or values belonging to data type sets. The internal state of an object consists of attributes, that may be references to objects and/or values. This graph (Figure 1 (a)) can be considered as a type graph for an object-based system, and therefore we will call it *object-based type graph*[1]. Note that a type graph models kinds of objects and links that may be present in an actual state of the system, but say nothing about the number of elements of each kind that must be present at a particular state.

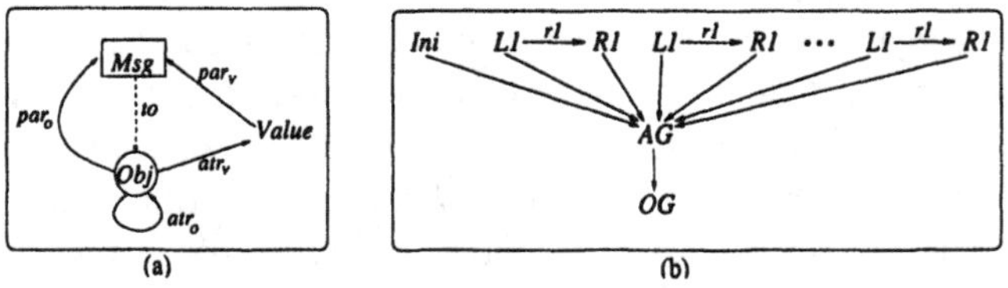

Figure 1. (a) Object-Based Type Graph *OG* (b) Object-Based Graph Grammar

[1]Formally, the carrier set *Value* will be used to label the edges of a graph, that is, the edges par_V and atr_V are actually loops on vertices *Msg* and *Obj*, respectively.

For each specific object-based system we may have various types of objects and messages that are relevant for that application. Thus, to build a specification for an object-based system using graph grammars one must first define what we call the *application type-graph*. This graph must be typed over the object-based model type-graph. The resulting structure of a object-based graph grammar is illustrated in Figure 1 (a) [2].

Object-based Application Graph. A typed graph G over a type graph T is a total graph morphism $type : G \to T$ (this morphism has three components, one function to map vertices, one to map edges and one algebra homomorphism to map the algebra of the instance graph to the one of the type graph). An object-based application graph is a graph typed over the object-based type graph where messages have only one destination.

Object-based Graph. An object-based graph G is a graph typed over an object-based application type graph where: *i)* Each message has all defined parameters; *ii)* An object does not have two times the same attribute.

Rules specify the behavior of the system in terms of local state changes. The left-hand side of the rule specifies a pattern that must be present in some state for the rule to be applied; the right-hand side shows the effect of the application of the rule; and the mapping from left- to right-hand side describes deletion (items that are not mapped), creation (items that are not in the range of the mapping) and preservation (items that are mapped). For an object-based graph grammar we will only allow rules that consume an element of type message, i.e., each rule represents a reaction to the kind of message that was consumed. Moreover, only one message may be consumed at a time by each rule. Note that the system may have many rules that specify reactions to the same kind of message (non-determinism), and that many rules may be applied in parallel if their triggers (messages) are present at an actual state. Many messages may be generated in reaction to one message. To make sure that a rule may be applied whenever its trigger is found in the actual state graph we will require that whenever a message appears in a graph, it has exactly all specified arguments and one destination.

Given object-based graphs $G1 = (GI1, type1, T)$ and $G2 = (GI2, type2, T)$, a morphism $g : G1 \to G2$ is a graph homomorphism $g : GI1 \to GI2$ that preserves types, that is, $type2 \circ g = type1$. This graph homomorphism may be partial on vertices and edges, but must be total on the algebra component. The compatibility condition required by the partial graph homomorphism is that the graph structure is preserved for the items in the domain of definition of the homomorphism, that is, preserved itens can not change types.

Object-based Rule. A morphism $r : L \to R$ between object-based graphs is a ***(object-based) rule*** iff L and R are finite r is injective, there is exactly one

[2]Formally, this structure can be defined as a doubly-typed graph grammar (see [16], [3] for the formal definitions). One of the advantages of defining explicitly the model type-graph within the specification is to ease the comparison among specifications with respect to different model graphs (once we relate the model graphs, the relationships among the specifications can be obtained automatically).

message vertex m in L, and this vertex is deleted by the rule. In this case, m is called ***trigger*** of r. Moreover, all attributes appearing in L must also appear in R (maybe with different values).

Object-based Graph Grammar. An ***object-based graph grammar*** is a tuple $GG = (AG, GI, R)$ where AG, the **type** of the grammar, is a finite object-based graph, GI is a finite graph typed over AG, called the ***initial graph*** of the grammar, and ***Rules*** is a finite set of object-based rules typed over AG.

The behavior of a graph grammar is based on the notion of derivation step, that is, rule application. A rule is applicable if all the items in its left-hand side are found in the graph representing the current state. The application of the rule performs the deletions and creations according to what is specified in the rule. Here we follow the (Single-Pushout) Algebraic Approach to Graph Gramamrs [15, 6]. The graph representing the state after the application of a rule $r : L \rightarrow R$ can be obtained as follows:

Occurrence. Given a rule $r : L \rightarrow R$ and a graph G, an occurrence for r in G is a total (typed) graph morphism.

Rule application. The graph representing the state after the application of a rule $r : L \rightarrow R$ can be obtained as follows: i) insert in the graph representing the state all items that are in R and not in L; ii) Remove from the resulting graph, the items that are in L and not in R, and also all dangling edges, such that the result is a graph. This construction can be formally defined as the pushout in the category of (typed) graphs and partial graph morphisms [6].

The semantics of a graph grammar can be defined as the class of all computations that can be performed using the rules of the grammar starting with the initial state. These computations may be sequential or concurrent, giving raise to sequential and concurrent semantic models. As here we do not have to reason about parallelism, we will stick to this sequential model (where parallelism is described by interleaving). Note that, if we would have chosen a true concurrency semantical model, like unfolding [16] or concurrent derivations [10], the set of reachable graphs would be the same, as well as the set of rules that may be applied in each state. What changes is that true concurrency models allow more derivations to occur in parallel (interleaving is not equivalent to true concurrency in graph grammars [12] due to the hability to preserve items).

2. Definition of the Problem RAGG

To define when a task has been completed in a system we can use the generation of a message (end-message), or the fact that some procedure has been called. In graph grammars, these two would correspond to the application of a rule: the rule that generates a message, or a rule that starts some procedure. Thus, we can use the knowledge about the fact that a rule has been applied to describe when a task has been completed. In order to determine if a rule can be applied, we must verify if it is possible to generate the message that enables this rule, and if the atributes used by rule may reach the necessary values. The problem RAGG is the problem of application of a rule in graph grammars, to define it we use the notation of Garey and Johnson [9].

Definition 2.1 *RAGG Problem*
Instance: Object-based Graph Grammar $GG = (T, GI, R)$, and a rule $r \in R$.
Question: Is there a sequential derivation of GG in which the rule r is applied?

Definition 2.2 *RAGG-m Problem*
Instance: Object-based Graph Grammar $GG = (T, GI, R)$, and a rule $r \in R$.
Question: Is there a sequential derivation of GG in which the rule r is applied m times?

These two problems will be studied in many contexts, with graph grammars with differents restrictions. Let us define some restrictions:
1) The rules have no attributes;
2) In left-hand side of a rule there exists exactly one message (restriction of object-based system);
3) The attributes values vary over a finite set;
4) The attributes values vary over the $\{T, F\}$ set;
5) Each attribute can change its status at most one time in the derivation;
6) Each rule can change the status of at most one attribute;
7) The status of an attribute can be changed only by one rule.

Graph grammars with restriction 1 were studied in [14], where it was shown that RAGG-m is decidable (considering restrictions 1 and 2). The cost was calculated (number of necessary rule applications to reach the m executions of r). The algorithm that calculates the cost has exponential complexity, however it was not proven that there is no polynomial algorithm nor that the problem is NP-Hard or NP-Complete. The same problem but with restrictions 2, 4, 5, 6 and 7 will be analyzed in the section 3 (Theorem 3). Restriction 2 is typical for object-based system: each action is triggered by one message. Restrictions 3 and 4 are fineteness assumptions (actually we can enconde grammars having restriction 3 into grammars having restriction 4). Restriction 5 will be used to prove termination. Although it seems to be very restrictive for practical applications, in the conclusion we discuss a way to weaken this condition. Restrictions 6 and 7 are only used to allow easier proofs, they actually do not restrict the expression power of the grammar. Any grammar can be translated into a grammar satisfying 6 and 7.

3. Analysis of RAGG-m and RAGG Problems

In this section the complexity of the problems RAGG and RAGG-m will be investigated. We will reduct the satisfiability problem (SAT) into RAGG-m. The resulting grammar has many of the restrictions defined in Sect. 2.

Definition 3.1 Reduction SAT$\propto$RAGG-m. [3]

3

A problem L is NP-Hard if and only if Satisfiability reduces to L[9], that is, there is a deterministic polynomial time algorithm that transforms each instance of SAT into and instance of L, preserving the answer. This transformation (reduction) is denoted by $\propto$.

Let $I_{SAT} = (U, C)$, with variable set $U = \{u_1, u_2, ..., u_n\}$ and clause set $C = \{c_1, c_2, ..., c_m\}$, be an instance of the SAT problem. The instance I_{SAT} reduces to the instance $I_{RAGG-m} = (G, k, r)$ of RAGG, where:

- *$G = (T, GI, R)$ is a graph grammar having T as type graph, GI as initial graph (Figure 2 (a) and (b), respectively), where $Bool = \{F, T\}$, and the rules set R is the union of the following rule sets:*

  Group AV: To each variable $u_i, i = 1, ..., n$ we define two rules u_i^+ and u_i^-, representing the two possible value attributions to this variable (true or false). The scheme of rules of this group are depicted in Figure 3.

  Group SV: Is composed by the rule described in Figure 4 (a), named r.

  Group CVV: To each variable u_i present in clause c_j there is a rule named $c_j u_i^+$ as shown in Figure 4 (b). Analogously, to each variable u_i present in clause c_j there is a rule named $c_j u_i^-$ (Figure 4 (c)).

- *m is the number of clauses in I_{SAT}*
- *r is the rule shown in Figure 4 (a)*

Lemma 3.1 *Let $GG = (T, GI, R)$ be as defined in Def. 3.1. For each $i = 1..n$, exactly one of the following sets of messages will be generated: $\{c_1 : u_i^+, ..., c_m : u_i^+\}$ or $\{c_1 : u_i^-, ..., c_m : u_i^-\}$ (representing the attribution of T/F to variable u_i, respectively). Each of these sets contains m messages (one for each clause in C). It means that each variable will have the same attribution in all clauses.*

Proof. To each variable u_i, either rule u_i^+ or rule u_i^- can be applied (because both delete the message $u_i?$) generating for each clause one message $c_j u_i^+$ or $c_j u_i^-$, respectively.

Lemma 3.2 *To each $j \in \{1, .., m\}$, at most one rule from $\{c_j u_1^+, c_j u_1^-, \ldots, c_j u_n^+, c_j u_n^-\}$ can be applied, and therefore at most m SAT messages will be generated.*

Proof. If rule $c_j u_i^+$ is applied to generate a message *SAT*, the value of the corresponding attribute $c_j Set$ must have been false and set to true. There is no rule that changes the status of $c_j Set$ to true again.

Now, we can prove that, when we have an affirmative answer to instance *SAT*, we have an affirmative answer to correspondent instance of *RAGG-m*.

Theorem 1 *There is a Yes answer to the I_{SAT} of SAT iff there is a Yes answer to the $I_{RAGG\text{-}m}$ instance of the RAGG-m problem.*

Proof. ($\Rightarrow$)Suppose $I_{SAT} = (U, C)$ gives a Yes to SAT problem, i.e. there is a truth value attribution *atr* to the variables of U that makes all clauses in C true. We must find a sequential derivation σ of GG in which the rule r is executed m times. This derivation will be composed by $n + m + m$ derivation steps, as described below:

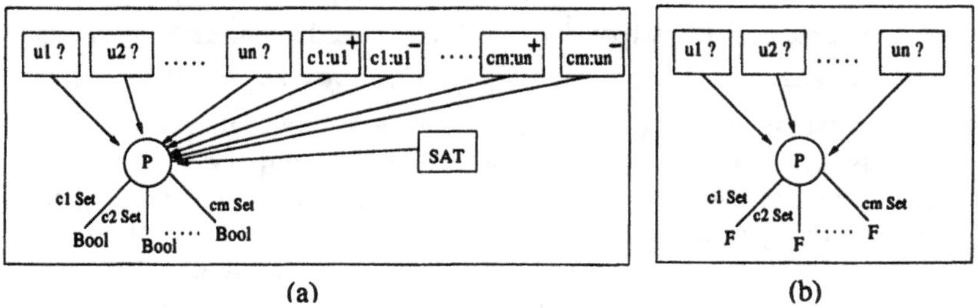

Figure 2. (a) Type Graph T (b)Initial Graph GI

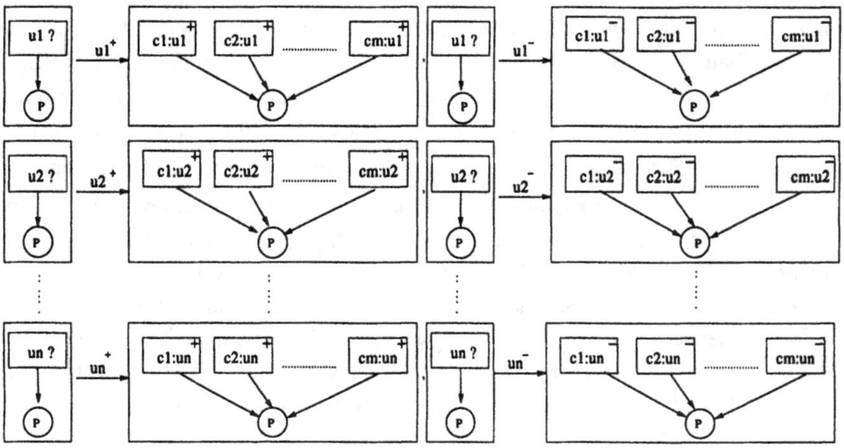

Figure 3. Rules Group AV

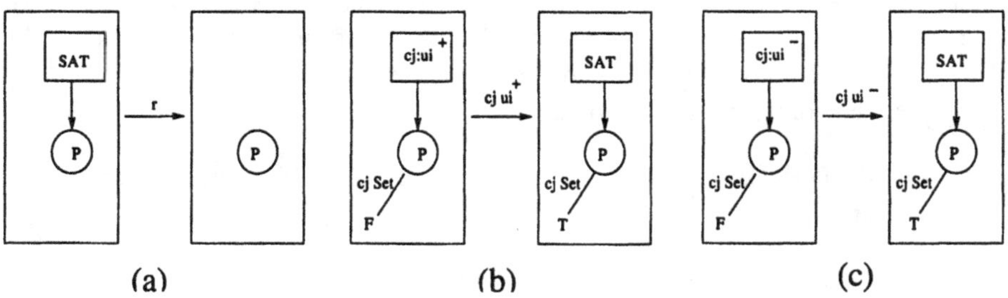

Figure 4. (a) SV: Rule r (b) CVV: Rule $c_j u_i^+$ (c) CVV: Rule $c_j u_i^-$

1..n steps: If atr assigns true to u_i, then consider the derivation where the rule u_i^+ is applied. This will have the effect that messages $\{c_1 : u_i^+, \ldots, c_m : u_i^+\}$ will be generated and the trigger $\boxed{u_i?}$ will be deleted (Lemma 3.1). Analogously, if atr assigns false to u_i, messages $\{c_1 : u_i^-, \ldots, c_m : u_i^-\}$ will be generated.

$n+1..n+m$ steps: For each clause $c_j \in C$, at least one of its components is true (because atr satisfies all clauses). Let u_i be that component. In this case, rule $c_j u_i^+$ can be applied because u_i^+ must have been generated in step i, generating one SAT message for clause j. Moreover, Lemma 3.2 assures that no other SAT message will ever be generated for this clause. The same would happen if $\neg u_i$ is the component that makes the clause true. Therefore, as the attribution atr makes all clauses true, we would generate m SAT messages.

$n+m+1..n+m+m$ steps: As we have m SAT messages, rule r can be applied m times.

($\Leftarrow$) Let σ be a sequential derivation with m occurrences of rule r. For this to happen, m messages SAT must have been generated (because each application of r consumes one message SAT). That means that m messages of set CVV must have been applied before (because these are the only rules that generate SAT messages). Lemma 3.2 assures that, for all j, at most one rule of $\{c_j u_1^+, c_j u_1^-, \ldots, c_j u_n^+, c_j u_n^-\}$ can be applied. This implies that each of the m SAT messages must have been generated for a different c_j. Moreover, for each u_i, either $\boxed{c_j : u_i^+}$ or $\boxed{c_j : u_i^-}$ is generated (Lemma 3.1). Consider the following attribution atr: if $c_j u_i^+$ was applied in σ then $atr(u_i) = true$; if $c_j u_i^-$ was applied in σ then $atr(u_i) = false$, for all i. If this attribution is well-defined, it would make all m clauses c_j true because if u_i is in c_j and $atr(u_i) = true$, the clause is true. The same holds for $\neg u_i$. This attribution is well-defined because rules $c_j u_i^+$ and $c_j u_i^-$ are mutually exclusive. Moreover, if $c_j u_i^+$ happens, there can be no other rule $c_l u_i^-$ that have been applied (this would mean that both $\boxed{u_i^+}$ and $\boxed{u_i^-}$ have been applied, and this is not possible – Lemma 3.1).

Theorem 2 *The reduction $\propto$ can be done in polynomial time.*

Proof. The initial graph GI has n messages and m attributes, that is $n+1$ vertices and $n+m$ edges and the carrier set of the attribute algebra has two elements (T and F), and therefore it can be constructed in polynomial time steps. The total number of rules is: in group AV $2n$; in group CVV $m * n$ rules, at most; in group SV, only one rule. Each rule in AV has exactly $m+3$ vertices and $m+1$ edges. Each rule in CVV has 4 vertices and 4 edges. And the r rule has 3 vertices and one edge. Thus, I_{RAGG} can be constructed in polynomial time.

Theorem 3 (NP-Hardness) *The problem RAGG-m with restrictions 2, 4, 5, 6 and 7 is NP-Hard.*

Proof. Follows directly from theorems 1 and 2.

Corollary 3.1 *Changing only the rule SV in definition 3.1 adding on the left-hand side the attributes $c_1Set, \ldots, c_mSet$, all set to T, we prove thatRAGG is NP-Hard (with the same restrictions of the theorem).*

Theorem 4 *The RAGG problem with restrictions 2 and 3 is NP-Hard.*

Proof. It is enough to adapt the definition 3.1 placing the counter with value m in the intial graph, decreasing one in each rule $c_j u_i^+$ and $c_j u_i^-$ and condition your application by $m \neq 0$ and changed the r rule, placing the attribute *cont* with value zero in its right-hand side.

To prove that this problem is decidable we will define an algorithm that the builds a tree of all possible derivations that generate the necessary conditions for the execution of r. These derivations will first set the necessary attribute values, and then generate the trigger message (without changing the attribute set by the first step). To describe this algorithm, we will divide the rules of a grammar into classes, according to the attributes they change.

Definition 3.2 *Given a grammar $GG = (T, GI, R)$ and $r \in R$. We can partition the rules of GG in four classes: A, B, C and D. A is the class of rules that modify attributes present in left-hand side of r to values that disable the application of r and T_A is the set of these attributes. B is the class of rules that modify attributes present in left-hand side of r to values that enable the application of r and T_B contains this attributes. C is the class of rules that modify the attributes not present in left-hand side of r and T_C contains this attributes. And D is the class of rules that does not modify attributes and T_D is empty.*

This means that, in order to apply rule r, the rules of A can not be applied, the rules of B must be applied and the rules of C and D can or not be applied. Let $G_{\text{-}A}$ be denote the GG grammar without the rules of A class.

Definition 3.3 *Causal sequence: Let $s = < r_1, r_2, ... >$ be a sequence of rules in $G_{\text{-}A}$, and t_i the attribute changed by r_i rule, if r_i modify some attribute. Then s is a Causal sequence if to each r_i, $< r_1, r_2, ..., r_i >$ is a sequence of application of rules from GI that generates the conditions to r_i to be applied.*

Now, consider the $G_{\text{-}A\text{-}M}$ the grammar equal to $G_{\text{-}A}$, but without messages. The proofs of lemmas 3.3, 3.4 and 3.5 are straightforward and were omited.

Lemma 3.3 *If there is a causal sequence $< r_1, r_2, ..., r_k >$ from rules of $B \cup C$ containing all the rules of B and possible some of C then r can be applied in the $G_{\text{-}A\text{-}M}$ grammar.*

Lemma 3.4 *If r is not applicable in $G_{\text{-}A}$ then r is not applicable in GG.*

Lemma 3.5 *If there are derivations $T1 : GI \to GF$ and $T2 : GF \to GM$, where GF satisfies all the attributes present in left-hand side of r and $T2$ change no attributes necessary to execution of r and does not consume the necessary messages for r, then r is applicable in GM.*

Definition 3.4 *Consider all causal sequences $< r_1, ..., r_k >$ satisfying Lemma 3.3 with the same subsequence of rules of B. From these sequences and the*

initial graph GI, the tree illustrated in Figure 5 is built. Many derivations can be constructed based on these sequences, the upper indices identify these variations. Each leave vertex is the graph resulting from the application of the $d^j = d^j_1; \ldots; d^j_k$ derivation that preceeds it, where each d^j_i is a derivation that contains no rule that modifies attributes and where the last applied rule is r_i.

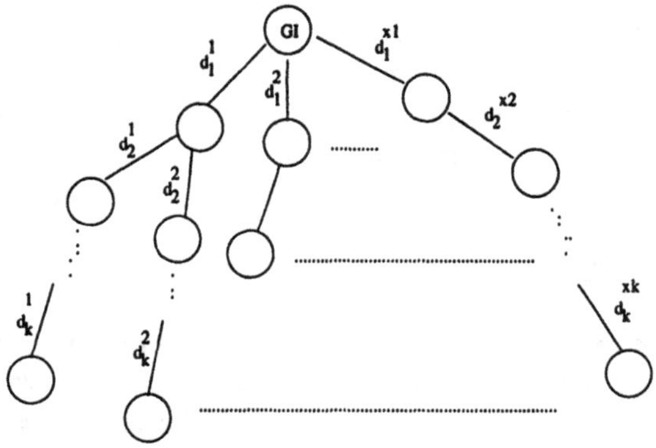

Figure 5. Tree of Derivation Sequence

Theorem 5 *The RAGG problem with restrictions 2, 4, 5, 6 and 7 is decidable.*

Proof. Let $GG = (T, GI, R)$ be a graph grammar with the restrictions 2, 4, 5, 6 and 7 and $r \in R$. Let A, B, C, D be defined from GG and r, by definition 3.2. In case of $C = \varnothing$, consider the follow algorithm: For each sequence $< r_1, ..., r_k >$ that satisfies Lemma 3.3 do the steps i) and ii) below. If this sequence does no exist, r is not applicable in GG (Lemma 3.4). This problem is obviously decidable because the sequence set to test is finite.
i) Build the derivation tree of Figure 5. Each d_i is a derivation is a gramar without attributes and therefore we can calculate [14] if r_i is applicable and which (finite) derivation creates the needed conditions to apply r_i;
ii) Verify, for each leaf of the tree, if r is applicable in the grammar (T, GI', R'), where GI' is the leaf graph removing the attributes; R' is the rule set D of GG. For a grammar without attributes, the problem is decidable [14].

Using Lemma 3.5, under the condition of $C = \varnothing$, this algorithm decides if r is or not applicable in GG. As the number of sequences $< r_1, ..., r_k >$ satisfying Lemma 3.3 is finite the problem is decidable.

The condition $C = \varnothing$ corresponds to the case where rule r reads all the attributes. When this condition is not true we must replaced r by rules that read all the attributes, considering all the possible assignments to the attributes not read by r. If one of these rules is applicable in the new graph grammar, r will be applicable in GG. As the number of these rules is finite, it is possible to decide whether r is applicable in GG or not.

4. Conclusion

This paper investigates the complexity of the problems *RAGG* and *RAGG-m* in object-based graph grammars with some restrictions. The problems stayed in different complexities classes depending on the considered restrictions. For graph grammars with restrictions 2, 4, 5, 6 and 7, the problem *RAGG* was proven decidable (theorem 5) and NP-Hard (theorem 3.1). Adapting the reduction of definition 3.1 we can prove that, with restrictions 2 and 3, the *RAGG* problem is also NP-Hard.

Graph Grammars are a generalization of Petri Nets. The complexity of verification of properties have been exaustive investigated for Petri Nets. One of the most investigated ones is the reachability problem. This problem is in different classes of complexities depending on the characteristics of the net. Reachability in Conflict-free Petri Nets is in P, for Petri Nets without cycles it is in NP-Complete class and for some extended Petri Nets like with inhibitor arcs the problem is undecidable [7, 8]. It is to expect that for grammar graphs the same occurs, the corresponding problem will be in different classes depending on the particularties of the investigated grammar.

In Formal Languages, given a grammar G, the properties usually investigated are "does $w \in L(G)$?" or "$L(G) \neq \varnothing$?". The former corresponds to the reachability problem that was already discussed above. The latter is not very useful because it would mean to verify if a system can evolve (at least one step). When using a graph grammar to represent a concurrent system a more interest property is to know if a r rule can be applied. This property in Context Free Grammar (*CFG*) corresponds to find out if a non-terminal is reachable. This test is usually done in the phase of simplification of grammar, the algorithm is simple and of polynomial degree. A graph grammar without attributes in which rules have an unique message in their left-hand sides is like the *CFG*. The messages of the type graph are the variables (non terminals) and the messages in the initial graph are the initial symbols. It is therefore to hope that the problem of verifying if a rule is applicable is a problem in P (has algorithm of polynomial order) for this particular case of graph grammar. Yet, the problem to know whether a rule can be applied m times does not have easy solution, even in *CFG*.

Although this paper has theoretical constributions, for the practice we still have to find conditions that are not too restrictive for system specification. The ideal solution would be that these restrictions should be enough to put the problem in a more promising complexity class, like P, allowing for efficient verification. On the other hand, the restrictions presented here could be weakened to detect the limits of decidability of this problem. Concerning this topic and disregarding restriction 1 (that is too restrictive for practical applications and was not imposed in the results presented in this paper) and 2 (that is usually required for practical applications), we can see that the restrictions that really matter for the complexity and decidability results are restrictions 3 and 5, that state that the attribute sets are finite and attributes may change only once in a derivation, respectively. Finiteness is a typical requirement when thinking of verification problems. Restriction 5 was used to assure the termination of the process. It could probably be weakened to imposing the existence of a well-founded partial order on the sets of attributes that change during a derivation, and requiring that each rule that changes such attributes sets them to values closer to the bottom (or top) of this partial order. In this sense, the application of the last rule (r) would mean the end of the execution of the desired task (this rule could change attributes disregarding the partial order, to put the system in some initial state again).

References

[1] Agha, G. *Actors: a model for concurrent computation in distributed systems,* MIT Press, 1986.

[2] Corradini, A., Montanari, U. and Rossi, F. *Graph processes,* Fundamentae Informatica, vol. 26, no. 3-4, 1996, pp. 241–265.

[3] Dotti, F. and Ribeiro, L., *Specification of mobile code systems using graph grammars.* In S. Smith and C. Talcott, editors, *Formal Methods for Open Object-based Systems IV*, pages 45–64. Kluwer Academic Publishers, 2000.

[4] H. Ehrig, G. Engels, H.-J. Kreowski, and G. Rozenberg, editors, *Handbook of Graph Grammars and Computing by Graph Transformation, Volume 2: Applications, Languages and Tools*, World Scientific, 1999.

[5] H. Ehrig, M. Pfender and H. Schneider, *Graph grammars: an algebraic approach*,Proc. IEEE Conference SWAT'73, 1973, pp. 167–180.

[6] H. Ehrig, R. Heckel, M. Korff, M. Löwe, L. Ribeiro, A. Wagner and A. Corradini, *Algebraic approaches to graph transformation II: Single pushout approach and comparison with double pushout approach*, in *The Handbook of Graph Grammars, vol. 1: Foundations*, World Scientific, 1997, pp. 247–312.

[7] J. Esparza and M. Nielsen. *Decidability Issues for Petri Nets-a survey*, J. Inform. Process. Cybernet. EIK 30 (1994) 3, 143-160.

[8] J. Esparza, *Decidability and Complexity of Petri Net Problems-An Introduction*, Lecture Notes in Computer Science 1491, Springer, 1998, pp. 376-428.

[9] M. R. Garey & D. S. Johnson, *Computers and Intractability: a guide to the theory of NP-completeness*, W. H. Freeman, San Francisco, 1979.

[10] M. Korff, *Generalized graph structures with application to concurrent object-oriented systems*, Ph.D. thesis, Technical University of Berlin, 1995.

[11] Korff, M. and Ribeiro, L., *Formal relationships between graph grammars and Petri nets*, Lecture Notes in Computer Science 1073, Springer, 1996, pp. 288–303.

[12] M. Korff and L. Ribeiro, *True concurrency = interleaving + weak conflict*, Eletronic Notes in Theoretical Computer Science, vol. 14, 1998.

[13] M. Löwe, M. Korff and A. Wagner, *An algebraic framework for the transformation of attributed graphs*, Term Graph Rewriting: Theory and Practice, John Wiley & Sons, 1993, pp. 185–199.

[14] A. Loreto, L. Ribeiro and L. Toscani, *Complexity Analysis of Reactive Graph Grammars*, Revista de Informtica Terica e Aplicada, vol. VII(2000). Porto Alegre, Instituto de Informtica, UFRGS, pp.109-128.

[15] M. Löwe, *Algebraic approach to single-pushout graph transformation*, Theoretical computer Science, vol. 109, 1993, 181-224.

[16] L. Ribeiro, *Parallel composition and unfolding semantics of graph grammars*, Ph. D. Thesis. Technical University of Berlin, 1996.

COVERAGE OF IMPLEMENTATIONS BY SIMULATING SPECIFICATIONS

Hana Chockler and Orna Kupferman
School of Engineering and Computer Science
Hebrew University
Jerusalem 91904, Israel
{ hanac,orna } @cs.huji.ac.il

Abstract In formal verification, we verify that an implementation is correct with respect to a specification. When verification succeeds and the implementation is proven to be correct, there is still a question of how complete the specification is, and whether it really covers all the behaviors of the implementation. In this paper we study coverage for simulation-based formal verification, where both the implementation and the specification are modelled by labeled state-transition graphs, and an implementation $\mathcal{I}$ satisfies a specification $\mathcal{S}$ if $\mathcal{S}$ simulates $\mathcal{I}$. Our measure of coverage is based on small modifications we apply to $\mathcal{I}$. A part of $\mathcal{I}$ is covered by $\mathcal{S}$ if the mutant implementation in which this part is modified is no longer simulated by $\mathcal{S}$. Thus, "mutation coverage" tells us which parts of the implementation were actually essential for the success of the verification. We describe two algorithms for finding the parts of the implementation that are covered by $\mathcal{S}$. The first algorithm improves a naive algorithm that checks the mutant implementations one by one by exploiting the significant overlaps among the mutant implementations. The second algorithm is symbolic, and it improves a naive symbolic algorithm by reducing the number of variables in the OBDDs involved. In addition, we compare our coverage measure with other approaches for measuring coverage.

Keywords: Model Checking, Simulation, Coverage.

Introduction

In *formal verification*, we verify the correctness of a finite-state implementation with respect to a desired behavior by checking whether a labeled state-transition graph that models the implementation satisfies a specification of this behavior, expressed in terms of a temporal logic formula or a more abstract labeled state-transition graph. Beyond being fully-automatic, an additional attraction of formal verification tools is their ability to accompany a negative answer to the correctness query by a counterexample to the satisfaction of the specification in the implementation (for a survey, see Clarke et al., 1999). On the other hand, when the answer to the correctness query is positive, most verification tools terminate with no further information to the user. Since a positive answer means that the implementation is correct with respect to the specification, this seems like a reasonable policy. In the last few years, however, there has been growing awareness of the importance of suspecting the implementation of containing an error also in the case verification succeeds. The main justification of such suspects are possible errors in the modeling of the implementation or of the behavior, and possible incompleteness in the specification.

There are various ways to look for possible errors in the modeling of the implementation or the behavior. One direction is to detect *vacuous satisfaction* of the specification [Beer et al., 1997; Kupferman and Vardi, 1999], where cases like antecedent failure [Beaty and Bryant, 1994] make parts of the specification irrelevant to its satisfaction. For example, the specification $\varphi = AG(req \rightarrow AF\,grant)$ is vacuously satisfied in an implementation in which *req* is always **false**. A similar direction is to check the validity of the specification. Clearly, a specification that is valid or is vacuously satisfied suggests some problem. It is less clear how to check completeness of the specification. Indeed, specifications are written manually, and their completeness depends entirely on the competence of the person who writes them. The motivation for such a check is clear: an erroneous behavior of the implementation can escape the verification efforts if this behavior is not captured by the specification. In fact, it is likely that a behavior not captured by the specification also escapes the attention of the designer, who is often the one to provide the specification.

In simulation-based verification techniques, *coverage metrics* are used in order to reveal states that were not visited during the testing procedure (i.e., not "covered" by this procedure); see [Ho and Horowitz, 1996; Bergmann and Horowitz, 1999; Fallah et al., 1999] and others. These metrics are a useful way of measuring progress of the verification process. However, the same intuition cannot be applied to formal verification, as the process of formal verification may visit all states regardless their essence to the success of the verification process. We can say that in testing, a state is "uncovered" if it is not essential to the success of the testing procedure. A similar idea can be applied to formal verification, where a state is defined as "uncovered" if its labeling is not essential to the success of the verification process. This approach was first suggested by Hoskote et al., 1999. Low coverage can point to several problems. One possibility is that the specification is not complete enough to fully describe all the possible behaviors of the implementation. Then, the output of a coverage check is helpful in completing the specification. Another possibility is that the implementation contains redundancies. Then, the output of the coverage check is helpful in simplifying the implementation.

There are two different approaches to coverage in model checking, where the specification is given as a temporal logic formula. One approach, introduced in Katz et al., 1999, states that a well-covered implementation should closely resemble the *reduced tableau* of its specification. Thus the coverage criteria of Katz et al. are based on the analysis of the differences between the implementation and the tableau of its specification. In the full version we discuss this approach in more detail. Another approach, introduced in Hoskote et al., 1999, is to check the influence of small changes in the implementation on the satisfaction of the specification. This approach is inspired by the definition of *mutation coverage* in simulation-based verification [Dill, 1998]. For a given implementation, we can consider a set of *mutants*, each representing one small change in the original implementation. The specification *covers* a mutation in an implementation if it is not satisfied in the corresponding mutant. Formally, for an implementation $\mathcal{I}$, modeled as a labeled state-transition graph, a state w in $\mathcal{I}$, and an *observable signal* q, the *mutant implementation* $\tilde{\mathcal{I}}_{w,q}$ is obtained from $\mathcal{I}$ by flipping the value of q in w (the signal q corresponds to a Boolean variable that is **true** in w if w is labeled with q and is **false** otherwise; when we say that we flip the value of q, we mean that we switch the value of this variable). For a specification φ that is satisfied in $\mathcal{I}$ and an observable signal q, a state w of $\mathcal{I}$ is q-covered by φ if $\tilde{\mathcal{I}}_{w,q}$ does not satisfy φ. Indeed, this indicates that the value of q in w is crucial for the satisfaction of φ in $\mathcal{I}$. It is easy to see that for each observable signal, the set of q-covered states can be computed by a naive algorithm that performs model checking of φ in $\tilde{\mathcal{I}}_{w,q}$ for each state w of I. The naive algorithm, however, is very expensive, and is useless for practical applications.

The approach of Hoskote et al., 1999 is followed by Chockler et al., 2001b, where two alternatives to the naive algorithm are presented for specifications in the branching time temporal logic CTL. The first algorithm is symbolic and computes the set of pairs $\langle w, w' \rangle$ such that flipping the value of q in w' falsifies φ in w. The second algorithm improves the naive algorithm by exploiting overlaps in the many mutant implementations that we need to check. The "mutant approach" is also taken in Chockler et al., 2001a, which studies coverage by specifications that are given as formulas in the linear temporal logic LTL or by automata on infinite words. Chockler et al. suggest alternative definitions of coverage, which suit better the linear case, and presents two algorithms for LTL and automata-based specifications. Both algorithms can be relatively easily implemented on top of existing model-checking tools.

In this paper we study coverage in design and verification methods in which the specification is given as a labeled state-transition graph. Consider an implementation and a specification. Both describe possible behaviors of the system, but the specification is more abstract than the implementation [Abadi and Lamport, 1991]. This approach, of representing both specifications and implementation as labeled state-transition graphs, suggests a top-down method for design development, called *hierarchical refinement* [Lam and Shankar, 1984; Kurshan, 1994]: starting with a highly abstract specification, we construct a sequence of behavior descriptions, each of which refers to its predecessor as a specification, and is thus less abstract than the predecessor. At each stage the current implementation is verified to satisfy its specification. Verifying that an intermediate implementation satisfies its specification leads to detection of errors in the design as soon as they are introduced. Likewise, measuring coverage of an intermediate implementation with respect to its specification would lead to early detection of low coverage.

There are several ways of defining what it means for an implementation $\mathcal{I}$ to satisfy a specification $\mathcal{S}$. The two main ones are *trace-based* and *tree-based*. The former requires each computation of $\mathcal{I}$ to correlate with some computation of $\mathcal{S}$, and the latter requires each computation tree embodied in $\mathcal{I}$ to correlate with some computation tree embodied in $\mathcal{S}$. The simplest definition of such correlation is equivalence with respect to the variables joint to $\mathcal{I}$ and $\mathcal{S}$, as the implementation is typically defined over a wider set of variables, reflecting the fact that it is more concrete than the specification. With this interpretation, trace-based verification corresponds to *trace containment* [Kurshan, 1994], and tree-based verification corresponds to *simulation* [Milner, 1971].

Simulation has several theoretically and practically appealing properties. First, since the definition of simulation is local, checking whether $\mathcal{S}$ simulates $\mathcal{I}$ can be done efficiently [Milner, 1980; Henzinger et al., 1995] and a witnessing relation for simulation can be computed symbolically [McMillan, 1993; Henzinger et al., 1995]. Second, simulation implies trace containment, whose checking for nondeterministic specifications is PSPACE-complete [Meyer and Stockmeyer, 1972]. The computational advantage is so compelling as to make simulation useful also to researchers that favor the linear approach to specification: in automatic verification, simulation is widely used as a sufficient condition for trace containment [Cleaveland et al., 1993]; in manual verification, trace containment is most naturally proved by exhibiting local witnesses such as simulation relations or refinement mappings (a restricted form of simulation relations) [Lamport, 1983; Lynch and Tuttle, 1987; Lynch, 1996].

We apply mutation-based coverage to simulation and suggest efficient algorithms to measure coverage in simulation. As in Hoskote et al., 1999, for an implementation $\mathcal{I}$, a state w in $\mathcal{I}$, and an observable signal q, we say that w is *q-covered* by a specification $\mathcal{S}$ if $\tilde{\mathcal{I}}_{w,q}$ is not simulated by $\mathcal{S}$. Intuitively, w is q-covered by $\mathcal{S}$ if flipping the value of q in w creates a behavior that is not permitted by $\mathcal{S}$. As in the context of model checking, the naive algorithm computes coverage by executing a simulation computation algorithm $|W|$ times, once for each mutant implementation. We suggest two algorithms that improve the naive algorithm. Our algorithms are built on top of algorithms that compute the simulation relation. The first algorithm is built on top of the enumerative simulation algorithm of Henzinger et al., 1995. The time complexity of the algorithm of Henzinger et al. is $O(m'n + mn')$, where m, n, and m', n' are the sizes of transition relations and state spaces of the implementation and specification. To the best of our knowledge, this is the best time complexity known for the problem [1]. Our algorithm exploits similarities between the mutant implementations, and has an average running time of $O((m'n + mn')\log n)$, while in the worst case its complexity does not exceed the complexity of the naive algorithm, which is $O((m'n + mn')n)$. The second algorithm is symbolic, and it computes, given an implementation $\mathcal{I}$ with state space W and a specification $\mathcal{S}$ with state space W', the following ternary relation.

$$C = \{\langle w, v, w' \rangle : w, v \in W, w' \in W', \text{ and } w' \text{ simulates } w \text{ in } \tilde{\mathcal{I}}_{v,q}\}.$$

Thus, a triplet $\langle w, v, w' \rangle$ is in C iff w' simulates w in the mutant implementation $\tilde{\mathcal{I}}_{v,q}$, obtained from $\mathcal{I}$ by flipping the value of q in v. In particular, a state v is q-covered iff there exists an initial state w_0 of $\mathcal{I}$ such that for all initial states w_0' of $\mathcal{S}$ we have $\langle w_0, v, w_0' \rangle$ not in C, in which case $\mathcal{S}$ does not simulate $\tilde{\mathcal{I}}_{v,q}$. A naive implementation of Milner's fixed-point expression for simulation requires $2(n + n')$ OBDD variables [Milner, 1980]. It has been recently shown in [Katz et al., 1999; Katz, 2001] how *early quantification* and *variable interleaving* in the OBDD can be used in order to reduce the number of required variables to $2n$. Similarly, a naive implementation of the fixed-point expression with which the relation C is computed requires $4n + 2n'$ OBDD variables. We show how early quantification and variable interleaving can be used also here, reducing the number of required variables to 3γ, where $\gamma = \max\{n, n'\}$.

Often, the designer is sufficiently familiar with the implementation and the specification to suspect that specific parts of the implementation are not covered by a specification. In such cases, it makes sense to replace the above described coverage algorithms by algorithms that get as input a set $MUT \subseteq W \times AP$ of mutations with respect to which coverage should be checked. A pair $\langle w, q \rangle \in MUT$ corresponds to the mutant implementation $\tilde{\mathcal{I}}_{w,q}$. Again, a naive algorithm checks the corresponding mutant implementations one by one, and is more complex than simulation in a factor of $|MUT|$. We show that our improved algorithms can be applied also in this case. The enumerative algorithm is more complex than simulation only in a factor of $\log |MUT|$, and the symbolic algorithm requires 3γ variables, with $\gamma = \max\{n, n', |MUT|\}$. In fact, the above described algorithms can be viewed as a special case where $MUT = W \times \{q\}$, for an observable signal q.

Due to the lack of space, we omit some proofs and technical details from this version. A full version can be found in the authors' URL.

1. Preliminaries

We model systems by *labeled state transition graphs*. Formally, a system S is a tuple $S = \langle AP, W, R, W_0, L \rangle$, where AP is a set of atomic propositions, W is a set of states, $R \subseteq W \times W$ is a total transition relation, W_0 is a set of initial states, and $L : W \times AP \to \{\mathbf{true}, \mathbf{false}\}$ is a labeling function that maps a state w and an atomic proposition p to the value of p in w. We use $L(w)$ to denote the set $\{p : p \in AP$ and $L(w, p) = \mathbf{true}\}$. For a state w, we denote by *pre*(w) the set of direct predecessors of w in the system, and by *post*(w) the set of direct successors of w in the system. Formally, *pre*$(w) = \{v \in W : R(v, w)\}$, and *post*$(w) = \{v \in W : R(w, v)\}$.

Consider an implementation $\mathcal{I} = \langle AP, W, R, W_0, L \rangle$, and a specification $\mathcal{S} = \langle AP', W', R', W_0', L' \rangle$. For technical convenience, we assume that $AP = AP'$; thus, the implementation and the specification are defined over the same set of atomic propositions[2]. A binary relation $B \subseteq W \times W'$ is a *simulation* (of $\mathcal{I}$ by $\mathcal{S}$) if for all $\langle w, w' \rangle \in B$ the following conditions hold:

1 $L(w) = L'(w')$.

2 For each u such that $R(w, u)$ there exists u' such that $R'(w', u')$ and $B(u, u')$.

It is easy to see that the union of two simulations is a simulation. Consequently, the maximal simulation between $\mathcal{I}$ and $\mathcal{S}$, denoted $\mathcal{B}$, is the union of all simulations of $\mathcal{I}$ by $\mathcal{S}$. We say that $w' \in W'$ *simulates* $w \in W$ if $\langle w, w' \rangle \in \mathcal{B}$. We say that $\mathcal{S}$ *simulates* $\mathcal{I}$, or, equivalently, $\mathcal{I}$ *is simulated by* $\mathcal{S}$ (denoted $\mathcal{I} \leq \mathcal{S}$), if for every $w_0 \in W_0$ there exists $w_0' \in W_0'$ that simulates w_0. Intuitively, it means that $\mathcal{S}$ has more behaviors than $\mathcal{I}$. In fact, every $\forall$CTL* formula that is satisfied in $\mathcal{S}$ is satisfied also in $\mathcal{I}$ [Browne et al., 1988; Grumberg and Long, 1994].

For the implementation $\mathcal{I}$, a state $w \in W$, and an atomic proposition $q \in AP$, the *mutant implementation* $\tilde{\mathcal{I}}_{w,q} = \langle AP, W, R, W_0, \tilde{L}_{w,q} \rangle$ is obtained from $\mathcal{I}$ by flipping the value of q in w. Formally, $\tilde{L}_{w,q}(w, q) = \neg L(w, q)$, and for all $\langle v, p \rangle \neq \langle w, q \rangle$, we have $\tilde{L}_{w,q}(v,p) = L(v,p)$. Consider a specification $\mathcal{S}$ such that $\mathcal{I} \leq \mathcal{S}$ and an atomic proposition $q \in AP$. We say that w is *q-covered* by $\mathcal{S}$ if $\tilde{\mathcal{I}}_{w,q} \not\leq \mathcal{S}$. Intuitively, w is q-covered by $\mathcal{S}$ if flipping the value of q in w creates a behavior that is not permitted by $\mathcal{S}$.

2. Enumerative Approach

In this section we describe an efficient algorithm for computing the set of q-covered states. Our algorithm is based on the enumerative algorithm of Henzinger et al., 1995 for simulation computation. Consider an implementation $\mathcal{I} = \langle AP, W, R, W_0, L\rangle$ and a specification $\mathcal{S} = \langle AP, W', R', W_0', L'\rangle$ such that $\mathcal{I} \leq \mathcal{S}$. Given $q \in AP$, we compute the set of states of $\mathcal{I}$ that are q-covered by $\mathcal{S}$. When q is clear from the context, we omit it from our notations.

We first describe the algorithm of Henzinger et al. The maximal simulation relation $\mathcal{B} \subseteq W \times W'$ is the set of all pairs $\langle w, w'\rangle \in W \times W'$ such that w' simulates w. That is, the labeling of w is equal to the labeling of w', and for each successor u of w there is a successor u' of w' such that u' simulates u. By the definition, $\mathcal{B}$ is the greatest fixed point of the equation

$$\mathcal{B} = \{\langle w, w'\rangle : \mathcal{B}_0(w, w') \wedge \forall u\, \exists u' : [R(w,u) \rightarrow R'(w',u') \wedge \mathcal{B}(u,u')]\},$$

where

$$\mathcal{B}_0 = \{\langle w.w'\rangle : L(w) = L'(w')\}.$$

An alternative way to compute the relation $\mathcal{B}$ is to compute, for each $w \in W$, the set $sim(w)$ of states of $\mathcal{S}$ that simulate w. The algorithm of Henzinger et al. starts with the maximal possible candidate for a simulation set for each w, namely, the set of all states in W' with the same labeling, and it repeatedly reduces the sets until it reaches a fixed point: as long as there is a state w, a successor u of w, and a state $w' \in sim(w)$ such that there is no successor of w' in $sim(u)$, the set $sim(w)$ should be reduced by removing w'. A straightforward implementation of such a fixed-point calculation is presented in the procedure *Schematic_Similarity* in Figure 1 and has time complexity $O(mm'n^2n')$. Henzinger et al. improve this algorithm in the following way. For each $w \in W$, the algorithm maintains two sets of states: $oldsim(w)$ and $sim(w)$. The set $oldsim(w)$ is the candidate for the simulation set for w that was computed in the previous iteration, and $sim(w)$ is the reduction of $oldsim(w)$ that is computed in this iteration. A fixed-point is reached when $sim(w) = oldsim(w)$ for all $w \in W$, thus no further reduction is possible. The reduction of $sim(w)$ is based on the same observation as in the straightforward algorithm: if a state w' is removed from the simulation set of w, then the predecessors of w' that have no other successors in the simulation set of w should be removed from the simulation sets of predecessors of w. An important observation that leads to the complexity of $O(m'n + mn')$ is that a state w' can be removed from the simulation set of w at most once during the algorithm. In addition, the algorithm uses a data structure that allows to compute the size of $post(v') \cap sim(u)$ for all $v' \in W'$ and $u \in W$ in constant time. The efficient version of the algorithm is described in the procedure *Efficient_Similarity* in Figure 1.

The naive approach for coverage runs the algorithm of Henzinger et al. for $\tilde{\mathcal{I}}_{v,q}$, for all $v \in W$. The complexity of this is n times the complexity of the algorithm of Henzinger et al, which is $O((m'n + mn')n)$. A better approach is to use the fact

```
procedure Schematic_Similarity:
  for all w ∈ W do
    sim(w) = {w' ∈ W' : L(w) = L'(w')}
  od;
  while there are w, v ∈ W, and w' ∈ W' such that
      v ∈ post(w), w' ∈ sim(w), and post(w') ∩ sim(v) = ∅ do
    sim(w) = sim(w) \ {w'}
  od;
  return sim.

procedure Efficient_Similarity:
  for all w ∈ W do
    oldsim(w) := W'; sim(w) := {w' ∈ W' : L(w) = L'(w')};
    rem(w) := pre(oldsim(w)) \ pre(sim(w));
  od;
  while there exists w ∈ W such that rem(w) ≠ ∅ do
    for all u ∈ pre(w) do
      for all u' ∈ rem(w) ∩ sim(u) do
        sim(u) := sim(u) \ {u'};
        for all v' ∈ pre(u') do
          if post(v') ∩ sim(u) = ∅ then rem(u) := rem(u) ∪ {v'} fi
        od
      od
    od
    oldsim(w) := sim(w); rem(w) := ∅;
  od;
  return sim.
```

Figure 1. The similarity algorithm of Henzinger et al., 1995.

that for all $w, v \in W$, the implementations $\tilde{\mathcal{I}}_{w,q}$ and $\tilde{\mathcal{I}}_{v,q}$ differ only slightly (that is, only in labeling of two states). Thus there is room for hope that the simulation computation for $\tilde{\mathcal{I}}_{w,q}$ and $\tilde{\mathcal{I}}_{v,q}$ is also almost the same. In order to explain our approach, we introduce the notion of *incomplete simulation*. Let X be the set of variables $\{x_w : w \in W\}$. For a subset of states $S \subseteq W$, the *incomplete labeling function* $L_S : W \times AP \to \{\mathbf{true}, \mathbf{false}\} \cup X$ maps a pair $\langle w, p \rangle$ to $L(w,p)$ if $w \notin S$ or $p \neq q$, and to x_w if $w \in S$ and $p = q$. As in the definition of L, we use $L_S(w)$ as a shortcut for the set $\{p : p \in AP \text{ and } L_S(w,p) = \mathbf{true}\}$. For two states $w \in W$ and $w' \in W'$, and a set $S \subseteq W$, we say that $L_S(w) = L'(w')$ if for every atomic proposition p, either $L_S(w,p) = L'(w',p)$, or $L_S(w,p)$ is a variable. For a set of states $S \subseteq W$ we define the implementation $\mathcal{I}_S = \langle AP, W, R, W_0, L_S \rangle$ as $\mathcal{I}$ with the incomplete labeling function L_S. Let $sim_S : W \to 2^{W'}$ denote the maximal simulation relation from $\mathcal{I}_S$ to $\mathcal{S}$. Also, for $w \in W$, let $\widetilde{sim}_w : W \to 2^{W'}$ denote the maximal simulation from $\tilde{\mathcal{I}}_{w,q}$ to $\mathcal{S}$.

Consider a state $w \in W$ and two sets $S_1 \subseteq S_2 \subseteq W$. It is easy to see that $sim_{S_1}(w) \subseteq sim_{S_2}(w)$. Indeed, the set $\{w' : w' \in W' \text{ and } L_{S_1}(w) = L'(w')\}$ is contained in the set $\{w' : w' \in W' \text{ and } L_{S_2}(w) = L'(w')\}$, and both simulation sets are computed from the above sets using the same monotonic fixed-point expression. In particular, when $S_1 = \emptyset$, we have that $sim(w) \subseteq sim_S(w)$ for all $w \in W$ and $S \subseteq W$.

Let $S_1 \subset S_2 \subseteq W$ be two sets of states of $\mathcal{I}$. Assume that we have computed sim_{S_2} and now we wish to compute sim_{S_1}. We claim that the computation of sim_{S_1} can be done using the algorithm of Henzinger et al. with the following modification: for each $w \in W$, we initialize the set $oldsim_{S_1}(w)$ to $sim_{S_2}(w)$ and the set $sim_{S_1}(w)$ to $sim_{S_2}(w) \cap \{w' \in W' : L'(w') = L_{S_1}(w)\}$. In other words, in the initialization of $oldsim_{S_1}(w)$ and $sim_{S_1}(w)$, we intersect the sets initialized in [Henzinger et al., 1995] with the set $sim_{S_2}(w)$. Formally, consider the procedure *Efficient_Incomplete_Similarity* described in Figure 2. The procedure gets two parameters: $S \subseteq W$, and a simulation function $sim' : W \to 2^{W'}$. It differs from *Efficient_Similarity* only in the initialization stage: when S is not a singleton, the procedure computes the simulation relation by initializing sim and $oldsim$ with respect to L_S and sim'. When $S = \{w\}$, the procedure computes the simulation relation by initializing sim and $oldsim$ with respect to $\tilde{L}_{w,q}$ and sim'.

procedure *Efficient_Incomplete_Similarity*(S, sim'):
  for all $w \in W$ **do**
    $oldsim(w) := sim'(w)$;
    if $S \neq \{w\}$ **then**
      $sim(w) := \{w' \in W' : L_S(w) = L'(w')\} \cap sim'(w)$;
    else
      $sim(w) := \{w' \in W' : \tilde{L}_{w,q}(w) = L'(w')\} \cap sim'(w)$;
    $rem(w) := pre(oldsim(w)) \setminus pre(sim(w))$;
  od;
... % continues as in *Efficient_Similarity*.

Figure 2. Incomplete similarity algorithm

When $sim' = sim_{S'}$ for $S \subseteq S'$, the tighter initialization does not effect the correctness of the procedure. Formally, we have the following.

Lemma 1 *Let $S_1 \subseteq S_2$ be two subsets of W.*

- *If $|S_1| > 1$, then Efficient_Incomplete_Similarity(S_1, sim_{S_2}) returns sim_{S_1} (from $\mathcal{I}_{S_1}$ to $\mathcal{S}$).*
- *If $S_1 = \{v\}$, then Efficient_Incomplete_Similarity(S_1, sim_{S_2}) returns $\widetilde{sim}_v$ (from $\tilde{\mathcal{I}}_{v,q}$ to $\mathcal{S}$).*

We are now ready to describe our algorithm. The algorithm is based on a stepwise computation of the simulation relation sim from $\mathcal{I}$ to $\mathcal{S}$. In the first step, we compute incomplete simulation sim_W from $\mathcal{I}_W$ to $\mathcal{S}$. Note that sim_W refers to the labels of all the atomic propositions except q, and is very likely (a likelihood that increases for large sets of atomic propositions) to be much tighter than the initial candidate used in *Efficient_Similarity*. Consider a partition of W into two equal sets, W_1 and W_2. Our algorithm essentially works as follows. For all the mutant implementations $\tilde{\mathcal{I}}_{w,q}$ such that $w \in W_1$, the states in W_2 maintain their original labeling. Therefore, we start by computing incomplete simulation from $\mathcal{I}_{W_1}$ to $\mathcal{S}$; that is, we compute simulation with a labeling function that does not rely on the values of q in states in W_1. We end up with the function sim_{W_1}. Then, we continue and partition the set W_1 into two equal sets, W_{11} and W_{12}, and calculate incomplete simulation from $\mathcal{I}_{W_{11}}$ to $\mathcal{S}$. The important point is that we can start the computation of $sim_{W_{11}}$ from sim_{W_1}. Thus, we have to reduce the current candidate only with respect to information that involves the values of q in W_{12}. In a similar way, we compute incomplete simulation from $\mathcal{I}_{W_2}$ by $\mathcal{S}$, and then partition the set W_2 into two equal sets W_{21} and W_{22}, and compute incomplete simulation from $\mathcal{I}_{W_{21}}$ and $\mathcal{I}_{W_{22}}$ to $\mathcal{S}$. Here, we can start the computation of $sim_{W_{21}}$ and $sim_{W_{22}}$ from sim_{W_2}. Thus, as we go deeper in the recursion described above, we perform less work. The depth of the recursion is bounded by $\log|W|$. As we shall analyze exactly below, the work in depth i amounts in average to performing $1/2^i$ of the work required for computing the full simulation relation. Hence the $O((m'n + mn')\log n)$ complexity.

In the full version, we describe the algorithm and analyze its complexity in detail.

Mutant vector. As discussed in the introduction, often it is helpful to allow the designer to specify a set $MUT \subseteq W \times AP$ of mutations with respect to which coverage should be checked. Each pair $\langle w, q\rangle$ in MUT represents the mutant implementation $\tilde{\mathcal{I}}_{w,q}$. The algorithm above can be viewed as a special case where $MUT = W \times \{q\}$. It is easy to extend the algorithm to the more general case as follows. Given MUT, let $X_{MUT} = \{x_{w,q} : \langle w, q\rangle \in MUT\}$ be a set of variables that correspond to possible mutations. The incomplete labeling function L_{MUT} agrees with L for all $\langle w, q\rangle \notin MUT$ and is $x_{w,q}$ for $\langle w, q\rangle \in MUT$. In each step we randomly divide the set of variables into two equal subsets, and assign half of the variables their original values. Then we compute the incomplete simulation for this assignment. When the set of variables becomes a singleton $\{x_{w,q}\}$, we assign to $x_{w,q}$ the complementary value (that is, flip the value of q in w), and compute the simulation function from the mutant implementation $\tilde{\mathcal{I}}_{w,q}$ to $\mathcal{S}$. The number of steps in the algorithm is $O(\log|MUT|)$. By the same considerations detailed for the special case, this leads to an average time complexity of $O((m'n + mn')\log|MUT|)$.

3. Symbolic Approach

In this section we present an algorithm that symbolically computes the set of q-covered states. Note that the naive approach, which executes a symbolic algorithm $|W|$ times for all mutant implementations, is no longer symbolic, as it requires explicit enumeration of the state space. The algorithm we present in this section is symbolic, and it computes the relation $\mathcal{C}$ that is defined as follows.

$$\mathcal{C} = \{\langle w, v, w' \rangle : w, v \in W, w' \in W', \text{ and } w' \text{ simulates } w \text{ in } \tilde{\mathcal{I}}_{v,q}\}.$$

Then, v is q-covered by $\mathcal{S}$ if there is $w_0 \in W_0$ such that for all $w'_0 \in W'_0$, we have $\langle w_0, v, w'_0 \rangle \notin \mathcal{C}$. Several symbolic simulation algorithms are described in the literature [McMillan, 1993; Henzinger et al., 1995; Katz et al., 1999]. We build our coverage algorithm on top of the straightforward symbolic implementation of Milner's fixed-point expression for simulation (see $\mathcal{B}$ in Section 2). The reason for this, as we elaborate below, is the small number of OBDD variables that are needed in this approach. It is not hard to see that the relation $\mathcal{C}$ is the greatest fixed point of the following equation.

$$\mathcal{C} = \{\langle w, v, w' \rangle : \mathcal{C}_0(w, v, w') \wedge \forall u \, \exists u' : [R(w, u) \rightarrow R'(w', u') \wedge \mathcal{C}(u, v, u')]\},$$

where $\mathcal{C}_0 = \{\langle w, v, w' \rangle : \tilde{L}_{v,q}(w) = L'(w')\}$.

Thus, the calculation of $\mathcal{C}$ is very similar to that of $\mathcal{B}$, only that the state v affects the labels that are compared in $\mathcal{C}_0$. The straightforward symbolic implementation of the above fixed point involves OBDDs with $6n$ variables, where $n = \max(|W|, |W'|)$. We show how to reduce the number of OBDD variables to $3n$. In order to do so, we use *early quantification* and *variable interleaving* in the OBDDs. Both techniques are used in [Katz et al., 1999] (see also [Katz, 2001]) in order to reduce the number of OBDD variables required for computing the simulation relation $\mathcal{B}$ from $4n$ to $2n$.

We first explain the techniques in more detail. In early quantification, we try to push existential quantification inside in order to quantify out variables as soon as possible. Early quantification is traditionally used in *conjunctive partitioning* [Chen and Bryant, 1998; Yang, 1999] and is based on the property that sub-expressions can be moved out of the scope of an existential quantifier if they do not depend on any of the variables being quantified. In [Katz et al., 1999], early quantification is used for computing simulation as follows. Recall that $\mathcal{B}$ is the greatest fixed point of the expression

$$\mathcal{B} = \{\langle w, w' \rangle : \mathcal{B}_0(w, w') \wedge \forall u \, \exists u' : [R(w, u) \rightarrow R'(w', u') \wedge \mathcal{B}(u, u')]\},$$

whose calculation involves OBDDs with $4n$ variables. By early quantification of u', we get

$$\mathcal{B} = \{\langle w, w' \rangle : \mathcal{B}_0(w, w') \wedge \forall u \, [R(w, u) \rightarrow \exists u' \text{ such that } R'(w', u') \wedge \mathcal{B}(u, u')]\}.$$

A naive implementation of the new fixed-point involves OBDDs with $3n$ variables. Indeed, the variables of u are introduced only after these of u' are quantified out. In order to reduce the number of variables further, Katz et al. order the variables in the OBDDs so that the variables of a binary relation $f(x, y)$ are interleaved: the variables of x are in the even levels of the OBDD for f and these of y are in the odd levels. Then, Katz et al. define two new operations on OBDD: *comp* and *comp_odd*. These operations compute "exist ... and" as one operation. Formally,

$$comp(f(x, y), g(x, z)) = \exists x (f(x, y) \wedge g(x, z)),$$

and

$$comp\_odd(f(y, x), g(z, x)) = \exists x (f(y, x) \wedge g(z, x)).$$

When the variables of f and g interleaved as described above, the implementation of *comp* and *comp_odd* can proceed in levels, where the corresponding element of x is quantified simultaneously in the OBDDs of f and g. The resulting OBDD refers to the variables in y and z only, so we stay with an OBDD with $2n$ variables. For the detailed implementation of the operations see [Katz, 2001].

Now $\mathcal{B}$ can be calculated using only $2n$ variables, by finding the greatest fixed point of the expression

$$\mathcal{B} = \{\langle w, w'\rangle : \mathcal{B}_0(w, w') \wedge \neg comp\_odd(R(w, u), \neg comp\_odd(R'(w', u'), \mathcal{B}(u, u')))\}.$$

Using the same ideas, we apply early quantification to the fixed-point expression for $\mathcal{C}$ and get

$$\mathcal{C} = \{\langle w, v, w'\rangle : \mathcal{C}_0(w, v, w') \wedge \forall u\ [R(w, u) \rightarrow \exists u' : R'(w', u') \wedge \mathcal{C}(u, v, u')]\}.$$

Then, we define two new operations on OBDDs, *(3:0)-c* and *(3:2)-c*, as follows.

$$\textit{(3:0)-c}(f(x, y), g(x, z, u)) = \exists x (f(x, y) \wedge g(x, z, u)),$$

and

$$\textit{(3:2)-c}(f(y, x), g(u, z, x)) = \exists x (f(y, x) \wedge g(u, z, x)).$$

The operations assume that the variables in the OBDDs of a ternary relation $g(x, z, u)$ are interleaved: the variables of x are in the 0 mod 3 levels, these of z are in the 1 mod 3 levels, and these of u are in the 2 mod 3 levels. Then, as in the case of binary relations, the existential quantification can be done in levels, with one pass on the OBDDs of f and g, and with only $3n$ variables.

Then, the relation $\mathcal{C}$ is the greatest fixed point of the expression

$$\mathcal{C} = \{\langle w, v, w'\rangle : \mathcal{C}_0(w, v, w') \wedge \neg \textit{(3:2)-c}(R(w, u), \neg \textit{(3:2)-c}(R'(w', u'), \mathcal{C}(u, v, u')))\},$$

which can be calculated with $3n$ variables.

Mutant vector. Given a mutant vector $MUT \subseteq W \times AP$, the above algorithm can be adjusted to calculate symbolically which of the mutants are covered by the specification. Note that MUT can be given in some symbolic way (in particular, the above algorithm handles the case where $MUT = W \times \{q\}$ for some observable signal q), in which case it may be crucial to avoid an explicit enumeration of its members.

For a mutant $\lambda = \langle w, q\rangle \in MUT$, let $\tilde{L}_\lambda$ be the labeling function with q flipped in w, and let $\tilde{\mathcal{I}}_\lambda$ be the corresponding mutant implementation. We would like to calculate the relation

$$\mathcal{C} = \{\langle w, \lambda, w'\rangle : w \in W, \lambda \in MUT, w' \in W', \text{ and } w' \text{ simulates } w \text{ in } \tilde{\mathcal{I}}_\lambda\}.$$

The relation $\mathcal{C}$ is the greatest fixed point of the expression

$$\mathcal{C} = \{\langle w, \lambda, w'\rangle : \mathcal{C}_0(w, \lambda, w') \wedge \forall u\ \exists u' : [R(w, u) \rightarrow R'(w', u') \wedge \mathcal{C}(u, \lambda, u')]\},$$

where

$$\mathcal{C}_0 = \{\langle w, \lambda, w'\rangle : \lambda \in MUT \text{ and } \tilde{L}_\lambda(w) = L'(w')\}.$$

As in the algorithm above, we can rewrite $\mathcal{C}$ so that the quantification on u' is pushed inside. Unlike in the algorithm above, here the members of the triplets are not only states of $\mathcal{I}$ and $\mathcal{S}$ but also members of MUT. Accordingly, we define $n = \max\{|W|, |W'|, |MUT|\}$. Now, we can interleave the variables of w, λ, and w' as in the algorithm above, and calculate $\mathcal{C}$ as the greatest fixed point of the expression

$$\mathcal{C} = \{\langle w, \lambda, w' \rangle : \mathcal{C}_0(w, \lambda, w') \wedge \neg (3{:}2)\text{-}c(R(w,u), \neg (3{:}2)\text{-}c(R'(w',u'), \mathcal{C}(u, \lambda, u'))),$$

which can be calculated with $3n$ variables.

4. Discussion

We defined mutation-based coverage for implementations and specifications given by labeled state-transition graphs and described two algorithms for computing the set of states covered by the specification. The general idea of the algorithms is similar to the idea used in [Chockler et al., 2001b] for mutation-based coverage in model-checking. The technical details, however, are different and nontrivial: in the enumerative algorithm, the overlaps between the mutant implementations lead to tighter candidates for simulation to start with[3]. In the symbolic approach, we addressed the problem of reducing the number of variables in the OBDDs involved, an issue that is not referred to in [Chockler et al., 2001b], where a naive implementation requires only $2n$ variables. In addition, we show how the ideas in [Chockler et al., 2001b] can be extended to handle a given vector of mutations.

Our work brings together the "mutant-based approach" of [Hoskote et al., 1999] and the "simulation approach" of [Katz et al., 1999]. As in [Hoskote et al., 1999], coverage is measured with respect to mutant implementations. As in [Katz et al., 1999], conformance to specification is checked by simulation. In the full version, we discuss the relation between the two approaches in detail and show that the criteria defined in [Katz et al., 1999] for measuring coverage are orthogonal to our coverage measure.

Notes

1. The algorithm in Henzinger et al., 1995 is presented for the computation of the simulation relation in the same system, yet it can be easily adjusted to compute the simulation between an implementation and its specification.

2. By replacing a set $L(w) \in 2^{AP}$ by the set $L(w) \cap AP'$, all our algorithms and results are valid also for the case $AP \supset AP'$.

3. Alternatively, one could have followed the game-theoretic approach to simulation, show how incomplete simulation shrinks the game graph, and apply results from circuit complexity about shrinkage. This, less direct, approach would have been very similar to the approach taken in [Chockler et al., 2001b].

References

Abadi, M. and Lamport, L. (1991). The existence of refinement mappings. *TCS*, 82(2):253–284.

Beaty, D. and Bryant, R. (1994). Formally verifying a microprocessor using a simulation methodology. In *31st DAC*, pp. 596–602. IEEE Comp. Soc.

Beer, I., Ben-David, S., Eisner, C., and Rodeh, Y. (1997). Efficient detection of vacuity in ACTL formulas. In *9th CAV*, *LNCS* 1254, pp. 279–290.

Bergmann, J. and Horowitz, M. (1999). Improving coverage analysis and test generation for large designs. In *IEEE ICCAD*, pp. 580–584.

Browne, M., Clarke, E., and Grumberg, O. (1988). Characterizing finite Kripke structures in propositional temporal logic. *TCS*, 59:115–131.

Chen, Y. A. and Bryant, R. (1998). Verification of floating point adders. In *10th CAV, LNCS* 1427, pp. 488–499.

Chockler, H., Kupferman, O., Kurshan, R., and Vardi, M. (2001a). A practical approach to coverage in model checking. In *13th CAV, LNCS* 2102, pp. 66–78.

Chockler, H., Kupferman, O., and Vardi, M. (2001b). Coverage metrics for temporal logic model checking. In *7th TACAS, LNCS* 2031, pp. 528 – 542.

Clarke, E.M., Grumberg O., and Peled D. (1999). Model Checking. MIT Press.

Cleaveland, R., Parrow, J., and Steffen, B. (1993). The concurrency workbench: A semantics-based tool for the verification of concurrent systems. *ACM TOPLAS*, 15:36–72.

Dill, D. (1998). What's between simulation and formal verification? In *35st DAC*, pp. 328–329.

Fallah, F., Ashar, P., and Devadas, S. (1999). Simulation vector generation from HDL descriptions for observability enhanced-statement coverage. In *36th DAC*, pp. 666–671.

Grumberg, O. and Long, D. (1994). Model checking and modular verification. *ACM TOPLAS*, 16(3):843–871.

Henzinger, M., Henzinger, T., and Kopke, P. (1995). Computing simulations on finite and infinite graphs. In *36th FOCS*, pp. 453–462.

Ho, R. and Horowitz, M. (1996). Validation coverage analysis for complex digital designs. In *ICCAD*, pp. 146–151.

Hoskote, Y., Kam, T., Ho, P.-H., and Zhao, X. (1999). Coverage estimation for symbolic model checking. In *36th DAC*, pp. 300–305.

Katz, S. (2001). Techniques for increasing coverage of formal verification. M.Sc. Thesis, The Technion, Israel.

Katz, S., Geist, D., and Grumberg, O. (1999). "Have I written enough properties ?" a method of comparison between specification and implementation. In *10th CHARME, LNCS* 1703, pp. 280–297.

Kupferman, O. and Vardi, M. (1999). Vacuity detection in temporal model checking. In *10th CHARME, LNCS* 1703, pp. 82–96.

Kurshan, R. (1994). *Computer Aided Verification of Coordinating Processes*. Princeton Univ. Press.

Lam, S. and Shankar, A. (1984). Protocol verification via projection. *IEEE TSE*, 10:325–342.

Lamport, L. (1983). Specifying concurrent program modules. *ACM TOPLAS*, 5:190–222.

Lynch, N. (1996). *Distributed algorithms*. Morgan Kaufmann.

Lynch, N. A. and Tuttle, M. (1987). Hierarchical correctness proofs for distributed algorithms. In *6th PODC*, pp. 137–151.

McMillan, K. (1993). *Symbolic Model Checking*. Kluwer Academic Publishers.

Meyer, A. and Stockmeyer, L. (1972). The equivalence problem for regular expressions with squaring requires exponential time. In *13th IEEE SSAT*, pp. 125–129.

Milner, R. (1971). An algebraic definition of simulation between programs. In *2nd IJCAI*, pp. 481–489. British Computer Society.

Milner, R. (1980). *A Calculus of Communicating Systems, LNCS* 92. Springer Verlag, Berlin.

Yang, B. (1999). *Optimizing Model Checking Based on BDD Characterization*. PhD thesis, School of Computer Science, Carnegie Mellon University.

TQL ALGEBRA AND ITS IMPLEMENTATION (EXTENDED ABSTRACT)

Giovanni Conforti, Orlando Ferrara, and Giorgio Ghelli
Università di Pisa
{confor, ferrara, ghelli}@di.unipi.it

Abstract TQL is a query language for semi-structured data. TQL binding mechanism is based upon the *ambient logic*. This binding mechanism is the key feature of TQL, but its implementation is far from obvious, being based on a logic which includes "difficult" operators such as negation, universal quantification, recursion, and new tree-related operators. In [6] an "implementation model" is presented, here we first extend it with tree operations, hence obtaining an algebra for the full TQL language. Then we shortly describe the evaluation techniques that we exploit in the actual implementation.

1. Introduction

TQL is a query language for semi-structured data based on the *tree logic*, an enriched subset of the ambient logic defined in [7, 5]. The *tree logic* is a logic to define sets of trees. It can be naturally used to express types and constraints over semistructured data. As a consequence, problems as subtyping, constraint implication, constraint satisfiability, can all be expressed and investigated as the validity (or satisfiability) of some class of TQL formulae. TQL uses the tree logic as its matching mechanism; as a consequence, more problems, such as query correctness and query containment (and their combinations), become special cases of the validity problem. The high expressivity of the logic allows us to express complex types, constraints, and queries, giving us, for types and constraints, an expressive power that is higher than the one of other proposals [12, 4]. This unified framework for types, constraints, and queries is a central aim of the TQL project, but its further discussion is out of the scope of this paper.

In this paper we describe the foundations of some focal aspects of our implementation of the TQL evaluator. The implementation goes through five steps: *source level rewriting* of the TQL query into a normal form; *translation* of the TQL query into a term of the TQL algebra; *logical optimization* of the algebraic term into a more efficient form; *execution* of the algebraic term. We are still designing a *physical optimization* phase, where queries will be rewritten taking physical information into account. For reasons of space, we cannot go through all those phases, but we will focus on the most original aspects, that are the TQL algebra, and the algorithms to implement its operations.

The major constributions of this paper are: (i) the definition of the TQL algebra, an algebra of operators over trees and over tables (i.e. *relations*) of trees, where both the trees and the tables may be infinite, and the translation of TQL into the TQL algebra; (ii) the description of the implementation of the TQL algebra; the crucial problems we solve are: the finite representation of the infinite tables that arise during evaluation of TQL, and the algorithms used to implement operators such as negation and universal quantification.

2. TQL by examples

Consider the following bibliography, where, informally, *a*[*F*] represents a piece of data labelled *a* with contents *F* (the data model will be fully defined in the next section); *F* is empty, or is a collection of similar pieces of data, separated by "|". When *F* is empty, we can omit the brackets, so that, for example, *Darwen*[ ] can be written as *Darwen*. In this paper we consider a data model where the content *F* is unordered.

The bibliography below consists of a set of references all labeled *book*. Each entry contains a number of *author* fields, a *title* field, and possibly other fields.

BOOKS= *book*[*author*[*Date*] | *title*[*DB*] | *publisher*[*Addison-Wesley*]] |
book[*author*[*Date*] | *author*[*Darwen*] | *title*[*Foundation for Future DB*]
| *year*[*2000*] | *pages*[*608*]] |
book[*author*[*Abiteboul*] | *author*[*Hull*] | *author*[*Vianu*]
| *title*[*Foundation of DB*] | *publisher*[*Addison-Wesley*] | *year*[*1994*]]

Suppose we want to find all the books in *BOOKS* where one author is *Date*; then we can write the following query (hereafter $\mathcal{X}$ and x are variables and everything else is a constant; in the concrete syntax, variable names begin with a $ character):

$$\textit{from} \quad BOOKS \models .book[\mathcal{X}], \quad \mathcal{X} \models .author[Date] \quad \textit{select} \quad text[\mathcal{X}]$$

The query consists of a list of *matching expressions* contained between *from* and *select*, and a *reconstruction expression*, following *select*. The matching expressions bind $\mathcal{X}$ with every piece of data that is reachable from the root *BOOKS* through a *book* path, and such that a path *author* goes from $\mathcal{X}$ to *Date*; the answer is *text*[*author*[*Date*] | *title*[*DB*] | ...] | *text*[*author*[*Date*] | *author*[*Darwen*] | ...], i.e. the first two books in the database, with the outer *book* rewritten as *text*. The operator $.book[\mathcal{X}]$ is actually an abbreviation for $book[\mathcal{X}] \mid \mathbf{T}$. The $BOOKS \models book[\mathcal{X}] \mid \mathbf{T}$ statement means: *BOOKS* can be split in two parts, one that satisfies $book[\mathcal{X}]$, the other one that satisfies **T**. Every piece of data satisfies **T** (*True*), while only an element *book*[...] satisfies $book[\mathcal{X}]$; hence, $BOOKS \models book[\mathcal{X}] \mid \mathbf{T}$ means: there is an element $book[\mathcal{X}]$ at the top level of *BOOKS*.

In TQL a matching expression is actually a logic expression, combining matching-like and classical logical operators. For example, the following query combines path-expression-like logical operators and classical logical operators ($\forall$, $\Rightarrow$) to get schema information out of the data source. It retrieves the tags appearing into each book.

$$\textit{from} \quad BOOKS \models \forall\mathcal{X}(.book[\mathcal{X}] \Rightarrow .book[\mathcal{X} \wedge .x[\mathbf{T}]]) \quad \textit{select} \quad tag[x]$$

The query can be read as: get $tag[x]$ for those labels x such that, for each book $book[\mathcal{X}]$, x is the tag of one of the elements of the book. Observe how the free variable x carries information from the binder to the result. The same property is expressed below using negation, as 'there exists no *book* where x is not a sub-tag. For more examples, see [6, 10].

$$\textit{from} \quad BOOKS \models \neg\ .book[\neg\ .x[\mathbf{T}]] \quad \textit{select} \quad tag[x]$$

3. TQL data model

Every query, and every piece of data, in TQL denotes an *information tree.* An information tree (over a label set Λ) is an unordered tree whose edges are labelled over Λ (i.e. $a[\,b[\,]|c[\,]\,] \mid a[\,]$ would be a tree with two edges labelled by a carrying $b[\,]|c[\,]$ and the empty tree as children). We allow infinitely branching trees in the formalization, but we do not support them in the implementation. Formally, information trees are nested multisets of label-tree pairs:

Definition *For a given set of* labels Λ, *the set* $\mathcal{IT}$ *of information trees over* Λ, *ranged over by* I, *is the smallest collection such that: (a) the empty multiset,* $\{\}$, *is in* $\mathcal{IT}$*; we will use* $\mathbf{0}$ *as a notation for* $\{\}$*; (b) if* m *is in* Λ *and* I *is in* $\mathcal{IT}$ *then the singleton multiset* $\{\langle m, I\rangle\}$ *is in* $\mathcal{IT}$*; we will use* $m[I]$ *as a notation for* $\{\langle m, I\rangle\}$*; (c)* $\mathcal{IT}$ *is closed under multiset union* $\biguplus_{j\in J} M(j)$ *where* $M \in J \rightarrow \mathcal{IT}$*; we will use* $\mathrm{Par}_{j\in J}\ M(j)$ *as a notation for* $\biguplus_{j\in J} M(j)$*, and* $I \mid I'$ *for* $I \uplus I'$.

In examples and discussions, we will often abbreviate $m[\mathbf{0}]$ as $m[]$, or as m. We assume that Λ includes the disjoint union of each basic data type of interest.

4. TQL Syntax and Semantics

We give here only a synthetic definition of the language; for a complete formal exposition see [6], for an informal one see [10].

In the syntax below, A and B denote formulas of the tree logic, Q denotes queries, and the symbol $\sim$ denotes a binary operator belonging to a fixed set of label comparison operators, such as $=$, $\leq$, closed under negation.

In a query $f(Q)$, the function f is chosen from a fixed set $\boldsymbol{f}_{set}$ of functions of type $\mathcal{IT} \rightarrow \mathcal{IT}$. $\boldsymbol{f}_{set}$ includes functions such as *count*, that returns the information tree $n[\mathbf{0}]$ when is applied to an information tree with n elements. To simplify notation, we skip the distinction between functions and their syntactical representation.

In a formula A, variables that are not bound by $\exists x$, $\exists \mathcal{X}$, or $\mu\xi$, are *free* in A, and they are used (as $\mathcal{X}$ and x in previous examples) to pass information from the binder ($Q \models A$) to the query result. Thus, in *from* $Q \models A$ *select* Q' all the label and tree variables that are free in A are (by definition) bound in the scope Q' (i.e., they score as bound variables when we consider the whole from-select expression). In the syntax below, we write E_v whenever the variable v is bound in the scope E, as in $\exists \mathcal{X}.A_{\mathcal{X}}$ or in *from* $Q \models A$ *select* $Q'_{FV(A)}$. Finally, a binder $Q \models A$ is only well formed when no recursion variable ξ is free in A.

Hereafter, we use $\boldsymbol{A}_{set}$ to denote the set of all formulae A, $\boldsymbol{x}_{set}$ the set of label variables x, $\boldsymbol{\mathcal{X}}_{set}$ the set of tree variables $\mathcal{X}$ and $\boldsymbol{\xi}_{set}$ the set of recursive variables ξ, and similarly for any other syntactic entity in Section 5.

TQL syntax

$$
\begin{array}{lll}
L & ::= & n \mid x \\
A, B & ::= & \mathbf{0} \mid L[A] \mid A|B \mid \mathbf{T} \mid \neg A \mid A \wedge B \mid \mathcal{X} \mid \exists x.A_x \mid \exists \mathcal{X}.A_{\mathcal{X}} \mid L \sim L' \mid \xi \mid \mu\xi.A_\xi \\
Q & ::= & \textit{from } Q \models A \textit{ select } Q'_{FV(A)} \mid \mathcal{X} \mid \mathbf{0} \mid Q|Q' \mid L[Q] \mid f(Q)
\end{array}
$$

A formula $\mu\xi.A_\xi$ is well formed when ξ only appears positively in A_ξ.

The interpretation of a formula A, i.e. the set of all information trees that satisfy A, is only defined with respect to a pair of valuations ρ and δ that give a value to the free variables of A. The valuation ρ maps label variables x to labels (elements of Λ) and tree variables $\mathcal{X}$ to information trees, while δ maps recursion variables ξ to sets of information trees. This interpretation is defined by the map $[\![A]\!]_{\rho,\delta}$, as specified in the table below.

To simplify the notation in the comparison case, we define the ρ's extension ρ^+ by fixing $\rho^+(n) = n$ for each $n \in \Lambda$ and $\rho^+(x) = \rho(x)$; hence, we can express in one line all the four cases of label comparison.

Tree Logic: formulas as sets of information trees

$$
\begin{array}{llll}
[\![\mathbf{0}]\!]_{\rho,\delta} =_{def} \{\mathbf{0}\} & & [\![L[A]]\!]_{\rho,\delta} =_{def} \{\rho^+(L)[I] \mid I \in [\![A]\!]_{\rho,\delta}\} \\
[\![\mathbf{T}]\!]_{\rho,\delta} =_{def} \mathcal{IT} & & [\![A \mid B]\!]_{\rho,\delta} =_{def} \{I \mid I' \mid I \in [\![A]\!]_{\rho,\delta}, I' \in [\![B]\!]_{\rho,\delta}\} \\
[\![\neg A]\!]_{\rho,\delta} =_{def} \mathcal{IT} \setminus [\![A]\!]_{\rho,\delta} & & [\![A \wedge B]\!]_{\rho,\delta} =_{def} [\![A]\!]_{\rho,\delta} \cap [\![B]\!]_{\rho,\delta} \\
[\![\exists x.A]\!]_{\rho,\delta} =_{def} \bigcup_{n \in \Lambda} [\![A]\!]_{\rho[x \mapsto n],\delta} & & [\![\exists \mathcal{X}.A]\!]_{\rho,\delta} =_{def} \bigcup_{I \in \mathcal{IT}} [\![A]\!]_{\rho[\mathcal{X} \mapsto I],\delta} \\
[\![\mathcal{X}]\!]_{\rho,\delta} =_{def} \{\rho(\mathcal{X})\} & & [\![L \sim L']\!]_{\rho,\delta} =_{def} \text{if } \rho^+(L) \sim \rho^+(L') \text{ then } \mathcal{IT} \text{ else } \emptyset \\
[\![\xi]\!]_{\rho,\delta} =_{def} \delta(\xi) & & [\![\mu\xi.A]\!]_{\rho,\delta} =_{def} \bigcap \{S \subseteq \mathcal{IT} \mid S \supseteq [\![A]\!]_{\rho,\delta[\xi \mapsto S]}\}
\end{array}
$$

We say that an information tree I satisfies a formula A with respect to ρ, δ, and write $I \models_{\rho,\delta} A$, when $I \in [\![A]\!]_{\rho,\delta}$. The definition above can be read, in terms of satisfaction with respect to ρ, δ, as follows. $\mathbf{0}$ is only satisfied by the information tree $\mathbf{0}$. $L[A]$ is satisfied by $m[I]$, if $m = \rho^+(L)$ and I satisfies A.$\mathbf{T}$ is satisfied by any I. $A' \mid A''$ is satisfied by I iff there exist I' and I'' such that $I' \mid I'' = I$ (where $\mid$ is multiset union) and I' satisfies A', and I'' satisfies A''. $\neg A$ is classical negation: it is satisfied by I iff I does not satisfy A. $A' \wedge A''$ is satisfied by I iff I satisfies both A' and A''. I satisfies $\exists x.A$ iff there exists some value n for x such that I is in $[\![A]\!]_{\rho[x \mapsto n],\delta}$. Here $\rho[x \mapsto n]$ denotes the valuation that maps x to n and otherwise coincides with ρ. $[\![L \sim L']\!]_{\rho,\delta}$ is the set $\mathcal{IT}$ if the comparison holds (w.r.t. ρ), else it is the empty set. $\mu\xi.A$ is satisfied by I iff I satisfies $A\{\xi \leftarrow \mu\xi.A\}$. Formally, $[\![\mu\xi.A]\!]_{\rho,\delta}$ is the least fix-point (with respect to set inclusion) of the function that maps any set of information trees S to $[\![A]\!]_{\rho,\delta[\xi \mapsto S]}$; the function is monotonic since any path from ξ to its binder is required to contain an even number of negations.

Valuations are the "pattern matching" mechanism of our query language; for example, $m[n[\mathbf{0}]]$ is in $[\![x[\mathcal{X}]]\!]_{\rho,\delta}$ if ρ maps x to m and $\mathcal{X}$ to $n[\mathbf{0}]$. We call *binding process* the process of finding all possible ρ's such that $I \in [\![A]\!]_{\rho,\delta}$. The implementation of the binding process is the core of the TQL processor.

The semantics of a query is defined in the following table. A query is evaluated with respect to an *input valuation* ρ, initialized with bindings for all the reachability roots of the database and also used to pass information from the surrounding *from-select* clauses.

from-select is the interesting case. Here, the sub-query Q' is evaluated once for each valuation ρ' obtained by the binding process for the formula A and the tree Q, that is once for each valuation ρ' that extends the input valuation ρ ($\rho' \supseteq \rho$) and such that $[\![Q]\!]_\rho \in [\![A]\!]_{\rho',\epsilon}$ (δ is initialized with the empty valuation ϵ since no recursion variable can be free in A); all the resulting trees are then combined using the | operator.

Query semantics

$[\![0]\!]_\rho =_{def} \mathbf{0}$ $\qquad$ $[\![\mathcal{X}]\!]_\rho =_{def} \rho(\mathcal{X})$ $\qquad$ $[\![m[Q]]\!]_\rho =_{def} m[[\![Q]\!]_\rho]$

$[\![x[Q]]\!]_\rho =_{def} \rho(x)[[\![Q]\!]_\rho]$ $\qquad$ $[\![f(Q)]\!]_\rho =_{def} f([\![Q]\!]_\rho)$ $\qquad$ $[\![Q \mid Q']\!]_\rho =_{def} [\![Q]\!]_\rho \mid [\![Q']\!]_\rho$

$[\![\mathit{from}\ Q \models A \mathit{select}\ Q']\!]_\rho =_{def} Par_{\rho' \in \{\rho' \mid\ dom(\rho')=dom(\rho)\cup FV(A),\ \rho' \supseteq \rho,\ [\![Q]\!]_\rho \in [\![A]\!]_{\rho',\epsilon}\}} [\![Q']\!]_{\rho'}$

As usual, negation allows us to derive useful 'dual' logical operators, such as universal quantification and disjunction. In [9] we describe the semantics and implementation of such operators and of *path formulas*, the derived logical operators that allow the programmer to retrieve information found at the end, or in the middle, of any path described by a regular expression over labels.

5. TQL Algebra

As happens with any declarative query language, TQL queries are translated into an algebraic form before execution. They are translated into terms of *TQL Algebra*, an algebra with two main sorts, tables and information trees, that are used to translate, respectively, binders and queries. Binder translation performs a "semantic inversion": it transforms the operators of the tree logic, whose terms denote functions from a valuation to a set of trees, into algebraic operators, that receive a tree and return a set of valuations (a *table*). For example, a formula $x[\mathbf{T}]$, that denotes the function $\lambda\rho.\ \{\rho(x)[I] \mid I \in \mathcal{IT}\}$, is translated into the table expression *if* $\mathcal{Q} = y[\mathcal{Y}]$ *then* $\{(x \mapsto y)\}$ *else* $\mathbf{0}$ that (informally) for each tree denoted by $\mathcal{Q}$, returns the set containing the valuation $(x \mapsto m)$ if $\mathcal{Q} = m[I]$ for some m, I, and the empty table otherwise.

TQL Algebra has been defined as a tool to translate TQL but is quite natural and general. The table operators are essentially the standard relational operators [1], generalized to infinite tables and to admit $\mathcal{IT}$ as a domain. The only new operators are the two operators, *if* and $\bigcup$, that are needed to build a table depending on the structure of the input information tree, and two more that are used to define and apply recursive functions. The tree expressions exactly mirror operators used to build an information tree, plus the operator that, mirroring the behavior of from-select, uses a table to build a tree.

In this section we will present the syntax and semantics of TQL Algebra; in the next sections we will show how TQL is translated into the algebra, and how the algebra is implemented.

5.1. Algebra Sorts and their Semantics

The example above shows how the TQL variables x and $\mathcal{X}$ become, in the algebra, the field names of the *rows* of the algebraic tables (i.e., the column names), while new algebraic variables (y, $\mathcal{Y}$) are introduced. Hence, in this section, the term *variable* will refer to the algebraic variables y, $\mathcal{Y}$, while x and $\mathcal{X}$ will be called *row field names*. Hereafter, the metavariable V will stand for either a field name $\mathcal{X}$, whose *universe* $U(\mathcal{X})$ is defined to be the set $\mathcal{IT}$ of all information trees, or a field name x, whose *universe* $U(x)$ is defined to be the set Λ of all labels. The metavariable $\mathbf{V}$ will range over schemas, i.e. finite sequences $V_1, \ldots, V_n$.

The query algebra is based on four sorts: a sort of *row expressions*, ranged over by $\mathcal{R}$ or $\mathcal{R}^{\mathbf{V}}$, a sort of *label expressions*, ranged over by $\mathcal{L}$, a sort of *table expressions*, ranged over by $\mathcal{T}$ and $\mathcal{T}^{\mathbf{V}}$, and a sort of *tree expressions*, ranged over by $\mathcal{Q}$.

A row expression $\mathcal{R}^{\mathbf{V}}$ denotes a *row* (or *valuation*) over $\mathbf{V}$, that is a function that maps each $V \in \mathbf{V}$ to an element of $U(V)$ (such as $(x \mapsto m,\ \mathcal{X} \mapsto m[\mathbf{0}])$, if $\mathbf{V} = \{x, \mathcal{X}\}$). $1^{\mathbf{V}}$ will denote the set of all rows having schema $\mathbf{V}$.

A table expressions $\mathcal{T}^{\mathbf{V}}$ denotes a finite or infinite table with schema $\mathbf{V}$, that is a set of rows over $\mathbf{V}$. A table expression is used to represent the evaluation of a TQL binding operation $Q \vDash A$; this evaluation returns a set of valuations with the same schema. Hence, $\mathcal{P}(1^{\mathbf{V}})$ denotes the set of all tables with schema $\mathbf{V}$. The set $\mathcal{P}(1^{\mathbf{V}})$ contains two special tables: $0^{\mathbf{V}}$, the *empty* table with schema $\mathbf{V}$, and $1^{\mathbf{V}}$, the *full* table with schema $\mathbf{V}$. Finally, a label expression $\mathcal{L}$ denotes an element of Λ, and a tree expression $\mathcal{Q}$ denotes an element of $\mathcal{IT}$.

5.2. Syntax

The syntax is presented in the table below. The algebra variables are r, y, $\mathcal{Y}$, M. Pedices are used to specify where variables are bound: for example, in *letrec* $M = \lambda\mathcal{Y}.\ \mathcal{T}_{M,\mathcal{Y}}$ *in* $\mathcal{T}'_M$, M is bound in both $\mathcal{T}_{M,\mathcal{Y}}$ and $\mathcal{T}'_M$, while $\mathcal{Y}$ is bound in $\mathcal{T}_{M,\mathcal{Y}}$ only.

As shown in the table below, the TQL Algebra has two forms of row expressions: the variable row expression $r^{\mathbf{V}}$, and the concatenation of two row expressions. Row variables arise during the translation of *from* $Q \models A$ *select* Q' queries, and range over the valuations obtained by the binding $Q \models A$; in the algebra, a row variable is bound by the operator $Par_{r \in \mathcal{T}}\ \mathcal{Q}_r$.

The TQL Algebra has three label expressions. $\mathcal{R}(x)$ extracts a label field x from $\mathcal{R}$; m is a label constant; y is a label variable, bound by the *if* operator.

The TQL Algebra has three operators to build one-row tables, that are $\{\mathcal{R}^{\mathbf{V}}\}$, $\{(x \mapsto \mathcal{L})\}$, and $\{(\mathcal{X} \mapsto \mathcal{Q})\}$: $\{\mathcal{R}^{\mathbf{V}}\}$ denotes a table, with schema $\mathbf{V}$, only containing the row denoted by $\mathcal{R}^{\mathbf{V}}$. $\{(x \mapsto \mathcal{L})\}$ and $\{(\mathcal{X} \mapsto \mathcal{Q})\}$ both denote a table with one row and one column only, mapping, respectively, x to the label denoted by $\mathcal{L}$, $\mathcal{X}$ to the denotation of $\mathcal{Q}$.

Then we have six table operators: universe (denoting the full table $1^{\mathbf{V}}$), union, cartesian product, projection, complement, and restriction, each carrying schema information. They correspond to standard operations of relational algebra. Restriction $\sigma^{\mathbf{V}}_{\mathcal{C} \sim \mathcal{C}'} \mathcal{T}^{\mathbf{V}}$ is subtle, since each argument $\mathcal{C}$ and $\mathcal{C}'$ of the comparison may be either a label $\mathcal{L}$ or a field name x. When at least one argument is a field name x, then x must appear in the schema $\mathbf{V}$ of $\mathcal{T}^{\mathbf{V}}$, and restriction is used to select a subset of the rows of $\mathcal{T}^{\mathbf{V}}$, depending on the value of their x field. In the special case when both arguments are label expressions, restriction returns either the whole $\mathcal{T}^{\mathbf{V}}$, if the comparison succeeds, or an empty table, if the comparison fails (evaluates to false).

Then, the table algebra contains two operators that analyze a tree and build a table according to its structure; the first (*if*) analyzes the vertical structure $m[I]$ of a singleton information tree, and the second analyzes the horizontal structure of an information tree, by evaluating an expression $\mathcal{T}_{\mathcal{Y}',\mathcal{Y}''}$ for each horizontal partition $\mathcal{Y}' \mid \mathcal{Y}''$ of the information tree denoted by $\mathcal{Q}$.

Finally, the table algebra has two operators used to translate recursive formulas: *letrec* $M = \lambda\mathcal{Y}.\ \mathcal{T}_{M,\mathcal{Y}}$ *in* $\mathcal{T}'_M$ computes the least fix-point of the monotone function $\lambda M.(\lambda\mathcal{Y}.\ \mathcal{T}_{M,\mathcal{Y}})$, in the space of functions from trees to tables, while $M(\mathcal{Q})$ applies such a fix-point to a tree.

The tree algebra reflects the TQL operators used to build trees. The essential difference is that here $\mathcal{X}$ does not denote a variable but the name of a field in the row ρ, while we have a new metavariable $\mathcal{Y}$, ranging over the tree variables. A variable $\mathcal{Y}$ is bound by the *if*, $\bigcup$, and *letrec* operators.

Query algebra, primitive operators:

$\mathcal{R}^{\mathbf{V}}$ ::=	row expression
$r^{\mathbf{V}}$	variable row expression ($r^{\mathbf{V}} \in \boldsymbol{r}_{set}$)
$\mathcal{R}'^{\mathbf{V}'}; \mathcal{R}''^{\mathbf{V}''}$	row concatenation ($\mathbf{V}' \cap \mathbf{V}'' = \emptyset,\ \mathbf{V}' \cup \mathbf{V}'' = \mathbf{V}$)
$\mathcal{L}$ ::=	label expression
y	label variable ($y \in \boldsymbol{y}_{set}$)
m	label
$\mathcal{R}^{\mathbf{V}}(x)$	field extraction from the row ($x \in \mathbf{V}$)
$\mathcal{C}$ ::=	comparison argument in the restriction operator
x	row field name ($x \in \boldsymbol{x}_{set}$)
$\mathcal{L}$	label expression

$\mathcal{T}^{\mathbf{V}} ::=$	table expression
$\{\mathcal{R}^{\mathbf{V}}\}$	one-row table
$\{(x \mapsto \mathcal{L})\}$	singleton: one column/one row ($\mathbf{V} = \{x\}$)
$\{(\mathcal{X} \mapsto \mathcal{Q})\}$	singleton ($\mathbf{V} = \{\mathcal{X}\}$)
$\mathbf{1}^{\mathbf{V}}$	universe: every row over $\mathbf{V}$
$\mathcal{T}^{\mathbf{V}} \cup^{\mathbf{V}} \mathcal{T}'^{\mathbf{V}}$	binary union
$\mathcal{T}'^{\mathbf{V}'} \times^{\mathbf{V}',\mathbf{V}''} \mathcal{T}''^{\mathbf{V}''}$	cartesian product ($\mathbf{V}' \cap \mathbf{V}'' = \emptyset$, $\mathbf{V}' \cup \mathbf{V}'' = \mathbf{V}$)
$\Pi_{\mathbf{V}}^{\mathbf{V}'} \mathcal{T}^{\mathbf{V}'}$	projection ($\mathbf{V} \subseteq \mathbf{V}'$)
$Co^{\mathbf{V}}(\mathcal{T}^{\mathbf{V}})$	complement
$\sigma^{\mathbf{V}}_{\mathcal{C}\sim\mathcal{C}'} \mathcal{T}^{\mathbf{V}}$	restriction
$\mathit{if}\ \mathcal{Q} = y[\mathcal{Y}]\ \mathit{then}\ \mathcal{T}^{\mathbf{V}}_{\mathcal{Y},y}\ \mathit{else}\ \mathcal{T}'^{\mathbf{V}}$	test for $y[\mathcal{Y}]$
$\bigcup^{\mathbf{V}}_{\{\mathcal{Y}' \mid \mathcal{Y}'' = \mathcal{Q}\}} \mathcal{T}_{\mathcal{Y}',\mathcal{Y}''}$	union of $\mathcal{T}_{\mathcal{Y}',\mathcal{Y}''}$ for decompositions $\mathcal{Y}' \mid \mathcal{Y}''$ of $\mathcal{Q}$
$\mathit{letrec}\ M^{\mathbf{V}'} = \lambda\mathcal{Y}.\ \mathcal{T}'^{\mathbf{V}'}_{M^{\mathbf{V}'},\mathcal{Y}}$ $\mathit{in}\ \mathcal{T}^{\mathbf{V}}_{M^{\mathbf{V}'}}$	recursive definition of a function ($M^{\mathbf{V}'} \in \boldsymbol{M}_{set}$) from trees ($\mathcal{Y}$) to tables; $M^{\mathbf{V}'}$ appears positively in $\mathcal{T}'^{\mathbf{V}'}$
$M^{\mathbf{V}}(\mathcal{Q})$	application of a recursive function to a tree
$\mathcal{Q} ::=$	tree expression
$Par_{r^{\mathbf{V}} \in \mathcal{T}^{\mathbf{V}}}\ \mathcal{Q}_{r^{\mathbf{V}}}$	union of $\mathcal{Q}_{r^{\mathbf{V}}}$ computed once for each $r^{\mathbf{V}}$ of $\mathcal{T}^{\mathbf{V}}$
$\mathcal{Y}$	tree variable ($\mathcal{Y} \in \boldsymbol{\mathcal{Y}}_{set}$)
$\mathcal{R}^{\mathbf{V}}(\mathcal{X})$	field extraction from the row $\mathcal{R}$ ($\mathcal{X} \in \mathbf{V}$)
$\mathbf{0}$	empty tree
$\mathcal{Q} \mid \mathcal{Q}'$	binary union
$\mathcal{L}[\mathcal{Q}]$	singleton tree
$f(\mathcal{Q})$	predefined function application

5.3. Semantics of TQL Algebra

Algebra expressions are evaluated with respect to an environment e, that associates each free label variable with a label, each free tree variable with an information tree, each free recursive variable with a function from trees to tables, and each row variable with a row of the right type. Formally, the type $\boldsymbol{e}_{sem}$ of e is defined as follows, where $(T \overset{\mathrm{P}}{\to} U) \times (T' \overset{\mathrm{P}}{\to} U')$ is the type of partial functions mapping T to U and T' to U', and $dom(f) = K$ means that $f : K \to K'$.

$$\begin{aligned}
\boldsymbol{\mathcal{R}}_{sem} &= (\boldsymbol{x}_{set} \overset{\mathrm{P}}{\to} \Lambda) \times (\boldsymbol{\mathcal{X}}_{set} \overset{\mathrm{P}}{\to} \mathcal{IT}) \\
\boldsymbol{\mathcal{R}}^{\mathbf{V}}{}_{sem} &= \{\rho \mid \rho \in \boldsymbol{\mathcal{R}}_{sem},\ dom(\rho) = \mathbf{V}\} \\
\boldsymbol{\mathcal{T}}_{sem} &= \cup_{\mathbf{V}} \mathcal{P}(1^{\mathbf{V}}) \\
\boldsymbol{e}'_{sem} &= (\boldsymbol{y}_{set} \overset{\mathrm{P}}{\to} \Lambda) \times (\boldsymbol{\mathcal{Y}}_{set} \overset{\mathrm{P}}{\to} \mathcal{IT}) \times (\boldsymbol{M}_{set} \overset{\mathrm{P}}{\to} \mathcal{IT} \overset{\mathrm{P}}{\to} \boldsymbol{\mathcal{T}}_{sem}) \times (\boldsymbol{r}_{set} \overset{\mathrm{P}}{\to} \boldsymbol{\mathcal{R}}_{sem}) \\
\boldsymbol{e}_{sem} &= \{e \mid e \in \boldsymbol{e}'_{sem},\ \forall r^{\mathbf{V}} \in dom(e).\ e(r^{\mathbf{V}}) \in \boldsymbol{\mathcal{R}}^{\mathbf{V}}{}_{sem}\}
\end{aligned}$$

The type of the semantic function $\lambda e.\lambda\_.\ [\![\_]\!]_e$ is: $\boldsymbol{e}_{sem} \to ((\boldsymbol{\mathcal{R}}_{set} \overset{\mathrm{P}}{\to} \boldsymbol{\mathcal{R}}_{sem}) \times (\boldsymbol{\mathcal{L}}_{set} \overset{\mathrm{P}}{\to} \Lambda) \times (\boldsymbol{\mathcal{Q}}_{set} \overset{\mathrm{P}}{\to} \mathcal{IT}) \times (\boldsymbol{\mathcal{T}}_{set} \overset{\mathrm{P}}{\to} \boldsymbol{\mathcal{T}}_{sem}))$. Most of the algebra semantics is straightforward. The crucial point is the information flow among the different sorts: tables depend on tree expressions when a singleton table is built, and when a tree analysis operation (*if* or $\biguplus$) is performed. Information flows from table expressions to trees when $Par_{r \in \mathcal{T}}\ \mathcal{Q}_r$ is evaluated, and it flows through the row variable r.

letrec involves the computation of a minimal fixpoint, that is well-defined since the function mapping f' to $\lambda t : \mathcal{IT}.\llbracket \mathcal{T} \rrbracket_{e[\mathcal{Y} \mapsto t][M^{\mathbf{V}} \mapsto f']}$ is monotone, because $M^{\mathbf{V}}$ only appears positively inside $\mathcal{T}$. We report here a couple of cases, while the full table is in [9].

The semantics of some table and tree expressions:

$$
\begin{array}{lcl}
\llbracket \sigma^{\mathbf{V}}_{x \sim \mathcal{L}} \mathcal{T}^{\mathbf{V}} \rrbracket_e & = & \{\rho \mid \rho \in \llbracket \mathcal{T}^{\mathbf{V}} \rrbracket_e,\ \rho(x) \sim \llbracket \mathcal{L} \rrbracket_e\} \\
\llbracket \mathit{letrec}\ M^{\mathbf{V}} = \lambda \mathcal{Y}.\ \mathcal{T}\ \mathit{in}\ \mathcal{T}' \rrbracket_e & = & \mathit{let}\ f = \mathit{minfix}_{f'}(\lambda t : \mathcal{IT}.\llbracket \mathcal{T} \rrbracket_{e[\mathcal{Y} \mapsto t][M^{\mathbf{V}} \mapsto f']}) \\
 & & \quad \mathit{in}\ \llbracket \mathcal{T}' \rrbracket_{e[M^{\mathbf{V}} \mapsto f]} \\
\llbracket M^{\mathbf{V}}(\mathcal{Q}) \rrbracket_e & = & e(M^{\mathbf{V}})(\llbracket \mathcal{Q} \rrbracket_e) \\
\llbracket Par_{r \in \mathcal{T}}\ \mathcal{Q} \rrbracket_e & = & Par_{\rho \in \llbracket \mathcal{T} \rrbracket_e}\ \llbracket \mathcal{Q} \rrbracket_{e[r \mapsto \rho]}
\end{array}
$$

5.4. Derived Operators

As for TQL, in the actual system there are many more algebraic operators that can be defined in terms of the primitive ones. In the translation of base operators of TQL (Section 6) the only derived operators we use are some variants of *if* and a generalized *natural join* operator $\mathcal{T}^{\mathbf{V}} \bowtie \mathcal{T}'^{\mathbf{V}'}$ (defined in [9]).

6. Translation of TQL into TQL Algebra

The translation of a formula A is the kernel of the translation problem. By definition, a formula defines a function from a substitution to a set of trees, but we want to transform it into an algebraic expression which, applied to a tree, yields a set of substitutions.

In detail, we assume that the algebraic expressions $\mathcal{Q}$ and $\mathcal{R}$, that compute the database Q and the input substitution ρ, are given. We define now the translation $\llbracket A \rrbracket_{\mathcal{Q},\mathcal{R}^{\mathbf{V}},\epsilon}$ of a formula A as an algebraic expression that computes the set of all substitutions ρ' such that $\llbracket Q \rrbracket_\rho \models_{(\rho;\rho'),\epsilon} A$, i.e. the rows that are passed from the binder $Q \models A$ to the *select* branch of a query. In order to deal with recursive formulae, we have to add a third parameter $\gamma : \boldsymbol{\xi}_{set} \xrightarrow{\mathrm{P}} \boldsymbol{M}_{set}$ that maps logical recursive variable to the algebraic ones. For each γ, $\hat{\gamma}$ describes its schema, i.e. $\hat{\gamma}(\xi) = \mathbf{V} \Leftrightarrow \gamma(\xi) = M^{\mathbf{V}}$, for some M.

The translation $\llbracket A \rrbracket_{\mathcal{Q},\mathcal{R}^{\mathbf{V}},\gamma}$ depends on a function $\mathcal{S}(A, \mathbf{V}, \hat{\gamma})$ that computes the schema of $\llbracket A \rrbracket_{\mathcal{Q},\mathcal{R}^{\mathbf{V}},\gamma}$ (essentially, it subtracts $\mathbf{V}$ from $FV(A)$, but recursion is subtle; see [9]).

We shortly describe some cases of the binder translation algorithm. $\mathbf{T}$ returns the table $\mathbf{1}^{\emptyset}$, the table that, when joined to any other table, does not exclude any row. A clause $Q \models \mathbf{0}$ returns the table $\mathbf{1}^{\emptyset}$ if Q is $\mathbf{0}$, and the table $\mathbf{0}^{\emptyset}$ otherwise. The operator $\wedge$ is interpreted by table join. The translation $\llbracket \mathcal{X} \rrbracket_{\mathcal{Q},\mathcal{R}^{\mathbf{V}},\gamma}$ depends on which variables are already bound by $\mathcal{R}^{\mathbf{V}}$: if $\mathcal{X}$ belongs to $\mathbf{V}$, then the resulting expression must have an empty schema, hence can only be $\mathbf{1}^{\emptyset}$ or $\mathbf{0}^{\emptyset}$ (depending on whether $\mathcal{Q}$ coincides with $\mathcal{R}^{\mathbf{V}}(\mathcal{X})$); otherwise, the expression must denote a table mapping $\mathcal{X}$ to $\mathcal{Q}$. $\llbracket n[A] \rrbracket_{\mathcal{Q},\mathcal{R}^{\mathbf{V}},\gamma}$ denotes a table $\mathbf{0}^{S(A,\mathbf{V},\hat{\gamma})}$ if the information tree denoted by $\mathcal{Q}$ does not match $n[I']$, otherwise it denotes the same table as $\llbracket A \rrbracket_{\mathcal{Y},\mathcal{R}^{\mathbf{V}},\gamma}$, evaluated in an environment that binds $\mathcal{Y}$ to I'. If x is in $\mathbf{V}$, the translation $\llbracket x[A] \rrbracket_{\mathcal{Q},\mathcal{R}^{\mathbf{V}},\gamma}$ goes essentially as in the previous case. Otherwise, the table denoted by $\llbracket A \rrbracket_{\mathcal{Y},\mathcal{R}^{\mathbf{V}},\gamma}$ has to be joined with the one mapping x to the label y (table join corresponds to conjunction). $\llbracket \neg A \rrbracket_{\mathcal{Q},\mathcal{R}^{\mathbf{V}},\gamma}$ is the table with all the rows that do not satisfy $\llbracket A \rrbracket_{\mathcal{Q},\mathcal{R}^{\mathbf{V}},\gamma}$. Comparison translation uses the abbreviation $\mathcal{R}^{\mathbf{V}}_{+}(L)$, which stands for $\mathcal{R}^{\mathbf{V}}(L)$ if $L \in \mathbf{V}$, and for L otherwise (when L is a variable not in $\mathbf{V}$, or a constant). $\llbracket A \mid B \rrbracket_{\mathcal{Q},\mathcal{R}^{\mathbf{V}},\gamma}$ is executed by considering every possible decomposition $I' \mid I''$ of the denotation of $\mathcal{Q}$, and by collecting all rows that satisfy both $I' \models A$ and $I'' \models B$. The nesting $\bigcup(\_ \bowtie \_)$ of the algorithm corresponds to the nesting "**there exists** a decomposition $I' \mid I''$ such that **both** $I' \models A$ **and** $I'' \models B$ hold". For existential quantification, $\llbracket \exists \mathcal{X}.\ A \rrbracket_{\mathcal{Q},\mathcal{R}^{\mathbf{V}},\gamma}$ can be computed as $\prod_{FV(A)\setminus\{\mathcal{X}\}} \llbracket A \rrbracket_{\mathcal{Q},\mathcal{R}^{\mathbf{V}},\gamma}$, because an information tree I belongs to $[\![\exists \mathcal{X}.A]\!]_{\rho,\delta}$ if, for some I', $I \in [\![A]\!]_{\rho[\mathcal{X} \mapsto I'],\delta}$.

Observe that the translation actually depends only on the shape of A and on the schema $\mathbf{V}$ of $\mathcal{R}^{\mathbf{V}}$, while $\mathcal{Q}$, $\mathcal{R}^{\mathbf{V}}$ and γ are only 'plugged' somewhere, without ever analyzing their shape.

The translation of recursion is the trickiest bit. In $\llbracket A \rrbracket_{\mathcal{Y},\mathcal{R}^{\mathbf{V}},\gamma[\xi \mapsto M]}$, the M variable corresponds to ξ, hence it means 'here you evaluate the translation of A again'. However, in general, you have to evaluate it against a different tree, since some of the logical operations (and their algebraic counterparts) 'walk' inside the input database; for example, $m[I] \models m[A]$ is reduced to $I \models A$, changing both the formula and the model ($m[A] \to A$, $m[I] \to I$). For this reason, the translation process $\llbracket A \rrbracket_{\mathcal{Q},\mathcal{R}^{\mathbf{V}},\gamma}$ analyzes A and produces a translation by keeping track, at any time, of the 'current tree expression' $\mathcal{Q}$. Therefore, the translation of the recursion body A is performed parametrically with respect to the actual input tree ($\lambda \mathcal{Y}.\llbracket A \rrbracket_{\mathcal{Y},\mathcal{R}^{\mathbf{V}},\gamma[\xi \mapsto M]}$), and, whenever ξ is met, the corresponding M (i.e. $\gamma(\xi)$) is applied to the current input tree $\mathcal{Q}$.

Binder and query translation

$$
\begin{array}{lll}
\llbracket \mathbf{T} \rrbracket_{\mathcal{Q},\mathcal{R}^{\mathbf{V}},\gamma} =_{def} \mathbf{1}^{\emptyset} & & \\
\llbracket \mathbf{0} \rrbracket_{\mathcal{Q},\mathcal{R}^{\mathbf{V}},\gamma} =_{def} \mathit{if}\ \mathcal{Q} = \mathbf{0}\ \mathit{then}\ \mathbf{1}^{\emptyset}\ \mathit{else}\ \mathbf{0}^{\emptyset} & & \\
\llbracket A \wedge B \rrbracket_{\mathcal{Q},\mathcal{R}^{\mathbf{V}},\gamma} =_{def} \llbracket A \rrbracket_{\mathcal{Q},\mathcal{R}^{\mathbf{V}},\gamma} \bowtie^{S(A,\mathbf{V},\hat{\gamma}),S(B,\mathbf{V},\hat{\gamma})} \llbracket B \rrbracket_{\mathcal{Q},\mathcal{R}^{\mathbf{V}},\gamma} & & \\
\llbracket \mathcal{X} \rrbracket_{\mathcal{Q},\mathcal{R}^{\mathbf{V}},\gamma} =_{def} \mathit{if}\ \mathcal{Q} = \mathcal{R}(\mathcal{X})\ \mathit{then}\ \mathbf{1}^{\emptyset}\ \mathit{else}\ \mathbf{0}^{\emptyset} & & \text{if } \mathcal{X} \in \mathbf{V} \\
\llbracket \mathcal{X} \rrbracket_{\mathcal{Q},\mathcal{R}^{\mathbf{V}},\gamma} =_{def} \{(\mathcal{X} \mapsto \mathcal{Q})\} & & \text{if } \mathcal{X} \notin \mathbf{V} \\
\llbracket n[A] \rrbracket_{\mathcal{Q},\mathcal{R}^{\mathbf{V}},\gamma} =_{def} \mathit{if}\ \mathcal{Q} = n[\mathcal{Y}]\ \mathit{then}\ \llbracket A \rrbracket_{\mathcal{Y},\mathcal{R}^{\mathbf{V}},\gamma}\ \mathit{else}\ \mathbf{0}^{S(A,\mathbf{V},\hat{\gamma})} & & \\
\llbracket x[A] \rrbracket_{\mathcal{Q},\mathcal{R}^{\mathbf{V}},\gamma} =_{def} \mathit{if}\ \mathcal{Q} = \mathcal{R}(x)[\mathcal{Y}]\ \mathit{then}\ \llbracket A \rrbracket_{\mathcal{Y},\mathcal{R}^{\mathbf{V}},\gamma}\ \mathit{else}\ \mathbf{0}^{S(A,\mathbf{V},\hat{\gamma})} & & \text{if } x \in \mathbf{V} \\
\llbracket x[A] \rrbracket_{\mathcal{Q},\mathcal{R}^{\mathbf{V}},\gamma} =_{def} \mathit{if}\ \mathcal{Q} = y[\mathcal{Y}] & & \text{if } x \notin \mathbf{V} \\
\qquad \mathit{then}\ \{(x \mapsto y)\} \bowtie^{\{x\},S(A,\mathbf{V},\hat{\gamma})} \llbracket A \rrbracket_{\mathcal{Y},\mathcal{R}^{\mathbf{V}},\gamma} & & \\
\qquad \mathit{else}\ \mathbf{0}^{S(x[A],\mathbf{V},\hat{\gamma})} & & \\
\llbracket \neg A \rrbracket_{\mathcal{Q},\mathcal{R}^{\mathbf{V}},\gamma} =_{def} Co^{S(A,\mathbf{V},\hat{\gamma})}(\llbracket A \rrbracket_{\mathcal{Q},\mathcal{R}^{\mathbf{V}},\gamma}) & & \\
\llbracket L \sim L' \rrbracket_{\mathcal{Q},\mathcal{R}^{\mathbf{V}},\gamma} =_{def} \sigma^{S(L \sim L',\mathbf{V},\hat{\gamma})}_{\mathcal{R}^{\mathbf{V}}_{+}(L) \sim \mathcal{R}^{\mathbf{V}}_{+}(L')} \mathbf{1}^{S(L \sim L',\mathbf{V},\hat{\gamma})} & &
\end{array}
$$

$$\llbracket A \mid B \rrbracket_{\mathcal{Q},\mathcal{R}^{\mathbf{V}},\gamma} =_{def} \bigcup^{S(A|B,\mathbf{V},\hat{\gamma})}_{\{\mathcal{Y}'|\mathcal{Y}''=\mathcal{Q}\}} (\llbracket A \rrbracket_{\mathcal{Y}',\mathcal{R}^{\mathbf{V}},\gamma} \bowtie^{S(A,\mathbf{V},\hat{\gamma}),S(B,\mathbf{V},\hat{\gamma})} \llbracket B \rrbracket_{\mathcal{Y}'',\mathcal{R}^{\mathbf{V}},\gamma})$$

$$\llbracket \exists x.\ A \rrbracket_{\mathcal{Q},\mathcal{R}^{\mathbf{V}},\gamma} =_{def} \prod^{S(A,\mathbf{V},\hat{\gamma})}_{S(A,\mathbf{V},\hat{\gamma})\setminus\{x\}} \llbracket A \rrbracket_{\mathcal{Q},\mathcal{R}^{\mathbf{V}},\gamma}$$

$$\llbracket \exists \mathcal{X}.\ A \rrbracket_{\mathcal{Q},\mathcal{R}^{\mathbf{V}},\gamma} =_{def} \prod^{S(A,\mathbf{V},\hat{\gamma})}_{S(A,\mathbf{V},\hat{\gamma})\setminus\{\mathcal{X}\}} \llbracket A \rrbracket_{\mathcal{Q},\mathcal{R}^{\mathbf{V}},\gamma}$$

$$\llbracket \mu\xi.A \rrbracket_{\mathcal{Q},\mathcal{R}^{\mathbf{V}},\gamma} =_{def} \mathit{letrec}\ M^{S(\mu\xi.A,\mathbf{V},\hat{\gamma})} = \lambda\mathcal{Y}.\llbracket A \rrbracket_{\mathcal{Y},\mathcal{R}^{\mathbf{V}},\gamma[\xi\mapsto M]}\ \mathit{in}\ M^{S(\mu\xi.A,\mathbf{V},\hat{\gamma})}(\mathcal{Q})$$

$$\llbracket \xi \rrbracket_{\mathcal{Q},\mathcal{R}^{\mathbf{V}},\gamma} =_{def} \gamma(\xi)(\mathcal{Q})$$

$$\llbracket \mathbf{0} \rrbracket_{\mathcal{R}^{\mathbf{V}}} =_{def} \mathbf{0} \qquad \llbracket \mathcal{X} \rrbracket_{\mathcal{R}^{\mathbf{V}}} =_{def} \mathcal{R}^{\mathbf{V}}(\mathcal{X})$$

$$\llbracket m[Q] \rrbracket_{\mathcal{R}^{\mathbf{V}}} =_{def} m[\llbracket Q \rrbracket_{\mathcal{R}^{\mathbf{V}}}] \qquad \llbracket x[Q] \rrbracket_{\mathcal{R}^{\mathbf{V}}} =_{def} \mathcal{R}^{\mathbf{V}}(x)[\llbracket Q \rrbracket_{\mathcal{R}^{\mathbf{V}}}]$$

$$\llbracket f(Q) \rrbracket_{\mathcal{R}^{\mathbf{V}}} =_{def} f(\llbracket Q \rrbracket_{\mathcal{R}^{\mathbf{V}}}) \qquad \llbracket Q \mid Q' \rrbracket_{\mathcal{R}^{\mathbf{V}}} =_{def} \llbracket Q \rrbracket_{\mathcal{R}^{\mathbf{V}}} \mid \llbracket Q' \rrbracket_{\mathcal{R}^{\mathbf{V}}}$$

$$\llbracket \mathit{from}\ Q \models A\ \mathit{select}\ Q' \rrbracket_{\mathcal{R}^{\mathbf{V}}} =_{def} \mathit{Par}_{r^{\mathbf{V}'} \in \llbracket A \rrbracket_{\mathcal{Q},\mathcal{R}^{\mathbf{V}},\epsilon}} \llbracket Q' \rrbracket_{\mathcal{R}^{\mathbf{V}};r^{\mathbf{V}'}} \quad \text{where } \mathcal{Q} = \llbracket Q \rrbracket_{\mathcal{R}^{\mathbf{V}}}$$

The following theorem states the query translation correctness. The core of the proof is the binder translation correctness statement [9], needed in the *from-select* case.

Theorem 1 *Let* $Q \in \boldsymbol{Q}_{set}$, $e \in \boldsymbol{e}_{sem}$, *and* $\mathcal{R}^{\mathbf{V}} \in \boldsymbol{\mathcal{R}}^{\mathbf{V}}_{set}$. *Then:*

$$FV(\mathcal{R}^{\mathbf{V}}) \subseteq dom(e),\ FV(Q) \subseteq \mathbf{V} \ \Rightarrow\ \llbracket Q \rrbracket_{e(\mathcal{R}^{\mathbf{V}})} = \lfloor\!\lfloor \llbracket Q \rrbracket_{\mathcal{R}^{\mathbf{V}}} \rfloor\!\rfloor_e$$

7. Implementing the Algebra

The essential problem we have to face is the representation of infinite tables and trees. Our solution is not complete; we actually implement a finite representation of infinite tables, but with the following limitations: (i) we only deal with finite trees; if the user runs a query that would need to evaluate a sub-query an infinite number of times, the evaluation is aborted; (ii) we do not implement general comparison between label expressions, but only equality comparison (and its negation) when at most one of the two compared label expressions is an unbound label variable, and full comparison when none of the two label expressions is an unbound label variable; (iii) recursion evaluation may loop forever; guarded recursion (when the recursive variable is separated from its definition by a $L[\_]$ operator) is safe, however, and it seems to be expressive enough for most purposes, including all queries that use path formulas.

A finite information tree is simply represented by an implementation of nested multi-sets; we keep the semantic notation ($\mathbf{0}$, $|$, $m[I]$) for the implementation of information tree operators and of labels (m).

The implementation of tables is the interesting part, since we have to represent infinite tables, and complex operations over them. A table is represented by a structure called *disjunctive constraint*, closely related to proposals in the field of constraint databases ([14], [15]). The constraint algebra we define here, however, does not seem to have been studied before.

A *constraint* is, essentially, a table with schema $\mathbf{V}$ where each cell contains either a value or the finite representation of a cofinite set. For example, an infinite table containing $[(x \mapsto l), (\mathcal{X} \mapsto m[0])]$ and $[(x \mapsto m), (\mathcal{X} \mapsto I)]$ for any I but $l[0]$, $m[0]$ would be represented by the following constraint:

x	$\mathcal{X}$
$\{l\}$	$\{m[0]\}$
$\{m\}$	$\overline{\{l[0], m[0]\}}$

i.e.: $\{\ [x{:=}\{l\}, \mathcal{X}{:=}\{m[0]\}],\ [x{:=}\{m\}, \mathcal{X}{:=}\overline{\{l[0], m[0]\}}]\}$

Formally, a *simple constraint* R is a tuple of sets S_i, each labelled with a different variable V_i, written as $[V_1 := S_1, \ldots, V_n := S_n]$, and such that: each V_i is either a label variable or a tree variable, and S_i is, respectively, a set of labels or a set of information trees; each S_i is either a singleton or the *cofinite* complement $\overline{P}$ of a finite set P. The set $dom(R) =_{def} \{V_i\}^{i \in I}$ is the domain of the simple constraint. Each simple constraint R defined on a domain $\mathbf{V}$ represents a *set* of rows $\rho^{\mathbf{V}}$; in detail $R = [V_1{:=}S_1, \ldots, V_n{:=}S_n]$ represents the set of all rows ρ over $\mathbf{V}$ that satisfy the constraint:

$$[\![R]\!] =_{def} \{\rho \mid \rho \in 1^{dom(R)},\ \rho(V_1) \in S_1 \wedge \ldots \wedge \rho(V_n) \in S_n\}$$

A *disjunctive constraint* $T^{\mathbf{V}}$ (or simply *constraint*) is a set of simple constraints, each with the same domain $\mathbf{V}$. It represents the union of all the sets of rows represented by its simple constraints.

Given this model, for each operator *op* defined on tables in $\mathcal{T}_{set}$ we have to define (at least) an implementation **op** that works on disjunctive constraints. In [9] we describe them, in particular we describe original and effective algorithms for complement and coprojection (the dual operator of projection).

8. Related Works and Conclusions

There are many algebras dealing with semi-structured data and XML [11, 8, 3, 16, 13, 2], but only some of them have a documented implementation [8, 11, 13]. These algebras operate on trees and tables of trees too, although some of them represent tables as trees or forests.

However, due to the specific, logic-based, nature of TQL, none of the other algebras has the operators we need to support our language. Namely, TQL Algebra is the only one that supports: (i) *logical complement* operator, that is a complement that is not defined on the active domain of the DBMS but on the infinite sets Λ and $\mathcal{IT}$; (ii) dual operators, such as *co-projection*, that allows the translation of universal quantification even in presence of free variables; (iii) specific fix-point operators to deal with horizontal and vertical recursion; (iv) iterators allowing analysis of the horizontal structure of a forest.

Finally, due to formal approach we have taken, our algebra is the only one where the correctness of the language translation has been proved.

The work on design and implementation of TQL is far from finished. At the language level, we are currently working on (i) extensions of the language to deal with order and with trees having a superimposed graph structure; (ii) adding a type and constraint system to the language; (iii) defining a TQL sublanguage that, by exhibiting a lower expressive power, may be implemented on more standard algebras.

At the implementation level, we are working towards the design of better persistent data structures and physical operators, endowed with a cost model, to allow cost-based physical optimization. The current TQL system is available at `http://tql.di.unipi.it/tql`.

References

[1] S. Abiteboul, R. Hull, and V. Vianu. *Foundations of Databases*. Addison-Wesley, 1995.

[2] D. Beech, A. Malhotra, and M. Rys. A formal data model and algebra for XML. Communication to the W3C, September 1999.

[3] Catriel Beeri and Yariv Tzaban. SAL: An algebra for semistructured data and XML. In *WebDB (Informal Proceedings)*, pages 37–42, 1999.

[4] P. Buneman, S. Davidson, W. Fan, C. Hara, and W. Tan. Keys for XML. In *Proc. of International World Wide Web Conference, WWW10*, May 2001.

[5] L. Cardelli. Describing semistructured data. *SIGMOD Record, Database Principles Column*, 2002. To appear.

[6] L. Cardelli and G. Ghelli. A query language based on the ambient logic. In *Proc. of European Symposium on Programming (ESOP), Genova, Italy*, April 2001.

[7] L. Cardelli and A. D. Gordon. Anytime, anywhere: Modal logics for mobile ambients. In *Proc. of POPL*. ACM Press, 2000.

[8] S. Cluet, C. Delobel, J. Siméon, and K. Smaga. Your mediators need data conversion. In *Proc. of ACM SIGMOD*, 1998.

[9] G. Conforti, O. Ferrara, and G. Ghelli. TQL Algebra and its Implementation. Working Draft available at http://tql.di.unipi.it/tql. Full version.

[10] G. Conforti, G. Ghelli, A. Albano, D. Colazzo, P. Manghi, and C. Sartiani. The Query Language TQL. In *Proc. of WebDB*, 2002. To appear.

[11] P. Fankhauser, M. Fernández, A. Malhotra, M. Rys, J. Siméon, and P. Wadler. XQuery 1.0 Formal Semantics, June 2001. W3C Working Draft.

[12] B.C. Pierce H. Hosoya. XDuce: A typed XML processing language (preliminary report). In *Proc. of Workshop on the Web and Data Bases (WebDB)*, 2000.

[13] H. V. Jagadish, L. V. S. Lakshmanan, D. Srivastava, and K. Thompson. TAX: A Tree Algebra for XML. In *Proceedings of DBPL'01*, 2001.

[14] P. Kanellakis. Tutorial: Constraint programming and database languages. In *Proc. of the 14th PODS*, pages 46–53. ACM Press, 1995.

[15] Peter Z. Revesz. Safe query languages for constraint databases. *ACM Transactions on Database Systems*, 23(1):58–99, March 1998.

[16] C. Sartiani and A. Albano. Yet another query algebra for XML data. In *Proc. of IEEE IDEAS*, 2002.

MODEL CHECKING BIRTH AND DEATH

Dino Distefano, Arend Rensink, Joost-Pieter Katoen
Faculty of Computer Science, University of Twente
P.O. Box 217, 7500 AE Enschede, The Netherlands
E-mail: {ddino, rensink, katoen}@cs.utwente.nl

Abstract This paper proposes Allocational Temporal Logic ($\mathcal{A}\ell\ell$TL) as a formalism to express properties concerning the dynamic allocation (birth) and de-allocation (death) of entities, such as the objects in an object-based system. The logic is interpreted on History-Dependent Automata, extended with a symbolic representation for certain cases of unbounded allocation. The paper also presents a simple imperative language with primitive statements for (de)allocation, with an operational semantics, to illustrate the kind of behaviour that can be modelled. The main contribution of the paper is a tableau-based model checking algorithm for $\mathcal{A}\ell\ell$TL, along the lines of Lichtenstein and Pnueli's algorithm for LTL.

1. Introduction

One of the aspects of computation that state-of-the-art model checking does not deal with very well is that of dynamic *allocation* and *deallocation* (birth and death) of entities. This is especially true if the number of entities is not known beforehand, or even unbounded. Though there are now calculi (such as the π-calculus [15]) that can express the generation of fresh names, as well as models (such as History-Dependent automata [16]) that can describe both the birth and the death of entities, what has been missing so far is a logic where these concepts are captured as primitives; a logic that should be as fundamental to reasoning about dynamic allocation as standard propositional logic is to reasoning about a fixed state space.

An attempt to formulate such a logic is presented in this paper. Called *allocational temporal logic* ($\mathcal{A}\ell\ell$TL), it has the following features: *(i)* Entity variables x, y, interpreted by a mapping to the entities existing (i.e., *alive*) in a given state. The interpretation is *partial*: a variable not mapped onto an existing entity stands for an entity that has *died*. *(ii)* Entity equations $x = y$ (where x, y are entity variables), asserting that x and y refer to the same entity. This cannot hold if either x or y has died; hence the entity equations express a *partial equivalence* of entity variables (symmetric and transitive, but not reflexive). *(iii)* Entity quantification $\exists x.\phi$, which holds in a given state if ϕ holds for some interpretation of x, provided that x is alive. *(iv)* A predicate x new to express that the entity referred to by x is *fresh*, i.e., newly born. In addition, $\mathcal{A}\ell\ell$TL has the standard LTL temporal operators.

The logic is interpreted over *high-level allocational Büchi automata* (HABA) which extend HD-automata [16] with a predicate for the *unboundedness* of (the number of entities in) a state, and with a (generalised) Büchi acceptance condition.

Together with the logic $\mathcal{A}\ell\ell$TL, the main contribution of this paper is that the model-checking problem for $\mathcal{A}\ell\ell$TL is shown to be decidable. In particular, we present a tableau-based model-checking algorithm that decides whether a given $\mathcal{A}\ell\ell$TL-formula holds for a given HABA. Our algorithm extends the tableau-based algorithm for LTL [14]. To the best of our knowledge, this yields the first approach to effectively model-check models with an unbounded number of entities. This is of particular interest to e.g. the verification of object-oriented systems in which the number of objects is typically not known in advance and may be unbounded. Furthermore, reasoning about (de)allocation of fresh names is relevant also in relation to *privacy* and *locality* as discussed in, e.g., [1, 5, 15].

Organisation of the paper. This paper introduces the logic (Section 2) and HABA (Section 3), as well as a simple imperative language, featuring statements for the allocation and deallocation of entities, with an operational semantics in terms of HABA (Section 4). The latter provides an intuition about the sort of behaviour that HABA can model. The main contribution of the paper is the proof of $\mathcal{A}\ell\ell$TL model checking property (Section 5). A discussion about related and future work completes the paper (Section 6). Proofs are reported in the long version of this paper [9].

2. Allocational temporal logic

Syntax. Let *LVar* be a countable set of logical variables ranged over by x, y, z, and *Ent* be a countable set of entities ranged over by $e, e', e_1, \ldots$. Allocational Temporal Logic ($\mathcal{A}\ell\ell$TL) is an extension of propositional LTL [17] that allows existential quantification over logical variables that can denote entities, or may be undefined. For $x \in LVar$, the syntax of $\mathcal{A}\ell\ell$TL is defined by the following grammar:

$$\phi ::= x \ \mathsf{new} \mid x \ \mathsf{dead} \mid x = x \mid \exists x.\phi \mid \neg\phi \mid \phi \vee \phi \mid \mathsf{X}\phi \mid \phi \,\mathsf{U}\, \phi$$

The operators have the following intuitive meaning. Formula x new holds if the entity denoted by x is new in the current state. Formula x dead holds if the entity denoted by x has died. Formula $x = y$ holds if variables x and y denote the same entity in the current state; $x = x$ is violated if x is undefined, i.e., if x does not denote any entity. $\exists x.\phi$ is valid in the current state if there exists an entity for which ϕ holds if assigned to x. X (next) and U (until) are the standard LTL operators. We denote $x \neq y$ for $\neg(x = y)$, x alive for $\neg(x \ \mathsf{dead})$, and x old for x alive $\wedge \neg(x \ \mathsf{new})$. The other boolean connectives and temporal operators F (eventually) and G (always) are standard.

Semantics. An *allocational sequence* σ is an infinite sequence of sets of entities $E_0 E_1 E_2 \cdots$ where $E_i \subseteq Ent$, for $i \in \mathbb{N}$. Let $\sigma^i = E_i E_{i+1} \cdots$. For given σ, E_i^σ denotes the set of entities in the i-th state of σ. The semantics of $\mathcal{A}\ell\ell$TL-formulae is defined by a satisfaction relation $\sigma, N, \theta \models \phi$ where σ is an allocational sequence, $N \subseteq E_0^\sigma$ is the set of entities that is initially new, and $\theta : LVar \rightharpoonup Ent$ is a partial valuation of the free variables in ϕ. Let N_i^σ denote the set of new entities in state i, i.e., $N_0^\sigma = N$ and $N_{i+1}^\sigma = E_{i+1}^\sigma \backslash E_i^\sigma$, and let $\theta_i^\sigma : LVar \rightharpoonup Ent$ denote the valuation at state i, where $\theta_i^\sigma(x) = \theta(x)$ if $\theta(x) \in E_k^\sigma$ for all $k \leqslant i$, and is undefined otherwise.

The condition avoids that contradictions like $\exists x.\mathsf{X}(x \text{ dead} \Rightarrow \mathsf{X}x \text{ alive})$ are fulfilled. Note that once a logical variable is mapped to an entity, then this association remains along σ unless the entity dies, i.e., is deallocated. Thereafter, although the entity may be reallocated, the logical variable remains undefined. The satisfaction relation $\models$ is defined as follows:

$$
\begin{array}{lll}
\sigma, N, \theta \models x \text{ new} & \text{iff} & x \in \text{dom}(\theta) \text{ and } \theta(x) \in N \\
\sigma, N, \theta \models x \text{ dead} & \text{iff} & x \notin \text{dom}(\theta) \\
\sigma, N, \theta \models x = y & \text{iff} & x, y \in \text{dom}(\theta) \text{ and } \theta(x) = \theta(y) \\
\sigma, N, \theta \models \exists x.\phi & \text{iff} & \exists e \in E_0^\sigma : \sigma, N, \theta\{e/x\} \models \phi \\
\sigma, N, \theta \models \neg\phi & \text{iff} & \sigma, N, \theta \not\models \phi \\
\sigma, N, \theta \models \phi \vee \psi & \text{iff} & \text{either } \sigma, N, \theta \models \phi \text{ or } \sigma, N, \theta \models \psi \\
\sigma, N, \theta \models \mathsf{X}\phi & \text{iff} & \sigma^1, N_1^\sigma, \theta_1^\sigma \models \phi \\
\sigma, N, \theta \models \phi \,\mathsf{U}\, \psi & \text{iff} & \exists i : (\sigma^i, N_i^\sigma, \theta_i^\sigma \models \psi \text{ and } \forall j < i : \sigma^j, N_j^\sigma, \theta_j^\sigma \models \phi).
\end{array}
$$

Here, $\theta\{e/x\}$ is defined as: $\theta\{e/x\}(x) = e$ and $\theta\{e/x\}(y) = \theta(y)$ for $y \neq x$.

Example 2.1. Properties concerning dynamic allocation and de-allocation can be formalised in $\mathcal{A}\ell\ell$TL. For example, formula $\mathsf{G}(\forall x.\forall y.\forall z.(x = y \vee x = z \vee y = z))$ asserts that the number of entities that are alive never exceeds 2, while $\mathsf{G}((\mathsf{F}\exists x.x \text{ new}) \wedge \forall x.\mathsf{X}(x \text{ alive}))$ states that the number of entities that are alive grows unboundedly. As a more involved example,

$$x \text{ alive} \,\mathsf{U}\, \exists y.(y \text{ new} \wedge (x \text{ alive} \,\mathsf{U}\, \exists z.(z \text{ new} \wedge y \neq z \wedge x \text{ alive})))$$

states that before x is deallocated, two new entities will be allocated. Note that formulas like $\mathsf{G}(x \text{ dead} \Rightarrow \mathsf{X}(x \text{ dead}))$, stating that entities cannot be allocated once they are de-allocated, and $\mathsf{X}(x \text{ dead} \vee x \text{ old})$ are tautologies.

Folded allocational sequences. In $\mathcal{A}\ell\ell$TL-formulae, entities can only be addressed through logical variables and valuations of variables (i.e., entities) can only be compared in the same state. These observations allow a reallocation (re-denomination) of entities from one state to its next state, as long as this is done injectively. For $E, E' \subseteq Ent$, a *reallocation* λ from E to E' is a partial injective function $\lambda : E \rightharpoonup E'$. A *folded allocational sequence* is an infinite alternating sequence $E_0\lambda_0E_1\lambda_1\cdots$, where λ_i is a reallocation from E_i to E_{i+1} for $i \geqslant 0$. We write λ_i^σ for the reallocation function of σ in state i. Note that for folded allocational sequence σ, $N_0^\sigma = N$, and $N_{i+1}^\sigma = E_{i+1}^\sigma \backslash \text{cod}(\lambda_i^\sigma)$. Similarly, $\theta_0^\sigma = \theta$ and $\theta_{i+1}^\sigma = \lambda_i^\sigma \circ \theta_i^\sigma$. Thus, entity e is considered to be deallocated if $e \notin \text{dom}(\lambda)$. Using these adapted definitions of N and θ, a satisfaction relation for $\mathcal{A}\ell\ell$TL can be defined in terms of folded allocational sequences precisely in the same way as above. The two kinds of sequences are equivalent models for $\mathcal{A}\ell\ell$TL-formulae [9]. The use of reallocations yields a local notion of entity identity that in turn allows minimisation of models [16].

3. High-level Allocational Büchi automata

In this section, we introduce an extension of (generalised) Büchi automata. High-level Allocational Büchi automata (HABA) generate folded allocational sequences and are inspired by History-Dependent automata [16]. HABA are basically Büchi automata where to each state a set of entities is associated. These entities, in turn, serve as valuation of logical (entity) variables.

Let $\infty \notin Ent$ be a special, distinguished entity, called *black hole*. Its role will become clear later on. We denote $E^\infty = E \cup \{\infty\}$ for arbitrary $E \subseteq Ent$. Furthermore, for $E, E_1 \subseteq Ent$, a ∞-*reallocation* is a partial function $\lambda : E^\infty \rightharpoonup E_1^\infty$ such that $\lambda(e) = \lambda(e') \neq \infty \Rightarrow e = e'$ for all $e, e' \in E$ and $\infty \in \text{dom}(\lambda) \Rightarrow \lambda(\infty) = \infty$. That is, λ is injective when mapping away from ∞ and preserves ∞.

Definition 3.1. A *High-level Allocational Büchi Automaton* (HABA) $\mathcal{H}$ is a tuple $\langle X, Q, E, \rightarrow, I, \mathcal{F}\rangle$ with

- $X \subseteq LVar$ a finite set of logical variables;
- Q a (possibly infinite) set of states;
- $E : Q \rightarrow 2^{Ent} \times \mathbb{B}$, a function that associates to each state $q \in Q$ a finite set E_q of entities and a predicate B_q which holds iff there is a bounded number of entities in q.
- $\rightarrow \subseteq Q \times (Ent^\infty \rightharpoonup Ent^\infty) \times Q$, such that for $q \rightarrow_\lambda q'$, λ is an ∞-reallocation from E_q^∞ to $E_{q'}^\infty$ with (i) $\infty \in \text{dom}(\lambda)$ iff $E_q = (E, \text{ff})$ and $E_{q'} = (E', \text{ff})$, and (ii) $\infty \in \text{cod}(\lambda) \Rightarrow E_{q'} = (E', \text{ff})$.
- $I : Q \rightharpoonup 2^{Ent} \times (X \rightharpoonup Ent)$ a partial function yielding for every initial state $q \in \text{dom}(I)$ an *initial valuation* (N, θ), where $N \subseteq E_q$ is a finite set of entities, and $\theta : X \rightharpoonup E_q$ is a partial valuation of the variables in X;
- $\mathcal{F} \subseteq 2^Q$ a set of sets of accept states.

We write $q \rightarrow_\lambda q'$ for $(q, \lambda, q') \in \rightarrow$. We adopt the generalised Büchi acceptance condition, i.e, $\rho = q_0\lambda_0 q_1\lambda_1 q_2 \cdots$ is a *run* of HABA $\mathcal{H}$ if $q_i \rightarrow_\lambda q_{i+1}$ for all $i \in \mathbb{N}$ and $|\{i | q_i \in F\}| = \omega$ for all $F \in \mathcal{F}$. Predicate B_q holds in state q iff the number of entities in q is bounded (denoted $\lceil q \rceil$). An unbounded state q (denoted $\lfloor q \rfloor$), possesses the distinguished entity ∞ that represents all entities that may be added to q. High-level state q thus represents all possible (concrete) states obtained from q by adding a finite number of entities to E_q. If a transition to state q' maps entities onto the black hole ∞, these entities cannot be distinguished anymore from there on. Moreover, if $q \rightarrow_\lambda q'$, entities in the black hole are either preserved (if $\lfloor q' \rfloor$), or are destroyed (if $\lceil q' \rceil$). The black hole thus allows to abstract from the identity of entities if these are not relevant anymore. The initial valuation (N, θ) associated to an initial state facilitates the generation of models for $A\ell\ell$TL-formulae. This is shown in the following definition that formalises the correspondence between runs of the HABA and folded allocational sequences.

Definition 3.2. A run $\rho = q_0\lambda_0 q_1\lambda_1 \cdots$ of HABA $\mathcal{H} = \langle X, Q, E, \rightarrow, I, \mathcal{F}\rangle$ *generates* an allocation triple (σ, N, θ), where $\sigma = E_0\lambda_0^\sigma E_1\lambda_1^\sigma \cdots$ is a folded allocational sequence, if there is a *generator*, i.e., a family of functions $\phi_i : E_i \rightarrow E_{q_i}^\infty$ satisfying for all $i \geqslant 0$:

1. $\forall e, e' \in E_i.\ (\phi_i(e) = \phi_i(e') \neq \infty \Rightarrow e = e')$
2. $\forall e \in E_{i+1}.\ (\phi_{i+1}(e) = \infty \Rightarrow e \in \text{cod}(\lambda_i^\sigma))$
3. $\lceil q_i \rceil \Rightarrow (\forall e \in E_i : \phi_i(e) \neq \infty)$
4. $\lambda_i \circ \phi_i = \phi_{i+1} \circ \lambda_i^\sigma$
5. $E_{q_i} \subseteq \text{cod}(\phi_i)$
6. $I(q_0) = (\phi_0(N), \phi_0 \circ \theta)$

In the previous definition, notice the difference between ∞-reallocations λ_i of HABA transitions and reallocations λ_i^σ of folded allocational sequence σ. Let runs($\mathcal{H}$) denote the set of runs of $\mathcal{H}$ and $\mathcal{L}(\mathcal{H}) = \{(\sigma, N, \theta) \mid \exists \rho \in \text{runs}(\mathcal{H}) : \rho \text{ generates } (\sigma, N, \theta)\}$.

Example 3.3. The picture just below depicts a HABA with $X = \{x, y\}$. Squares denote bounded states, (large) circles denote unbounded states, small circles denote entities, and accept states have a double boundary. Here for simplicity we assume $|\mathcal{F}| = 1$. Dashed arrows indicate ∞-reallocations. In initial states, dotted lines represent θ, and filled circles denote new entities. In q_1, variable x denotes (old) entity e_1, while y is undefined. Entity e_3 in state q_2 represents the same entity as e_1 in q_1, while e_1 (in q_2) represents a new entity. Run $q_1\lambda_{12}(q_2\lambda_{22})^\omega$ generates sequences where the initial entity dies after the second state, while the new entity created in the second state will be alive forever. After the second state, at every step, a new entity is created and it will be alive only in one state. Run $q_1\lambda_{14}(q_4\lambda_{44})^\omega$ generates sequences where the entity in the initial state dies immediately. Once q_4 is reached, a new entity e_3 is created at every step, and in this run thus the number of entities grows unboundedly.

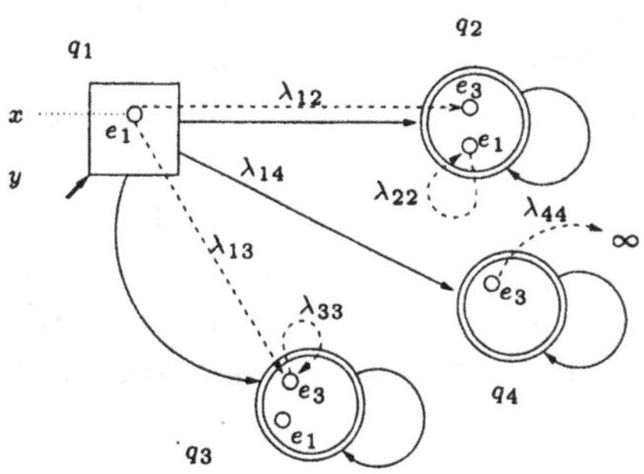

4. Programming allocation and deallocation

This section introduces a simple programming language $\mathcal{L}$ capturing the essence of allocation and deallocation. It is used for providing an intuition about the setup and the sort of behaviour that can be modelled by HABA. The operational semantics for $\mathcal{L}$ is defined using HABA as underlying model.

Syntax. For $PVar$ a set of program variables with $v, v_i \in PVar$ and $PVar \cap LVar = \varnothing$, the set of statements of $\mathcal{L}$ is given by:

$$
\begin{array}{lcl}
(p \in) \mathcal{L} & ::= & \mathsf{decl}\ v_1, \ldots, v_n : (s_1 \parallel \cdots \parallel s_k) \\
(s \in) Stat & ::= & \mathsf{new}(v) \mid \mathsf{del}(v) \mid v := v \mid \mathsf{skip} \mid s; s \mid \mathsf{if}\ b\ \mathsf{then}\ s\ \mathsf{else}\ s\ \mathsf{fi} \\
 & & \mid \mathsf{while}\ b\ \mathsf{do}\ s\ \mathsf{od} \\
(b \in) Bexp & ::= & v = v \mid b \vee b \mid \neg b
\end{array}
$$

A program p is a parallel composition of a finite number of statements preceded by the declaration of a finite number of global variables. new(v) creates (i.e., allocates) a new entity that will be referred to by the program variable v. The old value of v is lost. Thus, if v is the only variable that refers to entity e, say, then after the execution of new(v), e cannot be referenced anymore. In particular, e cannot be deallocated anymore. In other words, there is no automatic garbage collection. del(v) destroys (i.e., deallocates) the entity associated to v, and makes v undefined. The assignment $v := w$ passes the reference held by w (if any) to v. Again, the entity v was referring to might become unreferenced (for ever). Sequential composition, while, skip, and conditional statement have the standard interpretation. For the sake of simplicity, new and del create and destroy, respectively, a single entity only; generalisations in which several entities are considered simultaneously can be added in a straightforward manner.

Example 4.1. The following program, where $g(i) = (i+1) \bmod 4$, models the implementation of a naive solution to the dining philosopher problem:

$$\begin{array}{lcl} DPhil & \equiv & \mathsf{decl}\ v_1, v_2, v_3, v_4 : Ph_1 \parallel Ph_2 \parallel Ph_3 \parallel Ph_4)\ \text{where} \\ Ph_i & \equiv & \mathsf{while\ tt\ do\ if}\ (v_i\ \mathsf{alive} \wedge v_{g(i)}\ \mathsf{alive})\ \mathsf{then} \\ & & \quad \mathsf{del}(v_i);\ \mathsf{del}(v_{g(i)});\ \mathsf{new}(v_i);\ \mathsf{new}(v_{g(i)})\ \mathsf{else\ skip\ fi} \\ & & \mathsf{od} \end{array}$$

The variables v_i and $v_{g(i)}$ represent the left and the right chopstick of philosopher Ph_i, respectively. If v_i and $v_{g(i)}$ are defined[1], then the chopsticks are on the table. Taking the chopsticks from the table is represented by destroying the corresponding entities, while putting the chopsticks back on the table is modelled by creating new entities. Some properties that can possibly be satisfied by this program, stated in $\mathcal{A}\ell\ell$TL, are: $\mathsf{FG}(\forall x.\forall y.(x = y))$, expressing that eventually there is only one chopstick on the table (an inconsistency), or $\mathsf{G}(\forall x.\forall y.\forall z.(x\ \mathsf{old} \wedge y\ \mathsf{old} \wedge z\ \mathsf{old} \wedge (x = y \vee x = z \vee y = z)))$, expressing that among the philosophers there exists a greedy impostor who always eats and never thinks (an unfair computation).

Operational semantics. A (symbolic) semantics of our example language is given in terms of HABA where entities are represented by a partial partition of a subset of $PVar$; that is, the set E of entities is of the form $\{X_1, \ldots, X_n\}$ with $X_i \subseteq PVar$ and $X_i \cap X_j = \varnothing$ (for $i \neq j$). Note that we do not require $\bigcup_i X_i = PVar$ which would make it a full partitioning. Variable v is defined iff $v \in X_i$ for some i. Then, v refers to the entity represented by the set X_i. Otherwise, v is undefined. Using this approach, there is no need to represent (in a state) a mapping from the set of program variables onto the entities.

Let Par denote the compound statements, i.e., $r (\in Par) ::= s \mid r \parallel s$. The semantics of $p = \mathsf{decl}\ v_1, \ldots, v_n : (s_1 \parallel \cdots \parallel s_k)$ is the HABA $\mathcal{H}_p = \langle \varnothing, Q, E, \rightarrow, I, \mathcal{F} \rangle$ where

- $Q \subseteq Par \times 2^{2^{PVar}}$, i.e., a state $q = (r, E)$ consists of a compound statement and a set of entities; we have $\lfloor q \rfloor$ iff $\varnothing \in E$ (i.e., we represent the black hole by $\varnothing$).
- $E(r, E') = E' \backslash \{\varnothing\}$ and $I(s_1; \mathsf{skip} \parallel \cdots \parallel s_k; \mathsf{skip}, \varnothing) = (\varnothing, \varnothing)$;
- $\rightarrow$ is the smallest relation defined by the rules in Table 1 such that for $r, E \rightarrow_\lambda r', E'$ we have $\varnothing \in E \Rightarrow \varnothing \in \mathrm{dom}(\lambda)$.
- let $\widehat{F}_i = \{(s'_1 \parallel \cdots \parallel s'_k, E) \in Q \mid s'_i = \mathsf{skip} \vee s'_i = \mathsf{while}\ b\ \mathsf{do}\ s\ \mathsf{od}; s''\}$ and $\widetilde{F}_i = \{(s'_1 \parallel \cdots \parallel s'_k, E) \in Q \mid s'_i = \mathsf{skip} \vee s'_i = s; \mathsf{while}\ b\ \mathsf{do}\ s\ \mathsf{od}; s''\}$; then $\mathcal{F} = \{\widehat{F}_i \mid 0 < i \leqslant k\} \cup \{\widetilde{F}_i \mid 0 < i \leqslant k\}$.

---

[1]Notice that v dead iff $\neg(v = v)$. Again, here v alive stands for $\neg(v\ \mathsf{dead})$.

Table 1. Operational rules for the semantics of $\mathcal{L}$.

$$\frac{}{v := w, E \rightarrow_{\lambda} \mathsf{skip}, \{X_i \backslash \{v\} | w \notin X_i\} \cup \{X_i \cup \{v\} | w \in X_i\}} \; \lambda : X_i \mapsto \begin{cases} X_i \backslash \{v\} & \text{if } w \notin X_i \\ X_i \cup \{v\} & \text{otherwise} \end{cases}$$

$$\frac{}{\mathsf{new}(v), E \rightarrow_{\lambda} \mathsf{skip}, \{X_i \backslash \{v\} | X_i \in E\} \cup \{\{v\}\}} \; \lambda(X_i) = X_i \backslash \{v\}$$

$$\frac{v \in X_i}{\mathsf{del}(v), E \rightarrow_{\lambda} \mathsf{skip}, (E \backslash \{X_i\})} \; \lambda : X_j \mapsto \begin{cases} X_j & \text{if } j \neq i \\ \bot & \text{otherwise} \end{cases} \qquad \frac{s_1, E \rightarrow_{\lambda} s_1', E'}{s_1; s_2, E \rightarrow_{\lambda} s_1'; s_2, E'}$$

$$\frac{}{\mathsf{while}\ b\ \mathsf{do}\ s\ \mathsf{od}, E \rightarrow_{id} \mathsf{if}\ b\ \mathsf{then}\ s; \mathsf{while}\ b\ \mathsf{do}\ s\ \mathsf{od}\ \mathsf{else}\ \mathsf{skip}\ \mathsf{fi}, E} \qquad \frac{}{\mathsf{skip}; s_2, E \rightarrow_{id} s_2, E}$$

$$\frac{1 \leqslant j \leqslant k \;\wedge\; s_j, E \rightarrow_{\lambda} s_j', E'}{s_1 \parallel \cdots \parallel s_j \parallel \cdots \parallel s_k, E \rightarrow_{\lambda} s_1 \parallel \cdots \parallel s_j' \parallel \cdots \parallel s_k, E'} \qquad \frac{}{\parallel \mathsf{skip}, E \rightarrow_{id} \parallel \mathsf{skip}, E}$$

$$\frac{\mathcal{V}(b)(E)}{\mathsf{if}\ b\ \mathsf{then}\ s_1\ \mathsf{else}\ s_2\ \mathsf{fi}, E \rightarrow_{id} s_1, E} \qquad \frac{\neg\mathcal{V}(b)(E)}{\mathsf{if}\ b\ \mathsf{then}\ s_1\ \mathsf{else}\ s_2\ \mathsf{fi}, E \rightarrow_{id} s_2, E}$$

A few remarks are in order. $\mathcal{H}_p$ has a single initial state $s_1; \mathsf{skip} \parallel \cdots \parallel s_k; \mathsf{skip}$, where each sequential component is terminated by a skip statement. The set of accept states for the i-th sequential component consists of all states in which the component has either terminated ($s_i = \mathsf{skip}$) or is processing a loop (which could be infinite).

The condition on $\varnothing$ in the definition of $\rightarrow$ can be seen as a kind of "preservation law" of the black hole. In fact, once a state explodes into an unbounded one, the black hole generated by this explosion will last forever. Note that in Def. 3.1 this is not always the case. The semantics of the boolean expressions is given by the function $\mathcal{V} : \mathit{Bexp} \times 2^{2^{PVar}} \rightarrow \mathbb{B}$ defined by $\mathcal{V}(v = w, E) = \mathsf{tt}$ if $\exists X_i \in E : v, w \in X_i$ and false otherwise, $\mathcal{V}(b_1 \vee b_2, E) = \mathcal{V}(b_1, E) \vee \mathcal{V}(b_2, E)$, and $\mathcal{V}(\neg b, E) = \neg\mathcal{V}(b, E)$. Note that $\parallel$ skip is a shorthand for skip $\parallel \ldots \parallel$ skip. Whenever entity X_i is not referenced by any program variable, the state will become unbounded. Entity X_i will then be mapped by λ onto $\varnothing$ (recall that in the special case of $\mathcal{H}_p$ we represent ∞ by $\varnothing$), which can be viewed as a "black hole" collecting every non-referenced entity. These entities share the property that they cannot be deallocated anymore, thus they will have the same future, i.e, they will be "floating" in the black hole *ad infinitum*.

Although, there may be an unbounded number of entity creations, for the semantics defined in this section we have the following result:

Theorem 4.2. For any $p \in \mathcal{L}$: $\mathcal{H}_p$ is finite state.

In [9] it is shown that $|Q_{\mathcal{H}_p}|$ is exponential in the number of sequential components of p and super-exponential in $|PVar|$.

5. Model-checking $\mathcal{A}\ell\ell$TL

In this section, we define an algorithm for model-checking $\mathcal{A}\ell\ell$TL-formulae against a HABA. The algorithm extends the tableau method for LTL [14] to $\mathcal{A}\ell\ell$TL.

We will evaluate $\mathcal{A\ell\ell}$TL-formulae on states of a HABA by mapping the free variables of the formula to entities of the state. It should be clear that, in principle, any such mapping resolves all basic propositions. In turn, the basic propositions determine the validity of arbitrary formulae. There are, however, two obstacles to this principle, the first of which is slight and the other more difficult to overcome.

- It is not always uniquely determined whether or not an entity is fresh in a state. Our model allows states in which a given entity is considered fresh when arriving by one incoming transition (since it is not in the codomain of the reallocation associated with that transition), but not when arriving by another (the entity is the image of an entity in the previous state). This obstacle is dealt with by *duplicating* the states where such an ambiguity exists.
- For variables (of the formula in question) that are mapped onto entities in the black hole, entity equations are not resolved, since it is not clear whether the variables are mapped to distinct entities that have imploded into the black hole, or to the same one. To deal with this obstacle, we introduce an intermediate layer in the evaluation of the formula on the state. This additional layer consists of a *partial partitioning* of the free variables; that is, a set of nonempty, disjoint subsets of the set of all free variables. An entity equation is then resolved by the question whether the equated variables are in the same partition. It is the partitions, rather than the individual variables, that are mapped to the entities.

Assumptions. The duplication proposed above to overcome the first of these obstacles is straightforward; we will not work it out in more detail in this paper (see [9] for details). In the remainder, we assume that the necessary duplication has been carried out already: that is, we will assume that for every state $q \in Q$ there is an associated set $N_q \subseteq E_q$ that contains the entities that are *new in* q; i.e., such that

a) $q' \rightarrow_\lambda q$ implies $E_q \setminus \mathrm{cod}(\lambda) = N_q$ b) $I(q) = (N, \theta)$ implies $N = N_q$.

Note that, because of b), we can henceforth assume that I has just θ as its image — the component N is now uniquely associated with q. Another assumption needed below is that every quantified variable actually appears free in the subformula; that is, we only consider formulae $\exists x.\phi$ for which $x \in fv(\phi)$. Note that this imposes no real restriction, since $\exists x.\phi$ is equivalent to $\exists x.(x \text{ alive} \wedge \phi)$.

Valuations. A valuation of a formula in a given state is an interpretation of the free variables of the formula as entities of the state. Such an interpretation establishes the validity of at least the *atomic propositions* within the formula, i.e., the sub-formulae of the form $x = y$ (which holds if x and y are interpreted as the same entity) and x new (which holds if x is interpreted as a fresh entity).

Definition 5.1 (valuations). Let $E \subseteq Ent^\infty$. An *E-valuation* is a triple (ϕ, Ξ, Ψ) where ϕ is an $\mathcal{A\ell\ell}$TL-formula and

- Ξ is a partial partitioning of $fv(\phi)$; that is, $\Xi = \{X_1, \ldots, X_n\}$ such that $\varnothing \subset X_i \subseteq fv(\phi)$ for $1 \leq i \leq n$ and $X_i \cap X_j = \varnothing$ for $1 \leq i < j \leq n$ (but not necessarily $\bigcup_n X_n = fv(\phi)$, which would make it a *full* partitioning).

- $\Theta\colon \Xi \rightarrow E$ is a function mapping the partitions of Ξ to E, such that Θ is injective where it maps away from ∞ — i.e., $\Theta(X_i) = \Theta(X_j) \neq \infty \Rightarrow i = j$.

This is easily lifted to the states of a HABA: (ϕ, Ξ, Θ) is a q-valuation (for some $q \in Q_{\mathcal{H}}$) if it is an E_q-valuation (if $\lceil q \rceil$) or E_q^∞-valuation (if $\lfloor q \rfloor$). We write $V_q(\phi)$, ranged over by v, to denote the set of q-valuations of ϕ, and V_q to denote the set of *all* q-valuations. We denote the components of a valuation v as $(\phi_v, \Xi_v, \Theta_v)$.

A technicality: below we will need to restrict partial partitioning Ξ and mappings Θ of a valuation (ϕ, Ξ, Θ) to subformulae of ϕ, which means restricting the underlying sets of (free) variables upon which Ξ and Θ are built to those of that subformula. For this purpose, we define $\Xi \restriction \psi = \{X \cap fv(\psi) \mid X \in \Xi, X \cap fv(\psi) \neq \varnothing\}$ and $\Theta \restriction \psi = \{(X \cap fv(\psi), \Theta(X)) \mid X \in \mathrm{dom}(\Theta), X \cap fv(\psi) \neq \varnothing\}$.

The *atomic proposition valuations* of a state q of a HABA are those q-valuations of basic propositions of $\mathcal{A}\ell\ell$TL (i.e., freshness predicates and entity equations) that make the corresponding properties true.

Definition 5.2. Let $\mathcal{H}$ be a HABA and let $q \in Q_{\mathcal{H}}$ be arbitrary. The *atomic proposition valuations* of q are defined by the set $AV_q \subseteq V_q$ of all triples (ϕ, Ξ, Θ) for which one of the following holds:

- $\phi = \mathsf{tt}$;
- $\phi = (x = y)$, and $x, y \in X$ for some $X \in \Xi$;
- $\phi = (x \text{ new})$, and $x \in X \in \Xi$ implies $\Theta(X) \in N_q$.

Closure. Along the lines of [14], we associate to each state q of a HABA sets of q-valuations, specifically aimed at establishing the validity of a given formula ϕ. For this purpose, we first collect all $\mathcal{A}\ell\ell$TL-formulae whose validity is possibly relevant to the validity of a given formula ϕ into the so-called *closure of* ϕ.

Definition 5.3. Let ϕ be an $\mathcal{A}\ell\ell$TL-formula. The *closure* of ϕ, $CL(\phi)$, is the smallest set of formulae (identifying $\neg\neg\psi$ with ψ) such that:

- $\phi, \mathsf{tt}, \mathsf{ff} \in CL(\phi)$;
- $\neg\psi \in CL(\phi)$ iff $\psi \in CL(\phi)$;
- if $\psi_1 \vee \psi_2 \in CL(\phi)$ then $\psi_1, \psi_2 \in CL(\phi)$;
- if $\exists x.\psi \in CL(\phi)$ then $\psi \in CL(\phi)$;
- if $\mathsf{X}\psi \in CL(\phi)$ then $\psi \in CL(\phi)$;
- if $\neg\mathsf{X}\psi \in CL(\phi)$ then $\mathsf{X}\neg\psi \in CL(\phi)$;
- if $\psi_1 \,\mathsf{U}\, \psi_2 \in CL(\phi)$ then $\psi_1, \psi_2, \mathsf{X}(\psi_1 \,\mathsf{U}\, \psi_2) \in CL(\phi)$.

Since valuations map (sets of) variables of a given formula to entities, possibly to the black hole, it is important to know how many of these variables have to be taken into account at the most. This is obviously bounded by the number of variables occurring (free or bound) in ϕ, but in fact we can be a little more precise: the number is given by $K(\phi)$ defined as $K(\phi) = \max\{|fv(\psi)| \mid \psi \in CL(\phi)\}$.

The interesting case for the model checking construction is when one or more variables are indeed mapped to the black hole. Among other things, we will then have to make sure that sufficiently many entities of the state have imploded into the black hole to meet the demands of the valuation. For this purpose, we introduce the *black number* of a function, which is the number of entities that that function maps (implodes) into the black hole. For an arbitrary set A and (partial) mapping $\alpha: A \rightharpoonup Ent^{\infty}$ we define $\Omega(\alpha) = |\{a \in A \mid \alpha(a) = \infty\}|$.

Tableau graph. We now construct a graph that will be the basis of the model checking algorithm. The nodes of this graph, called *atoms* after [14], are built from states of a HABA, valuations of formulae from the closure, and a bound on the black number.

Definition 5.4. Given a HABA $\mathcal{H}$ and an $\mathcal{A}\ell\ell$TL-formula ϕ, an *atom* is a triple (q, D, k) where $q \in Q_{\mathcal{H}}$, $D \subseteq \{v \in V_q(\psi) \mid \psi \in CL(\phi), \Omega(\Theta_v) \leq k\}$ and $k \leq K(\phi)$ if $\lfloor q \rfloor$ or $k = 0$ if $\lceil q \rceil$, such that for all $v = (\psi, \Xi, \Theta) \in V_q$ with $\psi \in CL(\phi)$ and $\Omega(\Theta) \leq k$:

- if $v \in AV_q$, then $v \in D$;
- if $\psi = \neg\psi'$, then $v \in D$ iff $(\psi', \Xi, \Theta) \notin D$;
- if $\psi = \psi_1 \vee \psi_2$, then $v \in D$ iff $(\psi_i, \Xi \restriction \psi_i, \Theta \restriction \psi_i) \in D$ for $i = 1$ or $i = 2$;
- if $\psi = \exists x.\psi'$, then $v \in D$ iff there exists a $(\psi', \Xi', \Theta') \in D$ such that $\Xi = \Xi' \restriction \psi$, $\Theta = \Theta' \restriction \psi$ and $x \in \bigcup \Xi'$;
- if $\psi = \neg \mathsf{X}\psi'$, then $v \in D$ iff $(\mathsf{X}\neg\psi', \Xi, \Theta) \in D$;
- if $\psi = \psi_1 \mathsf{U} \psi_2$, then $v \in D$ iff either $(\psi_2, \Xi \restriction \psi_2, \Theta \restriction \psi_2) \in D$, or both $(\psi_1, \Xi \restriction \psi_1, \Theta \restriction \psi_1) \in D$ and $(\mathsf{X}\psi, \Xi, \Theta) \in D$.

The set of all atoms for a given formula ϕ constructed on top of $\mathcal{H}$ is denoted $A_{\mathcal{H}}(\phi)$, ranged over by A, B. We denote the components of an atom A by (q_A, D_A, k_A).

Definition 5.5. The *tableau graph* for a HABA $\mathcal{H}$ and an ATL-formula ϕ, denoted $G_{\mathcal{H}}(\phi)$, consists of vertices $A_{\mathcal{H}}(\phi)$ and edges $\rightarrow \subseteq A_{\mathcal{H}}(\phi) \times (Ent^{\infty} \rightharpoonup Ent^{\infty}) \times A_{\mathcal{H}}(\phi)$ determined by:

$$
\begin{aligned}
(q, D, k) \rightarrow_{\lambda} (q', D', k') \quad \text{iff} \quad & q \rightarrow_{\lambda} q', \\
& \forall \mathsf{X}\psi \in CL(\phi): (\mathsf{X}\psi, \Xi, \Theta) \in D \Leftrightarrow (\psi, \Xi, \lambda \circ \Theta) \in D', \\
& k' = \begin{cases} \min(K(\phi), k + \Omega(\lambda)) & \text{if } \lfloor q' \rfloor \\ 0 & \text{if } \lceil q' \rceil. \end{cases}
\end{aligned}
$$

Note that if $\mathcal{H}$ is finite-state, then $G_{\mathcal{H}}(\phi)$ can be effectively constructed: the set of atoms is finite for every given state. A *path* through a tableau graph is an infinite sequence of states and transitions, starting at an initial state of the HABA and satisfying the acceptance condition of the HABA, such that all "until"-subformulae in any of the atoms are satisfied somewhere further down the sequence.

Definition 5.6. An *allocational path* in $G_{\mathcal{H}}(\phi)$ is an infinite sequence $\pi = (q_0, D_0, k_0)$ λ_0 (q_1, D_1, k_1) λ_1 $\cdots$ such that:

1 $q_0\lambda_0 q_1\lambda_1\cdots \in \mathsf{runs}(\mathcal{H})$;

2 for all $i \geq 0$, $(q_i, D_i, k_i) \rightarrow_{\lambda_i} (q_{i+1}, D_{i+1}, k_{i+1})$;

3 for all $i \geq 0$ and all $(\psi_1 \mathsf{U} \psi_2, \Xi, \Theta) \in D_i$, there exists a $j \geq i$ such that $(\psi_2, \Xi \restriction \psi_2, \lambda_{j-1} \circ \cdots \lambda_i \circ (\Theta \restriction \psi_2)) \in D_j$.

Given an allocational path π in $G_{\mathcal{H}}(\phi)$ of this form, we say that π *fulfills* ϕ if the underlying run $\rho = q_0\lambda_0 q_1\lambda_1\cdots$ generates an allocation triple (σ, N, θ) with a generator $(h_i)_{i\in\mathbb{N}}$ such that $k_0 = \min(K(\phi), \Omega(h_0))$ and $\sigma, N, \theta \models \phi$. If ϕ is clear from the context, we call π a *fulfilling path.* Furthermore, if there exists $(\sigma, N, \theta) \in \mathcal{L}(\mathcal{H})$ such that $\sigma, N, \theta \models \phi$ we say that ϕ is $\mathcal{H}$-satisfiable.

This sets the stage for the main results. We first state the correspondence between the fulfilment of a formula by a path and the presence of that formula in the initial atom of the path. For a partition interpretation Θ let $\overline{\Theta}\colon fv(\phi) \rightharpoonup Ent^{\infty}$ (*flattening* of Θ) be defined as $\overline{\Theta}\colon x \mapsto \Theta(X)$ if $x \in X \in \mathrm{dom}(\Theta)$.

Proposition 5.7. A path π in $G_{\mathcal{H}}(\phi)$ fulfills ϕ if and only if there exists $(\phi, \Xi, \Theta) \in D_0$ (for some Ξ, Θ) such that $I_{\mathcal{H}}(q_0) = \overline{\Theta}$.

Furthermore, there is a correspondence between the satisfiability of a formula in the HABA and the existence of a fulfilling path in the tableau graph.

Proposition 5.8. ϕ is $\mathcal{H}$-satisfiable iff there exists a path in $G_{\mathcal{H}}(\phi)$ that fulfills ϕ.

From now on we can (almost) rely on standard theory (see [14]). The first observation is that a tableau graph can have infinitely many different paths, therefore looking for a fulfilling path for ϕ is still not an effective method for model checking. We need the following definitions.

A subgraph $G' \subseteq G_{\mathcal{H}}(\phi)$ is *self-fulfilling* if every node A in G' has at least an outgoing edge and for every $(\psi_1 \mathsf{U} \psi_2, \Xi, \Theta) \in D_A$ there exists a node $B \in G'$ s.t.

- $A = A_0 \rightarrow_{\lambda_0} A_1 \rightarrow_{\lambda_1} \cdots \rightarrow_{\lambda_{i-2}} A_{i-1} \rightarrow_{\lambda_{i-1}} A_i = B$
- $(\psi_2, \Xi \restriction \psi_2, \lambda_{i-1} \circ \cdots \lambda_0 \circ (\Theta \restriction \psi_2)) \in D_B$.

A *prefix* in $G_{\mathcal{H}}(\phi)$ is a sequence $A_0 \rightarrow_{\lambda_0} A_1 \rightarrow_{\lambda_1} \cdots \rightarrow_{\lambda_{i-2}} A_{i-1} \rightarrow_{\lambda_{i-1}} A_i$ such that A_0 is an initial atom (i.e., $q_{A_0} \in I_{\mathcal{H}}$) and A_i is in a self-fulfilling subgraph.

Let $Inf(\pi)$ denote the set of nodes that appear infinitely often in the path π. $Inf(\pi)$ is a strongly connected subgraph (SCS). We can prove the following implications:

Proposition 5.9. π is a fulfilling path in $G_{\mathcal{H}}(\phi) \Rightarrow Inf(\pi)$ is a self-fulfilling SCS of $G_{\mathcal{H}}(\phi)$.

Proposition 5.10. Let $G' \subseteq G_{\mathcal{H}}(\phi)$ be self-fulfilling SCS such that

- there exists a fulfilling prefix of G' starting at an initial atom A with $(\phi, \Xi, \Theta) \in D_A$ such that $I_{\mathcal{H}}(q_A) = \overline{\Theta}$;
- for all $F \in \mathcal{F}_{\mathcal{H}} : F \cap \{q | (q, D, k) \in G'\} \neq \varnothing$;

Then there exists a path π in $G_{\mathcal{H}}(\phi)$ that fulfils ϕ and such that $Inf(\pi) = G'$.

Finally, we present the main result of the paper:

Theorem 5.11. For any HABA $\mathcal{H}$ and formula ϕ, it is decidable whether or not ϕ is $\mathcal{H}$-satisfiable.

The complexity of the algorithm is double exponential in $|\phi|$, polynomial in $|Q_{\mathcal{H}}|$ (conjecture) and in the largest number of entities (in a state). A detailed analysis can be found in [9].

6. Related and future work

History-dependent automata. History-dependent (HD) automata [16] are the main inspiration for HABAs. An HD-automaton is an automaton where states, transitions and labels are equipped with a set of local names that can be created dynamically. HD-automata represent an adequate model for history-dependent formalisms such as the π-calculus. Reallocation of entities in HABA resembles the reallocation of names in HD-automata. The novelty introduced in HABAs is the black hole abstraction. This key feature allows us to deal with a possibly unbounded number of entities.

Spatial logic. Related to $\mathcal{A}\ell\ell$TL, concerning properties of freshness, is the Spatial Logic (SL) [4, 3]. SL is defined for the Ambient Calculus and has modalities that refer to space as well as time. Freshness can be identified in SL using a special quantifier, and has a somewhat different interpretation than in $\mathcal{A}\ell\ell$TL. In SL "fresh" means distinct from any name used in the formula and in the model satisfying it. If there is a fresh name, there are infinitely many of them. In contrast, in $\mathcal{A}\ell\ell$TL, if an entity is fresh it means that the entity is used in the current state and did not appear previously. This conceptual difference has several consequences. For instance, there exist non-contradictory $\mathcal{A}\ell\ell$TL-formulae where more than one distinct fresh entity is identified in the same state. Another difference between SL and $\mathcal{A}\ell\ell$TL concerns quantification. In SL, quantification is over a fixed (countable) set of names, whereas in $\mathcal{A}\ell\ell$TL, quantification ranges over entities that are alive in the current state. This set is not fixed from state to state. Therefore, e.g., $\forall x.\mathrm{X}\phi$ is not equivalent to $\mathrm{X}\forall x.\phi$.

Tableau-based methods. There are basically two approaches to model-checking temporal logics: the automata-theoretic approach (for LTL [19] and CTL [10, 13]) and the tableau method. Tableaux are typically used for the solution of more general problems, like satisfiability. For model checking, the tableau approach was first developed for CTL [6, 2]. Our algorithm is based on the tableau method for LTL reported in [14].

Model-checking and logics for object-oriented systems. Model-checking tools for object-oriented systems are becoming more and more popular, but the property specification formalisms are not tailored towards properties over objects (such as allocation and de-allocation). Bandera [7] is a model checker for Java that uses abstract interpretation and program slicing to yield compact state spaces. Another model checker for Java is Java PathFinder [12]. JPF employs garbage collection in order to obtain a finite state space. Dynamic creation of objects is only supported to a limited extent (the number of created objects must be bounded). The verification of (only) *safety* properties for systems with an unbounded number of objects is recently reported in [20]. Opposed to our approach which always provides correct answers, this approach may report *false negatives*. Apart from these tool-oriented approaches, several temporal logics for object-oriented systems have been defined [18, 11, 8], that, however, do not support primitives for the birth and death of objects.

Future work. In the future we plan to investigate the use of HABA-like models for the definition of the semantics of more realistic OOP languages. The first step would be the definition of automata where entities can reference each other. Another (long term) open research question that needs further investigation is satisfiability of $\mathcal{A}\ell\ell$TL. Similarly, it would be interesting to develop a proof theory for $\mathcal{A}\ell\ell$TL, as well as, to explore a possible embedding of LTL in $\mathcal{A}\ell\ell$TL.

References

[1] M. Abadi, A. Gordon. A calculus for cryptographic protocols: The spi calculus. *Inf. & Comp.* 148(1): 1-70, 1999.

[2] M. Ben-Ari, A. Pnueli, Z. Manna. The temporal logic of branching time. *Acta Inf.* 20(3):207–226,1983.

[3] L. Caires, L. Cardelli. A spatial logic for concurrency (part I). In *TACS'01*, LNCS 2255:1–37, Springer, 2001.

[4] L. Cardelli, A. Gordon. Logical properties of name restriction. In *TLCA'01*, LNCS 2044:46–60, Springer, 2001.

[5] L. Cardelli, A. Gordon. Mobile ambients. In *FoSSaCS'98*, LNCS 1378:140–155, Springer, 1998.

[6] E. Clarke, E. Emerson. Design and synthesis of synchronization skeletons using branching time temporal logic. In *Workshop on Logics of Programs*, LNCS 131:52–71, Springer, 1981.

[7] J. Corbett, M. Dwyer, J. Hatcliff, C. Pasareanu, Robby, S. Laubach, H. Zheng. Bandera: Extracting finite-state models from Java source code. In *ICSE'00* pp. 439–448, IEEE CS Press, 2000.

[8] D. Distefano, J.-P. Katoen, A. Rensink. On a temporal logic for object-based systems. In *FMOODS'00*, pp. 305–326, Kluwer, 2000.

[9] D. Distefano, A. Rensink J.-P. Katoen. Model checking dynamic allocation and deallocation. Technical report TR-01-40, University of Twente, 2002. Available on line at `http://fmt.cs.utwente.nl/~ddino/papers/DSRK01-report.ps.gz`

[10] E. A. Emerson. Automata, tableaux and temporal logics. In *Logic of Programs*, LNCS 193:79–88, Springer, 1985.

[11] J. Fiadeiro, T. Maibaum. Verifying for reuse: foundations of object-oriented system verification. In *Theory and Formal Methods*, pp. 235–257, 1995.

[12] K. Havelund, T. Pressburger. Model checking Java programs using Java PathFinder. *Int. J. on Software Tools for Technology Transfer*, 2(4):366–381, 2000.

[13] O. Kupferman, M. Y. Vardi, P. Wolper. An automata-theoretic approach to branching-time model checking. *J. of the ACM*, 47(2):312–360, 2000.

[14] O. Lichtenstein, A. Pnueli. Checking that finite state concurrent programs satisfy their linear specification. In *POPL'85*, pp. 97–107, ACM Press, 1985.

[15] R. Milner, J. Parrow, D. Walker. A calculus of mobile processes. *Inf. & Comp.* 100(1):1-77, 1992.

[16] U. Montanari, M. Pistore. An introduction to history-dependent automata. *Electr. Notes in Th. Comp. Sci.*, 10, 1998.

[17] A. Pnueli. The temporal logic of programs. In *FOCS'77*, pp. 46–57, IEEE CS Press, 1977.

[18] A. Sernadas, C. Sernadas, J.F. Costa. Object specification logic. *J. of Logic & Computation*, 5(5):603–630, 1995.

[19] M. Y. Vardi, P. Wolper. An automata-theoretic approach to automatic program verification. In *LICS'86*, pp. 332–344, IEEE CS Press, 1986.

[20] E. Yahav. Verifying safety properties of concurrent Java programs using 3-valued logic. In *POPL 2001*, pp. 27-40 ACM Press, 2001.

PHANTOM TYPES AND SUBTYPING

Matthew Fluet and Riccardo Pucella
Department of Computer Science
Cornell University
{fluet,riccardo}@cs.cornell.edu

Abstract We investigate a technique from the literature, called the phantom types technique, that uses parametric polymorphism, type constraints, and unification of polymorphic types to model a subtyping hierarchy. Hindley-Milner type systems, such as the one found in ML, can be used to enforce the subtyping relation. We show that this technique can be used to encode any finite subtyping hierarchy (including hierarchies arising from multiple interface inheritance). We then formally demonstrate the suitability of the phantom types technique for capturing subtyping by exhibiting a type-preserving translation from a simple calculus with bounded polymorphism to a calculus embodying the type system of ML.

1. Introduction

It is well known that traditional type systems, such as the one found in Standard ML [10], with parametric polymorphism and type constructors can be used to capture program properties beyond those naturally associated with a Hindley-Milner type system [9]. For concreteness, let us review a simple example, due to Leijen and Meijer [8]. Consider a type of atoms, either booleans or integers, that can be easily represented as an algebraic datatype:

```
datatype atom = I of int | B of bool
```

There are a number of operations that we may perform on such atoms (see Figure 1(a)). When the domain of an operation is restricted to only one kind of atom, as with `conj` and `double`, a run-time check must be made and an error or exception reported if the check fails.

One aim of static type checking is to reduce the number of run-time checks by catching type errors at compile time. Of course, in the example above, the ML type system does not consider `conj (mkI 3, mkB true)` to be ill-typed; evaluating this expression will simply raise a run-time exception.

If we were working in a language with subtyping, we would like to consider integer atoms and boolean atoms as distinct subtypes of the general type of atoms and use these subtypes to refine the types of the operations. Then the type system would report a type error in the expression `double (mkB false)` at compile time. Fortunately, we can write the operations in a way that utilizes the ML type system to do just this. We

```
fun mkI (i:int):atom = I (i)
fun mkB (b:bool):atom = B (b)

fun toString (v:atom):string =
  (case v
    of I (i) => Int.toString (i)
     | B (b) => Bool.toString (b))
fun double (v:atom):atom =
  (case v
    of I (i) => I (i * 2)
     | _ => raise Fail "type mismatch")
fun conj (v1:atom,
          v2:atom):atom =
  (case (v1,v2)
    of (B (b1), B (b2)) => B (b1 andalso b2)
     | _ => raise Fail "type mismatch")
```

(a) Unsafe operations

```
fun mkI (i:int):int atom = I (i)
fun mkB (b:bool):bool atom = B (b)

fun toString (v:'a atom):string =
  (case v
    of I (i) => Int.toString (i)
     | B (b) => Bool.toString (b))
fun double (v:int atom):int atom =
  (case v
    of I (i) => I (i * 2)
     | _ => raise Fail "type mismatch")
fun conj (v1:bool atom,
          v2:bool atom):bool atom =
  (case (v1,v2)
    of (B (b1), B (b2)) => B (b1 andalso b2)
     | _ => raise Fail "type mismatch")
```

(b) Safe operations

Figure 1

change the definition of the datatype to the following:

```
datatype 'a atom = I of int | B of bool
```

and constrain the types of the operations (see Figure 1(b)). We use the superfluous type variable in the datatype definition to encode information about the kind of atom. (Because instantiations of this type variable do not contribute to the run-time representation of atoms, it is called a *phantom type*.) The type *int atom* is used to represent integer atoms and *bool atom* is used to represent boolean atoms. Now, the expression `conj (mkI 3, mkB true)` results in a compile-time type error, because the types *int atom* and *bool atom* do not unify. (Observe that our use of `int` and `bool` as phantom types is arbitrary; we could have used any two types that do not unify to make the integer versus boolean distinction.) On the other hand, both `toString (mkI 3)` and `toString (mkB true)` are well-typed; `toString` can be used on any atom. This is the essence of the technique explored in this paper: using a free type variable to encode subtyping information and using an ML-like type system to enforce the subtyping. This "phantom types" technique, where user-defined restrictions are reflected in the constrained types of values and functions, underlies many interesting uses of type systems [14, 12, 2, 13, 6, 8, 5, 11, 1].

The main contributions of this paper are to exhibit a general encoding of subtyping hierarchies and to give one formalization of the use of the phantom types technique. We present a type-preserving translation from a calculus with subtyping to a calculus with let-bounded polymorphism. The kind of subtyping that can be captured turns out to be an interesting variant of bounded polymorphism [3], with a very restricted subsumption rule.

This paper is structured as follows. In the next section, we describe a simple recipe for deriving an interface enforcing a given subtyping hierarchy. The interface is parameterized by an encoding, via phantom types, of the subtyping hierarchy. In Section 3, we focus on a simple encoding for hierarchies. In Section 4, we extend the recipe to capture a limited form of bounded polymorphism. In Section 5, we formally define the kind of subtyping captured by our encodings by giving a simple calculus with subtyping and showing that our encodings provide a type-preserving translation to a variant of the Damas-Milner calculus, embodying the essence of the ML type system. We conclude with some problems inherent to the approach and a consideration of future work. Due to space considerations, proofs of our results, a more involved discussion of the encodings in Section 3, as well as the full typing rules for the formalization in Section 5 have been deferred to the full paper.

2. From subtyping to polymorphism

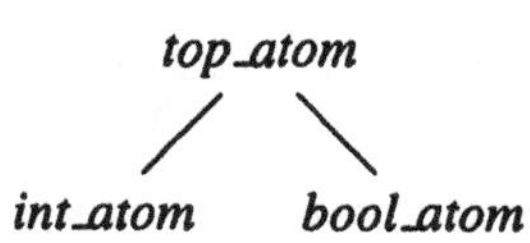

Figure 2

The example in the introduction has the following features: an underlying primitive type of values (the original type *atom*), a set of operations, and "implicit" subtypes that correspond to the sensible domains of the operations. The subtyping hierarchy corresponding to the example is given in Figure 2. The subtyping hierarchy is modeled by assigning a type to every implicit subtype in the hierarchy. For instance, integer atoms with implicit subtype *int_atom* are encoded by the ML type *int atom*. The appropriate use of polymorphic type variables in the type of an operation indicates the maximal type in the domain of the operation. For instance, the operation `toString` has the conceptual type *top_atom* $\rightarrow$ *string* which is encoded by the ML type *'a atom* $\rightarrow$ *string*. The key observation is the use of type unification to enforce the subtyping hierarchy: an *int atom* can be passed to a function expecting an *'a atom*, because these types unify.

We consider the following problem. Given an abstract type τ_p, a subtyping hierarchy, and an implementation of τ_p and its operations, we wish to derive a "safe" ML signature which uses phantom types to encode the subtyping and a "safe" implementation from the "unsafe" implementation. We will call the elements of the subtyping hierarchy *implicit types* and talk about *implicit subtyping* in the hierarchy. All values share the same underlying representation and each operation has a single implementation that acts on this underlying representation. The imposed subtyping captures restrictions that arise because of some external knowledge about the semantics of the operations; intuitively, it captures a "real" subtyping relationship that is not exposed by the abstract type.

We first consider deriving the safe interface. The new interface defines a type $\alpha\ \tau$ corresponding to the abstract type τ_p. The type variable α will be used to encode implicit subtype information. We require an encoding $\langle\sigma\rangle$ of each implicit type σ in the hierarchy; this encoding should yield a type in the underlying ML type system, with the property that $\langle\sigma_1\rangle$ unifies with $\langle\sigma_2\rangle$ if and only if σ_1 is an implicit subtype of σ_2. An obvious issue is that we want to use unification (a symmetric relation) to capture subtyping (an asymmetric relation). The simplest approach is to use two encodings $\langle\cdot\rangle_C$ and $\langle\cdot\rangle_A$ defined over all the implicit types in the hierarchy. A *value* of implicit type σ will be assigned a type $\langle\sigma\rangle_C\ \tau$. We call $\langle\sigma\rangle_C$ the *concrete* subtype encoding of σ, and we assume that it uses only ground types (i.e., no type variables). In order to restrict the domain of an operation to the set of values in any implicit subtype of σ, we use $\langle\sigma\rangle_A$, the *abstract* subtype encoding of σ. In order for the underlying type system to enforce the subtype hierarchy, we require the encodings $\langle\cdot\rangle_C$ and $\langle\cdot\rangle_A$ to be

```
signature ATOM = sig
  type atom
  val int : int -> atom
  val bool : bool -> atom
  val toString : atom -> string
  val double : atom -> atom
  val conj : atom * atom -> atom
end
```

(a) Unsafe signature

```
signature SAFE_ATOM = sig
  type 'a atom
  val int : int -> ⟨int⟩C atom
  val bool : bool -> ⟨bool⟩C atom
  val toString : ⟨top⟩A atom -> string
  val double : ⟨int⟩A atom -> ⟨int⟩C atom
  val conj : ⟨bool⟩A atom * ⟨bool⟩A atom -> ⟨bool⟩C atom
end
```

(b) Safe signature

Figure 3

respectful by satisfying the following property:

$$\text{for all } \sigma_1 \text{ and } \sigma_2,\ \langle\sigma_1\rangle_C \text{ matches } \langle\sigma_2\rangle_A \text{ iff } \sigma_1 \leq \sigma_2.$$

For example, the encodings used in the introduction are respectful:

$$\begin{array}{rclrcl} \langle \mathit{top\_atom}\rangle_A & = & \mathit{'a\ atom} & \langle \mathit{top\_atom}\rangle_C & = & \mathit{unit\ atom} \\ \langle \mathit{int\_atom}\rangle_A & = & \mathit{int\ atom} & \langle \mathit{int\_atom}\rangle_C & = & \mathit{int\ atom} \\ \langle \mathit{bool\_atom}\rangle_A & = & \mathit{bool\ atom} & \langle \mathit{bool\_atom}\rangle_C & = & \mathit{bool\ atom} \end{array}$$

The utility of the phantom types technique relies on being able to find respectful encodings for subtyping hierarchies of interest.

To allow for matching, the abstract subtype encoding will introduce free type variables. Since in a Hindley-Milner type system, a type cannot contain free type variables, the abstract encoding will be part of the larger type scheme of some polymorphic function operating on the value of implicit subtypes. This leads to some restrictions on when we should constrain values by concrete or abstract encodings. We will restrict ourselves to using concrete encodings in all covariant type positions, and using abstract encodings in most contravariant type positions. We will return to this issue in Section 5.

Consider again the example from the introduction. Assume we have encodings $\langle\cdot\rangle_C$ and $\langle\cdot\rangle_A$ for the hierarchy and a structure `Atom` implementing the "unsafe" operations, with the signature given in Figure 3(a). Deriving an interface using the recipe above, we get the safe signature given in Figure 3(b).

We must now derive a corresponding "safe" implementation. We need a type $\alpha\ \tau$ isomorphic to τ_p such that the type system considers $\tau_1\ \tau$ and $\tau_2\ \tau$ equivalent iff τ_1 and

```
structure SafeAtom1 :> SAFE_ATOM = struct
  type 'a atom = Atom.atom
  val int = Atom.int
  val bool = Atom.bool
  val toString = Atom.toString
  val double = Atom.double
  val conj = Atom.conj
end
```

(a) Opaque signature

```
structure SafeAtom2 : SAFE_ATOM = struct
  datatype 'a atom = C of Atom.atom
  fun int (i) = C (Atom.int (i))
  fun bool (b) = C (Atom.bool (b))
  fun toString (C v) = Atom.toString (v)
  fun double (C v) = C (Atom.double (v))
  fun conj (C b1, C b2) = C (Atom.conj (b1,b2))
end
```

(b) Datatype declaration

Figure 4

τ_2 are equivalent. (Note that this requirement precludes the use of type abbreviations of the form `type` $\alpha\ \tau = \tau_p$, which define constant type functions.) We can then constrain the types of values and operations using $\langle\sigma\rangle_C\ \tau$ and $\langle\sigma\rangle_A\ \tau$. In ML, the easiest way to achieve this is to use an abstract type at the module system level, as shown in Figure 4(a). The use of an opaque signature is critical to get the required behavior in terms of type equivalence. The advantage of this method is that there is no overhead.

In a language without abstract types at the module level, another approach is to wrap the primitive type τ_p using a datatype declaration

$$\texttt{datatype 'a}\ \tau = \texttt{C of}\ \tau_p$$

The type $\alpha\ \tau$ behaves as required, because the datatype declaration defines a generative type operator. However, we must explicitly convert primitive values to and from $\alpha\ \tau$ to witness the isomorphism. This yields the implementation given in Figure 4(b).

We should stress that the "safe" interface must ensure that the type $\alpha\ \tau$ is abstract—either through the use of opaque signature matching, or by hiding the value constructors of the type. Otherwise, it may be possible to create values that do not respect the subtyping invariants enforced by the encodings. Similarly, the use of an abstract subtype encoding in a covariant type position can lead to violations in the subtyping invariants.

We now have a way to derive a safe interface and implementation, by adding type information to a generic, unsafe implementation. In the next section, we show how to construct respectful encodings $\langle\cdot\rangle_C$ and $\langle\cdot\rangle_A$ by taking advantage of the structure of the subtyping hierarchy.

3. Encoding subtyping hierarchies

The framework presented in the previous section relies on having concrete and abstract encodings of the implicit subtypes in the subtyping hierarchy with the property that unification of the results of the encoding respects the subtype relation. In this section, we describe one general construction for such encodings.

We first consider a particular lattice that will be useful in our development. Recall that a lattice is a hierarchy where every set of elements has both a least upper bound and a greatest lower bound. Given a finite set S, we let the *powerset lattice* of S be the lattice of all subsets of S, ordered by inclusion, written $(\wp(S), \subseteq)$. We now exhibit an encoding of powerset lattices.

Let n be the cardinality of S and assume an ordering $s_1, \ldots, s_n$ on the elements of S. We encode subset X of S as an n-tuple type, where the ith entry expresses that $s_i \in X$ or $s_i \notin X$. First, we introduce a datatype definition:

```
datatype 'a z = Z
```

(The name of the datatype constructor is irrelevant, because we will never construct values of this type.) The encoding of an arbitrary subset of S is given by:

$$\langle X\rangle_C = (t_1, \ldots, t_n) \quad \text{where } t_i = \begin{cases} \texttt{unit} & \text{if } s_i \in X \\ \texttt{unit z} & \text{otherwise} \end{cases}$$

$$\langle X\rangle_A = (t_1, \ldots, t_n) \quad \text{where } t_i = \begin{cases} \alpha_i & \text{if } s_i \in X \\ \alpha_i\ \texttt{z} & \text{otherwise} \end{cases}$$

Note that $\langle\cdot\rangle_A$ requires every type variabe α_i to be a fresh type variable, unique in its context. This ensures that we do not inadvertently refer to any type variable bound in the context where we are introducing the abstractly encoded type.

As an example, consider the powerset lattice of $\{1,2,3,4\}$, which encodes into a four-tuple. We can verify, for example, that the concrete encoding for $\{2\}$, namely *(unit z, unit, unit z, unit z)*, unifies with the abstract encoding for $\{1,2\}$, namely *(*α_1*,* α_2*,* α_3 *z,* α_4 *z)*. On the other hand, the concrete encoding of $\{1,2\}$ does not unify with the abstract encoding of $\{2,3\}$.

The main reason we introduced powerset lattices is the fact that any finite hierarchy can be embedded in the powerset lattice of a set S. It is a simple matter, given a hierarchy H' embedded in a hierarchy H, to derive an encoding for H' given an encoding for H. Let $inj(\cdot)$ be the injection from H' to H witnessing the embedding and let $\langle\cdot\rangle_{C_H}$ and $\langle\cdot\rangle_{A_H}$ be the encodings for the hierarchy H. Deriving an encoding for H' simply involves defining $\langle\sigma\rangle_{C_{H'}} = \langle inj(\sigma)\rangle_{C_H}$ and $\langle\sigma\rangle_{A_{H'}} = \langle inj(\sigma)\rangle_{A_H}$. It is straightforward to verify that if $\langle\cdot\rangle_{C_H}$ and $\langle\cdot\rangle_{A_H}$ are respectful encodings, so are $\langle\cdot\rangle_{C_{H'}}$ and $\langle\cdot\rangle_{A_{H'}}$. By the result above, this allows us to derive an encoding for an arbitrary finite hierarchy.

We have presented a strategy for obtaining respectful encodings, which is sufficient for the remainder of this paper. However, there are encodings for specific hierarchies that are in general more efficient than their embedding in a powerset lattice, for instance, the encoding for tree hierarchies found in [6]. We discuss such encodings and address the issue of encoding extensibility in the full paper.

4. Towards bounded polymorphism

As mentioned in Section 3, the handling of type variables is somewhat delicate. If we allow common type variables to be used across abstract encodings, then we can capture a form of *bounded polymorphism* as in $F_{<:}$ [3]. Bounded polymorphism *à la* $F_{<:}$ is a typing discipline which extends both parametric polymorphism and subtyping. From parametric polymorphism, it borrows type variables and universal quantification; from subtyping, it allows one to set bounds on quantified type variables. For example, one can guarantee that the argument and return types of a function are the same and a subtype of σ, as in $\forall\alpha \leq \sigma.\alpha \rightarrow \alpha$. Similarly, one can guarantee that two arguments have the same type that is a subtype of σ, as in $\forall\alpha \leq \sigma.(\alpha \times \alpha) \rightarrow \sigma$. Notice that neither function can be written in a language that supports only subtyping.

Returning to the example from the introduction, consider adding natural numbers as a subtype of integers, so that *nat_atom* is a subtype of *int_atom*. Using bounded polymorphism, we can assign to `double` the reasonable type $\forall\alpha \leq int\_atom.\alpha \rightarrow \alpha$. However, bounded polymorphism has its limitations. One reasonable type for a `plus` operation is $\forall\alpha \leq int\_atom.\alpha \times \alpha \rightarrow \alpha$ where the same kind of atom is required for both arguments. In order to add an integer and a natural number we need a function `toInt` (operationally, an identity function) to coerce the type of the natural number to that of an integer.

We can adapt our "recipe" from Section 2 to types of the form $\forall\beta \leq \sigma_1.(\beta \times \sigma_2) \rightarrow \beta$. Let the "safe" interface use types of the form $\alpha\ \tau$. Since β stands for a subtype of σ_1, we let $\phi_\beta = \langle\sigma_1\rangle_A$, the abstract encoding of the bound. We then translate the type as we did in Section 2, but replace occurrences of the type variable β by ϕ_β instead of applying $\langle\cdot\rangle_A$ repeatedly, thereby sharing the type variables introduced by $\langle\sigma_1\rangle_A$. Hence, we get the type $\phi_\beta\ \tau \times \langle\sigma_2\rangle_A\ \tau \to \phi_\beta\ \tau$. In fact, we can further simplify the process by noting that we can "pull out" all the subtyping into bounded polymorphism. If a function expects an argument of any implicit subtype of σ, we can introduce a fresh type variable for that argument and bound it by σ. For example, the type above can be rewritten as: $\forall\beta \leq \sigma_1, \gamma \leq \sigma_2.(\beta \times \gamma) \to \beta$.

Unfortunately, this technique does not generalize to full $F_{<:}$. For example, we cannot encode bounded polymorphism where the bound on a type variable uses a type variable, such as a function `f` with type $\forall\alpha \leq \sigma, \beta \leq \alpha.\alpha \times \beta \to \alpha$. Encoding this type as $\phi_\alpha\ \tau \times \phi_\beta\ \tau \to \phi_\alpha\ \tau$ where $\phi_\alpha = \langle\sigma\rangle_A$ and $\phi_\beta = \langle\alpha\rangle_A$ fails, because we have no definition of $\langle\alpha\rangle_A$. Essentially, we need a different encoding of β for each instantiation of α at each application of `f`, something that cannot be accommodated by a single encoding of the type at the definition of `f`.

Likewise, we cannot encode first-class polymorphism, such as a function g with type $\forall\alpha \leq \sigma_1.\alpha \to (\forall\beta \leq \sigma_2.\beta \to \beta)$. Applying the technique yields a type $\phi_\alpha\ \tau \to \phi_\beta\ \tau \to \phi_\beta\ \tau$ where ϕ_α and ϕ_β contain free type variables. A Hindley-Milner style type system requires quantification over these variables in prenex position, which doesn't match the intuition of the original type. In fact, because we are translating into a language with prenex polymorphism, we can only capture bounded polymorphism that is itself in prenex form.

In other words, we cannot account for the general subsuption rule found in $F_{<:}$. Instead, we require all subtyping to occur at type application. This is the real motivation for the simplification above which "pulls out" all subtyping into bounded polymorphism. By introducing type variables for each argument, we move the resolution of the subtyping to the point of type application (when we instantiate the type variables).

These two restrictions impose one final restriction on the kind of subtyping we can encode. Consider a higher-order function `h` with type $\alpha \leq (\sigma_1 \to \sigma_2).\alpha \to \sigma_2$. What are the possible encodings of the bound $\sigma_1 \to \sigma_2$ that allow subtyping? Clearly encoding the bound as $\langle\sigma_1\rangle_C\ \tau \to \langle\sigma_2\rangle_C\ \tau$ does not allow any subtyping. Encoding the bound as $\langle\sigma_1\rangle_A\ \tau \to \langle\sigma_2\rangle_A\ \tau$ or $\langle\sigma_1\rangle_A\ \tau \to \langle\sigma_2\rangle_C\ \tau$ leads to an unsound system. (Consider applying the argument function to a value of type $\sigma_0 \geq \sigma_1$, which would typecheck in the encoding, because $\langle\sigma_0\rangle_C$ unifies with $\langle\sigma_1\rangle_A$ by the definition of a respectful encoding.) However, we can soundly encode the bound as $\langle\sigma_1\rangle_C\ \tau \to \langle\sigma_2\rangle_A\ \tau$. This corresponds to a subtyping rule on functional types that asserts $\tau_1 \to \tau_2 \leq \tau_1 \to \tau_2'$ iff $\tau_2 \leq \tau_2'$.

Despite these restrictions, the phantom types technique is still a viable method for encoding subtyping in a language like ML. All of the examples of phantom types found in the literature satisfy these restrictions. In practice, one rarely needs first-class polymorphism or complicated dependencies between the subtypes of function arguments, particularly when implementing a safe interface to existing library functions.

5. A formalization

There are subtle issues regarding the kind of subtyping that can be captured using phantom types. In this section, we clarify the picture by exhibiting a typed calculus with a suitable notion of subtyping that can be faithfully translated into a language such as ML, via a phantom types encoding. The idea is simple: to see if an interface can be implemented using phantom types, first express the interface in this calculus in such a way that the program type-checks. If it is possible to do so, our results show that a translation using phantom types exists. The target of the translation is a calculus embodying the essence of ML, essentially the calculus of Damas and Milner [4], a predicative polymorphic λ-calculus.

Let us first introduce the source calculus, $\lambda^{\mathrm{DM}}_{<:}$, also a variant of the Damas-Milner calculus, but with a very restricted notion of subtyping, and allowing multiple types for constants. We assume a partially ordered set $(T, \leq)$ of basic types. The types and prenex quantified type schemes of $\lambda^{\mathrm{DM}}_{<:}$ are as follows:

$$\begin{array}{rcl} \tau & ::= & t \mid \alpha \mid \tau_1 \to \tau_2 \\ \sigma & ::= & \forall \alpha_1 <: \tau_1, \ldots, \alpha_n <: \tau_n . \tau \end{array}$$

(where $t \in T$). Furthermore, we make a syntactic restriction that precludes the use of type variables in the bounds of quantified type variables.

An important aspect of our calculus, at least for our purposes, is the constants that we allow. We distinguish between two types of constants: basic constants and primitive operations. Basic constants, taken from a set C_b, are constants representing values of basic types $t \in T$. We suppose a function $\pi_b : C_b \to T$ assigning a basic type to every basic constant. The primitive operations, taken from a set C_p, are operations acting on constants and returning constants.[1] Rather than giving primitive operations polymorphic types, we assume that the operations can have multiple types, which encode the allowed subtyping. The primitive operation `double` in our example would get the types *int_value* $\to$ *int_value* and *nat_value* $\to$ *nat_value*. We suppose a function π_p assigning to every constant $c \in C_p$ a set of types $\pi_p(c)$, each type a functional type of the form $t \to t'$ (for $t, t' \in T$).

Our expression language is again typical:

$$\begin{array}{rcl} e & ::= & c \mid \lambda x{:}\tau.e \mid e_1\, e_2 \mid x \mid p\,[\tau_1, \ldots, \tau_n] \mid \textbf{let } x = p \textbf{ in } e \\ p & ::= & x \mid \Lambda \alpha_1 <: \tau_1, \ldots, \alpha_n <: \tau_n . e \\ v & ::= & c \mid \lambda x{:}\tau.e \\ E & ::= & [\,] \mid E\, e \mid v\, E \mid E\,[\tau_1, \ldots, \tau_n] \mid \textbf{let } x = E \textbf{ in } e \end{array}$$

(where $c \in C_b \cup C_p$). The operational semantics are given using a standard rewriting system. The basic reductions are

$$\begin{array}{rcl} (\lambda x{:}\tau.e)\, v & \longrightarrow_{<:} & e\{v/x\} \\ (\Lambda \alpha_1 <: \tau_1, \ldots, \alpha_n <: \tau_n.e)\,[\tau_1', \ldots, \tau_n'] & \longrightarrow_{<:} & e\{\tau_1'/\alpha_1, \ldots, \tau_n'/\alpha_n\} \\ \textbf{let } x = v \textbf{ in } e & \longrightarrow_{<:} & e\{v/x\} \\ c_1\, c_2 & \longrightarrow_{<:} & c_3 \quad \text{iff } \delta(c_1, c_2) = c_3 \end{array}$$

where $\delta : C_p \times C_b \rightharpoonup C_p$ is a partial function defining the result of applying a primitive operation to a basic constant. This reduction extends to contexts via the rule:

$$E[e_1] \longrightarrow_{<:} E[e_2] \text{ iff } e_1 \longrightarrow_{<:} e_2$$

[1] For simplicity, we will not deal with higher-order functions here—they would simply complicate the formalism without bringing any new insight. Likewise, allowing primitive operations to act on and return tuples of values is a simple extension of the formalism presented here.

As previously noted, we only allow primitive operations to be monotyped. However, we can easily use the fact that they can take on many types to write polymorphic wrappers. Returning to the `double` example, we can write a polymorphic wrapper $\Lambda\alpha <: \mathit{int\_value}.\lambda x : \alpha.$`double` x to capture the expected behavior. We will see shortly that this function is well-typed.

The typing rules for $\lambda^{\mathrm{DM}}_{<:}$ are the standard Damas-Milner typing rules, modified to account for subtyping. Subtyping is given by a judgment $\Delta \vdash_{<:} \tau_1 <: \tau_2$, and is derived from the subtyping on the basic types. The interesting rules are:

$$\frac{t_1 \leq t_2}{\Delta \vdash_{<:} t_1 <: t_2} \qquad \frac{\Delta \vdash_{<:} \tau_2 <: \tau_2'}{\Delta \vdash_{<:} \tau_1 \rightarrow \tau_2 <: \tau_1 \rightarrow \tau_2'}$$

Notice that subtyping at higher types only involves the result type. The typing rules are given by judgments $\Delta; \Gamma \vdash_{<:} e : \tau$ for monotypes and $\Delta; \Gamma \vdash_{<:} p : \sigma$ for type schemes. The rule for primitive operations is interesting:

$$\frac{\forall \tau_1' <: \tau_1, \ldots, \tau_n' <: \tau_n \quad \tau\{\tau_1'/\alpha_1, \ldots, \tau_n'/\alpha_n\} \in \pi_p(c)}{\Delta, \alpha_1 <: \tau_1, \ldots, \alpha_n <: \tau_n; \Gamma \vdash_{<:} c : \tau} \quad \left(\begin{array}{c} c \in C_p, \\ FV(\tau) = \langle \alpha_1, \ldots, \alpha_n \rangle \end{array} \right)$$

The syntactic restriction on type variable bounds ensures that each τ_i has no type variables, so each $\tau_i' <: \tau_i$ is well-defined. The rule captures the notion that any subtyping on a primitive operation through the use of bounded polymorphism is in fact realized by the "many types" interpretation of the operation.

Subtyping occurs at type application:

$$\frac{\Delta; \Gamma \vdash_{<:} p : \forall \alpha_1 <: \tau_1, \ldots, \alpha_n <: \tau_n.\tau \quad \Delta \vdash_{<:} \tau_1' <: \tau_1 \quad \cdots \quad \Delta \vdash_{<:} \tau_n' <: \tau_n}{\Delta; \Gamma \vdash_{<:} p\,[\tau_1', \ldots, \tau_n'] : \tau\{\tau_1'/\alpha_1, \ldots, \tau_n'/\alpha_n\}}$$

As discussed in the previous section, there is no subsumption in the system: subtyping must be witnessed by type application. Hence, there is a difference between the type $t_1 \rightarrow t_2$ (where $t_1, t_2 \in T$) and $\forall \alpha <: t_1.\alpha \rightarrow t_2$; namely, the former does not allow any subtyping. The restrictions of Section 4 are formalized by prenex quantification and the syntactic restriction on type variable bounds.

Clearly, type soundness of the above system depends on the definition of δ over the constants. We say that π_p is sound with respect to δ if for all $c_1 \in C_p$ and $c_2 \in C_b$, we have $\vdash_{<:} c_1\ c_2 : \tau$ implies that $\delta(c_1, c_2)$ is defined and $\pi_b(\delta(c_1, c_2)) = \tau$. This definition ensures that any application of a primitive operation c_1 to a basic constant c_2 yields exactly one value $\delta(c_1, c_2)$ at exactly one type $\pi_b(\delta(c_1, c_2)) = \tau$. This leads to the following conditional type soundness result for $\lambda^{\mathrm{DM}}_{<:}$:

Theorem 1 *If π_p is sound with respect to δ, $\vdash_{<:} e : \tau$, and $e \longrightarrow_{<:} e'$, then $\vdash_{<:} e' : \tau$ and either e' is a value or there exists e'' such that $e' \longrightarrow_{<:} e''$.*

Our target calculus, $\lambda^{\mathrm{DM}}_{\mathsf{T}}$, is meant to capture the appropriate aspects of ML that are relevant for the phantom types encoding of subtyping. Essentially, it is the Damas-Milner calculus [4] extended with a single type constructor T. Formally,

$$\begin{array}{lcl}
\tau & ::= & \alpha \mid \tau_1 \rightarrow \tau_2 \mid \mathsf{T}\,\tau \mid 1 \mid \tau_1 \times \tau_2 \\
\sigma & ::= & \forall \alpha_1, \ldots, \alpha_n.\tau \\
e & ::= & c \mid \lambda x : \tau.e \mid e_1\, e_2 \mid p[\tau_1, \ldots, \tau_n] \mid x \mid \mathbf{let}\ x = p\ \mathbf{in}\ e \\
v & ::= & c \mid \lambda x : \tau.e \\
p & ::= & x \mid \Lambda \alpha_1 \ldots, \alpha_n.e \\
E & ::= & [\,] \mid E\,e \mid v\,E \mid E\,[\tau_1, \ldots, \tau_n] \mid \mathbf{let}\ x = E\ \mathbf{in}\ e
\end{array}$$

The operational semantics (via a reduction relation $\longrightarrow_{\mathsf{T}}$) and most typing rules (via a judgment $\Delta;\Gamma \vdash_{\mathsf{T}} e : \tau$) are standard. As before, we assume that we have constants C_b and C_p and a function δ providing semantics for primitive applications. Likewise, we assume that π_b and π_p provide types for constants, with the same restrictions. The typing rule for primitive operations in $\lambda_{\mathsf{T}}^{\mathsf{DM}}$ is similar to the corresponding rule in $\lambda_{<:}^{\mathsf{DM}}$. Given two types τ and τ' in $\lambda_{\mathsf{T}}^{\mathsf{DM}}$, we define their unification $\mathit{unify}(\tau,\tau')$ to be a sequence of bindings $\langle(\alpha_1,\tau_1),(\alpha_2,\tau_2),\ldots\rangle$ in depth-first, left-to-right order of appearance of $\alpha_1,\ldots,\alpha_n$ in τ, or $\emptyset$ if τ' is not a substitution instance of τ. Given a type τ in $\lambda_{\mathsf{T}}^{\mathsf{DM}}$, we define $FV(\tau)$ to be the sequence of free type variables appearing in τ, in depth-first, left-to-right order.

$$\frac{\begin{array}{c}\forall\tau' \in \pi_b(C_b) \text{ with } \mathit{unify}(\tau_1,\tau') = \langle(\alpha_1,\tau_1'),\ldots,(\alpha_n,\tau_n'),\ldots\rangle \\ (\tau_1 \to \tau_2)\{\tau_1'/\alpha_1,\ldots,\tau_n'/\alpha_n\} \in \pi_p(c)\end{array}}{\Delta,\alpha_1,\ldots,\alpha_n;\Gamma \vdash_{\mathsf{T}} c : \tau_1 \to \tau_2} \quad \left(\begin{array}{c} c \in C_p, \\ FV(\tau_1 \to \tau_2) = \\ \langle\alpha_1,\ldots,\alpha_n\rangle \end{array}\right)$$

Again, this rule captures our notion of "subtyping through unification" by ensuring that the operation is defined at every basic type that unifies with its argument type. Our notion of soundness of π_p with respect to δ carries over and we can again establish a conditional type soundness result:

Theorem 2 *If π_p is sound with respect to δ, $\vdash_{\mathsf{T}} e : \tau$, and $e \longrightarrow_{\mathsf{T}} e'$, then $\vdash_{\mathsf{T}} e' : \tau$ and either e' is a value or there exists e'' such that $e' \longrightarrow_{\mathsf{T}} e''$.*

Note that the types $\mathsf{T}\ \tau$, 1, and $\tau_1 \times \tau_2$ have no corresponding introduction and elimination expressions. We include these types for the exclusive purpose of constructing the phantom types used by the encodings. We could add other types to allow more encodings, but these suffice for the lattice encodings of Section 3.

Thus far, we have a calculus $\lambda_{<:}^{\mathsf{DM}}$ embodying the notion of subtyping that interests us and a calculus $\lambda_{\mathsf{T}}^{\mathsf{DM}}$ capturing the essence of the ML type system. We now establish a translation from the first calculus into the second using phantom types to encode the subtyping, showing that we can indeed capture that particular notion of subtyping in ML. Moreover, we show that the translation preserves the soundness of the types assigned to constants, thereby guaranteeing that if the original system was sound, the system obtained by translation is sound as well.

We first describe how to translate types in $\lambda_{<:}^{\mathsf{DM}}$. Since subtyping is only witnessed at type abstraction, the type translation realizes the subtyping using the phantom types encoding of abstract and concrete subtypes. The translation is parameterized by an environment ρ associating every (free) type variable with a type in $\lambda_{\mathsf{T}}^{\mathsf{DM}}$ representing the abstract encoding of the bound.

$$\begin{array}{rcl}
\mathcal{T}[\![\alpha]\!]\rho & = & \rho(\alpha) \\
\mathcal{T}[\![t]\!]\rho & = & \mathsf{T}\ \langle t\rangle_C \\
\mathcal{T}[\![\tau_1 \to \tau_2]\!]\rho & = & \mathcal{T}[\![\tau_1]\!]\rho \to \mathcal{T}[\![\tau_2]\!]\rho \\
\mathcal{T}[\![\forall\alpha_1 <: \tau_1,\ldots,\alpha_n <: \tau_n.\tau]\!]\rho & = & \\
\multicolumn{3}{r}{\forall\alpha_{11},\ldots,\alpha_{1k_1},\ldots,\alpha_{n1},\ldots,\alpha_{nk_n}.\mathcal{T}[\![\tau]\!]\rho[\alpha_i \mapsto \tau_i^A]} \\
\multicolumn{3}{r}{\text{where } \tau_i^A = \mathcal{A}[\![\tau_i]\!] \text{ and } FV(\tau_i^A) = \langle\alpha_{i1},\ldots,\alpha_{ik_i}\rangle}
\end{array}$$

If ρ is empty, we will simple write $\mathcal{T}[\![\tau]\!]$. To compute the abstract and concrete encodings of a type, we define:

$$\begin{array}{rclcrcl}
\mathcal{A}[\![t]\!] & = & \mathsf{T}\ \langle t\rangle_A & \quad & \mathcal{C}[\![t]\!] & = & \mathsf{T}\ \langle t\rangle_C \\
\mathcal{A}[\![\tau_1 \to \tau_2]\!] & = & \mathcal{C}[\![\tau_1]\!] \to \mathcal{A}[\![\tau_2]\!] & & \mathcal{C}[\![\tau_1 \to \tau_2]\!] & = & \mathcal{C}[\![\tau_1]\!] \to \mathcal{C}[\![\tau_2]\!]
\end{array}$$

Note that the syntactic restriction on type variable bounds ensures that $\mathcal{A}$ and $\mathcal{C}$ are always well-defined, as they will never be applied to type variables. Furthermore, observe that the above translation depends on the fact that the type encodings $\langle t\rangle_C$ and $\langle t\rangle_A$ are expressible in the $\lambda_{\mathsf{T}}^{\mathrm{DM}}$ type system using $\top$, 1, and $\times$.

We extend the type transformation $\mathcal{T}$ to type contexts Γ in the obvious way:

$$\mathcal{T}[x_1 : \tau_1, \ldots, x_n : \tau_n]\rho \quad = \quad x_1 : \mathcal{T}[\tau_1]\rho, \ldots, x_n : \mathcal{T}[\tau_n]\rho$$

Finally, if we take the basic constants and the primitive operations in $\lambda_{<:}^{\mathrm{DM}}$ and assume that π_p is sound with respect to δ, then the translation can be used to assign types to the constants and operations such that they are sound in the target calculus. We first extend the definition of $\mathcal{T}$ to π_b and π_p in the obvious way:

$$\begin{array}{rcl} \mathcal{T}[\pi_b] & = & \pi_b' \text{ where } \pi_b'(c) = \mathcal{T}[\pi_b(c)] \\ \mathcal{T}[\pi_p] & = & \pi_p' \text{ where } \pi_p'(c) = \{\mathcal{T}[\tau] \mid \tau \in \pi_p(c)\} \end{array}$$

We can further show that the translated types do not allow us to "misuse" the constants in $\lambda_{\mathsf{T}}^{\mathrm{DM}}$:

Theorem 3 *If π_p is sound with respect to δ in $\lambda_{<:}^{\mathrm{DM}}$, then $\mathcal{T}[\![\pi_p]\!]$ is sound with respect to δ in $\lambda_{\mathsf{T}}^{\mathrm{DM}}$.*

We can now define the translation of expressions via a translation of typing derivations, $\mathcal{E}$, taking care to respect the types given by the above type translation. We note that the translation below only works if the concrete encodings being used do not contain free type variables. Again, the translation is parameterized by an environment ρ, as in the type translation.

$$\begin{array}{rcl} \mathcal{E}[\Delta; \Gamma \vdash_{<:} x : \tau]\rho & = & x \\ \mathcal{E}[\Delta; \Gamma \vdash_{<:} c : \tau]\rho & = & c \\ \mathcal{E}[\Delta; \Gamma \vdash_{<:} \lambda x{:}\tau'.e : \tau]\rho & = & \lambda x{:}\mathcal{T}[\tau]\rho.\mathcal{E}[e]\rho \\ \mathcal{E}[\Delta; \Gamma \vdash_{<:} e_1\, e_2 : \tau]\rho & = & (\mathcal{E}[e_1]\rho)\, \mathcal{E}[e_2]\rho \\ \mathcal{E}[\Delta; \Gamma \vdash_{<:} \mathbf{let}\ x = p\ \mathbf{in}\ e : \tau]\rho & = & \mathbf{let}\ x = \mathcal{E}[p]\rho\ \mathbf{in}\ \mathcal{E}[e]\rho \\ \mathcal{E}[\Delta; \Gamma \vdash_{<:} p[\tau_1, \ldots, \tau_n] : \tau]\rho & = & \end{array}$$

$$(\mathcal{E}[p]\rho)[\tau_{11}, \cdots, \tau_{1k_1}, \cdots, \tau_{n1}, \cdots, \tau_{nk_n}]$$
$$\text{where } \mathcal{B}[p]\Gamma = (\alpha_1, \tau_1^B), \ldots, (\alpha_n, \tau_n^B)\rangle \text{ and } \tau_i^A = \mathcal{A}[\tau_i^B]$$
$$\text{and } FV(\tau_i^B) = \langle \alpha_{i1}, \ldots, \alpha_{ik_i}\rangle \text{ and } \tau_i^T = \mathcal{T}[\tau_i]\rho$$
$$\text{and } \mathit{unify}(\tau_i^A, \tau_i^T) = \langle(\alpha_{i1}, \tau_{i1}), \ldots, (\alpha_{ik_i}, \tau_{ik_i}), \ldots\rangle$$

$$\begin{array}{rcl} \mathcal{E}[\Delta; \Gamma \vdash_{<:} x : \sigma]\rho & = & x \\ \mathcal{E}[\Delta; \Gamma \vdash_{<:} \Lambda\alpha_1 <: \tau_1, \ldots, \alpha_n <: \tau_n.e : \sigma]\rho & = & \end{array}$$

$$\Lambda\alpha_{11}, \ldots, \alpha_{1k_1}, \ldots, \alpha_{n1}, \ldots, \alpha_{nk_n}.\mathcal{E}[e]\rho[\alpha_i \mapsto \tau_i^A]$$
$$\text{where } \tau_i^A = \mathcal{A}[\tau_i] \text{ and } FV(\tau_i^A) = \langle \alpha_{i1}, \ldots, \alpha_{ik_i}\rangle$$

Again, if ρ is empty, we simply write $\mathcal{E}[\![e]\!]$. The function $\mathcal{B}$ returns the bounds of a type abstraction, using the environment Γ to resolve variables. It is defined as follows:

$$\begin{array}{rcl} \mathcal{B}[x]\Gamma & = & \langle(\alpha_1, \tau_1), \ldots, (\alpha_n, \tau_n)\rangle \\ & & \text{where } \Gamma(x) = \forall\alpha_1 <: \tau_1, \ldots, \alpha_n <: \tau_n.\tau \\ \mathcal{B}[\Lambda\alpha_1 <: \tau_1, \ldots, \alpha_n <: \tau_n.e]\Gamma & = & \langle(\alpha_1, \tau_1), \ldots, (\alpha_n, \tau_n)\rangle \end{array}$$

We use $\mathcal{B}$ and *unify* to perform unification "by hand." In most programming languages, type inference performs this automatically.

We can verify that this translation is type-preserving:

Theorem 4 *If* $\vdash_{<:} e : \tau$, *then* $\vdash_{\mathsf{T}} \mathcal{E}[\![\vdash_{<:} e : \tau]\!] : \mathcal{T}[\![\tau]\!]$.

Theorem 4 is interesting in that it shows that the translation, in a sense, captures the right notion of subtyping, particularly when designing an interface. Given a set of constants making up the interface, suppose we can assign types to those constants in $\lambda^{\mathsf{DM}}_{<:}$ in a way that gives the desired subtyping; that is, we can write type correct expressions of the form $\Lambda\alpha <: t.\lambda x : \alpha.c\ x$ with type $\forall\alpha <: t.\alpha \rightarrow \tau$. In other words, the typing π_p is sound with respect to the semantics of δ. By Theorem 1, this means that $\lambda^{\mathsf{DM}}_{<:}$ with these constants is sound and we can safely use these constants in $\lambda^{\mathsf{DM}}_{<:}$. In particular, we can write the program:

$$
\begin{array}{l}
\textbf{let } f_1 = \Lambda\alpha <: t_{i_1}.\lambda x : \alpha.c_1\ x \textbf{ in} \\
\quad\vdots \\
\textbf{let } f_n = \Lambda\alpha <: t_{i_n}.\lambda x : \alpha.c_n\ x \textbf{ in} \\
e
\end{array}
$$

By Theorem 4, the translation of the above program executes without run-time errors. Furthermore, by Theorem 3, the phantom types encoding of the types of these constants are sound with respect to δ in $\lambda^{\mathsf{DM}}_{\mathsf{T}}$. Hence, by Theorem 2, $\lambda^{\mathsf{DM}}_{\mathsf{T}}$ with these constants is sound and we can safely use these constants in $\lambda^{\mathsf{DM}}_{\mathsf{T}}$. Therefore, we can replace to the body of the translated program with an arbitrary $\lambda^{\mathsf{DM}}_{\mathsf{T}}$ expression that type-checks in that context and the resulting program will still execute without run-time errors. Essentially, the translation of the let bindings corresponds to a "safe" interface to the primitives; programs that use this interface in a type-safe manner are guaranteed to execute without run-time errors.

6. Conclusion

Essentially, the phantom types technique uses the definition of type equivalence in ML to encode information in a free type variable of a type. Unification can then be used to enforce a particular structure on the information carried by two such types. In this paper, we have focused on encoding subtyping information. We also showed how to extend the techniques we developed to encode a form of prenex bounded polymorphism, with subsumption occuring only at type application. It goes without saying that this approach to encoding subtyping is not without its problems from a practical point of view. As the encodings in this paper show, the types involved can become quite large. Type abbreviations can help simplify the presentation of concrete types, but for abstract encodings, which require type variables, those type variables must appear in the interface.

We also note that the source language of Section 5 provides only a lower bound on the power of phantom types. For example, one can use features of the *specific* encoding used to further constrain operations [13]. One can also capture programming invariants associated with user-defined datatypes [7]. An interesting direction for future work is formalizing these additional applications of the phantom types technique.

Acknowledgments

We have benefitted from discussions with Greg Morrisett and Dave MacQueen. John Reppy pointed out the work of Burton. Stephanie Weirich, Vicky Weissman, and Steve Zdancewic provided helpful comments on an early draft of this paper. Thanks also to the anonymous referees. The second author was partially supported by ONR grant N00014-00-1-03-41.

References

[1] M. Blume. No-Longer-Foreign: Teaching an ML compiler to speak C "natively". In *Electronic Notes in Theoretical Computer Science*, volume 59. Elsevier Science Publishers, 2001.

[2] F. Burton. Type extension through polymorphism. *ACM Transactions on Programming Languages and Systems*, 12(1):135–138, January 1990.

[3] L. Cardelli, S. Martini, J. C. Mitchell, and A. Scedrov. An extension of System F with subtyping. *Information and Computation*, 109(1–2):4–56, 1994.

[4] L. Damas and R. Milner. Principal type-schemes for functional programs. In *Conference Record of the Ninth Annual ACM Symposium on Principles of Programming Languages*, pages 207–212. ACM Press, 1982.

[5] C. Elliott, S. Finne, and O. de Moor. Compiling embedded languages. In *Workshop on Semantics, Applications, and Implementation of Program Generation*, 2000.

[6] S. Finne, D. Leijen, E. Meijer, and S. Peyton Jones. Calling hell from heaven and heaven from hell. In *Proceedings of the 1999 ACM SIGPLAN International Conference on Functional Programming*, pages 114–125. ACM Press, 1999.

[7] S. Kahrs. Red-black trees with types. *Journal of Functional Programming*, 11(3):425–432, 2001.

[8] D. Leijen and E. Meijer. Domain specific embedded compilers. In *Proceedings of the Second Conference on Domain-Specific Languages (DSL'99)*, pages 109–122, 1999.

[9] R. Milner. A theory of type polymorphism in programming. *Journal of Computer and Systems Sciences*, 17(3):348–375, 1978.

[10] R. Milner, M. Tofte, R. Harper, and D. MacQueen. *The Definition of Standard ML (Revised)*. The MIT Press, Cambridge, Mass., 1997.

[11] F. Pessaux and X. Leroy. Type-based analysis of uncaught exceptions. In *Conference Record of the Twenty-Sixth Annual ACM Symposium on Principles of Programming Languages*, pages 276–290. ACM Press, 1999.

[12] D. Rémy. Records and variants as a natural extension of ML. In *Conference Record of the Sixteenth Annual ACM Symposium on Principles of Programming Languages*, pages 77–88. ACM Press, 1989.

[13] J. H. Reppy. A safe interface to sockets. Technical memorandum, AT&T Bell Laboratories, 1996.

[14] M. Wand. Complete type inference for simple objects. In *Proceedings of the 2nd Annual IEEE Symposium on Logic in Computer Science*, 1987.

ON THE WEAKEST FAILURE DETECTOR FOR NON-BLOCKING ATOMIC COMMIT*

Rachid Guerraoui
Petr Kouznetsov
Distributed Programming Laboratory
Swiss Federal Institute of Technology in Lausanne

Abstract This paper addresses the question of the weakest failure detector to solve the Non-Blocking Atomic Commit (NBAC) problem in an asynchronous system. We define the set $\mathcal{A}$ of timeless failure detectors which excludes failure detectors that provide information about global time but includes all known meaningful failure detectors such as $\diamond\mathcal{S}$, $\diamond\mathcal{P}$ and $\mathcal{P}$ [2]. We show that, within $\mathcal{A}$, the weakest failure detector for NBAC is $?\mathcal{P} + \diamond\mathcal{S}$.

1. Introduction

Problem. *Non-Blocking Atomic Commit (NBAC)* is a typical *agreement* problem in distributed computing [9]. To ensure the atomicity of a distributed transaction, the processes must agree on a common outcome: *commit* or *abort*. Every process that does not crash during the execution of the algorithm (i.e., a correct process), should eventually decide on an outcome without waiting for crashed processes to recover.

More precisely, the NBAC problem consists for a set of processes to reach a common *decision*, *commit* or *abort*, according to some initial votes of the processes, *yes* or *no*, such that the following properties are satisfied: *(1) Agreement:* No two processes decide differently; *(2) Termination:* Every correct process eventually decides; *(3) A-Validity: Abort* is the only possible decision if some process votes *no*; and *(4) C-Validity: Commit* is the only possible decision if all processes are correct and vote *yes*.

In this paper, we discuss the solvability of the problem in a *crash-stop asynchronous message-passing* model of distributed computing. Informally, the model is one in which processes exchange messages through reliable commu-

*This work is partially supported by the Swiss National Science Foundation (project number 510-207).

nication channels, processes can fail by crashing, and there are no bounds on message transmission time and relative processor speeds.

Background. It is well-known that many fundamental agreement problems in distributed computing, in particular, the well-known Consensus problem, cannot be solved deterministically in an asynchronous system that is subject to even a single crash failure [3]. In Consensus, the processes need to decide on one out of two values, 0 or 1, based on proposed values, 0 or 1, so that, in addition to the *Agreement* and *Termination* properties of NBAC, the following *Validity* property holds: A value decided must be a value proposed.

To circumvent the impossibility of Consensus, Chandra and Toueg [2] introduced the notion of *failure detector*. Informally, a failure detector is a distributed oracle that gives (possibly incorrect) hints about the crashes of processes. Each process has access to a local *failure detector module* that monitors other processes in the system. In [2], it is shown that a rather weak failure detector $\diamond\mathcal{S}$, called the *eventually strong failure detector*, is sufficient to solve Consensus in an asynchronous system with a majority of correct processes. It is shown in [1] that and failure detector that solves Consensus can emulate $\diamond\mathcal{S}$: hence $\diamond\mathcal{S}$ is *the weakest* failure detector to solve the problem. In other words, $\diamond\mathcal{S}$ encapsulates the exact information about failures needed to solve Consensus in a system with a majority of correct processes.

Like Consensus, NBAC does not admit a deterministic solution in an asynchronous system even in the face of a single failure. In this paper we focus on the question of the weakest failure detector to solve NBAC.

Conjecture: $?\mathcal{P} + \diamond\mathcal{S}$. Guerraoui introduced in [6] the *anonymously perfect* failure detector $?\mathcal{P}$ and showed that $?\mathcal{P}$ is *necessary* to solve NBAC. Each module of $?\mathcal{P}$ at a given process outputs either the empty set or the identifier of the process. When the failure detector module of $?\mathcal{P}$ at a process p_i outputs p_i, we say that p_i *detects a crash*. $?\mathcal{P}$ satisfies the following properties: *Anonymous Completeness:* If some process crashes, then there is a time after which every correct process permanently detects a crash, and *Anonymous Accuracy:* No crash is detected unless some process crashes.

In other words, $?\mathcal{P}$ correctly detects that *some* process has crashed, but does not tell *which* process has actually crashed. An algorithm that transforms Consensus into NBAC using $?\mathcal{P}$ is presented in [6]. Since $\diamond\mathcal{S}$ is sufficient to solve Consensus in a system with a majority of correct processes, $?\mathcal{P} + \diamond\mathcal{S}$ is sufficient to solve NBAC in this environment. It is also shown in [6] that $?\mathcal{P} + \diamond\mathcal{S}$ is strictly weaker than the Perfect failure detector $\mathcal{P}$. Evidently, $\diamond\mathcal{S}$ is strictly weaker than $?\mathcal{P} + \diamond\mathcal{S}$.

The conjecture we want to prove is that $?\mathcal{P} + \diamond\mathcal{S}$ is the weakest failure detector to solve NBAC (with a majority of correct processes). To show this, we need to prove that any algorithm that solves NBAC can be used to emulate $\diamond\mathcal{S}$.

Assumptions. If we consider the overall universe of failure detectors defined in [2], $\diamond\mathcal{S}$ is not necessary to solve NBAC (i.e., our conjecture is not true). Indeed, Guerraoui introduced in [6] a *stillborn* failure detector, denoted by $\mathcal{B}$, that solves NBAC and cannot be transformed into $?\mathcal{P} + \diamond\mathcal{S}$ [6]. More precisely, $\mathcal{B}$ ensures that every initial crash is immediately detected by every process p_i, so that p_i can safely decide *abort* without synchronizing with others processes.

More generally, any failure detector that tells *the time* when a failure occurred can solve NBAC without employing Consensus. Consider a failure detector $\mathcal{B}[\alpha]$, such that at each process p_i, $\mathcal{B}[\alpha]$ outputs a singleton $\top$ until some time t_i. At time t_i, if some process is crashed at time α, the failure detector module outputs p_i. Otherwise, after t_i, $\mathcal{B}[\alpha]$ behaves like $\mathcal{P}$ (the perfect failure detector). It can be easily shown that $\mathcal{B}[\alpha]$ is not transformable into $\diamond\mathcal{S}$, although it solves NBAC as follows: each process p_i decides *abort* whenever its failure detector module outputs p_i instead of $\top$, otherwise p_i runs the 3PC algorithm [9]. However, $\mathcal{B}[\alpha]$ reports the exact time when a failure occurred, which can be provided only through the global time source. We question ourselves what happens if we rule out *time-based* failure detectors like $\mathcal{B}[\alpha]$: among the remaining failure detectors, is $?\mathcal{P} + \diamond\mathcal{S}$ indeed the weakest to solve NBAC?

Contributions. This paper shows that the answer is "yes", i.e., our initial conjecture is true under the assumption that failure detectors do not reveal time.

We define a new class $\mathcal{A}$ of *timeless* failure detectors (restricting the original universe of failure detectors of [2]) that excludes time-based failure detectors like $\mathcal{B}[\alpha]$, but includes all known failure detectors like $\mathcal{P}$, $\diamond\mathcal{S}$ and $?\mathcal{P}$. Informally, a timeless failure detector module is not able to provide information about *when* exactly (in the sense of the global time) failures have occurred.

We show that in $\mathcal{A}$, $\diamond\mathcal{S}$ is *necessary* to solve NBAC. That is, any failure detector of $\mathcal{A}$ that solves NBAC can emulate $\diamond\mathcal{S}$. To show this, we extend the technique used in [1] to prove that $\diamond\mathcal{S}$ is necessary to solve Consensus [1]. This extension is not trivial. Given that no information about time can be provided by a timeless failure detector, for any execution scenario, we construct an *imaginary* run that helps eventually deduce valuable information about correct processes in the system and emulate $\diamond\mathcal{S}$.

$?\mathcal{P}+\diamond\mathcal{S}$ is shown to be the weakest failure detector within $\mathcal{A}$ to solve NBAC with a majority of correct processes. As a corollary of our result, we show that in a system equipped with timeless failure detectors, NBAC is strictly harder than Consensus. Roughly speaking, in the class $\mathcal{A}$ of timeless failure detectors, the difference between the problems is exactly captured by $?\mathcal{P}$.

Roadmap. Section 2 defines our system model. Section 3 presents the class $\mathcal{A}$ of timeless failure detectors. Section 4 gives a brief reminder of the technique of [1] and discusses its applicability to the NBAC problem. Section 5 proves formally that, within $\mathcal{A}$, $\diamond\mathcal{S}$ is necessary to solve NBAC. Section 5 draws as corollaries that $\diamond\mathcal{S}+?\mathcal{P}$ is the weakest failure detector to solve NBAC and that NBAC is strictly harder than Consensus in $\mathcal{A}$. Section 6 concludes the paper by discussing some related work.

2. Model

We consider in this paper a crash-prone asynchronous message passing model augmented with the failure detector abstraction. We recall here what in the model is needed to state and prove our results. More details on the model can be found in [2].

System. We assume the existence of a global clock to simplify the presentation. This is actually a fictional device: the processes do not have *direct* access to it (timing assumptions are captured within failure detectors). We take the range $\mathcal{T}$ of the clock output values to be the positive real axis: $[0, +\infty)$. We consider an *asynchronous* distributed system in which there is no time bound on message delay, clock drift, or the time necessary to execute a step [3]. The system consists of a set of n processes $\Pi = \{p_1, .., p_n\}(n > 0)$. Every pair of processes is connected by a reliable communication channel.

Failures and failure patterns. Processes are subject to *crash* failures. A *failure pattern* F is a function from the global time range $\mathcal{T}$ to 2^Π, where $F(t)$ denotes the set of processes that have crashed by time t. Once a process crashes, it does not recover, i.e., $\forall t < t' : F(t) \subseteq F(t')$. We define $correct(F) = \Pi - \cup_{t\in\mathcal{T}} F(t)$ to be the set of *correct* processes. A process $p \notin F(t)$ is said to be *up* at time t. A process $p \in F(t)$ is said to be *crashed* (or *incorrect*) at time t. We do not consider Byzantine failures: a process either correctly executes the algorithm assigned to it, or crashes and stops executing any action forever. An *environment* $\mathcal{E}$ is a set of possible failure patterns. By default, we consider here environments of the form $\mathcal{E}_f$ in which up to f processes can fail. We assume that there is at least one correct process: $0 \leq f < n$.

Failure detectors. A *failure detector history H with range $\mathcal{R}$* is a function from $\Pi \times \mathcal{T}$ to $\mathcal{R}$. $H(p,t)$ is the output of the failure detector module of process p at time t. A *failure detector* $\mathcal{D}$ is a function that maps each failure pattern F to a *set* of failure detector histories $\mathcal{D}(F)$ with range $\mathcal{R}_\mathcal{D}$. $\mathcal{D}(F)$ denotes the set of possible failure detector histories permitted by $\mathcal{D}$ for the failure pattern F.

Every process p_i has a failure detector module $\mathcal{D}_i$ that p_i queries to obtain information about the failures in the system. Typically, this information includes the set of processes that a process currently suspects to have crashed.[1]

[1]In [1], failure detectors can output values from an arbitrary range. In determining the weakest failure detector for NBAC, we do not make any assumption a priori on the range of a failure detector.

Among the failure detectors defined in [2], we consider the following ones, each defined by a *completeness* and an *accuracy* property:

Perfect ($\mathcal{P}$): strong completeness (i.e., every incorrect process is eventually suspected by *every* correct process) and strong accuracy (i.e., no process is suspected before it crashes);

Eventually strong ($\diamond\mathcal{S}$): strong completeness and eventual weak accuracy (i.e., there is a time after which one correct process is never suspected).

For any failure pattern F, $\mathcal{P}(F)$, $\diamond\mathcal{S}(F)$ and $?\mathcal{P}(F)$ denote the sets of *all* histories satisfying the corresponding properties.

Algorithms. We model the set of asynchronous communication channels as a message buffer which contains messages not yet received by their destinations. An algorithm A is a collection of n (possibly infinite state) deterministic automata, one for each of the processes. $A(p_i)$ denotes the automaton running on process p_i. In each step of A, process p_i performs atomically the following three actions: (1) p_i chooses *non-deterministically* a single message addressed to p_i from the message buffer, or a null message, denoted λ; (2) p_i queries and receives a value from its failure detector module; (3) p_i changes its state and sends a message to a single process according to the automaton $A(p_i)$, based on its state at the beginning of the step, the message received in the receive action, and the value that p_i sees in the failure detector query action.

Configurations, schedules and runs. A *configuration* defines the current state of each process in the system and the set of messages currently in the message buffer. Initially, the message buffer is empty. A step (p_i, m, d, A) of an algorithm A is uniquely determined by the identity of the process p_i that takes the step, the message m received by p_i during the step (m might be the null message λ), and the failure detector value d seen by p_i during the step. We say that a step $e = (p_i, m, d, A)$ *is applicable* to the current configuration if and only if $m = \lambda$ or m is a message from the current message buffer destined to p_i. $e(C)$ denotes the unique configuration that results when e is applied to C. A *schedule* S of algorithm A is a (finite or infinite) sequence of steps of A. $S_\top$ denotes the empty schedule. We say that a schedule S *is applicable to a configuration* C if and only if (a) $S = S_\top$, or (b) $S[1]$ is applicable to C, $S[2]$ is applicable to $S[1](C)$, etc. For a finite schedule S applicable to C, $S(C)$ denotes the unique configuration that results from applying S to C.

A *partial run of algorithm A in an environment $\mathcal{E}$ using a failure detector $\mathcal{D}$* is a tuple $R = \langle F, H_\mathcal{D}, I, S, T\rangle$ where $F \in \mathcal{E}$ is a failure pattern, $H_\mathcal{D} \in \mathcal{D}(F)$ is a failure detector history, I is an initial configuration of A, S is a *finite* schedule of A, and $T \subseteq \mathcal{T}$ is a *finite* list of increasing time values (indicating when each step S occurred) such that $|S| = |T|$, S is applicable to I, and for all $t \leq |S|$, if $S[t]$ is of the form (p_i, m, d, A) then: (1) p_i has not crashed by time $T[t]$, i.e., $p_i \notin F(T[t])$ and (2) d is the value of the failure detector module $\mathcal{D}_i$ at time $T[t]$, i.e., $d = H_\mathcal{D}(p_i, T[t])$.

A *run of algorithm A in an environment $\mathcal{E}$ using a failure detector $\mathcal{D}$* is a tuple $R = \langle F, H_{\mathcal{D}}, I, S, T\rangle$, where S is an *infinite* schedule of A and $T \subseteq \mathcal{T}$ is an *infinite* list of increasing time values indicating when each step S occurred. In addition to satisfying the properties (1) and (2) of a partial run, run R should guarantee that (3) every correct process in F takes an infinite number of steps in S and eventually receives every message sent to it (this conveys the reliability of the communication channels).

Weakest failure detector. A *problem* is a set of runs (usually defined by a set of properties that these runs should satisfy). We say that a failure detector $\mathcal{D}$ *solves problem M in an environment $\mathcal{E}$* if there is an algorithm A, such that all the runs of A in $\mathcal{E}$ using $\mathcal{D}$ are in M (i.e., they satisfy the properties of M).

Let $\mathcal{D}$ and $\mathcal{D}'$ be any two failure detectors and $\mathcal{E}$ be any environment. If there is an algorithm $T_{\mathcal{D}'\to\mathcal{D}}$ that emulates $\mathcal{D}$ with $\mathcal{D}'$ in $\mathcal{E}$ ($T_{\mathcal{D}'\to\mathcal{D}}$ is called a *reduction* algorithm), we say that $\mathcal{D}$ *is weaker than $\mathcal{D}'$ in $\mathcal{E}$*, or $\mathcal{D} \preceq_{\mathcal{E}} \mathcal{D}'$. If $\mathcal{D} \preceq_{\mathcal{E}} \mathcal{D}'$ but $\mathcal{D}' \not\preceq_{\mathcal{E}} \mathcal{D}$ we say that $\mathcal{D}$ *is strictly weaker than $\mathcal{D}'$ in $\mathcal{E}$*, or $\mathcal{D} \prec_{\mathcal{E}} \mathcal{D}'$.[2] Note that $T_{\mathcal{D}'\to\mathcal{D}}$ does not need to emulate *all* histories of $\mathcal{D}$; it is required that all the failure detector histories it emulates be histories of $\mathcal{D}$.

We say that a failure detector $\mathcal{D}$ is *the weakest failure detector to solve a problem M in environment $\mathcal{E}$* if two conditions are satisfied: (1) Sufficiency: $\mathcal{D}$ solves M in $\mathcal{E}$, and (2) Necessity: if a failure detector $\mathcal{D}'$ solves M in $\mathcal{E}$ then $\mathcal{D} \preceq_{\mathcal{E}} \mathcal{D}'$.

We say that *problem M is harder than problem M' in environment $\mathcal{E}$*, if any failure detector $\mathcal{D}$ solving M in $\mathcal{E}$ solves also M' in $\mathcal{E}$. Respectively, *M is strictly harder than M' in $\mathcal{E}$*, if M is harder than M' in $\mathcal{E}$ and there exists a failure detector $\mathcal{D}'$ that solves M' (in $\mathcal{E}$) but not M.

3. Timeless failure detectors

This section introduces a new class $\mathcal{A}$ of *timeless* failure detectors. Intuitively, a timeless failure detector module is not able to provide information about *when* exactly (in the sense of the global time) failures have occurred.

We denote by F_0 the *failure-free* failure pattern: $\forall t \in \mathcal{T}, F_0(t) = \varnothing$. For any $F \in \mathcal{E}_f$ and $\delta \in \mathcal{T}$, we introduce the failure pattern F_δ, such that, for all $t \in \mathcal{T}$:

$$F_\delta(t) = \begin{cases} \varnothing & \text{if} \quad t < \delta \\ F(t-\delta) & \text{if} \quad t \geq \delta \end{cases}$$

Thus, for a failure that occurs at time t in F, the corresponding failure in F_δ occurs at $t + \delta$, and no failure occurs before time δ in F_δ. Note that $\forall \delta \in \mathcal{T}$: $correct(F) = correct(F_\delta)$. That is, a process is correct in F if and only if it is correct in F_δ. Thus, for any f, if $F \in \mathcal{E}_f$, then $F_\delta \in \mathcal{E}_f$.

[2]Later we omit $\mathcal{E}$ in $\prec_{\mathcal{E}}$ and $\preceq_{\mathcal{E}}$ when there is no ambiguity on the environment $\mathcal{E}$.

Formally, the class $\mathcal{A}$ consists of all (timeless) failure detectors $\mathcal{D}$, such that:

$$\begin{array}{l} \exists H_0 \in \mathcal{D}(F_0),\ \forall \delta \in \mathcal{T} \\ \forall F \in \mathcal{E}_f,\ \forall H \in \mathcal{D}(F), \\ \exists H_\delta \in \mathcal{D}(F_\delta) : \forall p_i \in \Pi,\ \forall t \in \mathcal{T}, \\ H_\delta(p_i,t) = \begin{cases} H_0(p_i,t) & \text{if} \quad t < \delta \\ H(p_i, t-\delta) & \text{if} \quad t \geq \delta \end{cases} \end{array} \tag{1}$$

It follows from (1) that, for any failure detector $\mathcal{D} \in \mathcal{A}$ and $\delta \in \mathcal{T}$, if a failure occurred at time t_0 and is reported by a module of $\mathcal{D}$ at time t_1, then a failure that occurred at $t_0 + \delta$ could be reported *in the same way* at $t_1 + \delta$: the process does not know when exactly the failure occurred. This captures our idea that failure detectors of class $\mathcal{A}$ provide no information about the time when failures occur.

We restrict our scope from the original universe of failure detectors [2] to $\mathcal{A}$. Note that $\mathcal{P}$, $\diamond\mathcal{S}$ and $?\mathcal{P}$ are timeless. With $?\mathcal{P}$, for instance, a process can detect the very fact that some process has crashed, but there is no way to acquire the actual time at which the failure occurred.

Examples of histories output by a failure detector $\mathcal{D}$ from $\mathcal{A}$ are depicted in Figure 1. The system consists of two processes p_1 and p_2. At any time, $\mathcal{D}_i$ outputs a set of processes suspected by p_i ($i = 1, 2$). Assume that (a) p_2 crashes at time t_1 and p_1 detects the failure at time t_2. Then, by the definition of $\mathcal{A}$, if (b) p_2 crashes at time $t_1 + d$, then there exists a history of $\mathcal{D}$ in which p_1 detects the failure at time $t_2 + d$.

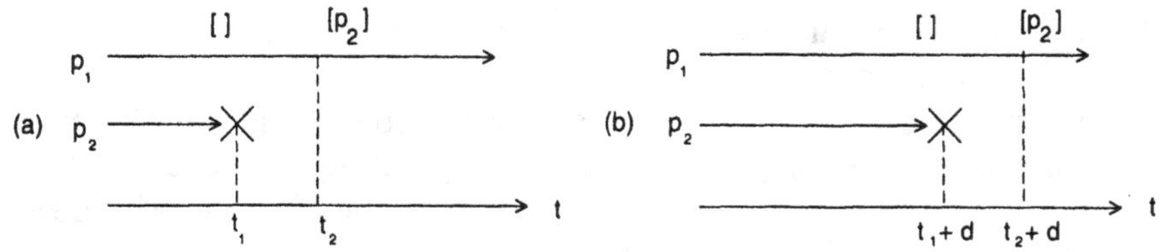

Figure 1. Examples of timeless failure detector histories.

Consider the failure detector $\mathcal{B}[\alpha]$ mentioned in the introduction: each module of $\mathcal{B}[\alpha]$ outputs $\top$ or a subset of the processes in Π ($\mathcal{R}_{\mathcal{B}[\alpha]} = \top \cup 2^\Pi$). Formally, $\mathcal{B}[\alpha]$ is defined as follows:

$$\begin{array}{l} \forall F \in \mathcal{E}_f, \forall H \in B[\alpha](F), \\ \exists H_\mathcal{P} \in \mathcal{P}(F), \\ \forall p_i \in \Pi, \exists t_i \in \mathcal{T}, \forall t \in \mathcal{T} : \\ H(p_i,t) = \begin{cases} \top & \text{if } t < t_i \\ p_i & \text{if } t \geq t_i \ \wedge\ F(\alpha) \neq \varnothing \\ H_\mathcal{P}(p_i,t) & \text{if } t \geq t_i \ \wedge\ F(\alpha) = \varnothing \end{cases} \end{array}$$

(The Stillborn failure detector $\mathcal{B}$ is a particular case of $\mathcal{B}[\alpha]$ with $\alpha = 0$ and $t_i = 0, \forall p_i \in \Pi$.) Clearly, $\mathcal{B}[\alpha]$ does not belong to $\mathcal{A}$. Indeed, consider a failure pattern F in which only one process is crashed at time α. Take a corresponding history $H \in B[\alpha](F)$. For every process p_i, there is a time t_i, such that, for any $t \geq t_i$, $H(p_i, t) = p_i$. Now consider a failure pattern F_δ in which no process is crashed at time α (the condition holds for all $\delta > \alpha$). Since failure detector module of p_i behaves now like $\mathcal{P}_i$ (i.e., the perfect failure detector), its own identity p_i is never output. Thus, $\mathcal{B}[\alpha] \notin \mathcal{A}$.

4. Proof technique

Before proving that, in $\mathcal{A}$, $\diamond\mathcal{S}$ is necessary to solve NBAC, we briefly recall here the technique used in [1] to prove that $\diamond\mathcal{S}$ is necessary to solve Consensus and we discuss the applicablity of this technique to NBAC.

The weakest failure detector to solve Consensus. A failure detector, denoted Ω ($\diamond\mathcal{S} \preceq \Omega$) is defined in [1]. The output of Ω_i is a single process p_j, that p_i currently *trusts*, i.e., that p_i considers to be correct ($\mathcal{R}_\Omega = \Pi$). For each failure pattern F, $\Omega(F)$ is the set of all failure detector histories H_Ω, such that there is a time after which all correct processes always trust the same correct process. Clearly, $\Omega \in \mathcal{A}$.

Let $\mathcal{E}$ be any environment, $\mathcal{D}$ be any failure detector that solves Consensus in $\mathcal{E}$, and $Consensus_\mathcal{D}$ be any Consensus algorithm that uses $\mathcal{D}$. The algorithm $T_{\mathcal{D}\to\Omega}$ that transforms $\mathcal{D}$ into Ω in $\mathcal{E}$ works as follows.

Fix an arbitrary run of $T_{\mathcal{D}\to\Omega}$ using $\mathcal{D}$, with failure pattern $F \in \mathcal{E}$ and failure detector history $H_\mathcal{D} \in \mathcal{D}(F)$. All processes periodically query their failure detector $\mathcal{D}$ and exchange information about the values of $H_\mathcal{D}$ that they see in this run. Using this information, the processes construct a directed acyclic graph (DAG) that represents a "sampling" of failure detector values in $H_\mathcal{D}$ and causal relationships between the values. By periodically sending its current version of the DAG to all processes, and incorporating all the DAGs that it receives into its own DAG, every correct process constructs ever increasing finite approximations of the same infinite limit DAG G.

The DAG G can be used to simulate runs of $Consensus_\mathcal{D}$ with failure pattern F and failure detector history $H_\mathcal{D}$. These runs *could have happened* if processes were running $Consensus_\mathcal{D}$ instead of $T_{\mathcal{D}\to\Omega}$. If we simulate all possible runs of $Consensus_\mathcal{D}$ applied to the DAG G with all possible initial configurations I, we obtain a *simulation forest*: a tree for each initial configuration.

Thus, the infinite DAG G induces an infinite simulation forest Υ of runs of $Consensus_\mathcal{D}$ with failure pattern F and failure detector history $H_\mathcal{D} \in \mathcal{D}(F)$. From the properties of the Consensus problem, it follows that Υ comprises schedules corresponding to the runs of Consensus in which every correct process decides 0 and runs in which every correct process decides 1. This allows to design a deterministic algorithm that identifies a process p^* that is correct in F, namely a process whose step defines which decision is going to be taken by the rest of correct processes in the descending schedules.

Although the simulation forest Υ is infinite and cannot be computed by any process, there exists a *finite subforest* of Υ that gives sufficient information to identify p^*. Thus, there is a time after which, every correct process running $T_{\mathcal{D}\to\Omega}$ obtains a reference p^*. In other words, $T_{\mathcal{D}\to\Omega}$ emulates Ω.

NBAC: a hard nut. As we discussed for $\Diamond\mathcal{S}$, Ω is not at all necessary to solve NBAC (as shown in [6]). Thus, the technique of [1] cannot be directly applied to show the necessity of Ω for NBAC. For instance, if a module of the stillborn failure detector $\mathcal{B}$ outputs $\top$, then there is an initial failure in the system and NBAC is trivially solved by deciding *abort* at every correct process. There is no way to identify *correct* processes and, thus, no algorithm $T_{\mathcal{B}\to\Omega}$ is possible.

However, even if we exclude failure detectors like $\mathcal{B}[\alpha]$ by considering timeless failure detectors only (i.e., focusing on class $\mathcal{A}$), we are still not able to apply the technique of [1]. Indeed, let $\mathcal{D}$ be any failure detector that solves NBAC in an environment $\mathcal{E}$. Consider a run of an NBAC algorithm using $\mathcal{D}$ in a failure pattern F, such that $F(0) \neq \varnothing$ (some process is initially crashed). Clearly, no process can decide *commit* (no matter which failure detector history $H_{\mathcal{D}}$ is output by $\mathcal{D}$). The only decision a correct process can take is *abort* (otherwise, the A-Validity property of NBAC would be violated, since some $p \in F(0)$ could have voted *no*). In this case, the corresponding simulation forest Υ does not bring any valuable information about failures to identify a correct process (no matter which failure detector history $H_{\mathcal{D}}$ is output by $\mathcal{D}$).

Fortunately, thanks to the very nature of timeless failure detectors, we can modify the original DAG G in order to fetch a valuable information about correct processes of F. The details are presented in Section 5.

5. Necessary condition

This section shows the necessity of Ω to solve NBAC using failure detectors in class $\mathcal{A}$. To this end, we present a reduction algorithm $T_{\mathcal{D}\to\Omega}$ transforming any failure detector $\mathcal{D} \in \mathcal{A}$ that solves NBAC into Ω. A corollary of our result is that $\Diamond\mathcal{S}$ (which is weaker than Ω) is necessary to solve NBAC (using $\mathcal{A}$), and hence $?\mathcal{P} + \Diamond\mathcal{S}$ is the weakest failure detector within $\mathcal{A}$ to solve NBAC.

Nice runs and nice DAGs. Let $\mathcal{D}$ be any failure detector in $\mathcal{A}$ and $NBAC_{\mathcal{D}}$ be any NBAC algorithm using $\mathcal{D}$. From now on, we denote by $e = (p, m, d)$ a step of process p executing $NBAC_{\mathcal{D}}$.

Let F_0 be the failure-free pattern and H_0 be the history from $\mathcal{D}(F_0)$, such that the condition (1) in Section 3 for $\mathcal{D}$ holds with H_0. Let I be any initial configuration in which all processes vote *yes*. Due to the properties of NBAC, there exists a partial run $R_0 = \langle F_0, H_0, I, S_0, T_0\rangle$ of $NBAC_{\mathcal{D}}$ comprising a finite number of steps in which every process decides *commit*.

Taking the *nice run* R_0 as a basis, we can now construct a *nice DAG* (directed acyclic graph) G_0 induced by the failure-free pattern F_0. For any step $e = (p, m, d)$ in S_0, we create a vertex $[p, d, k]$ of G_0, where $k - 1$ is the number of steps of p in S_0 preceding e. For any steps $e_1 = (p_1, m_1, d_1)$ and $e_2 = (p_2, m_2, d_2)$ in S_0, such that e_1 precedes e_2 in S_0, we create a corresponding edge in $[p_1, d_1, k_1] \rightarrow [p_2, d_2, k_2]$ in G_0. This means that p_2 queried its failure detector for the k_2-th time *after* p_1 queried its failure detector for the k_1-th time.

Constructing a DAG. Let F be any failure pattern from $\mathcal{E}_f$ and $H \in \mathcal{D}(F)$ and assume that the same nice DAG G_0 is initially available to all processes. Consider a run R of $T_{\mathcal{D}\rightarrow\Omega}$. Processes periodically query their failure detector $\mathcal{D}$ and exchange information about the values of $H \in \mathcal{D}(F)$ that they see in the current run. Using this information, every process p constructs an *imaginary* DAG G_p, in which the real samples of H are assumed to be seen *after* all the values of H_0 presented in G_0. That is, every time a process p sees a failure detector value d, (1) a new vertex $[p, d, k]$ is added to G_p, such that $k = k_0 + k_R$, where k_0 is the number of steps of p in S_0 and k_R is the number of times p queried its failure detector module so far, and (2) a new edge from every vertex of G_p to $[p, d, k]$ is added. As a result, every correct process p maintains an ever growing graph $G_p(t)$, such that $G_p(t) \rightarrow_{t\rightarrow\infty} G$ for some infinite DAG G. Note that G contains a sampling of the failure detector history H corresponding to a *real* failure pattern F ($H \in \mathcal{D}(F)$ as well as of some *imaginary* history $H_0 \in \mathcal{D}(F_0)$, where F_0 is the failure-free pattern.

We say that an infinite DAG $\tilde{G}$ is a *sampling DAG of history* $\tilde{H} \in \mathcal{D}(\tilde{F})$ if the following properties are satisfied:

1 The vertices of $\tilde{G}$ are of the form $[p, d, k]$ where $d = \tilde{H}(p, t)$ for some t and $k \in \mathbb{N}$.

2 If $[p_1, d_1, k_1] \rightarrow [p_2, d_2, k_2]$ is an edge of $\tilde{G}$ and $d_1 = \tilde{H}(p_1, t_1)$ and $d_2 = \tilde{H}(p_2, t_2)$, then $t_1 < t_2$.

3 If $[p, d, k]$,$[p, d', k']$ are vertices of $\tilde{G}$ and $k < k'$ then $[p, d, k] \rightarrow [p, d', k']$ is an edge of $\tilde{G}$.

4 $\tilde{G}$ is transitively closed.

5 Let V be any finite subset of vertices of $\tilde{G}$ and p be any correct process. There is $d \in \mathcal{R}_\mathcal{D}$ and $k \in \mathbb{N}$, such that for every vertex $v \in V$, $v \rightarrow [p, d, k]$ is an edge of $\tilde{G}$.

The following lemma precisely captures the relationship between the real failure pattern F and the DAG G constructed by $T_{\mathcal{D}\rightarrow\Omega}$ (the proof is omitted due to the space limitations and it can be found in [7]):

Lemma 1 *There exists $\delta \in \mathcal{T}$ and a failure detector history $H_\delta \in \mathcal{D}(F_\delta)$ such that G is a sampling DAG of H_δ.*

Thus, G represents a sample of a failure detector history H_δ that *could be* seen if the failure pattern was F_δ. Note that even if a process p is initially crashed in F, G contains the samples of its failure detector module output. However, the number of vertices of the form $[p, \cdot, \cdot] \in G$ is finite, thus, p cannot be considered to be correct in F_δ. In other words, a crashed process in F cannot appear to be correct in F_δ.

Tags and decision gadgets. Lemma 1 allows us to use G to simulate some of the runs of $NBAC_{\mathcal{D}}$ in the failure pattern F_δ. Take an initial configuration I of $NBAC_{\mathcal{D}}$ in which every process votes *yes*. The set of simulated schedules of $NBAC_{\mathcal{D}}$ that are compatible with some path of G and are applicable to I can be organized as a tree Υ: paths in this tree represent simulated schedules of $NBAC_{\mathcal{D}}$ with initial configuration I. The fact that $G_0 \subset G$ guarantees that there exists a schedule in Υ in which every process decides *commit*.

Following [1], we assign a set of tags (*abort* or *commit*) to each vertex of the simulation tree Υ induced by G. Vertex S of tree Υ gets tag k if and only if it has a descendant S' (possibly $S = S'$) such that some correct process has decided k in $S'(I)$. A vertex of Υ is *monovalent* if it has only one tag, and *bivalent* if it has both tags (following the terminology of [3]).

Still following [1], we also introduce the notion of *decision gadgets* and *deciding processes* and show that any deciding process in Υ is correct. Informally, a decision gadget is a vertex S of Υ having exactly two monovalent leaves: one *abort*-valent and one *commit*-valent. In turn, a deciding process of S is a process whose step defines the decision taken by a descendant of S. The following lemma gives a condition of the existence of at least one decision gadget in Υ:

Lemma 2 *If correct$(F) \neq \Pi$ (F is not failure-free), then Υ has a decision gadget.*

Proof: Let $p \notin correct(F)$. There exists a *finite* schedule E in Υ containing only steps of correct processes such that all correct processes have decided in in $E(I)$ (Lemma 10 of [1]). Since E contains no step of process p, no information is available about its initial vote, and the decision value must be *abort* (otherwise the A-Validity property is violated). From the way the simulation tree is constructed, it follows that Υ contains a schedule in which *commit* is decided. Thus the initial configuration of Υ is bivalent. By Lemma 18 of [1], Υ has at least one decision gadget (and hence a deciding process). □

Reduction algorithm. Now we are ready to define a rule according to which the same correct process is eventually identified by every correct process. Every process p periodically updates and tags a simulation tree Υ_p induced by G_p with the initial configuration I in which all processes vote *yes*. If there exists a decision gadget in Υ_p, then $T_{\mathcal{D}\rightarrow\Omega}$ outputs the deciding process of the smallest decision gadget of Υ_p (since the set of vertices of Υ_p is countable, we can easily impose a rule to define the smallest decision gadget in it), otherwise $T_{\mathcal{D}\rightarrow\Omega}$ outputs p_1. Note that for any correct process p, $G = \lim_{t\rightarrow\infty} G_p(t)$ and thus $\Upsilon = \lim_{t\rightarrow\infty} \Upsilon_p(t)$.

Theorem 3 *There exists a process $p^* \in correct(F)$, such that, for every correct process p, there is a time after which $T_{\mathcal{D}\to\Omega}$ outputs p^*, forever.*

Proof: Consider $\delta \in \mathcal{T}$ and failure pattern F_δ, such that G is a sampling DAG for some $H_\delta \in \mathcal{D}(F_\delta)$. Note that $correct(F) = correct(F_\delta)$. Two cases are possible:

(1) F, and thus F_δ are failure-free. Then all vertices in Υ are monovalent and the reduction algorithm forever outputs $p_1 \in correct(F)$.

(2) F, and thus F_δ are not failure-free. By Lemma 2, Υ has a deciding process. Let p^* be the deciding process of the smallest decision gadget. Since ever growing simulation trees $\Upsilon_p(t)$ of all correct processes p tend to Υ, there exists t_0 such that $\forall t > t_0, \forall p \in correct(F)$, the decision process of the smallest decision gadget is p^*. Thus, $\forall t > t_0$ all correct processes p have $output_p = p^*$. By Lemma 21 of [1], the deciding process is correct in F. Thus $T_{\mathcal{D}\to\Omega}$ outputs the identity of a correct process. □

Theorem 4 *For any environment $\mathcal{E}_f$, if a failure detector $\mathcal{D} \in \mathcal{A}$ can be used to solve NBAC in $\mathcal{E}_f$, then $\mathcal{D} \succeq_{\mathcal{E}_f} \Omega$.*

From the facts that $?\mathcal{P}$ is necessary to solve NBAC in any environment [6], $\Diamond\mathcal{S}$ is weaker than Ω [1], and Theorem 4, the following result holds:

Theorem 5 *For any environment $\mathcal{E}_f$, if a failure detector $\mathcal{D} \in \mathcal{A}$ solves NBAC in $\mathcal{E}_f$, then $\mathcal{D} \succeq_{\mathcal{E}_f} ?\mathcal{P} + \Diamond\mathcal{S}$.*

From the theorem above and [6], we have:

Corollary 6 *$?\mathcal{P} + \Diamond\mathcal{S}$ is the weakest among timeless failure detectors to solve NBAC in any environment $\mathcal{E}_f$ with $f < \lceil \frac{n}{2} \rceil$.*

From [1], [6] and Corollary 6, we have:

Corollary 7 *For any environment $\mathcal{E}_f$ with $0 < f < \lceil \frac{n}{2} \rceil$ in a system augmented with timeless failure detectors, NBAC is strictly harder than Consensus.*

6. Concluding remarks

Sabel and Marzullo showed in [8] that $\mathcal{P}$ is the weakest failure detector to solve the Leader Election problem within a specific class of failure detectors. They focus on failure detectors that output sets of suspected processes and satisfy the following symmetry property: if a process detects a failure erroneously, then any process can detect a failure erroneously an arbitrary number of times. The requirement is rather strong: for instance, it excludes all failure detectors that make a finite number of mistakes. The approach is somewhat similar to ours. We also defined a subset $\mathcal{A}$ of the overall universe of failure detectors [2] in which $?\mathcal{P} + \Diamond\mathcal{S} \prec \mathcal{P}$ is shown to be the weakest to solve our NBAC problem.

The class of *symmetric* failure detectors of [8] and our class $\mathcal{A}$ of timeless failure detectors are however incomparable.

Fromentin, Raynal and Tronel stated in [4] that $\mathcal{P}$ is the weakest failure detector to solve NBAC. Guerraoui [6] pointed out that [4] assumes NBAC to be solved among any subset of the processes in the system and showed that $\mathcal{P}$ is not the weakest failure detector to solve NBAC without that assumption. In this paper, we make a step further showing that a failure detector $?\mathcal{P}+\diamond\mathcal{S} \prec \mathcal{P}$ is the weakest to solve NBAC in a wide class $\mathcal{A}$ of timeless failure detectors (provided an environment with a majority of correct processes). Thus, in this environment, NBAC is strictly harder than Consensus (which is not true in general [5, 6]). The question of the weakest failure detector to solve NBAC without assuming a majority of correct processes is open for future research.

References

[1] T. D. Chandra, V. Hadzilacos, and S. Toueg. The weakest failure detector for solving consensus. *Journal of the ACM*, 43(4):685–722, March 1996.

[2] T. D. Chandra and S. Toueg. Unreliable failure detectors for reliable distributed systems. *Journal of the ACM*, 43(2):225–267, March 1996.

[3] M. J. Fischer, N. A. Lynch, and M. S. Paterson. Impossibility of distributed consensus with one faulty process. *Journal of the ACM*, 32(3):374–382, April 1985.

[4] E. Fromentin, M. Raynal, and F. Tronel. On classes of problems in asynchronous distributed systems with process crashes. In *Proceedings of the IEEE International Conference on Distributed Systems (ICDCS)*, pages 470–477, 1999.

[5] R. Guerraoui. On the hardness of failure-sensitive agreement problems. *Information Processing Letters*, 79(2):99–104, June 2001.

[6] R. Guerraoui. Non-blocking atomic commit in asynchronous distributed systems with failure detectors. *Distributed Computing*, 15:17–25, January 2002.

[7] R. Guerraoui and P. Kouznetsov. On the weakest failure detector for non-blocking atomic commit. Technical report, School of Computer and Communication Sciences, Swiss Institute of Technology in Lausanne, 2002. Available at http://icwww.epfl.ch/publications/list.php.

[8] L. S. Sabel and K. Marzullo. Election vs. consensus in asynchronous systems. Technical report, Cornell University, Ithaca, NY, TR95-1488, 1995.

[9] D. Skeen. Nonblocking commit protocols. In *ACM SIGMOD International Conference on Management of Data*, pages 133–142. ACM Press, May 1981.

COMBINING COMPUTATIONAL EFFECTS: COMMUTATIVITY AND SUM

Martin Hyland
Department of Mathematics
University of Cambridge
Cambridge CB3 0WB, England
M.Hyland@dpmms.cam.ac.uk

Gordon Plotkin and John Power*
Laboratory for the Foundations of Computer Science
University of Edinburgh
King's Buildings, Edinburgh EH9 3JZ, Scotland
gdp@dcs.ed.ac.uk, ajp@dcs.ed.ac.uk

Abstract We seek a unified account of modularity for computational effects, using the notion of enriched Lawvere theory, together with its relationship with strong monads, to reformulate Moggi's paradigm for modelling computational effects. Effects qua theories are then combined by appropriate bifunctors (on the category of theories). We give a theory of the commutative combination of effects, which in particular yields Moggi's side-effects monad transformer. And we give a theory for the sum of computational effects, which in particular yields Moggi's exceptions monad transformer.

Keywords: Computational effects, enriched Lawvere theories, commutative combination, sum.

Introduction

We seek a unified account of modularity for computational effects. More precisely, we seek a mathematical theory that supports the combining of computational effects such as nondeterminism, probabilistic nondeterminism, side-effects, exceptions, interactive input/output, and continuations. Ideally, we should like to develop mathematical operations, together with associated rel-

*This work has been done with the support of EPSRC grant GR/M56333, a British Council grant and the COE budget of Japan.

evant theory. There is more than one such operation as, for example, the combination of side-effects and nondeterminism is of a different nature to the combination of side-effects and exceptions, and, further, as one is sometimes interested in different ways to combine even the same pair of effects. So we seek to find and develop what we expect will be a small number of computationally and mathematically natural such combining operations. We do not address continuations at all in this paper. But, with that caveat, the paper is devoted to two such ways of combining effects: that of combining them commutatively, as we shall see holds for combining side-effects with all effects we know other than exceptions; and that of taking their sum, as we shall see holds for combining exceptions with all other effects.

In order to give such operations, we first need a unified way to model the various computational effects individually. In this, we start by following Eugenio Moggi, who, in [15, 17], gave a unified category theoretic account of computational effects, which he called notions of computation. He modelled each effect by means of a strong monad T on a base category C with finite products. The monads corresponding to the effects listed above are given by a powerdomain [18], a probabilistic powerdomain [9, 10], and the monads $(S \times -)^S$, $-+E, TX = \mu Y.(O \times Y + Y^I + X)$, and R^{R^-} respectively [15, 16, 17], assuming C has appropriate additional structure; the set S of *states* is typically analysed as V^{Loc} where V is a set of *values* and *Loc* is a set of *locations*. Moggi's unified approach has proved useful, especially in functional programming [2].

Strong monads in hand, we first seek a binary operation that, to each pair of strong monads (T, T'), yields a new strong monad $T \otimes T'$, such that, in the case where $T' = (S \times -)^S$, the monad $T \otimes T'$ is $T(S \times -)^S$, which computational experience tells us is the natural combination of side-effects with all effects we know other than exceptions. So we ask: can we give a mathematical theory yielding such an operation on a pair of strong monads? Modulo a few side conditions, the answer is yes; we make fundamental use of the correspondence between strong monads and a generalised notion of Lawvere theory in order to provide it [23]. That correspondence is computationally natural and is already implicit in our previous work on computational effects [20, 21, 22]. We are unaware of any direct justification for the existence of $T \otimes T'$.

Here, we are following an algebraic programme that shifts focus away from monads to the study of natural operations that yield the required effects (see [20, 21] for other recent work along these lines), with the monads then corresponding to natural theories for these operations [22]. For instance, rather than emphasise the side-effects monad $(S \times -)^S$, we emphasise the operations *lookup* and *update* associated with side-effects, and the equations that relate them [22]. In the case where $S = V^{Loc}$, *lookup* can be considered as a *Loc*-indexed family of V-ary operations, and *update* as a $Loc \times V$-indexed family; the idea is that $lookup_l(x)$ proceeds with x_v if the contents of l is v and $update_{\langle l,v \rangle}(y)$ proceeds with y, having updated l with v. Again, rather than emphasise the powerdomain P, we emphasise the operation of nondeterministic choice $\vee$ with its equations for associativity, symmetry, and idempotence [7, 19].

This change in emphasis, supported by the correspondence between strong monads and enriched Lawvere theories in [23] (and see the expository [25]), is computationally natural for all the examples of computational effects listed above except for continuations [22]; in that case one can still make a formal change in emphasis, but it seems computationally unnatural, and we believe continuations should be treated separately.

Having reformulated our account of computational effects in terms of enriched Lawvere theories, we can reformulate our question to read: is there a mathematical theory yielding an operation that to each pair (L, L') of enriched Lawvere theories, gives a new enriched Lawvere theory $L \otimes L'$, such that, if L' is the enriched Lawvere theory associated with side-effects, the new enriched Lawvere theory corresponds to $T(S \times -)^S$, where L corresponds to T? The answer is yes, it is remarkably natural, and, in various guises, forms of it have existed since the 1960's [4, 13]. It is known as the ***tensor product*** of theories and simply amounts to taking the operations of both theories and demanding that they commute with each other, while retaining the equations of both. For instance, in the case where $S = V^{Loc}$, combining side-effects with nondeterminism, if there were three values, one would have the equation

$$lookup_l(x_1 \vee y_1, x_2 \vee y_2, x_3 \vee y_3) = lookup_l(x_1, x_2, x_3) \vee lookup_l(y_1, y_2, y_3)$$

In a functional language with references and nondeterminism this would induce the program equivalence:

$$\textbf{let } x \textbf{ be } !y \textbf{ in } (M \textbf{ or } N) \equiv (\textbf{let } x \textbf{ be } !y \textbf{ in } M) \textbf{ or } (\textbf{let } x \textbf{ be } !y \textbf{ in } N)$$

where $!M$ is the dereferencing operator and M **or** N is non-deterministic choice. There is a similar commutation equation for *update* and $\vee$, with a corresponding induced program equivalence. A recent reference for mathematical theory that supports this construction is [8], for which this is a leading example.

Having studied the commutative combination of effects, we turn to their sum. The natural combination of side-effects with exceptions is not their commutative combination. Exceptions combines with all other computational effects by taking the sum of the two theories: one has the operations for exceptions together with all operations for the other effect subject to all its equations, with no further equations. We shall show that the sum of theories yields Moggi's exceptions monad transformer, taking a monad T to the monad $T(- + E)$.

Of course, one typically combines more than two effects, so the operations we define may be used several times. For instance, to combine partiality, side-effects and nondeterminism, one can first combine partiality and semilattices by sum, then combine the result with side-effects by commutativity; similarly for partiality, side-effects and interactive input/output.

The published work most closely related to ours is that of Moggi and Cenciarelli on monad transformers. They defined a monad transformer to be a function from the set of strong monads on a category C with finite products to itself [2, 3]. The monad transformer for side-effects takes a monad T to the monad $T(S \times -)^S$, assuming C is cartesian closed. To model the combination of nondeterminism with side-effects, one would apply the side-effects monad transformer to a powerdomain P, yielding the monad $P(S\times -)^S$. So the resulting monad agrees with ours, as it must, but we have an associated mathematical theory: the question we pose could equally be posed to ask how one might derive the side-effects monad transformer from the side-effects monad *qua* monad, but the work on monad transformers to date has not answered that. Moreover, our work involves no asymmetry: there seems no a priori reason why the combination of side-effects with nondeterminism should be achieved by applying the side-effects monad transformer to the nondeterminism monad rather than vice-versa. And in the case of exceptions, the side-effects monad transformer does not even give the required result for the usual interpretation of the combination.

There is also relevant unpublished work by Paul Levy. He has observed that the sum of any monad T with that for exceptions $- + E$ is given by $T(- + E)$. He has also defined a notion of commutative combination of monads and shown that $T(S \times -)^S$ is the commutative combination of T and $(S \times -)^S$ with that definition. His construction does not exist for all pairs of monads, and he has not developed accompanying theory; but when the monads have rank, his definition agrees with ours.

Other than the work on monad transformers, the other main attempt we know to account for the combination of side-effects with other computational effects has been the development of dyads [24] which amount to a decomposition of the side-effects monad into strictly more primitive structure. Dyads come equipped with a notion of Kleisli category, in which one may model the computational λ-calculus, and have been integrated with *Freyd*-structure, which models a delicate feature of contexts arising with side-effects or exceptions, where the order of evaluation is crucial. The relationship between the two notions remains to be investigated.

The paper is organised as follows. We do not have space here to explain the general enriched setting. So in each section we investigate the unenriched case, which largely amounts to the situation where computational effects are modelled in *Set*, then we briefly remark on its enrichment, especially to ωCpo (whose objects are ω-cpos, partial orders with least upper bounds of increasing ω-chains, and whose morphisms are continuous functions, i.e., maps of partial orders that preserve the least upper bounds). In Section 1, we describe the relationship between monads and Lawvere theories, and explain how the latter appear in our leading examples. In Section 2, we explore the commutative combination of Lawvere theories. In Section 3, we show that the commutative combination of side-effects with any other Lawvere theory gives the outcome we seek. And in Section 4, we develop a theory for the sum of Lawvere theories and explain how this gives rise to the exceptions monad transformer.

A clear omission from this paper is the study of distributivity: this seems to be the main other way in which computational effects combine. One has distributivity of one set of operations over the other in combining nondeterminism with probabilistic nondeterminism [14], and one has distributivity of each set of operations over the other in combining internal and external nondeterminism [6]. Another important question concerns the combination of effects with local state [22] (this paper only concerns global state). In [22] local state is specified using an additional operation *block* together with additional equations. But it is unclear yet how best to integrate it with enriched Lawvere theories, let alone consider combinations with other effects. We have also not considered the combination of interactive input/output with other effects in this paper, but we believe that sum is the operation of primary interest there. Finally, we have not considered the relationship of all these effects with that of continuations; this will be substantially different as the continuations monad does not have a rank.

1. Monads and Lawvere theories

For simplicity of exposition, we start by restricting our attention to the base category Set. The side-effects monad is then the monad $(S \times -)^S$ for a set of states S. We impose the natural restriction that S is a countable set, and in case $S = V^{Loc}$ restrict V to be countable and Loc to be finite (when dealing with local state with its need for unboundedly many locations, one can use a presheaf semantics [22]).

The side-effects monad is then of *countable rank*, which means that, in a precise sense, it is of bounded size [11]. The category of monads with countable rank is equivalent to the category of countable Lawvere theories as we shall outline. So, in principle, the side-effects monad can equally be seen as a countable Lawvere theory, and that turns out to be a computationally natural way in which to see it. All our mathematics generalises to arbitrary rank, but, as all our examples are of countable rank, we naturally restrict our exposition to that case.

Let $\aleph_1$ denote a skeleton of the category of countable sets and all functions between them. So $\aleph_1$ has an object for each natural number n and an object for $\aleph_0$. Up to equivalence, $\aleph_1$ is the free category with countable coproducts on 1. So, in referring to $\aleph_1$, we implicitly make a choice of the structure of its countable coproducts.

Definition 1 *A* countable Lawvere theory *consists of a small category L with countable products and a strict countable-product preserving identity-on-objects functor $I : \aleph_1^{op} \longrightarrow L$. A* model *of a countable Lawvere theory L in any category C with countable products is a countable-product preserving functor $M : L \longrightarrow C$.*

For any countable Lawvere theory L and any category with countable products C, we thus have the category $Mod(L, C)$ of models of L in C. There is a canonical forgetful functor $U : Mod(L, C) \longrightarrow C$, and, when $C = Set$, this forgetful functor has a left adjoint, exhibiting $Mod(L, Set)$ as equivalent to the category T_L-Alg for the induced monad T_L on Set.

Conversely, given a monad T with countable rank on Set, the category $Kl(T)^{op}_{\aleph_1}$ determined by restricting $Kl(T)$ to the objects of $\aleph_1$ is a countable Lawvere theory L_T, and the functor from T-Alg to $Mod(L_T, Set)$ induced by the restriction is an equivalence of categories. The following theorem appears in enriched form, as we ultimately require it, in [23].

Theorem 1 *The construction sending a countable Lawvere theory L to T_L together with that sending a monad T with countable rank to L_T induce an equivalence of categories between the category of countable Lawvere theories and the category of monads with countable rank on Set. Moreover, the comparison functor exhibits an equivalence of the categories $Mod(L, Set)$ and T_L-Alg.*

So, in principle, the side-effects monad can be described as a countable Lawvere theory. The usual way in which to define countable Lawvere theories is by means of sketches, with the Lawvere theory given freely on the sketch. To give a sketch amounts to giving operations and equations, here the operations being allowed to be of countable arity: Barr and Wells' book [1] treats sketches in loving detail. A sketch, and hence the countable Lawvere theory, corresponding to the side-effects monad is essentially given in [22] and is easy to describe:

Example 1 *The countable Lawvere theory L_S for side-effects (when $S = V^{Loc}$) is the free countable Lawvere theory generated by operations lookup : $V \longrightarrow Loc$ and update : $1 \longrightarrow Loc \times V$ subject to the seven natural equations listed in [22], four of them specifying interaction equations for lookup and update and three of them specifying commutation equations. Note the use of the targets Loc and $Loc \times V$ to handle indexing at the Lawvere theory level.*

Proposition 1 *[22] For any category C with countable products and countable coproducts, the canonical comparison functor from $Mod(L_S, C)$ to T-Alg is an equivalence of categories, where T is the monad on C defined by $TX = (\Sigma_S X)^S$.*

Example 2 *Ignoring partiality, the countable Lawvere theory L_N for (binary) nondeterminism is the countable Lawvere theory freely generated by a binary operation $\vee : 2 \longrightarrow 1$ subject to equations for associativity, commutativity and idempotence, i.e., the countable Lawvere theory for a semilattice.*

Example 3 *The countable Lawvere theory L_P for (binary) probabilistic nondeterminism is that freely generated by $[0, 1]$-many binary operations $+_r : 2 \longrightarrow 1$ subject to the equations for associativity, commutativity and idempotence in [5].*

Example 4 *The countable Lawvere theory $L_{I/O}$ for interactive input/output is the free countable Lawvere theory generated by operations read : $I \longrightarrow 1$ and write : $1 \longrightarrow O$, where I is a countable set of* inputs *and O of* outputs. *So, interactive input/output is more directly modelled by the countable Lawvere theory than by the corresponding monad $TX = \mu Y.(O \times Y + Y^I + X)$.*

Example 5 *The countable Lawvere theory L_E for exceptions is the free countable Lawvere theory generated by the operation raise : $0 \longrightarrow E$, where E is a countable set of* exceptions.

Of course, *Set* is not the category of primary interest in denotational semantics. One is more interested in ωCpo, and variants, in order to model recursion. The relationship between countable Lawvere theories and countable monads extends without fuss to one between countable enriched Lawvere theories and countable strong monads on such categories [23].

Example 6 *The countable Lawvere ωCpo-theory $L_\perp$ for partiality is the theory freely generated by a nullary operation $\perp: 0 \longrightarrow 1$ subject to the condition that there is an inequality*

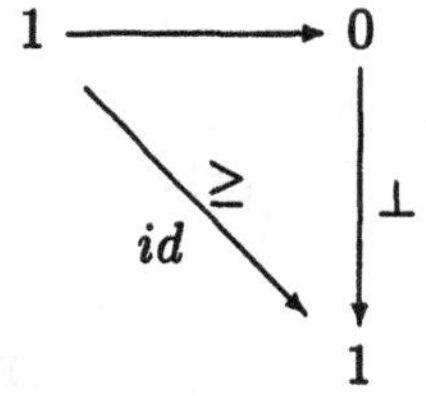

where the unlabelled map is the unique map determined because 0 *is the terminal object of $V_{\aleph_1}^{op}$. A model of $L_\perp$ in ωCpo is exactly an ω-cpo with least element.*

We have already introduced a countable Lawvere theory L_P corresponding to a powerdomain: it is the countable Lawvere theory for a semilattice. We overload notation a little here by also using the notation L_P to denote the countable Lawvere ωCpo-theory for a semilattice: the generators and equations are the same, but the ωCpo-theory has more objects as there are countably presentable ω-cpos other than flat ones, and these additional objects generate additional maps. But the countable Lawvere ωCpo-theory for a semilattice is still just the free countable Lawvere ωCpo-theory on the countable Lawvere theory for a semilattice.

This definition allows us to make immediate reference to the sum of effects that we shall define later. Using the terminology we shall define, we can therefore describe the countable Lawvere ωCpo-theory for nondeterminism.

Example 7 *The countable Lawvere ωCpo-theory for nondeterminism is given by the sum of the countable Lawvere ωCpo-theories L_N for a semilattice and $L_\perp$ for partiality.*

The combination of partiality with other effects is typically given by sum. But that is not always the case: the combination with side-effects is given by taking the commutative combination, which we define in the next section.

Another non-trivial example of a computationally natural countable Lawvere ωCpo-theory is given by probabilistic nondeterminism [5, 9, 10, 22]. This includes L_P and $L_\perp$ as well as an infinitary axiom.

2. The commutative combination of effects

In this section, we define the commutative combination $L \otimes L'$ of countable Lawvere theories L and L' and develop mathematical theory in support of the definition of this tensor product.

The category $\aleph_1$ not only has countable coproducts, but also has finite products, which we denote by $A \times A'$. The object $A \times A'$ may also be seen as the coproduct of A copies of A', so, given an arbitrary map $f' : A' \longrightarrow B'$ in a countable Lawvere theory, it is immediately clear what we mean by the morphism $A \times f' : A \times A' \longrightarrow A \times B'$. We define $f \times A'$ by conjugation, and, in the following, we suppress the canonical isomorphisms.

Definition 2 *Given countable Lawvere theories L and L', the countable Lawvere theory $L \otimes L'$ is defined by the universal property of having maps of countable Lawvere theories from L and L' to $L \otimes L'$, with commutativity of all operations of L with respect to all operations of L', i.e., given $f : A \longrightarrow B$ in L and $f' : A' \longrightarrow B'$ in L', we demand commutativity of the diagram*

$$\begin{array}{ccc} A \times A' & \xrightarrow{A \times f'} & A \times B' \\ {\scriptstyle f \times A'} \downarrow & & \downarrow {\scriptstyle f \times B'} \\ B \times A' & \xrightarrow[B \times f']{} & B \times B' \end{array}$$

Theorem 2 *The construction $\otimes$ extends canonically to a symmetric monoidal structure on the category of countable Lawvere theories. Moreover, for any small category C with countable products, there is a coherent equivalence of categories between $Mod(L \otimes L', C)$ and $Mod(L, Mod(L', C))$.*

Example 8 *Letting L_S be the countable Lawvere theory for side-effects, if C has countable products and countable coproducts, we have seen that $Mod(L_S, C)$ is equivalent to the category T-Alg for the monad $TX = (\Sigma_S X)^S$ on C. For any countable Lawvere theory L, the category $Mod(L, Set)$ is always complete and cocomplete, so has countable products and countable coproducts. So, by the theorem, $Mod(L_S \otimes L, Set)$ is equivalent to T-Alg for $TX = (\Sigma_S X)^S$ taken as a monad on $Mod(L, Set)$.*

3. The commutative combination of side-effects with other effects

Here, we study the commutative combination of side-effects with other computational effects in more detail. Our central result is as follows.

Theorem 3 *Let L_S denote the countable Lawvere theory for side-effects (where $S = V^{Loc}$) and let L denote any countable Lawvere theory. Then the monad $T_{L_S \otimes L}$ is isomorphic to $T_L(S \times -)^S$.*

Proof 1 *We have seen in preceding sections that $Mod(L_S, Mod(L, Set))$ is equivalent to T-Alg, where T is the monad on $Mod(L, Set)$ given by $TX = (\Sigma_S X)^S$. The category $Mod(L, Set)$ is equivalent to T_L-Alg. So we denote the canonical adjunction by $F_L \dashv U_L : Mod(L, Set) \longrightarrow Set$. Right adjoints preserve products, left adjoints preserve coproducts, and a coproduct*

$\Sigma_Y X$ in Set is given by $Y \times X$. So the monad $T_{L_S \otimes L}$, which, by our main theorem, is the monad determined by the composite forgetful functor from T-Alg to Set, must be given by $T_{L_S \otimes L}X = U_L(\Sigma_S F_L X)^S = T_L(S \times X)^S$ as required.

We do not require rank, in particular countability, for this result. We could define a notion of theory that does not involve a rank, retain a correspondence with strong monads, and make the commutative combination of the theorem, but the general theory becomes more complicated because commutative combinitions of such theories do not always exist.

This result shows that, under the hypotheses of the theorem, our theory of the commutative combination of computational effects agrees with Moggi's definition of the side-effects monad transformer. In particular, this accounts for the interaction between side-effects and nondeterminism, and in doing so, the theory yields not just an object of values for the combination but a description of natural operations and the way in which they interact with each other, and it follows immediately from the definition of $\otimes$ that one does not lose any of the equations for either nondeterminism or side-effects with which one began. It is also interesting to note that the side-effects theory for $S = V^{Loc}$ is the *Loc*-fold tensor product of the side-effects theory for $S = V$.

4. The sum of effects

Finally, we turn to the sum of effects, our leading example being given by the combination of exceptions with all other computational effects we have considered, such as side-effects, nondeterminism, and interactive input/output.

Theorem 4 *[12, 23] The category of countable Lawvere theories is cocomplete.*

Theorem 5 *Given a set E, if L_E is the countable Lawvere theory for E nullary operations, and if L is any countable Lawvere theory, then T_{L_E+L} is given by the theory $T_L(- + E)$.*

Proof 2 *The category $T_L(- + E)$-Alg is isomorphic to T'_L-Alg, where T'_L is the monad on $(- + E)$-Alg determined by lifting T_L, using the canonical distributive law of $- + E$ over T_L. By direct calculation, one can see that the latter category is in turn isomorphic to $(T_L + (- + E))$-Alg: a T'_L-algebra consists of a set X together with E elements of X and a T_L-structure on X, i.e., a $(T_L + (- + E))$-algebra.*

This result explains how the exceptions monad transformer, sending a monad T to the composite $T(- + E)$, arises: take the disjoint union of the two sets of operations and retain the equations for T. And it provides our usual theory of coproducts, such as its associativity and commutativity, and its interaction with other operations.

Definition 3 *Given a countable Lawvere theory L and a category C with countable products, denote by $Mod^*(L, C)$ the identity-on-objects/fully faithful factorisation of the forgetful functor $U : Mod(L, C) \longrightarrow C$.*

Theorem 6 *There is a natural equivalence between* $Mod^*(L + L', C)$ *and* $Mod^*(L, Mod^*(L', C))$.

As in previous sections, the analysis of this section all enriches without fuss, with the sum again being the correct operation in the enriched setting.

References

[1] M. Barr and C. Wells, *Category Theory for Computing Science*, Prentice-Hall, 1990.

[2] N. Benton, J. Hughes, and E. Moggi, *Monads and Effects*, APPSEM '00 Summer School, 2000.

[3] P. Cenciarelli and E. Moggi, *A Syntactic Approach to Modularity in Denotational Semantics*, CWI Technical Report, 1983.

[4] P. J. Freyd, Algebra-Valued Functors in General and Tensor Products in Particular, *Colloq. Math. Wroclaw* Vol. 14, pp. 89–106, 1966.

[5] R. Heckmann, Probabilistic Domains, in *Proc. CAAP '94*, LNCS, Vol. 136, pp. 21-56, Berlin: Springer-Verlag, 1994.

[6] M. C. B. Hennessy, *Algebraic Theory of Processes*, Cambridge, Massachusetts: MIT Press, 1988.

[7] M. C. B. Hennessy and G. D. Plotkin, Full Abstraction for a Simple Parallel Programming Language, in *Proc. MFCS '79* (ed. J. Beĉvář), LNCS, Vol. 74, pp. 108-120, Berlin: Springer-Verlag, 1979.

[8] J. M. E. Hyland and A. J. Power, Pseudo-Closed 2-Categories and Pseudo-Commutativities, *J. Pure Appl. Algebra*, to appear.

[9] C. Jones, *Probabilistic Non-Determinism*, Ph.D. Thesis, University of Edinburgh, Report ECS-LFCS-90-105, 1990.

[10] C. Jones and G. D. Plotkin, A Probabilistic Powerdomain of Evaluations, in *Proc. LICS '89*, pp. 186–195, Washington: IEEE Press, 1989.

[11] G. M. Kelly, *Basic Concepts of Enriched Category Theory*, Cambridge: Cambridge University Press, 1982.

[12] G. M. Kelly and A. J. Power, Adjunctions whose Counits are Coequalizers, and Presentations of Finitary Enriched Monads, *J. Pure Appl. Algebra*, Vol. 89, pp. 163–179, 1993.

[13] E. G. Manes, *Algebraic Theories*, Graduate Texts in Mathematics, Vol. 26, New York: Springer-Verlag, 1976.

[14] M. W. Mislove, Nondeterminism and Probabilistic Choice: Obeying the Laws, in *International Conference on Concurrency Theory*, pp. 350–364, URL: http://www.math.tulane.edu/ mwm, 2000.

[15] E. Moggi, Computational Lambda-Calculus and Monads, in *Proc. LICS '89*, pp. 14–23, Washington: IEEE Press, 1989.

[16] E. Moggi, *An Abstract View of Programming Languages*, University of Edinburgh, Report ECS-LFCS-90-113, 1989.

[17] E. Moggi, Notions of Computation and Monads, *Inf. and Comp.*, Vol. 93, No. 1, pp. 55–92, 1991.

[18] G. D. Plotkin, A Powerdomain Construction, *SIAM J. Comput.* Vol. 5, No. 3, pp. 452–487, 1976.

[19] G. D. Plotkin, *Domains*, URL: http://www.dcs.ed.ac.uk/home/gdp, 1983.

[20] G. D. Plotkin and A. J. Power, Adequacy for Algebraic Effects, in *Proc. FOSSACS 2001* (eds. F. Honsell and M. Miculan), LNCS, Vol. 2030, pp. 1–24, Berlin: Springer-Verlag, 2001.

[21] G. D. Plotkin and A. J. Power, Semantics for Algebraic Operations (extended abstract), in *Proc. MFPS XVII* (eds. S. Brookes and M. Mislove), ENTCS, Vol. 45, Amsterdam: Elsevier, 2001.

[22] G. D. Plotkin and A. J. Power, Notions of Computation Determine Monads, in *Proc. FOSSACS 2002* (eds. M. Neilsen and U. Engberg), LNCS, Vol. 2303, pp. 342–356, Berlin: Springer-Verlag, 2002.

[23] A. J. Power, Enriched Lawvere Theories, in *Theory and Applications of Categories*, pp. 83–93, 2000.

[24] A. J. Power and E. P. Robinson, Modularity and Dyads, in *Proc. MFPS XV* (eds. S. Brookes, A. Jung, M. Mislove and A. Scedrov), ENTCS Vol. 20, Amsterdam: Elsevier, 1999.

[25] E. Robinson, *Variations on Algebra: Monadicity and Generalisations of Equational Theories*, to appear, URL: http://www.dcs.qmul.ac.uk/ edmundr/publications.html, 2001.

OPTIMAL-REACHABILITY AND CONTROL FOR ACYCLIC WEIGHTED TIMED AUTOMATA*

Salvatore La Torre
University of Pennsylvania &
Università degli Studi di Salerno
sallat@dia.unisa.it

Supratik Mukhopadhyay
University of Pennsylvania
supratik@saul.cis.upenn.edu

Aniello Murano
Università degli Studi di Salerno
murano@dia.unisa.it

Keywords: Timed Automata, Control Synthesis, Optimization.

Abstract Weighted timed automata extend timed automata with costs on both locations and transitions. In this framework we study the optimal reachability and the optimal control synthesis problems for the automata with acyclic control graphs. This class of automata is relevant for some practical problems such as some static scheduling problems or air-traffic control problems. We give a nondeterministic polynomial time algorithm to solve the decision version of the considered optimal reachability problem. This algorithm matches the known lower bound on the reachability for acyclic timed automata, and thus the problem is NP-complete. We also solve in doubly exponential time the corresponding control synthesis problem.

*The first and the second authors were supported in part by the NSF award CCR99-70925, by the SRC award 99-TJ-688, and by the DARPA ITO Mobies award F33615-00-C-1707. The first and the third authors were supported in part by the MURST in the framework of the project "Metodi Formali per la Sicurezza" (MEFISTO). The first author was also supported by the NSF ITR award.

Introduction

Timed automata were introduced by Alur and Dill [2] to model real-time systems, that is, systems interacting with physical processes and whose correct behavior crucially depends upon real-time considerations. A timed automaton is a finite automaton augmented with a finite set of real-valued *clocks*. Transitions are enabled according the current *location* and the current clock values. In a transition, clocks can be instantaneously reset, and the value of a clock is exactly the time elapsed since the last time it was reset. A basic decision problem on timed automata is *reachability*. Given an automaton A, the reachability problem can be stated as the problem of deciding if there exists a run of A from a location s to a location t. Reachability for timed automata is known to be PSPACE-complete [2], and can be used in many contexts, such as the verification of *safety properties* ("nothing bad will eventually happen") or the *emptiness problem* ("is the language accepted by a given automaton empty?").

Designing reactive systems often requires each component to be viewed as an open system, that is, a system which interacts with its environment and whose behavior depends on its current state as well as the behavior of the environment. In open systems, a central problem is *controller synthesis*: "given a specification φ and a *plant* P, we want to determine a *controller* for P such that φ is satisfied". For example, a typical specification in an air-traffic management problem is to require that a controller must try to prevent aircraft collisions. A plant P is described by all its possible behaviors and a controller for P is then a system that can observe the state of P and issue a *control action* to influence its behavior. In an air-traffic management problem, control actions might consist of establishing priorities among aircrafts, or forcing an aircraft to modify its trajectory. Specifications in control synthesis problems can be given using different formalisms. In a reachability control synthesis, the specification is usually expressed by a set of states with the meaning that the desired controller must drive the plant to these states despite of the behavior of the environment. Reachability control synthesis problems on timed automata are considered in [5, 7, 15, 17]. LTL specifications for rectangular hybrid systems are studied in [14]. LTL and CTL specifications for timed automata are investigated using an automata-theoretic approach in [11]. Also specifications given as timed automata [10] and TCTL formulas [12] have been recently addressed.

Associating with a run of a timed automaton a performance measure, it is possible to compare runs, and thus, search for an optimal run connecting two locations. The most natural performance measure for timed automata is clearly time. Time-optimal reachability was first considered in [9], where the problem of computing lower and upper bounds on time delays in timed automata was solved. Minimum-time reachability is also considered in [18] and is shown to be NP-complete acyclic timed automata in [3]. The related problem for open systems is the synthesis of a time-optimal controller which is shown to be decidable in [5]. We recall that, it is known that the related decision problem is EXPTIME-complete [14, 15]. Besides time, other cost functions have been considered. Given a timed automaton, a weight w can be associated with each

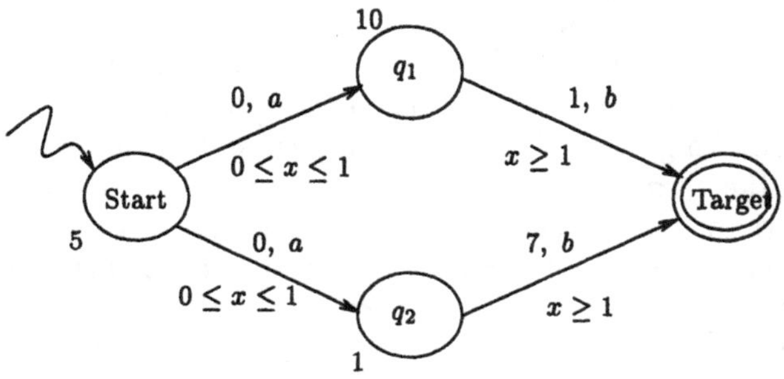

Figure 1. A weighted timed automaton.

location q such that w gives the cost of a unit of time spent in q. On such a model, given a cost interval I and two states s and t, the decision problem "is t reachable from s at a cost $c \in I$?" (***duration-bounded reachability***) is addressed in [1]. A ***weighted timed automaton*** extends a timed automaton with costs on both locations and transitions[1]. On this model the optimal-reachability problem was independently addressed in [4] and [8].

The techniques, used in [4, 8] to solve the optimal-reachability problem, do not extend to solving the corresponding optimal-synthesis problem. In fact, the linearity of the cost function implies that optimal runs can be found among those having only transitions taken in one of the following cases: as soon as the corresponding guard holds true, right before the corresponding guard holds false, or when at least a clock variable has an integer value. In the optimal-synthesis problem we consider this property does not hold. To see this consider the weighted timed automaton in Figure 1. In the automaton, nondeterminism is used to model the choices of the environment and the symbols labeling the edges are the actions that can be taken by the controller. Denoting by t the time spent in location *Start*, the minimum cost to reach *Target* through location q_1 is given by the infimum value of $(11-5t)$ for $t \in [0,1]$ and the minimum cost to reach *Target* through q_2 is the infimum of $(4t+8)$ for $t \in [0,1]$. In this example, the controller is only allowed to select the time at which a transition from *Start* is taken (since both transitions are labeled by a) and the environment selects instead the transition. Thus the best choice for the controller is t^* such that $\max\{11-5t, 4t+8\}$ is minimum, that is, $t^* = \frac{1}{3}$. In case we consider the optimal-reachability problem on the same automaton (i.e., the controller is allowed to select the actual transition in presence of nondeterminism), an optimal run goes through location q_1 with the first transition taken at time 1.

In this paper we are concerned with both the optimal reachability and the optimal control synthesis for acyclic weighted timed automata. There are several interesting problems that can be modeled by acyclic automata such as some static scheduling problems and air-traffic control problems. Recently the *job-shop scheduling problem* [6] has been modeled as minimum-time reachability on acyclic timed automata. Considering different costs for each task to schedule is realistic and can be captured by a weighted timed automaton. We show that the optimal reachability problem for acyclic weighted timed automata is in FP^{NP}. We recall that in the general case, the known upper bound is EXPTIME[4]. We give an algorithm that combines binary search along with a solution to the corresponding decision problem: "is there a run r from s to t whose cost is not larger than c?" (*cost-bounded reachability*). We prove that the cost-bounded reachability can be solved in nondeterministic polynomial time, and since reachability is NP-hard for acyclic timed automata [3], we get that this problem is NP-complete. We also solve the optimal synthesis problem for acyclic weighted timed automata in doubly exponential time. Our solution consists of reducing this problem to deciding some particular kind of first-order logic formulas of the theory of reals with addition and order. The translation causes an exponential blow-up and the satisfiability of such formulas can be decided in exponential time. We recall that in general the first-order theory of reals with addition and order is decidable in doubly exponential time [13].

The rest of the paper is organized as follows. In Section 1 we recall the definition of weighted timed automata and introduce our notation. Our solution to the optimal reachability problem for acyclic weighted timed automata is given in Section 2. In Section 3 the optimal control synthesis problem for acyclic weighted timed automata is solved. Finally, in Section 4 we give our conclusions.

1. Weighted timed automata

A timed automaton models a real-time system. We assume that there is a central (real-valued) clock, and the model can use a finite set of *clock variables* (also said simply *clocks*) along with timing constraints to check the satisfaction of timing requirements. Each clock can be seen as a chronograph synchronized with the central clock, thus it can be read or set to zero (reset): after a reset, a clock restarts automatically. In each automaton, timing constraints are expressed by clock constraints. Let C be a set of clocks, the set of clock constraints $\Xi(C)$ contains:

- $x \leq y+c$, $x \geq y+c$, $x \leq c$ and $x \geq c$ $\forall x, y \in C$ and for a rational number c; we call such constraints *atomic* clock constraints;
- $\neg\delta$ and $\delta_1 \wedge \delta_2$ where $\delta, \delta_1, \delta_2 \in \Xi(C)$.

Furthermore, a *clock interpretation* is a mapping $\nu : C \longrightarrow \mathbb{R}_+$. We denote by $\vec{0}$ the clock interpretation mapping each clock to 0. If ν is a clock interpretation, λ is a set of clocks and d is a real number, we denote with $\nu(\lambda, d)$ the clock interpretation that for each clock $x \in \lambda$ gives 0 and for each clock $x \notin \lambda$ gives the value $\nu(x) + d$.

Definition 1 *A* timed automaton A *is a tuple* $(\Sigma, \mathcal{Q}, \mathcal{Q}_0, C, \Delta, inv)$ *where:*

- Σ *is a finite set of symbols (the alphabet);*
- $\mathcal{Q}$ *is a finite set of locations;*

- *$Q_0 \subseteq Q$ is the set of initial locations;*
- *C is a finite set of n clock variables;*
- *Δ is a finite subset of $Q \times \Sigma \times \Xi(C) \times 2^C \times Q$ (edges);*
- *$inv : Q \longrightarrow \Xi(C)$ maps each location q to its invariant $inv(q)$.*

A *state* of a timed automaton A is a pair $\langle q, \nu \rangle$ where $q \in Q$ and $\nu \in \mathbb{R}_+^n$. An *initial state* is a pair $\langle q_0, \vec{0} \rangle$ where $q_0 \in Q_0$ is an initial location. The semantics of a timed automaton is given by a transition system over the set of states. The transitions of this system are divided into *discrete steps* and *time steps.* A discrete step is $\langle q, \nu \rangle \xrightarrow{\sigma} \langle q', \nu' \rangle$ where $(q, \sigma, \delta, \lambda, q') \in \Delta$, ν satisfies δ, $\nu' = \nu(\lambda, 0)$, and ν' satisfies $inv(q')$. A time step is $\langle q, \nu \rangle \xrightarrow{d} \langle q, \nu' \rangle$ where $\nu' = \nu + d$, $d \geq 0$, and $\nu + d'$ satisfies $inv(q)$ for all $0 \leq d' \leq d$. A *step* is $\langle q, \nu \rangle \xrightarrow{\sigma, d} \langle q', \nu' \rangle$ where $\langle q, \nu \rangle \xrightarrow{d} \langle q, \nu'' \rangle$ and $\langle q, \nu'' \rangle \xrightarrow{\sigma} \langle q', \nu' \rangle$, for some $\nu'' \in \mathbb{R}^n$. A *timed sequence* (σ, τ) over the alphabet Σ is such that $\sigma \in \Sigma^*$, $\tau \in \mathbb{R}_+^*$, and $|\sigma| = |\tau|$. The sequence τ is called a *time sequence.* In a timed sequence, each symbol σ_i at input is associated with a positive real number τ_i, which expresses (except for the first symbol σ_1) the time which has elapsed since the symbol σ_{i-1} was at input. Time τ_1 represents instead the time at which the symbol σ_1 appears at input assuming that time is 0 when the computation starts. A run r of a timed automaton A on a timed sequence (σ, τ), where $\sigma = \sigma_1 \dots \sigma_k$ and $\tau = \tau_1 \dots \tau_k$, is a finite sequence $\langle q_0, \nu_0 \rangle \xrightarrow{\sigma_1, \tau_1} \langle q_1, \nu_1 \rangle \xrightarrow{\sigma_2, \tau_2} \dots \xrightarrow{\sigma_k, \tau_k} \langle q_k, \nu_k \rangle$. We say that r starts at q_0 and ends at q_k. We denote the set of A runs by Run(A). A timed automaton is *acyclic* if its control graph is acyclic.

A *weighted timed automaton* is a timed automaton A with cost functions:

- $J_s : \Delta \longrightarrow \mathbb{N}$ (*switch cost*), and
- $J_d : Q \longrightarrow \mathbb{N}$ (*duration cost*).

Given a run r of A, let $e_1, \dots, e_k$ be the sequence of transitions taken in r and $q_0, \dots, q_k$ be the sequence of visited locations. We associate to r the following costs:

- $J_s(r) = \sum_{i=1}^{k} J_s(e_i)$, and
- $J_d(r) = \sum_{i=0}^{k-1} t_i \cdot J_d(q_i)$.

The total cost associated with a run r is then $J(r) = J_s(r) + J_d(r)$. As an example of a weighted timed automaton consider the one modeling a simple air traffic control system given in Figure 2. A different air traffic control system modeled by an acyclic weighted timed automaton is described in [16].

Example 1 Consider the timed transition system in Figure 2. It models a scenario in which two aircrafts send a landing request to a control tower, and

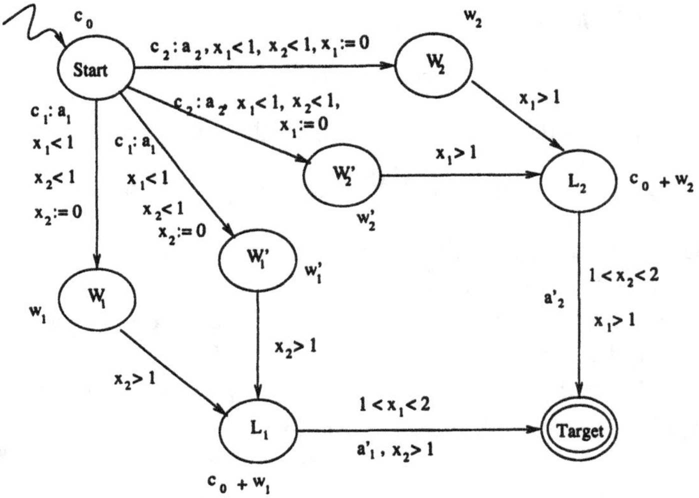

Figure 2. An air-traffic control problem.

our goal is to allow both the aircrafts to land safely and at minimum cost. Safety requires that only one aircraft at a time must be acknowledged for landing, thus there are two possible choices: aircraft 1 waits for aircraft 2 to land (action a_1), or vice-versa (action a_2). There are costs c_1 and c_2 relative to actions a_1 and a_2 respectively, that correspond to the cost to pay for forcing respectively aircraft 1 and aircraft 2 to wait. Moreover, the cost of waiting may vary according to different environmental situations. We consider two possible scenarios for each taken action. If action a_1 is taken, depending on the environment the cost of waiting is either w_1 or w_1' per time unit. Analogously, we have costs w_2 and w_2' relatively to action a_2. Since it is realistic to reduce the amount of time a runway stays unused, we penalize this event by a cost c_0 per time unit. Finally, we assume that the landing of each aircraft takes at least time 1 since the related acknowledgment was issued by the control tower. Actions a_1' and a_2' correspond to the landing of aircrafts 1 and 2, respectively, after they are forced to wait.

2. Optimal reachability

Given an automaton A, the *reachability problem* can be stated as the problem of deciding if there exists a run of A from a location s to a location t. Here we study a related optimization problem on acyclic weighted timed automata. Given a weighted timed automaton A, a source location src and a target location trg, the *optimal cost* J^* to reach trg from src is the infimum of $J(r)$ over all the runs r from $\langle src, \vec{0}\rangle$ to a state $\langle trg, \nu\rangle$, for some clock interpretation ν. If location trg is reachable from $\langle s, \vec{0}\rangle$, the optimal cost is well defined while a run matching this cost may not exist. Thus the *optimal reachability problem* can be stated as the problem of determining a run of optimal cost, if there exists one, and an infinite family of runs parameterized on ξ and containing for any $\xi > 0$ a run r such that $J(r) < J^* + \xi$, otherwise. In this section we give an algorithm to determine the optimal cost J^* given an acyclic weighted timed automaton, and two locations src and trg. Clearly a solution to the optimal reachability problem can be constructed by tracking back the transitions leading to the optimal cost.

In relation to the optimal reachability problem we consider the following *cost-bounded reachability problem*: "given a positive real c and a weighted timed automaton A, is there a run r from the state $\langle src, \vec{0}\rangle$ to a location trg such that $J(r) \leq c$?" We prove that this problem is NP-complete for acyclic weighted timed automata.

In the rest of the section we assume that the constants in the clock constraints of the considered automata are natural numbers. All the obtained results extend to the general case in the following way. Let u be the minimum common denominator among all the constants used in a timed automaton. We use $\frac{1}{u}$ instead of 1 as time unit and thus all the results rescale according to this.

Denote by $c_{\max}$ the largest constant used in the clock constraints incremented by 1, by $J_d^{\max}$ the largest duration cost, by $J_s^{\max}$ the largest switch cost, and by M the longest path in the control graph of a timed automaton. We have the following result.

Lemma 1 *Given two locations src and trg of a acyclic weighted timed automaton A, if trg is reachable from* $\langle src, \vec{0}\rangle$ *then there exists a run r such that* $J(r) \leq M \cdot (J_s^{\max} + J_d^{\max} \cdot c_{\max})$.

Proof. We observe that on any run from $\langle src, \vec{0}\rangle$ to a state in location trg there are at most M transitions. Moreover, any transition guard either is satisfied within time $c_{\max}$ or stays unsatisfied forever. Thus the lemma holds. ■

The above result gives a simple polynomial-time reduction from the reachability problem for acyclic timed automata.

Theorem 1 *The cost-bounded reachability problem for acyclic weighted timed automata is NP-complete.*

Proof. NP-hardness is consequence of the NP-hardness of the reachability problem for acyclic timed automata [3] and Lemma 1. To prove NP-membership we use an algorithm that nondeterministically guesses a linear program of linear size and solves it. Since all the variables involved in the guessed linear program are real-valued, this algorithm runs in nondeterministic polynomial time. We sketch the algorithm in the following. For the sake of simplicity we assume that the control graph of the automaton we consider does not have multiple edges connecting two given locations. We denote by $q_0, \ldots, q_m$ the locations of the weighted timed automaton, where q_0 is the source location and q_m is the target location, and by $e_{i,j}$ the transition from a location q_i to a location q_j. Observe first that given a boolean formula φ and a formula φ' which is a conjunction of literals containing exactly a literal for each variable in φ, there exists a linear-time algorithm to check if φ' implies φ (this is equivalent to evaluate φ on a given variable assignment). Denote by $\delta_{i,j}$ and $\lambda_{i,j}$ respectively the clock constraint and the reset associated to a transition $e_{i,j}$. Let $\varphi_1, \dots, \varphi_k$ be the atomic clock constraints in $\delta_{i,j}$. Given a sequence of bits $\beta = b_1 \dots b_k$, let φ'_h be φ_h, if $b_h = 1$, and $\neg\varphi_h$, otherwise. We define $\Delta^{\beta}_{i,j}$ be $\bigwedge_h \varphi'_h$, if $\bigwedge_h \varphi'_h$ implies $\delta_{i,j}$, and the constant FALSE, otherwise. Moreover, we define a variable $a_{i,j}$ which evaluates 1, if transition $e_{i,j}$ is taken, and 0 otherwise. We use a variable $\nu_{i,j}$ to store the clock interpretation after that transition $e_{i,j}$ is taken, and a variable $t_{i,j}$ to store the time spent in location q_i before taking $e_{i,j}$. For a given location q_i we denote by $succ(i)$ the set $\{j \mid e_{i,j} \text{ is a transition}\}$ and by $pre(i)$ the set $\{j \mid e_{j,i} \text{ is a transition}\}$. Denote by $l_i = \sum_{j \in succ(i)} a_{i,j}$ for $i = 0, \dots, m-1$ and $l_m = \sum_{j \in pre(m)} a_{j,m}$. The objective of the program is to minimize the function $\sum_{i,j} a_{i,j}\, J_s(e_{i,j}) + \sum_{i,j} l_i\, J_d(q_i)\, t_{i,j}$ according to the following constraints:

1 $\sum_{j \in succ(0)} a_{0,j} = 1$ (exactly a transition must be taken from q_0);

2 $\sum_{j \in pre(i)} a_{j,i} = \sum_{h \in succ(i)} a_{i,h}$ (conservation property: if q_i is reached, just a transition can be taken);

3 $\sum_{j \in pre(m)} a_{j,m} = 1$ (the target location must be reached);

4 $\nu_{0,j} = \vec{0}(\lambda_{0,j}, t_{0,j})$, and for $i > 0$, $\nu_{i,j} = \sum_{k \in pre(i)} \nu_{k,i}(\lambda_{i,j}, t_{i,j})\, a_{k,i}$ (resets);

5 $a_{0,j} = 0 \vee \Delta^{\beta}_{0,j}(\vec{0} + t_{0,j})$, and for $i > 0$, $a_{i,j} = 0 \vee \Delta^{\beta}_{i,j}(\sum_{k \in pre(i)}(\nu_{k,i}\, a_{k,i}) + t_{i,j})$ (time constraints).

We claim that the above program gives a nondeterministic polynomial-time algorithm to solve the cost-bounded reachability problem. To see this, observe that a valuation of all $a_{i,j}$'s fulfilling the above constraints 1, 2, and 3 selects a path from q_0 to q_m. Moreover, providing a valuation for any sequence β, for any transition $e_{i,j}$, we have either an empty region (if some $e_{i,j}$ has been selected and sequence chosen for it does not correspond to a conjunctive formula implying $\delta_{i,j}$) or a linear program P such that the optimal solution corresponds to the optimal cost of a run corresponding to the selected path. Since any of such P contains only real-valued variables, the optimal solution can be computed in polynomial time. Moreover, the size of any P is linear in the size of the given automaton. Thus, we have proved NP membership for the cost-bounded reachability problem. ∎

The nondeterministic polynomial-time algorithm solving the cost-bounded reachability problem can be used to obtain a nondeterministic polynomial-time algorithm to solve the optimal reachability problem.

Theorem 2 *The optimal reachability problem for acyclic weighted timed automata is in FP^{NP}.*

Proof. Since the cost function is linear in the time, we have that, under the assumption that all constants in the automaton are integers, the optimal cost is also an integer. The optimal cost can be computed simply by combining a binary search in the interval from 0 to $M \cdot (J_s^{\max} + J_d^{\max} \cdot c_{\max})$, with an algorithm solving the cost-bounded reachability problem. By Theorem 1 this last step can be done in nondeterministic polynomial-time. Since the algorithm for computing the optimum cost, as well as the optimum run, calls a polynomial number of times the nondeterministic polynomial time procedure, we get that the total algorithm is in FP^{NP}. ■

3. Optimal Control

We model the controller synthesis problem for timed automata as a *timed game* of a *controller* against the *environment*, and then the synthesis of a controller corresponds to determine a winning strategy. In our settings, the game is modeled as a nondeterministic timed automaton, where actions represent the choices of the controller and the nondeterminism is used to model the possible choices of the *environment* for a given action taken by the controller[2]. We use a special action denoted by ε to capture the case that the controller is *idle* and the environment is moving. When both players are idle, a time step is taken. In the following we will denote by Σ^ε the set of actions including the idle action ε. A *play* of a timed game is constructed in the following way. At each time, a player declares how long it will wait idling until its next choice. At the time one of the players or both move, both the players are allowed to redeclare their next move and the time they will issue it. That is, if a player moves before the other, the latter is allowed to change its former decision. A play is represented by a run of the automaton. Formally, a *timed game* is a tuple (A, src, trg) where

- $A = (\Sigma^\varepsilon, \mathcal{Q}, \mathcal{Q}_0, C, \Delta, inv)$ is a timed automaton,
- $src \in \mathcal{Q}$ is the *source location* and $trg \in \mathcal{Q}$ is the *target location.*

A *weighted timed game* is defined in the same way by considering a weighted timed automaton instead of a timed automaton. In the following we can assume that A has no outgoing transition from the target location trg. The goal of our computations is to reach trg. Given a run $r = \langle q_0, \nu_0\rangle \xrightarrow{\tau_1,\sigma_1} \langle q_1, \nu_1\rangle \xrightarrow{\tau_2,\sigma_2} \ldots \xrightarrow{\tau_k,\sigma_k} \langle q_k, \nu_k\rangle$, we denote by r_i the run $\langle q_0, \nu_0\rangle \xrightarrow{\tau_1,\sigma_1} \langle q_1, \nu_1\rangle \xrightarrow{\tau_2,\sigma_2} \ldots \xrightarrow{\tau_i,\sigma_i} \langle q_i, \nu_i\rangle$. A *strategy* is a function $\mathcal{F} : Plays(\mathcal{F}) \longrightarrow \mathbb{R} \times \Sigma$, where $Plays(\mathcal{F}) \subseteq \mathrm{Run}(A)$, $\langle q_0, \nu_0\rangle = \langle src, \vec{0}\rangle \in Plays(\mathcal{F})$ and for any $r = \langle q_0, \nu_0\rangle \xrightarrow{\tau_1,\sigma_1} \langle q_1, \nu_1\rangle \xrightarrow{\tau_2,\sigma_2} \ldots \xrightarrow{\tau_k,\sigma_k} \langle q_k, \nu_k\rangle$ belonging to $Plays(\mathcal{F})$, it holds that for $i = 1, \ldots, k-1$, either $\mathcal{F}(r_i) = (\tau_{i+1}, \sigma_{i+1})$ or $\mathcal{F}(r_i) = (d, \sigma)$, $\tau_{i+1} < d$ and $\sigma_{i+1} = \varepsilon$. In other words a strategy gives the moves of the controller on each play which is "consistent" with the strategy and the case $\sigma_{i+1} = \varepsilon$ corresponds to a move of the environment taken before the next declared move of the controller according to a strategy $\mathcal{F}$. A run is *maximal* if it ends in a state from which it will be possible only to have time steps. A strategy $\mathcal{F}$, is *winning* if all maximal runs $r \in Plays(\mathcal{F})$ end in a state of the target location. Given a winning strategy $\mathcal{F}$, the cost of $\mathcal{F}$ is defined as $J(\mathcal{F}) = \sup_{r \in Plays(\mathcal{F})} J(r)$. The optimal cost for a winning strategy is defined as

$$J^* = \inf_{\mathcal{F}} J(\mathcal{F})$$

and there exists an optimal winning strategy if and only if there exists $\mathcal{F}^*$ such that

$$J(\mathcal{F}^*) = J^*.$$

If there exists a winning strategy, the optimal cost is always well defined while an optimal winning strategy may not exist. When this is the case, it is possible to determine a family of strategies parameterized on ξ such that for any $\xi > 0$ there exists a strategy in this family such that its cost J is such that $J \leq J^* + \xi$. In the following we refer as an optimal solution of a weighted timed game to either a winning strategy, if one exists, or to such a family of winning strategies, otherwise. We focus on the following optimization problem: "given a weighted timed game, determine if there exists a winning strategy, and in this case, the cost of an optimal winning strategy".

In the sequel, we will refer to the controller as the protagonist and to the environment as the adversary. Let $\mathcal{C}(q,\nu)$ be the cost of an optimal winning strategy for the protagonist when the game starts at the state $\langle q,\nu\rangle$. For the target location trg, $\mathcal{C}(trg,\nu) = 0$ for all $\nu \in \mathbb{R}^n$. The cost $\mathcal{C}(q,\nu)$ of an optimal strategy for any state with location q, different from trg, is computed as described below.

Given a $t \geq 0$, the cost incurred by the protagonist taking an action σ after time t is the maximum cost over all the transitions on σ enabled at time t and all the transitions on ε enabled at any $0 \leq t' < t$. While the protagonist aims to minimize the cost of a strategy, the adversary will chose the transitions that will maximize it. Let $\mathcal{E}_\sigma$ be the set of σ transitions and E be the set of ε transitions (adversary transitions) that are enabled at time t. Then we have the following recurrence:

$$\mathcal{C}(q,\nu) = \inf_{t\geq 0} \max \begin{cases} \min_{\sigma\in\Sigma} \max_{e\in\mathcal{E}_\sigma} (J_d(q)\cdot t + J_s(e) + \mathcal{C}(q',\nu(\lambda,t))) \\ \sup_{0\leq t'\leq t} \max_{e\in E} (J_d(q)\cdot t' + J_s(e) + \mathcal{C}(q'',\nu(\lambda',t'))) \end{cases} \tag{1}$$

In order to compute the cost of an optimal strategy for a game starting at state $\langle q,\nu\rangle$, we need to solve the above equation. In the rest of this section, we outline an algorithm for solving it. Notice that Equation 1 is defined inductively and involves max, min, sup, and inf operators. Since these operators are definable in first order logic and the graph is acyclic, it is possible to describe Equation 1 in terms of a formula in the first order theory of reals with addition and order. Moreover, this theory is decidable and admits quantifier elimination.

More precisely, let $p(Q,\vec{X},C)$ be an $(n+2)$-ary predicate defined over the set of locations, the set of clock values and the set of costs corresponding to strategies adopted by the protagonist, with $p(q,\nu,c)$ standing for the fact that the cost of an optimal strategy starting from the state $\langle q,\nu\rangle$ is c. In the rest of this section we will use upper case letters to denote variables and lower case letters to denote constants; also, whenever we omit quantifiers on variables, we will assume that they are universally quantified. Observe that for the target location trg, $p(trg, \nu, 0)$ is true for all clock values $\nu \in \mathbb{R}^n$. For any other location q, different from trg, we briefly sketch how to define a first order logic formula that describes Equation 1, and leave the details to the full version of this paper. The first order logic formula corresponding to q is given by[3]:

$$p(q, \vec{X}, C) \longleftarrow \exists T \geq 0.\, \varphi(q, \vec{X}, T, C),$$

where the first order logic formula φ corresponds to $\mathcal{C}(q, \nu)$. From Equation 1:

$$\begin{aligned} \varphi(q, \vec{X}, T, C) \quad \equiv \quad & \varphi_{lb}(q, \vec{X}, T, C) \\ & \wedge \forall C'.(\varphi_{lb}(q, \vec{X}, T, C') \longrightarrow C \geq C'), \end{aligned}$$

where the formula φ_{lb} holds true for any C expressing a lower bound on the costs of winning strategies from $\langle q, \vec{X}\rangle$. We define φ_{lb} as:

$$\varphi_{lb}(q, \vec{X}, T, C) \quad \equiv \quad \forall C'.\forall T'.((T' \geq 0 \wedge \psi_1(q, \vec{X}, T', C')) \longrightarrow C' \geq C).$$

In the above formula, the first order formula ψ_1 corresponds to $g(t)$ such that $\mathcal{C}(q, \nu) = \inf_{t \geq 0} g(t)$ corresponds to Equation 1. We define ψ_1 as:

$$\begin{aligned} \psi_1(q, \vec{X}, T, C) \quad \equiv \quad & (\psi_2(q, \vec{X}, T, C) \vee \psi_3(q, \vec{X}, T, C)) \\ & \wedge (\forall C'.\psi_2(q, \vec{X}, T, C') \longrightarrow C \geq C') \\ & \wedge (\forall C'.\psi_3(q, \vec{X}, T, C') \longrightarrow C \geq C'), \end{aligned}$$

where the first order formulas ψ_2 and ψ_3 correspond respectively to terms $\min_{\sigma \in \Sigma} \max_{e \in \mathcal{E}_\sigma} (J_d(q) \cdot t + J_s(e) + \mathcal{C}(q', \nu(\lambda, t)))$ and $\sup_{0 \leq t' \leq t} \max_{e \in E}(J_d(q) \cdot t' + J_s(e) + \mathcal{C}(q'', \nu(\lambda', t')))$ in Equation 1. We give below the definition of ψ_3, and the first order formula ψ_2 can be defined similarly:

$$\begin{aligned} \psi_3(q, \vec{X}, T, C) \quad \equiv \quad & \exists T''.0 \leq T'' \leq T \wedge \psi_{ub}(q, \vec{X}, T'', C) \\ & \wedge \forall C'.(\psi_{ub}(q, \vec{X}, T'', C') \longrightarrow C' \geq C), \end{aligned}$$

where the first order formula ψ_{ub} holds true for any upper bound C on the costs of winning strategies from $\langle q, \vec{X}\rangle$. We define this last formula as:

$$\psi_{ub}(q, \vec{X}, T, C) \quad \equiv \quad \forall T'.\forall C'.((0 \leq T' \leq T \wedge \psi_4(q, \vec{X}, T', C')) \longrightarrow C \geq C'),$$

where the first order formula ψ_4 corresponds to $\max_{e \in E}(J_d(q) \cdot t' + J_s(e) + \mathcal{C}(q'', \nu(\lambda', t')))$ in Equation 1. We define ψ_4 as:

$$\begin{aligned} \psi_4(q, \vec{X}, T, C) \quad \equiv \quad & \textstyle\bigvee_{e \in E} \alpha(q, \vec{X}, e, T, C) \\ & \wedge \textstyle\bigwedge_{e \in E}(\forall C'.\alpha(q, \vec{X}, e, T, C') \longrightarrow C \geq C'), \end{aligned}$$

where E is the set of ε moves enabled at location q. The first order subformula α above corresponds to the expression $(J_d(q) \cdot t' + J_s(e) + \mathcal{C}(q'', \nu(\lambda', t')))$ in Equation 1 and is defined as follows:

$$\begin{aligned} \alpha(q, \vec{X}, e, T, C) \quad \equiv \quad & \exists C'.\, p(q', \vec{X}(\lambda, T), C') \wedge C = C' + J_d(q) \cdot T + J_s(e) \\ & \wedge \delta[\vec{X} \leftarrow \vec{X} + T], \end{aligned}$$

where $e = (q, \varepsilon, \delta, \lambda, q')$, and $\delta[\vec{X} \leftarrow \vec{X} + T]$ is the formula obtained by replacing all free occurrences of $\vec{X}$ in δ by $\vec{X} + T$.

Expanding the definition of φ, for each location q, different from t, we obtain a formula $p(q, \vec{X}, C) \longleftarrow \varphi$ such that q does not occur in φ. Since the control graph of the timed automaton is acyclic, a forward chaining starting from the fact $p(trg, \vec{X}, 0)$ for the target node trg will terminate. The length of the chaining is bounded by the size of the control graph of the automaton. At each step in the chaining, we eliminate the quantifiers occurring in the formula on the right-hand-side of the implication. Notice that at each step of the deduction the quantifier depth is constant. Hence at each step the quantifier elimination can be done in exponential time in the size of the formula (the Ferrante-Rackoff [13] algorithm runs in exponential time for such formulas). Since in the worst case the size of each formula is exponential in the size of the automaton, the complexity of the deduction is doubly exponential. Thus the following theorem holds.

Theorem 3 *The optimal control problem for weighted acyclic timed automata can be solved in doubly exponential time.*

4. Conclusions

In this paper we have solved the optimal reachability and the optimal control synthesis for acyclic weighted timed automata. We have proved that the considered optimal reachability problem is FP^{NP}. Moreover, we give a doubly exponential upper bound for the optimal control synthesis problem for acyclic automata. Proving a better upper bound or a lower bound matching our solution is an open problem. It is also an open problem to find a solution to this problem for general weighted timed automata. Our solution strongly exploits the acyclicity of the control graph and thus does not allow a direct extension to the general case.

Notes

1. In the literature, this model is also known as *linearly priced timed automaton*.

2. Games are usually defined in a symmetric way with respect to each of the players. In optimal control synthesis, the interest is focused on the winning strategies of the controller.

3. We write a formula φ as $\varphi(\beta)$ to denote that the expression β possibly occurs in φ.

References

[1] R. Alur, C. Courcoubetis, and T.A. Henzinger. Computing accumulated delays in real-time system. In *Proc. of the Fifth International Conference on Computer-Aided Verification, CAV'93*, LNCS 697, pages 181 – 193. Springer, 1993.

[2] R. Alur and D.L. Dill. A theory of timed automata. *Theoretical Computer Science*, 126:183 – 235, 1994.

[3] R. Alur, S. La Torre and S. Mukhopadhyay. *Subclasses of Timed Automata with NP-complete Reachability Problem.* CIS Department, University of Pennsylvania, 2001. URL:"`http://www.cis.upenn.edu/~latorre/Papers/ata.ps.gz`".

[4] R. Alur, S. La Torre, and G. J. Pappas. Optimal paths in weighted timed automata. In *Proc. of the 4th International Workshop on Hybrid Systems: Computation and Control, HSCC'01*, LNCS 2034, pages 49 – 62. Springer, 2001.

[5] E. Asarin and O. Maler. As soon as possible: Time optimal control for timed automata. In *Proc. of the 2nd International Workshop on Hybrid Systems: Computation and Control*, LNCS 1569, pages 19 – 30. Springer, 1999.

[6] Y. Abdedaim and O. Maler. Job-shop scheduling using timed automata. In *Proc. of the 13th Intern. Conference on Computer Aided Verification, CAV'01*, LNCS 2102, pages 478 –492, 2001.

[7] E. Asarin, O. Maler, and A. Pnueli. Symbolic controller synthesis for discrete and timed systems. In *Proc. of the 2nd International Workshop on Hybrid Systems*, LNCS 999, pages 1 – 20. Springer, 1995.

[8] G. Behrman, T. Hune, A. Fehnker, K. Larsen, P. Pettersson, R. Romijn, and F. Vaandrager. Minimum-cost reachability for priced timed automata. In *Proc. of the 4th International Workshop on Hybrid Systems: Computation and Control, HSCC'01*, LNCS 2034, pages 147–161. Springer, 2001.

[9] C. Courcoubetis and M. Yannakakis. Minimum and maximum delay problems in real-time systems. In *Proc. of the 3rd International Conference on Computer Aided Verification*, LNCS 575, pages 399 – 409. Springer, 1991.

[10] D. D'Souza and P. Madhusudan. Timed control synthesis for external specifications. In *Proc. of the 19th Annual Symposium on Theoretical Aspects of Computer Science, STACS'02*, LNCS 2285, pages 571 – 582. Springer, 2002.

[11] M. Faella, S. La Torre, and A. Murano. Automata-theoretic decision of timed games. In *Proc. of the 3rd Intern. Workshop on Verification, Model Checking, and Abstract Interpretation, VMCAI'02*, LNCS 2294, pages 94 –108. Springer, 2002.

[12] M. Faella, S. La Torre, and A. Murano. Dense Real-time Games. In *Proc. of the 17th Annual IEEE Symposium on Logic in Computer Science, LICS'02*, IEEE Computer Society Press, 2002.

[13] J. Ferrante and C. Rackoff. A decision procedure for the first order theory on real addition with order. *SIAM Journal of Computing*, 4(1):69 – 76, 1975.

[14] T. Henzinger, B. Horowitz, and R. Majumdar. Rectangular hybrid games. In *Proc. of the 10th International Conference on Concurrency Theory, CONCUR'99*, LNCS 1664, pages 320 – 335, 1999.

[15] T. Henzinger and P. Kopke. Discrete-time control for rectangular hybrid automata. *Theoretical Computer Science*, 221(1–2):369–392, 1999.

[16] K. G. Larsen, G. Behrman, E. Brinksma, A. Fehnker, T. Hune, P. Petersson, and J. Romijn. As cheap as possible: Efficient cost-optimal reachability for priced timed automata. In *Proc. of the 13th International Conference on Computer Aided Verification, CAV'01*, LNCS, pages 493–505. Springer, 2001.

[17] O. Maler, A. Pnueli, and J. Sifakis. On the synthesis of discrete controllers for timed systems. In *Proc. of the 12th Annual Symposium on Theoretical Aspects of Computer Science, STACS'95*, LNCS 900, pages 229 – 242. Springer, 1995.

[18] P. Niebert, S. Tripakis, and S. Yovine. Minimum-time reachability for timed automata. In *Proc. of the 8-th IEEE Mediterranean Conference on Control and Automation*, 2000.

SUBSTRUCTURAL VERIFICATION AND COMPUTATIONAL FEASIBILITY*

Daniel Leivant
Indiana University

Abstract We refer to the intrinsic theories of [14, 17], a generic framework for uncoded reasoning about equational programs. In particular, a natural notion of provable functions corresponds to the provably recursive functions of Peano Arithmetic and similar systems. A natural-deduction formulation of these systems map directly, via a Curry-Howard morphism, to terms of the simply typed lambda calculus with recurrence, with a termination proof for a function f mapping to a representation of f.

In [16] we showed that natural structural restrictions on derivations correspond to major complexity classes. When induction is restricted to positive formulas, a generalization of Σ_1^0 formulas, exactly the primitive recursive functions are provable. When only a "predicative" form of induction is allowed we obtain the Kalmar elementary functions. The combination of both restrictions yields the functions computable in polynomial time.

We show here that induction over arbitrary formulas does not add new provable functions if we disallow in derivations the closing of multiple data-complex assumptions. This significantly extends the class of proofs that can be accepted as "feasible mathematics."

We also show that if multiple closing of data-complex assumptions is only prohibited when above distinct premises of implication elimination, then the provable functions are precisely the functions computable in polynomial space.

Keywords: Implicit computational complexity, proof theory, substructural proofs, intrinsic theories, program verification, ramified induction, equational programs, program termination, feasibility, polynomial time, polynomial space, elementary functions, typed lambda calculi.

*Research partially supported by NSF grant CCR-0105651.

1. BACKGROUND

1.1. Intrinsic theories

In [14, 17] we introduced a verification methodology for equational programs, dubbed *intrinsic theories.* For each inductively-generated data system C one uses a skeletal theory $\mathbf{IT}(C)$, whose axioms are merely data-introduction axioms, i.e. the closure of data under the basic constructors, and data-elimination, i.e. induction schemas for the data types. To keep this condensed presentation uncluttered, we focus on the term algebra most relevant to computational complexity, namely the algebra $\mathbb{W}$ generated from the constant ε and the unary constructors $\mathbf{0}$ and $\mathbf{1}$, i.e. essentially the set $\{0,1\}^*$. The axioms of $\mathbf{IT}(\mathbb{W})$ are then the data-introduction axioms, which we supplement with a destructor rule, and write as inference rules:

$$\frac{}{\mathsf{W}(\varepsilon)} \qquad \frac{\mathsf{W}(\mathsf{t})}{\mathsf{W}(\mathsf{it})} \qquad \frac{\mathsf{W}(\mathsf{it})}{\mathsf{W}(\mathsf{t})} \qquad (i=0,1)$$

and data-elimination, i.e. the induction schema

$$\frac{\mathsf{W}(t) \quad \varphi[\varepsilon] \quad \varphi[z]\rightarrow\varphi[\mathbf{0}z] \quad \varphi[z]\rightarrow\varphi[\mathbf{1}z]}{\varphi[t]}$$

with z not free in open assumptions.[1] A *degenerated form* of data-elimination is

$$\frac{\mathsf{W}(t) \quad \varphi[\varepsilon] \quad \varphi[\mathbf{0}x] \quad \varphi[\mathbf{1}x]}{\varphi[t]}$$

i.e. reasoning by cases.

Throughout the paper we refer to provability in *intuitionistic logic.*[2] Moreover, we greatly simplify the discussion by referring to the fragment of logic without disjunction and $\exists$.

1.2. Provable equational programs

We refer to equational programs over $\mathbf{IT}(\mathbb{W})$. Each such program consists of a finite set P of equations between terms, where the terms are built from variables, the constructors ε, $\mathbf{0}$ and $\mathbf{1}$, and program function-identifiers. One identifier is singled out as the program's principal identifier. If $\mathbf{f}$ is the principal identifier of P, we say that $(P,\mathbf{f})$ *computes* a function f over $\mathbb{N}$ if $f(\vec{n}) = m$ exactly when the formal equation $\mathbf{f}(\vec{n}) = m$ is derived from P in equational

[1]See [17] for the generic rules. It is also natural to consider *separation axioms,* which guarantee that the denotation of all ground terms are distinct; for $\mathbb{N}$ these are Peano's third and fourth axioms, $\forall x.\ \mathrm{s}x \neq \mathbf{0}$ and $\forall x,y.\ \mathrm{s}x = \mathrm{s}y \rightarrow x = y$. However, these axioms have no effect on the provability of programs, as defined below; see [16].

[2]Indeed, the calibration of proofs' computational contents by structural conditions is more problematic and less rewarding when classical logic is used; compare [16, §3.3].

logic. A program P with principal r-ary function identifier $\mathbf{f}$, is *provable* (over a given logic $\mathbf{L}$) if

$$\mathbf{IT}(\mathbb{W}),\ \forall P,\ \mathbf{W}(x_1)\ldots\mathbf{W}(x_r) \vdash \mathbf{W}(\mathbf{f}(\vec{x}))$$

where $\forall P$ is the universal closure of the conjunction of P, and provability is in $\mathbf{L}$.

Two examples of provable programs, which will be of use later, are addition and multiplication over $\mathbb{W}$, defined by $+\varepsilon x = x$, $+\mathbf{i}yx = \mathbf{i}(+yx)$, $*x\varepsilon = \varepsilon$, and $*x(\mathbf{i}y) = +x(*xy)$ $(i = 0,1)$. Here are derivations for these two functions, where we use double-bars for the contraction of trivial steps, and display generically the induction cases for the successor functions $\mathbf{0}$ and $\mathbf{1}$.

$$\frac{\mathbf{W}(y) \qquad \dfrac{\dfrac{\forall x.\ +\varepsilon x = x}{x = +\varepsilon x} \quad \mathbf{W}(x)}{\mathbf{W}(+\varepsilon x)} \qquad \dfrac{\dfrac{\dfrac{\forall P}{\mathbf{i}(+yz) = +(\mathbf{i}y)z} \quad \dfrac{\overset{(1)}{\mathbf{W}(+yz)}}{\mathbf{W}(\mathbf{i}(+yz))}}{\mathbf{W}(+(\mathbf{i}y)z)}}{\mathbf{W}(+yz)\to\mathbf{W}(+(\mathbf{i}y)z)}\,(1)}{\mathbf{W}(+xy)}$$

$$\frac{\mathbf{W}(y) \qquad \dfrac{\dfrac{\forall P}{\varepsilon = *x\varepsilon} \quad \mathbf{W}(\varepsilon)}{\mathbf{W}(*x\varepsilon)} \qquad \dfrac{\dfrac{\dfrac{\forall P}{+x(*xz) = *x(\mathbf{i}z)} \quad \begin{matrix}\mathbf{W}(x) \quad \overset{(1)}{\mathbf{W}(*xz)}\\ \mathcal{D}\\ \mathbf{W}(+x(*xz))\end{matrix}}{\mathbf{W}(*x(\mathbf{i}z))}}{\mathbf{W}(*xz)\to\mathbf{W}(*x(\mathbf{i}z))}\,(1)}{\mathbf{W}(*xy)}$$

where $\mathcal{D}$ is the derivation above for addition, with $*xz$ substituted for the free occurrences of y.

1.3. Morphism to λ-terms

Let λ_1 be the (Church-style) simply-typed lambda calculus defined as follows. The *types* are generated from base types ι (for elements of $\mathbb{W}$) and θ (a unit type), using the binary type operations $\to$ and $\times$. We call arrow-free types *positive*. For all type τ we identify all of $\theta\times\tau$, $\tau\times\theta$, and $\theta\to\tau$ with τ; also, we identify $\tau\to\theta$ with θ. In each case we say that the shorter form is a contraction of the longer one, and we say that τ' is the *contracted form of* τ if τ' is obtained from τ by successive contractions, and cannot be contracted further (i.e. is either θ or free of θ).

We omit parentheses when in no danger of ambiguity, modulo the proviso that $\times$ binds stronger than $\to$, and then that $\to$ and $\times$ associate to the right. For example, $\iota\to\iota\times\iota\to\iota$ abbreviates $\iota\to((\iota\times\iota)\to\iota)$. We call a type *positive* if its contracted form is free of $\to$.

For each type τ we posit an unbounded stock of *variables* of type τ, x_i^τ (we omit the type superscript when convenient). Terms are generated from the variables using λ-abstraction, type-correct application, pairing (written $\langle E_0, E_1 \rangle$), and type-correct projection (written $\pi_i E$, $i = 0$ or 1). The corresponding types are defined as usual. We write $\langle E_0, \ldots, E_m \rangle$ for $\langle E_0, \langle E_1, \cdots, \langle E_{m-1}, E_m \rangle \cdots \rangle \rangle$. The computational rules are β-reduction and projection-reduction. We write $E \Rightarrow E'$ (and say that E converts to E') if E' arises by replacing in E a subterm F by its reductum.

The *typed lambda calculus over* $\mathbb{W}$, $\lambda_1(\mathbb{W})$, is the extension of λ_1 with constants $*$ of type θ, ε of type ι, **0**, **1** and p of type $\iota \to \iota$, B (branching) of type $(\iota, \iota \to \iota, \iota \to \iota, \iota) \to \iota$, and for each type τ R_τ of type $\tau \to (\tau \to \tau) \to \iota \to \tau$. The reduction rules of λ_1 are augmented with:

$$
\begin{array}{rcll}
\mathsf{p}(\mathbf{it}) & \to & \mathbf{t} & (i = 0, 1) \\[6pt]
\mathsf{B}\mathbf{t}_\epsilon \mathbf{t}_0 \mathbf{t}_1 \varepsilon & \to & \mathbf{t}_\epsilon & \\
\mathsf{B}\mathbf{t}_\epsilon \mathbf{t}_0 \mathbf{t}_1 (\mathbf{iw}) & \to & \mathbf{t}_i(\mathbf{w}) & (i = 0, 1) \\[6pt]
\mathsf{R}\mathbf{t}_\epsilon \mathbf{t}_0 \mathbf{t}_1 \varepsilon & \to & \mathbf{t}_\epsilon & \\
\mathsf{R}\mathbf{t}_\epsilon \mathbf{t}_0 \mathbf{t}_1 (\mathbf{iw}) & \to & \mathbf{t}_i(\mathsf{R}\mathbf{t}_\epsilon \mathbf{t}_0 \mathbf{t}_1 \mathbf{w}) &
\end{array}
$$

We define a mapping κ from $\mathbf{IT}(\mathbb{W})$ formulas to types of $\lambda_1(\mathbb{W})$:[3]

$$
\begin{array}{rcll}
\kappa(E) & = & \theta & \text{if } E \text{ is an equation} \\
\kappa(\mathsf{W}(\mathbf{t})) & = & \iota & \\
\kappa(\varphi_0 \wedge \varphi_1) & = & \kappa(\varphi_0) \times \kappa(\varphi_1) & \\
\kappa(\varphi_0 \to \varphi_1) & = & \kappa(\varphi_0) \to \kappa(\varphi_1) & \\
\kappa(\forall x \varphi) & = & \kappa(\varphi) &
\end{array}
$$

Thus, κ extracts from a formula φ the type $\kappa\varphi$ of its "computational contents." We extend κ to a Curry-Howard mapping from derivations $\mathcal{D}$ of $\mathbf{IT}(\mathbb{W})$ to to terms $\kappa\mathcal{D}$ of $\lambda_1(\mathbb{W})$. If $\mathcal{D}$ derives a formula φ from labeled assumptions $\overset{(\ell_i)}{\psi_i}$ then $\kappa\mathcal{D}$ will be a term of type $\kappa\varphi$, with free variables $x_{\ell_i}^{\kappa\psi_i}$. The definition of κ is given in a table at the end of the paper.[4] The mapping κ is a homomorphism with respect to reductions:[5] if $\mathcal{D}$ reduces to $\mathcal{D}'$, then $\kappa\mathcal{D}'$ is either identical to $\kappa\mathcal{D}$ or is obtained from it by a reduction in $\lambda_1(\mathbb{W})$.

THEOREM 1 [14, 17, 16] *(1) If $\mathcal{D}$ is a proof of* $\mathbf{IT}(\mathbb{W})$ *for* $(P, \mathbf{f})$ *then* $\kappa\mathcal{D}$ *is a program of* $\lambda_1(\mathbb{W})$ *for the function computed by* $(P, \mathbf{f})$.

[3]The oblivion of this mapping to terms and first order quantifiers was first used in [13]. The unit type was first used in this context in [17].

[4]The definition of κ for data rules (i.e. those referring to N) is shorter when the latter are formulated as axioms, but we prefer to keep them as as inference rules, which offer a more streamlined proof theoretic treatment of normalization, as well as of structural conditions on induction.

[5]See [17] for the definition of reduction for the data-rules.

(2) The provable functions of **IT**(𝕎) *are precisely (modulo canonical codings) the provably-recursive functions of Peano Arithmetic.*

(3) The functions provable in **IT**(𝕎) *with induction for positive formulas are the primitive recursive functions.* ⊣

Part (1) is a useful tool here and in similar formalism: it enables one to focus attention on the computationally relevant aspects of proofs, namely those that are coded in the corresponding $\lambda_1(\mathbb{W})$-terms. For example, the fact that every function provable using positive induction is primitive recursive immediately follows from (1), since such proofs map under κ to terms of $\lambda_1(\mathbb{W})$ with first-order recurrence. Similarly, (2) follows from (1) by [7].

1.4. Predicative induction

Ed Nelson [20] and others have noticed that first order arithmetic has an impredicative ingredient, responsible for the admission of unfeasible functions such as exponentiation. This implicit impredicativity is clearly identified in the induction rule of **IT**(𝕎): from $\mathbf{W}(\mathbf{t})$ one derives $\varphi[\mathbf{t}]$ in which $\mathbf{W}$ itself may occur. Viewing induction as delineating 𝕎 (or — similarly — ℕ), is therefore circular.

Intrinsic theories are useful not only for articulating this impredicativity, but also for addressing it. One such method is ramification [14], which is analogous to the ramification of second order logic to break its impredicativity [21]. Combinatorially speaking, ramification of induction blocks an exponential explosion of proof normalization by preventing that the major premise of induction depend on an induction hypothesis of another induction. A more direct blocking was defined in [16]. Call a labeled assumption φ (in a natural deduction derivation $\mathcal{D}$) a ***working assumption*** if it is closed in $\mathcal{D}$. Call a formula φ ***data significant*** if the contracted form of $\kappa\varphi$ is not θ. Call an instance of induction ***predicative*** if its major (i.e. leftmost) premise does not depend on an open data-siginficant working assumption. A derivation is ***predicative*** if all non-degenerated instances of induction are predicative.

THEOREM **2** [16] *A function over* 𝕎 *is computable in polynomial time iff it is computed by an equational program* $(P, \mathbf{f})$ *which s provable in* **IT**(𝕎) *by a predicative derivation with induction for data-positive formulas.*

The proof uses the homomorphism κ: if $\mathcal{D}$ is a derivation of $(P, \mathbf{f})$ with only predicative instances of induction, and all induction formulas data-positive, then $\kappa\mathcal{D}$ is a $\lambda_1(\mathbb{W})$-term with $\mathbf{R}_\tau$ for positive τ only, where all instances of recurrence are ***predicative***, i.e. with no variable both free in the recurrence argument and bound in $\kappa\mathcal{D}$.

2. SOLITARY INFERENCES AND POLY-TIME

2.1. Multiple uses of assumptions

We illustrate the power of induction for non-positive formulas by proving a function over $\mathbb{N}$ of exponential growth rate, namely the function $e(x,y) = 2^x + y$ defined by the program consisting of the equations $\mathbf{e}(0,y) = \mathbf{s}y$, $\mathbf{e}(\mathbf{s}x,y) = \mathbf{e}(x,\mathbf{e}(x,y))$. Here is a natural deduction proof for this program, in $\mathbf{IT}(\mathbb{N})$, the intrinsic theory for $\mathbb{N}$ analogous to $\mathbf{IT}(\mathbb{W})$.

$$\dfrac{\mathsf{N}(x) \quad \dfrac{\dfrac{\dfrac{\mathsf{N}(y)}{\mathsf{N}(\mathsf{s}y)}}{\mathsf{N}(\mathsf{e}(0,y))}}{\forall y\,(\mathsf{N}(y)\to\mathsf{N}(\mathsf{e}(0,y)))} \quad \dfrac{\dfrac{\dfrac{\dfrac{\forall y\,(\mathsf{N}(y)\to\mathsf{N}(\mathsf{e}(u,y)))}{\mathsf{N}(\mathsf{e}(u,y))\to\mathsf{N}(\mathsf{e}(u,\mathsf{e}(u,y)))} \quad \dfrac{\dfrac{\forall y\,(\mathsf{N}(y)\to\mathsf{N}(\mathsf{e}(u,y)))}{\mathsf{N}(y)\to\mathsf{N}(\mathsf{e}(u,y))} \quad \mathsf{N}(y)}{\mathsf{N}(\mathsf{e}(u,y))}}{\mathsf{N}(\mathsf{e}(u,\mathsf{e}(u,y)))}}{\mathsf{N}(\mathsf{e}(\mathsf{s}u,y))}}{\forall y\,(\mathsf{N}(y)\to\mathsf{N}(\mathsf{e}(\mathsf{s}u,y)))}}{\forall y\,(\mathsf{N}(y)\to\mathsf{N}(\mathsf{e}(x,y)))}\ ind$$

The fact that the induction formula here is not positive is exploited by using it twice, and applying the outcome of one use to the outcome of the other. The question arrises then: is the culprit for exponential growth-rate the very complexity of the induction formula, or merely the duplicate use made of it? The answer is the latter.

2.2. Solitary assumption-classes

Theorem 2 shows that powerful restrictions on proof methods yield poly-time complexity. We show that these restrictions can be relaxed without yielding additional provable functions (though possibly proving additional *programs* for such functions). We start by presenting the main idea in its simplest form.

A labeled-assumption in a derivation $\mathcal{D}$ is *solitary* if it has at most one formula-occurrence.[6] A derivation is *strictly-solitary* if all its assumption classes are solitary.

If $\mathcal{D}$ is a strictly-solitary derivation, then the $\lambda_1(\mathbb{W})$-term $\kappa\mathcal{D}$ has the property that every λ-abstraction closes at most one variable-occurrence. Let M be such a term, and suppose that M reduces to M'. The two salient properties of the reduction are: (1) all λ-abstraction in M' again close at most one variable-occurrence; and (2) the size of M' is smaller than the size M (though the height of M' may be roughly double that of M).

PROPOSITION **3** *A function over $\mathbb{W}$ is computable in polynomial time iff it is computed by an equational program $(P, \mathbf{f})$ which is provable in $\mathbf{IT}(\mathbb{W})$ by a strictly-solitary derivation $\mathcal{D}$.*

[6]Recall that an assumption-class is the set of commonly-labeled assumptions in a natural deduction derivation, closed jointly by some inference.

Notice that in the derivation $\mathcal{D}$ for a program $(P, \mathbf{f})$ the assumptions $\mathbf{W}(x_i)$ are not closed, and therefore may have multiple occurrences. Also note that we make no stipulation about the complexity of induction formulas or their predicative use.

Proof Outline. The proof of Theorem 2 shows that all poly-time functions have a program with a strictly-solitary derivation $\mathcal{D}$.

To prove the converse, consider a strictly-solitary derivation $\mathcal{D}$ for a program $(P, \mathbf{f})$ computing the function f. Then $M \equiv \kappa\mathcal{D}$ is a $\lambda_1(\mathbb{W})$-term that computes f, and in which no λ-abstraction closes more than one variable occurrence. Consider a term Mw, where $w \in \mathbb{W}$. By induction on M and secondary induction on $|w|$ it is easy to see, using properties (1) and (2) above, that the reduction sequence of Mw to its normal form has length polynomial in $|w|$. ⊣

2.3. Solitary inferences and poly-time

Proposition 3 shows that restricting induction to positive formulas can be traded for a prohibition of assumption multiplicity. The class of provable functions is poly-time in either case. While this result is potentially beneficial in some cases, it seems that multiple invocation of assumptions is, in fact, used and needed in actual proofs far more frequently than induction over data-complex formulas. Fortunately, we can combine the advantages of both approaches. Call a formula data-complex if the contracted form of $\kappa\varphi$ contains $\rightarrow$. Call a derivation $\mathcal{D}$ *solitary* if all *data-complex* assumption classes are solitary.

THEOREM 4 *A function f is poly-time iff it is provable by a solitary and predicative derivation $\mathcal{D}$.*

Note that induction is permitted here over all formulas. If induction is restricted to positive formulas, then $\mathcal{D}$, which w.l.o.g. can be assumed normal, has no data-complex assumption classes, so Theorem 2 is a special case of Theorem 4.

Proof Outline. The proof of Proposition 3 shows that every poly-time function is provable as required.

For the converse, consider a derivation $\mathcal{D}$ as above. The term $\kappa\mathcal{D}$, which by Theorem 1(1) defines f, has only predicative instances of recurrence, and no higher-type variable is multiply-closed by a λ-abstraction in $\kappa\mathcal{D}$. In [15, Lemma 3.12] we showed that such terms define poly-time functions. ⊣

2.4. Origins of solitary deductions

It has been known for long that allowing resources to be invoked only once is related to poly-time computation. For instance, linear logic leads to poly-time [6, 5], second-order existential database queries are poly-time if the matrix is Horn [12, 8], monotone inductive definitions (where each object is inserted only once) define exactly the poly-time queries over finite structures, ramified recurrence with parameters does not lead out of poly-time if only one parameter is used [1], Turing machines operating in poly-space accept exactly the poly-time languages if non-blank tape-cells cannot be reused, etc.

Continuing in this vein, Martin Hofmann [9, 10, 11] has developed a linear-type ramified functional calculus that defines exactly the poly-time non-size-increasing functions, even if recurrence is used at all finite types. This has been further refined in [3]. Independently we showed that allowing abstracted higher order functions to be used only once in λ-recurrence terms, yield exactly poly-time [15].

Proof theoretic characterizations of poly-time that build on linearity have also appeared recently, among others in [2, 22]. The main advantage of our present result is that it does not require an overlay of syntactic machinery on the formulas; the proofs we consider are all proofs in intrinsic theories (whose syntax is very simple), and the structural properties that they satisfy can be automatically checked. This yields a transparent machinery for certifying program feasibility.

Since poly-time has rather simple proof-theoretic characterizations (e.g. [16] above), the main motivation of restricted-multiplicity conditions is the attempt to permit induction for all formulas, thereby providing the user of the formalism (human or automated) with a larger arsenal. Consequently, it is self-defeating to abandon in the process other methods, in particular when these are important and natural. For instance, taken in isolation, restricted multiplicity disallows a direct and simple proof of the squaring function! Indeed, one would wish to combine the advantages of various approaches, rather than piling up the hurdles to using them.

There is a trade-off, of course, in our avoiding the explicit use of linear and other resource-control operators. Our combinatorial conditions may be viewed as corresponding to the use of such operators at the outer level of reasoning, whereas more explicit resource-control operators, as in [10], might be used to convey more subtle interactions between parts of proofs. Examples illustrating such gains are yet to be developed, however.

3. POLY-SPACE

3.1. Weakly-solitary derivations

Returning to our example above of a derivation for the functions 2^x+y, we can further ask: is the culprit for exponential growth-rate the very duplicate use of the assumption, or only the particular setting where one use is applied to another, across an instance of implication elimination? In other words, what would happen if we allow duplicate uses of data-complex working assumptions, as long as they are not ancestors of distinct premises of implication elimination? Interestingly, the provable functions are then precisely the functions computable in polynomial space.

Call an assumption-class in a natural deduction $\mathcal{D}$ *weakly-solitary* if no two of its members are ancestors of distinct premises of an instance of implication elimination. A derivation $\mathcal{D}$ is *weakly-solitary* if every data-complex assumption-class in $\mathcal{D}$ is weakly-solitary.

We show below that a function $f : \mathbb{W}^* \to \{0,1\}$ is provable by predicative and weakly solitary derivations iff it is computable in polynomial space. Recall that a function $g : \mathbb{W}^* \to \mathbb{W}$ is computable by a Turing machine in polynomial space iff the associated bit-function

$$g'(\vec{x}, y) =_{\rm df} \text{the bit of } g(\vec{x}) \text{ at address } y$$

is computable in PSpace.[7] However, our proof does not apply directly to functions g as above, since it relies on the characterization of PSpace as alternating PTime [4]. With this in mind, we will prove the following.

THEOREM 5 *A function $f : \mathbb{W}^* \to \{0,1\}$ is provable in* $\mathbf{IT}(\mathbb{W})$ *by a predicative and weakly-solitary derivation iff it is in PSpace.*

3.2. From PSpace to provability

PROPOSITION 6 *Every function $f : \mathbb{W}^* \to \{0,1\}$ computable in polynomial space is provable in* $\mathbf{IT}(\mathbb{W})$ *by a predicative and weakly-solitary derivation.*

Proof Outline. The proof uses a proof-theoretic analog of the technique of [18, 19]. By [4] a boolean-valued function computable in PSpace is computable in polynomial time by an alternating Turing machine, which w.l.o.g. has branching degree 2. The latter is computable by composing a polynomial function (generating the computation clock) with a function defined by parameterized recurrence of the form:

$$\begin{aligned} f(\varepsilon, \vec{x}) &= g_\epsilon(\vec{x}) \\ f(\mathbf{c}t, \vec{x}) &= g_s(\vec{x}, f(t, \vec{h}_0(\vec{x})), f(t, \vec{h}_1(\vec{x}))) \end{aligned}$$

The intent is that $f(t, \vec{x})$ is the acceptance status returned by the given alternating machine, when in configuration $\vec{x}$, and provided with $|t|$ computation steps along each branch. The functions h_i return the two subsequent configuration, and are defined explicitly without use of recurrence (other than definition by cases). The function g_s returns the conjunction or disjunction of its last two arguments, depending on the state whose code is part of $\vec{x}$.

It is easy to see that the equational program described above is provable, using induction for the formula

$$\varphi[t] \;\; \equiv_{\rm df} \;\; \forall \vec{x}.\mathbf{W}(\vec{x}) \to \mathbf{W}(\mathbf{f}(t, \vec{x}))$$

The formula is used twice in the induction step, with $\vec{x}$ instantiated once to $\vec{h}_0(\vec{x})$ and once to $\vec{h}_1(\vec{x})$. The two instances are combined using only basic operations, without use of implication elimination or induction. ⊣

[7] More precisely, g' has co-domain $\{0, 1, \emptyset\}$, and returns $\emptyset$ if the address y exceeds the length of $f'(\vec{x})$.

3.3. From Provability to PSpace

We complete the proof of Theorem 5 by showing that every function f provable by a predicative and weakly-solitary derivation is in PSpace.

Suppose $\mathcal{D}$ is a predicative and weakly-solitary derivation. The term $M \equiv \kappa\mathcal{D}$ has the following properties: (1) No recurrence argument has a free variable bound in M; and (2) M is *weakly-solitary* in the following sense: If EF is a subterm of M, then no variable of higher type occurs free in both E and F. Call a $\boldsymbol{\lambda}_1(\mathbb{W})$-term $\lambda\vec{x}.M[\vec{x}]$ *input-driven* if the recurrence arguments in M are all variables out of the list $\vec{x}$. Note that this is a stronger condition than (1) above.

LEMMA 7 *Every function represented in $\boldsymbol{\lambda}_1(\mathbb{W})$ by a predicative and weakly-solitary term is the composition of functions represented in $\boldsymbol{\lambda}_1(\mathbb{W})$ by input-driven and weakly-solitary terms.*

See [15, Lemma 2.2] for a proof. The proof of Theorem 5 is now concluded by the following.

LEMMA 8 *If a function f is representable in $\boldsymbol{\lambda}_1(\mathbb{W})$ by a weakly-solitary and predicative term, then it is computable in polynomial space.*

Proof Outline. By Lemma 7 it suffices to consider the case where f is representable by a weakly-solitary and input-driven term $\lambda\vec{x}.F$. We may further assume w.l.o.g. that the recurrence arguments in F are *distinct* variables out of the list $\vec{x}$. Let $y_1, \ldots, y_q$ be the x_i's used for higher order recurrence, and $z_1, \ldots, z_k$ the x_i's used for recurrence in $\nrightarrow$-free types. Given $\mathbf{y}_1, \ldots, \mathbf{y}_q, \mathbf{z}_1, \ldots, \mathbf{z}_k \in \mathbb{W}$, consider the term $F^* = \{\vec{\mathbf{y}}, \vec{\mathbf{z}}/\vec{y}, \vec{z}\}F$. By the condition above we can independently unfold each higher-type recurrence in F^*, finally yielding some term M. Clearly, M is of polynomial height (and exponential size).

Let m the maximal number of occurrences of higher-type variables that are bound by a λ-abstraction in F. Consider the symbolic parse-tree T_M of M, with the root at the bottom. Each node in $T(M)$ has below it the junctures of recurrence-unfolding on F^*. Each such juncture corresponds to one choice out of (at most) m positions, for a λ-abstracted variable. Thus, to each node there corresponds a "reduction address" in $\{0, \ldots, m-1\}^h$, where $h \leq$ the height of M. Call a node N in T_M *relevant* to node N' if the reduction address of N is a subsequence (not necessarily strict) of the reduction address of N'. It can then be seen that: (a) Each node has only polynomially many relevant nodes relevant to it; and (b) The behavior of each node under reductions can be affected only by nodes relevant to it (here the definition of weakly-solitary terms is crucial). Consider now a reduction sequence on M that eliminates higher-type redexes. Since M is itself weakly-solitary, no node of T_M can be duplicated on the same branch of any redex term. It follows that for each term M' along the reduction sequence, the parse-tree $T(M')$ is the "horizontal union" of (perhaps exponentially many) subtrees of polynomial size and height: each such subtree corresponds to a reduction address in M.

Consequently, the entire reduction sequence of M can be computed in polynomial-space, leading to an input-driven and predicative term whose only redexes are for positive-type reductions. This can now be normalized in space polynomial in the height of the term, which is itself polynomial in the size of the input. The final normal form is the value of F^*, i.e. the value of the function f for input $\mathbf{y}_1, \ldots, \mathbf{y}_q, \mathbf{z}_1, \ldots, \mathbf{z}_k \in \mathbb{W}$. ⊣

References

[1] A. Beckmann and A. Weiermann. A term rewriting characterization of the polytime functions and related complexity classes. *Archive for Mathematical Logic*, 36:11–30, 1996.

[2] Stephen Bellantoni and Martin Hofmann. A new feasible arithmetic. *Journal for Symbolic Logic*, 2001.

[3] Stephen J. Bellantoni, Karl-Heinz Niggl, and Helmut Schwichtenberg. Higher type recursion, ramification and polynomial time. *Annals of Pure and Applied Logic*, 104 (1-3):17–30, 2000.

[4] A. Chandra, D. Kozen, and L. Stockmeyer. Alternation. *Journal of the ACM*, 28:114–133, 1981.

[5] Jean-Yves Girard. Light linear logic. *Information and Computation*, 143, 1998.

[6] Jean-Yves Girard, Andre Scedrov, and Philip Scott. Bounded linear logic: A modular approach to polynomial time computability. *Theoretical Computer Science*, 97:1–66, 1992.

[7] Kurt Gödel. Über eine bisher noch nicht benutzte erweiterung des finiten standpunktes. *Dialectica*, 12:280–287, 1958.

[8] E. Grädel. Capturing Complexity Classes by Fragments of Second Order Logic. *Theoretical Computer Science*, 101:35–57, 1992.

[9] Martin Hofmann. A mixed modal/linear lambda calculus with applications to bellantoni-cook safe recursion. In *Proceedings of CSL'97*, pages 275–294. Springer-Verlag LNCS 1414, 1998.

[10] Martin Hofmann. Linear types and non-size-increasing polynomial time computation. In *Proceedings of LICS'99*, pages 464–473. IEEE Computer Society, 1999.

[11] Martin Hofmann. Safe recursion with higher types and bck-algebra. *Annals of Pure and Applied Logic*, 104 (1-3):113–166, 2000.

[12] Daniel Leivant. Descriptive characterizations of computational complexity. In *Second Annual Conference on Structure in Complexity Theory*, pages 203–217, Washington, 1987. IEEE Computer Society Press. Revised in *Journal of Computer and System Sciences*, 39:51–83, 1989.

[13] Daniel Leivant. Contracting proofs to programs. In P. Odifreddi, editor, *Logic and Computer Science*, pages 279–327. Academic Press, London, 1990.

[14] Daniel Leivant. Intrinsic theories and computational complexity. In D. Leivant, editor, *Logic and Computational Complexity*, LNCS, pages 177–194, Berlin, 1995. Springer-Verlag.

[15] Daniel Leivant. Applicative control and computational complexity. In J. Flum and M. Rodriguez-Artalejo, editors, *Computer Science Logic (Proceedings of the*

Thirteenth CSL Conference, pages 82–95, Berlin, 1999. Springer Verlag (LNCS 1683).

[16] Daniel Leivant. Termination proofs and complexity certification. In Naoki Kobayashi and Benjamin C. Pierce, editors, *Theoretical Aspects of Computer Software (TACS 2001)*, Springer LNCS 2215, pages 183–200, 2001.

[17] Daniel Leivant. Intrinsic reasoning about functional programs I: first order theories. *Annals of Pure and Applied Logic*, 114:117–153, 2002.

[18] Daniel Leivant and Jean-Yves Marion. Ramified recurrence and computational complexity IV: Predicative functionals and poly-space. *Information and Computation*. To appear.

[19] Daniel Leivant and Jean-Yves Marion. Predicative functional recurrence and poly-space. In M.Bidoit and M. Dauchet, editors, *Theory and Practice of Software Development*, LNCS 1214, pages 369–380, Berlin, 1997. Springer-Verlag.

[20] Edward Nelson. *Predicative Arithmetic.* Princeton University Press, Princeton, 1986.

[21] Kurt Schütte. *Proof Theory.* Springer-Verlag, Berlin, 1977.

[22] Helmut Schwichtenberg. An arithmetic for polynomial-time computation. Submitted for publication, 2002.

$\mathcal{D}$	$\kappa\mathcal{D}$
$\begin{matrix}(\ell)\\ \psi\end{matrix}$ (labeled assumption)	$x_\ell^{\kappa\psi}$ (ℓ-th variable of type $\kappa\psi$)
$\dfrac{\begin{matrix}\mathcal{D}_0 & \mathcal{D}_1\\ \varphi_0 & \varphi_1\end{matrix}}{\varphi_0 \wedge \varphi_1}$	$\langle \kappa\mathcal{D}_0, \kappa\mathcal{D}_1 \rangle$
$\dfrac{\begin{matrix}\mathcal{D}_0\\ \varphi_0 \wedge \varphi_1\end{matrix}}{\varphi_i}$	$\pi_i \kappa\mathcal{D}_0$
$\dfrac{\begin{matrix}(\ell)\\ \psi\\ \mathcal{D}_0\\ \varphi\end{matrix}}{\varphi \to \psi}\ (\ell)$	$\lambda x_\ell^{\kappa\psi}.\, \kappa\mathcal{D}_0$
$\dfrac{\begin{matrix}\mathcal{D}_0 & \mathcal{D}_0\\ \varphi \to \psi & \varphi\end{matrix}}{\psi}$	$(\kappa\mathcal{D}_0)(\kappa\mathcal{D}_1)$
$\dfrac{\begin{matrix}\mathcal{D}_0\\ \varphi[z]\end{matrix}}{\forall x\, \varphi[x]}$	$\kappa\mathcal{D}_0$
$\dfrac{\begin{matrix}\mathcal{D}_0\\ \forall x\, \varphi[x]\end{matrix}}{\varphi[\mathbf{t}]}$	$\kappa\mathcal{D}_0$
$\mathbf{t} = \mathbf{t}$	$*$
$\dfrac{\begin{matrix}\mathcal{D}_0 & \mathcal{D}_1\\ \mathbf{t} = \mathbf{t}' & \varphi[\mathbf{t}]\end{matrix}}{\varphi[\mathbf{t}']}$	$\kappa\mathcal{D}_1$
$\mathbf{W}(\varepsilon)$	$\mathbf{0}$
$\dfrac{\begin{matrix}\mathcal{D}_0\\ \mathbf{W}(\mathbf{t})\end{matrix}}{\mathbf{W}(\mathbf{i}\mathbf{t})}\quad (i = 0, 1)$	$\mathbf{i}\kappa\mathcal{D}_0$
$\dfrac{\begin{matrix}\mathcal{D}_0\\ \mathbf{W}(\mathbf{i}\mathbf{t})\end{matrix}}{\mathbf{W}(\mathbf{t})}\quad (i = 0, 1)$	$\mathbf{p}\kappa\mathcal{D}_0$
$\dfrac{\begin{matrix}\mathcal{P} & \mathcal{D}_\epsilon & \mathcal{D}_0 & \mathcal{D}_1\\ \mathbf{W}(\mathbf{t}) & \varphi[\varepsilon] & \varphi[z] \to \varphi[\mathbf{0}z] & \varphi[z] \to \varphi[\mathbf{1}z]\end{matrix}}{\varphi[\mathbf{t}]}$	$\mathbf{R}_{\kappa\varphi}(\kappa\mathcal{D}_\epsilon)(\kappa\mathcal{D}_0)(\kappa\mathcal{D}_1)(\kappa\mathcal{P})$

AN IMPROVED SYSTEM OF INTERSECTION TYPES FOR EXPLICIT SUBSTITUTIONS

Dan Dougherty
Department of Mathematics and Computer Science, Wesleyan University
Middletown, CT 06459 USA
ddougherty@wesleyan.edu

Stephane Lengrand and Pierre Lescanne
Ecole Normale Superieure de Lyon
46, Alle d'Italie, 69364 Lyon 07, FRANCE
{Stephane.Lengrand,Pierre.Lescanne}@ens-lyon.fr

Abstract We characterize those terms which are strongly normalizing in a composition-free calculus of explicit substitutions by defining a suitable type system using intersection types. The key idea is the notion of *available* variable in a term, which is a generalization of the classical notion of free variable.

1. Introduction

An explicit substitutions calculus is a refinement of traditional λ-calculus in which substitution is not treated as a meta-operation on terms but rather as an operation of the calculus itself. The inspiration for such a study is the observation that, in the presence of variable-binding, substitution is a complex operation to define and to implement, so that making substitutions explicit leads to a more pertinent analysis of the correctness and efficiency of compilers, theorem provers, and proof-checkers. Abadi, Cardelli, Curien, and Lévy defined the first calculus of explicit substitutions in [Abadi et al., 1991].

A fundamental property of classical typed lambda-calculi is strong normalization: no term admits an infinite reduction sequence. Melliès [Melliès, 1995] made the somewhat surprising discovery that strong normalization fails even for simply-typed terms of the calculi of [Abadi et al., 1991] and [Curien et al., 1996].

Given the central place that strong normalization occupies in the theory and application of classical lambda calculus it is important to study this property in systems of explicit substitutions. Melliès' result exploits the existence of a *composition* operator on substitutions, so there are two obvious and complementary research directions. The first is to define classes of reduction strategies in the original calculus which support strong normalization; a notable example of work in this area is that of Eike Ritter [Ritter, 1999]. The second direction is to investigate calculi in which substitutions are explicit but composition is absent; the current paper is part of this effort.

Composition-free calculi of explicit substitutions have been studied in [Lescanne, 1994, Bloo and Rose, 1995, Kamareddine and Ríos, 1997, Bloo and Geuvers, 1999, Benaissa et al., 1996] among others. Here we work in the composition-free calculus $\lambda\mathbf{x}$ [Bloo and Rose, 1995] and the calculus $\lambda\mathbf{x}_{gc}$ obtained by adding explicit garbage collection to $\lambda\mathbf{x}$.

In previous work [Dougherty and Lescanne, 2001, Dougherty and Lescanne, pear] we explored some reduction properties of this system using intersection types. Working with the natural generalization of the classical type systems we were able to characterize the sets of normalizing and head-normalizing terms in terms of typability. But it was shown in [Dougherty and Lescanne, 2001] that the naive generalization of the classical system did not characterize the strongly normalizing terms. Typable terms were strongly normalizing but the converse fails.

Example 1 Let S be the term $\lambda u.uu$. Consider the terms

$$M_1 \equiv ((\lambda y.z)xx)\langle x = S\rangle \longrightarrow M_2 \equiv z\langle y = xx\rangle\langle x = S\rangle$$

(The syntax and reduction rules of the calculus are given in section 2.) The term M_2 is readily seen to be strongly normalizing. But M_2 is not typable in the system $\mathcal{D}$ of [Dougherty and Lescanne, 2001]: it is obtained from the (non-SN, hence untypable) term M_1 by contracting a β-redex, and such a contraction does not change the typing behavior of terms under $\mathcal{D}$. Finding a type system characterizing the strongly normalizing terms was left as an open problem in [Dougherty and Lescanne, 2001].

Main results. In this paper we solve the aforementioned problem: we define an extension $\mathcal{E}$ of system $\mathcal{D}$ which types precisely the strongly normalizing terms. Furthermore when a universal type ω is added the resulting system $\mathcal{E}_\omega$ satisfies the same theorems as those in [Dougherty and Lescanne, 2001] characterizing the weakly normalizing, head normalizing, and solvable terms. Our claim, then, is that the system presented here — with or without a universal type — is a robust type system appropriate for analyzing reduction properties in explicit substitutions calculi.

The key insight for the solution is a new notion, that of *available* variable occurrence in a term (Definition 3). This is a refinement of the notion of free variable and is the key to extending $\mathcal{D}$. As a corollary of our approach we are able to define a somewhat more general notion of garbage collection than has been studied in the literature of λ**x** and show that adding a reduction for garbage-collection does not change the set of strongly normalizing terms.

2. The calculus λx

Definition 2 The set Λx of terms with explicit substitutions is defined as:

$$M, N \;:=\; x \mid \lambda x.M \mid M\,N \mid M\langle x = N\rangle$$

One defines the notions of free and bound variable occurrences in a term as usual. But it turns out that in the presence of explicit substitutions a refinement of the notion of free variable, called ***available*** variable occurrence, is key.

Definition 3 The set of *free* variables $FV(M)$ is the same as in [Dougherty and Lescanne, 2001] and the set of ***available*** variables $AV(M)$ in a term M is

$$\begin{cases} AV(x) & := & \{x\} \\ AV(\lambda x.M) & := & AV(M)\backslash\{x\} \\ AV(M\,N) & := & AV(M) \cup AV(N) \\ AV(M\langle x = N\rangle) & := & (AV(M)\backslash\{x\}) \cup AV(N) & \text{if } x \in AV(M) \\ AV(M\langle x = N\rangle) & := & AV(M) & \text{if } x \notin AV(M) \end{cases}$$

For pure terms the notions of freeness and availability coincide. But availability differs from freeness in that the available variables of $M\langle x = N\rangle$, where x is not available in M, are exactly those of M, whereas the free variables in any case are those of M <u>and</u> of N. The intuition is that x is not available just when the term N disappears in the course of fully applying the substitutions in $M\langle x = N\rangle$.

Further discussion of the motivation for defining available variable occurrences will be given after we present our type system. For now we can observe, referring to Example 1, that in the term $z\langle y = xx\rangle$ the variable x is free, but is not available.

>From the definitions of $AV(M)$ and $FV(M)$, an easy induction over the structure of M shows that the available variable occurrences in a term are a subset of the free variable occurrences.

In what follows we consider terms up to a α-conversion. Moreover, when one chooses a representative in a term, one does that in such a way that the Barendregt convention [Barendregt, 1984] is fulfilled: *no variable occurs both free and bound.* Since available variables are free it follows that we may assume that no variable occurs both available and bound.

As usual we often treat *contexts,* terms $C[\;]$ with a designated variable $[\;]$ called a *hole*; terms can be "grafted" into the hole with variable-capture permitted (see [Barendregt, 1984]).

Definition 4 (λx and λx_{gc}) We identify the following reduction rules on λx terms.

$$
\begin{array}{llll}
(\lambda x.M)\,A & \longrightarrow & M\langle x = A\rangle & \mathsf{B} \\
(M\,N)\langle x = A\rangle & \longrightarrow & M\langle x = A\rangle\,N\langle x = A\rangle & \mathsf{App} \\
(\lambda y.M)\langle x = A\rangle & \longrightarrow & \lambda y.(M\langle x = A\rangle) & \mathsf{Abs} \\
x\langle x = A\rangle & \longrightarrow & A & \mathsf{VarI} \\
y\langle x = A\rangle & \longrightarrow & y & \mathsf{VarK} \\
M\langle x = A\rangle & \longrightarrow & M \quad \text{if } x \notin AV(M) & \mathsf{gc}
\end{array}
$$

The notion of reduction λx is obtained by deleting the rule gc, and the notion of reduction λx_{gc} is obtained by deleting the rule VarK. The rule gc is called "garbage collection", as it removes useless substitutions.

In contrast with the classical λ-calculus we are considering a rewrite system with several rules, which in fact interact with each other in interesting ways. For example there is a *critical pair* formed by the rules B and App, which is responsible for much of the complexity in analyzing the theory.

Definition 5 (Reduction) Let $l \longrightarrow r$ be a reduction rule; we refer to an instantiation $s(l)$ of l as a *redex*. A (unconstrained) *reduction* is determined by a redex occurrence in a term $C[s(l)]$ and gives rise to the ordered pair $C[s(l)] \longrightarrow C[s(r)]$.

We write $\mathcal{SN}$ for the set of strongly normalizing terms under λx, and $\mathcal{SN}_{\mathsf{gc}}$ for the corresponding set under λx_{gc}.

The notion of garbage collection in this paper is more liberal than that originally defined by Bloo and Rose [Bloo and Rose, 1995] and treated in [Dougherty and Lescanne, 2001]: here we define "garbage" in terms of available occurrences rather than free occurrences. Our results will imply that a term is strongly normalizing under λx_{gc} if and only if it is strongly normalizing under λx (a result shown directly in [Bloo and Rose, 1995] for their notion of garbage-collection). The following easy observations will be useful later. They apply to each of λx and λx_{gc}.

Lemma 6 *If $M\langle x = N\rangle$ is not a redex then $M \equiv M'\langle y = N'\rangle$. In particular $M\langle x = N\rangle$ is never a normal form.*

We will also need the fact that if we omit rule B then the resulting reduction is strongly normalizing, so that an infinite derivation contains infinitely many applications of rule B. To prove that strongly normalizing terms are typable, we will induct over the reduction relation, and so we will want to show that the converse of the reduction relation preserves typability. Of course this is not true in full generality, and we restrict attention to reductions following a certain strategy, a *leftmost-outermost* strategy. As discussed in [Dougherty and Lescanne, 2001] the notion of leftmost reduction is not as straightforward as in classical λ-calculus, in particular the notion there is a non-deterministic strategy. The strategy defined below is a deterministic restriction. It makes sense for each of $\lambda\mathbf{x}$ and $\lambda\mathbf{x}_{gc}$, although we make use of it only in Section 4, where we consider only $\lambda\mathbf{x}$.

Definition 7 For any term not in normal form, the *leftmost-outermost* strategy reduces the leftmost-outermost redex, where the leftmost-outermost redex of M, written $\mathsf{lo}(M)$ is

M	if M is a redex
$\mathsf{lo}(N_1)$	if $M \equiv N_1\ N_2$ where N_1 is not a normal form
$\mathsf{lo}(N_2)$	if $M \equiv N_1\ N_2$ where N_1 is a normal form
$\mathsf{lo}(N)$	if $M \equiv N\langle x = A\rangle$
$\mathsf{lo}(N)$	if $M \equiv \lambda x.N$

3. The system $\mathcal{E}$ of intersection types

Definition 8 The set of *types* is inductively defined as

$$\tau_1, \tau_2 \ := \ \sigma \mid \tau_1 \cap \tau_2 \mid \tau_1 \to \tau_2$$

The standard ordering $\leq$ on types is the smallest transitive and reflexive relation such that

$$\tau_1 \cap \tau_2 \ \leq \ \tau_1 \qquad \tau_1 \cap \tau_2 \ \leq \ \tau_2 \qquad \text{if } \sigma \leq \tau_1 \text{ and } \sigma \leq \tau_2 \text{ then } \sigma \leq \tau_1 \cap \tau_2$$

Definition 9 An *environment* is an assignment from variables to types, where each individual assignment is written $(x : \tau)$. Environments are partially ordered as follows.

$$\Gamma \leq \Gamma' \quad \text{iff} \quad (x : \tau') \in \Gamma' \Longrightarrow (\exists\tau)\ (x : \tau) \in \Gamma \text{ and } \tau \leq \tau'$$

Definition 10 A *judgment* is a triple consisting of an environment Γ, a term M, and a type τ. A judgment is *derivable* in system $\mathcal{E}$, denoted $\Gamma \vdash M : \tau$, if this form can be derived by the rules of inference given in Table 1. A term M is *typable* if for some Γ and τ, $\Gamma \ \vdash \ M : \tau$ is derivable.

The innovation in the type system here is the presence of the rule drop. The type system of [Dougherty and Lescanne, 2001] had no such rule: the point of view taken there was that a closure $M\langle x = N\rangle$ should always have the same typing behavior as the B-redex $(\lambda x.M)N$ which yields it. This is a plausible strategy since B-reduction involves no (immediate) erasing of subterms, even when x is not free in M; and indeed the resulting system — in the presence of a universal type — yields the expected characterizations of head-normalizing and leftmost-normalizing terms. But as we have seen in Example 1 this system failed to provide a characterization of the strongly normalizing terms. This example makes clear that we must allow the type system to distinguish between certain B-redexes and their contractions.

Perhaps one's first instinct is to note that in Example 1 the input variable of the B-redex in M_1 does not occur free in the function body (*i.e.*, we have a "K-redex" in classical λ-calculus). This suggests modifying the cut-rule to obtain

$$\text{start}\ \frac{}{\Gamma \vdash x : \sigma}\ (x : \sigma) \in \Gamma$$

$$\to \mathsf{I}\ \frac{\Gamma, x : \sigma \vdash M : \tau}{\Gamma \vdash \lambda x.M : \sigma \to \tau} \qquad \to \mathsf{E}\ \frac{\Gamma \vdash M : \sigma \to \tau \quad \Gamma \vdash N : \sigma}{\Gamma \vdash M\,N : \tau}$$

$$\text{cut}\ \frac{\Gamma, x : \sigma \vdash M : \tau \quad \Gamma \vdash A : \sigma}{\Gamma \vdash M\langle x = A\rangle : \tau}$$

$$\text{drop}\ \frac{\Gamma \vdash M : \tau \quad A \text{ typable}}{\Gamma \vdash M\langle x = A\rangle : \tau}\ x \notin AV(M)$$

$$\cap\text{-I}\ \frac{\Gamma \vdash M : \tau_1 \quad \Gamma \vdash M : \tau_2}{\Gamma \vdash M : \tau_1 \cap \tau_2} \qquad \cap\text{-E}\ \frac{\Gamma \vdash M : \tau_1 \cap \tau_2}{\Gamma \vdash M : \tau_i}\ i \in \{1,2\}$$

Table 1. Typing rules for $\mathcal{E}$.

one which, when typing $M\langle x = N\rangle$ with x not free in M, relaxes the typing hypothesis for N to merely ask that it be typable under *some* environment. This seems particularly appropriate since it echoes the hypotheses of the Subject Expansion Theorem in treatments of intersection types for classical λ-calculus. But such a rule doesn't work: it is still too restrictive. For example, the reader can easily check that the term $M_2' \equiv x\langle y = xx\rangle\langle x = S\rangle$ cannot be typed in such a system, but is clearly strongly normalizing. This last example should motivate our notion of *available* variable occurrence and the corresponding typing rule drop.

A good exercise at this point is to check that the terms M_2 and M_2' can be typed in system $\mathcal{E}$. On another hand, notice that rule cut has no side condition, therefore when $x \notin AV(M)$ and $\Gamma \vdash A : \sigma$, one can freely use cut or drop.

The following are some elementary properties of the type system.

Lemma 11

1. *If $x \notin AV(M)$, then for all types σ, $\Gamma \vdash M : \tau$ if and only if $\Gamma, (x : \sigma) \vdash M : \tau$.*

2. *If $\tau \leq \tau'$ then $(x : \tau) \vdash x : \tau'$*

3. *If $\Gamma \leq \Gamma'$ and $\Gamma' \vdash M : \tau$ then $\Gamma \vdash M : \tau$*

Adding a universal type to $\mathcal{E}$

The system $\mathcal{E}$ is obtained from the system $\mathcal{D}$ of [Dougherty and Lescanne, 2001] by adding the rule drop. The system $\mathcal{D}_\omega$ is the extension of $\mathcal{D}$ obtained by adding a universal type ω; in [Dougherty and Lescanne, 2001] characterizations of the head-normalizing and leftmost-normalizing terms of λx were obtained in terms of typability in $\mathcal{D}_\omega$.

The main result of this paper is that typability in system $\mathcal{E}$ serves to characterize the strongly-normalizing terms of λx, and therefore that the rule drop captures an important aspect of reduction in explicit substitutions calculi. But an important question to raise at this point is whether the addition of rule drop behaves well in the presence of a universal type. In particular we may ask whether the normalization theorems of [Dougherty and Lescanne, 2001] still hold in the presence of drop.

Definition 12 The type system $\mathcal{E}_\omega$ is obtained from system $\mathcal{E}$ by adding the type constant ω and the judgement $\Gamma \vdash M : \omega$ as an axiom.

Since system $\mathcal{D}_\omega$ is a subsystem of $\mathcal{E}_\omega$ it is clear that the following results follow from the corresponding results for $\mathcal{D}_\omega$. *1. If M is head normalizing then M is typable in system $\mathcal{E}_\omega$ with a non-trivial type. 2. If M is normalizing then M is typable in system $\mathcal{E}_\omega$ with a type not involving ω.* The following theorem is sufficient to establish the converses of these results.

Theorem 13 *Suppose $\Gamma \vdash M : \tau$ in system $\mathcal{E}_\omega$. Then $\Gamma \vdash M : \tau$ in system $\mathcal{D}_\omega$ as well.*

For completeness we state here the results relating reduction properties of terms with their typing properties in system $\mathcal{E}_\omega$. The results follow from those in [Dougherty and Lescanne, 2001] together with Theorem 13.

The following definitions are due to Cardone and Coppo [Cardone and Coppo, 1990]: A type is *proper* if it has no positive occurrence of ω. A type is *trivial* if it can be generated by the following rules: (i) ω is trivial, (ii) If σ is trivial and θ is any type, then $\theta \to \sigma$ is trivial, (iii) If σ and τ are trivial, then $\sigma \cap \tau$ is trivial.

Theorem 14 *Let M be a closed term. The following are equivalent.*

1 *M is typable with a non-trivial type in system $\mathcal{E}_\omega$.*

2 *M is head-normalizing in the calculus $\lambda\mathrm{x}_{gc}$.*

Theorem 15 *Let M be a closed term. The following are equivalent.*

1 *M is typable in system $\mathcal{E}_\omega$ with a type not involving ω.*

2 *M is leftmost-normalizing in the calculus $\lambda\mathrm{x}_{gc}$.*

4. Typing strongly normalizing terms

In this section "reduction" will always mean "λx reduction," that is, we do not consider garbage collection. Our goal is to prove, by induction over the leftmost-outermost strategy, that strongly normalizing terms are typable.

Definition 16 When M is not strongly normalizing we set $h(M) := \infty$. Otherwise we define $h(M) := \max\{h(N)+1 \mid M \longrightarrow N\}$.

This definition makes sense since the reduction $\longrightarrow$ is finitely-branching, so that a term M which is strongly normalizing under $\longrightarrow$ will have only finitely many N such that $M \twoheadrightarrow N$, and the definition of $h(M)$ involves taking the maximum of a finite set. The height $h(M)$ of a term M is the length of the longest derivation to normal form, and in particular the height of a normal form is 0. Note that $M \longrightarrow N \Rightarrow h(M) > h(N)$ and $M = C[N] \Rightarrow h(M) \geq h(N)$.

Normal forms in $\Lambda\mathsf{x}$ are the same as in classical λ-calculus, and the type system $\mathcal{E}$ is an extension of the standard system of intersection types for classical λ-calculus. The following proposition is thus an immediate consequence of the classical result.

Proposition 17 *If M is a normal form then M is typable in system $\mathcal{E}$.*

4.1. From restricted judgments to general judgments

In Section 4.2 we show that if $M \longrightarrow N$ by a leftmost-outermost reduction, we assign a type to M built from the type assigned to N. If the last rule of the typing tree is not an intersection rule, then it is directly determined by the structure of N. If in every of those cases we can assign the same type to M, then we can do it for every type, whether or not the last rule is $\cap$-I or $\cap$-E. This will allow us, as we treat the various cases for N, to avoid explicitly considering the situation when the last typing rule is an intersection rule.

4.2. Subject expansion

It is convenient to identify a general property we will refer to throughout this section as we induct over the height of terms:

$$\boxed{M \in \mathcal{SN} \text{ and } (\forall P \in \Lambda\mathsf{x})\ h(P) < h(M) \Longrightarrow P \text{ is typable} \qquad \mathbb{P}(M)}$$

We wish to prove that for any term whose leftmost-outermost redex is $s(l)$, if it reduces to a typable term, then the term itself is typable. For such a term, we consider the context $C[\,]$ such that the term is $C[s(l)]$. The proof lies on a structural induction on contexts $C[\,]$.

For this induction to work, we need a somewhat stronger statement.

- A term is assigned the same type as this of the term obtained by contracting its leftmost-outermost redex, except when the rule is B and the term is an abstraction.
- This assignment is made in the same environment, except when the rule is B, in which case the environment is more constrained.

Notice that the initial case of the induction concerns terms whose root can be reduced (so the type is preserved, since such terms are not abstractions) and is treated in the following Lemma. The proof is omitted for lack of space.

Lemma 18 (Root reduction) *Given a rule $l \longrightarrow r$ and an instance $s(l)$ of l, assume $\mathbb{P}(s(l))$.*

$$\Gamma \vdash s(r) : \tau \Rightarrow \Gamma \vdash s(l) : \tau \text{ if the rule is not (B)}$$

$$\Gamma \vdash s(r) : \tau \Rightarrow \exists\, \Gamma' \leq \Gamma \mid \Gamma' \vdash s(l) : \tau \text{ if the rule is (B)}$$

Lemma 19 (Leftmost-outermost reduction) *Let $l \longrightarrow r$ be a rule and let $s(l)$ be an instance of l. Let $C[\,]$ be a context such that $s(l)$ is the leftmost-outermost redex of $C[s(l)]$. Assume $\mathbb{P}(s(l))$ and $\Gamma \vdash C[s(r)] : \tau$.*

If the rule is not (B) : $\Gamma \vdash C[s(l)] : \tau$.

If the rule is (B) : $\exists\, \Gamma' \leq \Gamma \mid \begin{cases} \exists \tau' \mid \Gamma' \vdash C[s(l)] : \tau' & \text{if } C[\,] = \lambda x.C'[\,] \\ \Gamma' \vdash C[s(l)] : \tau & \text{if } C[\,] \neq \lambda x.C'[\,] \end{cases}$

In the above proof it is important to notice that the reduction *is* the leftmost-outermost one. Indeed this way we escape difficult cases, namely when the term to reduce is a closure or when $C[\,] = C'[\,]\, N$ and we have a B-redex.

Theorem 20 *If $M \in \mathcal{SN}$ then M is typable.*

Proof. By induction on the height of terms: if M is a normal form, we are done. If not, then it can be reduced to a term N by a leftmost-outermost reduction. By induction every P such $h(P) < h(M)$ is typable, thus $\mathbb{P}(M)$ is fulfilled. Especially $h(N) < h(M)$, hence N is typable. Then, using Lemma 19, we get M is typable. ///

5. Characterization of strongly normalizing terms

In this section "reduction" will always mean "λx_{gc} reduction," that is, we allow garbage collection. Our goal is to prove that typable terms are strongly normalizing (we use $\mathcal{SN}_{gc}$ to refer to the set of such terms). As described in the introduction, a consequence of this result and the result of the previous section is the fact that garbage collection does not change the set of strongly normalizing terms. See Theorem 26.

We only slightly modify the proof made in [Dougherty and Lescanne, 2001], in changing FV to AV when required namely in definition sat-gc.

Definition 21 A set $\mathcal{S}$ is $\mathcal{X}$-saturated (or saturated if there is no ambiguity about the set $\mathcal{X}$), if it is closed under the rules of inference in Table 2.

Lemma 22 *$\mathcal{SN}_{gc}$ is $\mathcal{SN}_{gc}$-saturated.*

$$\text{sat-B}\ \frac{B\langle x = A\rangle\ \boldsymbol{T}}{(\lambda x.B)A\ \boldsymbol{T}} \qquad \text{sat-I}\ \frac{A\langle z = \boldsymbol{S}\rangle\ \boldsymbol{T}}{x\langle x = A\rangle\langle z = \boldsymbol{S}\rangle\ \boldsymbol{T}}$$

$$\text{sat-Abs}\ \frac{(\lambda y.B\langle x = A\rangle)\ \langle z = \boldsymbol{S}\rangle\ \boldsymbol{T}}{(\lambda y.B)\langle x = A\rangle\langle z = \boldsymbol{S}\rangle\ \boldsymbol{T}}$$

$$\text{sat-App}\ \frac{(U\langle x = A\rangle)(V\langle x = A\rangle)\ \langle z = \boldsymbol{S}\rangle\ \boldsymbol{T}}{(UV)\langle x = A\rangle\ \langle z = \boldsymbol{S}\rangle\ \boldsymbol{T}}$$

$$\text{sat-comp}\ \frac{M\langle y = Q\rangle\langle x = P\langle y = Q\rangle\rangle\langle z = \boldsymbol{S}\rangle\ \boldsymbol{T}}{M\langle x = P\rangle\langle y = Q\rangle\langle z = \boldsymbol{S}\rangle\ \boldsymbol{T}}$$

$$\text{sat-gc}\ \frac{N\langle z = \boldsymbol{S}\rangle\ \boldsymbol{T} \qquad A \in \mathcal{X}, \quad x \notin AV(N)}{N\langle x = A\rangle\langle z = \boldsymbol{S}\rangle\ \boldsymbol{T}}$$

Table 2. $\mathcal{X}$-saturated sets

Proof. The proof relies of Corollary 3.6 of [Dougherty and Lescanne, 2001], which itself relies on Lemma 3.5 there. But in this lemma, FV can be changed safely into AV, since the statement $x_i \notin FV(M_i)$ is used for insuring that rule gc can be applied, but in our formulation of gc, FV has been precisely changed into AV. ///

Definition 23 We define $\mathcal{S}_\tau$for each type τ as

- $\mathcal{S}_t := \mathcal{SN}_{\mathsf{gc}}$ for each type variable t.
- $\mathcal{S}_{\tau\cap\sigma} := \mathcal{S}_\tau \bigcap \mathcal{S}_\sigma$.
- $\mathcal{S}_{\sigma\to\tau} := \{F \in \Lambda\mathsf{x} \mid (\forall A \in \mathcal{S}_\sigma)\ (F\ A) \in \mathcal{S}_\tau\}$.

Lemma 24 *Then for any type τ, $\mathcal{S}_\tau \subset \mathcal{SN}_{gc}$, and $\mathcal{S}_\tau$ is $\mathcal{SN}_{gc}$-saturated.*

Proof. The proof is entirely done in [Dougherty and Lescanne, 2001] by structural induction on types. Although the first statement is completely independent, the initial case of the second one relies on the former lemma. ///

Theorem 25 (Soundness theorem for $\mathcal{SN}_{\mathsf{gc}}$) *For any terms $M, A_1, \ldots, A_n$, suppose*

- $(x_1 : \sigma_1), \ldots, (x_n : \sigma_n) \vdash M : \tau$
- $\forall i \in [1, n], A_i \in \mathcal{S}_{\sigma_i}$

- $\forall i \in [1,n], \forall j \geq 0, x_{i+j} \notin AV(A_i)$

Then $M\langle x_1 = A_1\rangle \ldots \langle x_n = A_n\rangle \in \mathcal{S}_\tau$.

Proof. The proof in [Dougherty and Lescanne, 2001] consists in a structural induction on the typing tree of M. Most of it need not to be modified (the weakening of the hypothesis -AV instead of FV- is balanced by the change in the definition of saturated sets), we only have to proceed with a new case due to the addition of the drop rule.

Let $\Gamma := (x_1 : \sigma_1), \ldots, (x_n : \sigma_n)$ and assume $\Gamma \vdash M\langle x = A\rangle : \tau$ comes by the drop rule from $\Gamma \vdash M : \tau$ and $\Gamma' \vdash A : \sigma$ for some Γ' and σ.

Applying the induction hypothesis to $\Gamma \vdash M : \tau$ we get $M\langle x_1 = A_1\rangle \ldots \langle x_n = A_n\rangle \in \mathcal{S}_\tau$. Applying the induction hypothesis to $\Gamma' \vdash A : \sigma$ and using Lemma 24, we get $A \in \mathcal{SN}_{gc}$. Since $\mathcal{S}_\tau$ is $\mathcal{SN}_{gc}$-saturated, we can apply rule sat-gc which yields $M\langle x = A\rangle\langle x_1 = A_1\rangle \ldots \langle x_n = A_n\rangle \in \mathcal{S}_\tau$. ///

Theorem 26 *The following are equivalent.*

1 M *is typable in system* $\mathcal{E}$.

2 $M \in \mathcal{SN}_{gc}$.

3 $M \in \mathcal{SN}$.

Proof. Part 1 implies part 2 by Theorem 25 together with Lemma 24. Clearly 2 implies 3. Theorem 20 yields 3 implies 1. ///

The fact that garbage-collection does not change the set of strongly normalizing terms was originally established by Rose [Rose, 1996] for the slightly more restricted original notion of garbage-collection.

6. The type system of Dezani and van Bakel

As mentioned in the introduction, Dezani and van Bakel [van Bakel and Dezani-Ciancaglini, 2002] have independently found a typing characterization of the strongly normalizing terms. The innovation in their typing system also involves a new rule for typing closures $M\langle x = N\rangle$, but rather than attending directly to the way x occurs in M, as we do, they focus on whether M can be typed in an environment not binding x. Specifically, they use the following rule in place of our drop (let us write $\vdash_{DvB}$ for typability in the system of Dezani and van Bakel).

$$\text{K-cut} \quad \frac{\Gamma \vdash_{DvB} M : \tau \qquad \Delta \vdash_{DvB} N : \sigma}{\Gamma \vdash_{DvB} M\langle x = N\rangle : \tau} \quad x \text{ not in dom}(\Gamma)$$

They prove that typability in their system characterizes the strongly normalizing terms, so clearly their system types the same terms as ours. In fact the system of Dezani and van Bakel is equivalent to ours in the strong sense that it types the same terms as ours with the same types.

Theorem 27 $\Gamma \vdash M : \tau$ *if and only if* $\Gamma \vdash_{DvB} M : \tau$.

7. Conclusions and future work

We have defined a new intersection-types system $\mathcal{E}$ for terms of the explicit substitutions calculus $\lambda\mathbf{x}$ and shown that typability in $\mathcal{E}$ characterizes strong normalization. We defined a new notion of garbage-collection and proved that a term is strongly normalizing in the core calculus if and only if it is strongly normalizing in the presence of garbage collection. Using results from [Dougherty and Lescanne, 2001] we can show that the system $\mathcal{E}_\omega$ obtained by adding a universal type smoothly characterizes the weakly normalizing terms and the head-normalizing, or solvable terms. We can also give a direct proof of equivalence with a different system, found independently by Dezani and van Bakel, which also characterizes strong normalization in $\lambda\mathbf{x}$.

Future work. Intersection types have long been known to be a robust tool for exploring properties of classical λ-terms: Krivine's book [Krivine, 1993] has many examples of this; recent work includes [Bucciarelli et al., 1999, Kfoury and Wells, 1999, Dezani-Ciancaglini et al., 2000, Davies and Pfenning, 2000, Ghilezan, 2001]. There is much more work to be done in applying intersection types to calculi of explicit substitutions. One intriguing idea is to attempt to better understand the reduction properties of calculi with substitution-composition with the help of these type systems. Another, largely unexplored, area of investigation is semantics for explicit substitutions calculi: of course intersection types have proven to be a very fruitful tool for studying semantics of the classical λ-calculus. For lack of space we removed the bibliography which can found in the full version at `http://www.ens-lyon.fr/~plescann/PUBLICATIONS/avail.ps`.

References

Abadi, M., Cardelli, L., Curien, P.-L., and Lévy, J.-J. (1991). Explicit substitutions. *Journal of Functional Programming*, 1(4):375–416.

Barendregt, H. P. (1984). *The Lambda-Calculus, its syntax and semantics*. Studies in Logic and the Foundation of Mathematics. Elsevier Science Publishers B. V. (North-Holland), Amsterdam. Second edition.

Benaissa, Z., Briaud, D., Lescanne, P., and Rouyer-Degli, J. (1996). $\lambda\upsilon$, a calculus of explicit substitutions which preserves strong normalisation. *Journal of Functional Programming*, 6(5):699–722.

Bloo, R. and Geuvers, J. H. (1999). Explicit substitution: on the edge of strong normalization. *Theoretical Computer Science*, 211:375 – 395.

Bloo, R. and Rose, K. H. (1995). Preservation of strong normalisation in named lambda calculi with explicit substitution and garbage collection. In *CSN '95—Computing Science in the Netherlands*, pages 62–72, Koninklijke Jaarbeurs, Utrecht.

Bucciarelli, A., Lorenzis, S. D., Piperno, A., and Salvo, I. (1999). Some computational properties of intersection types. In *14th Symposium on Logic in Computer Science (LICS'99)*, pages 109–118, Washington - Brussels - Tokyo. IEEE.

Cardone, F. and Coppo, M. (1990). Two extension of Curry's type inference system. In Odifreddi, P., editor, *Logic and Computer Science*, volume 31 of *APIC Series*, pages 19–75. Academic Press, New York, NY.

Curien, P.-L., Hardin, T., and Lévy, J.-J. (1996). Confluence properties of weak and strong calculi of explicit substitutions. *Journal of the ACM*, 43(2):362–397.

Davies, R. and Pfenning, F. (2000). Intersection types and computational effects. In *Proceedings of the ACM Sigplan International Conference on Functional Programming (ICFP-00)*, volume 35.9 of *ACM Sigplan Notices*, pages 198–208, N.Y. ACM Press.

Dezani-Ciancaglini, M., Honsell, F., and Motohama, Y. (2000). Compositional characterization of lambda -terms using intersection types. In *Mathematical Foundation of Computer Science*, volume 1893 of *Lecture Notes in Computer Science*, pages 304–314. Springer-Verlag. extended abstract.

Dougherty, D. and Lescanne, P. (2001). Reductions, intersection types, and explicit substitutions (extended abstract). In Abramsky, S., editor, *TLCA 2001—5th Int. Conf. on Typed Lambda Calculus and Applications*, volume 2044 of *Lecture Notes in Computer Science*, pages 121–135, Krakow, Poland. Springer-Verlag.

Dougherty, D. and Lescanne, P. (to appear). Reductions, intersection types, and explicit substitutions. *Mathematical Structures in Computer Science.*

Ghilezan, S. (2001). Full intersection types and topologies in lambda calculus. *JCSS: Journal of Computer and System Sciences*, 62(1):1–14.

Kamareddine, F. and Ríos, A. (1997). Extending a lambda-calculus with explicit substitution which preserves strong normalisation into a confluent calculus on open terms. *Journal of Functional Programming*, 7(4):395–420.

Kfoury, A. J. and Wells, J. B. (1999). Principality and decidable type inference for finite-rank intersection types. In ACM, editor, *POPL '99. Proceedings of the 26th ACM SIGPLAN-SIGACT on Principles of programming languages, January 20–22, 1999, San Antonio, TX*, ACM SIGPLAN Notices, pages 161–174, New York, NY, USA. ACM Press.

Krivine, J.-L. (1993). *Lambda calculus, types and models.* Ellis Horwood.

Lescanne, P. (1994). From $\lambda\sigma$ to $\lambda\upsilon$: a journey through calculi of explicit substitutions. In Boehm, H.-J., editor, *POPL '94—21st Annual ACM Symposium on Principles of Programming Languages*, pages 60–69, Portland, Oregon. ACM.

Melliès, P.-A. (1995). Typed λ-calculi with explicit substitution may not terminate. In Dezani, M., editor, *TLCA '95—Int. Conf. on Typed Lambda Calculus and Applications*, volume 902 of *Lecture Notes in Computer Science*, pages 328–334, Edinburgh, Scotland. Springer-Verlag.

Ritter, E. (1999). Characterising explicit substitutions which preserve termination. In *TLCA'99, Int. Conf. on Typed Lambda Calculus and Applications*, volume 1581 of *Lecture Notes in Computer Science*, pages 325–339. Springer-Verlag.

Rose, K. (1996). *Operational Reduction Models for Functional Programming Languages.* PhD thesis, DIKU, Universitetsparken 1, DK-2100 København Ø. DIKU report 96/1.

van Bakel, S. and Dezani-Ciancaglini, M. (2002). Characterizing strong normalization for explicit substitutions. In *Latin American Theoretical INformatics - LATIN.* to appear.

ABOUT COMPOSITIONAL ANALYSIS OF π–CALCULUS PROCESSES*

Fabio Martinelli
Istituto per le Applicazioni Telematiche - C.N.R., Pisa, Italy.
Fabio.Martinelli@iat.cnr.it

Abstract We set up a logical framework for the compositional analysis of finite π–calculus processes. In particular, we extend the partial model checking techniques developed for value passing process algebras to a nominal calculus, i.e. the π–calculus. The logic considered is an adaptation of the ambient logic to the π–calculus. As one of the possible applications, we show that our techniques may be used to study interesting security properties as confidentiality for (finite) π–calculus processes.

Keywords: π-calculus, partial model checking, ambient logic, confidentiality.

1. Introduction

The π–calculus (Milner et al., 1992) is a compact and expressive language for describing concurrent systems. This calculus is suitable for describing processes whose communication topology may change during the computation. Processes communicate by performing sending or receiving actions on channels. Actions may be performed only on channels whose name is known by the process. Thus, the notion of name plays a central role in this calculus. Processes can send and receive names. So, if P sends the name n to Q then also Q can communicate on the channel n. Consider the following two terms:

$$\begin{array}{rcl} P & \doteq & c\langle n\rangle \\ Q & \doteq & c(y).y\langle m\rangle \end{array}$$

denoting two π–calculus processes. The process P emits on the channel c the name n (although note that both c and n are names in the π–calculus). The process Q is willing to receive a name on the channel c and after it emits the

*This is an extended abstract of (Martinelli, a). Work partially supported by Microsoft Research Europe (Cambridge); by MIUR project "MEFISTO"; by MIUR project " Tecniche e strumenti software per l'analisi della sicurezza delle comunicazioni in applicazioni telematiche di interesse economico e sociale"; by CNR project "Strumenti, ambienti ed applicazioni innovative per la società dell'informazione" and finally by CSP with the project "SeTAPS".

name m on that channel. The parallel composition of P and Q, i.e. $P \mid Q$, evolves in the process $n\langle m\rangle$ through a reduction, i.e. an internal communication, between P and Q. This fact is represented as $P \mid Q \longrightarrow n\langle m\rangle$. Thus, the process Q now is able to communicate on the channel n. Another interesting feature is the possibility for processes to create new local names, i.e. names which no other process can refer to. Consider the process P' defined as

$$P' \doteq \nu n(c\langle n\rangle \mid n(y))$$

The idea is that n is a name different from all the others outside the restriction ν. Note that also restricted (or private) names can be communicated. When this happens, the scope of the restriction changes (*scope extrusion*) by including also the receiving process, e.g.:

$$P' \mid Q = \nu n(c\langle n\rangle \mid n(y)) \mid Q \longrightarrow \nu n(n(y) \mid n\langle m\rangle)$$

This models n is a private name of P' and Q after the reduction.

In this paper, we are interested in extending the compositional analysis techniques called partial model checking (e.g., see Andersen, 1995; Larsen and Xinxin, 1991) to the π–calculus. Basically, suppose we want to verify that a system $P \mid Q$ enjoys a property expressed by a formula A of a certain logic. Then, we can simply study if one of the two components, say Q, satisfies a property A' which encodes the necessary and sufficient conditions on Q s.t. $P \mid Q$ enjoys A.

There are several verification problems where the compositional analysis provided by partial model checking is particularly useful. In particular, we consider here the analysis of security protocols. The verification scenario for such protocols is to check whether the protocol participants are able to successfully complete their assigned roles even in the presence of an enemy which tries to interfere with the execution (e.g., see Focardi et al., 2000). Let P be the process describing the behavior of honest agents of the protocol. The enemy could be whatever process one may specify in a given language, say X, possibly enjoying certain initial assumptions (e.g., the set of messages it knows). Thus, by following (Martinelli, b), we can state security properties as:

$$\forall X \quad P \mid X \models A$$

and then apply partial model checking techniques to reduce such a verification problem to a validity one, i.e.:

$$\forall X \quad X \models A'$$

which may be faced by using standard results of logic. So far, this idea has been applied to the analysis of several security properties for systems which may be described through variants of the CCS process algebra, and properties expressed with modal logics as the Hennessy-Milner one or the μ–calculus.

In this paper, as logic for describing the process properties, we adopt a restriction of the *ambient* logic developed in (Cardelli and Gordon, 2000; Cardelli and Gordon, 2001) to the π–calculus. The reason is that this logic is suitable for reasoning about free and restricted names. Furthermore, this logic is defined in terms of ***structural congruence*** between processes. This equivalence relation takes into account the spatial structure of processes, e.g. how many parallel processes are running and how these are related by the scope of restriction operators. To the best of our knowledge, this is the first attempt to develop a partial model checking analysis for a nominal calculus, and moreover, for a logic with operators which can express also the spatial structure of processes.

The techniques we develop here, even though present some restrictions to their application, are powerful enough to study interesting properties of the π–calculus. In particular, we obtain an effective method for the verification of confidentiality properties for finite π–calculus processes, i.e. if a (restricted) name is leaked to the external environment.

Organization of the paper. In Section 2, we briefly recall the asynchronous (finite) π–calculus. In Section 3, we introduce the logic we use, i.e. a restriction of the ambient logic to the π–calculus. Section 4 presents the partial model checking techniques. In Section 5 we show how to apply partial model checking techniques to study security properties, in particular the so–called Dolev-Yao confidentiality. In Section 6, we discuss about some further work.

2. Asynchronous π–calculus

In this section we briefly recall some basic concepts about the asynchronous π–calculus.

Given a countable set of names $\mathcal{N}$ (ranged over by $a, b, \ldots, n, m, \ldots$) the set of π–calculus processes is defined through the following BNF grammar:

$$P, Q ::= \mathbf{0} \mid a\langle n\rangle \mid a(n).P \mid \nu n\, P \mid P \,|\, Q$$

The name n is said bound in the terms $(\nu n)P$ and $a(n).P$. The set $fn(P)$ of free names of P is defined as usual.

We give an intuitive explanation of the operators of the calculus:

- $\mathbf{0}$ is the stuck process that does nothing.
- $a\langle n\rangle$ is the output process. Briefly, it denotes a communication on the channel a of the name n. Note that channel names can be communicated.
- $a(n).P$ is the input construct. A name is received on the channel a and its value is substituted to the free occurrences of the name n.
- $(\nu n)P$ is the name restriction. The idea is that n is a local name of P.
- $P \,|\, Q$ is the parallel composition of two processes P and Q.

We also define the ***structural congruence*** as follows. Let $\equiv$ be the least congruence relation over processes closed under the following rules:

1 $P \equiv Q$, if P is obtained through α–conversion from Q.
2 $P \,|\, \mathbf{0} \equiv P$;
3 $P \,|\, Q \equiv Q \,|\, P$;

4 $P\,|(Q\,|\,R) \equiv (P\,|\,Q)\,|\,R$;
5 $\nu n\mathbf{0} \equiv \mathbf{0}$;
6 $\nu n \nu m P \equiv \nu m \nu n P$;
7 $\nu n \nu n(P) \equiv \nu n(P)$;
8 $\nu n(a\langle m\rangle) \equiv a\langle n\rangle$, if $n \notin \{a, m\}$;
9 $\nu n(a(m).P) \equiv a(m).\nu n(P)$, if $n \notin \{a, m\}$;
10 $\nu n(P\,|\,Q) \equiv P\,|\,\nu n Q$ if $n \notin fn(P)$.

For convenience, we often write $\nu N(P)$ for $\nu n_1..\nu n_j(P)$ where $N = \{n_1, \ldots, n_j\}$. When N is empty, we assume that $\nu N(P) = P$. (We do not loose information by considering $\nu N(P)$ instead of $\nu n_1..\nu n_j(P)$ because of the rules on structural congruence.)

We give the *reduction* semantics for the asynchronous π–calculus. Processes communicate among them by exchanging messages. An internal communication (or *reduction*) of the process P is denoted by $P \longrightarrow P'$. We have the following rules for calculating the reduction relation between processes:

$$\frac{}{a\langle n\rangle\,|\,a(m).P \longrightarrow P[n/m]} \qquad \frac{P \equiv Q, Q \longrightarrow Q', Q' \equiv P'}{P \longrightarrow P'}$$

$$\frac{P \longrightarrow P'}{P\,|\,Q \longrightarrow P'\,|\,Q} \qquad \frac{P \longrightarrow P'}{\nu n(P) \longrightarrow \nu n(P')}$$

where $P[n/m]$ denotes the process P where all the free occurrences of m are replaced with n.

3. A logic on π–calculus

In this section we describe the logic we use to express properties of π–processes. This is a restriction of the ambient logic of Cardelli and Gordon, 2000; Cardelli and Gordon, 2001 to π–calculus[1]. The syntax of formulas is as follows:

$$\begin{array}{rcl} A & ::= & \mathbf{T} \mid \neg A \mid A_1 \vee A_2 \mid \eta\langle\eta'\rangle A \mid \eta(\eta')A \mid \bigcirc A \mid A \cdot_R \eta \mid \eta \cdot_R A \mid \\ & & \mathbf{0} \mid A\,|\,B \mid A \rhd B \mid \forall x A \end{array}$$

where $\eta(\eta')$ are variables $x \in \mathcal{V}$ or names $n \in \mathcal{N}$.

The logic permits us to express both temporal and spatial properties of processes; moreover it allows to treat with restricted names in a convenient way. The logic, besides the usual constants and operators of propositional logic, has three modalities for expressing the temporal behavior of processes:

- $\eta\langle\eta'\rangle A$ (*output*). This formula expresses a process may send the name η' on the channel η and then it satisfies A.

[1]Recently in (Caires and Cardelli, 2001), Cardelli and Caires adapted several concepts of the ambient logic to the π–calculus by adding also recursion. Clearly, the logic we use here is also a restriction of the one of Cardelli and Caires (although we adopt slightly different input/output modalities).

- $\eta(\eta')A$ (*input*). This formula expresses that a process may receive the name η' on the channel η and then it satisfies A.
- $\bigcirc A$ (*reduction*). This formula expresses that a process performs a reduction (an internal communication) and then it satisfies A[2].

Moreover, the logic permits us to represent the spatial structure of processes, in particular:

- $\mathbf{0}$ (*zero*). This formula requires that the process is structurally equivalent to $\mathbf{0}$.
- $A \mid B$ (*composition*). This formula expresses that the process is (or better is structurally equivalent to) a composition of two processes. One of them satisfies A, while the other satisfies B.
- $A \triangleright B$ (*adjunct of the composition*). This formula expresses that the composition of the process with whatever process satisfying A enjoys the formula B.

But, the main feature of this logic is its treatment of the (restricted) names. The logic uses two operators for managing names.

- $A{\cdot}_R n$ (*hiding*). This formula expresses that a process, after the restriction of the name n enjoys A.
- $n \cdot_R A$ (*revelation*). This formula expresses that a process is equivalent to another one under the restriction of n. After that the restriction is removed, the resulting process enjoys A.

We have also a universal quantifier $\forall x A$, which may be used to state that a property A always holds when one substitutes any name for the variable x.

The truth relation $\models$ for the logic is inductively defined in Tab. 1. As usual, we define $\exists x A$ as $\neg \forall x \neg A$, $A \wedge B$ as $\neg(\neg A \vee \neg B)$, $A \Longrightarrow B$ as $\neg A \vee B$ and $\mathbf{F} = \neg \mathbf{T}$. Let $Names(A)$ be the set of all names appearing in a formula A.

We give some examples of the usage of the logic for expressing properties of π–calculus processes.

Example 1 *To express that a process P emits a restricted name, we can check whether $P \models \exists y \exists x(\neg(y \cdot_R T) \wedge x \cdot_R y\langle x \rangle)$. The revelation operator picks a restricted name of P, if any, call it x and then checks if x is emitted on some channel y. Note, that we can reveal a name x only if x is not a free name of the process. Indeed, the formula $x \cdot_R \mathbf{T}$ states that x is not free in a process (and thus it can be revealed). Similarly, if we are not able to reveal y, it means that y is free in P.* ∎

[2] In (Cardelli and Gordon, 2001), a different operator is used, namely $\Diamond$, whose semantics is similar to $\bigcirc A$ where the reduction relation is replaced with its reflexive and transitive closure. Thus, all the reachable processes through finite sequences of reductions are inspected instead of the ones reachable with only one reduction step. However, when dealing with finite π–calculus processes, the $\Diamond$ modality may often be equivalently expressed through the $\bigcirc$ one plus the disjunction operator (e.g., see Section 5).

$P \models \mathbf{T}$		For all P
$P \models A \vee B$	iff	$P \models A$ or $P \models B$
$P \models \neg A$	iff	Not $P \models A$
$P \models a\langle n\rangle A$	iff	$P \equiv a\langle n\rangle \mid P'$ and $P' \models A$
$P \models a(n)A$	iff	$P \equiv \nu N(a(m).P' \mid P'')$, with $a, n \notin N$, and $\nu N(P'[n/m] \mid P'') \models A$
$P \models \bigcirc A$	iff	$P \longrightarrow P'$ and $P' \models A$
$P \models n \cdot_R A$	iff	$P \equiv \nu n P'$ and $P' \models A$
$P \models A \cdot_R n$	iff	$\nu n P \models A$
$P \models \mathbf{0}$	iff	$P \equiv \mathbf{0}$
$P \models A \mid B$	iff	$P \equiv Q' \mid Q''$ and $Q' \models A, Q'' \models B$
$P \models A \triangleright B$	iff	$\forall P' \models A$ we have $P \mid P' \models B$
$P \models \forall x A$	iff	$\forall n \in \mathcal{N}$ we have $P \models A[n/x]$

Table 1. Formal semantics of the logic.

Example 2 *The adjunct of the composition is an interesting operator whose definition involves the quantification over processes. It is interesting to note that, through this operator, it is possible to encode in the logic, the schema for defining the security properties given in the introduction. In particular, we can encode:* $\forall X \quad P \mid X \models B$ *as* $P \models \mathbf{T} \triangleright B$ ■

4. Compositional analysis of processes

In this section we provide a technique to reason compositionally about the satisfaction of the properties of the logic.

Our aim is to find the necessary and sufficient condition, expressed by a formula A', on the process X s.t.

$$P \mid X \models A \text{ iff } X \models A'$$

Thus, instead of reasoning about the whole system, we can directly work on one (or more) of its (parallel) components.

Example 3 *To show how this works consider, for instance, a process* $P = n\langle n'\rangle$ *and a formula* $A = n\langle n'\rangle\mathbf{T}$. *Then, from the definition of the truth relation, we have* $P \mid X \models A$ *iff there exists* P' *s.t.* $P \mid X \equiv n\langle n'\rangle \mid P'$ *and* $P' \models \mathbf{T}$. *Note that* $P \mid X \equiv n\langle n'\rangle \mid P'$ *iff* P' *is* X, *or* $P' \equiv P \mid X'$ *and* $X \equiv n\langle n'\rangle \mid X'$. *Thus, the former case imposes no conditions on* X, *i.e.* X *could be whatever process; the latter one imposes that* $X \models n\langle n'\rangle\mathbf{T}$. *By putting together the conditions on* X *we get* $X \models \mathbf{T} \vee n\langle n'\rangle\mathbf{T}$, *which is equivalent to require that* $X \models \mathbf{T}$. *Indeed, the process* P *is enough to satisfy* A. *The situation slightly changes if we take* $A = n\langle n''\rangle$, *with* $n'' \neq n'$. *In this case, the condition on* X *is* $X \models n\langle n''\rangle$, *since* P *cannot contribute to the satisfaction of the formula* A. ■

We must face a specific problem directly related with the restriction operator of the π–calculus and its scope extrusion mechanism. Indeed, the process $P \mid X$ may perform a reduction which depends on a communication of a private name of P (resp. of X) to X (resp. P). Thus, we may have $P \mid X \longrightarrow \nu n(P' \mid X')$. So, the evaluation context is changed. Note that this situation does not arise for CCS-like process algebras (e.g., see Andersen, 1995), where the channel restriction is a static operator, i.e. it does not change its scope, whereas in the π–calculus is dynamic. Thus, we perform the partial model checking of more general contexts, as $\nu N(P \mid (\_))$ where N could be possibly empty.

Another specific problem that we encounter in this study is that the definition of the semantics of both the π–calculus and the logic heavily depend on the structural congruence. Instead, the previously defined frameworks for partial model checking usually rely on Labeled Transition Systems (LTSs) (e.g., see Andersen, 1995; Larsen and Xinxin, 1991). Note that it is possible to give a semantics of the π–calculus in terms of LTSs, however the spatial operators of the logic require the notion of structural congruence (while for the others it is possible to give a semantics through a suitable notion of LTSs). Thus, we develop the partial model checking techniques in a framework completely based on structural congruence.

We introduce some auxiliary lemmas that show how it is possible to decompose processes in several formats, up to structural equivalence. Most of them are straightforwardly derived from the results about structural congruence in Engelfriet and Gelsema, 1999.

Given a name n, we can find only a finite number of processes P', up to structural equivalence, such that P is structurally equivalent to the restriction of n in P'.

Lemma 1 *Given a process P, we can effectively compute a finite set of processes $res(P,n)$ s.t.:*

1. *$P \equiv \nu n P'$ implies that there exists $P'' \in res(P,n)$ s.t. $P'' \equiv P'$;*
2. *$P'' \in res(P,n)$ implies $P \equiv \nu n P''$.* ∎

Given a process $a\langle n\rangle$, we can find only a finite number of processes P', up to structural equivalence, such that P is structurally equivalent to the composition of the output process $a\langle n\rangle$ with P'. This gives us all the possible continuations of the process P after an output.

Lemma 2 *Given a process P, we can effectively compute a finite set of processes $C^o(a\langle n\rangle, P)$ s.t.:*

1. *$P \equiv a\langle n\rangle \mid P'$ implies that there exists $P'' \in C^o(a\langle n\rangle, P)$ s.t. $P'' \equiv P'$;*
2. *$P'' \in C^o(a\langle n\rangle, P)$ implies $P \equiv a\langle n\rangle \mid P''$.* ∎

Given a process $a\langle n\rangle$, we can find only a finite number of processes P', up to structural equivalence, such that P is structurally equivalent to the composition of the output process $a\langle n\rangle$ with P' under the restriction of n. This gives us all the possible continuations of the process P after the output of a restricted name.

Lemma 3 *Given a process P, we can effectively compute a finite set of processes $C^{ro}(a\langle n\rangle, P)$ s.t.:*

1. *$P \equiv \nu n(a\langle n\rangle \mid P')$, with $a \neq n$, implies that there exists $P'' \in C^{ro}(a\langle n\rangle, P)$ s.t. $P'' \equiv P'$;*
2. *$P'' \in C^{ro}(a\langle n\rangle, P)$ implies $P \equiv \nu n(a\langle n\rangle \mid P'')$, with $a \neq n$.* ■

A process P after the receiving on a channel a of a name n may be only finitely decomposed, up to structural equivalence, as the restriction on a set of channels different from a and n of a composition of two processes s.t. one is the residual after the communication. This gives us all the possible continuations after the reception of the value n.

Lemma 4 *Given a process P, we can effectively compute a finite set of triples $C^{i}(a(-), n, P)$ s.t.:*

1. *$P \equiv \nu N(a(m).P' \mid P'')$, with $a, n \notin N$, implies that there exists $(N_1, P_1', P_1'') \in C^{i}(a(-), n, P)$ s.t. $N \cap fn(a(m).P' \mid P'') = N_1, P'[n/m] \equiv P_1'$ and $P'' \equiv P_1''$;*
2. *$(N_1, P_1', P_1'') \in C^{i}(a(-), n, P)$ implies $P \equiv \nu N_1(a(m).P' \mid P'')$, with $a, n \notin N_1$, $P_1' = P'[n/m]$ and $P_1'' = P''$.* ■

A process P may be finitely represented as the composition of pairs of processes, up to structural equivalence.

Lemma 5 *Given a process P, we can effectively compute a finite set of pairs $C^{comp}(P)$ s.t.:*

1. *$P \equiv P' \mid P''$, implies that there exists $(P_1', P_1'') \in C^{comp}(P)$ s.t. $P' \equiv P_1'$ and $P'' \equiv P_1''$;*
2. *$(P_1', P_1'') \in C^{comp}(P)$ implies $P \equiv P_1' \mid P_1''$.* ■

Note that we can study the fact that a process performs a reduction, by considering its possible decompositions. In particular, the following lemma states the possible decompositions on P and Q s.t. $\nu N(P \mid Q) \longrightarrow R$.

Lemma 6 *We have that $\nu N(P \mid Q) \longrightarrow R$ iff one of the following cases holds:*

1. *$P \longrightarrow P'$ and $R \equiv \nu N(P' \mid Q)$;*
2. *$P \equiv a\langle n\rangle \mid P', Q \equiv \nu N''(a(m).Q' \mid Q'')$, with $a, n \notin N''$, and $R \equiv \nu N(P' \mid \nu N''(Q'[n/m] \mid Q''))$;*
3. *$P \equiv \nu n(a\langle n\rangle \mid P'), Q \equiv \nu N''(a(m).Q' \mid Q'')$, with $a, n \notin N'', a \neq n, n \notin fn(Q)$ and $R \equiv \nu(N \cup \{n\})(P' \mid \nu N''(Q'[n/m] \mid Q''))$;*
4. *$Q \longrightarrow Q'$ and $R \equiv \nu N(P \mid Q')$;*
5. *$Q \equiv a\langle n\rangle \mid Q', P \equiv \nu N''(a(m).P' \mid P'')$, with $a, n \notin N''$, and $R \equiv \nu N(Q' \mid \nu N''(P'[n/m] \mid P''))$;*
6. *$Q \equiv \nu n(a\langle n\rangle \mid Q'), P \equiv \nu N''(a(m).P' \mid P'')$, with $a, n \notin N'', a \neq n, n \notin fn(P)$ and $R \equiv \nu N \cup \{n\}(Q' \mid \nu N''(P'[n/m] \mid P''))$.* ■

$$
\begin{array}{lcl}
\mathbf{T}/\!/_{P,N,\phi} & \doteq & \mathbf{T} \\
(A \vee B)/\!/_{P,N,\phi} & \doteq & A/\!/_{P,N,\phi} \vee B/\!/_{P,N,\phi} \\
(\neg A)/\!/_{P,N,\phi} & \doteq & \neg(A/\!/_{P,N,\phi}) \\
(a\langle n\rangle A)/\!/_{P,N,\phi} & \doteq & A_1 \vee A_2,\ \text{where} \\
 & & A_1 \doteq a\langle n\rangle(A/\!/_{P,N,\phi}) \quad \text{if } \{a,n\} \subseteq \phi \\
 & & A_2 \doteq \bigvee_{P' \in C^{o}(a\langle n\rangle,P)} A/\!/_{P',N,\phi} \\
(a(n)A)/\!/_{P,N,\phi} & \doteq & A_1 \vee A_2,\ \text{where} \\
 & & A_1 \doteq a(n)(A/\!/_{P,N,\phi\cup\{n\}}) \quad \text{if } a \in \phi \\
 & & A_2 \doteq \bigvee_{(N',P',P'') \in C^{i}(a(-),n,P)} A/\!/_{\nu N'(P'|P''),N,\phi} \\
(\bigcirc A)/\!/_{P,N,\phi} & \doteq & \bigvee_{P':P \longrightarrow P'} A/\!/_{P',N,\phi} \\
 & \vee & \bigvee_{a\in\phi} \bigvee_{n \in fn(P)} \bigvee_{P' \in C^{o}(a\langle n\rangle,P)} a(n)(A/\!/_{P',N,\phi\cup\{n\}}) \\
 & \vee & \bigvee_{a\in\phi} \bigvee_{P' \in C^{ro}(a\langle n'\rangle,P)} a(n')(A/\!/_{P',N\cup\{n'\},\phi\cup\{n'\}}) \\
 & \vee & \bigcirc(A/\!/_{P,N,\phi}) \\
 & \vee & \bigvee_{a,n\in\phi} \bigvee_{(N_1,P',P'') \in C^{i}(a(-),n,P)} a\langle n\rangle(A/\!/_{\nu N_1(P'|P''),N,\phi}) \\
 & \vee & \bigvee_{a\in\phi} \bigvee_{(N_1,P',P'') \in C^{i}(a(-),n',P)} \\
 & & \quad n' \cdot_R a\langle n'\rangle(A/\!/_{\nu N_1(P'|P''),N\cup\{n'\},\phi\cup\{n'\}}) \\
 & & \text{where } n' \text{ is s.t. } n' \notin fn(P) \cup \phi \cup N \cup Names(A) \\
(A \cdot_R n)/\!/_{P,N,\phi} & \doteq & (A[n'/n])/\!/_{P,N\cup\{n\},\phi} \\
 & & \text{where } n' \text{ is s.t. } n' \notin fn(P) \cup \phi \cup N \cup Names(A) \\
(n \cdot_R A)/\!/_{P,N,\phi} & \doteq & A_1 \vee A_2 \vee A_3,\ \text{where} \\
 & & A_1 \doteq n \cdot_R \mathbf{T} \wedge (\bigvee_{n' \in N}(A[n'/n])/\!/_{P,N\setminus\{n'\},\phi}) \ \text{if}\, n \notin fn(P)\setminus N \\
 & & A_2 \doteq n \cdot_R (A/\!/_{P,N,\phi\cup\{n\}}) \ \text{if } n \notin fn(P) \cup N \\
 & & A_3 \doteq n \cdot_R \mathbf{T} \wedge (\bigvee_{P' \in res(P,n)} A/\!/_{P',N,\phi}) \quad \text{if } n \notin fn(P) \cup N \\
\mathbf{0}/\!/_{P,N,\phi} & \doteq & \mathbf{0} \quad \text{if } P \equiv \mathbf{0} \\
A \mid B/\!/_{P,N,\phi} & \doteq & \bigvee_{(P',P'') \in C^{comp}(P), fn(P') \cap N = \emptyset} \\
 & & (fn^{\neg}(N) \wedge A/\!/_{P',\emptyset,\phi\setminus N}) \mid (B/\!/_{P'',N,\phi}) \vee \\
 & & (fn^{\neg}(N) \wedge B/\!/_{P',\emptyset,\phi\setminus N}) \mid (A/\!/_{P'',N,\phi}) \\
 & & \text{where } fn^{\neg}(\{n_1,\ldots,n_k\}) \doteq \bigwedge_{l \in \{1,\ldots,k\}} n_l \cdot_R \mathbf{T} \\
(A' \rhd B)/\!/_{P,N,\phi} & \doteq & A' \rhd (B/\!/_{P,N,\phi\cup N_1}) \\
 & & \text{where } A' = A \wedge fn^{\subseteq}(N_1)
\end{array}
$$

Table 2. Partial model checking function for the context $\nu N(P \mid (\_))$. The "else" branch in the definition of auxiliary formulas A_i, with $i = 1..3$, is always $\neg\mathbf{T}$.

We make some assumptions that help us to make more tractable the partial model checking problem. In particular, we consider only components X whose set of free names is fixed *a priori*. Moreover, we consider only formulas where the adjunct of the composition has the following format $A \wedge fn^{\subseteq}(N) \rhd B$, where $fn^{\subseteq}(N)$ is a short-cut for $\forall x(x \cdot_R \mathbf{T} \vee \bigvee_{n \in N} x = n)$. Basically, $fn^{\subseteq}(N)$ is satisfied by a process P iff $fn(P)$ is contained in N. We also require that the quantification is only present within the definition of $fn^{\subseteq}(N)$. Call $\mathcal{L}_{\setminus\forall}$ such sub-logic.

We have now the technical notions to state the main result of this paper.

Proposition 1 *Let A be a formula in $\mathcal{L}_{\setminus\forall}$. Consider a finite set of names ϕ, a context $\nu N(P\,|(\_))$, with $Names(A) \cap N = \emptyset$, and process X with $fn(X) \subseteq \phi$. Then, we have:*

$$\nu N(P \mid X) \models A \text{ iff } X \models A//_{P,N,\phi}$$

where $A//_{P,N,\phi}$ is the formula defined in Tab. 2. ■

Requiring that the names of the formula A are not in N is not restrictive. Indeed, we can simply rename each name of A in N with a fresh one and then perform the analysis w.r.t. this new formula (e.g., see Lemma 2.3 of Cardelli and Gordon, 2001).

Remark 1 *Note that with the previous proposition we are able to reduce some model checking problems for the logic to validity checking problems. For instance, checking that $P \models A' \triangleright B$ holds, where A' is equivalent to require that the free names of the process are contained in ϕ, can be reduced to checking that $A' \Longrightarrow B//_{P,\emptyset,\phi}$ is valid. Indeed, $P \models A' \triangleright B$ iff for all X with $fn(X) \subseteq \phi$ we have $P\,|\,X \models B$; by partial model checking we obtain that X must satisfy $B//_{P,\emptyset,\phi}$. On the other hand, note that validity problems may be encoded as model checking problems. For instance, in order to establish whether or not A is valid, one can simply check if $\mathbf{0} \models \mathbf{T} \triangleright A$ holds. By definition, $\mathbf{0} \models \mathbf{T} \triangleright A$ iff $\forall P \models \mathbf{T}$ we have $\mathbf{0}\,|\,P \models A$. So, we have $\mathbf{0}\,|\,P \models A$ iff $P \models A$.* ■

5. An application to confidentiality analysis

The results of the previous section, even if deal only with a subset of the logic, are strong enough to prove interesting properties as the confidentiality one.

Consider a protocol P, which runs in a hostile environment X. We may be interested to study whether a name of P remains confined among the agents of P or it is leaked to the outside environment. (This form of confidentiality is sometimes called *Dolev-Yao* secrecy.) This leakage may be represented as the sending on an open channel, say *pub*, of the confidential value, say v.

Definition 1 *Given a context $\nu N(P\,|\,\_)$ and a process X, with $v \in fn(P), pub \in fn(X) \setminus N, v \notin fn(X) \cup N$, we say that v is leaked to X, if $\nu N(P\,|\,X) \longrightarrow^* Q\,|\,pub\langle v\rangle$. If a name v is not leaked, we say it is confidential (w.r.t. X).* ■

Remark 2 *Similarly, we could define the leakage of a restricted name n of $P \equiv \nu n P'$, with $n \in fn(P')$ to a process X. But, this kind of property may be treated as an instance of the previous definition, i.e. as the leakage of the free name n of P' to a process X' s.t. $n \notin fn(X')$.* ■

Note that if it could be possible to fix an upper bound to the number of possible interactions between whatever intruder and the system P then it would be possible to express in the logic the confidentiality property. For a while, assume that such bound is n. Then, the formula

$$leaked_v^n = \vee_{1 \leq i \leq n} \bigcirc^i pub\langle v\rangle$$

may be used to state that v is leaked by P (where $pub\langle v\rangle\mathbf{T}$ is abbreviated as $pub\langle v\rangle$) and $\neg leaked_v^n$ that v is confidential. We can prove that such bound actually exists. First, we give an estimation for the maximal length of possible interactions of a system P with its environment. Let $ml(P)$ be defined as:

$$ml(\mathbf{0}) = 0; \quad ml(\nu n P) = ml(P); \quad ml(a\langle n\rangle) = 1; \quad ml(a(n).P) = 1 + ml(P); \\ ml(P \mid P') = ml(P) + ml(P')$$

Thus, $\nu N(P \mid X) \longrightarrow^*$ may consists of at most $ml(P)$ interactions between P and X plus the interactions internal to the process X. Consider the situation where X does not contribute to the computation with internal actions, so:

$$\begin{array}{rl} \nu N(P \mid X) \longrightarrow^* Q \mid pub\langle v\rangle & \text{iff} \\ \nu N(P \mid X) \overbrace{\longrightarrow \ldots \longrightarrow}^{n'} Q \mid pub\langle v\rangle \quad \text{with } n' \leq ml(P) & \text{iff} \\ \nu N(P \mid X) \models leaked_v^{ml(P)} & \end{array}$$

It is possible to build a process X' s.t. if $\nu N(P \mid X) \longrightarrow^* Q \mid pub\langle v\rangle$ then also $\nu N(P \mid X') \longrightarrow^* Q' \mid pub\langle v\rangle$ but X' does not perform any internal reduction.

Lemma 7 *If $\nu N(P \mid X) \longrightarrow^* Q \mid pub\langle v\rangle$ then we can find a process X' s.t. $\nu N(P \mid X') \longrightarrow^* Q' \mid pub\langle v\rangle$, with $fn(X) = fn(X')$ and X' during this computation does not perform any internal reduction.* ■

Note also that we can consider as intruders only processes X s.t. $fn(X) \subseteq (fn(P) \cup \{pub\}) \setminus \{v\}$.

Lemma 8 *Assume that $\nu N'(P \mid X) \longrightarrow^* Q \mid pub\langle v\rangle$, with $pub, v \notin N'$ and $v \notin fn(X)$. Then, there exists X', with $fn(X') \subseteq (fn(P) \cup \{pub\}) \setminus \{v\}$, s.t.:*

$$\nu N'(P \mid X') \longrightarrow^* Q \mid pub\langle v\rangle$$ ■

Moreover, we can restrict ourselves to consider the analysis of the formula $leaked_v^{ml(P)}$ only for contexts $\nu fn(P)(P \mid X)$, where $fn(X) \subseteq fn(P) \cup \{pub\}$. Thus, we can study whether the value $v \in fn(P)$ is confidential in P, by requiring that there is no process X, with $fn(X) \subseteq (fn(P) \cup pub) \setminus \{v\}$, s.t.:

$$P \mid X \models leaked_v^{ml(P)}$$

By partial model checking, we can find

$$F = leaked_v^{ml(P)} //_{P,\emptyset,(fn(P)\cup\{pub\})\setminus\{v\}}$$

that is satisfiable by some process (whose set of free names is contained in $(fn(P) \cup pub) \setminus \{v\}$) if and only if v can be leaked. Alternatively, v is confidential iff F is not satisfiable. Note that the formula obtained after the partial model checking only consists of logical constants, disjunctions, revelations, outputs, inputs and reductions. A satisfiability procedure for such formulas exists.

Lemma 9 *Let A be a formula which consists only of logical constants, disjunctions, inputs, outputs, revelations and reductions. Then, the problem of* *establishing whether or not there exists a process X, with $fn(X) \subseteq \phi$, s.t. $X \models A$ is decidable.* ∎

Thus, as a simple application of our theory we have that the confidentiality analysis for our processes is decidable.

Proposition 2 *The confidentiality analysis for finite π–calculus processes is decidable.* ∎

It is worthy noticing that, from results in (Marchignoli and Martinelli, 1999; Martinelli, c), several authentication properties can be encoded as properties of the intruder knowledge, and ultimately as confidentiality properties. Thus our results may be used to deal also with authentication properties of finite π–calculus processes.

Remark 3 *In Remark 1, it has been shown how by using $\triangleright$ operator one can encode validity checking problems as model checking ones. This feature makes the model checking problem for the full logic rather difficult. Indeed, in (Charatonik et al., 2001), where the model checking problem for the ambient logic have been studied, no results have been given about fragments with the $\triangleright$ operator. By means of partial model checking and by Lemma 9, we are able to give a decision procedure for a small class of properties defined in the logic with this operator. In particular, we are able to perform the model checking of formulas like $fn^{\subseteq}(N) \triangleright B$ where N is a finite set of names and B is a formula which consists only of logical constants, disjunctions, inputs, outputs, revelations, hidings and reductions. The key point is that the partial model checking of the formula B does not introduce other operators (moreover the hidings are removed)*[3]. ∎

6. Further work

In this paper, we performed some preliminary steps towards a compositional analysis framework for a nominal calculus, i.e. the π–calculus. So far, we have considered only finite π–calculus processes and a simple logic without recursion and without universal quantification. This clearly limits the range of application of such techniques. For dealing with universal quantification one could try to resort to infinite conjunctions or to a symbolic semantics for the π–calculus. A recent work (Caires and Cardelli, 2001) shows that the interplay between new name generation and recursion is rather complex and interesting. Whether it is possible to extend the partial model checking techniques to a calculus and a logic with some form of recursion deserves further investigation. However, we argue that the results in this paper and the semantics of Cardelli and Caires for the logic with recursion may be considered as building blocks for such a study. In particular, we plan to consider the partial model checking problem for finite-control processes which may have infinite behavior but whose

[3] Actually, the partial model checking for revelation introduces some conjunctions with simple revelations. However, these can be simply treated by modifying the satisfiability procedure defined in the proof of Lemma 9.

model checking problem (actually for a different logic) may be solved (see Dam, 1996). On the other hand, some preliminary investigations show that it is possible to extend the same ideas applied in this paper to other nominal calculi such as the spi–calculus (Abadi and Gordon, 1999) and *ambient* calculus (Cardelli and Gordon, 1998).

Acknowledgments. We wish to thank Andy Gordon for encouraging comments about this work. Thanks are also due to the anonymous referees for their help for the presentation of the paper.

References

Abadi, M. and Gordon, A. D. (1999). A calculus for cryptographic protocols: The spi calculus. *Information and Computation*, 148(1):1–70.

Andersen, H. R. (1995). Partial model checking (extended abstract). In *Proc. of LICS*, pages 398–407. IEEE Computer Society Press.

Caires, L. and Cardelli, L. (2001). A spatial logic for concurrency (Part I). In *Proc. of TACS*, volume 2215 of *LNCS.*

Cardelli, L. and Gordon, A. (1998). Mobile ambients. In *Proc. of FOSSACS*, volume 1378 of *LNCS*, pages 140–155.

Cardelli, L. and Gordon, A. D. (2000). Anytime, anywhere: Modal logics for mobile ambients. In *Proc. of POPL*, pages 365–377. ACM Press.

Cardelli, L. and Gordon, A. D. (2001). Logical properties of name restriction. In *Proc. of TCLA 2001*, volume 2044 of *LNCS*, pages 46–60. Springer.

Charatonik, W., Dal Zilio, S., Gordon, A. D., Mukhopadhyay, S., , and Talbot, J.-M. (2001). The complexity of model checking mobile ambients. In *Proc. of FoSSaCS*, volume 2030 of *LNCS*, pages 52–167. Springer.

Dam, M. (1996). Model checking mobile processes. *Information and Computation*, 129(1):35–51.

Engelfriet, J. and Gelsema, T. (1999). Multisets and structural congruence of the π-calculus with replication. *Theoretical Computer Science*, 211(1–2):311–337.

Focardi, R., Gorrieri, R., and Martinelli, F. (2000). Non interference for the analysis of cryptographic protocols. In *Proc. of ICALP*, volume 1853 of *LNCS*, pages 354–372.

Larsen, K. G. and Xinxin, L. (1991). Compositionality through an operational semantics of contexts. *Journal of Logic and Computation*, 1(6):761–795.

Marchignoli, D. and Martinelli, F. (1999). Automatic verification of cryptographic protocols through compositional analysis techniques. In *Proc. of TACAS*, volume 1579 of *LNCS.*

Martinelli, F. About compositional analysis of π–calculus processes. Technical Report IAT-B4-019, December 2001. Full version of this paper.

Martinelli, F. Analysis of security protocols as open systems. Technical Report IAT-B4-006, July 2001. Accepted for publication on TCS (under minor revisions).

Martinelli, F. Encoding several security properties as properties of the intruder's knowledge. Technical Report IAT-B4-020, December 2001. Submitted.

Milner, R., Parrow, J., and Walker, D. (1992). A calculus of mobile processes. *Information and Computation*, 100(1):1–77.

A RANDOMIZED DISTRIBUTED ENCODING OF THE π-CALCULUS WITH MIXED CHOICE*

Catuscia Palamidessi and Oltea Mihaela Herescu
Dept. of Comp. Sci. and Eng., The Pennsylvania State University
{catuscia,herescu}@cse.psu.edu

Abstract We consider the problem of encoding the π-calculus (more precisely, the version of the π-calculus with mixed choice) into the asynchronous π-calculus via a *uniform* translation preserving a *reasonable* semantics. Although it has been shown that this is not possible with an exact encoding, we suggest a randomized approach using a probabilistic extension of the asynchronous π-calculus, and we show that our solution is correct with probability 1 under any proper adversary wrt a notion of testing semantics. This result establishes the basis for a distributed and symmetric implementation of mixed choice which, differently from previous proposals in literature, does not rely on assumptions on the relative speed of processes and is robust to attacks of *proper* adversaries.

1. Introduction

In [15] it has been shown that the π-calculus is strictly more expressive than the asynchronous π-calculus, in the sense that it is not possible to encode the first into the latter in a *uniform* way while preserving a *reasonable* semantics. Uniform essentially means homomorphic wrt the parallel and the renaming operators, and reasonable means sensitive to livelocks (divergencies) and to the visible actions. This result is due to the fact that in the π-calculus we can define an algorithm for solving the leader election problem in a *symmetric* network (i.e. a network of isomorphic processes), while this is not possible in the asynchronous π-calculus.

The picture changes if we decide to be content with an encoding that does not preserve *exactly* a reasonable semantics, but preserves it with probability 1. To this purpose, we consider as target language a probabilistic extension of the asynchronous π-calculus, π_{pa} ([4]), based on the probabilistic automata of Segala and Lynch ([22]). The characteristic of this model is that it distinguishes between probabilistic and nondeterministic behavior. The first is associated with the random choices of the process, while the second is related to the arbitrary decisions of an external scheduler. This separation allows us to reason about adverse conditions, i.e. schedulers that "try to sabotage" the encoding, by forcing the translated processes to loop. We propose an

*Work supported by the NSF-POWRE grant EIA-0074909.

encoding that is robust wrt a large class of adversary schedulers: they can make use of all the information about the state and the history of the system, including the result of the past random choices of the processes. The only assumption we need is that the scheduler treats the output action of the asynchronous π-calculus "properly", i.e. as a message that should eventually become available to the reader.

In order to prove the correctness of the encoding we develop a π_{pa} extension of the notion of testing semantics ([14, 2]). This semantics is sensitive to divergencies and deadlocks, hence it is "reasonable" in the sense of [15]. We will show that our encoding is correct in the sense that translated processes preserve, under any proper adversary, and with probability 1, the may and must conditions with respect to each translated observer. There have been other notions of testing semantics developed for probabilistic automata or similar systems, see [8, 7, 21], however, those notions are formalized as orderings among probabilistic processes, and as such they would not be suitable to formulate the correctness of the encoding, which needs to be stated as a correspondence between processes of different kind (non-probabilistic, π, and probabilistic, π_{pa}). It is worth noting that we could not use bisimulation, barbed bisimulation, or coupled simulation either, not even in their weak and asynchronous versions, because these semantics are on one hand "too concrete" for the kind of translation developed here (see Section 4 for more details), and, on the other hand, they are not sensitive to divergencies.

The interest in considering π_{pa} as target language lies on the fact that it can be implemented in a distributed way, i.e. without using centralized control or shared memory. In fact, like in the asynchronous π-calculus, the output actions are not allowed to have a continuation, hence they can be mapped naturally into asynchronous communication, which is the only form of communication available in a distributed architecture. A uniform translation of π_{pa} into a distributed Java-like machine is illustrated in [4]. The condition of uniformity on the encodings of π into π_{pa} and of π_{pa} into Java ensure that the distribution and symmetry are preserved, thus we can argue that our results provide an approach to the distributed and symmetric implementation of the π-calculus.

The distributed implementation of mixed choice, also known as *the binary interaction problem*, has been widely investigated, as well as the more general *multiway interaction problem*. Most of the proposed solutions are asymmetric, see for instance [10, 17, 23], and most of them rely on an ordering among the identifiers of the processes (or equivalently among the nodes of the connection graph). The only symmetric solutions that have been proposed are, not surprisingly, randomized ([3, 9, 19]). They also rely on assumptions about the relative speed of the processes during the phase in which the processes attempt to establish communication (or equivalently, on particular restrictions on the scheduler). The solution proposed in [12] can also be considered as belonging to the category of randomized approaches, although the randomization is not used explicitly by the process, but assumed implicitly in the scheduler. In the full version of this paper ([16]) we show that the algorithms of [3, 9, 19] do not work, when the processes proceed at independent speed, by giving an example of network and adversary for which any attempt of synchronization produces a livelock. The relation with [12] is discussed in Section 3.

To our knowledge, our proposal is the first symmetric solution to the binary interaction problem which makes no assumptions about the relative speed of the processes and it is robust wrt any proper adversary. We also regard as a pleasant feature of our encoding the fact that it does not require the fairness assumption on the scheduler. Most of the randomized algorithms for coordination of distributed processes do require fairness, including the one in [18], but the implementations of concurrent programming languages (for instance Java) usually do not guarantee a fair scheduling policy.

2. Preliminaries

In this section we recall the definition of the π-calculus with mixed choice, of the probabilistic asynchronous π-calculus, and of probabilistic automata.

2.1. The π-calculus with mixed choice

We consider the variant of the π-calculus presented in [20]. The main difference with the original version ([11]) is the absence of a matching operator, and the replacement of free choice with a construct for mixed choice. In our presentation, we will use recursion instead of the replication operator, as we find it more convenient for writing programs.

Consider a countable set of *channel names*, $x, y, \ldots$, and a countable set of *process names* $X, Y, \ldots$. The set of prefixes, $\alpha, \beta, \ldots$, and the set of π-calculus processes, $P, Q, \ldots$, are defined by the following syntax:

$$\begin{array}{llll} \textit{Prefixes} & \alpha & ::= & x(y) \mid \bar{x}y \mid \tau \\ \textit{Processes} & P & ::= & \sum_i \alpha_i.P_i \mid \nu x P \mid P \,|\, P \mid X \mid rec_X P \end{array}$$

Prefixes represent the basic actions of processes: $x(y)$ is the *input* of the (formal) name y from channel x; $\bar{x}y$ is the *output* of the name y on channel x; τ stands for any silent (non-communication) action.

The process $\sum_i \alpha_i.P_i$ represents guarded choice and it is usually assumed to be finite. We will use the abbreviations $\mathbf{0}$ (*inaction*) to represent the empty sum, $\alpha.P$ (*prefix*) to represent sum on one element only, and $P + Q$ for the binary sum. The symbols νx and $|$ are the *restriction* and the *parallel* operator, respectively. The process $rec_X P$ represents a process X defined as $X \stackrel{\text{def}}{=} P$, where P may contain occurrences of X (recursive definition). We assume that all occurrences of X in P are prefixed.

The operators νx and $y(x)$ are *x-binders*, i.e. in the processes $\nu x P$ and $y(x).P$ the occurrences of x in P are considered *bound*, with the usual rules of scoping. The *alpha-conversion* of bound names is defined as usual, and the renaming (or substitution) $P[y/x]$ is defined as the result of replacing all occurrences of x in P by y, possibly applying alpha-conversion to avoid capture.

The operational semantics is specified via a transition system labeled by *actions* $\mu, \mu' \ldots$, which represent either a prefix or the *bound output* $\bar{x}(y)$. This is introduced to model *scope extrusion*, i.e. the result of sending to another process a private (ν-bound) name. In the following, we will use *fn* and *bn* to denote the set of free and bound

$$\text{SUM} \quad \sum_i \alpha_i.P_i \xrightarrow{\alpha_j} P_j \qquad\qquad \text{RES} \quad \frac{P \xrightarrow{\mu} P'}{\nu y P \xrightarrow{\mu} \nu y P'} \quad y \notin names(\mu)$$

$$\text{OPEN} \quad \frac{P \xrightarrow{\bar{x}y} P'}{\nu y P \xrightarrow{\bar{x}(y)} P'} \quad x \neq y \qquad\qquad \text{PAR} \quad \frac{P \xrightarrow{\mu} P'}{P \mid Q \xrightarrow{\mu} P' \mid Q} \quad bn(\mu) \cap fn(Q) = \emptyset$$

$$\text{COM} \quad \frac{P \xrightarrow{\bar{x}y} P' \quad Q \xrightarrow{x(z)} Q'}{P \mid Q \xrightarrow{\tau} P' \mid Q'[y/z]} \qquad\qquad \text{CLOSE} \quad \frac{P \xrightarrow{\bar{x}(y)} P' \quad Q \xrightarrow{x(y)} Q'}{P \mid Q \xrightarrow{\tau} \nu y(P' \mid Q')}$$

$$\text{CONG} \quad \frac{P \equiv P' \quad P' \xrightarrow{\mu} Q' \quad Q' \equiv Q}{P \xrightarrow{\mu} Q}$$

Table 1. The late-instantiation transition system of the π-calculus.

names, respectively, in processes and actions. We will also use *names* to denote both kind of names.

In literature there are two main definitions for the transition system of the π-calculus: the *early* system and the *late* system. Here we choose to present the second one because it is more refined, hence more challenging for obtaining positive embedding results.

The rules for the late semantics are given in Table 1. The symbol $\equiv$ used in Rule CONG stands for *structural congruence*, a form of equivalence which identifies "statically" two processes and which is used to simplify the presentation. We assume this congruence to satisfy the associative monoid rules for $|$, alpha-conversion, $rec_X P \equiv P[rec_X P/X]$, and $(\nu x P) \mid Q \equiv \nu x(P \mid Q)$ if $x \notin fn(Q)$.

2.2. Probabilistic automata, adversaries, and executions

Asynchronous automata have been proposed in [22]. Here we consider a variant suitable for π_{pa}. The main difference is that we consider only discrete probabilistic spaces, and that the concept of deadlock is simply a node with no out-transitions.

A discrete probabilistic space is a pair (X, pb) where X is a set and pb is a function $pb : X \to (0,1]$ such that $\sum_{x \in X} pb(x) = 1$. Given a set Y, we define

$$Prob(Y) = \{(X, pb) \mid X \subseteq Y \text{ and } (X, pb) \text{ is a discrete probabilistic space}\}.$$

Given a set of states S and a set of actions A, a *probabilistic automaton* on S and A is a triple $(S, \mathcal{T}, s_0)$ where $s_0 \in S$ (initial state) and $\mathcal{T} \subseteq S \times Prob(A \times S)$. We call the elements of $\mathcal{T}$ *transition groups* (in [22] they are called *steps*). The idea behind this model is that the choice between two different groups is made nondeterministically and possibly controlled by an external agent, e.g. a scheduler, while the transition within the same group is chosen probabilistically and it is controlled internally (e.g. by a probabilistic choice operator). An automaton in which at most one transition group is allowed for each state is called *fully probabilistic*.

We define now the notion of execution of an automaton under a *scheduler*, by adapting and simplifying the corresponding notion given in [22]. A scheduler can be seen as a function which solves the nondeterminism of the automaton by selecting, at each moment of the computation, a transition group among all the ones allowed in the present state. Schedulers are sometimes called *adversaries*, thus conveying the idea of an external entity playing "against" the process. A process is *robust* wrt a certain class of adversaries if it gives the intended result for each possible scheduling imposed by an adversary in the class. Clearly, the reliability of an algorithm depends on how "smart" the adversaries of this class can be. We will assume that an adversary can decide the next transition group depending not only on the current state, but also on the whole history of the computation till that moment, including the random choices made by the automaton.

In the following, we assume M to be the probabilistic automaton $(S, \mathcal{T}, s_0)$.

We define $tree(M)$ as the tree obtained by unfolding the transition system, i.e. the tree with a root n_0 labeled by s_0, and such that, for each node n, if $s \in S$ is the label of n, then for each $(s, (X, pb)) \in \mathcal{T}$, and for each $(\mu, s') \in X$, there is a node n' child of n labeled by s', and the arc from n to n' is labeled by μ and $pb(\mu, s')$. We will denote by $nodes(M)$ the set of nodes in $tree(M)$, and by $state(n)$ the state labeling a node n.

An *adversary* for M is a function ζ that associates to each node n of $tree(M)$ a transition group among those which are allowed in $state(n)$. More formally, $\zeta : nodes(M) \to Prob(A \times S)$ is an adversary for M iff $\zeta(n) = (X, pb)$ implies $(state(n), (X, pb)) \in \mathcal{T}$.

The *execution tree* of M under an adversary ζ, denoted by $etree(M, \zeta)$, is the tree obtained from $tree(M)$ by pruning all the groups of transitions which are not selected by ζ. More formally, $etree(M, \zeta)$ is a fully probabilistic automaton $(S', \mathcal{T}', n_0)$, where $S' \subseteq nodes(M)$, n_0 is the root of $tree(M)$, and $(n, (X', pb')) \in \mathcal{T}'$ iff $X' = \{(\mu, n') \mid (\mu, state(n')) \in X\}$ and $pb'(\mu, n') = pb(\mu, state(n'))$, where $(X, pb) = \zeta(n)$.

An *execution fragment* ξ is any path (finite or infinite) from the root of $etree(M, \zeta)$. The notation $\xi \leq \xi'$ means that ξ is a prefix of ξ'. If ξ is $n_0 \xrightarrow[p_0]{\mu_0} n_1 \xrightarrow[p_1]{\mu_1} n_2 \xrightarrow[p_2]{\mu_2} \ldots$, the *probability* of ξ is defined as $pb(\xi) = \prod_i p_i$. If ξ is maximal, then it is called *execution*. We denote by $exec(M, \zeta)$ the set of all executions in $etree(M, \zeta)$.

We define now a probability on certain sets of executions, following a standard construction of Measure Theory. Given an execution fragment ξ, let $C_\xi = \{\xi' \in exec(M, \zeta) \mid \xi \leq \xi'\}$ (*cone* with prefix ξ). Define $pb(C_\xi) = pb(\xi)$. Let $\{C_i\}_{i \in I}$ be a countable set of disjoint cones (i.e. I is countable, and $\forall i, j.\, i \neq j \Rightarrow C_i \cap C_j = \emptyset$). Then define $pb(\bigcup_{i \in I} C_i) = \sum_{i \in I} pb(C_i)$. It is possible to show that pb is well defined, i.e. two countable sets of disjoint cones with the same union produce the same result for pb. We can also define the probability of an empty set of executions as 0, and the probability of the complement of a certain set of executions as the complement wrt 1 of the probability of the set. The closure of the cones wrt the empty set, the countable union, and the complementation generates what in Measure Theory is known as a σ-field.

2.3. The probabilistic asynchronous π-calculus

In this section we recall the definition of π_{pa} ([4]). This calculus is a probabilistic extension of the asynchronous π-calculus ([6]), whose fundamental feature is the absence of the output prefix construct, namely the continuation after an output action. This is why the calculus is called "asynchronous", although communication is modeled by handshaking. We consider a version of the asynchronous π-calculus which admits the input-guarded choice (see for instance [1]), in contrast to the original version which is choiceless. In [13] it is proved that this difference is irrelevant wrt expressiveness.

The novelty of π_{pa} is that each branch of the choice is associated with a probability. The grammar is as follows:

$$\begin{array}{llll} \textit{Prefixes} & \alpha & ::= & x(y) \mid \tau \\ \textit{Processes} & P & ::= & \bar{x}y \mid \sum_i p_i\alpha_i.P_i \mid \nu x P \mid P \mid P \mid X \mid rec_X P \end{array}$$

In the *probabilistic choice operator* $\sum_i p_i\alpha_i.P_i$, the p_i's represent positive probabilities, i.e. they satisfy $p_i \in (0,1]$ and $\sum_i p_i = 1$.

In order to give the formal definition of the probabilistic model for π_{pa}, it is convenient to introduce the following notation for representing transition groups: given a probabilistic automaton $(S, \mathcal{T}, s_0)$ and $s \in S$, we write

$$s\,\{\xrightarrow[p_i]{\mu_i} s_i \mid i \in I\}$$

iff $(s, (\{(\mu_i, s_i) \mid i \in I\}, pb)) \in \mathcal{T}$ and $\forall i \in I\ p_i = pb(\mu_i, s_i)$, where I is an index set. When I is not relevant, we will use the simpler notation $s\,\{\xrightarrow[p_i]{\mu_i} s_i\}_i$. We will also use the notation $s\,\{\xrightarrow[p_i]{\mu_i} s_i\}_{i:\phi(i)}$, where $\phi(i)$ is a logical formula depending on i, for the set $s\,\{\xrightarrow[p_i]{\mu_i} s_i \mid i \in I \text{ and } \phi(i)\}$.

The operational semantics of a π_{pa} process P is defined as a probabilistic automaton whose states are the processes reachable from P and the $\mathcal{T}$ relation is defined by the rules in Table 2. In order to keep the presentation simple, we assume that all branches in SUM are different, namely, if $i \neq j$, then $\alpha_i.P_i \not\equiv \alpha_j.P_j$. Furthermore, in RES and PAR we assume that all bound variables are distinct from each other, and from the free variables. We also assume that the transition groups that are interderivable using CONG are identified.

The SUM rule models the behavior of a choice process. Note that all possible transitions belong to the same group, meaning that the transition is chosen probabilistically by the process itself. RES models restriction on channel y: only the actions on channels different from y can be performed and possibly synchronize with an external process. The probability is redistributed among these actions. PAR represents the interleaving of parallel processes. All the transitions of the processes involved are made possible, and they are kept separated in the original groups. In this way we model the fact that the selection of the process for the next computation step is determined by a scheduler. In fact, choosing a group corresponds to choosing a process. COM models communication by handshaking. The output action synchronizes with

SUM $\sum_i p_i \alpha_i.P_i \ \{\xrightarrow[p_i]{\alpha_i} P_i\}_i$ OUT $\bar{x}y \ \{\xrightarrow[1]{\bar{x}y} 0\}$

$$\text{OPEN} \quad \frac{P\ \{\xrightarrow[1]{\bar{x}y} P'\}}{\nu y P\ \{\xrightarrow[1]{\bar{x}(y)} P'\}} \quad x \neq y \qquad \text{PAR} \quad \frac{P\ \{\xrightarrow[p_i]{\mu_i} P_i\}_i}{P \mid Q\ \{\xrightarrow[p_i]{\mu_i} P_i \mid Q\}_i}$$

$$\text{RES} \quad \frac{P\ \{\xrightarrow[p_i]{\mu_i} P_i\}_i}{\nu y P\ \{\xrightarrow[p'_i]{\mu_i} \nu y P_i\}_{i: y \notin fn(\mu_i)}} \quad \begin{array}{l}\exists i.\ y \notin fn(\mu_i) \text{ and} \\ \forall i.\ p'_i = p_i / \sum_{j: y \notin fn(\mu_j)} p_j\end{array}$$

$$\text{COM} \quad \frac{P\ \{\xrightarrow[1]{\bar{x}y} P'\} \quad Q\ \{\xrightarrow[p_i]{\mu_i} Q_i\}_i}{P \mid Q\ \{\xrightarrow[p_i]{\tau} P' \mid Q_i[y/z_i]\}_{i:\mu_i = x(z_i)} \cup \{\xrightarrow[p_i]{\mu_i} P \mid Q_i\}_{i:\mu_i \neq x(z_i)}}$$

$$\text{CLOSE} \quad \frac{P\ \{\xrightarrow[1]{\bar{x}(y)} P'\} \quad Q\ \{\xrightarrow[p_i]{\mu_i} Q_i\}_i}{P \mid Q\ \{\xrightarrow[p_i]{\tau} \nu y(P' \mid Q_i[y/z_i])\}_{i:\mu_i = x(z_i)} \cup \{\xrightarrow[p_i]{\mu_i} P \mid Q_i\}_{i:\mu_i \neq x(z_i)}}$$

$$\text{CONG} \quad \frac{P \equiv P' \quad P'\ \{\xrightarrow[p_i]{\mu_i} Q'_i\}_i \quad \forall i.\ Q'_i \equiv Q_i}{P\ \{\xrightarrow[p_i]{\mu_i} Q_i\}_i}$$

Table 2. The late-instantiation probabilistic transition system of the π_{pa}-calculus. In PAR we assume that if the argument of μ_i is bound then it does not occur free in Q.

all matching input actions of a partner, with the same probability of the input action. The other possible transitions of the partner are kept with the original probability as well. CLOSE is analogous to COM, the only difference is that the name being transmitted is private to the sender. OPEN works in combination with CLOSE like in the standard (asynchronous) π-calculus. The other rules, OUT and CONG, should be self-explanatory.

3. Encoding π into π_{pa}

In this section we define a uniform, compositional translation from π to π_{pa}. For the sake of simplicity we assume that the same channel cannot be used as both input and output guard in the same choice construct.

The main difficulty of course consists in encoding the choice operator. We follow an idea used by Nestmann in [12], which consists in associating a lock l, initially set to *true*, to each choice, and then launch a parallel process for each branch. A process P corresponding to an input branch will try to get both its lock (local lock, l) and the partner's lock (remote lock, r). When P succeeds, it tests the locks: if they are both *true* (meaning that P has won the competition) then P sets the locks to *false* so that all the other processes can *abort*, sends a positive acknowledgment (*true*) to the partner, and proceeds with its continuation. The partner also proceeds when it receives the positive acknowledgment. If the local lock is *false* then P aborts. If the remote lock is *false* then P tells the partner to abort by sending it a negative acknowledgment (*false*).

The problem with the algorithm in [12] is that processes might loop forever in the attempt to get both locks. If the initial situation is symmetric, then it is possible to define a scheduler (even a fair one) which always selects the processes in the same order, and never breaks the symmetry. In [12] it is assumed that the scheduler itself has a random behavior, i.e. it selects at random which process to execute next (and in a way totally independent from the history of the system). As argued in the introduction, we believe that it is important to consider stronger schedulers.

In order to make the algorithm robust with respect to every scheduler (under an assumption of "proper" behavior that will be explained later), we enhance it with a randomized choice made internally by the processes involved in the synchronization. The idea is similar to the one used by Lehmann and Rabin for solving the dining philosophers problem ([18]). The forks, in this case, are the locks. The idea is to let the process choose randomly the first lock, and wait on it until it becomes available. When the connection graph (where the forks are the nodes and the philosophers are the arcs) is a simple ring, like in the classic dining philosophers case, this algorithm is deadlock and livelock free with probability 1 under any fair adversary ([18]).

The problem of the mixed choice however still presents a complication: the connection graph may be rather complex, and in [5] it has been shown that the classic algorithm of Lehmann and Rabin does not work for general graphs. For instance, it does not work when a node is part of two or more cycles. In order to cope with this problem, we associate to each choice containing output guards an additional lock h. The processes corresponding to the input branches will first have to compete for the lock h of the partner. Thus, at most one output branch for each choice will be involved at a time in an interaction attempt. This ensures that there will be no connected cycles in the graph representing the interaction attempts, and we will see that this condition is sufficient for the correctness of the algorithm.

In the encoding we make use of some syntactic sugar: we assume polyadic communication (i.e. more than one parameter in the communication actions), boolean values **t** and **f** and an if-then-else construct, which is defined by the structural rules

$$\textit{if } \mathbf{t} \textit{ then } P \textit{ else } Q \equiv P \qquad \textit{if } \mathbf{f} \textit{ then } P \textit{ else } Q \equiv Q$$

As discussed in [13], these features can be translated into π_a. The encoding of π into π_{pa} is defined in Table 3. Note that all the operators are translated homomorphically except for the choice. In the encoding of the choice, l represents the principal lock (corresponding to a fork in the algorithm of Lehmann and Rabin), h represents the auxiliary lock (for ensuring that no more than one output branch for each choice will be involved simoultaneously in an interaction attempt). In the encoding of the input prefix, l represents the local principal lock, and r represents the remote principal lock. The name a is used to send an acknowledgment to the partner.

Note that in the encoding of the input-prefix the top-level choice, which represents the arbitrary choice of the first principal lock, is a blind choice ($1/2\,\tau \ldots + 1/2\,\tau \ldots$). This means that the process commits to a lock *before* knowing whether such lock is available. It can be proved that this commitment is essential for the termination of

$$
\begin{array}{rcl}
[\![\nu x P]\!] & = & \nu x [\![P]\!] \\
[\![P_1 \mid P_2]\!] & = & [\![P_1]\!] \mid [\![P_2]\!] \\
[\![X]\!] & = & X \\
[\![rec_X P]\!] & = & rec_X [\![P]\!] \\
\left[\!\!\left[\begin{array}{l} \sum_i \alpha_i.P_i \\ + \\ \sum_j \tau.Q_j \\ + \\ \sum_k \beta_k.R_k \end{array}\right]\!\!\right] & = & \nu l\,(\bar{l}\mathbf{t} \mid \nu h\,(\bar{h} \mid \prod_i [\![\alpha_i.P_i]\!]_{lh}) \mid \prod_j [\![\tau.Q_j]\!]_l \mid \prod_k [\![\beta_k.R_k]\!]_l) \\
[\![\bar{x}y.P]\!]_{rh} & = & \nu a\,(\bar{x}\langle r,a,h,y\rangle \mid a(b).\ \mathit{if}\ b\ \mathit{then}\ [\![P]\!]\ \mathit{else}\ \mathbf{0}) \\
[\![\tau.Q]\!]_l & = & l(b).(\bar{l}\mathbf{f} \mid\ \mathit{if}\ b\ \mathit{then}\ [\![Q]\!]\ \mathit{else}\ \mathbf{0}) \\
[\![x(y).R]\!]_l & = & rec_X\,(x(r,a,h,y).h.rec_Y(\ 1/2\ \tau.l(b_L).((1-\varepsilon)\ r(b_R).B + \varepsilon\ \tau.(\bar{l}b_L \mid Y)) \\
 & & \qquad + \\
 & & \qquad 1/2\ \tau.r(b_R).((1-\varepsilon)\ l(b_L).B + \varepsilon\ \tau.(\bar{r}b_R \mid Y))\ ) \\
\text{where} & & \\
B & = & \begin{array}{llll} & \mathit{if}\ b_L \wedge b_R & \mathit{then} & \bar{h} \mid \bar{l}\mathbf{f} \mid \bar{r}\mathbf{f} \mid \bar{a}\mathbf{t} \mid [\![R]\!] \\ \mathit{else} & \mathit{if}\ b_L & \mathit{then} & \bar{h} \mid \bar{l}\mathbf{t} \mid \bar{r}\mathbf{f} \mid \bar{a}\mathbf{f} \mid X \\ \mathit{else} & \mathit{if}\ \ b_R & \mathit{then} & \bar{h} \mid \bar{l}\mathbf{f} \mid \bar{r}\mathbf{t} \mid \bar{x}\langle r,a,h,y\rangle \\ & & \mathit{else} & \bar{h} \mid \bar{l}\mathbf{f} \mid \bar{r}\mathbf{f} \mid \bar{a}\mathbf{f} \end{array}
\end{array}
$$

Table 3. The encoding of π into π_{pa}. In the translation of the mixed choice, the α_i's represent output actions, and the β_k's represent input actions. ε stands for a real number in $[0, 1)$.

the algorithm. The distribution of the probabilities, on the contrary, is not essential for termination. However, this distribution affects the efficiency, i.e. how soon the synchronization protocol will converge. It can be proved that in this choice it is better to split the probability as evenly as possible, hence $1/2$ and $1/2$.

Once the process has obtained the first principal lock, the idea is that it should try to get the second one. If it succeeds, then it should test the locks and proceeds accordingly to the results of the tests as explained at the beginning of this section. Otherwise, it should release both locks and go back to the beginning of the inner loop, where it will make another random draw for selecting the first lock. This conditional behavior would need a priority choice to be expressed, namely a choice in which the first branch would always be selected whenever the corresponding guard is enabled. Such construct does not exist in the (asynchronous) π-calculus, and its introduction would make the semantics rather complicated (although it would be easy to implement it in a language like Java). To overcome the problem, we use a probabilistic choice $((1-\epsilon)\ldots+\epsilon\ldots)$ to approximate a priority choice. Of course, the smaller ϵ is, the tighter the approximation is.

4. Correctness of the encoding

In order to assess the correctness of the translation of π into π_{pa}, we consider a probabilistic extension of the notion of testing semantics proposed in [14, 2]. This extension has the advantage of being probabilistically "reasonable", i.e. sensitive to deadlocks and livelocks with non-null probability. Furthermore, in testing semantics all communications are internalized (except the one used by the observer to declare *success*), and this spares us from the problem, discussed in [12], which arises with semantics like bisimulation, barbed bisimulation, and coupled simulation, even in their weak and asynchronous versions. The kind of encoding that we use for choice cannot be correct wrt these semantics, due to their sensitiveness to the output capabilities. In fact, in the original process the output guards which are not chosen disappear after the choice is made. In the translation, however, a choice is mapped into the parallel composition of the branches, hence an output guard which is not able to interact with a partner will remain present even after some other branch wins the competition, thus causing the presence of a residual output barb. However these barbs are "garbage" by definition, not able to synchronize with any other process at this point (at least, not according to the synchronization protocol of the translated process), so they should not be counted. This sensitivity to the synchronization capabilities is exactly what testing semantics features, differently from bisimulation semantics.

Let us recall briefly the key concepts of the testing semantics for the π-calculus. An *observer* O is a π-calculus process able to perform a special action ω, denoting success. Usually ω is seen as an output action, but it does not really matter. We assume this action to be different from all those performed by tested processes. Given a π-calculus process P and an observer O, an *interaction* between P and O is a maximal (finite or infinite) sequence of τ transitions starting from $P \mid O$:

$$P \mid O = Q_0 \xrightarrow{\tau} Q_1 \xrightarrow{\tau} Q_2 \xrightarrow{\tau} \dots$$

Maximal means that the sequence is either infinite, or the last state is not able to make any further τ transition.

We say that P **may** O iff there exists an interaction such that $Q_i \xrightarrow{\omega}$ for some i. We say that P **must** O iff for every interaction there exists i such that $Q_i \xrightarrow{\omega}$. Finally, P is *testing equivalent* to Q, notation $P \simeq Q$, if for every observer O, P **may** O iff Q **may** O, and P **must** O iff Q **must** O.

In order to state the correctness of the embedding, we need to extend the notion of testing to the π_{pa}-calculus. We propose the following extension, which, we believe, captures the spirit of testing semantics.

4.1. Testing semantics for the π_{pa}-calculus

The natural extension to π_{pa} of the concept of interaction between a process P and an observer O is an execution starting from $P \mid O$, under some adversary ζ, and consisting only of arcs labeled by τ. An interaction is *successful* if it passes trough a state in which an ω step can be performed.

Our intended notion of successful must testing is that the probability that an interaction be successful is 1. To this end, we need to consider the probability of successful executions *relatively* to those executions which are interactions.

In the sequel we denote by νP the process $\nu x_1 \dots \nu x_n P$, where $x_1, \dots, x_n$ are all the free names occurring in P. With a slight abuse of notation, we denote the execution tree of the automaton generated by P under the adversary ζ as $etree(P, \zeta)$, and the set of its branches (executions) as $exec(P, \zeta)$.

Let P be a π_{pa} process and let O be a π_{pa} observer. An interaction ξ between P and O is an element of $exec(\nu(P|O), \zeta)$. Given an interaction ξ of the form:

$$\nu(P \mid O) = Q_0 \xrightarrow[p_0]{\tau} Q_1 \xrightarrow[p_1]{\tau} Q_2 \xrightarrow[p_2]{\tau} \dots,$$

we say that ξ is *successful* if there exist i and p such that $Q_i \xrightarrow[p]{\omega}$. We denote by $sexec(\nu(P|O), \zeta)$ the set $\{\xi \in exec(\nu(P|O), \zeta) \mid \xi \text{ is successful}\}$.

The following property is fundamental for defining our notion of testing for π_{pa}:

Proposition 1 *Given an adversary ζ, the set $sexec(\nu(P|O), \zeta)$ can be obtained as a countable union of disjoint cones.*

As a consequence of this proposition, the probability of $sexec(\nu(P|O), \zeta)$ is well defined (cfr. Section 2.2).

Definition 2 Let $\mathcal{A}$ be a class of adversaries. Let P, Q be π_{pa} processes and O be a π_{pa} observer.

- P $\mathbf{may}_{\mathcal{A}}$ O iff there exists an adversary $\zeta \in \mathcal{A}$ s.t. $pb(sexec(\nu(P|O), \zeta)) > 0$.
- P $\mathbf{must}_{\mathcal{A}}$ O iff for every adversary $\zeta \in \mathcal{A}$, $pb(sexec(\nu(P|O), \zeta)) = 1$.
- $P \simeq_{\mathcal{A}} Q$ iff for every O, P $\mathbf{may}_{\mathcal{A}}$ O iff Q $\mathbf{may}_{\mathcal{A}}$ O and P $\mathbf{must}_{\mathcal{A}}$ O iff Q $\mathbf{must}_{\mathcal{A}}$ O.

4.2. Correctness of the encoding wrt testing semantics

First of all, we need to make precise what class of adversaries our algorithm can cope with. Clearly, we wish this class to be as large as possible. Yet, we cannot allow just *any* adversary. The problem is related to the output actions: a malicious adversary that never schedules $\bar{l}b_L$ or $\bar{r}b_R$ in the definition of $[\![x(y).P]\!]_l$ will make it impossible for the process to get the lock and therefore will force it to loop forever.

In the intended meaning of the asynchronous π-calculus, however, these actions represent messages rather than processes. The idea is that they are "sent" when they reach the top-level in a parallel context, and are "received" when the handshaking with the corresponding input action takes place. Thus it is reasonable to assume that the scheduler will not delay forever the reception of a message, i.e. if an output action is in parallel with a process able to execute the corresponding input action, then the handshaking will eventually take place.

Definition 3 An adversary ζ for P is *proper* if, whenever P evolves into a process of the form $\nu x_1 \dots \nu x_k(P_1| \dots |P_n)$, in which one of the P_i's is an output action on one of the channels $x_1 \dots x_k$, if ζ selects infinitely often a parallel process ready to execute the corresponding input action, then P_i will eventually be scheduled for handshaking. Namely, P_i will be in the premise of a COM or CLOSE rule. We will denote by $\mathcal{P}$ the class of *proper* adversaries.

Note that the above definition is weaker than the notion of fair scheduler, which requires that *any* process which is ready infinitely often will eventually be scheduled for execution. Clearly, the fairness assumption would be sufficient for our encoding, however it is not necessary. This may seem surprising, since the solution to the dining philosophers proposed in [18] *requires* fairness. However, a careful analysis of the algorithm in [18] reveals that the fairness assumption is used *only* because a philosopher who has committed to a fork enters a *busy waiting* loop, and it remains in the loop until the fork becomes available. An unfair scheduler, hence, could keep scheduling always the same philosopher in a busy waiting loop, thus generating a livelock. If the busy wait is replaced by a suspension command (obliging the scheduler to select another process) then the fairness assumption is not be necessary ([16]).

It is important to note that π_{pa} (like most process algebra) has a suspension mechanism associated with the communication actions: if a process can proceed only by performing a handshaking, then the process will suspend until the partner is ready. Furthermore the semantics of π_{pa} ensures that a scheduler is obliged to select processes which are not suspended. Note that in $[\![x(y).P]\!]_l$ (Table 3) the acquisition of h (auxiliary lock) and of the first lock are done by input prefixes (with no alternatives) and therefore they will suspend if the locks are unavailable. It is easy to implement such suspension mechanism in a language like Java by using the `wait()` and `notify()` primitives.

Theorem 4 *For every π-calculus process P, and every observer O*

(i) P **may** O *iff* $[\![P]\!]$ $\mathbf{may}_{\mathcal{P}}$ $[\![O]\!]$

(ii) P **must** O *iff* $[\![P]\!]$ $\mathbf{must}_{\mathcal{P}}$ $[\![O]\!]$

An important ingredient of the proof of the above result is that, at any point of the execution of $[\![P]\!]$, in the graph representing the interaction attempts all cycles are disconnected (i.e. they are not connected to each other by any path). It has been shown in [5] that this is a necessary condition for the algorithm of [18] to be livelock-free, even under the fairness hypothesis.

The following corollary, which is an immediate consequence of the above theorem, states correctness in the standard process algebra sense.

Corollary 5 *For every π-calculus processes P and Q, if $[\![P]\!] \simeq_{\mathcal{P}} [\![Q]\!]$ then $P \simeq Q$.*

Note that the viceversa (full abstraction) does not hold: This is due to the fact that, if we allow arbitrary observers in π_{pa}, then we can distinguish $[\![P]\!]$ and $[\![Q]\!]$ by using observers which interact directly with their actions, i.e. without following the synchronization protocol enforced by the algorithm.

References

[1] Roberto M. Amadio, Ilaria Castellani, and Davide Sangiorgi. On bisimulations for the asynchronous π-calculus. *Theoretical Computer Science*, 195(2):291–324, 1998.

[2] Michele Boreale and Rocco De Nicola. Testing equivalence for mobile processes. *Information and Computation*, 120(2):279–303, 1995.

[3] N. Francez and M. Rodeh. A distributed abstract data type implemented by a probabilistic communication scheme. In *Proc. of FOCS*, pages 373–379, 1980.

[4] Oltea Mihaela Herescu and Catuscia Palamidessi. Probabilistic asynchronous π-calculus. In *Proc. of FOSSACS*, LNCS 1784, pages 146–160. Springer-Verlag, 2000.

[5] Oltea Mihaela Herescu and Catuscia Palamidessi. On the generalized dining philosophers problem. In *Proc. of PODC*, pages 81–89, 2001.

[6] Kohei Honda and Mario Tokoro. An object calculus for asynchronous communication. In *Proc. of ECOOP*, LNCS 512, pages 133–147. Springer-Verlag, 1991.

[7] Bengt Jonsson, Kim G. Larsen, and Wang Yi. Probabilistic extensions of process algebras. In J. Bergstra, A. Ponse, and S. Smolka, Eds., *Handbook of Process Algebras*. Elsevier. 2001.

[8] Bengt Jonsson and Wang Yi. Compositional testing preorders for probabilistic processes. In *Proc. of LICS*, pages 431–441, 1995. IEEE Computer Society Press.

[9] Yuh-Jzer Joung and Scott A. Smolka. Strong interaction fairness via randomization. *IEEE Transactions on Parallel and Distributed Systems*, 9(2):137–149, 1998.

[10] F. Knabe. A distributed protocol for channel-based communications with choice. *Computers and Artificial Intelligence*, 12(5):475–490, 1993.

[11] Robin Milner, Joachim Parrow, and David Walker. A calculus of mobile processes, I and II. *Information and Computation*, 100(1):1–40 & 41–77, 1992.

[12] Uwe Nestmann. What is a 'good' encoding of guarded choice? *Journal of Information and Computation*, 156:287–319, 2000.

[13] Uwe Nestmann and Benjamin C. Pierce. Decoding choice encodings. *Journal of Information and Computation*, 163:1–59, 2000.

[14] Rocco De Nicola and Matthew C. B. Hennessy. Testing equivalences for processes. *Theoretical Computer Science*, 34(1-2):83–133, 1984.

[15] Catuscia Palamidessi. Comparing the expressive power of the synchronous and the asynchronous π-calculus. To appear in *Mathematical Structures in Computer Science*. A preliminary version appeared in *Proc. of POPL'97*.

[16] Catuscia Palamidessi and Oltea Mihaela Herescu. A randomized distribuited encoding of the π-calculus with mixed choice. Internal Report. Postscript available at `http://www.cse.psu.edu/~catuscia/papers/Prob_impl/report.ps`.

[17] Joachim Parrow and Peter Sjodin. Multiway synchronization verified with coupled simulation. In *Proc. of CONCUR*, LNCS 630, pages 518–533. Springer-Verlag, 1992.

[18] Michael O. Rabin and Daniel Lehmann. On the advantages of free choice: A symmetric and fully distributed solution to the dining philosophers problem. In A. W. Roscoe, editor, *A Classical Mind: Essays in Honour of C.A.R. Hoare*, chapter 20, pages 333–352. Prentice Hall, 1994. An extended abstract appeared in the *Proc. of POPL'81*, pages 133-138.

[19] John H. Reif and Paul G. Spirakis. Real-time synchronization of interprocess communications. *ACM Trans. on Programming Languages and Systems*, 6(2):215–238, 1984.

[20] Davide Sangiorgi. π-calculus, internal mobility and agent-passing calculi. *Theoretical Computer Science*, 167(1,2):235–274, 1996.

[21] Roberto Segala. Testing probabilistic automata. In *Proc. of CONCUR*, LNCS 1119, pages 299–314, 1996. Springer-Verlag.

[22] Roberto Segala and Nancy Lynch. Probabilistic simulations for probabilistic processes. *Nordic Journal of Computing*, 2(2):250–273, 1995.

[23] Yih-Kuen Tsay and Rajive L. Bagrodia. Fault-tolerant algorithms for fair interprocess synchronization. *IEEE Trans. on Parallel and Distributed Systems*, 5(7):737–748, 1994.

ON REDUCTION SEMANTICS FOR THE PUSH AND PULL AMBIENT CALCULUS

Iain Phillips
Maria Grazia Vigliotti
Department of Computing, Imperial College, London
{ iccp,mgv98 } @doc.ic.ac.uk

Abstract Honda and Yoshida showed how to obtain a meaningful equivalence on processes in the asynchronous π-calculus using equational theories identifying insensitive processes. We apply their approach to a dialect of Cardelli and Gordon's Mobile Ambients. The version we propose here is the Push and Pull Ambient Calculus, where the operational semantics is no longer defined in terms of ambients entering and exiting other ambients, but in terms of being pulled and being pushed away by another ambient. Like the standard ambient calculus, this dialect has the computational power of the asynchronous π-calculus. We contend that the new calculus allows a better modelling of certain security issues.

1. Introduction

The Ambient Calculus (Mobile Ambients, or MA) has been introduced by Cardelli and Gordon as a new model for distributed mobile computation that would take into consideration the reality of the World Wide Web, where division into administrative domains requires handling the notions of both mobile code and authorisation [3]. Ambients, which are meant to represent administrative domains, have a tree structure possibly containing sub-ambients; the notion of access and mobility is captured by the operational semantics where processes equipped with the appropriate capability can freely enter or exit an ambient. The calculus has become in a short time very popular [4, 2, 1, 13] (just to cite a few papers); in particular a great deal of research has been done on types and security. Moreover different dialects have been proposed [10, 11].

On the semantics for the calculus there has been fairly modest progress. Some techniques for proving contextual equivalence have been developed [7], but in terms of bisimulation, the only successful definition has been developed for a dialect of MA, safe ambients with passwords (SAP) [11], and not for the standard calculus (to the best of our knowledge), despite different attempts [6, 15]. The techniques developed in [11] for SAP depend on the particular features of SAP; therefore these are not applicable to standard MA.

Another possible way to define equivalence is to work with *barbed bisimulation*, relying on the definition of reduction relation and an observational predicate, or *barb* [12, 9, 14]. However with barbed bisimulation one needs to use the power of contexts to get a congruence, and one must also commit a priori to a notion of barb, when in certain calculi there might be several possible choices.

Honda and Yoshida [9] have proposed a general formulation of process semantics which induces a canonical congruence based solely on the reduction relation and equational reasoning. Notice that the equivalence obtained is a congruence and we do not mention barbs. In this paper we present the results of applying Honda-Yoshida methods to a variant of MA. The starting point is the definition of *insensitive processes*, which are processes that do not interact with the environment. We then define sound theories. A theory is *sound* if it identifies insensitive and structurally equivalent processes, is consistent, and is preserved by contexts and by the reduction relation. We will see that these minimal requirements are enough to induce a meaningful equivalence, *contextual barbed bisimulation* over the class of processes. In the asynchronous π-calculus setting, contextual barbed bisimulation and barbed congruence coincides [5]. In our setting it remains future work to prove this.

Notice that we do not have to commit in advance to any kind of observation—any observable becomes a property with which the theory is equipped a posteriori. Using the Honda-Yoshida techniques, the observables are indeed the names of the top level ambients, confirming the original choice of barbs made by Cardelli and Gordon.

The Honda-Yoshida approach can be applied to standard MA, but it can be applied to other dialects. The particular dialect we present here is the Push and Pull Ambient Calculus (PAC). The philosophy of PAC is different from MA. The difference lies in the control of the movements of the ambient. In the pure AC, ambients are seen as agents that can enter and exit other administrative domains if equipped with the right capability, but which are not in control of their own boundaries. Other ambients might cross their boundaries. In the following example we show that an ambient called Client, willing to enter into a server (Server), can do that if it is in possession of the name of the Server.

$$\textbf{Client}\boxed{EnterServer} \mid \textbf{Server}\boxed{Program}$$

reduces to

$$\textbf{Server}\boxed{\textbf{Client}\boxed{EnteredServer} \mid Program}$$

Notice that in this setting, Client decides to enter Server, and Server cannot send Client away in the case that Client is a malicious agent. Similarly for exit. In the PAC setting, administrative domains have total control of their own boundaries, in the sense that an ambient can pull in other ambients and push away unwanted or dangerous agents.

The scenario above would be rewritten as follows:

$$\textbf{Client}\boxed{Program} \mid \textbf{Server}\boxed{PullClient \mid Program}$$

reduces to

$$\textbf{Server}\boxed{\textbf{Client}\boxed{Program} \mid PulledClient \mid Program}$$

The reduction relation is very similar to standard MA, but in this case the *control* of movements of the ambient lies in the server and not in the client. This new dialect is as powerful as the asynchronous π-calculus and therefore it is Turing complete. There does not seem to exist any obvious encoding from PAC into MA or vice versa.

The main advantage of this new dialect is that we can model interesting scenarios in distributed systems. For instance, consider the scenario of a bank's server with its own LAN, which is an administrative domain. A customer, which is a registered user, can gain access to some services offered by the bank. Even though the customer might not gain access to private or classified administrative sub-domains, it can nevertheless misbehave, for instance by consuming too much time or resources of the service and depriving other customers of the use of the same service. In the standard ambient world, unless the customer leaves, there is no way in which the main server can send away the problematic customer. However this is very easy to represent using the new primitives of PAC. The server deals with a request call from the customer, it allows the customer to gain access to the services, but if the customer misbehaves, the server can always usher the customer away and call another customer.

The main contribution of the paper is a novel approach to the semantics of ambients, in the unlabelled setting, using the techniques developed in [9]. We will obtain contextual bisimulation starting from a minimal set of equations involving insensitive processes. We also propose a different version of MA, where the capabilities are changed in such a way that the control of the movements of ambients lies within the boundaries of the administrative domain. We believe this version helps towards the modelling of problems of network security.

The rest of the paper is organised as follows: in Section 2 we will introduce the new primitives for the Push and Pull Ambient Calculus and its operational semantics based on the reduction relation. In Section 3 we will introduce the notion of insensitive processes that will play a crucial role for the main development of the paper. In Section 4 we will present sound theories and how to define a meaningful congruence based solely on equations; the main result is Theorem 1. In Section 5 we give the operational characterisation of the congruence as contextual bisimulation (Theorem 2) and in Section 6 we will see that our theory is independent from the definition of insensitive processes. In Section 7 we discuss an encoding of the asynchronous π-calculus into our calculus, the main result being Theorem 4. Finally we give conclusions and discuss future work.

2. Syntax and reductions of the Push and Pull Ambient Calculus (PAC)

In this section we are going to introduce the primitives for the Push and Pull Ambient Calculus. In comparison with the Mobile Ambients (MA) of Cardelli and Gordon [3], two different capabilities are introduced: $\mathbf{push}_m\, n$ and $\mathbf{pull}_m\, n$ instead of $\mathbf{in}\, n$ and $\mathbf{out}\, n$; the rest of the syntax remains unchanged. Capability $\mathbf{push}_m\, n$ is the same as the objective move $mv\ n\ out\ m$ briefly mentioned in [3], but $\mathbf{pull}_m\, n$ has no such counterpart. We assume there is an infinite set of names $\mathcal{N}$, ranged over by $n, m, \ldots$.

The set of processes $\mathcal{P}$ is given by the following grammar:

$$m\,[\mathbf{pull}_m\, n.P \mid Q] \mid n\,[R] \rightarrow m\,[P \mid Q \mid n\,[R]]$$

$$m\,[\mathbf{push}_m\, n.P \mid n\,[Q] \mid R] \rightarrow m\,[P \mid R] \mid n\,[Q]$$

$$\mathbf{open}\, n.Q \mid n\,[R] \rightarrow Q \mid R$$

$$\langle n \rangle \mid (m)P \rightarrow P\{n/m\}$$

$$\frac{P \rightarrow Q}{P \mid R \rightarrow Q \mid R} \qquad \frac{P \rightarrow Q}{(new\ \ n)P \rightarrow (new\ \ n)Q}$$

$$\frac{P \rightarrow Q}{n\,[P] \rightarrow n\,[Q]} \qquad \frac{P \equiv P' \qquad P' \rightarrow Q' \qquad Q' \equiv Q}{P \rightarrow Q}$$

Figure 1. Reduction Relation

$$P, Q ::= \mathbf{0} \mid !P \mid P|Q \mid (new\ \ n)P \mid n\,[P] \mid M.P \mid (n).P \mid \langle n \rangle$$

where M stands for the capabilities, which are the following:

$$M \quad ::= \quad \mathbf{push}_m\, n \mid \mathbf{pull}_m\, n \mid \mathbf{open}\, n$$

Intuitively $\mathbf{0}$ stands for the inactive process, $!P$ simulates recursion by spinning off copies of P, $P \mid Q$ is the parallel composition of two processes, $(new\ \ n)P$ (restriction) creates a new name n in P, $n\,[P]$ is the ambient n containing the active process P and $M.P$ is the process P guarded by the capability M. The meaning of the capabilities is intuitively the following: $\mathbf{open}\, n$ dissolves an ambient with name n; $\mathbf{pull}_m\, n$ and $\mathbf{push}_m\, n$, are primitives which can be exercised within an ambient named m only; $\mathbf{pull}_m\, n$ causes an ambient with name n to be pulled inside the current ambient named m; $\mathbf{push}_m\, n$ pushes an ambient with name n out of the current ambient named m. $(n)P$ will input any message in the top-level ambient, and $\langle n \rangle$ is an output with no continuation. This form of communication is more limited than in MA, where sequences of capabilities can be passed; we have adopted this formulation for simplicity, and because this is all we need for the encoding of the asynchronous π-calculus (Section 7). The set of free names of P is written $\mathsf{fn}(P)$ and is defined in the standard way, taking into account that the binding operators are restriction and input.

The *reduction relation* $P \rightarrow P'$, is defined in Figure 1. We use $\twoheadrightarrow$ for the reflexive and transitive closure of the reduction relation. Structural congruence $\equiv$ identifies processes that we do not want to tell apart for any semantic reason. It is defined in Figure 2. As ever, if $P \equiv Q$ then $\mathsf{fn}(P) = \mathsf{fn}(Q)$.

Contextual equivalence (also called may testing) equates two processes that admit the same observation in any context. In the following definition we are following Cardelli and Gordon. We can observe the name of a top level ambient, which does not fall in the scope of the restriction as the following definition expresses.

$$
\begin{array}{rcl}
P \mid \mathbf{0} & \equiv & P \\
P \mid Q & \equiv & Q \mid P \\
(P \mid Q) \mid R & \equiv & P \mid (Q \mid R) \\
(new\ \ n)\mathbf{0} & \equiv & \mathbf{0} \\
(new\ \ m)(new\ \ n)P & \equiv & (new\ \ n)(new\ \ m)P \\
(new\ \ n)(P \mid Q) & \equiv & P \mid (new\ \ n)Q \text{ if } n \notin \mathsf{fn}(P) \\
(new\ \ m)n\,[P] & \equiv & n\,[(new\ \ m)P] \text{ if } n \neq m \\
!P & \equiv & P \mid\ !P \\
!\mathbf{0} & \equiv & \mathbf{0}
\end{array}
$$

Figure 2. Structural Congruence

Definition 1 (Strong and weak barbs) *A process* P exhibits ambient n, *written as* $P \downarrow n$, *iff* $P \equiv (new\ \ p_1 \ldots p_n)(n\,[P'] \mid P'')$ *and* $n \notin \{p_1 \ldots p_n\}$. *The predicate* $P \Downarrow n$ *means* $P \twoheadrightarrow\downarrow n$.

A *context* is a process $\mathcal{C}\{\}$ with zero or more occurrences of a hole. $\mathcal{C}\{P\}$ denotes the result of filling each occurrence of a hole in $\mathcal{C}$ with a copy of P, allowing the possible capture of free names. We shall also need contexts with *different* holes, producing $\mathcal{C}\{P_1, \ldots, P_k\}$ in an obvious fashion.

Definition 2 *Processes* P, Q *are* contextually equivalent *(*$P \simeq Q$*) if for all names* n *and all contexts* $\mathcal{C}$, $\mathcal{C}\{P\} \Downarrow n$ *if and only if* $\mathcal{C}\{Q\} \Downarrow n$.

2.1. Examples

There is no obvious way to encode Cardelli and Gordon's MA in PAC (or vice versa). In this section we give some examples to show that in our calculus some basic computation is possible, such as choice. In Section 7 we shall see that we can encode the asynchronous π-calculus.

Cardelli and Gordon discuss a scenario of 'mobile agent authentication' [3]. They point out that a process at the top level inside an ambient is 'privileged', in that it can directly affect the movement of the ambient, and can open subambients. Care must be taken that unauthorised processes do not achieve a privileged position. Although PAC handles movement somewhat differently, the idea of top-level processes being privileged is equally true.

Cardelli and Gordon show how to model a top-level process which wishes to leave the surrounding ambient (named *home*) and then return to resume a top-level position. We can do this quite straightforwardly in PAC as follows:

$$home\,[(new\ \ n)(\mathbf{push}_{home}\, n.\mathbf{pull}_{home}\, n.\mathbf{open}\, n \mid n\,[P])]$$

A *firewall* in PAC is simply a restricted ambient $(new\ \ n)n\,[P]$. Thus a firewall prepared to admit trusted ambients named m is $(new\ \ n)n\,[!\mathbf{pull}_n\, m]$. This is rather simpler than in MA, precisely because we have chosen to let control reside with the admitting ambient rather than the entering ambient.

Finally we show how to encode choice. This was done by Cardelli and Gordon and ours is a modification of what is present in [3].

$$\begin{array}{rcl} n \Rightarrow P + m \Rightarrow Q & \stackrel{\text{df}}{=} & (new\ \ pqr) \\ & & (p\,[\mathbf{pull}_p\, n.\mathbf{push}_p\, n.(\mathbf{push}_p\, q \mid q\,[\mathbf{open}\, r.P])] \mid \\ & & (p\,[\mathbf{pull}_p\, m.\mathbf{push}_p\, m.(\mathbf{push}_p\, q \mid q\,[\mathbf{open}\, r.Q])]) \mid \\ & & \mathbf{open}\, q \mid r\,[\,]) \end{array}$$

In the presence of $n\,[R]$ and assuming that p, q, r are not free in P, Q, R, we have that:

$$n \Rightarrow P + m \Rightarrow Q \mid n\,[R] \twoheadrightarrow\simeq P \mid n\,[R]$$

3. Insensitive processes

An insensitive process is one which can never react with its environment. In the setting of the asynchronous π-calculus Honda and Yoshida [9] define the active names $\mathsf{an}(P)$ of a process P:

$$\begin{array}{rcl} \mathsf{an}(a(b).P) & = & \{a\} \\ \mathsf{an}(\overline{a}\langle b\rangle) & = & \{a\} \end{array}$$

(together with the clauses for replication, parallel, restriction). A name a is *active* in P if P is immediately able to engage in communication on channel a. A process P is *insensitive* if $\mathsf{an}(P') = \emptyset$ for all P' such that $P \twoheadrightarrow P'$.

It is rather harder to define active names for the Ambient Calculus. Clearly n is active in $n\,[P]$ as well as $\mathbf{open}\, n.P$. But we also have to allow that n is active in $m\,[\mathbf{pull}_m\, n.P]$.

Definition 3 *The set of* active names *is defined as follows:*

$$\begin{array}{rclrcl} \mathsf{an}(\mathbf{0}) & = & \emptyset & \mathsf{an}((new\ \ n)P) & = & \mathsf{an}(P) - \{n\} \\ \mathsf{an}(\mathbf{pull}_m\, n.P) & = & \{m, n\} & \mathsf{an}(P \mid Q) & = & \mathsf{an}(P) \cup \mathsf{an}(Q) \\ \mathsf{an}(\mathbf{push}_m\, n.P) & = & \{m, n\} & \mathsf{an}(!P) & = & \mathsf{an}(P) \\ \mathsf{an}(\mathbf{open}\, n.P) & = & \{n\} & \mathsf{an}((n)P) & = & \{\star\} \\ \mathsf{an}(n\,[P]) & = & \{n\} \cup (\mathsf{an}(P) - \{\star\}) & \mathsf{an}(\langle n\rangle) & = & \{\star\} \end{array}$$

Here $\star$ is a special 'name' representing possible communication at the top level.

It is straightforward that $\mathsf{an}(P) \subseteq \mathsf{fn}(P) \cup \{\star\}$ for any P; also that if $P \equiv Q$ then $\mathsf{an}(P) = \mathsf{an}(Q)$, and that if $P \downarrow n$ then $n \in \mathsf{an}(P)$. We are now ready to give the definition of insensitive processes.

Definition 4 (cf [9]) *A process P is* insensitive *if $\mathsf{an}(P') = \emptyset$ for all P' such that $P \twoheadrightarrow P'$. We let U, V range over insensitive processes.*

Example 1 *Examples of insensitive processes:*

- $(new\ \ n)(\mathbf{open}\, n.P)$;

- $(new\ m, n)(\mathbf{pull}_m\, n.P)$;
- $(new\ k, p_1 \dots p_n)k\,[P]$ *where* $\mathsf{fn}(P) \subseteq \{p_1 \dots p_n\}$

A *renaming* is a function $\sigma : \mathcal{N} \to \mathcal{N}$. We denote by $P\sigma$ the result of renaming the free names of P according to σ. We conventionally extend renamings to active names by letting $\sigma(\star) = \star$. Renamings can introduce extra reactions if they identify two distinct names. However, we can show that insensitive processes remain insensitive after renaming, which will be useful later on.

Proposition 1 *If U is insensitive and σ is a renaming, then $U\sigma$ is insensitive.* □

The following lemma tells us that insensitive processes never interact with the surrounding environment.

Lemma 1 *Let P be a process such that* $\mathsf{an}(P) = \emptyset$ *and let $\mathcal{C}\{\}$ be a context. If $\mathcal{C}\{P\} \to Q$ then either $Q \equiv \mathcal{C}\{P'\}$ where $P \to P'$, or $Q \equiv \mathcal{C}'\{P, P\sigma\}$ (some $\mathcal{C}', \sigma$) where $\mathcal{C}\{R\} \to \mathcal{C}'\{R, R\sigma\}$ (any R).* □

The statement of Lemma 1 has to involve contexts with renamings and two different holes because of examples such as $\langle m\rangle \,|\, !(n)P \to P\{m/n\} \,|\, !(n)P$.

4. Sound theories

Honda and Yoshida showed how to obtain an equivalence on processes starting from a very basic set of equations [9]. We are going to apply the same method to the PAC. The main idea is to define sound theories in general, and then consider the union of all such theories.

We shall consider equational theories, which are simply sets of pairs $\langle P, Q\rangle$ of processes, from which we can deduce equations between processes by the usual laws of equational reasoning. As usual, a theory is *consistent* if not all processes are identified in the theory.

We shall need to impose some conditions of *soundness* on theories in order that they capture behavioural equivalence. These conditions include the identification of insensitive processes and *reduction* closure, meaning that equality is preserved through reduction. We will see these minimal requirements are powerful enough to define a meaningful congruence over processes.

First we fix the notation. We let $\mathcal{E}$ range over theories. We will say that $\mathcal{E} \vdash P = Q$ if $P = Q$ is derivable in $\mathcal{E}$, and $\mathcal{E} \not\vdash P = Q$ otherwise. We will write $P = Q, P \neq Q$ when it is clear from the context which particular theory we are referring to.

Definition 5 *A theory is* reduction closed *if whenever $P = Q$ and $P \twoheadrightarrow P'$ then there exists a Q' such that $Q \twoheadrightarrow Q'$ and $P' = Q'$.*

Definition 6 *A theory is* sound *if:*

- *it contains structural congruence;*
- *it is consistent;*

- *it is reduction closed;*
- *it identifies all the insensitive processes.*

We shall deem two processes to be equivalent if they are equated in some sound theory. In order for this to make sense, we need to show that sound theories exist, and that the union of all sound theories is itself sound. We first establish that sound theories exist.

Definition 7 *Let $\mathcal{E}_{Ins}$ be the theory generated by $\equiv$ and identifying all the insensitive processes.*

We shall see that $\mathcal{E}_{Ins}$ is sound.

Definition 8 *An equational theory $\mathcal{E}$ preserves* strong (weak) barbs *if whenever $\mathcal{E} \vdash P = Q$, then for every name n, if $P \downarrow n$ then $Q \downarrow n$ (if $P \downarrow n$ then $Q \Downarrow n$).*

Lemma 2 *$\mathcal{E}_{Ins}$ preserves strong barbs.* □

Clearly a theory $\mathcal{E}$ that preserves strong (weak) barbs must be consistent since $\mathcal{E} \not\vdash n\,[\,] = \mathbf{0}$ (any n). Finally we are ready to prove the result:

Proposition 2 *$\mathcal{E}_{Ins}$ is sound.*

Proof By construction, $\equiv$ is included and insensitive processes are identified. Consistency follows from Lemma 2. So it remains to show reduction closure.

In the case of $\equiv$ it follows from the definition of $\rightarrow$. In the case of the laws identifying the insensitive processes, suppose that $\mathcal{C}\{U\} = \mathcal{C}\{V\}$. Then by Lemma 1 we have that if $\mathcal{C}\{U\}$ reacts (in one step) it is either $\mathcal{C}\{\ \}$ that reacts to produce $\mathcal{C}\{U, U\sigma\}$, which can be imitated by $\mathcal{C}\{V, V\sigma\}$ (note that $U\sigma$ and $V\sigma$ are insensitive by Proposition 1); or it is U which reacts, $U \rightarrow U'$, and U' is still insensitive, in which case $\mathcal{C}\{V\} \twoheadrightarrow \mathcal{C}\{V\}$. □

So far we have proved that there is at least one sound theory, namely $\mathcal{E}_{Ins}$, which is the *minimal* sound theory according to our definition. A sound theory in general may be more generous, in the sense that it may equate more processes; therefore it is not trivial or automatic that *any* sound theory $\mathcal{E}$ preserves weak barbs. Indeed this is the core of the paper, whose proof relies heavily on the fact that a sound theory is consistent, unlike the proof of Proposition 2, where consistency was a consequence of the preservation of the barbs. We need the following lemma, which holds for any sound theory.

Lemma 3 (Incompatible pairs [9]) *Let $\mathcal{E}$ be a sound theory. Then, $n\,[\,] \neq \mathbf{0}$, all n.*

Proof Let's assume by contradiction that for some m we have that $m\,[\,] = \mathbf{0}$. First of all we are going to prove that for all n, $n\,[\,] = \mathbf{0}$.

Now $(new\ \ m)(\mathbf{open}\, m.n\,[\,] \mid m\,[\,]) = (new\ \ m)(\mathbf{open}\, m.n\,[\,] \mid \mathbf{0})$, then:

$$(new\ \ m)(\mathbf{open}\, m.n\,[\,] \mid m\,[\,]) \rightarrow n\,[\,]$$

and $(new\ \ m)(\mathbf{open}\, m.n\,[\,]\mid \mathbf{0})$ is insensitive. By reduction closure we have that $n\,[\,] = \mathbf{0}$.

We now prove that $\mathcal{E}$ is inconsistent and thereby obtain a contradiction. Take any process P, and take m such that $m \notin \mathsf{fn}(P)$. We know from above that $m\,[\,] = \mathbf{0}$. So

$$(new\ \ m)(\mathbf{open}\, m.P \mid m\,[\,]) = (new\ \ m)(\mathbf{open}\, m.P \mid \mathbf{0})\,.$$

Moreover, because $(new\ \ m)(\mathbf{open}\, m.P \mid \mathbf{0})$ is insensitive we have

$$(new\ \ m)(\mathbf{open}\, m.P \mid m\,[\,]) = \mathbf{0}\,.$$

By reduction closure we have that $P = \mathbf{0}$, so that the theory is inconsistent. Contradiction! So $m\,[\,] \neq \mathbf{0}$ after all, for any m. □

The following result is very important, because it tells us that any sound theory is a posteriori equipped with observables.

Proposition 3 *Let $\mathcal{E}$ be a sound theory. Then it preserves weak barbs.*

Proof Assume $\mathcal{E} \vdash P = Q$ and $P \downarrow n$. Let $A = \mathsf{fn}(P) \cup \mathsf{fn}(Q)$. Take $m, k \notin A$. Now $(new\ \ A,k)k\,[P \mid \mathbf{open}\, n.(\mathbf{push}_k\, m \mid m\,[\,])] \twoheadrightarrow (new\ \ A,k)(k\,[P']) \mid m\,[\,]$. By reduction closure $(new\ \ A,k)k\,[Q \mid \mathbf{open}\, n.(\mathbf{push}_k\, m \mid m\,[\,])] \twoheadrightarrow Q'$ with

$$(new\ \ k)(k\,[P']) \mid m\,[\,] = Q'\,.$$

Since $(new\ \ A,k)(k\,[P'])$ is insensitive, $m\,[\,] = Q'$. If $\mathbf{open}\, n.(\mathbf{push}_k\, m \mid m\,[\,])$ is no longer a subprocess of Q' then immediately $Q \Downarrow n$. So suppose $\mathbf{open}\, n.(\mathbf{push}_k\, m \mid m\,[\,])$ is still inside Q'. Then $Q' \equiv (new\ \ A,k)k\,[Q'' \mid \mathbf{open}\, n.(\mathbf{push}_k\, m \mid m\,[\,])]$. But it is not hard to see that Q' is insensitive, giving $m\,[\,] = \mathbf{0}$. Contradiction of Lemma 3. So $Q \Downarrow n$. □

Theorem 1 *Let $\mathcal{E}_m$ be the union of all sound theories. Then $\mathcal{E}_m$ is sound.*

Proof It is easy to see that the union of any set of sound theories is reduction closed and contains the set of insensitive processes and structural congruence. For consistency we use Proposition 3. It is easy to see that the union of any set of sound theories also preserves weak barbs, and is therefore consistent. □

We will see in the next section that there is an operational characterisation of the maximal sound theory, namely contextual barbed bisimulation.

5. Contextual barbed bisimulation

In this section we are going to give a more operational definition of the union of all the sound theories. This goes under the name of *contextual barbed bisimulation.* This equivalence relation, that is a congruence, holds only between equal processes, giving an operational meaning to the syntactic relation $\mathcal{E} \vdash P = Q$ defined in the previous section.

Definition 9 *A symmetric relation $\mathcal{S}$ is a* contextual barbed bisimulation *if whenever $P\ \mathcal{S}\ Q$ then for any n and contexts $\mathcal{C}$:*

- *if $\mathcal{C}\{P\} \downarrow n$ then $\mathcal{C}\{Q\} \Downarrow n$;*
- *if $\mathcal{C}\{P\} \rightarrow P'$ then for some Q' we have $\mathcal{C}\{Q\} \twoheadrightarrow Q'$ and $P'\ \mathcal{S}\ Q'$*

Two processes are said to be contextual barbed equivalent *($P \approx Q$) if there exists a contextual barbed bisimulation such that $P\ \mathcal{S}\ Q$.*

By standard arguments, $\approx$ is the largest bisimulation, and $\equiv \subseteq \approx \subseteq \simeq$. We can show that $\approx$ coincides with the union of the sound theories as the following theorem states:

Theorem 2 (cf [9]) $\approx = \mathcal{E}_m$. □

6. Adequate sets of insensitive processes

We earlier defined insensitive processes via active names. It is natural to wonder how far our results on sound theories depend on the particular definition of active names. In this section we show that a quite minimal subset of insensitive processes suffices, implying that the results are quite robust. This set is defined independently of active names.

Given a set of insensitive processes T, a theory is *T-sound* if it

- contains structural congruence;
- is consistent;
- is reduction closed;
- identifies all the processes in T.

Clearly a sound theory is T-sound (for any T). The union of all T-sound theories, denoted $\mathcal{E}_m^T$, plainly includes $\mathcal{E}_m$. We shall say that a set T is 'adequate' if $\mathcal{E}_m^T = \mathcal{E}_m$.

Definition 10 *T_{Open} consists of the following processes:*

- $(new\ \ n)n\ [P]$ *where* $\mathsf{fn}(P) = \emptyset$
- $(new\ \ n)(\mathbf{open}\, n.P)$
- $(new\ \ k)k\ [(new\ \ n)(\mathbf{open}\, n.P \mid Q)]$ *where* $\mathsf{fn}(Q) \subseteq \{n\}$ *and* $Q \not\Downarrow n$

The heart of the matter when choosing sets of insensitive processes is whether we can prove the next proposition, that allows us to build the maximal sound theory.

Proposition 4 *Let $\mathcal{E}$ be a T_{Open}-sound theory. Then $\mathcal{E}$ preserves weak barbs.*

Proof Follow the proof of Proposition 3, noting that only the axioms in T_{Open} are used. □

Theorem 3 *T_{Open} is adequate: $\mathcal{E}_m^{T_{Open}} = \mathcal{E}_m$.*

Proof We know $\mathcal{E}_m \subseteq \mathcal{E}_m^{T_{Open}}$. So it is enough to show that $\mathcal{E}_m^{T_{Open}}$ is sound. $\mathcal{E}_m^{T_{Open}}$ is consistent by Proposition 4. It is reduction closed, being the union of reduction closed theories. It contains all insensitive processes since $\mathcal{E}_m \subseteq \mathcal{E}_m^{T_{Open}}$. □

7. Encoding the asynchronous π-calculus

In this section we show that the asynchronous π-calculus can be encoded in our version of the ambient calculus, which guarantees that the PAC has not only the computational power of the asynchronous π-calculus, but is also Turing complete.

In [3] Cardelli and Gordon gave an encoding of the asynchronous π-calculus [8] into their MA with communication. The idea is that for each communication channel m, an ambient named m is created. Processes wishing to either input or output on channel m enter ambient m, where they communicate. Then the input process leaves m in order to continue its execution. In order to effect this, both input and output processes need to be encased in an ambient name io which is known to the channel ambient, so that they can be opened up and enabled to communicate. Thus the encoding requires a new public name.

We present an encoding into the PAC which follows quite different lines. The idea is that an output on channel m is an ambient named m which contains a value. These output ambients exist in the 'ether' represented by the top level in the tree of ambients. Communication occurs when an output ambient is pulled in by a restricted ambient representing an input on m. It then dissolves its own enclosing ambient before continuing its execution. It will be the case that barbs on outputs (which are what are normally considered in the asynchronous π-calculus) correspond exactly to barbs for ambients. The encoding seems simpler than the one in [3].

The π-calculus is often augmented with the *matching* operator $[m = n]P$. Structural congruence is then augmented with the rule $[m = n]P \equiv P$. We can define a *derived* matching operator in PAC. The basic idea is that $\mathbf{open}\, m \mid n\,[\,]$ can react iff $m = n$. We insulate this potential reaction from the outside world inside a restricted ambient, so that no other processes can react with either of the two sides.

$$[m = n]P \overset{\text{df}}{=} (new\ \ cd)(c\,[\mathbf{open}\, m.\mathbf{push}_c\, d \mid n\,[\,] \mid d\,[\,]\,] \mid \mathbf{open}\, d.\mathbf{open}\, c.P)$$

Matching can also be simulated in standard MA.

The syntax of the asynchronous π-calculus with matching ($A\pi_=$) is as follows:

$$P, Q ::= \mathbf{0} \mid !P \mid P|Q \mid (new\ \ n)P \mid m(n).P \mid \bar{m}\langle n\rangle \mid [m = n]P$$

We omit the standard definitions of structural congruence and reduction. Barbs in the asynchronous π-calculus are only on outputs (not inputs). We shall need contextual barbed equivalence ($\approx$) and contextual equivalence ($\simeq$) on $A\pi_=$. We omit the definitions, which are much as for PAC.

Our encoding of $A\pi_=$ into PAC is as follows:

$$\begin{aligned}
[\![m(n).P]\!] &= (new\ \ l, r)\\
&\quad\ (\mathbf{open}\, l \mid r\,[(n).l\,[\mathbf{open}\, r.[\![P]\!]] \mid \mathbf{pull}_r\, m.\mathbf{open}\, m.\mathbf{push}_r\, l])\\
&\quad\ \text{where } l, r \notin \mathsf{fn}([\![P]\!]) = \mathsf{fn}(P)\\
[\![\bar{m}\langle n\rangle]\!] &= m\,[\langle n\rangle]\\
[\![\mathbf{0}]\!] &= \mathbf{0}\\
[\![(new\ \ n)P]\!] &= (new\ \ n)[\![P]\!]\\
[\![P \mid Q]\!] &= [\![P]\!] \mid [\![Q]\!]\\
[\![!P]\!] &= ![\![P]\!]\\
[\![[m = n]P]\!] &= [m = n][\![P]\!]
\end{aligned}$$

Note that each π reduction is encoded by six steps. The first of these steps pulls the output inside the input, and commits to the π reduction. The remaining steps then proceed in a deterministic fashion, and cannot be interfered with, since they take place inside restriction.

Proposition 5 *If $P \twoheadrightarrow Q$ holds in $A\pi_=$, then $[\![P]\!] \twoheadrightarrow\approx [\![Q]\!]$.* $\square$

Lemma 4 *Let $P \in A\pi_=$. If $[\![P]\!] \twoheadrightarrow R$ then there is an $A\pi_=$ process Q such that $P \twoheadrightarrow Q$ and $[\![Q]\!] \approx R$. Moreover, if $R \downarrow n$ then $Q \downarrow n$.* $\square$

Lemma 4 does not hold for Cardelli and Gordon's encoding, since the io ambients are not restricted, and are consumed in stages during execution, so that there are intermediate states which do not correspond under any reasonable notion of equivalence to the encoding of a π process.

Lemma 5 *Let $P \in A\pi_=$ and let n be a name. Then $P \Downarrow n$ iff $[\![P]\!] \Downarrow n$.* $\square$

Theorem 4 *Let P and Q be $A\pi_=$ processes.*

1. *If $[\![P]\!] \approx [\![Q]\!]$ then $P \approx Q$.*
2. *If $[\![P]\!] \simeq [\![Q]\!]$ then $P \simeq Q$.*

Proof By Lemmas 4 and 5. $\square$

The converse results do not hold, since in $A\pi_=$ we have $m(n).\bar{m}\langle n\rangle \approx \mathbf{0}$, whereas $[\![m(n).\bar{m}\langle n\rangle]\!]$ is not equivalent to $[\![\mathbf{0}]\!]$ even under $\simeq$.

We do not know whether Theorem 4 holds for Cardelli and Gordon's translation. However our method of proof would not be available in their setting, since we make use of Lemma 4.

8. Conclusions

In this paper we have presented the Push and Pull version of Mobile Ambients, that allows us to model some interesting problems in distributed settings, which seem hard (or impossible) to model in standard MA. We have presented a novel approach to the semantics, which consists of applying semantic techniques developed in [9]. The results presented here could be easily recast for standard MA, but it remains to see whether this approach would equally work for other dialects such as SAP [11], where new syntactical constructs are present.

Acknowledgments

We would like to thank Martin Berger, Cedric Fournet, Andy Gordon, Philippa Gardner, Kohei Honda, Sergio Maffeis and Andrew Phillips for useful discussions. We also thank the anonymous referees for helpful comments. Maria Grazia Vigliotti is supported by an *EPSRC* grant.

References

[1] T. Amtoft, A.J. Kfoury, and S.M. Pericas-Geertsen. What are polymorphically-typed ambients? In *Proceedings of ESOP'01*, volume 2028 of *Lectures Notes in Computer Science*, pages 206–220. Springer-Verlag, 2001.

[2] L. Cardelli, G. Ghelli, and A.D. Gordon. Mobility types for Mobile Ambients. In *Proceedings of ICALP'99*, volume 1644 of *Lectures Notes in Computer Science*, pages 230–239. Springer-Verlag, 1999.

[3] L. Cardelli and A.D. Gordon. Mobile ambients. In *Proceedings of FoSSaCS'98*, volume 1378 of *Lectures Notes in Computer Science*, pages 140–155. Springer-Verlag, 1998. Also Theoretical Computer Science, 240(1), 177–213, 2000.

[4] L. Cardelli and A.D. Gordon. Types for Mobile Ambients. In *Proceedings of POPL '99*, pages 79–92. ACM, January 1999.

[5] C. Fournet and G. Gonthier. A hierarchy of equivalences for asynchronous calculi. In *Proceedings of ICALP'98*, volume 1443 of *Lectures Notes in Computer Science*, pages 844–855. Springer-Verlag, 1998.

[6] A.D. Gordon and L. Cardelli. Technical annex. Available at: www.adg.com/, 1998.

[7] A.D. Gordon and L. Cardelli. Equational Properties of mobile ambients. In *Proceedings of FoSSaCS'99*, volume 1578 of *Lectures Notes in Computer Science*, pages 212–226. Springer-Verlag, 1999.

[8] K. Honda and M. Tokoro. An object calculus for asynchronous communication. In *Proceedings of ECOOP '91*, volume 512 of *Lectures Notes in Computer Science*, pages 133–147. Springer-Verlag, 1991.

[9] K. Honda and N. Yoshida. On the reduction-based process semantics. *Theoretical Computer Science*, 151:437–486, 1995.

[10] F. Levi and D. Sangiorgi. Controlling Interference for Ambients. In *Proceedings of POPL'00*, pages 352–364. ACM, 2000. Extended version available at: www-sop.inria.fr/meije/personnel/Davide.Sangiorgi.

[11] M. Merro and M. Hennessy. Bisimulation congruences in Safe Ambients. In *Proceedings of POPL'02*. ACM, 2002. To appear.

[12] R. Milner and D. Sangiorgi. Barbed Bisimulation. In *Proceedings of ICALP*, volume 623 of *Lectures Notes in Computer Science*, pages 685–695. Springer-Verlag, 1992.

[13] H.R. Nielson and F. Nielson. Validating firewalls in mobile ambients. In *Proceedings of CONCUR'99*, volume 1664 of *Lectures Notes in Computer Science*, pages 463–477. Springer-Verlag, 1999.

[14] D. Sangiorgi. *Expressing mobility in Process Algebra: First-Order and Higher-Order Paradigms*. PhD thesis, University of Edinburgh, 1993.

[15] M.G. Vigliotti. Transition Systems for the Ambient Calculus. Master's thesis, Imperial College, 1999.

SAFE DYNAMIC BINDING IN THE JOIN CALCULUS

Alan Schmitt
INRIA Rocquencourt
alan.schmitt@inria.fr

Abstract This paper presents an extension of the distributed Join Calculus with messages dynamically bound to definitions according to their location. New dynamic channels and new definitions for dynamic channels may be created at runtime. A dynamic message is rebound when the location containing the message migrates. A sound type system is introduced to guarantee that every dynamic message is bound to a definition. A longer version of this paper with additional examples is available [17].

1. Introduction

In a distributed world, channel-based communication usually involves senders and receivers that may reside in different locations. A receiver is a *definition*, an association between a channel name and a process that is spawned upon reception of a message on this channel. A message sent on a channel is routed to a location containing a definition for this channel. Thus the binding of senders to receivers is dependent upon the routing of messages. On the one hand, the binding may be *static*, in the sense that the routing does not depend on where the message originates, greatly simplifying remote communication. The distributed Join Calculus [10, 9] makes such a choice, by restricting the definition of a given channel to a single location. On the other hand, the routing may depend on where the message is. This *dynamic* binding models location-dependent behaviors. This paper enriches the distributed Join Calculus with channels whose messages are routed dynamically. Because of space constraints, we do not present the distributed Join Calculus. Readers who do not feel familiar with this calculus can find a tutorial in [11].

The JoCaml system [15], an implementation of the distributed Join Calculus, provides some form of dynamic binding to functions defined in the current runtime. Modeling this behavior is also one of our goals.

One requirement for our system is the ability to create new definitions for existing dynamic channels. However, when considering such a system of first class dynamic channels where definitions may be added and new channel names may be created at runtime, it is difficult to check manually that every message is bound to a definition. This *receptiveness* property should prevent mistakes like trying to send a message on a channel when it is not bound. This paper aims at such a safe calculus of local resources using a static type system.

An important design choice is the relation between the process being rebound to new definitions and the process triggering this rebinding. As in [6], we distinguish between *objective* and *subjective* rebinding. The rebinding is objective if it is triggered by a guard, and the (guarded) process that is being rebound is not active. On the opposite, rebinding is subjective when it involves running processes that are in the environment of the process triggering the rebinding. As subjective dynamic calculi have less control on the scope of the rebinding, they let the programmer modify the execution environment of running processes. However, this makes the receptiveness property significantly harder to prove statically, since the process being rebound is not explicitly available when typing the process triggering the rebinding.

Our extension adds dynamic channels and corresponding definitions. The definition bound to a dynamic channel at a given location is the closest definition of this channel in the enclosing locations. We say that a dynamic channel is *available* when there is such a definition. Since migration in the distributed Join Calculus is subjective, our calculus is a subjective dynamic calculus. As a location may redefine a channel already defined in an enclosing location, this model is an extension of the JoCaml model, where only *runtimes*—top level locations that do not migrate—provide definitions for dynamic channels.

The dynamic Join Calculus is designed to be implemented in a distributed setting. In the distributed Join Calculus, each static channel is defined in a single location. This property does not hold for dynamic channels, as a dynamic channel may be defined in several locations. However, the routing of a message on a dynamic channel is deterministic as there is only one closest enclosing location defining this channel. Moreover, we require the determination of the destination of a message to be local. This insures that there is no need for distributed synchronization.

In section 2 we present the syntax and semantics of the dynamic Join Calculus; in section 3 we present a type system that guarantees the presence of definitions, and we state its soundness in section 4; we conclude in section 5.

2. Syntax and semantics

The chemical syntax of the dynamic Join Calculus is the following:

$$
\begin{array}{rcl}
\mathcal{P} & ::= & \mathbf{0} \mid \mathcal{P} \mid \mathcal{P}' \mid n\langle \widetilde{n}_i \rangle; P \mid \mathsf{go}\ n; P \mid \nu\mathbf{n}.P \mid a.n\langle \widetilde{n_i} \rangle \\
\mathcal{D} & ::= & \top \mid \mathcal{D}, \mathcal{D}' \mid J \triangleright P \mid a[\mathcal{D} : \mathcal{P}]^{\Delta, I} \\
J & ::= & n\langle \widetilde{y} \rangle \mid J \mid J' \\
\mathcal{S} & ::= & \Omega \mid \mathcal{S} \parallel \mathcal{S}' \mid \mathcal{D} \vdash_{\varphi a}^{\Delta, I, F} \mathcal{P} \\
\Delta, I & ::= & \emptyset \mid \{\mathbf{n}\} \mid \{y\} \mid \Delta \cup \Delta
\end{array}
$$

The structural and reduction semantics are given in figure 1. We presuppose the existence of an infinite set of names ranged over by m, n, x. Location names are ranged over by a, b, c, dynamic channel names are ranged over by $\mathbf{m}, \mathbf{n}, \mathbf{x}$, static names are ranged over by $\mathtt{m}, \mathtt{n}, \mathtt{x}$, and variables are ranged over by u, y. We write $\widetilde{n}$ for possibly empty tuples of names. We write P (resp. D) for processes (resp. definitions) that do not contain any occurrence of resolved messages (of the form $a.n\langle\widetilde{n_i}\rangle$). We write $\mathcal{P}$ (resp. $\mathcal{D}$) for processes (resp. definitions) which may contain occurrences of resolved messages. Free names, received names, and defined names are defined as usual. A local definition **def** D **in** P binds within D and P the static channel names and location names defined in D. However, it does not bind the dynamic channel names defined in D. A restriction $\nu\mathbf{n}.P$ binds the dynamic name $\mathbf{n}$ in P. A reaction rule $J \triangleright P$ binds the received names of J in P. The formal definitions can be found in [17]. We also introduce the notion of *defined local names* (*dln*) as names that are defined in a given location, *defined static names* (resp. *defined dynamic names*) (*dsn*) (resp. *ddn*) as defined names that are static (resp. defined names that are dynamic).

Intuitively, a configuration consists of several concurrently running locations. Each location contains a multiset of definitions $\mathcal{D}$ and a multiset of running processes $\mathcal{P}$. As in the Join Calculus, locations are structured as a tree, and each location has a unique static name. Since the chemical semantics acts on a flat structure of running locations, the tree structure is reflected in the names of the running locations: a location has name φa if its name is a and if the path from the root of the location tree to this location is φ.

In order to account for the different routings of static and dynamic messages, we split the routing in two steps (much as in [12]). The first step, called the *name lookup* step, resolves the location where to route the message, and prepends this location to the message. The destination is the location containing the definition the message is bound to. For static channels, it is the unique location defining the channel; for dynamic channels, it is the closest enclosing location containing a definition of the channel. The second step is the *communication* step, it corresponds to the migration of the resolved message to its destination (rule COMM). We remark that, unlike [19], our semantics only focuses on name lookup and does not deal with the actual routing.

To show that the name lookup of dynamic channels is local, each running location bears a lookup function F from dynamic channels to the the name of the closest enclosing location defining the channel. This function is used in the dynamic name lookup rule (NL-DYN), where $\perp$ is the undefined location. The static name lookup rule (NL-STAT) resolves the unique location defining the channel. Since the name lookup step is local, we let the programmer define a continuation to unresolved messages that is spawned when the lookup occurs (rules NL-STAT and NL-DYN). However, the delivery and consumption of the message are asynchronous. In the following we may write $n\langle\widetilde{m}\rangle$ for $n\langle\widetilde{m}\rangle; \mathbf{0}$.

A definition has the form $n_1\langle\widetilde{y_1}\rangle \mid \ldots \mid n_k\langle\widetilde{y_k}\rangle \triangleright P$ where the n_i are the channel names, the $\widetilde{y_i}$ are the received names, and P is the guarded process. A channel name in a join pattern may either be of the form $\mathbf{n}_i$, if the channel

$$\frac{\mathcal{S} =_\alpha \mathcal{S}'}{\mathcal{S} \equiv \mathcal{S}'} \text{[STR-}\alpha\text{]} \qquad \frac{\forall \psi \in loc(\mathcal{S}), a \notin \psi \qquad G = Lookup(F, I, \Delta, a)}{a[\mathcal{D} : \mathcal{P}]^{\Delta, I} \vdash_{\varphi}^{\Delta', I', F} \;\equiv\; \vdash_{\varphi}^{\Delta', I', F} \;\|\; \mathcal{D} \vdash_{\varphi a}^{\Delta, I, G} \mathcal{P}} \text{[STR-LOC]}$$

$$\frac{dsn(D) \cap (bn(\mathcal{S}) \cup bn(\mathcal{D}, D) \cup bn(\mathcal{P} \mid P)) = \emptyset \qquad dln(D) \cap ddn(D) = \emptyset}{\mathcal{S} \;\|\; \mathcal{D} \vdash_{\varphi a}^{\Delta, I, F} \mathcal{P} \mid \mathbf{def}\ D\ \mathbf{in}\ P \longrightarrow \mathcal{S} \;\|\; \mathcal{D}, D \vdash_{\varphi a}^{\Delta, I, F} \mathcal{P} \mid P} \text{[DEF]}$$

$$\frac{\{\mathbf{n}\} \cap (bn(\mathcal{S}) \cup bn(\mathcal{D}) \cup bn(\mathcal{P} \mid P)) = \emptyset}{\mathcal{S} \;\|\; \mathcal{D} \vdash_{\varphi a}^{\Delta, I, F} \mathcal{P} \mid \nu\mathbf{n}.P \longrightarrow \mathcal{S} \;\|\; \mathcal{D} \vdash_{\varphi a}^{\Delta, I, F} \mathcal{P} \mid P} \text{[NU]}$$

$$\frac{\mathbf{n} \in dln(b)}{\vdash_{\varphi a}^{\Delta, I, F} \mathbf{n}\langle\widetilde{v}\rangle; P \longrightarrow \vdash_{\varphi a}^{\Delta, I, F} b.\mathbf{n}\langle\widetilde{v}\rangle \mid P} \text{[NL-STAT]}$$

$$\frac{F(\mathbf{n}) = b \wedge b \neq \bot}{\vdash_{\varphi a}^{\Delta, I, F} \mathbf{n}\langle\widetilde{v}\rangle; P \longrightarrow \vdash_{\varphi a}^{\Delta, I, F} b.\mathbf{n}\langle\widetilde{v}\rangle \mid P} \text{[NL-DYN]}$$

$$\frac{dom(\sigma_{rn}) = rn(J)}{J \triangleright P \vdash_{\varphi a}^{\Delta, I, F} a.J\sigma_{rn} \longrightarrow J \triangleright P \vdash_{\varphi a}^{\Delta, I, F} P\sigma_{rn}} \text{[JOIN]}$$

$$\vdash_{\varphi a}^{\Delta_a, I_a, F_a} b.n\langle\widetilde{v}\rangle \;\|\; \vdash_{\psi b}^{\Delta_b, I_b, F_b} \longrightarrow \vdash_{\varphi a}^{\Delta_a, I_a, F_a} \;\|\; \vdash_{\psi b}^{\Delta_b, I_b, F_b} b.n\langle\widetilde{v}\rangle \text{ [COMM]}$$

$$a[\mathcal{D} : \mathcal{P} \mid \mathrm{go}\ b; Q]^{\Delta_a, I_a} \vdash_{\varphi}^{\Delta, I, F} \;\|\; \vdash_{\psi b}^{\Delta_b, I_b, F_b} \longrightarrow \vdash_{\varphi}^{\Delta, I, F} \;\|\; a[\mathcal{D} : \mathcal{P} \mid Q]^{\Delta_a, I_a} \vdash_{\psi b}^{\Delta_b, I_b, F_b} \text{[GO]}$$

Figure 1. Semantics of the dynamic Join Calculus

is static, or $\mathbf{n}_i$ if it is dynamic, or y_i if it is a variable. A definition is triggered when among the running processes of the location there are messages on each of the n_i. These messages are consumed, and the guarded process is spawned, replacing the formal names (the received names) by the arguments of the messages using the substitution σ_{rn} (as described in the JOIN rule). Note that we use a slightly different JOIN rule: since only resolved messages may be consumed, we write $a.J$ for the join pattern where every message pattern has the prefix a (*i.e.* $a.(J \mid J') = a.J \mid a.J'$). New definitions are introduced using the **def** D **in** P construct, where the defined static names of D have scope D and P. New dynamic channels are introduced using the $\nu\mathbf{n}.P$ construct.

Locations, either folded or running, gather in the set Δ the dynamic channels they *define*, and in the set I the dynamic channels they *import* (*i.e.* the dynamic names they require to be defined in enclosing locations).

When a process go$(b); P$ is evaluated, the current location as well as all its sublocations migrate to location b (rule GO).

Some running location φa may be folded in its parent location φ (for subsequent migration, for instance) using rule STR-LOC. The first condition of this rule insures that there is no running sublocation of φa, in order to preserve the tree structure. Conversely, when unfolding a location, its lookup function needs to be computed using the operator *Lookup*, that takes the lookup function of the enclosing location and patches it to correspond to the current location.

Definition 2.1 (*Lookup* **operator**) *The operator Lookup takes the lookup function F of the enclosing location, the imported dynamic names I, the locally defined dynamic names Δ, and the name of the current location a, to create a lookup function $G = Lookup(F, I, \Delta, a)$ that associates the current location to locally defined dynamic names and the result of the enclosing lookup function for imported names (we write $\perp$ for the undefined location).*

We have:

$$G(n) = \begin{cases} a & n \in \Delta \\ F(n) & n \notin \Delta \wedge n \in I \\ \perp & n \notin \Delta \wedge n \notin I \end{cases}$$

We define α-conversion as in the Join Calculus (renaming of defined static names bound by a **def** and renaming of received names bound by join patterns), with the additional renaming of dynamic names bound by a ν operator.

In the following, we only consider a restricted class of processes: every location must have a unique name; every defined static name is defined in a single location; join patterns are linear, *i.e.* no defined name nor received name may occur more than once in a given join pattern; free and bound names are distinct ($fn(\mathcal{S}) \cap bn(\mathcal{S}) = \emptyset$). We call this last condition the *hygienic* condition.

The condition of rule DEF enforces the preservation of the hygienic condition. Since it must be true before the reduction, the defined names of D, that are free afterward, were bound. Thus they could not occur free in the initial configuration and the reduction cannot capture free names. The rule simply checks that these names are not bound in the final configuration, and that no defined local name of D is a dynamic name (well typed definitions satisfy this property). The hygienic condition is also enforced through rule NU.

In figure 1, only rules DEF and NU explicitly mention the context. In the other rules, the other running locations, definitions, and running processes in the locations involved in the reduction are left implicit.

The structural equivalence $\equiv$ is the smallest reflexive, symmetric and transitive relation generated by rules STR-α and STR-LOC, with the parallel operator "|" (resp. the definition composition operator ",") being associative, commutative and having $\mathbf{0}$ (resp. $\top$) as neutral element. The reduction relation $\rightarrow$ is the smallest relation generated by rules of figure 1 such that $\equiv\rightarrow\equiv \; \subseteq \; \rightarrow$. We recall that most of these rules include implicit contexts.

Discussion and examples of the dynamic Join Calculus are available in [17].

3. Safe dynamic binding

We describe a type system that allows only configurations where dynamic messages are bound to a dynamic definition in some enclosing location. This type system is similar to the one for the distributed Join Calculus. It uses the same generalization criterion as the one implemented in JoCaml and formalized in [12]. We use the following types:

$$\begin{array}{rcl} \tau & ::= & \widetilde{\tau} \mid \langle\tau\rangle_w^+ \mid \langle\tau\rangle_{\boldsymbol{\Delta}} \mid loc(\boldsymbol{\Delta}) \mid \alpha \\ \boldsymbol{\Delta}, \mathbf{I} & ::= & \emptyset \mid \{w\} \mid \boldsymbol{\Delta} \cup \boldsymbol{\Delta} \end{array} \qquad \begin{array}{rcl} w & ::= & \mathbf{n} \mid \delta \\ \sigma & ::= & \forall\widetilde{\alpha}\widetilde{\delta}.\tau \end{array}$$

$$\frac{}{\langle\tau\rangle^{+}_{w} \leq \langle\tau\rangle_{\{w\}}}\ [\textsc{Plus}] \qquad \frac{\tau' \leq \tau \quad \boldsymbol{\Delta} \subseteq \boldsymbol{\Delta}'}{\langle\tau\rangle_{\boldsymbol{\Delta}} \leq \langle\tau'\rangle_{\boldsymbol{\Delta}'}}\ [\text{N-}\textsc{Sub}] \qquad \frac{\boldsymbol{\Delta}' \subseteq \boldsymbol{\Delta}}{\mathit{loc}(\boldsymbol{\Delta}) \leq \mathit{loc}(\boldsymbol{\Delta}')}\ [\text{L-}\textsc{Sub}]$$

Figure 2. Subtyping rules

Name type variables δ occur in the types of processes guarded by a join pattern, and represent a dynamic channel name (as a name variable y represents a channel or location name).

Intuitively, locations provide dynamic definitions, either by directly defining them, or by requesting enclosing locations to define them. The type of a location is $\mathit{loc}(\boldsymbol{\Delta})$, where $\boldsymbol{\Delta}$ is the set of dynamic names that are available in the location. Thus, inside such a location a message sent on a dynamic name of $\boldsymbol{\Delta}$ is correct, whereas a message sent on any other dynamic name should be considered as incorrectly typed.

The type of dynamic channel names reflects the required dynamic definitions. A message on a channel of type $\langle\tau\rangle_{\boldsymbol{\Delta}}$ carries an argument of type τ, and requires the availability of definitions for the names of $\boldsymbol{\Delta}$. For instance, a dynamic channel $\mathbf{n}$ that does not carry any argument has type $\langle\rangle_{\{\mathbf{n}\}}$, since a message on such a channel requires a definition for $\mathbf{n}$ to be available. Since static channels do not require the presence of any dynamic definition in enclosing locations, their type is of the form $\langle\tau\rangle_{\emptyset}$ which is simply written $\langle\tau\rangle$.

Since channel names are first class values, it is possible to write a definition such as $\mathtt{send}\langle x\rangle \triangleright x\langle\rangle$, that receives a name and sends a message on it. Any use of send with a static name is correct, and using send with a dynamic name is correct only if the dynamic name is defined in an enclosing location. In order to represent this behavior, we say that the type of the argument of send is $\langle\rangle_{\boldsymbol{\Delta}}$ if $\boldsymbol{\Delta}$ is the set of available dynamic channels. Thus send has the type $\langle\langle\rangle_{\boldsymbol{\Delta}}\rangle$. As $\boldsymbol{\Delta}$ represents an upper bound of the definitions that may be used, we have an immediate notion of subtyping (written $\leq$) on dynamic channels: $\langle\tau\rangle_{\boldsymbol{\Delta}'} \leq \langle\tau\rangle_{\boldsymbol{\Delta}}$ if $\boldsymbol{\Delta}' \subseteq \boldsymbol{\Delta}$; a channel that may access fewer dynamic definitions is a subtype of a channel that may access more dynamic definitions. Thus a static channel has a type that is a subtype of any dynamic channel carrying the same type of arguments. The subtyping rule N-Sub of figure 2 also introduces contravariant subtyping on the argument type.

Since a location that provides more dynamic definitions may be used instead of one that provides fewer dynamic definitions, we introduce a notion of subtyping on location types, in rule L-Sub.

One important condition for insuring soundness of the type system is to forbid subtyping on dynamic names that are redefined. We write $\langle\tau\rangle^{+}_{w}$ for the type of these channels. Since a redefinable channel may be used instead of a plain dynamic channel for message sending, we introduce the Plus subtyping rule. In the following we consider $\leq$ to be the smallest reflexive transitive closure generated by the rules of figure 2.

A type scheme $\forall\widetilde{\alpha}\widetilde{\delta}.\tau$ is composed of generalized type variables $\widetilde{\alpha}$, generalized name type variables $\widetilde{\delta}$, and type τ. An instantiation of this type scheme, written $Inst(\forall\widetilde{\alpha}\widetilde{\delta}.\tau)$, is a type $\tau\theta$, where θ is a substitution from the type variables $\widetilde{\alpha}$ to types and from the name type variables $\widetilde{\delta}$ to dynamic name types w.

For soundness reasons, we do not allow polymorphic redefinable channel types. Similarly, we do not allow polymorphic location types. All other types are said to be *well formed.*

A *type environment* B (resp. a *type scheme environment* A or Γ) is an association map between names and types (resp. names and type schemes), where each name occurs at most once.

In the following, we call $ftv(\tau)$ the free type variables in τ, $fnv(\tau)$ the free name type variables in τ. We write $fv(\tau)$ for $ftv(\tau) \cup fnv(\tau)$. We also extend *ftv* and *fnv* to type environments and to type scheme environments.

In the following typing rules, we use a *generalization* operator $Gen(B, \Lambda, \Theta)$ defined as: $Gen(B, \Lambda, \Theta) = \bigcup_{m:\tau\in B}\{m : \forall\widetilde{\alpha}\widetilde{\delta}.\tau\}$, where $\widetilde{\alpha} = ftv(\tau) \setminus (\Lambda \cup \Theta)$ and $\widetilde{\delta} = fnv(\tau) \setminus (\Lambda \cup \Theta)$. The set Λ contains the names and type variables that occur in the typing environment (in typing rule DEF, Λ is $fv(\Gamma)$); the set Θ contains the names and type variables that may not be generalized because they are shared in a join pattern (as in [12]).

A typing judgment has one of the following forms: $\Gamma \Vdash \widetilde{n} : \widetilde{\tau}$, $\mathbf{\Delta}; \mathbf{I}; \Gamma \Vdash P$, $\mathbf{\Delta}; \mathbf{I}; \Gamma \Vdash \mathcal{D} :: B; \Delta_1; \Theta$, $\Gamma \Vdash \mathcal{S}$, or $\mathbf{\Delta}; \mathbf{I}; \Gamma \Vdash \mathcal{D} : \Delta_1$ where Γ is the type scheme environment; $\mathbf{\Delta}$ is a set of dynamic name types, the dynamic channels defined in the current location; $\mathbf{I}$ is a set of dynamic names, the dynamic channels imported by the current location; B gathers the types of the defined static names of $\mathcal{D}$; Δ_1 collects the dynamic channels locally defined by $\mathcal{D}$; Θ is a set of type variables and name type variables that cannot be generalized.

The four main typing rules to guarantee the presence of a dynamic definition for every dynamic message are MSG, LOC, SOUP-LOC, and GO. Rule MSG checks that the channel used does not require more dynamic definitions than the ones available locally (*i.e.* the ones specified in $\mathbf{\Delta}$ and $\mathbf{I}$). Rules LOC and SOUP-LOC are very similar, the former being more complex as it can occur in the process guarded by a join pattern. These rules check that the specified dynamic channels are defined and that the imported dynamic channels are available locally (in rule CONF for the SOUP-LOC case). The contents of the location are typed in an environment where the defined and imported dynamic channels are available (these rules modify $\mathbf{\Delta}$ and $\mathbf{I}$). Rule GO checks that the target of the migration provides at least the dynamic channels imported by the current location.

The two typing rules for messages MSG and R-MSG are very similar, and follow the different states of a message as it is first resolved, then sent to its destination. In these rules, the name on which the message is sent needs to satisfy the typing judgment $\Gamma \Vdash n : \langle\widetilde{\tau}\rangle_{\mathbf{\Delta}\cup\mathbf{I}}$. Because of subtyping on channel names, this judgment gives an upper bound on the dynamic names that the message may use, thus on the definitions accessed; this upper bound consists of the available dynamic definitions at this point, thus insuring that every

$$\frac{m : \forall\widetilde{\alpha}\widetilde{\delta}.\tau' \in \Gamma \qquad \tau = \mathit{Inst}(\forall\widetilde{\alpha}\widetilde{\delta}.\tau')}{\Gamma \Vdash m : \tau}\ [\textsc{Name}] \qquad\qquad \frac{\Gamma \Vdash m : \tau \qquad \tau \leq \tau'}{\Gamma \Vdash m : \tau'}\ [\textsc{Sub}]$$

$$\frac{\Gamma \Vdash m_i : \tau_i \text{ for } i \in [1..n]}{\Gamma \Vdash m_1, \ldots, m_n : \tau_1, \ldots, \tau_n}\ [\textsc{Tuple}]$$

$$\frac{\Gamma \Vdash n : \langle\widetilde{\tau}\rangle_{\mathbf{\Delta}\cup\mathbf{I}} \qquad \Gamma \Vdash \widetilde{m} : \widetilde{\tau} \qquad \mathbf{\Delta};\mathbf{I};\Gamma \Vdash P}{\mathbf{\Delta};\mathbf{I};\Gamma \Vdash n\langle\widetilde{m}\rangle; P}\ [\textsc{Msg}]$$

$$\frac{\Gamma \Vdash n : \langle\widetilde{\tau}\rangle_{\mathbf{\Delta}\cup\mathbf{I}} \qquad \Gamma \Vdash \widetilde{m} : \widetilde{\tau} \qquad \Gamma \Vdash a : \mathit{loc}(\emptyset) \qquad n \in \mathit{dln}(a)}{\mathbf{\Delta};\mathbf{I};\Gamma \Vdash a.n\langle\widetilde{m}\rangle}\ [\textsc{R-Msg}]$$

$$\frac{\begin{array}{c}\mathbf{\Delta};\mathbf{I};\Gamma + A \Vdash D :: B;\emptyset;\Theta \\ A = \mathit{Gen}(B, \mathit{fv}(\Gamma), \Theta) \qquad \mathbf{\Delta};\mathbf{I};\Gamma + A \Vdash P \qquad \mathit{dom}(A) \cap \mathit{dom}(\Gamma) = \emptyset\end{array}}{\mathbf{\Delta};\mathbf{I};\Gamma \Vdash \mathbf{def}\ D \mathbf{\ in\ } P}\ [\textsc{Def}]$$

$$\frac{\mathbf{\Delta};\mathbf{I};\Gamma \Vdash \mathcal{P}_1 \qquad \mathbf{\Delta};\mathbf{I};\Gamma \Vdash \mathcal{P}_2}{\mathbf{\Delta};\mathbf{I};\Gamma \Vdash \mathcal{P}_1 \mid \mathcal{P}_2}\ [\textsc{Par}] \qquad\qquad \frac{\Gamma \Vdash n : \mathit{loc}(\mathbf{I}) \qquad \mathbf{\Delta};\mathbf{I};\Gamma \Vdash P}{\mathbf{\Delta};\mathbf{I};\Gamma \Vdash \mathbf{go}\ n; P}\ [\textsc{Go}]$$

$$\frac{\mathbf{d} \notin \mathit{fn}(\Gamma) \qquad \mathbf{\Delta};\mathbf{I};\Gamma + \mathbf{d}:\langle\tau\rangle^+_{\mathbf{d}} \Vdash P}{\mathbf{\Delta};\mathbf{I};\Gamma \Vdash \nu\mathbf{d}.P}\ [\textsc{Nu}] \qquad\qquad \frac{}{\mathbf{\Delta};\mathbf{I};\Gamma \Vdash \mathbf{0}}\ [\textsc{Nil}]$$

$$\frac{\begin{array}{c}\mathbf{\Delta};\mathbf{I};\Gamma + (\widetilde{u_i} : \widetilde{\tau_i})^i + (\widetilde{y_j} : \widetilde{\tau_j})^j \Vdash P \\ \Gamma \Vdash \mathbf{x_i} : \langle\widetilde{\tau_i}\rangle \qquad \Gamma \Vdash m_j : \langle\widetilde{\tau_j}\rangle^+_{w_j} \qquad (\bigcup_i \widetilde{u_i} \cup \bigcup_j \widetilde{y_j}) \cap \mathit{dom}(\Gamma) = \emptyset \\ \forall(n:\tau),(n':\tau') \in \left(\{\mathbf{x_i} : \langle\widetilde{\tau_i}\rangle\} \cup \{m_j : \langle\widetilde{\tau_j}\rangle^+_{w_j}\}\right) . n \neq n' \implies \mathit{fv}(\tau) \cap \mathit{fv}(\tau') \subseteq \Theta\end{array}}{\mathbf{\Delta};\mathbf{I};\Gamma \Vdash (\mathbf{x_i}\langle\widetilde{u_i}\rangle)^i \mid (m_j\langle\widetilde{y_j}\rangle)^j \triangleright P :: (\mathbf{x_i} : \langle\widetilde{\tau_i}\rangle)^i; \bigcup_j \{m_j\}; \Theta}\ [\textsc{Join}]$$

$$\frac{}{\mathbf{\Delta};\mathbf{I};\Gamma \Vdash \top :: \emptyset;\emptyset;\Theta}\ [\textsc{Top}] \qquad \frac{\mathbf{\Delta};\mathbf{I};\Gamma \Vdash \mathcal{D}_1 :: B_1;\Delta_1;\Theta \qquad \mathbf{\Delta};\mathbf{I};\Gamma \Vdash \mathcal{D}_2 :: B_2;\Delta_2;\Theta}{\mathbf{\Delta};\mathbf{I};\Gamma \Vdash \mathcal{D}_1, \mathcal{D}_2 :: B_1 \oplus B_2;\Delta_1 \cup \Delta_2;\Theta}\ [\textsc{And}]$$

$$\frac{\begin{array}{c}\forall m_i \in \Delta \,.\, m_i : \langle\tau_i\rangle^+_{w_i} \in \Gamma \qquad \forall n_j \in I \,.\, n_j : \langle\tau_j\rangle^+_{w'_j} \in \Gamma \qquad \mathbf{\Delta}' = \bigcup_i w_i \quad \mathbf{I}' = \bigcup_j w'_j \\ \mathbf{\Delta}';\mathbf{I}';\Gamma \Vdash \mathcal{D} :: B;\Delta;\Theta \qquad \mathbf{\Delta}';\mathbf{I}';\Gamma \Vdash \mathcal{P} \qquad \Gamma \Vdash a : \mathit{loc}(\mathbf{\Delta}' \cup \mathbf{I}') \qquad \mathbf{I}' \subseteq (\mathbf{\Delta}'' \cup \mathbf{I}'')\end{array}}{\mathbf{\Delta}'';\mathbf{I}'';\Gamma \Vdash a[\mathcal{D} : \mathcal{P}]^{\Delta,I} :: B + a : \mathit{loc}(\mathbf{\Delta}' \cup \mathbf{I}');\emptyset;\Theta}\ [\textsc{Loc}]$$

$$\frac{\begin{array}{c}\{\varphi_i\} \text{ form a tree with root } b \qquad \forall\psi a \in \{\varphi_i\}.I_{\psi a} \subseteq \Delta_\psi \cup I_\psi \qquad I_b = \emptyset \\ \forall\psi a \in \{\varphi_i\}.F_{\psi a} = \mathit{Lookup}(F_\psi, I_{\psi a}, \Delta_{\psi a}, a) \qquad (\Gamma \Vdash \mathcal{D}_i \vdash^{\Delta_{\varphi_i}, I_{\varphi_i}, F_{\varphi_i}}_{\varphi_i} \mathcal{P}_i)^i\end{array}}{\Gamma \Vdash \prod_i (\mathcal{D}_i \vdash^{\Delta_{\varphi_i}, I_{\varphi_i}, F_{\varphi_i}}_{\varphi_i} \mathcal{P}_i)}\ [\textsc{Conf}]$$

$$\frac{\begin{array}{c}n \in \Delta \cup I \implies n = \mathbf{n} \wedge \mathbf{n} : \langle\tau\rangle^+_{\mathbf{n}} \in \Gamma \\ a : \mathit{loc}(\Delta \cup I) \in \Gamma \qquad \Delta;I;\Gamma \Vdash \mathcal{P} \qquad \Delta;I;\Gamma \Vdash \mathcal{D} : \Delta\end{array}}{\Gamma \Vdash \mathcal{D} \vdash^{\Delta,I,F}_{\varphi a} \mathcal{P}}\ [\textsc{Soup-Loc}]$$

$$\frac{\mathbf{\Delta};\mathbf{I};\Gamma \Vdash \mathcal{D} :: B;\Delta_{\mathcal{D}};\Theta \qquad A = \mathit{Gen}(B, \mathit{fv}(\Gamma), \Theta) \qquad A \subseteq \Gamma}{\mathbf{\Delta};\mathbf{I};\Gamma \Vdash \mathcal{D} : \Delta_{\mathcal{D}}}\ [\textsc{Chem-Def}]$$

Figure 3. Typing rules

dynamic message is bound to a definition. The R-MSG rule checks that the location specified in the prefix of the message is present with a location type in Γ (every location type is a subtype of $loc(\emptyset)$), and that the channel name is a defined local name of this location.

The typing rule DEF checks that no local dynamic name is defined in D ($\Delta_1 = \emptyset$), and also checks that the defined static names of D—which are the domain of B and A—do not clash with names of the typing environment. We remark that we use polymorphic recursion here, which drastically simplifies the subject reduction proof for the distributed Join Calculus.

Rule GO requires that the destination location has type $loc(\mathbf{I})$, where $\mathbf{I}$ is the set of imported dynamic channels. By definition of subtyping on location, this set is a lower bound, and any location providing more dynamic definitions may be the target of migration.

Rule NU introduces a new dynamic channel, which is monomorphic, and which has the type of dynamic channels sending messages on their own name. Since every dynamic channel created has a monomorphic redefinable type, every subsequent definition must exactly follow this type.

Rule JOIN is used to type one join pattern. The defined names of the join pattern are partitioned into two sets: the static names and the dynamic names. Static names $\mathtt{x_i}$ are given the type $\langle\widetilde{\tau_i}\rangle$, and are collected in the typing environment B. Dynamic names need to be present in Γ, with a redefinable type $\langle\widetilde{\tau_j}\rangle^{+}_{w_j}$, as they are redefined. Dynamic names are not written $\mathbf{m}_j$ since they may be variables bound by an enclosing join pattern. In evaluation context, we always have $m_j = w_j = \mathbf{m}_j$. They may however be different if the join pattern occurs in the guarded process of another join pattern that receives a dynamic name and then redefines it. All dynamic names are collected in the set of local dynamic names. The set of non generalized variables Θ is checked to be big enough: any type variable or name that is shared between two types cannot be generalized (as in [12]).

The typing rule AND uses the $\oplus$ operator in $B_1 \oplus B_2$, that requires names that are both in the domain of B_1 and B_2 to have the same type.

The rule LOC extracts from the typing environment the types associated to the names of Δ and I, in order to collect the dynamic name types associated to these channels. As in rule JOIN, in evaluation context the names are the same. They may be different if the typing rule is used to type the guarded process of a join pattern that receives a dynamic name that is redefined. The typing of the definition $\mathcal{D}$ must yield a set of local dynamic names Δ identical to the one declared in the location. The typing of $\mathcal{D}$ and $\mathcal{P}$ may use the available dynamic channels associated to the ones declared by the location being typed. This rule also checks that the imported names are available in the enclosing location.

The CONF rule checks that the running locations form a tree, all running locations import names that are available in the enclosing location (the root location does not import any name), every name lookup function is correctly computed, and all running locations are well typed. To do this, the SOUP-LOC rule is used. It first checks that all the names declared in Δ and I are dynamic channels (not variables), and that the type of these channels in Γ is the type of redefinable channels sending messages on their own name. It then types the definition (using rule CHEM-DEF) and process of the location, using the available dynamic channels declared by the location, and checks that the location is present with the correct type in the typing environment. This rule is similar to rule LOC, although simpler as running locations occur only in evaluation contexts. The CHEM-DEF rule types the definition and checks that the resulting generalization is present in the environment.

Typing examples are available in [17].

4. Type Soundness

To prove the soundness of our system, we first prove a subject reduction theorem, then we prove a progress property that insures that well-typed configurations do not go wrong. We first specify *well-formed* typing environments.

Definition 4.1 *A typing environment* Γ *is* well formed *if and only if its types are well formed and every type binding is of the form* $\mathbf{n} : \langle\tau\rangle^{+}_{\mathbf{n}}$, $\mathrm{n} : \forall\tilde{\alpha}\delta.\langle\tau\rangle$, *and* $a : loc(\mathbf{\Delta})$ *where* $\mathbf{\Delta}$ *contains no name type variable.*

We now prove that structural equivalence and reductions preserve typing.

Lemma 4.2 *Let* $\mathcal{S}$ *be a configuration and* $\Gamma \Vdash \mathcal{S}$ *a typing of this configuration where* Γ *is well-formed. If* $\mathcal{S} \equiv \mathcal{S}'$, *there is a well-formed* Γ' *such that* $\Gamma' \Vdash \mathcal{S}'$.

Theorem 1 (Subject reduction) *Let* $\mathcal{S}$ *be a configuration and* Γ *be a well formed environment. If* $\Gamma \Vdash \mathcal{S}$ *and* $\mathcal{S} \to \mathcal{S}'$, *then there exists a well-formed environment* Γ' *such that* $\Gamma' \Vdash \mathcal{S}'$.

We remark that Γ and Γ' need not be related, since reduction steps involve configurations including implicit contexts. We now prove that the NL-DYN step resolves messages to the closest enclosing location defining them (where *dyn*(b) is the set of dynamic names locally defined in b). This proof insures that our way of computing the name lookup function corresponds to our specification.

Lemma 4.3 *Let* $\Gamma \Vdash \mathcal{S}$ *be a typing derivation. For any dissolved location* φa *of* $\mathcal{S}$, *if we have* $F_{\varphi a}(n) = b \neq \bot$, *then we have* $n \in dyn(b)$, $\varphi a = \psi b \psi'$ *with* $\forall c \in \psi'.n \notin dln(c)$.

We define the notion of a *stuck configuration.*

Definition 4.4 *We say that a configuration* $\mathcal{S}$ *is* stuck *when one of the following is true: there is a message* $n\langle\tilde{m}\rangle$ *or* $a.n\langle\tilde{m}\rangle$ *in evaluation context where* n *is not a channel name; there is a* $n \in dn(\mathcal{S})$ *that is not a channel name or a location name; there is a process* go $n; P$ *where* n *is not a location name; there is a message* $n\langle\tilde{m}\rangle$ *or* $a.n\langle\tilde{m}\rangle$ *and a definition of* n *with different arities; there is a message* $\mathbf{n}\langle\tilde{m}\rangle$ *in evaluation context that cannot be reduced by rule* NL-DYN; *there is a message* $a.n\langle\tilde{m}\rangle$ *with* $n \notin dln(a)$.

We can now state that no well-typed configuration is stuck.

Theorem 2 (Progress) *Let S be a configuration and Γ a well-formed environment. If $\Gamma \Vdash \mathcal{S}$, then $\mathcal{S}$ is not stuck.*

Combining lemma 4.2, 4.3, and theorems 1 and 2, we prove that well-typed configurations cannot become stuck, thus dynamic messages are always sent to the closest enclosing location providing a definition and there is always such a location. The proofs are fairly complex, especially the substitution lemma, as channel names occur in the types. They are available in [16].

5. Future work, related work, and conclusion

One of our first priorities is the integration of dynamic channels into JoCaml. A form of dynamic binding at the module level is already present in JoCaml, but is difficult to use. The implementation would provide the programmers with easier management of local resources. As our system was designed with implementation in mind, this should not present any difficulty. In particular, it seems clear that maintaining the lookup function F (that resolves dynamic names to the closest enclosing location defining them) would be cheap because it does not require more locking than the one already present in JoCaml.

On the theoretical side, it is unfortunate that the construct $a.n\langle\widetilde{m}\rangle$, very similar to the message construct of [19], cannot be made available to programmers. To use such a construct soundly, it is necessary to check that n is indeed defined in location a, where a may be a variable bound by an enclosing join pattern. We have been designing such a type system, which is very similar to the one introduced in this paper, and we are finishing the soundness proofs.

Another extension is the design of a type reconstruction algorithm. Since our type system uses polymorphic recursion, which greatly simplifies the proof of subject reduction, a type reconstruction algorithm requires a different type system (see [17]). This issue is already present in the distributed Join Calculus. We are also modifying our system to use a constraint based type system to recover principal types, adapting the B(T) framework of [8] to this end.

In this paper, our strategy for handling a message on a dynamic channel name is to resolve in one step the closest enclosing location defining the channel, and then to send the message to this location in an other step. An alternate approach would simply consist in sending a message on a dynamic name that is not defined locally to the parent location, which would then deal with the message (this second incremental approach is similar to [6] and [18]). The two semantics yield two different behaviors but, surprisingly enough, the type system presented in this paper is also sound for the second system.

Part of this work could have been achieved in a distributed π Calculus. However, this would have been more subtle since the association between senders and receivers is more dynamic than in the Join Calculus, since receivers may disappear. It is therefore more complex to distinguish between deadlock freedom [1, 14] and availability of receivers. This implies that our type system is simpler as we do not guarantee deadlock freedom because of join patterns.

However, the resulting type system of [1] requires that names passed on channels cannot be used as receivers, unlike our dynamic channels as shown in [17].

Several works deal with resource control (as opposed to resource availability) in a π-calculus with localities and objective rebinding. In [13], localities have a flat structure, and processes may migrate objectively between localities. A type system insures that no agent may access a resource if it was not given the capability to do so. In the *local area calculus* [7], localities form a fixed hierarchy of levels that do not migrate. Channels have a level of operation, meaning that no communication on such a channel may cross the boundary of an higher level area. In the box-π calculus [18], localities also form a fixed hierarchy, and communication may cross only one locality boundary at a time. This calculus aims at controlling the flow of information between localities.

The higher order π-calculus of [21, 20] also deals with access control, by explicitly specifying for each input which resources may be accessed by the input process, distinguishing between read and write accesses. However, this calculus is objective, guarantees locality of resources instead of availability, and uses dependent types instead of polymorphism with name type variables.

The work on secrecy and groups [5] also deals with controlling access to some names through the creation of fresh groups, and the assignment of channels to these groups. However, the different intent (secrecy vs receptiveness) leads to different type systems. We use polymorphism to let dynamic names escape the scope in which they are created, to type more processes, whereas in [5] secrecy is achieved by checking that groups cannot escape their initial scope.

A very simple calculus of localized resources is the Ambient Calculus [6]. Recent works on the Ambient Calculus add type systems to analyze the behavior of ambients or enforce security policies. In [4], groups are introduced to refine mobility types of ambients in order to control the escape of capabilities. In [2], safe ambients are typed according to a security policy. The type system takes into account the capabilities that the ambient may acquire. Boxed ambients [3] drops the *open* capabilities of the Ambient Calculus, but allows communication between parent and children. A type system allows the analysis of ambient behavior, distinguishing local communications from communications with the context. In these ambient calculi, the emphasis is more on restricting the behavior of processes than on checking the availability of resources. Moreover, there is no notion of static names.

We have presented an extension of the distributed Join Calculus that features the creation of definitions accessible through dynamic channels that are rebound at migration. New definitions for a given channel may be created at runtime, and new dynamic channels may be generated as well. A type system that has the subject reduction property was presented, which guarantees the presence of a definitions for any dynamic channel that may be used.

Acknowledgments

We would like to thank Sylvain Conchon, Cédric Fournet, James Leifer, Jean-Jacques Lévy, François Pottier, and Didier Rémy for comments.

References

[1] R. M. Amadio, G. Boudol, and C. Lhoussaine. The receptive distributed pi-calculus (extended abstract). In *FST-TCS'99*, LNCS, 1999.

[2] M. Bugliesi and G. Castagna. Secure safe ambients. In *Proceedings of POPL '01*, pages 222–235. ACM Press, 2001.

[3] M. Bugliesi, G. Castagna, and S. Crafa. Boxed ambients. In *TACS'01*, LNCS, 2001.

[4] L. Cardelli, G. Ghelli, and A. D. Gordon. Ambient groups and mobility types. In *IFIP TCS 2000 (Sendai, Japan)*, LNCS. IFIP, Springer, Aug. 2000.

[5] L. Cardelli, G. Ghelli, and A. D. Gordon. Secrecy and group creation. In *CONCUR 2000 (University Park, PA, USA)*, LNCS. Springer, Aug. 2000.

[6] L. Cardelli and A. D. Gordon. Mobile ambients. In *Foundations of Software Science and Computational Structures*. Springer (LNCS), 1998.

[7] T. Chothia and I. Stark. A distributed calculus with local areas of communication. In *Proceedings of HLCL '00*, 2001.

[8] S. Conchon and F. Pottier. JOIN(X): Constraint-Based Type Inference for the Join-Calculus. In *Proceedings of ESOP'01*, LNCS, Apr. 2001.

[9] C. Fournet. *The Join-Calculus: a Calculus for Distributed Mobile Programming*. PhD thesis, Ecole Polytechnique, Palaiseau, Nov. 1998. INRIA, TU-0556.

[10] C. Fournet, G. Gonthier, J.-J. Lévy, L. Maranget, and D. Rémy. A calculus of mobile agents. In *Proceedings of CONCUR'96*, Aug. 1996. LNCS 1119.

[11] C. Fournet, J.-J. Lévy, and A. Schmitt. A distributed implementation of Ambients. Draft of long version, `http://join.inria.fr/ambients.html`, 1999.

[12] C. Fournet, L. Maranget, C. Laneve, and D. Rémy. Inheritance in the join calculus. In *FST-TCS'00*, LNCS, Dec. 2000.

[13] M. Hennessy and J. Riely. Resource access control in systems of mobile agents. *Information and Computation*, To appear.

[14] N. Kobayashi, S. Saito, and E. Sumii. An implicitly-typed deadlock-free process calculus. In *Proceedings of CONCUR 2000*, LNCS, Aug. 2000.

[15] F. Le Fessant. The JoCAML system prototype. Software and documentation available from `http://pauillac.inria.fr/jocaml`, 1998.

[16] A. Schmitt. Safe Dynamic Binding in the Join Calculus, Draft. Available from `http://pauillac.inria.fr/~aschmitt/publications.html`, 2001.

[17] A. Schmitt. Safe dynamic binding in the join calculus. Version of this paper with discussions and examples, available from `http://pauillac.inria.fr/~aschmitt/publications.html`, 2002.

[18] P. Sewell and J. Vitek. Secure composition of insecure components. In *Proceedings of CSFW 99 (Mordano, Italy)*, June 1999.

[19] A. Unyapoth and P. Sewell. Nomadic Pict: Correct communication infrastructure for mobile computation. In *Proceedings of POPL 2001 (London)*, Jan. 2001.

[20] N. Yoshida and M. Hennessy. Subtyping and locality in distributed higher-order processes. In *Proceedings CONCUR 99, LNCS no 1664*, 1999.

[21] N. Yoshida and M. Hennessy. Assigning types to processes (extended abstract). In *Fifteenth Annual IEEE Symposium on Logic in Computer Science*, 2000.

VECTORIAL LANGUAGES AND LINEAR TEMPORAL LOGIC

Olivier Serre
LIAFA, Université Paris VII
2, place Jussieu, case 7014
F-75251 Paris Cedex 05
serre@liafa.jussieu.fr

Abstract

Determining for a given deterministic complete automaton the sequence of visited states while reading a given word is the core of important problems with automata-based solutions, such as approximate string matching. The main difficulty is to do this computation efficiently, especially when dealing with very large texts. Considering words as vectors and working on them using vectorial (parallel) operations allows to solve the problem faster than in linear time using sequential computations.

In this paper, we show first that the set of vectorial operations needed by an algorithm representing a given automaton depends only on the language accepted by the automaton. We give precise characterizations of vectorial algorithms for star-free, solvable and regular languages in terms of the vectorial operations allowed. We also consider classes of languages associated with restricted sets of vectorial operations and relate them with languages defined by fragments of linear temporal logic.

Finally, we consider the converse problem of constructing an automaton from a given vectorial algorithm. As a byproduct, we show that the satisfiability problem for some extensions of linear-time temporal logic characterizing solvable and regular languages is PSPACE-complete.

Keywords: Parallel automata simulation, linear temporal logic and extensions

Introduction

Given a deterministic complete automaton and an input word, a classical question is to decide whether or not the automaton accepts the word. A more detailed information is the sequence of visited states while processing the word. Computing this sequence is the core of important problems such as approximate string matching [8]. An elegant way to solve this problem consists in simulating a dynamic programming approach by a finite, deterministic automaton [2] on the input sequence. However, approximate string matching is generally used on very long sequences (as genomic ones) and the natural algorithm, which is linear in the length of the input word, is not performing enough. A natural solution to accelerate the computation is to consider words as vectors and therefore to compute the sequence of visited states using vectorial operations, that can be efficiently achieved using parallelism [8].

In this paper, we are interested in vectorial algorithms, that were introduced and investigated by A. Bergeron and S. Hamel in [2, 3]. Such an algorithm computes the sequence of visited states while reading a word using a *constant* number (independent of the length of the word) of vectorial operations. The existence of an algorithm for a given automaton depends on the automaton and on the kind of vectorial operations we allow. The problem can also be studied from the language point of view: can we find a deterministic complete automaton recognizing a given language and an associated vectorial algorithm? We first show that the existence of a vectorial algorithm depends on the languages only. Then we exhibit a very tight connection between temporal logic operators and vectorial operations. This relation will motivate the notion of PTL-vectorial algorithm, where PTL stands for Past Temporal Logic. Moreover, we obtain an alternative proof of the equivalence between star-free languages and vectorial algorithms (Note that the inclusion of star-free languages in the class of PTL-vectorial languages was established in [3]. Hence, we show here that the converse also holds). Then, we describe extensions of vectorial algorithms, first to capture a larger subclass of regular languages, the solvable ones, and finally for the whole class of regular languages.

Finally, we consider the converse problem, that is we want to check for a given vectorial algorithm whether there exists an automaton associated with it. To solve this problem, we show how to decide the satisfiability of formulas belonging to extensions of linear temporal logic introduced in [1]. Our constructions are based on alternating automata. Proofs omitted in the paper can be found in [13].

1. Notation and Definitions

1.1. Vectorial Algorithms

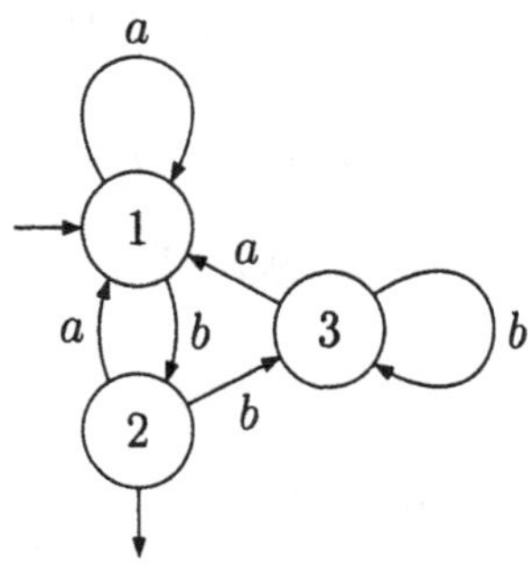

Throughout the paper, vectors are noted in bold characters (e.g. **u**) and are considered as words. Conversely, vectorial operations can be applied to words, considering them as vectors. Therefore, a word u is associated with a canonical vectorial representation **u** and a vector **v** is associated with a canonical word representation v.

Let $\mathcal{A} = (Q, A, \cdot, q_0, F)$ be a deterministic complete automaton, where A is a finite alphabet, Q the finite set of states of $\mathcal{A}$, q_0 the initial state, F the set of finite states and $\cdot$ the transition function $Q \times A \to Q$ of $\mathcal{A}$.

With each input vector $\mathbf{u} = a_1a_2\cdots a_m \in A^*$ we associate the output vector $\mathbf{r} = r_1r_2\cdots r_m \in Q^*$ representing the sequence of states reached reading $\mathbf{u}$ (we omit the leading initial state). Therefore, $\mathbf{u}$ and $\mathbf{r}$ have the same length. For instance, consider the automaton given in figure below. With the input vector $\mathbf{u} = bbaabbbababab$ we associate the output vector $\mathbf{r} = 2311233121212$. A *vectorial algorithm* for $\mathcal{A}$ consists of a sequence of vectorial operations of constant length (i.e., a straight-line expression of length which is independent on $\mathbf{u}$) computing $\mathbf{r}$ from $\mathbf{u}$.

Given a word $\mathbf{u} = a_1a_2\cdots a_m$, we consider for every letter $a \in A$, the boolean vector $(\mathbf{u} = \mathbf{a}) = b_1\cdots b_m$ where, for each i, $b_i = 1$ if $a_i = a$ an $b_i = 0$ otherwise. Hence, $(\mathbf{u} = \mathbf{a})$ is the characteristic boolean vector of the letter a in the word $\mathbf{u}$. Just as for words, for any state $q \in Q$, with an output vector $\mathbf{r} = r_1r_2\cdots r_m$ we associate the boolean vector $(\mathbf{r} = \mathbf{q}) = (r_1 = q)\cdots(r_m = q)$ that is, the characteristic vector of state q. For example, $(bbaabbbababab = \mathbf{a}) = 0011000101010$ and $(2311233121212 = \mathbf{1}) = 0011000101010$.

The sequence $(\mathbf{u} = \mathbf{a})_{a\in A}$ (respectively, the sequence $(\mathbf{r} = \mathbf{q})_{q\in Q}$) is an equivalent boolean representation for the input word $\mathbf{u}$ (respectively for the output vector $\mathbf{r}$). In order to work only with boolean vectors, the vectorial algorithms presented in this paper compute the sequence of characteristic vectors $(\mathbf{r} = \mathbf{q})_{q\in Q}$ from the sequence of characteristic vectors $(\mathbf{u} = \mathbf{a})_{a\in A}$.

Let Ω be a class of vectorial operations. Vector algorithms based on operations from Ω and on the bit-wise logical operations (combinations of $\vee$, $\wedge$ and $\neg$) are called *Ω-vectorial algorithms.* A deterministic complete automaton is called *Ω-vectorial* if there is an Ω-vectorial algorithm computing for every $\mathbf{u} \in A^*$ the sequence $(\mathbf{r} = \mathbf{q})_{q\in Q}$ from the sequence $(\mathbf{u} = \mathbf{a})_{a\in A}$. Finally, a language is *Ω-vectorial* if it is recognized by a deterministic complete Ω-vectorial automaton.

Given a class Ω of vectorial operations, we write VA$[\Omega]$ for the set of Ω-vectorial algorithms and V$\mathcal{L}[\Omega]$ for the set of Ω-vectorial languages.

Proposition 1 below shows that minimization preserves the property of being an Ω-vectorial automaton. Therefore a language is Ω-vectorial if and only if its minimal automaton is Ω-vectorial (by minimal automaton we always mean the minimal *complete* automaton). This property is very useful, because to decide whether or not a language is Ω-vectorial, it suffices to know how to decide whether or not a given automaton (the minimal one) is Ω-vectorial.

Proposition 1 *The property of being Ω-vectorial is preserve by automaton homomorphisms. Therefore if a deterministic complete automaton $\mathcal{A}$ is Ω-vectorial, then its minimal automaton $\mathcal{A}_{min}$ is also Ω-vectorial.*

1.2. Past Temporal Logic

Given a class Ω of vectorial operators our aim is to give a characterization of Ω-vectorial languages that allows us to decide whether or not a given language is Ω-vectorial. For this, we use characterizations in terms of past temporal logic PTL [6]. We first recall the syntax of PTL:

PTL-formulas are constructed inductively according to the following rules:

(1) For every $a \in A$, p_a is a PTL-formula.

(2) If φ_1 and φ_2 are PTL-formulas, so are $\varphi_1 \vee \varphi_2$, $\neg\varphi_1$, $\mathbf{Y}\varphi_1$ and $\varphi_1 \, \mathbf{S} \, \varphi_2$.

Semantics is defined by induction on the rules. Given a word $w \in A^+$ and an integer $n \in \{1, 2, \ldots, |w|\}$, we define that "$w$ satisfies φ at position n", denoted $(w, n) \models \varphi$, as follows:

(1) $(w, n) \models p_a$ if the nth letter of w is a.

(2) $(w, n) \models \varphi_1 \vee \varphi_2$ if $(w, n) \models \varphi_1$ or $(w, n) \models \varphi_2$.

(3) $(w, n) \models \neg\varphi_1$ if $(w, n) \not\models \varphi_1$.

(4) $(w, n) \models \mathbf{Y}\varphi_1$ if $n > 1$ and $(w, n-1) \models \varphi_1$.

(5) $(w, n) \models \varphi_1 \, \mathbf{S} \, \varphi_2$ if there exists $m \leq n$ such that $(w, m) \models \varphi_2$ and, for every k such that $m < k \leq n$, $(w, k) \models \varphi_1$.

With each PTL-formula φ, we associate the language L_φ of finite words satisfying φ as $L_\varphi = \{u \in A^+ \mid (u, |u|) \models \varphi\}$. Given a class Λ of logical operators we write TL$[\Lambda]$ for the set of temporal formulas in which modalities other than ones from Λ do not occur. A language $L \subseteq A^*$ is Λ-definable if there exists a formula $\varphi \in$ TL$[\Lambda]$ such that $L = L_\varphi$. The set of Λ-definable languages will be denoted by $\mathcal{L}[\Lambda]$.

1.3. Vectorial Languages and Temporal Logic

We will show that there exists a tight link between languages defined by logical conditions and languages defined by "equivalent" vectorial conditions. But, whereas logical satisfiability depends exclusively on the language, vectorial characterizations seem to be closely related with a specific automaton. Vectorial characterizations are stronger than logical characterizations because in order to have a vectorial algorithm for a given automaton one must be able to characterize *any* state, hence any language recognized by the automaton obtained by setting a given state as unique final state. For a logical formula one just needs to exhibit the set of final states needed for the given language.

But under some assumptions, logical fragments and vectorial fragments define the same class of languages. Let us be more explicit. Given a set Ω of vectorial operations and a set Λ of logical operators, we will say that Ω and Λ are *equivalent* if they verify the following conditions:

(1) To any vectorial algorithm Φ using only operations in Ω one can associate a formula φ using only operators in Λ such that for any word u and any positive integer i smaller than $|u|$, the i-th entry of the vector obtained by applying Φ to u is 1 if and only if $(u, i) \models \varphi$.

(2) To any formula φ using only operators in Λ, one can associate a vectorial algorithm Φ using only operators in Ω such that for any word $u = u_1 \cdots u_m$, the computation of the binary vector $\mathbf{v}_\varphi = v_1 \cdots v_m$, where $v_i = 1 \Leftrightarrow (u_1 \cdots u_i, i) \models \varphi$, is performed by the algorithm Φ.

Several fragments [4, 19, 20] and extensions [1] of temporal logic have been studied. Therefore, in order to characterize for a given class Ω of vectorial operations the class of Ω-vectorial languages, a solution consists in finding an equivalent fragment in temporal logic. We have to find a condition on two equivalent sets Ω and Λ to have $V\mathcal{L}[\Omega] = \mathcal{L}[\Lambda]$.

A class Λ of logical operators will be called *finally stable* if for every language L that belongs to $\mathcal{L}(\Lambda)$, any language recognized by an automaton obtained from the minimal automaton of L by letting some arbitrary state to be the unique final state, belongs to $\mathcal{L}(\Lambda)$.

For example, any set Λ such that $\mathcal{L}(\Lambda)$ is a variety of languages (see [9, 10] for the definition of variety) of languages is finally stable. Formally, if L is a language in $\mathcal{L}(\Lambda)$ and $\mathcal{A}$ its minimal automaton, any automaton $\mathcal{A}'$ obtained by modifying the final states of $\mathcal{A}$ recognizes a language L' of $\mathcal{L}(\Lambda)$ because the syntactic monoid of $\mathcal{A}'$ divides the syntactic monoid of $\mathcal{A}$.

The notion of final stability gives us the following lemma:

Lemma 1 *Let Ω be a class of vectorial operations and let Λ be an equivalent class of logical operators. Then Λ is finally stable if and only if* $V\mathcal{L}(\Omega) = \mathcal{L}(\Lambda)$.

2. A Vectorial Characterization of Star-Free Languages

To make our algorithms precise we have to state which vectorial operations are allowed. As in [3], we first consider a basic class of vectorial operations:

- Bit-wise logical operations such as $\vee$, $\wedge$, $\neg$ and the atomic formulas $(\mathbf{u} = \mathbf{a}) = (u_1 = a) \cdots (u_m = a)$ for each $a \in A$.
- Right shift: $\uparrow_i u_1 \cdots u_m = iu_1 \cdots u_{m-1}$, $i \in \{0, 1\}$.
- Binary addition between two vectors of same length: we perform the usual binary addition from left to right but we do not keep the highest bit (carry) if the length of the result exceeds the initial vectors' ones. For example $\uparrow_0 110101 + 101011 = 110100$.

Vectorial algorithms using only these operations are called in this paper *PTL-vectorial algorithms.* Recall that PTL stands for Past Temporal Logic, as we show that PTL-vectorial operations and PTL operators are equivalent:

Theorem 1 *The class $\{\uparrow_0, +\}$ of vectorial operation is equivalent to the class $\{\mathbf{Y}, \mathbf{S}\}$ of logical operators.*

The main point of the proof is to use the operator $\mathbf{S}$ in order to express the carry bit, and thus addition of vectors.

Corollary 1 *A regular language is PTL-vectorial if and only if it is star-free.*

3. Beyond Star-Freeness

3.1. Solvable Languages

An alternative proof for PTL-vectorial languages being star-free, is based on the following result:

Proposition 2 *Let* $\mathbf{u}$ *be a vector and let* Φ *be a PTL-vectorial algorithm. If* $\mathbf{u}$ *has period* $p \in \mathbb{N}$ *then the result* $\Phi(\mathbf{u})$ *of* Φ *applied on* $\mathbf{u}$ *has period* p.

Hence, to provide characterizations for families of regular languages that strictly contain the star-free languages we need to introduce vectorial operations, that do not preserve the period. For any integers k, l such that $0 \leq l < k$, we define the *modular operation* $S_{l,k}$ by

$$S_{l,k}(x_1 x_2 \cdots x_m) = (s_1 \cdots s_m) \text{ where } s_i = \begin{cases} 1 & \text{if } \sum_{j=1}^{i} x_j = l \pmod{k}, \\ 0 & \text{otherwise.} \end{cases}$$

Vectorial algorithms using only the PTL-vectorial operations plus the modular operations $S_{l,k}$ will be called *MTL-vectorial algorithms*. A language is an *MTL-vectorial language* if there is an MTL-vectorial algorithm for its minimal automaton.

Consider now the following extension of temporal logic, called modular temporal logic (MTL) [1]: by modular temporal logic we mean past temporal logic augmented with the unary operators $\mathrm{Mod}_{l,k}$ for integers $0 \leq l < k$. The new modular operators have the following natural semantics: given an MTL formula φ, we have $(u, i) \models \mathrm{Mod}_{l,k}\varphi$ if, there are l positions j, $1 \leq j \leq i$ (modulo k) such that $(u, j) \models \varphi$.

It was shown in [1], that MTL characterizes the variety of solvable languages. A regular language is called *solvable* if its syntactic monoid contains no non solvable group [15].

Theorem 2 ([1]) *A regular language is MTL-definable if and only if it is solvable.*

Therefore we have the following result:

Theorem 3 *The class* $\{\uparrow_0, +\} \cup \{S_{l,k} |\ 0 \leq l \leq k,\ k \in \mathbb{N}\}$ *of vectorial operations is equivalent to the finally stable class* $\{\mathbf{Y}, \mathbf{S}\} \cup \{\mathrm{Mod}_{l,k},\ 0 \leq l \leq k\}$ *of logical operators.*

We easily obtain now a vectorial characterization of solvable languages:

Corollary 2 *A regular language is MTL-vectorial if and only if it is solvable.*

3.2. Regular Languages

MTL-vectorial algorithms do not characterize all regular languages. Therefore, we propose an extension to MTL-vectorial algorithms, which we denote as *GTL-vectorial algorithms*, which captures all regular languages.

For this we consider an extension of modular temporal logic introduced in [1] and construct an equivalent class of vectorial operations. This extension of temporal logic is obtained by augmenting modular temporal logic with group temporal operators $\Gamma_{g,G}$ for any finite group G and any element $g \in G$. The operator $\Gamma_{g,G}$ always binds $|G| - 1$ formulas.

Let us now explain the semantics of $\Gamma_{g,G}$ for a given finite group G and an element $g \in G$. We first have to order the elements of the group G (this order will not be modified afterwards), say as $g_1, g_2, \ldots, g_q = id$. Let u be an element of A^+ and let $\varphi_1, \varphi_2, \ldots, \varphi_{q-1}$ be GTL-formulas. With each j, $1 \leq j \leq |u|$ we associate an element of G, denoted $\langle \varphi_1, \varphi_2 \ldots \varphi_{q-1} \rangle \langle u, j \rangle$, defined by:

$$\langle \varphi_1, \varphi_2, \ldots, \varphi_{q-1} \rangle \langle u, j \rangle = g_k$$

where $k = \min\{l \mid (u, j) \models \varphi_l\}$ with the convention that $\min \emptyset = q$.

Finally, we define $(u, i) \models \Gamma_{g,G} \langle \varphi_1, \varphi_2, \ldots, \varphi_{q-1} \rangle$ to mean that:

$$\prod_{j=1}^{i} \langle \varphi_1, \varphi_2, \ldots, \varphi_{q-1} \rangle \langle u, j \rangle = g$$

It was shown in [1] that a language is expressible in group temporal logic if and only if it is regular:

Theorem 4 ([1]) *A language is GTL-definable if and only if it is regular.*

Thus, to have a vectorial characterization of all regular languages it suffices to find vectorial operations equivalent to $\Gamma_{g,G}$ for all finite groups G. Let $G = \{g_1, g_2, \ldots, g_q = id\}$ be a finite group of cardinality q and let $\mathbf{v}_{g_1}, \mathbf{v}_{g_2}, \ldots, \mathbf{v}_{g_{q-1}}$ be $q-1$ vectors of same length m. With each position j, $1 \leq j \leq m$ we associate an element of G, denoted $\mathcal{G}(\langle \mathbf{v}_{g_1}, \mathbf{v}_{g_2}, \ldots, \mathbf{v}_{g_{q-1}} \rangle, G, j)$ and defined by:

$$\mathcal{G}(\langle \mathbf{v}_{g_1}, \mathbf{v}_{g_2}, \ldots, \mathbf{v}_{g_{q-1}} \rangle, G, j) = g_k$$

where $k = \min\{l \mid \mathbf{v}^j_{g_l} = 1\}$ ($\mathbf{v}^j_{g_l}$ representing the j-th bit of vector $\mathbf{v}_{g_l}$) with the convention that $\min \emptyset = q$.

Finally, we introduce the vectorial operator $P_{g,G}$ defined by:

$$P_{g,G}(\mathbf{v}_{g_1}, \mathbf{v}_{g_2}, \ldots, \mathbf{v}_{g_{q-1}}) = (s_1 \ldots s_m)$$

$$\text{with } s_i = \begin{cases} 1 & \text{if } \prod_{j=1}^{i} \mathcal{G}(\langle \mathbf{v}_{g_1}, \mathbf{v}_{g_2}, \ldots, \mathbf{v}_{g_{q-1}} \rangle, G, j) = g, \\ 0 & \text{otherwise.} \end{cases}$$

Remark 1 One can note that the modular operations are special cases of group operations that only use cyclic groups $(\mathbb{Z}/k\mathbb{Z},+)$, as we have $S_{l,k}(\mathbf{x}) = \Gamma_{l,(\mathbb{Z}/k\mathbb{Z},+)}(\mathbf{x},\mathbf{0},\ldots,\mathbf{0})$, for the elements of $(\mathbb{Z}/k\mathbb{Z},+)$ ordered as $(1,2,\ldots,k)$.

By construction we have the following result:

Theorem 5 *The class* $\{\uparrow_0,+\} \cup \{P_{g,G} \mid g \in G\}$ *of vectorial operations is equivalent to the finally stable class* $\{\mathbf{Y},\mathbf{S}\} \cup \{\Gamma_{g,G} \mid g \in G\}$ *of logical operators.*

Therefore, we obtain a vectorial characterization for regular languages:

Corollary 3 *A language is GTL-vectorial if and only if it is regular.*

4. Reconstructing an Automaton From a Vectorial Algorithm

In the preceding sections we determined a vectorial algorithm from a given automaton. We now consider the converse problem, that is we want to check for a given vectorial algorithm whether there exists a deterministic complete automaton associated with it (and determine an automaton, if this is the case). This question becomes interesting for instance when we modify a given vectorial algorithm (associated with a deterministic automaton) and we want to check afterwards that the new algorithm is equivalent to the old one. We will show that the complexity of this test is actually the same as testing the satisfiability of a GTL-formula i.e. PSPACE-complete.

Vectorial algorithms are associated with deterministic complete automata and therefore depend on the initial state, and not only on the underlying labeled graph structure of the given automaton. We will thus suppose that the initial state is part of the input.

To begin with, let us call a *valid* vectorial algorithm an algorithm for which there exists a corresponding deterministic complete automaton. Let us explain how to construct such an associated automaton. Let $A = \{a_1,\ldots,a_k\}$ be the alphabet of the automaton and let n be the number of states. We will identify the states with the integers $1\ldots n$. To compute an associated automaton $\mathcal{A}_\Phi$ from a given vectorial algorithm Φ we perform a depth-first search of $\mathcal{A}_\Phi$, that is we start from the initial state q_0 and compute the states that can be reached by reading a letter from q_0 and then we repeat this step with the new states found so far. We are done when we have explored all reachable states. With this method we explore all the transitions of the accessible part of the automaton. We just have to explain how to compute the reachable states from a given state. In our algorithm we maintain a vector, *state_direction*, giving for any encountered state q a word u leading from the initial state to q. Therefore, when considering a state q, and a letter a to compute the transition from q reading a we have to apply Φ to the word ua and consider the $|u|+1$ component of the result, denoted $\Phi^{|u|+1}(ua)$.

The algorithm for constructing $\mathcal{A}_\Phi$ is the following one:

Variables and initialization:
δ: $(n \times k)$-vector.
$new\_states = [1]$: last-in-first-out structure.
$known\_states = \{1\}$: Set structure.
$state\_direction = [\varepsilon, \varepsilon, \ldots, \varepsilon]$: n-vector.

Main loop:
While $new\_states \neq \emptyset$ **Do**
Let q =Delete element from new_states.
Let $u = state\_direction.(q)$.
Let $h = |u|$.
For $i = 1$ to k **Do**
Let $q' = \Phi^{h+1}(ua_i)$.
Let $\delta(q, i) = q'$.
If $q' \notin known\_states$ **Then**
Add q' to new_states and to $known\_states$.
Set $state\_direction.(q') = ua$.
End If.
End For
End While

Return δ

To test the validity of a given algorithm Φ we will first use the preceding algorithm to compute the automaton $\mathcal{A}_\Phi$ associated with Φ, if it is valid. If the algorithm does not work (that is if $\Phi^{h+1}(ua_i)$ is not defined for a given step of the algorithm) this implies that Φ is not valid. Otherwise we test the validity as stated in the theorem below. For any state q, let $L(q)$ denote the regular language defined by the logical formula obtained by translating the algorithm computing $(\mathbf{r} = \mathbf{q})$.

Theorem 6 *Let Φ be a PTL-vectorial algorithm and let $\mathcal{A}_\Phi$ be the deterministic complete automaton constructed by the algorithm above. Then Φ is a valid algorithm associated with $\mathcal{A}_\Phi$ if and only if:*

(1) *For any non reachable state q of $\mathcal{A}_\Phi$, we have $L(q) = \emptyset$.*

(2) *For any reachable state q, we have that $L(q) \neq \emptyset$. In addition, the following assertions are equivalent:*

   (i) *$L(q) = L(q_1)a_1 \cup \cdots \cup L(q_i)a_i \cup E_q$, where $E_q = \{\varepsilon\}$ if q is the initial state and $E_q = \emptyset$ otherwise. Moreover, a_j is a letter and each q_j is a reachable state.*

   (ii) *$\{(q_1, a_1), \ldots, (q_i, a_i)\}$ is exactly the set of the pairs (q_j, a_j) such that $q_j.a_j = q$ in $\mathcal{A}_\Phi$.*

We can now give a method to test the validity of a vectorial algorithm Φ:

(1) We apply the depth-first search algorithm described above to Φ. If the algorithm does not yield a deterministic automaton $\mathcal{A}_\Phi$, then Φ is not a valid algorithm and we can stop. Otherwise we go to the next step.

(2) We determine the reachable states and the non reachable states of the automaton $\mathcal{A}_\Phi$ constructed in the preceding step.

(3) For every non reachable state q we translate the associated component in Φ into a formula φ_q and test whether or not it can be satisfied. If φ_q is satisfiable for a non reachable-state q then Φ is not valid and we stop. Otherwise we go to the next step.

(4) For every reachable state q we determine the set $\{(q_1, a_1), \dots, (q_i, a_i)\}$ of the pairs (q_j, a_j) such that $q_j \cdot a_j = q$ in $\mathcal{A}_\Phi$ and we verify that $L(q) = L(q_1)a_1 \cup \dots \cup L(q_i)a_i \cup E_q$. To achieve this efficiently we can determine for every j a formula associated with $L(q_j)a_j$. It suffices to consider the formula $p_{a_j} \wedge \mathbf{Y}\varphi_{q_j}$ where φ_{q_j} is the translation of the component of Φ associated with q_j. Then, we can construct a PTL-formula for the language $L(q)\Delta[L(q_1)a_1 \cup \dots \cup L(q_i)a_i \cup E_q]$, where Δ holds for the symmetric difference, and verify that it cannot be satisfied, what is equivalent to the equality $L(q) = L(q_1)a_1 \cup \dots \cup L(q_i)a_i \cup E_q$. If the test does not fail, then Φ is valid and associated with $\mathcal{A}_\Phi$, otherwise Φ is not valid.

Let us now give the complexity of this algorithm. The most costly part of the algorithm is the satisfiability problems for GTL-formulas that appear in step (3) and (4) of the algorithm. This problem is known to be PSPACE-complete [18] for PTL-formulas. In fact this result can be extended to GTL:

Theorem 7 *Deciding whether an GTL-formula is satisfiable is a PSPACE-complete problem.*

Using this result we conclude that our algorithm works in polynomial space. In fact we can give a more precise result:

Theorem 8 *Deciding whether or not a GTL-vectorial algorithm is valid is a PSPACE-complete problem.*

Note that the same problem applied to some fragments of PTL-vectorial algorithms become simpler. For instance we have the following result for algorithms in VA($\uparrow_0$):

Theorem 9 *Deciding whether or not an algorithm in* VA($\uparrow_0$) *is valid is an NP-complete problem.*

5. Conclusion

Using vectorial algorithms we have given new characterizations of star-free languages (as the class of PTL-vectorial languages), of solvable languages (as the class of MTL-vectorial languages) and of regular languages (as the class of GTL-vectorial languages). However, even in the easiest case, that is for star-free languages, there is no general efficient method to compute an algorithm associated with a given language. Nevertheless, since vectorial languages are closely related with temporal logic this is not that surprising at all, as the computation of an algorithm associated with an automaton is at least as difficult as finding a temporal logic formula associated with a given language, which might be of exponential size with regard to the automaton.

We have characterized subsets of vectorial operations by equivalent sets of temporal logic operators.

It is interesting to note that vectorial algorithms provide a more detailed information about an automaton than logical formulas without any loss in computational complexity and in the complexity of the operators used in both models.

Finally, we have shown that deciding the validity of a GTL-vectorial algorithm is PSPACE-complete. As a byproduct we have obtained that the extension of LTL with group operators does not change the complexity of the satisfiability problem, which is still PSPACE-complete, and we have given an effective algorithm deciding this question.

Another interesting investigation is to study fragments of algorithms based on the set of PTL-vectorial operations. Here, we want to know which subset of star-free languages can be characterized by forbidding certain vector operations. One can show that these fragments are closely related with the fragments of past temporal logic as defined and characterized in [4, 19, 20].

Acknowledgments

I gratefully acknowledge the many helpful suggestions of Anca Muscholl during the preparation of the paper. I also wish to express my thanks to Jean-Eric Pin for suggesting many stimulating ideas, and to the anonymous referees for their remarks.

References

[1] A. Baziramwabo, P. McKenzie, and D. Therien. Modular temporal logic. In *14th Symposium on Logic in Computer Science (LICS'99)*, pages 344–351. IEEE, 1999.

[2] A. Bergeron and S. Hamel. Cascade decomposition are bit-vector algorithms. In *Proceedings of the Conference CIAA'01*, Lecture Notes in Computer Science. Springer Verlag, 2001, to appear. http://www.lacim.uqam.ca/~anne.

[3] A. Bergeron and S. Hamel. Vector algorithms for approximate string matching. *International Journal of Foundations of Computer Science*, 13(1):53–66, 2002.

[4] J. Cohen, D. Perrin, and J.-E. Pin. On the expressive power of temporal logic for finite words. *Journal of Computer and System Sciences*, 46:271–294, 1993.

[5] J.A. Kamp. *Tense Logic and the Theory of Linear Order.* Ph.d. thesis, University of California, Los Angeles, 1968.

[6] O. Lichtenstein, A. Pnueli, and L. Zuck. The glory of the past. In *Proceedings of the Conference on Logics of Programs (LICS'85)*, volume 193 of *Lecture Notes in Computer Science*, pages 196–218. Springer Verlag, 1985.

[7] R. McNaughton and S. Papert. *Counter-free Automata.* MIT Press, 1971.

[8] G. Myers. A fast bit-vector algorithm for approximate string matching based on dynamic programming. *Journal of the Association of Computing Machinery*, 46-3:395–415, 1999.

[9] J.-E. Pin. *Varieties of formal languages.* North Oxford, LondonPlenum, New-York, 1986. (Translation of Variétés de langages formels).

[10] J.-E. Pin. Syntactic semigroups. In G. Rozenberg and A. Salomaa, editors, *Handbook of formal languages*, volume 1, chapter 10, pages 679–746. Springer Verlag, 1997.

[11] A. Pnueli. The temporal logic of programs. In *18th IEEE Symposium Foundations of Computer Science (FOCS 1977)*, pages 46–57, 1977.

[12] M.P. Schützenberger and D. Perrin. On finite monoids having only trivial subgroups. *Information and Control*, 8:190–194, 1965.

[13] O. Serre. Vectorial languages an linear temporal logic: Version with proofs. http://www.liafa.jussieu.fr/~serre.

[14] A. P. Sistla and E. M. Clarke. The complexity of propositional linear temporal logics. *Journal of the Association for Computing Machinery*, 32(3):733–749, July 1985.

[15] H. Straubing. Families or recognizable sets corresponding to certain varieties of finite monoids. *Journal of Pure and Applied Algebra*, 15:305–318, 1979.

[16] H. Straubing. *Finite automata, formal logic, and circuit complexity.* Birkhäuser, 1994.

[17] D. Thérien. Classification of finite monoids: the language approach. *Theoretical Computer Science*, 14:195–208, 1981.

[18] M. Y. Vardi. An automata-theoretic approach to linear-temporal logic. In F. Moller and G. Birtwistle, editors, *Logics for concurrency*, number 1043 in Lecture Notes in Computer Science, pages 238–266. Springer, 1996.

[19] Th. Wilke. *Classifying discrete temporal properties.* Habilitation thesis, Kiel, Germany, 1998.

[20] Th. Wilke. Classifying discrete temporal properties. In *STACS 99*, number 1563 in Lecture Notes in Computer Science, pages 32–46, Berlin, 1999. Springer.

A BOUND ON ATTACKS ON AUTHENTICATION PROTOCOLS

Scott D. Stoller*
Computer Science Dept., SUNY at Stony Brook, Stony Brook, NY 11794-4400 USA

Abstract Authentication protocols are designed to work correctly in the presence of an adversary that can prompt honest principals to engage in an unbounded number of concurrent executions of the protocol. This paper establishes a bound on the number of protocol executions that could be useful in attacks. The bound applies to a large class of protocols, which contains versions of some well-known authentication protocols, including the Yahalom, Otway-Rees, and Needham-Schroeder-Lowe protocols.

1. Introduction

Many protocols are designed to work correctly in the presence of an adversary—hereafter called a penetrator—that can prompt honest principals to engage in an unbounded number of concurrent executions of the protocol. This paper focuses on authentication (including key establishment). Authentication protocols should satisfy at least two kinds of correctness requirements: *secrecy*, which states that certain values are not obtained by the penetrator, and *agreement*, which states, *e.g.*, that a principal's conclusion about the identity of a principal with whom it is communicating is never incorrect. Authentication protocols are short and look deceptively simple, but numerous flawed or weak protocols have been published. This attests to the importance of rigorous verification.

Allowing an unbounded number of concurrent protocol executions makes the number of reachable states unbounded, so automated verification using state-space exploration is not directly applicable. State-space exploration is feasible when small upper bounds are imposed on the size of messages and the number of protocol executions. Therefore, reduction theorems are needed, which show

*The author gratefully acknowledges the support of NSF under Grant CCR-9876058 and the support of ONR under Grants N00014-99-1-0358 and N00014-01-1-0109. This work was started while the author was at Indiana University in Bloomington. Email: stoller@cs.sunysb.edu

that if a protocol is correct in a system with certain bounds on these parameters, then the protocol is correct in the unbounded system as well.

Our reduction is formulated in the strand space model [13] but is relatively model-independent. A regular strand can be regarded as a thread that runs the program corresponding to one role (*e.g.*, initiator or responder) of the protocol and then terminates; thus, a regular strand corresponds to one execution of one role. Our reduction imposes three significant restrictions on protocols.

Shallow ciphertext restriction: the protocol does not use nested ciphertexts. This is easily checked by static analysis of the program, so we call it a *static restriction.* (This restriction can be relaxed; see Section 5.)

Bounded Support Restriction (BSR): in every history (*i.e.*, every possible behavior) of the system, each regular strand depends on at most a given number of regular strands. Correct authentication protocols are designed to involve only a small number of participants and hence typically satisfy BSR.

Revealed Genval Restriction (RGR): every genval revealed to the penetrator is revealed "directly", *i.e.*, the penetrator needs to perform at most one decryption on an intercepted message to obtain each genval.

The notion of dependence underlying BSR is a variant of Lamport's happened-before relation [5], modified to treat nonces and session keys—collectively called *generated values*, or *genvals* for short—appropriately. For example, if a genval g generated on strand s_1 appears in messages received by strand s_2 but only in contexts in which it could be replaced with a value generated by the penetrator, then g's presence in those messages does not cause s_2 to depend on s_1.

It seems difficult to develop static analyses to check BSR and RGR, so we call them *dynamic restrictions* and propose to check them during state-space exploration. Thus, we need reductions for them as well as for the correctness requirements. We prove: if a protocol satisfies the dynamic restrictions and correctness requirements when appropriate bounds are imposed on the number of regular strands in a history, then the protocol also satisfies the dynamic restrictions and correctness requirements without those bounds.

2. Related Work

Most existing techniques for automated verification of systems with unbounded numbers of processes, such as [3], are not applicable to authentication protocols, because they assume the set of values (equivalently, the set of local states of each process) is independent of the number of processes, whereas authentication protocols generate fresh nonces and session keys, so the set of values grows as the number of processes (strands) increases.

Roscoe and Broadfoot use data independence to bound the number of nonces that could be useful in attacks [10], assuming each honest principal participates in at most a given number of protocol executions at a time. Our reduction does not require such assumptions.

Lowe's reduction for authentication protocols [7] does not handle agreement requirements or known-key attacks and does not apply to the Otway-Rees [8], Yahalom [1], and Needham-Schroeder-Lowe (abbreviated NSL) [6] protocols, due to various restrictions.

Our reduction handles secrecy and agreement requirements, allows known-key attacks, and applies to some well-known protocols, including the Otway-Rees, Yahalom, and NSL protocols, after the Otway-Rees and Yahalom protocols have been modified slightly (in an obviously correctness-preserving way) to eliminate forwarding of ciphertexts, as in [7, 10].

Heather and Schneider's method [4] can efficiently (compared to state-space exploration) verify protocols for which a rank function exists. Currently, our method, unlike theirs, can verify secrecy properties for protocols that use temporary secrets, while their method, unlike ours, accommodates forwarded ciphertexts. In the absence of completeness results, it is unclear whether requiring BSR or requiring existence of a rank function is more restrictive.

The reduction in [12] is more general in some ways than this one, but it does not handle session keys, so it does not apply to most authentication protocols, especially if session keys are used to encrypt protocol messages, as in the Needham-Schroeder shared-key [1], Yahalom, and Kerberos protocols.

3. Model of Authentication Protocols

We adopt the strand space model [13], with minor modifications. We introduce simple languages for authentication protocols and correctness requirements, similar to the languages in [2] and [14], respectively.

3.1. Term, Directed Term, and Trace

The set of *primitive terms* is the union of the following five disjoint sets. (1) *Text* is a set of arbitrary non-cryptographic values, with a distinguished subset *Name* containing names of principals. (2) *Nonce* is a set of nonces. (3) Key_{sess} is a set of session keys. (4) $Key_{sym} = \{key(x, y) \mid x, y \in Name\}$ is a set of long-term symmetric keys; informally, $key(x, y)$ is intended to be shared by x and y. (5) $Key_{asym} = \{pubkey(x) \mid x \in Name\} \cup \{pvtkey(x) \mid x \in Name\}$ is a set of long-term asymmetric keys; $pubkey(x)$ and $pvtkey(x)$ represent x's public and private keys, respectively.

The set *Term* of terms is defined inductively as follows, where $Key = Key_{sym} \cup Key_{asym} \cup Key_{sess}$. (1) All primitive terms are terms. (2) If t and t' are terms and $k \in Key$, then $encr(t, k)$ (encryption of t with k, usually written $\{t\}_k$) and $pair(t, t')$ (pairing of t and t', usually written $t \cdot t'$) are terms.

The function $\mathrm{inv} \in Key \to Key$ maps each key to its inverse: decrypting $\{t\}_k$ with $\mathrm{inv}(k)$ yields t. For a symmetric key k, $\mathrm{inv}(k) = k$. We usually write $\mathrm{inv}(k)$ as k^{-1}. We assume perfect encryption.

Elements of $Nonce \cup Key_{sess}$ are called *generated values*, or *genvals* for short. Let $\mathrm{genvals}(t)$ be the set of genvals that occur in a term t. For $S \subseteq Term$, let $\mathrm{genvals}(S) = \bigcup_{t \in S} \mathrm{genvals}(t)$.

A *ciphertext* is a term whose outermost operator is *encr*. A term t' *occurs in the clear* in a term t if there is an occurrence of t' in t that is not in the scope of *encr*.

Let $|S|$ denote the size of a set S. Let $\mathrm{dom}(f)$ denote the domain of a function f. A sequence is a function from a finite prefix of the natural numbers to elements. Let $\mathrm{len}(\sigma)$ denote the length of a sequence σ. $\langle\!\langle a, b, \ldots \rangle\!\rangle$ denotes a sequence σ with $\sigma(0) = a$, $\sigma(1) = b$, and so on.

A *directed term* is $+t$ or $-t$, where t is a term. Positive and negative terms represent sending and receiving messages, respectively. Let $\pm\mathit{Term}$ denote the set of directed terms. For a directed term t, the *absolute value* of t, denoted $\mathrm{abs}(t)$, is t without its direction; for example, $\mathrm{abs}(-A) = A$. For $S \subseteq \pm\mathit{Term}$, let $\mathrm{abs}(S) = \{\mathrm{abs}(t) \mid t \in S\}$. We often refer to directed terms as terms.

A *trace* is a finite sequence of directed terms. Let $(\pm\mathit{Term})^*$ denote the set of traces.

3.2. Strand Space

A *strand space* is a function $\mathit{tr} \in \mathrm{dom}(\mathit{tr}) \to (\pm\mathit{Term})^*$, where $\mathrm{dom}(\mathit{tr})$ is an arbitrary set whose elements are called *strands* (think of them as "strand identifiers").

A *node* of tr is a pair $\langle s, i\rangle$ with $s \in \mathrm{dom}(\mathit{tr})$ and $0 \le i < \mathrm{len}(\mathit{tr}(s))$. Let $\mathcal{N}_{tr}$ denote the set of nodes of tr. We say that node $\langle s, i\rangle$ is on strand s. Let $\mathrm{nodes}_{tr}(s)$ denote the set of nodes on strand s in tr. Let $\mathrm{strand}(\langle s, i\rangle) = s$, $\mathrm{index}(\langle s, i\rangle) = i$, and $\mathrm{term}_{tr}(\langle s, i\rangle) = \mathit{tr}(s)(i)$. For $S \subseteq \mathcal{N}_{tr}$, let $\mathrm{strand}(S) = \{\mathrm{strand}(n) \mid n \in S\}$ and $\mathrm{term}_{tr}(S) = \{\mathrm{term}_{tr}(n) \mid n \in S\}$. If $\mathrm{term}_{tr}(n)$ is positive (or negative), we say that n is positive (or negative).

The local dependence relation on nodes is defined by: $n_1 \stackrel{lcl}{\to} n_2$ iff $\mathrm{strand}(n_1) = \mathrm{strand}(n_2)$ and $\mathrm{index}(n_2) = \mathrm{index}(n_1) + 1$.

A term t *originates* from a node $\langle s, i\rangle$ in tr iff $\langle s, i\rangle$ is positive, t is a subterm of $\mathrm{term}_{tr}(\langle s, i\rangle)$, and t is not a subterm of $\mathrm{term}_{tr}(\langle s, 0\rangle), \mathrm{term}_{tr}(\langle s, 1\rangle), \ldots,$ or $\mathrm{term}_{tr}(\langle s, i-1\rangle)$.

A term t *uniquely originates* from a node n in tr iff t originates from n in tr and not from any other node in tr. This is the strand space way of expressing freshness of genvals.

For symbols subscripted by a strand space, we elide the subscript when the strand space is evident from context.

3.3. Role and Protocol

Let *Param* be a set of parameters. The set of *parameterized terms* is defined like *Term* except with parameters as an additional base case.

A *role* r is a sequence of directed parameterized terms, with a type—*i.e.*, a set of allowed values—associated with each parameter, and with a subset of the parameters designated as uniquely-originated. Informally, parameters that represent genvals generated by r (and hence that first occur in r in a positive term) are so designated, to indicate that values of those parameters must be uniquely-originated. In examples, uniquely-originated parameters are underlined in the parameter list. Roles must also satisfy some well-formedness conditions, detailed in [11], notably that types may not contain ciphertexts. Let $r.x$ denote parameter x of role r. For example, the roles for the initiator and responder in the NSL protocol [6] are

$$\begin{array}{ll} \mathrm{Init}_{NSL}(i : \mathit{Name} \setminus \{P\},\, r : \mathit{Name}, & \mathrm{Resp}_{NSL}(i : \mathit{Name},\, r : \mathit{Name} \setminus \{P\}, \\ \quad \underline{ni} : \mathit{Nonce},\, nr : \mathit{Nonce}) = & \quad ni : \mathit{Nonce},\, \underline{nr} : \mathit{Nonce}) = \\ \langle\!\langle +\{ni \cdot i\}_{pubkey(r)}, & \langle\!\langle -\{ni \cdot i\}_{pubkey(r)}, \\ \quad -\{ni \cdot nr \cdot r\}_{pubkey(i)}, & \quad +\{ni \cdot nr \cdot r\}_{pubkey(i)}, \\ \quad +\{nr\}_{pubkey(r)}\rangle\!\rangle & \quad -\{nr\}_{pubkey(r)}\rangle\!\rangle. \end{array}$$

In both roles, parameters i and r hold the names of the initiator and responder, respectively. We exclude P from the type of $\mathrm{Init}_{NSL}.i$ and $\mathrm{Resp}_{NSL}.r$, because we interpret P as the name of a dishonest principal (the penetrator), and we interpret $\mathrm{Init}_{NSL}.i$ and $\mathrm{Resp}_{NSL}.r$ as the name of the principal executing the role, and all actions of the penetrator are represented by traces for penetrator roles, described in Section 3.4. $\mathrm{Init}_{NSL}.ni$ and $\mathrm{Resp}_{NSL}.nr$ are uniquely-originated, because they represent nonces generated by their respective roles.

A *genval parameter* is a parameter with type Key_{sess} or *Nonce*.

A *trace for role* r is a prefix of a trace obtained by substituting for each parameter x of r a term in the type of x.

A role r and a trace σ for r uniquely determine a mapping, denoted $\mathit{args}(r, \sigma)$, from the parameters of r that appear in $r(0), r(1), \ldots, r(\mathrm{len}(\sigma) - 1)$ to *Term*. For example, $\mathrm{dom}(\mathit{args}(\mathrm{Init}_{NSL}, \sigma_0)) = \{i, r, ni\}$ and $\mathit{args}(\mathrm{Init}_{NSL}, \sigma_0)(i) = B$.

A *protocol* is a set of roles. For example, the NSL protocol is $\Pi_{NSL} = \{\mathrm{Init}_{NSL}, \mathrm{Resp}_{NSL}\}$.

3.4. Penetrator

The penetrator model is parameterized by a set $\mathit{pik} \subseteq \mathit{Term}$, called the *penetrator's initial knowledge.* Typically, we assume there is a single dishonest principal, named P, and take $\mathit{pik} \supseteq \mathit{pik}_0$, where $\mathit{pik}_0 = \{\mathit{pvtkey}(P)\} \cup \{\mathit{pubkey}(x) \mid x \in \mathit{Name}\} \cup \{\mathit{key}(P, x), \mathit{key}(x, P) \mid x \in \mathit{Name} \setminus \{P\}\}$.

Known-key attacks are modeled by including in *pik* the absolute values of terms appearing in some executions of the protocol and the genvals generated during those executions.

$\Pi_P(\mathit{pik})$, the set of *penetrator roles* for initial knowledge *pik*, contains

$$\begin{array}{ll} \mathrm{Msg}(x : \mathit{Text} \cup \mathit{Nonce} \cup \mathit{Key}_{sess} \cup \mathit{pik}) = \langle\!\langle +x \rangle\!\rangle & \\ \mathrm{Pair}(x_1 : \mathit{Term}, x_2 : \mathit{Term}) = \langle\!\langle -x_1,\ -x_2,\ +x_1 \cdot x_2 \rangle\!\rangle & \\ \mathrm{Enc}(k : \mathit{Key}, x : \mathit{Term}) = \langle\!\langle -k,\ -x,\ +\{x\}_k \rangle\!\rangle & \\ \mathrm{Sep}_i(x_1 : \mathit{Term}, x_2 : \mathit{Term}) = \langle\!\langle -x_1 \cdot x_2,\ +x_i \rangle\!\rangle & \text{for } i \in \{1, 2\} \\ \mathrm{Dec}(k : \mathit{Key}, x : \mathit{Term}) = \langle\!\langle -k^{-1},\ -\{x\}_k,\ +x \rangle\!\rangle & \end{array}$$

A trace σ for a role r is *compromised* if it is running the protocol with the penetrator as a partner, specifically, if $\mathit{args}(r, \sigma)(x) = P$ for some parameter x of r with type *Name*.

3.5. System and History

A *system* is a pair $\langle \Pi, pik \rangle$. For example, $\mathcal{M}_{NSL} = \langle \Pi_{NSL}, pik_{NSL} \rangle$, where pik_{NSL} is a superset of pik_0 that also contains terms and genvals from one execution of Π_{NSL}.

A *history* of a system $\langle \Pi, pik \rangle$ is a tuple $h = \langle tr, \overset{msg}{\rightarrow}, role \rangle$, where tr is a strand space, $\overset{msg}{\rightarrow}$ is a binary relation on $\mathcal{N}_{tr}$ (read $n_1 \overset{msg}{\rightarrow} n_2$ as "n_1 is the sending of a message received at n_2"), and $role$ is a function from strands to roles (*i.e.*, $role \in \text{dom}(tr) \rightarrow (\Pi \cup \Pi_P(pik))$) such that: (1) for each negative node n_2, there exists a unique positive node n_1 such that $n_1 \overset{msg}{\rightarrow} n_2$ and $\text{abs}(\text{term}(n_1)) = \text{abs}(\text{term}(n_2))$; (2) the happened-before [5] (also called causal dependence) relation $\preceq_h$, defined to be the reflexive and transitive closure of $\overset{msg}{\rightarrow} \cup \overset{lcl}{\rightarrow}$, is well-founded and acyclic; (3) for all $s \in \text{dom}(tr)$, $tr(s)$ is a trace for $role(s)$; (4) for all $s \in \text{dom}(tr)$, for all $x \in \text{dom}(args(role(s), tr(s)))$, if parameter x is uniquely-originated and $tr(s)$ is uncompromised, then $args(role(s), tr(s))(x)$ is not in genvals(pik) and uniquely originates from $\langle s, i \rangle$, where i is the index of the first term in r that contains x.

If $role(s) = r$, then s is called a *strand for* r. If $role(s) \in \Pi$, then s is called a *regular strand*; otherwise, s is called a *penetrator strand*. Nodes on regular and penetrator strands are called *regular nodes* and *penetrator nodes*, respectively.

A system *satisfies* a predicate ϕ on histories iff all of its histories satisfy ϕ.

We sometimes use a history instead of a strand space as a subscript. For example, if $h = \langle tr, \overset{msg}{\rightarrow}, role \rangle$, we sometimes write $\mathcal{N}_h$ instead of $\mathcal{N}_{tr}$.

The set of predecessors of a node n in a history h is $\text{preds}_h(n) = \{n' \in \mathcal{N}_h \mid n' \preceq_h n \wedge n' \neq n\}$.

A set S of nodes is *backwards-closed* with respect to a binary relation R iff, for all nodes n_1 and n_2, if $n_2 \in S$ and $n_1 \; R \; n_2$, then $n_1 \in S$. Given a history $h = \langle tr, \overset{msg}{\rightarrow}, role \rangle$ of a system $\mathcal{M}$, a set S of nodes that is backward-closed with respect to $\preceq_h$ can be regarded as a history of $\mathcal{M}$, denoted $\text{nodesToHist}_h^{\mathcal{M}}(S)$, in a natural way.

3.6. Derivability

A term t is derivable (by the penetrator) from a set S of nodes of a history h of a system $\mathcal{M} = \langle \Pi, pik \rangle$, denoted $S \vdash_h^{\mathcal{M}} t$, if the penetrator can compute t from $\text{term}_h(S) \cup pik$, by performing encryption, decryption, pairing, and separation (*i.e.*, projection) operations, and by generating genvals that are not in $\text{uniqOrigRqrd}_h^{\mathcal{M}}(S)$, where $\text{uniqOrigRqrd}_h^{\mathcal{M}}(S)$ is the set of genvals g that originate from a node in S and are required to be uniquely originated in h (by item (4) in the definition of history). Similar derivability relations or functions have been considered by several researchers, *e.g.*, [9]. The new twist here is in the treatment of genvals.

3.7. Correctness Requirements

Genval Secrecy. Informally, genval secrecy says: the values of specified genval parameters are not revealed to the penetrator. Formally, a genval secrecy requirement for a system $\langle \Pi, pik \rangle$ is specified by a set of uniquely-originated genval parameters of Π. A history $h = \langle tr, \stackrel{msg}{\rightarrow}, role \rangle$ of a system $\mathcal{M}$ satisfies a genval secrecy requirement G iff, for every $r.x \in G$, for every uncompromised regular strand s for r, if $x \in \mathrm{dom}(args(role(s), tr(s)))$, then $\mathcal{N}_{tr} \not\vdash_h^{\mathcal{M}} args(role(s), tr(s))(x)$. For example, $\mathcal{M}_{NSL}$ satisfies the genval secrecy requirement $\{\mathrm{Init}_{NSL}.ni, \mathrm{Init}_{NSL}.nr, \mathrm{Resp}_{NSL}.ni, \mathrm{Resp}_{NSL}.nr\}$.

Agreement. Informally, agreement says: if some uncompromised strand executes a certain role to a certain point with certain arguments, then some strand must have executed a certain role to a certain point with certain arguments. An agreement requirement for a protocol Π has the form "$\langle r_1, len_1, xs_1 \rangle$ precedes $\langle r_2, len_2, xs_2 \rangle$", where $r_1 \in \Pi$, $r_2 \in \Pi$, and xs_1 and xs_2 are sequences of parameters of r_1 and r_2, respectively, such that $\mathrm{len}(xs_1) = \mathrm{len}(xs_2)$ and for $j \in \{1, 2\}$, every parameter in xs_j occurs in $r_j(0), r_j(1), \ldots, r_j(len_j - 1)$. A history $\langle tr, \stackrel{msg}{\rightarrow}, role \rangle$ of a system $\langle \Pi, pik \rangle$ satisfies that agreement requirement iff, if tr contains an uncompromised strand s_2 such that $role(s_2) = r_2$ and $\mathrm{len}(tr(s_2)) \geq len_2$, then tr contains a strand s_1 such that $role(s_1) = r_1$ and $\mathrm{len}(tr(s_1)) \geq len_1$ and the sequence of arguments of s_2 corresponding to parameters xs_2 equals the sequence of arguments of s_1 corresponding to parameters xs_1. For example, $\mathcal{M}_{NSL}$ satisfies the agreement requirement $\langle \mathrm{Resp}_{NSL}, 1, \langle\!\langle i, r, ni, nr \rangle\!\rangle \rangle$ precedes $\langle \mathrm{Init}_{NSL}, 1, \langle\!\langle i, r, ni, nr \rangle\!\rangle \rangle$.

4. Restrictions

Hereafter, we consider only systems $\langle \Pi, pik \rangle$ that satisfy the following static restrictions.

Shallow Ciphertext Restriction. In every term in every role in Π and in every term in pik, $encr$ does not occur in the scope of $encr$.

Unsent Long-Term Keys Restriction. In every parameterized term in every role of Π and in every term in $pik \setminus (Key_{sym} \cup Key_{asym})$, the operators key, $pubkey$, and $pvtkey$ occur only in the second argument of $encr$. This implies that long-term keys not in pik are not sent in messages.

4.1. Support

Informally, a set S' of nodes supports a set S of nodes if S' contains all of the nodes in S and all of the regular nodes on which nodes in S depend. Let $\mathcal{N}_h^{\mathrm{reg}}$ denote the set of regular nodes in history h of system $\mathcal{M}$. For a genval g that uniquely originates in a history h, let $\mathrm{origin}_h(g)$ denote the node from which g originates in h.

A set S' of nodes is a *support* for a set S of nodes in a history h of a system $\mathcal{M}$ if

Su1. $S \subseteq S' \subseteq \mathcal{N}_h$, and S' is backwards-closed with respect to $\overset{lcl}{\rightarrow}$.

Su2. Every received term is derivable from preceding terms, *i.e.*, for all negative nodes n in S', $\text{preds}_h(n) \cap S' \cap \mathcal{N}_h^{\text{reg}} \vdash_h^{\mathcal{M}} \text{term}_h(n)$.

Su3. For $g \in \text{genvals}(\text{term}_h(S')) \cap (\text{uniqOrigRqrd}_h^{\mathcal{M}}(\mathcal{N}_h) \setminus \text{uniqOrigRqrd}_h^{\mathcal{M}}(S'))$, g occurs in the clear in $\text{term}_h(\text{origin}_h(g))$. (Su3 is needed for Lemma 2.)

If S' is a support for S, we say that S' supports S. For a strand s, if S' supports nodes(s), we say that S' supports s. An algorithm for computing supports is in [11].

To illustrate the treatment of unique origination, consider the following history of a generic server-based authentication protocol that reveals at least one genval that originates from the initiator role; the Yahalom and Otway-Rees protocols are specific examples of this kind. Suppose s_I, s_R, and s_S are initiator, responder, and server strands, respectively, that interact without interference from the penetrator. Let n be a nonce that uniquely originates on s_I. The penetrator then behaves as an initiator, interacting with a responder strand s'_R and a server strand s'_S, except that the penetrator uses n instead of a fresh nonce. A support for s'_R or s'_S need *not* contain nodes on s_I. In that sense, s'_R and s'_S do not depend on s_I, even though the chain of messages that conveys n means that there is causal dependence between those nodes in the classical sense of Lamport [5]. Informally, that classical dependence can be ignored here because the penetrator could generate a nonce n' and replace n with n' in the terms of nodes on s'_R and s'_S. The careful treatment of unique origination in the definition of derivability allows such inessential classical dependencies to be ignored. If they were not ignored, few interesting protocols would satisfy BSR, defined in Section 4.2. The next lemma says that a support can be transformed into a history by adding only penetrator nodes.

Given a strand space tr, a strand $s \in \text{dom}(tr)$, and a set S of nodes of tr that is backwards-closed with respect to $\overset{lcl}{\rightarrow}$, S contains nodes on a prefix of $tr(s)$; let $\text{prefix}_{tr}(s, S)$ denote that prefix.

Lemma 1 *If S' is a support for S in a history $h = \langle tr, \overset{msg}{\rightarrow}, role\rangle$ of a system $\mathcal{M} = \langle \Pi, pik\rangle$, then there exists a history $h' = \langle tr', \overset{msg'}{\rightarrow}, role'\rangle$ of $\mathcal{M}$ such that*

$$(\forall s \in \text{strand}(S') : s \in \text{dom}(tr') \ \wedge\ tr'(s) = \text{prefix}_{tr}(s, S') \ \wedge\ role'(s) = role(s)) \\ \wedge (\forall s \in \text{dom}(tr') \setminus \text{strand}(S') : role'(s) \in \Pi_P(pik))$$

Proof: h' is constructed by combining nodes in S with histories that witness the derivability of terms, as required by Su2. Details are in [11]. ∎

Lemma 2 *Supports are compositional, i.e., if S'_0 and S'_1 support S_0 and S_1, respectively, in a history h of a system $\mathcal{M}$, then $S'_0 \cup S'_1$ supports $S_0 \cup S_1$ in history h of $\mathcal{M}$.*

System	Strand count f for BSR(f)			$\mathrm{DW_R}$	Total strands		
	f(Init)	f(Resp)	f(Srvr)		Init	Resp	Srvr
NSL	1	1	none	2	3	3	none
Yahalom	1	2	1	2	3	6	3
Otway-Rees	1	1	1	2	3	3	3

Figure 1. Results for some well-known authentication protocols. $\mathrm{DW_R}$ is defined in Section 5. The right part of the table gives the total number of strands for each role that need to be considered in a history to verify correctness requirements and dynamic restrictions (*cf.* Section 8).

Proof: The proof is straightforward. Details are in [11]. ∎

4.2. Bounded Support Restriction

A *strand count* for a protocol Π is a function from Π to the natural numbers. A set S of nodes *has strand count* f iff, for each role r, S contains nodes from exactly $f(r)$ strands for r. If $\mathcal{N}_h$ has strand count f, then we say that history h has strand count f. We define a partial ordering $\preceq_{SC}$ on strand counts for a protocol: $\preceq_{SC}$ is the pointwise extension of the usual ordering on numbers.

A history h satisfies the *bounded support restriction for strand count* f, abbreviated BSR(f), iff for each regular strand s in h, there exists a support for s in h with strand count at most f.

Figure 1 lists some systems and, for each system, a strand count f for which the system satisfies BSR(f). By Theorem 2 in Section 6, these results can be verified automatically through state-space exploration of histories with strand counts bounded by the values in the right part of the table. Our method is not currently implemented, so these results were proven by hand, which is not difficult. The proof for the NSL protocol appears in [11]; the other proofs are similar. Although we do not have formal guidelines for choosing a strand count f for verifying a given system, in practice, it appears that all correct and un-contrived authentication protocols satisfy BSR(f_2).

4.3. Revealed Genval Restriction

A node n *directly reveals* a term t in a history h of a system $\mathcal{M}$ iff n is a positive regular node and $\{n\} \vdash_h^{\mathcal{M}} t$. A history h of a system $\mathcal{M}$ satisfies the *revealed genval restriction* (RGR) if, for every genval $g \in \mathrm{uniqOrigRqrd}_h^{\mathcal{M}}(\mathcal{N}_{tr})$, if the penetrator learns g (*i.e.*, $\mathcal{N}_h \vdash_h^{\mathcal{M}} g$), then h contains a node that directly reveals g. RGR prevents genvals from being revealed to the penetrator indirectly, *e.g.*, by encrypting one genval with another and then revealing the latter genval. RGR helps us obtain a static bound on the dependence width (see Section 5). The NSL, Yahalom, and Otway-Rees protocols satisfy RGR. By Theorem 2 in Section 6, this can be verified automatically through state-space exploration of histories with the strand counts given in the right part of Figure 1. Currently. we proved these results by hand.

5. Dependence Width

Let r be a role of a system $\mathcal{M}$, and let i be the index of a negative parameterized term in r. An *instance* of $\langle r, i\rangle$ in a history h is a node $\langle s, i\rangle$ on a strand s for r. A *revealing set* for a term t at a node n in a history h of a system $\mathcal{M}$ is a set R of positive regular nodes of tr such that $R \cap \mathrm{preds}_h(n) \vdash_h^{\mathcal{M}} t$.

The dependence width of $\langle r, i\rangle$ in $\mathcal{M}$ is the maximum, over all histories h of $\mathcal{M}$ and all instances n of $\langle r, i\rangle$ in h, of $|R \setminus \mathrm{nodes}_h(\mathrm{strand}(n))|$, where R is a minimum-size revealing set for $\mathrm{term}_h(n)$ at n in h. For example, suppose h contains an instance $\langle s, i\rangle$ of $\langle r, i\rangle$ with $\mathrm{term}_h(\langle s, i\rangle) = g_1 \cdot \{g_2\}_k$. Suppose g_1 and g_2 were sent in the clear at positive regular nodes n_1 and n_2 not on s, respectively, and $k \in pik$, and no other regular nodes send or receive g_1 and g_2. Then n_1 and n_2 together reveal t, and no single node reveals t, so the dependence width of $\langle r, i\rangle$ in h is at least 2. If we suppose instead that n_1 is on s, then we would not count n_1 in the dependence width. Note that a support for $\{n\}$ would (in general) include nodes that n_1 and n_2 causally depend on; a revealing set for n does not. Dependence width is used in the proof of Theorem 2 (in Section 6) to bound the number of strands involved in a violation of BSR.

Nodes on strand(n) are not counted in the dependence width, because dependence width is designed to bound the size of the index set of the rightmost union in equation (3), and those nodes appear in $\mathrm{support}_{h_0}^{\mathcal{M}}(s_0)$ and hence are excluded from that index set.

The *dependence width* of a system $\mathcal{M}$ is the maximum, over all roles r of $\mathcal{M}$ and all negative parameterized terms $r(i)$ in r, of $\mathrm{DW}(\langle r, i\rangle, \mathcal{M})$.

The proof of Theorem 2 relies on an upper bound on the dependence width of a system. It is convenient to base this bound on the syntactic structure of the protocol. This is difficult if a protocol sends terms of the forms $\{g\}_{k_1}$, $\{k_1\}_{k_2}$, $\{k_2\}_{k_3}$, ..., $\{k_{i-1}\}_{k_i}$, k_i; in this case, a minimum-size revealing set for g might contain $i + 1$ nodes. RGR prohibits such behavior.

The *RGR dependence width* of $\langle r, i\rangle$ in $\mathcal{M}$, denoted $\mathrm{DW_R}(\langle r, i\rangle, \mathcal{M})$, is defined like $\mathrm{DW}(\langle r, i\rangle, \mathcal{M})$, except ignoring histories that do not satisfy RGR. The *RGR dependence width* of $\mathcal{M}$, denoted $\mathrm{DW_R}(\mathcal{M})$, is defined analogously.

Let genvalPar(r, i) be the set of genval parameters of role r that occur in the term $r(i)$. Let $\mathrm{genvalParClr}(r, i) = \{x \in \mathrm{genvalPar}(r, i) \mid x \text{ occurs in the clear in } r(i)\}$. Let genvalParClrBefore(r, i) be the set of genval parameters of r that occur in the clear in some $r(j)$ with $j < i$.

Theorem 1 *Let $\mathcal{M} = \langle \Pi, pik\rangle$ be a system satisfying the shallow ciphertext and unsent long-term keys restrictions. Let $r \in \Pi$. If $r(i)$ is negative and contains at most one occurrence of encr, then*

$$\mathrm{DW_R}(\langle r, i\rangle, \mathcal{M}) \le \max(|\mathrm{genvalPar}(r, i) \setminus \mathrm{genvalParClrBefore}(r, i)|, |\mathrm{genvalParClr}(r, i) \setminus \mathrm{genvalParClrBefore}(r, i)| + 1) \quad (1)$$

Proof: Consider how the penetrator learns each ciphertext and genval in term($\langle s,i\rangle$), where s is a strand for r, and sum the number of nodes involved in revealing each of them. Let $t = \text{term}(\langle s,i\rangle)$. The shallow ciphertext restriction ensures that each ciphertext is directly revealed by some node. RGR implies that each genval is directly revealed by some node. The first argument of max in (1) corresponds to the case in which the penetrator learns all of the genvals in t and then performs an encryption to compute the ciphertext (if any) in t; the second argument of max corresponds to the case in which the penetrator learns the genvals that occur in the clear in t and the ciphertext in t. A genval that occurs in the clear in a term before $\langle s,i\rangle$ on s does not contribute to the RGR dependence width of t, because nodes on s are not counted in the dependence width. This justifies subtracting genvalParClrBefore(r,i) in (1). ∎

Theorem 1 yields the bounds on RGR dependence width in Figure 1. A simple correctness-preserving transformation was applied to some of the protocols to satisfy the "at most one ciphertext" hypothesis; specifically, each parameterized term of the form $-\{t\}_k \cdot \{t'\}_{k'}$ was replaced with the sequence $-\{t\}_k, -\{t'\}_{k'}$.

Generalizing Theorem 1 to apply to terms containing multiple shallow ciphertexts is not difficult. Generalizing it to eliminate the shallow ciphertext restriction is also possible, thereby completely eliminating the need for this restriction. This requires extending the proof of Theorem 1 to consider values that are revealed by sequences of decryptions applied to nested ciphertexts.

6. Reduction for Dynamic Restrictions

For a strand count f and a system $\mathcal{M}$, define a strand count $\beta(f,\mathcal{M})$ by

$$\beta(f,\mathcal{M})(r) = \max(\{\mathrm{DW_R}(\mathcal{M})+1, 3\})f(r). \tag{2}$$

Theorem 2 *Let $\mathcal{M} = \langle \Pi, pik\rangle$ be a system satisfying the shallow ciphertext and unsent long-term keys restrictions. Let f be a strand count for Π. $\mathcal{M}$ satisfies BSR(f) and RGR iff all histories of $\mathcal{M}$ with strand count $\beta(f,\mathcal{M})$ do.*

Proof: (A more detailed proof is in [11].) The forward direction ($\Rightarrow$) of the "iff" follows immediately from the definitions. For the reverse direction ($\Leftarrow$), we prove the contrapositive, *i.e.*, we suppose there exists a history h of $\mathcal{M}$ that violates BSR(f) or RGR, and we construct a history of $\mathcal{M}$ with strand count at most $\beta(f,\mathcal{M})$ that violates the same property.

BSR(f) and RGR are safety properties satisfied by histories with zero nodes, so there exists a $\preceq_h$-minimal node n_0 such that (1) $\text{nodesToHist}_h^{\mathcal{M}}(\text{preds}_h(n_0))$ satisfies BSR(f) and RGR, and (2) $\text{nodesToHist}_h^{\mathcal{M}}(\text{preds}_h(n_0) \cup \{n_0\})$ violates BSR(f) or RGR.

Let $h_0 = \text{nodesToHist}_h^{\mathcal{M}}(\text{preds}_h(n_0))$. Let $s_0 = \text{strand}(n_0)$ and $i_0 = \text{index}(n_0)$. Note that $n_0 \notin \mathcal{N}_{h_0}$. For a history h' of $\mathcal{M}$ that satisfies $BSR(f)$, for a regular strand s of h', let $\text{support}_{h'}^{\mathcal{M}}(s)$ denote a support for s in h' that has strand count at most f and contains no penetrator nodes. Consider cases based on the sign of n_0.

Suppose n_0 is negative. n_0 cannot cause a violation of RGR, so n_0 causes a violation of BSR(f). Suppose $i_0 > 0$ (the proof for $i_0 = 0$ is similar). n_0 directly depends on $\langle s_0, i_0 - 1\rangle$ and on a revealing set R for term(n_0) at n_0. Let

$$S_1 = \{n_0\} \cup \text{support}_{h_0}^{\mathcal{M}}(s_0) \cup \bigcup_{n \in R \setminus \text{nodes}_{h_0}(s_0)} \text{support}_{h_0}^{\mathcal{M}}(\text{strand}(n)). \quad (3)$$

h_0 satisfies RGR, so Theorem 1 implies $|R \setminus \text{nodes}_{h_0}(s_0)| \leq \text{DW}_{\text{R}}(\mathcal{M})$. h_0 satisfies BSR(f), so each support in (3) has strand count at most f. n_0 is on s_0, so it does not increase the strand count of S_1. Thus, S_1 has strand count at most $\beta(f, \mathcal{M})$. It is easy to show that S_1 supports $\{n_0\}$ in h. Lemma 1 implies that S_1 can be transformed into a history h_1 of $\mathcal{M}$ by adding penetrator nodes. It is easy to show that h_1 violates BSR(f).

Suppose n_0 is positive. n_0 cannot cause a violation of BSR(f), so n_0 causes a violation of RGR in h. Let g_0 be a genval that is "indirectly" revealed jointly by n_0 and other nodes, causing a violation of RGR. Now perform a series of case analyses based on the decryption keys that the penetrator uses to obtain g_0. In each case, one can identify a set S_1 of nodes such that S_1 has strand count at most $\beta(f, \mathcal{M})$ and $S_1 \vdash_h^{\mathcal{M}} g_0$ and $\text{origin}_h(g_0) \in S_1$. Using Lemmas 1 and 2, one can show that S_1 can be transformed into a history h_1 of $\mathcal{M}$ by adding penetrator nodes. Furthermore, h_1 violates RGR. ∎

7. Reduction for Correctness Requirements

Given a strand count f for a protocol Π, define a strand count dbl(f) for Π by: $\text{dbl}(f)(r) = 2f(r)$.

Theorem 3 *Let $\mathcal{M} = \langle \Pi, pik\rangle$ be a system satisfying the shallow ciphertext and unsent long-term keys restrictions. Let f be a strand count for Π. Let ϕ be a genval secrecy or agreement requirement. Suppose all histories of $\mathcal{M}$ with strand count $\beta(f, \mathcal{M})$ satisfy BSR(f) and RGR. $\mathcal{M}$ satisfies ϕ iff all histories of $\mathcal{M}$ with strand count* dbl(f) *do.*

Proof: The proof is in [11]. It is similar in strategy to the proof of Theorem 2, but simpler. ∎

8. Bounds for Sample Protocols

The right part of Figure 1 contains the maximum of the bounds obtained from Theorems 2 and 3, *i.e.*, $\max(\beta(f, \mathcal{M}), \text{dbl}(f))$.

References

[1] Michael Burrows, Martín Abadi, and Roger Needham. A logic of authentication. *ACM Transactions on Computer Systems*, 8(1):18–36, February 1990.

[2] Iliano Cervesato, Nancy Durgin, Patrick Lincoln, John Mitchell, and Andre Scedrov. Relating strands and multiset rewriting for security protocol analysis. In Paul Syverson, editor, *Proc. 13th IEEE Computer Security Foundations Workshop*, pages 35–51. IEEE Press, 2000.

[3] Edmund M. Clarke, Orna Grumberg, and Somesh Jha. Verifying parameterized networks using abstractions and regular languages. In *Proc. Sixth Int'l. Conference on Concurrency Theory (CONCUR)*, 1995.

[4] James Heather and Steve Schneider. Towards automatic verification of authentication protocols on an unbounded network. In *Proc. 13th IEEE Computer Security Foundations Workshop (CSFW)*, July 2000.

[5] Leslie Lamport. Time, clocks, and the ordering of events in a distributed system. *Communications of the ACM*, 21(7):558–564, 1978.

[6] Gavin Lowe. Breaking and fixing the Needham-Schroeder public-key protocol using FDR. In *Proc. Workshop on Tools and Algorithms for The Construction and Analysis of Systems (TACAS)*, volume 1055 of *Lecture Notes in Computer Science*, pages 147–166. Springer, 1996.

[7] Gavin Lowe. Towards a completeness result for model checking of security protocols. *The Journal of Computer Security*, 7(2/3):89–146, 1999.

[8] Dave Otway and Owen Rees. Efficient and timely mutual authentication. *Operating Systems Review*, 21(1):8–10, January 1987.

[9] L. C. Paulson. The inductive approach to verifying cryptographic protocols. *The Journal of Computer Security*, 6(1/2):85–128, 1996.

[10] A. W. Roscoe and P. J. Broadfoot. Proving security protocols with model checkers by data independence techniques. *The Journal of Computer Security*, 7(2/3), 1999.

[11] Scott D. Stoller. A bound on attacks on authentication protocols. Technical Report 526, Computer Science Dept., Indiana University, July 1999. Revised April 2001. Also available at www.cs.sunysb.edu/~stoller/TR526.html .

[12] Scott D. Stoller. A bound on attacks on payment protocols. In *Proc. 16th Annual IEEE Symposium on Logic in Computer Science (LICS)*, pages 61–70. IEEE Press, June 2001.

[13] F. Javier Thayer Fábrega, Jonathan C. Herzog, and Joshua D. Guttman. Strand spaces: proving security protocols correct. *The Journal of Computer Security*, 7:191–230, 1999.

[14] Thomas Woo and Simon S. Lam. A semantic model for authentication protocols. In *Proc. 14th IEEE Symposium on Research in Security and Privacy*, pages 178–194. IEEE Press, 1993.

RESPONSIVE BISIMULATION

Xiaogang Zhang and John Potter
School of Computer Science and Engineering
University of New South Wales, Australia
{ xzhang,potter } @cse.unsw.edu.au

Abstract This paper introduces the responsive bisimulation, which treats local delays of incoming messages the same as external delays, as long as potential interference by competing receptors is avoided. By this bisimulation, the π-calculus process $k.m.P$, representing a lock k with a message receptor m, is equivalent to the process $(\nu\, n)(m.\bar{n}|k.n.P)$; the first process will delay messages externally, the second locally. Existing bisimulations distinguish between these processes. The responsive bisimulation is a congruence for the family of processes which model objects. It is useful for studying compositional synchronisation in such models.

1. Introduction

With the ability to directly model dynamic reference structures, process algebras such as the π-calculus ([Milner92, Milner99]) and its variations have been used to model concurrent objects ([Walker95, Jones93, Sangi96, Hüttel96, Zhang97]). Some researchers ([Schne97, Zhang98a, Zhang98b]) have also applied it to model compositional concurrent objects in the aspect-oriented programming style ([Aksit92, Holmes97]) attempting to avoid the inheritance anomaly [McHale94]. One of the important issues with such models is to identify the similarity between composed behaviours and the expected behaviour. There are many known bisimulation techniques available for various purposes. For example, the weak ground bisimulation and many others recognise equivalence of processes $(\nu\, n)(m.\bar{n}|n.P)$ and $m.P$, by ignoring the internal forwarding. This paper presumes familiarity with the π-calculus and its bisimulation relations.

Existing bisimulations distinguish between processes $(\nu\, n)(m.\bar{n}|k.n.P)$ and $k.m.P$, the former represents an object with a lock k and a forwarder (via n) for message m so that any delays in servicing a method call are local, whereas the latter process causes an external delay in reception of m. We want to treat these two processes as equivalent, since their behaviour in emitting responses to incoming messages is the same.

In this paper we propose the notion of *responsive bisimulation* to capture this kind equivalence by allowing the location of input delays to be ignored. This is different from asynchronous bisimulation which is insensitive to output delays and considers $(m.\bar{m}|P)$ to be the same as P; we detail the differences in Section 6. We can take alternative views of responsive bisimulation. First, from the target process, we can tolerate delays of incoming messages at input ports when determining the similarity between process evolution trees. Second, when testing a process, we can "localise" messages by buffering them in the environment, and make them only accessible to the target process. External observers cannot see these localised messages, so cannot mistake them as output from the target process. In Section 4 we will show these views are equivalent.

One of the major results of this paper is that the responsive bisimulation is a congruence for the family of processes which model objects, that is, the processes which uniquely own all the message receptors they use. All process combinators, except replication and parallel composition, always preserve responsive bisimulation. Replication preserves responsive bisimilarity for the family of processes which model objects, and parallel composition preserves it for an even larger family of processes.

Another interesting result, revealed in Proposition 20, is that, a persistently available receptor with an internal forwarder can be ignored.

The rest of the paper is structured as follows: Section 2 motivates this work; Section 3 briefly introduces the polar π-calculus and related notions; Section 4 defines responsive bisimulation; Section 5 gives some properties of the equivalence and other theoretical results; Section 6 discusses some further issues relating the responsive bisimulation with other notions; and Section 7 concludes the paper. Most proofs are omitted here in order to save space; interested readers can find them in [Zhang02b].

2. Background

In [Zhang98a] and [Zhang98b], the behaviour of a concurrent object is modelled using the π-calculus as the parallel composition of a process F, representing the object's functional behaviour with no constraint on its concurrent interactions, and a process C representing the constraints on the object's concurrent behaviour. For example, the functionality of a buffer object can be described by $F \stackrel{\text{def}}{=} !n_r(x).M_r\langle x\rangle | !n_w(x).M_w\langle x\rangle$, where $n_r(x).M_r\langle x\rangle$ and $n_w(x).M_w\langle x\rangle$ represent the behaviour of the read and write methods respectively; each of them can have unlimited invocations executing in parallel with no concern for any potential interference. To discipline those invocations, assume a synchronisation behaviour modelled by the control process $C \stackrel{\text{def}}{=} m_r(x).\bar{n}_r\langle x\rangle + m_w(x).\bar{n}_w\langle x\rangle$, where the choice operator in fact represents a mutual exclusion lock on those methods. Then the parallel composition of the two processes, $(\nu\ n)(C|F)$, will be weakly bisimilar to $R \stackrel{\text{def}}{=} m_r(x).M_r\langle x\rangle + m_w(x).M_w\langle x\rangle$, as expected. More complicated and generic method exclusion relations, besides mutual exclusion locks, can be simply modelled and composed in exactly the same way, once the κ-calculus ([Zhang02a]) is used. The κ-calculus is an extended calculus which welds the mobility power of the π-calculus with the synchronisation expressiveness of the algebra of exclusion ([Noble00]). In this paper the responsive bisimulation is presented in the polar π-calculus, an asynchronous π-calculus with polars, which is a simplified version of the κ-calculus and allows us to focus on those features essential for responsive bisimulation, without the full complexity of the κ-calculus. The following "real world" scenario illustrates the need for responsive bisimulation:

> In the mailroom of a commercial office building, the property manager uses internal mail to send bills to her tenants and collect payments. Each tenant has a locked mailbox, located either on the mailroom wall and accessible from outside of the mailroom by the tenant, or on the door of the tenant's suite to which a postman delivers mail from the mailroom. Suppose the manager only wants to know about the arrival of payments, and classify tenants who pay the bill on time as "good" tenants. She does not care whether the mailboxes are in the mailroom or on their doors. However, existing bisimulations distinguish between cases where the mailroom mailboxes are cleared or not, as the manager can observe this; however she cannot distinguish between mail being cleared or not from the door-based mailboxes. We aim to provide a model which is able to treat the mailroom and door-based mailboxes as equivalent.

Let us model this scenario. The process O_1 and O_2 illustrated in Figure 1 are composite processes representing two different versions of a mailbox locked with a key κ. The only difference between them is that O_1 has an extra "empty" control $Ctrl_e$ (the postman) which does nothing but forward whatever message received from channel m to the next control $Ctrl_l$ (the locked mailbox). The body (a tenant) of these two can always give the same response (a payment) if fed with the same message (a bill). If an unlocking signal is received via channel κ, both O_1 and O_2 can accept incoming messages and process them immediately. If some message arrives before the unlocking, O_1 will store it in an internal buffer (the door mailbox), but O_2 will leave the message in the external buffer (the mailroom) as it was. In both cases, the processing will only continue after unlocking.

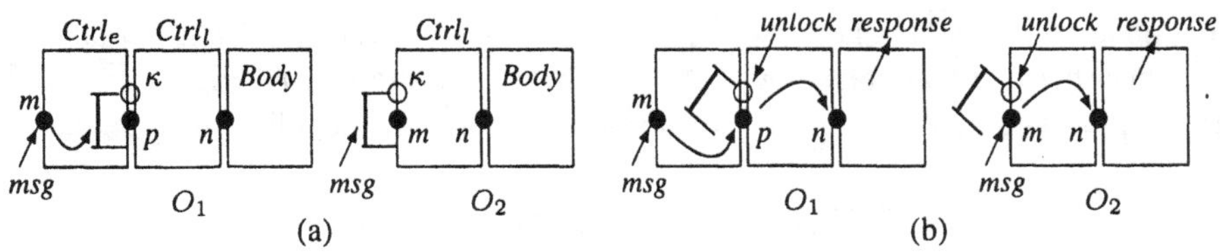

Figure 1. Two objects with the same responsive behavour

For a client (the manager) who is sending the message, the behaviour of the target object can be measured only by observing how it responds. Since the responses O_1 and O_2 can give are the same, their behaviour are identical in the client's eyes. However, this behavioural similarity cannot be captured by most of the known behavioural equivalences, since in some stage O_1 can perform an input action from the channel m while O_2 cannot. Even the weak barbed-equivalence, one of the weakest, is too strong for them, since $O_1|R$ and $O_2|R$ are not weakly barbed-bisimilar for some R, such as $R \stackrel{\text{def}}{=} \bar{m}\langle\tilde{v}\rangle$. The responsive bisimulation recovers this equivalence.

3. The polar π-calculus (π_p-calculus)

As in the asynchronous π-calculus ([Amadio96, Hüttel96]), output in the *polar π-calculus* (π_p-calculus) is non-blocking, and is not used as a prefix or choice point. Similar to [Odersky95], each name m, which can be considered as a reference to a communication channel, has an input polar $\grave{m}$ and an output polar $\acute{m}$, which in turn can be considered as the input port from, and the output port to, the channel m. The main difference from [Amadio96] and [Odersky95] is that in the π_p-calculus, only output polars can be transmitted through a communication channel. [Ravara97, Merro00] and others have adopted a similar restriction, but in the π_p-calculus this restriction is enforced syntactically. As a consequence, only output polar substitution can be caused by input prefix, while in [Odersky95] substitution may involve names with both polarities and in [Amadio96] it may affect both input and output usage of a name.

One of the advantages in using polars to enforce this restriction rather than using the implicit restriction as in [Ravara97] and [Merro00], is the simplification afforded in describing and proving some properties of bisimulations, such as bisimilarity between process P and Q being preserved by $\grave{m}(\tilde{x}).P$ and $\grave{m}(\tilde{x}).Q$ for all input prefixes $\grave{m}(\tilde{x})$.

The notion of polarised ports exists in many forms of communication. For example, an email account consists of a mailbox (input polar) and an email address (output polar). You may send the address to other people for them to send you message, but you can neither send your mailbox nor tell someone "receive my next email using this address". Moreover, if input polars can be transmitted, then the behaviour of a process will no longer be predictable from its structure. For object modelling, the prohibition on transmitting input polars respects the idea that object identity and method ownership cannot be transferred from one object to another (see Section 5).

The Syntax. Let $\mathcal{N}$ be the set of all names, ranged over by name expressions m, n, u, v and variables x, y. Let $\grave{\mathcal{N}} \stackrel{\text{def}}{=} \{\grave{n} : n \in \mathcal{N}\}$ and $\acute{\mathcal{N}} \stackrel{\text{def}}{=} \{\acute{n} : n \in \mathcal{N}\}$ be the sets of input polars and output polars respectively. Let polar expressions a, b and variable w range over the set of all polars, $\grave{\mathcal{N}} \cup \acute{\mathcal{N}}$. $\tilde{r}$ is the abbreviation for $r_1, r_2, \cdots, r_n$. The generic process terms P in the π_p-calculus are generated by the following grammar:

$$P ::= \acute{m}\langle\tilde{u}\rangle \mid (\nu\, \tilde{n})P \mid P_1|P_2 \mid !B \mid G$$

$$G ::= \mathbf{0} \mid B \mid (\nu\, \tilde{n})G \mid G_1+G_2, \qquad B ::= \grave{m}(\tilde{x}).P$$

The set of all actions a process may take is specified by

$$\alpha ::= \grave{m}(\tilde{u}) \mid (\nu\, \tilde{v})\acute{m}\langle\tilde{u}\rangle \mid \tau, \quad \text{where} \quad \tilde{v} \subseteq \tilde{u} \text{ and } m \notin \tilde{v}.$$

Here $\mathbf{0}$ is the inactive (terminated) process; $\acute{m}\langle\tilde{u}\rangle$ is the output action which sends output polars $\tilde{u}$ into the channel m; $(\nu\, \tilde{n})P$ binds the set of names $\tilde{n}$, and therefore both polars of each of those names, within the scope of P; $P_1|P_2$ indicates two processes run in parallel; B is an input-guarded process; $!B$ is replication and G is exclusive choice, both constructed from input-guarded processes.

As usual, auxiliary functions *fn*, *bn* and *n* are used to identify the sets of free, bound and all names, respectively, of a term or action. Specific functions to identify polars are also needed. For example, the function *fin* identifies those names whose input polar freely occurs in a term or action.

A formalised mapping between the π_p-calculus and the asynchronous π-calculus of [Amadio96], and that between the π_p-calculus and the polarised π-calculus of [Odersky95], can be found in [Zhang02b].

Table 1. Structural congruence rules for the polar π-calculus

Summation			
str-SUM1:	$P_1\|0 \equiv P_1$;	$P_1\|P_2 \equiv P_2\|P_1$;	$P_1\|(P_2\|P_3) \equiv (P_1\|P_2)\|P_3$
str-SUM2:	$G_1+0 \equiv G_1$;	$G_1+G_2 \equiv G_2+G_1$;	$G_1+(G_2+G_3) \equiv (G_1+G_2)+G_3$
Scope			
str-SCP1:	$(\nu\,\tilde{n})P \equiv P$, if $\tilde{n} \cap fn(P) = \emptyset$;		$(\nu\,\tilde{n})G \equiv G$, if $\tilde{n} \cap fn(G) = \emptyset$
str-SCP2:	$(\nu\,m)m\langle\tilde{y}\rangle \equiv 0$;		$(\nu\,m)\grave{m}(\tilde{x}).P \equiv 0$
str-SCP3:	$(\nu\,\tilde{m})(\nu\,\tilde{n})P \equiv (\nu\,\tilde{m},\tilde{n})P$;		$(\nu\,\tilde{m})(\nu\,\tilde{n})P \equiv (\nu\,\tilde{n})(\nu\,\tilde{m})P$
str-SCP4:	$(\nu\,\tilde{n})P_1\|P_2 \equiv (\nu\,\tilde{n})(P_1\|P_2)$,		if $\tilde{n} \cap fn(P_2) = \emptyset$
	$(\nu\,\tilde{n})G_1+G_2 \equiv (\nu\,\tilde{n})(G_1+G_2)$,		if $\tilde{n} \cap fn(G_2) = \emptyset$
str-REN:	$(\nu\,\tilde{n})P \equiv (\nu\,\tilde{m})(P\{\tilde{m}/\tilde{n}\})$,		if $\tilde{m} \cap fn(P) = \emptyset$

Table 2. Labelled transition rules for process terms in the polar π-calculus

tr-OUT:	$\dfrac{\cdot}{m\langle\tilde{u}\rangle \xrightarrow{m\langle\tilde{u}\rangle} 0_P}$; $\dfrac{P \xrightarrow{m\langle\tilde{u}\rangle} P',\quad m \notin \tilde{v}}{(\nu\,\tilde{v})P \xrightarrow{(\nu\,\tilde{v})m\langle\tilde{u}\rangle} P'}$	**tr-IN**:	$\dfrac{\cdot}{\grave{m}(\tilde{x}).P \xrightarrow{\grave{m}(\tilde{u})} P\{\tilde{u}/\tilde{x}\}}$
tr-RES:	$\dfrac{P \xrightarrow{\alpha} P',\quad \tilde{n} \cap fn(\alpha) = \emptyset}{(\nu\,\tilde{n})P \xrightarrow{\alpha} (\nu\,\tilde{n})P'}$;	**tr-REP**:	$\dfrac{B \xrightarrow{\grave{m}(\tilde{u})} P}{!B \xrightarrow{\grave{m}(\tilde{u})} P\|!B}$
tr-PARL:	$\dfrac{P \xrightarrow{\alpha} P'}{P\|Q \xrightarrow{\alpha} P'\|Q}$	**tr-CHOI**:	$\dfrac{G_1 \xrightarrow{\grave{m}(\tilde{u})} P}{G_1+G_2 \xrightarrow{\grave{m}(\tilde{u})} P}$
tr-INTL:	$\dfrac{P \xrightarrow{(\nu\,\tilde{v})m\langle\tilde{u}\rangle} P',\quad P' \xrightarrow{\grave{m}(\tilde{u})} P''}{(\nu\,m)P \xrightarrow{\tau} (\nu\,m)(\nu\,\tilde{v})P''}$	**tr-STRUC**:	$\dfrac{P_1' \equiv P_1,\quad P_1 \xrightarrow{\alpha} P_2,\quad P_2 \equiv P_2'}{P_1' \xrightarrow{\alpha} P_2'}$

The Semantics. The structural equivalences and labelled transitions are shown in Table 1 and Table 2. They are standard except the rule tr-INTL, which gives the meaning of an internal action τ. A variant of this rule can be written as:

$$\frac{P \xrightarrow{(\nu\,\tilde{v})m\langle\tilde{u}\rangle} P',\quad Q \xrightarrow{\grave{m}(\tilde{u})} Q',\quad \tilde{v} \cap fn(Q) = \emptyset}{(\nu\,m)(P|Q) \xrightarrow{\tau} (\nu\,m)(\nu\,\tilde{v})(P'|Q')}.$$

The name restriction $(\nu\,m)$ is necessary for preserving any bisimulation involving τ action, under the input prefixing $\grave{m}(\tilde{x}).P$ in which substitution can only occur for the output polar of the name m, rather than both polars.

Definition 1 *Weak transitions are defined as:*
$P \overset{\tau}{\Rightarrow} P'$ iff $P(\xrightarrow{\tau})^* P'$, where $*$ indicates zero or finitely many τ-transitions;
$P \overset{\alpha}{\Rightarrow} P'$ iff $P \overset{\tau}{\Rightarrow}\xrightarrow{\alpha}\overset{\tau}{\Rightarrow} P'$, where $\alpha \neq \tau$.

The reduction relation, a familiar concept in this literature, is defined in a non-standard way in the π_p-calculus:

Definition 2 *The strong and weak reduction relations are defined respectively as:*

$P \rightarrow P'$ *iff* $(\nu\, m)P \xrightarrow{\tau} (\nu\, m)P'$, *for some* m;
$P \Rightarrow P'$ *iff* $(\nu\, m)P \stackrel{\tau}{\Rightarrow} (\nu\, m)P'$, *for some* m.

Clearly, $P \xrightarrow{\tau} P'$ implies $P \rightarrow P'$, and $P \stackrel{\tau}{\Rightarrow} P'$ implies $P \Rightarrow P'$, and therefore, another variant of the rule tr-INTL can be written as:

$$\frac{P \xrightarrow{(\nu\ \tilde{v})\acute{m}\langle\tilde{u}\rangle} P',\ \ Q \xrightarrow{\grave{m}(\tilde{u})} Q',\ \ \tilde{v} \cap n(Q) = \emptyset}{P|Q \rightarrow (\nu\ \tilde{v})(P'|Q')},$$

which looks more like the usual communication rule for the π-calculus. Making the distinction between internal action and reduction is also necessary for the new bisimulation relation, as we will find out later.

Definition 3 *The strong commitments of process P are defined as:*

P can commit *the action* α, *denoted as* $P\downarrow\alpha$, *if* $P \xrightarrow{\alpha} P'$ *for some* P';
P can commit *on input polar* $\grave{m}$, *denoted as* $P\downarrow\grave{m}$, *if* $P\downarrow\grave{m}(\tilde{u})$;
P can commit *on output polar* $\acute{m}$, *denoted as* $P\downarrow\acute{m}$, *if* $P\downarrow(\nu\tilde{v})\acute{m}\langle\tilde{u}\rangle$;
The weak commitments $\Downarrow$, *are obtained by replacing* $\rightarrow$ *with* $\Rightarrow$ *and* $\downarrow$ *with* $\Downarrow$ *throughout.*

Definition 4 *Process* P' *is a* derivative *of* P, *if there exists some finite sequence of actions* $\alpha_1, \alpha_2, \cdots, \alpha_n$, *such that* $P \xrightarrow{\alpha_1}\xrightarrow{\alpha_2}\cdots\xrightarrow{\alpha_n} P'$.

4. Responsive bisimulation in the polar π-calculus

As pointed out at the end of Section 2, the barbed equivalence, even in its weak form, is too strong for compositional objects. We first review the version of [Amadio96] for an asynchronous π-calculus before leading on to responsive bisimulation. It is derived from the barbed bisimulation ([MilSan92, Sangi92]), a rather weak relation which traces the state changes of a process during the course of reductions and observes which channels are available for communication.

Definition 5 (Barbed bisimulation) *A symmetric relation* $\mathcal{S}$ *on P-terms is a (strong) barbed bisimulation if whenever* $P\mathcal{S}Q$ *then* $P\downarrow a$ *implies* $Q\downarrow a$ *for* $a \in \mathcal{N}$, *and* $P \rightarrow P'$ *implies* $\exists Q'$ *such that* $Q \rightarrow Q'$ *and* $P'\mathcal{S}Q'$.

Let $\sim_b$ *be the largest strong barbed bisimulation. The notion of weak barbed bisimulation* $\approx_b$ *is obtained by replacing the transition* $\downarrow$ *with* $\Downarrow$, *and* $\rightarrow$ *with* $\Rightarrow$ *throughout.*

Since barbed bisimulation cannot identify what messages are communicated, it is too rough to measure processes' behaviour.

Definition 6 (Barbed equivalence) *Let* R *be an arbitrary process, the strong and weak barbed equivalence are defined respectively as:* $P \simeq_b Q$ *if* $\forall R.(R|P \sim_b R|Q)$;
$P \approxeq_b Q$ *if* $\forall R.(R|P \approx_b R|P)$.

For our purposes, the weak barbed equivalence fails in at least three ways: (a) it cannot distinguish between a message sent out from the target process and a message

Table 3. Rules for localised messaging

Structural equivalence :

lStr-NULL	$[\grave{m}\langle\tilde{u}\rangle]0 \equiv 0$;	**lStr-LOC**	$(\nu\, m)[\grave{m}\langle\tilde{u}\rangle]P \equiv (\nu\, m)(\bar{m}\langle\tilde{u}\rangle\vert P)$
lStr-SUM2	$[\grave{m}\langle\tilde{u}\rangle][\grave{n}\langle\tilde{v}\rangle]P \equiv [\grave{n}\langle\tilde{v}\rangle][\grave{m}\langle\tilde{u}\rangle]P$;	**lStr-IND**	$([\grave{m}\langle\tilde{u}\rangle]P)\vert Q \equiv [\grave{m}\langle\tilde{u}\rangle](P\vert Q)$, if $m \notin fn(Q)$

Transition :

lTr-INTL2 $\dfrac{P \xrightarrow{\grave{m}(\tilde{u})} P'}{[\grave{m}\langle\tilde{u}\rangle]P \xrightarrow{\tau} P'}$ **lTr-INV** $\dfrac{P \xrightarrow{\alpha} P' \quad \alpha \neq \grave{m}(\tilde{u})}{[\grave{m}\langle\tilde{u}\rangle]P \xrightarrow{\alpha} [\grave{m}\langle\tilde{u}\rangle]P'}$

sent by another agent but buffered in the environment; (b) it cannot prevent input name clashes between the testing environment and the processes being tested; (c) it treats synchronisation actions occurring in public channels as single step reduction, and therefore cannot match them with incomplete synchronisations which have input delays. A different observation technique is needed. We begin with the $o\tau$-bisimulation similar to that in [Amadio96]:

Definition 7 ($o\tau$-bisimulation) *The $o\tau$-bisimulation is a symmetric relation $\mathcal{S}$ on processes, for which whenever $P\mathcal{S}Q$ then $P \xrightarrow{\alpha} P'$ implies $Q \xrightarrow{\alpha} Q'$ and $P'\mathcal{S}Q'$, where α is a non-input action and $bn(\alpha) \cap fn(Q) = \emptyset$.*

The weak $o\tau$-bisimulation is obtained by replacing $\xrightarrow{\alpha}$ with $\xRightarrow{\alpha}$ throughout. We denote the largest (strong) $o\tau$-bisimulation as $\sim_{o\tau}$, and the largest weak $o\tau$-bisimulation as $\approx_{o\tau}$.

The $o\tau$-bisimulation gives a measurement on processes' states by observing available reductions and output actions, but cannot determine how a process responds to incoming messages, since input actions are not observed. To determine responsive behaviours, we introduce a new term for specifying input messages.

Notation 8 *We extend the process syntax with an auxiliary P-term $[\grave{m}\langle\tilde{u}\rangle]P$, the localisation of the sent message $\bar{m}\langle\tilde{u}\rangle$ with process P. Properties for this term are shown in Table 3.*

The term $[\grave{m}\langle\tilde{u}\rangle]P$ is not for modelling processes, but only designed to express responsive bisimulation relations between processes. It couples process P with the message $\tilde{u}$ which is buffered in channel m and unobservable from outside, so that it cannot be mistaken by external observers as an output from P. However the output polar $\bar{m}$ is not hidden by this term. This issue will be discussed further in Section 6.

The rule lTr-INTL2 adds a new case for introducing a τ action. Unlike rule tr-INTL, the name restriction is not required here. However, since only the input polar, $\grave{m}$, of the channel name m is involved, the preservation of τ actions is maintained by input prefixing.

Definition 9 (Responsive equivalence) *Let $\mathcal{T}[.]$ be the responsive testing context with syntax $\mathcal{T} ::= [.] \;\big\vert\; [\grave{m}\langle\tilde{u}\rangle]\mathcal{T}$. We define strong and weak responsive equivalences as:*

$$P \simeq_r Q \;\; \textit{iff} \;\; \forall\mathcal{T}.(\mathcal{T}[P] \sim_{o\tau} \mathcal{T}[Q]); \qquad P \approxeq_r Q \;\; \textit{iff} \;\; \forall\mathcal{T}.(\mathcal{T}[P] \approx_{o\tau} \mathcal{T}[Q]).$$

This definition is clear about the meaning of equivalence in responsive behaviour, but is not so useful for proving equivalence, since it requires the exhaustive testing over the infinite set of responsive testing contexts. A more practical definition is the r1-bisimulation.

Definition 10 *A strong (or weak) r1-bisimulation is a strong (or weak, respectively) $o\tau$-bisimulation $\mathcal{S}$ such that whenever $P\mathcal{S}Q$ then $[\bar{m}\langle\tilde{u}\rangle]P\mathcal{S}[\bar{m}\langle\tilde{u}\rangle]Q$ for all $[\bar{m}\langle\tilde{u}\rangle]$.*

We denote the largest strong r1-bisimulation as $\sim_{r1}$, and the largest weak r1-bisimulation as $\approx_{r1}$.

It is easy to verify that $O_1 \approx_{r1} O_2$ holds for the example of Figure 1. The r1-bisimulation provides a test platform and measures behavioural equivalence from outside of target processes.

While responsive equivalence and r1-bisimulation provide a good base for describing similarities of responsive behaviours, it can tell little about why or when two processes may offer similar behaviours. For closer study, we need an inside view observing input actions.

Definition 11 (Responsive bisimulation) *A (strong) responsive bisimulation is a (strong) $o\tau$-bisimulation $\mathcal{S}$ such that whenever $P\mathcal{S}Q$ then $P \xrightarrow{\bar{m}(\tilde{u})} P'$ implies either $Q \xrightarrow{\bar{m}(\tilde{u})} Q'$ and $P'\mathcal{S}Q'$, or $Q \xrightarrow{\tau} Q'$ and $P'\mathcal{S}[\bar{m}\langle\tilde{u}\rangle]Q'$.*

The weak version is obtained by replacing transitions with weak transitions everywhere. We denote the largest strong and weak responsive bisimulation by $\sim_r$ and $\approx_r$ respectively. Clearly, $\sim_r \subseteq \approx_r$.

For the example of Figure 1, we can easily verify that $O_1 \approx_r O_2$. In general, the two views mentioned in the introduction are equivalent:

Lemma 12 *responsive bisimulation, r1-bisimulation and responsive equivalence all coincide: $\sim_r \equiv \sim_{r1} \equiv \simeq_r$ and $\approx_r \equiv \approx_{r1} \equiv \approxeq_r$.*

5. Properties of responsive bisimulation

We now investigate some formal properties of responsive bisimulation.

Lemma 13 *Responsive bisimulations are equivalences. That is, they are reflexive, symmetric and transitive.*

Proposition 14 *Responsive bisimulation is preserved by output polar substitution, input prefix, restriction, localisation and choice. That is, let $\mathcal{S}$ be either $\sim_r$ or $\approx_r$, then, $P\mathcal{S}Q$ implies*

1. $P\sigma\mathcal{S}Q\sigma$ *for all* $\sigma = \{\tilde{u}/\tilde{x}\}$;
2. $\bar{m}(\tilde{u}).P\mathcal{S}\bar{m}(\tilde{u}).Q$ *for all* $\bar{m}(\tilde{u})$;
3. $(\nu\,\tilde{v})P\mathcal{S}(\nu\,\tilde{v})Q$ *for all* $\tilde{v}$;
4. $[\bar{m}\langle\tilde{u}\rangle]P\mathcal{S}[\bar{m}\langle\tilde{u}\rangle]Q$ *for all* $[\bar{m}\langle\tilde{u}\rangle]$.

and, $G_1\mathcal{S}G_2$ implies $(G_1+G)\mathcal{S}(G_2+G)$ for all G.

A problem is apparent: the responsive bisimulation is not preserved by parallel composition in general. For example, in the Figure 1 we have $O_1 \approx_r O_2$, but

$(O_1|O_3) \not\approx_r (O_2|O_3)$ for $O_3 \stackrel{\text{def}}{=} \grave{m}.R$, because the occurrence of input polar $\grave{m}$ in O_3 has changed the ability of O_1 to receive messages on $\grave{m}$. However, the purpose of our study is object modelling, where the ownership of each input port should be unique. For example, the identity of an object is unique and each method of each object is also uniquely identified so that messages are delivered to their correct destination. In general, each input polar has a restricted scope (or ownership), and is never exported outside this scope. When responsive bisimulation is restricted to the object modelling domain, its preservation in parallel composition can be guaranteed. To show this, we formalise the restriction on input polars.

Definition 15 (Safe process) *A process P is safe for Env, and the environment Env is said to be safe for P, if $fin(P) \cap fin(Env) = \emptyset$.*

We may call P a safe process, when the behaviour of P is only considered within environments safe for P.

A process P is autonomous *if $fin(P) = \emptyset$.*

Lemma 16 *Evolution preserves process safety. That is, all derivatives of a process P safe for environment Env are safe for Env and its derivatives. An autonomous process and all its derivatives are safe in any system.*

When modelling objects, all method bodies can be considered as autonomous; even an object itself is initially autonomous until creation. After creation, input can only be performed via channels that were initially private.

Proposition 17 *Responsive bisimulation is preserved by parallel composition for safe processes: let $\mathcal{S}$ be either $\sim_r$ or $\approx_r$, then $P_1 \mathcal{S} P_2$ implies $(P_1|P)\mathcal{S}(P_2|P)$ for all P satisfying $(fin(P_1) \cup fin(P_2)) \cap fin(P) = \emptyset$.*

Proposition 18 *Replication preserves responsive bisimulations for autonomous processes. That is, let $\mathcal{S}$ be either $\sim_r$ or $\approx_r$, P_1 and P_2 be autonomous processes, then $P_1 \mathcal{S} P_2$ implies $(!\grave{n}(\tilde{x}).P_1)\mathcal{S}(!\grave{n}(\tilde{x}).P_2)$ for all input prefix $\grave{n}(\tilde{x})$.*

Corollary 19 *For autonomous processes, responsive bisimulations are congruences.*

An interesting result which demonstrates responsive equivalence of an object and a wrapped version of it, is :

Proposition 20 $P \approx_r (\nu\, n)(!\grave{m}(\tilde{x}).\acute{n}\langle\tilde{x}\rangle | P\{\grave{n}/\grave{m}\})$ *for all P and m, where $n \notin fn(P)$.*

6. Discussion

Privatise message versus privatise port. It is worth emphasising that the $[\grave{m}\langle\tilde{u}\rangle]$ in the term $[\grave{m}\langle\tilde{u}\rangle]P$ hides neither polar $\grave{m}$ nor $\acute{m}$, but the message $\tilde{u}$ which the input polar $\grave{m}$ is yet to consume. The output polar $\acute{m}$ remains public, and any message emitted via $\acute{m}$ by P itself must be considered as part of the observable behaviour of $[\grave{m}\langle\tilde{u}\rangle]P$. There is a difference between $[\grave{m}\langle\tilde{u}\rangle]P$ and $\acute{m}\langle\tilde{u}\rangle|P$: in the former, $[\grave{m}\langle\tilde{u}\rangle]$ is a buffered message arriving from the channel m, waiting for P or its derivatives to pick it up (though they may not); in the latter, the $\overline{m}\langle\tilde{u}\rangle$ is an outgoing message to be buffered in the channel m, and therefore is not distinguishable from an output of P.

The input polar $\grave{m}$ appearing in $[\grave{m}\langle\tilde{u}\rangle]$ will not be affected by an output polar substitution, just as one cannot change the destination address of mail after it is sent. Further intuition for the meaning of the term $[\grave{m}\langle\tilde{u}\rangle]P$ may be perceived from the immediate deduction from Proposition 20: $[\grave{m}\langle\tilde{u}\rangle]P \approx_r (\nu\, n)(!\grave{m}(\tilde{x}).\overline{n}\langle\tilde{x}\rangle|\overline{n}\langle\tilde{u}\rangle|P\{\grave{n}/\grave{m}\})$.

Delay of input versus delay of output. The *asynchronous bisimulation* $\sim_a$ (or $\approx_a$ for the weak version), is defined in [Amadio96] as an $o\tau$-bisimulation $\mathcal{S}$ for which whenever $P\mathcal{S}Q$ then $P \xrightarrow{\grave{m}(\tilde{u})} P'$ implies either $Q \xrightarrow{\grave{m}(\tilde{u})} Q'$ and $P'\mathcal{S}Q'$, or $Q \xrightarrow{\tau} Q'$ and $P'\mathcal{S}(\overline{m}\langle\tilde{u}\rangle|Q')$. It emphasises the possible delay of message output (or more precisely, delays during delivery), and considers message retransmission with the same communication channel as ignorable. In the view of OO systems, the delay in delivery is neither visible nor controllable for either sender or receiver. The responsive bisimulation, on the other hand, is quite natural for compositional objects since it concentrates on the delay of input, which is controllable by the receiver, and as pointed out by [McHale94] and [Zhang98b], can provide a synchronisation control point for the behavioural composition of objects. The $Ctrl_e$ and $Ctrl_l$ in Figure 1 are examples of these kind of controls.

Asynchronous and responsive bisimulations overlap, but neither contains the other. For example, the processes $\grave{m}.\overline{m}|P$ and P are weakly asynchronous bisimilar when $m \notin fin(P)$, but not weakly responsive bisimilar; while the processes $(\nu\, n)(\grave{m}.\overline{n}|\grave{k}.\grave{n}.P)$ and $\grave{k}.\grave{m}.P$ are weakly responsive but not asynchronous bisimilar. Details and proofs about the relation between responsive bisimulation and other well known bisimulations can be found in [Zhang02b].

7. Conclusion

This paper has presented the responsive bisimulation, which captures an equivalence between compositional concurrent objects, based on their response to inputs which may be delayed. For object systems, where input name clash can be eliminated, the responsive bisimulation is a congruence.

The responsive bisimulation can be understood in different ways. Apart from the view of "input delay", another view is that, when testing the behaviour of the target object, or black box, the only precondition we need to know is what messages have been provided to it, and the only postcondition we should examine is the response from the target object. We have proven these different views are equivalent.

With the responsive bisimulation, we can have a broader and more generic study of the behaviour of concurrent components, where existing bisimulations fail to give us the desired equivalence. Our approach enables us to establish a theory of concurrent objects with elegant compositional properties and provides a semantic basis for an extension to concurrent object-oriented programming languages.

References

[Aksit92] Mehmet Aksit and Lodewijk Bergmans. (1992). "Obstacles in Object-oriented Software Development", *Proc. of OOPSLA'92*, ACM SIGPLAN Notices vol. 27, pages 341–358, New York.

[Amadio96] Roberto M. Amadio, Ilaria Castellani and Davide Sangiorgi. (1996). "On Bisimulations for the Asynchronous π-calculus", in *Proc. of CONCUR'96*, LNCS vol. 1119, Springer Verlag.

[Holmes97] David Holmes, James Noble, John Potter. (1997). "Aspects of Synchronisation", in Christine Mingins, Roger Duke and Bertrand Meyer, editors, *Proc. of TOOLS 25* (TOOLS Pacific'97), pages 7–18, Melbourne, Australia.

[Hüttel96] Hans Hüttel and Josva Kleist. (1996). "*Objects as mobile processes*", Aalborg University.

[Jones93] Cliff B. Jones. (1993). "A π-calculus Semantics for an Object-based Design Notation", in E. Best, editor, *Proc. of CONCUR'93*, LNCS vol. 715, pages 158–172. Springer Verlag.

[McHale94] Ciaran McHale. (1994). "*Synchronisation in Concurrent, Object-oriented Languages: Expressive Power, Genericity and Inheritance*", PhD. Thesis, Department of Computer Science, Trinity college, University of Dublin, Ireland.

[Merro00] Massimo Merro, Josva Kleist and Uwe Nestmann. (2000). "Local π-Calculus at Work: Mobile Object as Mobile Processes", In *Proc. of IFIP TCS2000*, Sendai, Japan. LNCS vol. 1872, pages 390–408, Springe.

[Milner92] R. Milner, J. Parrow, D. Walker. (1992). "A Calculus of Mobile Process" (Parts I and II), *Journal of Information and Computation*, 100:1–77.

[MilSan92] Robin Milner and Davide Sangiorgi. (1992). "Barbed Bisimulation", in W. Kuich, editor, *Proceeding of 19th ICALP*, LNCS vol. 623, Springer Verlag.

[Milner99] Robin Milner. (1999). "*Communicating and Mobile Systems: the π-calculus*", Cambridge University Press.

[Noble00] James Noble and John Potter. (2000). "Exclusion for Composite Objects", In *Proc. of OOPSLA 2000*, Minneapolis, Minnesota USA, ACM press.

[Odersky95] Martin Odersky. (1995). "Polarized Name Passing", in *Proc. of FST& TCS'95*, Bangalore, India.

[Ravara97] António Ravara and Vasco T. Vasconcelos. (1997). Behavioural types for a calculus of concurrent objects. In C. Lengauer, M. Griebl, and S. Gorlatch, editors, *Proc. of Euro-Par'97*, LNCS 1300, pages 554–561. Springer-Verlag.

[Sangi92] Davide Sangiorgi. (1992). "*Expressing Mobility in Process Algebras: First-Order and Higher-Order paradigms*", PhD thesis, Computer Science Department, University of Edinburgh, UK.

[Sangi96] Davide Sangiorgi. (1996). "*An Interpretation of Typed Objects into Typed π-calculus*", INRIA Technical Report RR-3000.

[Schne97] Jean-guy Schneider and Markus Lumpe. (1997). "Synchronizing Concurrent Objects in the Pi-Calculus", *Proceedings of Langages et Modèles à Objets '97*, Roland Ducournau and Serge Garlatti (Ed.), pp.61–76, Hermes, Roscoff.

[Walker95] David Walker. (1995). "Objects in the π-Calculus", *Information and Computation*, 116(2):253–271.

[Zhang97] Xiaogang Zhang and John Potter. (1998). "Class-based models in π-calculus", in Christine Mingins, Roger Duke and Bertrand Meyer, editors, *Proc. of TOOLS 25* (TOOLS Pacific'97), Melbourne, Australia, 24th–27th November 1997, pages 238–251, IEEE Computing Society Press.

[Zhang98a] Xiaogang Zhang and John Potter. (1998). "*Compositional Concurrency Constraints for Object Models in π-calculus*", Technical Report C/TR-9804, Macquarie University, Sydney, Australia.

[Zhang98b] Xiaogang Zhang and John Potter. (1998). "A Composition Approach to Concurrent Objects", in Jian Chen, Mingshu Li, Christine Mingins and Bertrand Meyer, editors, *Proc. of TOOLS 27* (TOOLS Asia'98), Beijing, China, 22nd–25th September 1998, pages 116–126, IEEE Computing Society Press.

[Zhang02a] Xiaogang Zhang and John Potter. (20012). "*A Constraint Description Calculus for Compositional Concurrent Objects*", Technical report UNSW-CSE-TR-0204.

[Zhang02b] Xiaogang Zhang and John Potter. (2002). "*The Responsive Bisimulations in the polar π-calculus*", Technical report UNSW-CSE-TR-0203.

Author Index

GPSR Compliance
The European Union's (EU) General Product Safety Regulation (GPSR) is a set of rules that requires consumer products to be safe and our obligations to ensure this.

If you have any concerns about our products, you can contact us on

ProductSafety@springernature.com

In case Publisher is established outside the EU, the EU authorized representative is:

Springer Nature Customer Service Center GmbH
Europaplatz 3
69115 Heidelberg, Germany

www.ingramcontent.com/pod-product-compliance
Lightning Source LLC
LaVergne TN
LVHW011854200926
848484LV00002B/455
9781402071812